Wildlife Ecology and Management in Mexico

Perspectives on South Texas
Sponsored by Texas A&M University–Kingsville

This publication was made possible
by the generous funding support from
New Mexico State University,
Texas A&M University–Kingsville,
Aurelio Valdez Jr., and David A. Valdez.

Wildlife Ecology and Management in Mexico

Edited by

Raul Valdez and J. Alfonso Ortega-S.

Timothy E. Fulbright, General Editor

Texas A&M University Press • College Station

First edition

This paper meets the requirements of
ANSI/NISO Z39.48–1992 (Permanence of Paper).
Binding materials have been chosen for durability.
Manufactured in China through FCI Print Group

Library of Congress Cataloging-in-Publication Data

Names: Valdez, Raul, editor. | Ortega-S., J. Alfonso, editor.
Title: Wildlife ecology and management in Mexico / edited by Raul Valdez and J. Alfonso Ortega-S.
Other titles: Perspectives on south Texas.
Description: First edition. | College Station : Texas A&M University Press, [2019] | Series: Perspectives on south Texas | Includes bibliographical references and index.
Identifiers: LCCN 2018046155| ISBN 9781623497231 (book/ hardcover (printed case) : alk. paper) | ISBN 9781623497248 (e-book)
Subjects: LCSH: Wildlife conservation—Mexico. | Wildlife management—Mexico. | Animal ecology—Mexico.
Classification: LCC QL84.28 .W55 2019 | DDC 333.95/40972—dc23
LC record available at https://lccn.loc.go—v/2018046155

Cover photo by Santiago Gibert Isern.
Back cover photos courtesy of (clockwise): Sul Ross State University, Antony Challenger, Santiago Gibert Isern, Jorge G. Villarreal González, Cornell Lab of Ornithology, Marc Bartoskewitz, and Octavio C. Rosas Rosas.

A list of titles in this series is available at the end of the book.

Contents

Preface

Mexico is of great importance in the conservation of North America's wildlife. It is pivotally located at the transition between New World temperate and tropical regions and is a crucial element in the conservation and management of North American wildlife and the world's biodiversity. It ranks fifth in biodiversity among the nations of the world. It is particularly vital in providing stopover and wintering habitats for Canadian and American migratory birds and is a plant and faunal dispersal and migratory corridor. It is evident that the past and future of wildlife conservation and management in Mexico have had and will continue to have important effects on North American wildlife. A book that synthesizes wildlife conservation and management in Mexico is also an important reference work for Canadian and American wildlife biologists and natural resource managers because it can serve as the basis from which to integrate knowledge of Mexican wildlife populations and their habitats with those of the rest of North America. There are numerous books that are species specific, as well as many comprehensive treatises on game birds and mammals of North America. Most of these concern North American wildlife in temperate zones only, and none include the status, habitat use, and management of wildlife in Mexico.

The objectives of *Wildlife Ecology and Management in Mexico* are to (1) provide an updated text detailing the status, distribution, ecological relationships, habitat requirements, and habitat management of the most important game birds and mammals in Mexico; (2) provide an introduction to the diversity and management needs of Mexican wildlife; (3) provide wildlife biologists, natural resource administrators, land owners, and students with a comprehensive reference work on the wildlife of Mexico; (4) review the threats and challenges facing wildlife conservation in Mexico; and (5) present strategies for resolving wildlife management challenges based on an understanding of the political, economic, social, cultural, land governance, and ecological factors impacting the wildlife of Mexico.

A book relating to these topics was urgently needed. Since A. Starker Leopold's *Wildlife of Mexico,* which was published in 1959 and is now outdated and out of print, there has been no subsequent book on this topic. Leopold's book was based on field studies conducted principally in the 1940s, and there is much new published wildlife research and changes in the laws and policy of Mexican state and federal governments. Nonetheless, Leopold's book was a seminal treatise on Mexican wildlife and the only reference work for 60 years; it was translated into Spanish and served as a stimulus to a growing number of Mexican wildlife students and professionals to pursue wildlife research and conservation in Mexico.

The intended audience for this book is a broad sector of natural resource professionals, including private and public wildlife research scientists, veterinarians, administrators, private wildlife consultants, private landowners and communal landholders, natural resource agencies, nonprofit wildlife conservation organizations, the corporate sector, which has become an important sponsor of wildlife conservation in Mexico, and the general public. This book is also intended to aid land managers who lack a background in wildlife conservation, and to serve as a textbook and reference for Mexican wildlife educators and students. It is written using semitechnical terminology to make it usable to this broadly intended audience. The first chapter presents the basic principles of wildlife conservation in an environmental, ecological, social, and historical context for Mexico. The emphasis of the book is on the ecology and management of the major game birds and mammals, with additional chapters on the physical and biological setting of Mexico, wildlife laws and policy, wildlife as a public and economic resource, and the conservation and management challenges of wildlife in Mexico. We do not include exotics or the only reestablished big game species in Mexico, the bison, which occurs in a restricted, small population.

Wildlife ecology and conservation in Mexico have made great strides in the last 50 years. Numerous students with master's and doctoral degrees in wildlife and conservation ecology have graduated from Mexican, American, and European universities and have afforded Mexico much-needed expertise in natural resource conservation and educational endeavors. These professionals have been productive in furthering wildlife research and conservation in Mexico. It is evident by the chapter authors, many of which are Mexican biologists, that Mexico now has a solid foundation of expertise in wildlife conservation and management. However, an updated comprehensive book was still needed that could serve as a reference to the profession for current knowledge and literature. The current wildlife literature relating to Mexico is scattered in numerous technical publications and books and much of it is in English; however, it is increasingly being authored by Mexican biologists. A literature cited section is included at the end of each chapter to allow readers to readily refer to specific pertinent chapter topic references. This book should provide a framework and stimulus to further wildlife conservation in Mexico at scientific, educational, administrative, private, and communal ownership levels.

Both of us have extensive experience with wildlife in Mexico. We have conducted research and supervised graduate students who have carried out field research in Mexico. It has been a pleasure and a gratifying experience to work with Mexican professionals and students. Much of our fieldwork in Mexico has been conducted in conjunction with Mexican wildlife professionals and colleagues whose wealth of knowledge has stimulated and improved our work and who, equally importantly, have made us feel at home and have provided many hours of camaraderie and true friendship. We are optimistic for the future of wildlife in Mexico. A rural land and wildlife stewardship philosophy is becoming widespread. It is especially encouraging to know that the values of wildlife are increasingly being recognized and that numerous wildlife scientific and management programs have been initiated throughout the nation. Many private and communal owners view wildlife as an asset and an integral part of the natural resources of Mexico. The basic premise of this book is that it will enable individuals with an altruistic interest in the wildlife of Mexico to foster a stronger commitment to its sustainable management.

Raul Valdez
J. Alfonso (Poncho) Ortega-S.

Acknowledgments

We thank the College of Agricultural, Consumer, and Environmental Sciences at New Mexico State University, Las Cruces, New Mexico, and the Caesar Kleberg Wildlife Research Institute at Texas A&M University–Kingsville for their support during the editing of this book. Shannon Davies, director of Texas A&M University Press, encouraged our efforts and was instrumental in completing this book. A special thanks to Eahsan Shahriary for his tireless assistance in various capacities to make this book a reality. We thank the following individuals for reviewing chapters: Keith Arnold (retired), Texas A&M University–College Station; Richard Buchholz, University of Mississippi; Randy DeYoung, Caesar Kleberg Wildlife Research Institute; Jack Clinton Eitniear, Center for the Study of Tropical Birds; Jonathan Escobar Flores, Instituto Tecnológico de Durango; Shawn Gray, Sul Ross State University; Joe Haug (retired), Wisconsin Department of Natural Resources; James R. Heffelfinger, Arizona Game and Fish Department; Scott G. Hereford, US Fish and Wildlife Service; Bill Johnson, US Fish and Wildlife Service; Ray Lee, Ray Lee LLC; Heather A. Mathewson, Tarleton State University; Alicia Melgoza Castillo, Universidad Autónoma de Chihuahua; German D. Mendoza, Universidad Autónoma Metropolitana; Kevin Monteith, Wyoming Cooperative Fish and Wildlife Research Unit, University of Wyoming; Ryan O'Shaughnessy, Sul Ross State University; Nicole T. Quintana, New Mexico Department of Game and Fish; Steve Stringham, Bear Viewing Associates; Luis A. Tarángo Arámbula, Colegio de Postgraduados; Ron Thompson, Primero Conservation; Oscar A. Villarreal Espino Barros, Benemérita Universidad Autónoma de Puebla; Thomas Wayne Schwertner, Tarleton State University; Gerardo Bezanilla, Universidad Autónoma de Chihuahua. We sincerely thank our wives, Griselda Sánchez de Valdez and Caty Sánchez de Ortega, for their patience and support during the editing of this book.

Acronyms and Abbreviations

AMMAC (Asociación Mexicana de Mastozoología)

ANGADI (Asociación Nacional de Ganaderos Diversificados Criadores de Fauna)

ANP (Áreas Naturales Protegidas)

AOU (American Ornithologists' Union)

BLM (Bureau of Land Management)

CEDUA (Centro de Estudios Demográficos, Urbanos y Ambientales)

CEMEX (Cementos Mexicanos)

CIPAMEX (Sociedad para el Estudio y Conservación de las Aves en México)

CNOG (Confederación Nacional de Organizaciones Ganaderas)

CONABIO (Comisión Nacional para el Conocimiento y Uso de la Biodiversidad)

CONACyT (Consejo Nacional de Ciencia y Tecnología)

CONAGUA (Comisión Nacional del Agua)

CONANP (Comisión Nacional de Áreas Naturales Protegidas)

COTECOCA (Comisión Técnico Consultiva de Coeficientes de Agostadero)

DAFCP (Departamento Autónomo Forestal y de Caza y Pesca)

DFC (Dirección Forestal y de Caza)

DGFFS (Dirección General de Flora y Fauna Silvestres)

DGFS (Dirección General de Fauna Silvestre)

DGVS (Dirección General de Vida Silvestre)

DIPARES (Dirección General de Parques, Reservas y Áreas Ecológicas Protegidas)

DOF (Diario Oficial de la Federación)

ECOSUR (El Colegio de la Frontera Sur)

EUNA (Editorial Universidad Nacional)

FAO (Food and Agriculture Organization of the United Nations)

FCE (Fondo de Cultura Económica)

FEDECAVE (Federación Nacional de Caficultores de Venezuela)

FIRA (Fideicomisos Instituidos en Relación con la Agricultura)

FUDECI (Fundación para el Desarrollo de las Ciencias Físicas, Matemáticas y Naturales)

FUNEP (Fundação de Apoio a Pesquisa, Ensino e Extensão)

GCR (Greater Calakmul Region)

GEF (Global Environment Facility)

GTZ (Gesellschaft für Technische Zusammenarbeit)

IMERNAR (Instituto Mexicano de Recursos Naturales Renovables)

INE (Instituto Nacional de Ecología)

INEGI (Instituto Nacional de Estadística y Geografía)

IUCN (International Union for Conservation of Nature)

LFPA (Ley Federal de Protección al Ambiente)

LFPCCA (Ley Federal para Prevenir y Controlar la Contaminación Ambiental)

LGEEPA (Ley General del Equilibrio Ecologíco y Protección al Ambiente)

LGVS (Ley General de Vida Silvestre)

NAWCA (North American Wetlands Conservation Act)

NOM (Normas Oficiales Mexicanas)

OIMT (Organización Internacional de las Maderas Tropicales)

PACE (Programas de Acción para la Conservación de Especies)

PCVSDP (Programa de Conservación de la Vida Silvestre y Diversificación Productiva en el Sector Rural)

PIMVS (Predios o Instalaciones para el Manejo de Vida Silvestre)

PNUMA (Programa de las Naciones Unidas para el Medio Ambiente)

PREP (Programa para la Conservación de Especies Prioritarias)

PROCER (Programa de Conservación de Especies en Riesgo)

PROFAUNA (Protección de la Fauna Mexicana)

PROFEPA (Procuraduria Federal de Protección al Ambiente)

RCS (Ramsar Convention Secretariat)

SAF (Secretaría de Agricultura y Fomento)

SAG (Secretaría de Agricultura y Ganadería)

SAGARHPA (Secretaria de Agricultura, Ganadería, Recursos Hidráulicos, Pesca y Acuacultura)

SAGARPA (Secretaría de Agricultura, Ganadería, Desarrollo Rural, Pesca y Alimentación)

SARH (Secretaría de Agricultura y Recursos Hidráulicos)

SE (Subsecretaría de Ecología)

SEDESOL (Secretaría de Desarrollo Social)

SEDUE (Secretaría de Desarrollo Urbano y Ecología)

SEMARNAP (Secretaría de Medio Ambiente, Recursos Naturales y Pesca)

SEMARNAT (Secretaría de Medio Ambiente y Recursos Naturales)

SFFS (Subsecretaría Forestal y de Fauna Silvestre)

SNIARN (Sistema Nacional de Información Ambiental y de Recursos Naturales)

SNIB (Sistema Nacional de Información sobre Biodiversidad)

SRFC (Secretaría de Recursos Forestales y de Caza)

SUMA (Sistema de Unidades para la Conservación de la Vida Silvestre)

UAAAN (Universidad Autónoma Agraria Antonio Narro)

UABC (Universidad Autónoma de Baja California)

UACAM (Universidad Autónoma de Campeche)

UAEH (Universidad Autónoma del Estado de Hidalgo)

UAEM (Universidad Autónoma del Estado de Morelos)

UAM (Universidad Autónoma Metropolitana)

UANL (Universidad Autónoma de Nuevo León)

UAT (Universidad Autónoma de Tamaulipas)

UMA (Unidades para la Conservación, Manejo y Aprovechamiento Sustentable de la Vida Silvestre)

UMSNH (Universidad Michoacana de San Nicolás de Hidalgo)

UNAM (Universidad Nacional Autónoma de México)

UNDP (United Nations Development Program)

UNESCO (United Nations Educational, Scientific and Cultural Organization)

US (United States of America)

USDI (United States Department of the Interior)

USFWS (United States Fish and Wildlife Service)

Wildlife Ecology and Management in Mexico

Historical and Conservation Perspectives of Wildlife in Mexico

1

Raul Valdez

The United Mexican States encompass an area of 1,972,000 km^2 (761,393 mi^2) and comprise 31 states and Ciudad de México (formerly a federal district). Mexico is bounded on the north by the United States of America (US), with a boundary that extends 3,115 km, and on the south by Guatemala and Belize. The Gulf of Mexico borders the country on the east and the Pacific Ocean borders the west, and the Tropic of Cancer divides Mexico into almost equal north and south halves. The northern border region of Mexico constitutes the world's largest human migration corridor, with an alarming average annual population growth rate of 2.24%. Mexico is also the world's largest and most populous Spanish-speaking nation, with a population of 120 million in 2017, of which about 22% is rural. There are 196,328 villages of fewer than 2,500 inhabitants. Mexico is one of the world's most culturally diverse nations, with 291 spoken languages. Mexico City, the capital, has a population of 21 million and is the most populous city in the Western Hemisphere (Blouet and Blouet 2015). Mexico is the second-largest buyer of US goods, and the US is Mexico's top export market; bilateral trade reached US$584 billion in 2015 (Wilson 2017).

The merging of the New World temperate and tropical regions in Mexico is one of the major reasons for Mexico's high climatic and biological diversity. It is the fourteenth-largest country in the world and one of the 17 countries identified as megadiverse. Mexico's unique zoogeographical position as a transition zone between the Nearctic and Neotropical realms, its large size, its diversity of terrestrial, freshwater, and marine habitats, its phytophysiographical complexity, and its biological interconnectivity with the rest of North America establish Mexico as a crucial component in the conservation and management of North American wildlife. It is a major center for plant origins and domestication as well as a plant and faunal dispersal corridor, and it is noted for its large number of endemics (Valdez et al. 2006). As a faunal corridor, for example, it is an important wintering area for several species of bats, monarch butterflies, and temperate North American nesting birds; 51% of US and Canadian bird species spend six to nine months a year in Mexico (McNeely et al. 1990). For a detailed review of Mexico's ecosystems, biomes, and vegetation types, see chapter 2.

Mexico's rich and bountiful biodiversity is the basis for economic, aesthetic, and cultural values that transcend national boundaries. Mexico's biodiversity provided the progenitors for the domestication of some of the world's most important foods, including maize or Mexican corn (the world's most-produced grain crop), several species of beans, and the domesticated turkey, plus other basic foods. These agricultural products (chili, cacao bean [source of chocolate], tomato, vanilla, avocado, and pumpkin and other squashes) are now components of numerous international cuisines. More importantly, billions of the world's human population rely on these foods to provide a major portion of their daily nutrient intake. Mexico's varied indigenous heritage, ancient civilizations, archaeological splendors, stimulating human-dominated rural and metropolitan attractions, awe-inspiring natural landscapes with both biological and geological components, extensive sandy beaches in tropical settings, and distinctive cuisines have created alluring attractions for one of the world's most important venues for international tourism.

Mexico is also internationally recognized for having some of the highest wildlife diversity. It ranks fifth in biodiversity among the world's 196 nations. In vertebrates alone, Mexico has 1,107 bird species (more than the combined species total of Canada and the continental US), 544 mammals (ranks second in the world), 361 amphibians, 804 reptiles (ranks second in the world), and more than 450 freshwater fish species (Llorente-Bousquets and Ocegueda 2008,

Ceballos et al. 2014). Unfortunately, its wildlife diversity has been historically neglected at all levels of resource management planning during most of the modern period, and it has been ignorantly regarded as a resource without significant economic value, with no recognition of its ecological importance. This was not the case for Mexico's ancient cultures, who related culturally and spiritually to wildlife and depended on a multitude of wild plants and animals as primary sources of food, commercial products (e.g., live wildlife and by-products such as pelts, skins, feathers, pets, flowers, lumber, and artisanal products), construction materials, and medicines. Plants and animals were also used for religious purposes (fig. 1.1). During the modern era, Mexico's wildlife still has economic value and will ultimately have to provide the stimulus to ensure the sustainable management of its wildlife populations. Fortunately, Mexico is making progress in its efforts to manage its wildlife resources in economically and environmentally viable programs (Valdez et al. 2006, Martinez et al. 2006). In fact, wildlife may serve as the basis for initiating natural resource conservation programs and may thus assist in alleviating rural poverty.

Figure 1.1. The jaguar was revered in Mesoamerican cultures. Associated with power and war, it also had a religious role.

Wildlife conservation and management in Mexico are rapidly evolving, and people are increasingly recognizing the monetary significance of wildlife. Land managers are now engaged in initiating and incorporating economically viable game and nongame wildlife programs in land stewardship enterprises throughout Mexico, in some instances with the assistance of government-sponsored programs. There is growing recognition that Mexico's wildlife resources can be a critical component in the rural economic development of the country. Wildlife conservation policy decisions in Mexico can have impacts on individuals, communities, the nation, and even the world because of Mexico's status as a megadiverse country and because its crucial ecosystems provide habitats for migratory North American wildlife, including birds, bats, and butterflies. A growing cadre of wildlife professionals are enthusiastically pursuing wildlife research and management dedicated to the sustainable management of wildlife populations and their habitats. In addition, a growing number of university wildlife professors are actively engaged in developing and supervising graduate student research and producing competent wildlife professionals who are being integrated into natural resource management programs (fig. 1.2). Equally important, professionals in other disciplines are recognizing the value of wildlife in economic development programs and are incorporating wildlife in integrated natural resource planning. Wildlife enterprises can be highly profitable, but this has been realized only within the last 40

Figure 1.2. The wild turkey is one of several species that have been reestablished by releasing wild trapped animals in areas where they have been extirpated. Photo by Luis A. Tarango Arámbula.

years in northern Mexico (Sonora, Sinaloa, Coahuila, Nuevo León, and Tamaulipas) and has been largely neglected in central and southern Mexico. Because of its economic value, wildlife is now incorporated into agricultural, rangeland, and forestry programs throughout Mexico. The commodity value (including goods and services) of wildlife in some cases exceeds the economic value of other land uses such as livestock enterprises. At the local level, communities in Mexico continue to depend on wildlife as an important source of meat protein, and wildlife continues to have commercial value (see chapter 3) (Valdez et al. 2006, Santos-Fita et al. 2012).

Historical Ecology of Mexico

Historical ecology focuses on the interface of humans, the environment, and their past intertwined effects. It is the study of human interactions with the biota and abiota that have determined human and wildlife distribution and abundance, as well as the effects of humans on landscapes through time. Abiotic environmental effects, such as droughts, hurricanes, and long-term climate change, have altered human and animal population abundance. Humans have evolved strategies to adapt to environmental regimes, but when these have proved unsustainable, human populations have declined because degraded habitats have become less productive for human needs. These human-altered habitats have in turn affected the abundance of wildlife and have often resulted in degraded, less productive ecosystems (Redman 1999, Mann 2005).

Humans populated Mexico approximately 12,500 years ago, when mammoths were still present and hunted. The initial hunter-gatherers in Mexico's varied ecoregions brought dogs with them that assisted in hunting, as well as effective projectile weapons and habitat management techniques, fire being the most significant. They channeled resources through human-centered food webs and in the process altered ecosystems and reduced biodiversity. However, subsistence hunting had only locally significant environmental effects because of the dispersed nature and low densities of human populations. Watersheds were intact and provided large areas for aquifer recharge with undisturbed vegetation and minimal effects on biogeochemical cycles. Throughout the pre-Columbian period, Mexican cultures lacked metal implements and domesticated ungulates for traction and transportation; they depended on stone and wooden tools and human labor without wheeled vehicles. This limited the extensive development of agriculture away from population centers. Wildlife provided food, medicinal and artisanal products, tools, adornments, and pets, and it had commercial, ceremonial, and religious value. However, the domestication of plants, which was initiated about 7,500 years ago in small food plots and was later intensively developed by the Maya and Aztec cultures, among others, required significantly sophisticated technologies to maintain the extensive agroecosystems on which the increasing human populations, concentrated in permanent villages, depended. The enlarged agricultural land transformations caused decreases in native biota but also benefited species that proliferated in human-dominated landscapes. The magnified, concentrated human populations became ever more vulnerable to environmental disturbance such as climatic change, which in turn affected food production and human abundance, sometimes drastically. Regional overpopulation, resulting in intensive land conversion and habitat loss for wildlife, often had deleterious, long-term effects on the native biota.

Mayas

The Maya culture (3500 BP–1100 AD) of Central America, one of Mexico's most successful civilizations, had a population of perhaps 10 million at its peak and began to clear forests extensively about 3000 BP. The people used a variety of advanced farming methods, in conjunction with silvicultural, water, and soil conservation measures (Simonian 1995). Their major food production was based on a slash-and-burn system of cleared tropical forest patches (milpas or cornfields) intensively planted with corn, beans, and squash for several years, followed by a long-term fallow period (with abandoned cultivated forest tracts) to restore soil fertility and productivity. The fallow period allowed a resurgence of diversified wild plants that attracted wild ungulates, rodents, and birds, among others, which flourished because of the food and cover provided by the successional vegetation stages. In turn, these animal species were important sources of meat protein, especially in the absence of domesticated animals. Families had home crop and flower gardens, which also attracted wildlife. Wildlife that damaged planted fields were hunted to mitigate crop damage; some of the species considered harmful were also consumed. The interspersed, uncleared tropical forest and cleared-forest patches, in various successional stages, formed a mosaic of source-sink resilient habitats that resulted in increased biodiversity and a sustainable, bountiful anthropogenic ecosystem (Toledo et al. 2003, Anderson and Medina Tzuc 2005). These disturbed vegetative communities also provided favorable diverse habitats for hundreds of migratory bird species from temperate North America wintering in the tropics. Indeed, native broadleaf forests in multiple stages of succession support more species of migratory birds in the Neotropics (Petit et al. 1995). Moreover, modern native rural communities, especially in impoverished areas, continue to practice subsistence hunting of medium-sized and large animals in habitat mosaics of croplands, fallow fields, forests, and wetlands (Santos-Fita et al. 2012).

The Classic Maya civilization ended about 1,400 years ago, probably because of overpopulation and environmental degradation caused by severe regional soil erosion, exacerbated by periods of drought and shorter fallow intervals. This in turn resulted in reduced agricultural productivity, extensive deforestation, and extirpation of wildlife populations. The drastic decline of the human population initiated a regrowth of forests and initially an increase in the abundance of wildlife because of reduced hunting pressure, curtailed farming, and the restructuring and compositional changes of plant communities that provided habitats favorable to a host of tropical wildlife species (Dunning and Beach 2000, Hodell et al. 2000, Peters 2000). However, in the long term, the reduced habitat variability resulting from the regrowth of forests favored species more adapted to mature forest communities.

Aztecs

The Aztec civilization (1300–1521), the largest Mexican pre-Columbian empire, dominated most of central and southern Mexico. Centered in the Basin of Mexico (now Mexico City), it had a population of perhaps several million. The basin consisted of five shallow lakes, which supported multitudes of waterfowl (23 species), shorebirds, and other waterbirds, as well as abundant fishes, amphibians, arthropods, and algae that were exploited for food. The wild turkey (*Meleagris gallopavo*) and Muscovy duck (*Cairina moschata*) were the only indigenous domesticated vertebrates. The surrounding uplands supported a diverse vegetation of pine, oak, and juniper communities, grasslands, and xerophilous scrub with abundant populations of wild turkey and white-tailed deer (*Odocoileus virginianus*), which were exploited commercially (Ezcurra et al. 1999). Hernán Cortés, commander of the conquering Spanish army, described the astonishing menagerie of Montezuma, ruler of the empire, who had a vast collection of wild mammals and marine and freshwater birds for his pleasure and consumption (Pagden 1971). The empire was based on the subjugation of other cultures, which included the appropriation of natural resources with an extensive market economy of tribute (taxes) and trade domination. The open market in Mexico City, supplied by thousands of boats, was one of the largest in the world, with a multitude of products, including wild and domesticated plants and animals. The ruling elite prized the feathers of the resplendent quetzal (*Pharomachrus mocinno*) for adornment and ceremonial purposes. The area surrounding the lake was densely populated and, because of intensive agriculture and exploitation of ecosystems, could not supply sufficient natural resources to sustain the human population, which in turn had to depend on supplies from distant sources in the empire. Extensive areas of land were converted to agriculture, resulting in widespread environmental disturbances, but vast areas throughout Mexico remained minimally impacted. The reasons for this included low human densities distant from population centers, the absence of domesticated ungulates as a protein source and a source of manure to fertilize crops, a limited transportation system, and lack of technology to extract resources on a large scale (Challenger 1998).

Colonial period

The Spanish conquest (1521) of the Aztec Empire was catastrophic for the biota of Mexico and deleterious to the relationships between Native Americans and wildlife established over millennia. The seismic scale of environmental changes affected the behavior, interspecific interactions, population dynamics, and evolutionary processes of native species, which were suddenly confronted with extraordinarily transformed landscapes. Spaniards rejected indigenous land-use knowledge systems and introduced intensified European farming practices, favoring European wheat over maize, and brought cattle, horses, donkeys, sheep, goats, and pigs. Mismanagement of livestock was common and resulted in biodiversity loss from land conversion, overgrazing, and soil compaction and erosion. Oxen and horses were used to pull metal-tipped plows to farm previously untilled soil, facilitating the removal of native plant communities. While the native fauna was extirpated over greater areas, agriculturally adaptable new food sources, especially Old World fruits (e.g., bananas, mangoes, oranges, pears, and grapes), and numerous invasive forbs and grasses grew abundantly. The introduction of firearms added to the local decimation of wildlife. The death of several million Native Americans—largely the result of introduced diseases—tempered the initial ecological shock of the environmentally destructive onslaught. This decline in population in turn limited land development and natural resource extraction because of the lack of labor. The native population did not begin to increase until after the mid-1600s. Eventually, the Spaniards established military, cultural, and religious control over most of Mexico and initiated even greater agricultural developments at the expense of native ecosystems. This was exacerbated by an ever-increasing human population. Raw materials used as trade commodities, especially precious metals, were economic priorities. Imperial policy emphasized the search for botanical discoveries that had commercial value, such as indigo, a plant that produced a coveted blue vegetable dye. Silver mining, the major source of Spain's colonial revenue, required the felling of extensive forests for building materials and fuel to process mined minerals, which in turn released deleterious wastes, mercury being the most toxic, during ore production that resulted in polluted air and water. Livestock were one of the most transformative biotic elements imposed on New World ecosystems; unmanaged horses, cattle, pigs, and sheep thrived on American forage, which had not previously been grazed by domestic ungulates. The uncontrolled increase of domestic sheep in central Mexico led to the loss of vegetative cover, increased soil erosion, invasion of woody plants and less nutritious forage, and the eventual decline of sheep numbers because of degraded rangelands. Feral domestic ungulates added to the problem. However, remote areas were still relatively unexploited and retained their biodiversity (Crosby 1972, 1986; Challenger 1998; Melville 1994, 2000).

Postcolonial period

After Mexico gained independence from Spain in 1821, the nation became economically weak and agricultural development slowed. It was not until the latter part of the nineteenth century that exports from Mexico increased significantly, and the colonial pattern of mercantile exploitation of natural resources resumed. Foreign investments enabled the extraction of lumber and the large-scale production of henequen, sugar, rubber, and coffee, among other items, resulting in the felling of extensive native forests in southern Mexico without concern for wildlife. In arid northern Mexico, large private landholdings devoted to cattle enterprises degraded rangelands. As ranching in tropical areas became widespread, it required extensive forest clearing. Global and internal demands for and profitable returns from minerals, lumber, petroleum, and gas dictated Mexico's natural resource policies. Also, the emphasis on increased production of agricultural commodities based on damming rivers, clearing native vegetation, using aquifers unsustainably, and making extensive use of fertilizers and synthetic pesticides resulted in soil, water, and air pollution, devastating ecosystems. Mexico's population increased from 30 million in the 1950s to 100 million in the 1990s, much of it in larger cities with unconstrained urban sprawl, pollution, and extensive biodiversity loss. Economic and social exigencies led to government-sponsored programs of uncontrolled natural resource extraction and the establishment of human settlements in unexploited areas to meet the growing needs of the expanding population (fig. 1.3). Nonetheless, millions of Mexicans continued to live in poverty. Wildlife was considered an expendable resource without economic value; the result was a steady decrease in biodiversity (Boyer 2012, Wright 2012). Water mismanagement, an especially critical challenge, also contributed to ecosystem degradation (Arreguín Cortés et al. 2011). The major threats to sustainable use of wildlife continue to be deforestation, mismanagement of livestock resulting in rangeland resource degradation, unregulated mining and agricultural enterprises, drainage of wetlands, dam construction, industrial pollution, and illegal exploitation of native plant and animal resources (Challenger 1998, Geist and Lambin 2002, Stoleson et al. 2005, Valdez et al. 2006, Echeverría et al. 2007, Cappock et al. 2017).

Figure 1.3. Habitat loss is one of the factors that contribute the most to species extinctions, as in the case of converting critical habitat for the endangered aplomado falcon into farmland. Photo by Gill Sorg.

Concepts of Wildlife Conservation

A wildlife biologist is a natural resource professional who applies the tenets of social science and the scientific principles of ecology, genetics, and ethology (study of comparative animal behavior) to the sustainable management of native wildlife resources. "Wildlife" refers to wild animals—those that are not domesticated or tamed. However, vertebrates (mammals, birds, reptiles, amphibians, and fishes) usually take precedence because they include animals that are hunted (game species) and fished and that have direct economic value. Also, the public can relate to vertebrates culturally and emotionally (especially charismatic wildlife such as jaguars and hummingbirds that people consider fascinating and delightful). Nevertheless, invertebrates such as spiders, scorpions, and some insects are becoming popular in the pet trade. Nongame species (those used nonconsumptively) have aesthetic value, and although not hunted, they can have great economic value because of the price people are willing to pay to view these species, and the lucrative prices they and their by-products command in the legal and, unfortunately, illegal wildlife trade.

Wildlife managers are also concerned with nonnative wildlife, including domestic animals, because they may compete with wild animals for forage and space, deleteriously impact habitats, and have the potential to transmit diseases, so that monetary resources that could be spent on native wildlife have to be diverted to feral and exotic (nonnative introduced) species, as in control programs (Mungall 2000, Teer 2003). Feral animals, such as dogs and cats, may be the most significant predators in urban and suburban areas (Gehrt et al. 2010). Examples of exotics in Mexico include the English sparrow (*Passer domesticus*), Chinese ring-necked pheasant (*Phasianus colchicus*), and several species of ungulates (Álvarez-Romero et al. 2010).

Objectives of Wildlife Conservation

Wildlife is a renewable natural resource, as are forests and range plants. If managed in a sustainable manner, they can provide aesthetic enjoyment and can be harvested for many generations. Wildlife conservation is concerned primarily with managing native wildlife populations by maintaining and restoring the productivity of native wildlife habitats (Mangun 1992). The structure, diversity, and ecological functioning of habitats are themselves dependent on the ecological services of wildlife. For example, wild species serve as disseminators of pollen, seeds, and other propagules. Carnivores and other species can help prevent herbivorous species from growing to levels that have harmful impacts on vegetation structure and composition (Bleich 2012). As metaphorically phrased by A. Challenger (personal communication): "In this and other aspects, wildlife is not merely the icing on the ecological layer cake, but is the filling that holds the layers together." Wildlife conservation involves research, education (public relations and extension work), administration (logistics), law enforcement (making people obey regulations), and management (making it work). The objectives of wildlife conservation are to (1) manage native wildlife populations and habitats sustainably, (2) conserve the genetic diversity of native species, (3) maintain healthy ecosystems and restore degraded habitats, (4) gain economic benefit from wildlife, (5) reduce material damage from wildlife, (6) integrate wildlife management programs with those of other natural resources, and (7) incorporate the social and economic concerns and needs of stakeholders in the planning and implementation of wildlife conservation programs.

Wildlife management involves practices that will result in a sustained yield of wild animals for recreational, subsistence, commercial, and ecological restoration purposes. This requires the application of ecological research in manipulating wildlife populations and habitats (Randel et al. 2012). Equally important is the involvement of wildlife managers in human social, economic, and political activities that impact wildlife populations. Integrating human dimensions in wildlife policy and planning is crucial to successful wildlife management (Peterson and Rodriguez 2012, Decker et al. 2013). Wildlife biologists have become increasingly involved in managing and considering the impacts of endangered and nongame species, wildlife diseases including zoonoses (diseases transmitted from animals to humans) (Peterson and Ferro 2012, Conover and Vail 2015), invasive alien organisms (Clout and Williams 2009), climate change, ecosystem and agroecosystem integrity (Bohlen and House 2009), human-wildlife conflicts (Conover 2001, Thompson and Cassaigne 2017), and urban wildlife management (Adams 2016, McCleery et al. 2012). Wildlife biologists view all species as having inherent ecological value and potential genetic and economic value. Restoration of wildlife habitats is an important but complicated aspect of wildlife management that involves knowledge of species life histories, habitat requirements including plant community composition and structure, and present and historical abundance and distribution data, among other things (Morrison 2013). Wildlife management includes sustainably integrating urban and rural landscapes in ways that retain native biodiversity. The economic and social aspects of sustainable use are equally important. Most natural resource problems require managers to resolve the human social and economic concerns that arise in managing natural resources. Wildlife biologists promote ecologically sustainable use and development of wildlife resources in conjunction with promoting and incorporating justifiable economic, social, and political strategies.

Value of Wildlife

Wildlife increases income opportunities and land values for landowners. Recreational use of rangeland, forest, and wildlife resources in conjunction with livestock enterprises, such as multispecies grazing systems that integrate wild and domestic ungulates, can provide additional sources of income to Mexican landowners. Economically viable wildlife strategies can readily be integrated with agricultural, range, and forestry management programs. Wildlife can provide an economic safety net because it can continue to provide necessary income when other forms of production may be economically or environmentally stressed. For example, livestock production during drought conditions may have to be drastically curtailed, but wildlife can continue to be an economically viable resource, an important consideration in Mexico because of the vast areas of arid and semiarid rangelands. Wildlife also requires few economic inputs; livestock require fencing, supplemental feed and water, veterinary assistance, hired herders, equipment maintenance, and transport to market. Diversification programs have been especially successful in northern Mexico where dove, turkey, deer, wild sheep, and wild duck sport-hunting enterprises have greatly added to landowner incomes, resulting in an appreciation for wildlife (fig. 1.4).

The qualitative values of wildlife are equally important; these include recreational values (nonconsumptive, such as viewing and photography, in contrast to consumptive, such as sport and commercial hunting and trapping), as well as scientific, educational, emotional, psychological, and cultural (including religious) values, among others. Nonconsumptive activities such as ecotourism can be a

Figure 1.4. The economic value of sport hunting, such as for desert bighorn sheep, can in some instances surpass the economic value of other forms of land use. Photo by Bighorn Institute.

significant source of income to rural communities and landowners. Unfortunately, the commercial (tangible) value of wildlife must be quantified, either directly as a resource, or indirectly as ensuring vital and economically valuable ecosystem services, before modern society will justify wildlife's continued existence (see chapter 3). Refer to Bezaury Creel (2009) for the value of goods and services derived from national protected areas in Mexico.

Wildlife Conservation and Management Policy in Mexico

Modern wildlife conservation and management policy in Mexico is based on the realities of economic, political, social, and land tenure systems. There are three major land tenure types: federal, private, and communal. The principal communal lands are *ejidos* and *comunidades*. Comunidades are Indian communal landholdings, whereas in ejidos, communal properties are distributed to groups of individual peasants, but land ownership resides with the community and there is collective administration and management of these lands and their resources, including wildlife and forests. Most of these lands are farmed. Originally, communal properties could not be sold and remained the property of the federal government, but regulatory changes in 1991 permitted their sale. However, ejidos remain the second-largest form of land tenure in Mexico; there are 28,000 ejidos, encompassing about half of the rural lands in Mexico and occupied by over three million communal landowners and their families (Harvey 1996). About 80% of the forests in Mexico are managed in ejidos or indigenous communities (Bray and Wexler 1996). To retain the right to exploit them, communal property owners must manage resources productively, but they are rarely managed sustainably because of a lack of federal funding for modern management technology and economic incentives such as credits or income for conservation practices. There is widespread poverty in rural areas. Wildlife has been particularly neglected because it has not been considered an economically viable resource (Valdez et al. 2006).

Mexico's modern federal conservation policy, initiated in 1997, is incorporated in the Program of Wildlife Conservation and Rural Sector Production Diversification. It integrates environmental, economic, social, and legal strategies in wildlife conservation planning, while promoting broader societal participation and creating economic incentives. Key stakeholders are involved in management decisions, which promotes participatory conservation (Instituto Nacional de Ecología 2000, Valdez et al. 2006). This program focuses on the conservation and recovery of priority species and the creation of a system of wildlife management units that emphasizes wildlife conservation and its sustainable use. Priority is placed on threatened or endangered, umbrella, or charismatic species of animals and plants and those that possess cultural or economic value (DOF 1999, Instituto Nacional de Ecología 2000). The Secretariat of Environment and Natural Resources (SEMARNAT) is the major federal government department responsible for the protection, restoration, and sustainable use of natural resources, principally fauna and vegetation. Refer to chapter 21 for a review of the federal agencies in Mexico with natural resource administrative and management responsibilities.

The most innovative feature of the 1997 policy was the creation of wildlife management units (officially termed Management Units for Conservation and Sustainable Use of Wildlife [UMAs]). The objective of UMAs was to increase economic incentives for the sustainable management of wildlife by facilitating the integration of wildlife management programs in livestock, forestry, and agricultural schemes. Wildlife (including plants) in UMAs can be used for research, recreational, and environmental education programs, game farms, and commercialization of wildlife, which can be marketed through regulated laws. UMAs can vary in size, depending on management size and economic viability; they must be registered and approved by the Federal Office of Wildlife and must also include a management plan. The management plan must include censusing and monitoring methods, species-specific use criteria, harvest verification, and measures to prevent illegal use. SEMARNAT has initiated a subsidy program to encourage the establishment and improvement of units (DGVS 2015). By 2013, there were 12,000 registered units, and these have been increasing at the rate of 5% per year (Avila-Foucat and Pérez-Campuzano 2015). UMAs are a source of income diversification (ecotourism, sport hunting, commercialization of wildlife by-products), food security (subsistence hunting, captive breeding), and conservation of wildlife and habitats (sustainable use of natural resources). Refer to chapter 21 for a detailed review of wildlife laws, policy, and administration in Mexico, and to Sisk et al. (2008), García-Marmolejo et al. (2008), and chapter 21 for an appraisal of UMAs.

The current wildlife policy is successful because it includes rural ejido communities and private landowners in determining the utilization of wildlife. Rural communities, particularly, are motivated to participate in community-based conservation projects and to benefit economically from the commercialization of wildlife. Landowners and communities have benefited from sport-hunting enterprises because foreign and native hunters are willing to pay high prices for trophy big game mammals. Some livestock enterprises in northern Mexico now prioritize wildlife because of the lucrative earnings from hunting permits for deer and wild sheep (see chapters 3, 17, and 20), and because livestock production is economically unreliable in arid regions. Wildlife in ejidos and private holdings outside protected areas should be incorporated into coordinated wildlife management programs to maintain and restore unprotected species and habitats and to avoid isolating existing protected areas (Valdez et al. 2006). Private landowners and community stakeholders have the greatest potential to sustainably manage and protect wildlife given the limited ability of the government to fund law enforcement and invest in wildlife conservation. It is essential that corporate and nongovernmental sectors also be integrated in land conservation programs because of their involvement in purchasing and managing wildlife habitats, initiating land restoration, and conducting educational programs. Academic institutions are actively involved in wildlife research and can facilitate the establishment of long-term wildlife and habitat monitoring programs and determine the efficacy of conservation programs. Some state governments have assumed leading administrative roles in managing wildlife and will have ever-increasing participation. Refer to Avila-Foucat and Pérez-Campuzano (2015) for the socioeconomic and land use practices that influence the location and development of UMAs, and to chapter 21 for the role of state governments.

Mexico's natural protected areas (ANPs), most of which are inhabited by rural communities with specific limitations on land use and resource extraction, are also important in providing habitats for resident native and migratory wildlife. ANPs include biosphere reserves, national parks, national monuments, sanctuaries, areas for the protection of natural resources, and others specifically for fauna and vegetation. ANPs are valued for the ecosystem services they provide. Because they encompass habitats with a rich fauna, wildlife conservation is often a priority (Martinez 2003). However, natural resource conservation in some ANPs is jeopardized because most are inhabited by increasing populations of native and rural communities, and some form of natural resource exploitation is usually allowed regardless of its sustainability. ANPs encompass 25.4 million ha (62.8 million ac), or 12.9% of the land area of Mexico (CONANP 2014).

Biodiversity, Endangered Species, and Human Population Concerns

Biodiversity refers to the variety within and among organisms and is quantified in terms of genetic variation within species and the number of species or kinds of living organisms inhabiting an area, the number of communities and other biotic assemblages, and the variety of ecosystem functions and structures. Major factors contribute to biodiversity loss in Mexico, including rapid human population growth and associated poverty, lack of education, failure to enforce wildlife protection, and ineffectual government policies. Direct human impacts include subsistence hunting, illegal trade in wildlife and wildlife products, vegetation extraction, and unmanaged industrial and agricultural development that results in unsustainable resource exploitation, increasing waste products, habitat loss, degraded habitats, and extirpation of species (Stedman-Edwards 2000, Stoleson et al. 2005).

Biodiversity loss is one of the most urgent problems confronting wildlife biologists (Primack et al. 2001a, Primack 2014). Biodiversity protection and restoration are essential not only to the survival of wildlife, but also to human societies. The sustainable use of natural resources should take priority over unrestrained economic growth. Mexico's bountiful biodiversity provides a wealth of ecosystem goods and services. One study, in need of being updated, estimated the environmental value of forests alone in Mexico at US$4 billion per year, or US$80 per ha ($32 per ac) (Adger et al. 1995), which is likely an underestimate. Mexican game animals include 40 species of mammals and 55 species of birds, some of which have been locally extirpated outside managed areas. Forty-four percent of the species and subspecies of mammals in Mexico are threatened or at risk of extinction (Ceballos et al. 2014). Mexico's biodiversity should motivate both the internal and foreign tourism sectors to spur rural development, generate nonagricultural-based employment and income, and galvanize greater public awareness of, and support for, the conservation and sustainable management of wildlife. In Mexico, as in other areas of the Northern Hemisphere, biodiversity is greatest in the southern tropical latitudes and gradually decreases toward the northern temperate region.

Endangered species are in danger of extinction throughout all or a portion of their range. Major causes of endangerment are habitat loss and fragmentation resulting from human-induced environmental changes such as agricultural and industrial development; these factors affect both mammals and birds. Fragmentation is an urgent concern. Habitat or movement corridors are essential to ecological processes and the survival of wildlife populations, allowing dispersion of organisms and gene flow from one habitat patch to another and facilitating the colonization of unoccupied habitats (Anderson and Jenkins 2006, Primack 2014). In some cases, international cooperation is necessary for species that cross national boundaries, such as black bears (*Ursus americanus*) along the northern border of Mexico (see chapter 12), large carnivores (see chapter 13), and migrating monarch butterflies (*Danaus plexippus*). Additional factors contributing to endangerment are interspecific interactions, especially those associated with introduced species, and human overexploitation associated with the harvest, collection, or commercial trade of species. Invasive species can cause the extinction of vulnerable endemic species and alter the composition and structure of native communities. Commercial exploitation and collection are particularly detrimental to snails, reptiles, crustaceans, amphibians, charismatic invertebrates (butterflies, other large insects, and tarantulas), and plants (see chapter 3).

Unfortunately, several species of vertebrates in Mexico are extinct. One of the world's most impressive birds, the imperial woodpecker, was the world's largest woodpecker (56–61 cm, 22–24 in) and was endemic to Mexico, but it became extinct because of human depredations (Snyder et al. 2009). The California condor, the largest raptor in North America, was extirpated (eliminated from a portion of its distributional range) in Mexico but was recently reestablished in northern Baja California (Walters et al. 2010). Extirpated larger mammalian species in Mexico include the bison, grizzly bear, wolf, black-footed ferret, monk seal, and sea otter. In Mexico, as in the US, wolves and grizzly bears were eradicated by government and landowner programs directed toward the elimination of predators that preyed on domestic animals (Leopold 1959; Brown 1983, 1985; Ceballos et al. 2014). Many species were extirpated in states where they once occurred, such as desert bighorn sheep in Chihuahua, pronghorn in San Luis Potosí, and black bears in Durango (Medellín et al. 2005), but some species have been successfully restored, including black bears (see chapter 12), wild turkeys (see chapter 9), and desert bighorn sheep (see chapter 20).

Farming and livestock enterprises are the principal sources of land degradation in Mexico. These practices have resulted in the clearing of millions of hectares of native plant cover. The harmful environmental impacts of domestic animals—principally goats, sheep, cattle, horses, and intensively produced pigs and poultry—are a direct result of indiscriminate farming, ranching, and forestry practices. Further habitat loss, fragmentation, and disturbance accrue from dam, canal, fence, and road construction. Soil and water pollution result from agricultural runoff. Land degradation also results from mining, logging, urban expansion, unregulated sport and commercial hunting, unregulated outdoor recreation, introduction of exotic organisms, unregulated oil and gas exploration and extraction, and usurpation of natural watering sources by humans and domestic, feral, and exotic animals (Challenger 1998, Búrquez and Yrízar 2006). Road construction can be particularly detrimental to forest wildlife because it increases accessibility by illegal and commercial hunters and facilitates the transport of wild animal meats and by-products to markets (Primack et al. 2001b, Primack 2014).

Human population growth is the ultimate source of the accelerating socioeconomic, political, and environmental problems confronting natural resource managers. Mexico's population now exceeds 120 million, and the world's population has exceeded 7.5 billion. By 2050, the world's population is projected to reach 9.7 billion, and that of Mexico 165 million. The population of the northern Mexican border states alone exceeds 20 million (Wilson and Lee 2013). The continuously growing human population places greater demands on already overburdened natural resources and often leads to politically expedient decisions without regard for sustainable use. Most of Mexico's poverty is concentrated in rural areas, where many people are engaged in agriculture, and on the peripheries of large urban areas, where there is small-scale agriculture and backyard production of food, including poultry, rabbits, and swine for subsistence and commercial purposes. Overpopulation and overconsumption, along with poverty, result in environmental degradation such as: wildlife habitat destruction and fragmentation; illegal trade in wildlife; deforestation; soil erosion; intensification of natural disasters (storms, flooding, and landslides); loss of biodiversity; soil, air, and water pollution; excessive groundwater extraction; and ultimately a lower standard of living for human societies (Ezcurra et al. 1999, Stoleson et al. 2005, Primack 2014). Water overutilization is a particularly critical issue in Mexico because of the growing population and demands of industry and agriculture; this is of binational urgency in the northern border states. Economic development is especially critical in the rural sector, where the greatest poverty exists and where accelerated habitat degradation is occurring (Velasco Molina 2000, Ocampo Ruiz 2007). The failure of the federal government to incorporate natural resource conservation concerns in economic and social development programs is reflected in the present deteriorated environmental conditions in Mexico (Cuevas 2003).

Exotics

Although the foremost obligation of wildlife managers is to native fauna and plants, they must also focus on exotic species because of the potentially negative impacts exotics can have on native species. In Mexico and the US, some big game ranching enterprises have incorporated exotic ungulates. Although these animals are initially established in enclosures, there is always the danger that they will escape and become invasive species. Exotic large mammals in Mexico that have established wild populations include feral pigs (Solís-Cámara 2009); European wild boar (Weber et al. 2006); European red deer (*Cervus elaphus*); four other exotic deer (Gallina and Escobedo-Morales 2009); the aoudad, an ungulate native to northern Africa (Mungall 2000); and numerous other introduced species (Weber et al. 2006, Álvarez Romero et al. 2010). The nilgai (*Boselaphus tragocamelus*), an antelope native to India and a potential reservoir of bovine babesiosis, a cattle disease, was introduced in Texas in the 1920s, increased to 30,000 animals, and dispersed into Coahuila, Mexico (Cardenas-Canales et al. 2011). The aoudad is widespread in the northern Sierra Madre Oriental, where it could deleteriously impact the reestablishment of desert bighorn sheep (see chapter 20). Exotic small mammals such as house rats and mice, cats, and dogs, together with invasive plants, have become the organisms of greatest concern to ecologists and land managers, especially regarding their destructive impacts in island ecosystems (Gehrt et al. 2010). Relative to aquatic systems, exotic fish species such as tilapia (*Oreochromis* spp.), which have escaped from fish farms, have caused enormous ecological disruption of Mexico's freshwater ecosystems, in some cases displacing native species and altering habitat quality (Primack et al. 2001b, Miller et al. 2005, Primack 2014).

Historical Aspects of Wildlife Conservation in Mexico

In the nineteenth and early twentieth centuries, Mexican hunters wrote some of the earliest popular books describing the game birds and mammals of Mexico (Blázquez 1868, López and López 1911). A. Starker Leopold (1959) wrote the first comprehensive treatise on Mexican game birds and mammals, based on field surveys he conducted principally in the 1940s. This seminal study included descriptive life histories and provided foundational knowledge as well as a stimulus for wildlife conservation in Mexico. In contrast, in the US, several wildlife ecology and management textbooks were published between 1933 and 1960, and several others in subsequent years (Thomas 2007). These were in addition to popular books, which made the public aware of declining wildlife populations (Seton 1898, Hornaday 1913).

Soon after these publications, all states in the US established wildlife agencies, and most states established fish and wildlife curricula in universities that produced undergraduate and graduate fish and wildlife ecologists and managers. These wildlife professionals were hired by state and federal agencies, initially concentrating on providing outdoor recreation opportunities for hunters and fishermen and management information to landowners. State agencies and the US Fish and Wildlife Service provided billions of dollars in research funding that resulted in a voluminous production of technical and popular literature. Wildlife professionals in the US established a national society (the Wildlife Society) in 1936 and initiated the publication of a scientific journal (*Journal of Wildlife Management*) in 1937 to disseminate wildlife research. The public developed a wide interest in wildlife conservation, leading to the creation of numerous wildlife advocacy groups, several of which were established by hunters, such as Ducks Unlimited, the National Wildlife Federation, and the National Wild Turkey Federation (Brown 2013). Increased public pressure on legislative bodies, initiated by hunters, resulted in legislation to protect wildlife at national and state levels, establish protected wildlife areas, and fund the enforcement of federal and state laws (Grosse 1997, Thomas 2007).

In Mexico, the social, economic, and political support necessary for a viable and robust wildlife program did not develop because of socioeconomic factors and disastrous governmental policies for natural resource management. The educational system did not have a strong natural history component, and government at all levels failed to recognize the economic significance of wildlife. In addition, governmental planning failed to integrate wildlife in other natural resource development strategies. Also, socioeconomically and politically urgent problems, including widespread poverty, limited governmental investment in natural resource programs. The noninvolvement of the middle class in sport hunting hindered the development of prohunting advocacy groups and the necessary political support of the urban populace for widespread conservation programs and legislation. Sport hunting was not widespread among Mexico's middle class for a number of reasons, including the absence of public hunting areas, unmanaged and depleted wildlife populations, lack of wildlife law enforcement, and administrative hurdles such as gun ownership restrictions. Consequently, wildlife conservation did not attain widespread citizen appeal and economic importance (Leopold 1959, Simonian 1995, Valdez et al. 2006). Nonetheless, wildlife conservation in

Mexico has developed rapidly and successfully since the 1980s (see chapter 21).

Professional Challenges for Wildlife Biologists in Mexico

Wildlife professionals need an educational background with a broad knowledge base and communication (including interpersonal) skills. Knowledge also accrues with experience. Wildlife managers need to analyze problems from an interdisciplinary perspective. Highly complex and synergistic wildlife management challenges require that professionals work in teams and communicate with specialists in areas of expertise other than their own (Organ 2013). This is not possible in most cases in Mexico because of the need for more trained professionals. Also, our knowledge of the life histories and habitat requirements of Mexican wildlife is still rudimentary for many species and precludes the development of wildlife management strategies based on sound ecological knowledge. Wildlife management in northern Mexico is relatively advanced because the broad knowledge base developed in the US can be applied to shared wildlife species (Demarais and Krausman 2000, Feldhamer et al. 2003). However, most Mexican wildlife species in semitropical and tropical ecosystems, where the greatest number of species occurs, are virtually unstudied (Weber et al. 2006). For example, there are 6 quail species in the US, but 15 species in Mexico (see chapter 7). Most quail species in Mexico occur principally in semitropical and tropical regions, but only one of the species in these ecosystems has been studied. This diversity presents management challenges because of the specific habitat requirements of each species in a community. The species and ecosystem diversity of Mexico places a greater responsibility on Mexican biologists. This requires continual personal effort on their part to increase their knowledge and keep up with the most current wildlife ecology research and management applications.

Future of the Wildlife Profession in Mexico

Wildlife managers are members of a rewarding and challenging profession. Wildlife management is progressing rapidly in Mexico, with expanding opportunities to serve society and conserve the nation's enormous biodiversity. Mexican wildlife biologists, principally in university positions, are making important contributions to several scientific fields, including general ecology, restoration ecology, conservation biology, historical ecology, genetics, ethology, and wildlife diseases, and they are an increasingly vital component in the development of natural resource conservation programs (fig. 1.5). They widely disseminate their research findings to other profes-

Figure 1.5. Wildlife biologists in Mexico conduct research on a wide variety of species, including velvety tree ants, whose larvae are important commercially but have not been managed sustainably. Photo by Luis A. Tarango Arámbula.

sionals, principally through professional journals, books, and regional publications, and are recognized and respected nationally and internationally for their scientific contributions. Mexican wildlife biologists are knowledgeable and dedicated individuals who use a holistic approach (emphasizing the importance of the total biosphere while recognizing the interdependence of its abiotic and biotic components) that emphasizes the integration of all aspects of natural resource management, including the socioeconomic and political. They recognize the urgent need to develop sustainable and economically viable wildlife enterprises in the rural sector to alleviate poverty and curtail the further degradation and loss of wildlife habitats in Mexico.

Leopold (1959) expressed concerns about the ineffectual federal wildlife policy that still prevails in Mexico (lack of federal funding to promote integrated wildlife conservation and research, and inadequate law enforcement and protected areas for wildlife); nonetheless, there are reasons to be optimistic. A growing number of Mexican wildlife ecologists and managers are highly knowledgeable and passionately dedicated to preventing the loss of Mexico's biodiversity. They are fully aware of the ecologically and socioeconomically disastrous consequences of further deterioration and loss of the nation's biodiversity, and the important and challenging tasks they must undertake to conserve and manage the wildlife resources of Mexico. They are willing to actively share their knowledge and passion for the wildlife of Mexico by developing partnerships with experts in other disciplines, policy makers, and the public to maximize natural resource conservation efforts in Mexico. There is a growing job market in the private, communal, and government sectors. Several universities now teach courses in wildlife ecology, and there is a growing need to produce more university-trained wildlife biologists to meet present and future market needs. This will also require additional educators with expertise in wildlife conservation.

The foremost incentive for wildlife professionals in Mexico is the opportunity to experience Mexico's vast plant and animal diversity and the nation's spectacular landscapes, with their attendant temperate and tropical forests, grasslands, aquatic habitats, and deserts amid myriad elevational and climatic gradients. It is gratifying and rewarding for wildlife biologists to devote their professional lives to the adventurous pursuit of knowledge and service in one of the world's richest nations in cultural and wildlife resources. We can be confident that they will continue to make every effort to ensure that future generations of Mexican citizens will have the opportunity to enjoy and benefit from their bountiful wildlife resources.

Literature Cited

Adams, C. E. 2016. Urban wildlife management. Third edition. CRC Press, Boca Raton, Florida, USA.

Adger, W. N., K. Brown, R. Cervigini, and D. Moran. 1995. Total economic value of forests in Mexico. Ambio 24:286–296.

Álvarez Romero, J., R. A. Medellín, A. Oliveras de Ita, H. Goméz de Silva, and O. Sánchez. 2010. Animales exóticos en México: una amenaza para la biodiversidad. CONABIO, México, D. F., México.

Anderson, A., and C. N. Jenkins. 2006. Applying nature's design: corridors as a strategy for biodiversity conservation. Columbia University Press, New York, New York, USA.

Anderson, E. N., and F. Medina Tzuc. 2005. Animals and the Maya. University of Arizona Press, Tucson, USA.

Arreguín Cortés, F. I., M. López Pérez, and H. Marengo Mogollón. 2011. Mexico's water challenges for the 21st century. Pages 21–38 *in* U. Oswald Spring, editor. Water resources in Mexico: scarcity, degradation, stress, conflicts, management, and policy. Springer, Berlin, Germany.

Avila-Foucat, V. S., and E. Pérez-Campuzano. 2015. Municipality socioeconomic characteristics and the probability of occurrence of wildlife management units in Mexico. Environmental Science and Policy 45:146–153.

Bezaury Creel, J. E. 2009. El valor de los bienes y servicios que las áreas naturales protegidas proveen a los mexicanos. The Nature Conservancy and CONANP, México, D. F., México.

Blázquez, P. 1868. El cazador mexicano o el arte de la caza en México y en sus relaciones con la historia natural. Privately printed, Puebla, México.

Bleich, V. C. 2012. Managing rangelands for wildlife. Pages 74–94 *in* N. J. Silvy, editor. The wildlife techniques manual: management. Seventh edition. Volume 2. Johns Hopkins University Press, Baltimore, Maryland, USA.

Blouet, B. W., and O. M. Blouet. 2015. Latin America and the Caribbean: a systematic and regional survey. Seventh edition. John Wiley and Sons, Hoboken, New Jersey, USA.

Bohlen, P. J., and G. House, editors. 2009. Sustainable agroecosystem management: integrating ecology, economics, and society. CRC Press, Boca Raton, Florida, USA.

Boyer, C. R. 2012. The cycles of Mexican environmental history. Pages 1–21 *in* C. R. Boyer, editor. A land between waters: environmental histories of modern Mexico. University of Arizona Press, Tucson, USA.

Bray, D. B., and M. B. Wexler. 1996. Forest policies in Mexico. Pages 217–228 *in* R. Randall, editor. Changing structure in Mexico. M. E. Sharpe, Armonk, New York, New York, USA.

Brown, D. E., editor. 1983. The wolf in the Southwest: the making of an endangered species. University of Arizona Press, Tucson, USA.

Brown, D. E. 1985. The grizzly in the Southwest. University of Oklahoma Press, Norman, USA.

Brown, R. D. 2013. The history of wildlife conservation in North America. Pages 6–23 *in* P. R. Krausman and J. W. Cain III, editors. Wildlife management and conservation: contemporary principles and practices. Johns Hopkins University Press, Baltimore, Maryland, USA.

Búrquez, A., and A. M. Yrízar. 2006. Conservación, transformación del paisaje y biodiversidad en el noroeste de México. Pages 85–110 *in* K. Oyama and A. Castillo, coordinators. Manejo, conservación y restauración de recursos naturales en México. Siglo Veintiuno, México, D. F., México.

Cappock, D. L., et al. 2017. Rangeland systems in developing nations: conceptual advances and societal implications. Pages 569–641 *in* D. D. Briske, editor. Rangeland systems: processes, management and challenges. Springer, Dordrecht, Germany.

Cardenas-Canales, E. M., et al. 2011. Nilgai antelope in northern Mexico as a possible carrier for cattle fever ticks and *Babesia bovis* and *Babesia bigemina*. Journal of Wildlife Diseases 47:777–779.

Ceballos, G., J. Arroyo-Cabrales, R. A. Medellín, L. Medrano González, and G. Oliva. 2014. Diversity and conservation. Pages 1–44 *in* G. Ceballos, editor. Mammals of Mexico. Johns Hopkins University Press, Baltimore, Maryland, USA.

Challenger, A. 1998. Utilización y conservación de los ecosistemas terrestres de México: pasado, presente y futuro. CONABIO, Instituto de Biología UNAM, and Agrupación Sierra Madre, México, D. F., México.

Clout, M. N., and P. A. Williams. 2009. Invasive species management: a handbook of principles and techniques. Oxford University Press, Oxford, United Kingdom.

CONANP (Comisión Nacional de Áreas Naturales Protegidas). 2014. http://www.conanp.gob.mx/english.php. Accessed 16 Mar 2016.

Conover, M. R. 2001. Resolving human-wildlife conflicts: the science of wildlife damage control. Lewis Publisher, Boca Raton, Florida, USA.

Conover, M. R, and R. M. Vail. 2015. Human diseases from wildlife. CRC Press, Boca Raton, Florida, USA.

Crosby, A. W. 1972. The Columbian exchange: biological and cultural consequences of 1492. Greenwood Press, Westport, Connecticut, USA.

Crosby, A. W. 1986. Ecological imperialism: the biological expansion of Europe, 900–1900. Cambridge University Press, New York, New York, USA.

Cuevas, M. de L. N. 2003. La sustentabilidad como estrategia de desarrollo en la frontera norte de México. Pages 163–179 *in* H. M. Cappello, coordinator. Nuevos paradigmas sobre la frontera Estados Unidos-México: problemas asociados a una larga transición. UNAM, Cuernavaca, México.

Decker, D. J., S. J. Riley, and W. F. Siemer. 2013. Human dimensions of wildlife management. Pages 34–50 *in* P. R. Krausman and J. W. Cain III, editors. Wildlife management and conservation: contemporary principles and practices. Johns Hopkins University Press, Baltimore, Maryland, USA.

Demarais, S., and P. R. Krausman, editors. 2000. Ecology and management of large mammals in North America. Prentice Hall, Upper Saddle River, New Jersey, USA.

DGVS. 2015. Lineamientos para otorgar subsidios para el fomento a la conservación y aprovechamiento sustentable de la vida silvestre nativa en UMA en zonas y comunidades rurales de la República Mexicana. Ejercicio 2015. SEMARNAT, México, D. F., México.

DOF. 1999. Acuerdo de la SEMARNAP por el que se crea el comité técnico consultivo nacional para la recuperación de especies prioritarias. DOF, México, D. F., México.

Dunning, N., and T. Beach. 2000. Stability and instability in prehispanic Maya landscapes. Pages 179–202 *in* D. L. Lentz, editor. Imperfect balance: landscape transformation in the precolumbian Americas. Columbia University Press, New York, New York, USA.

Echeverría, C., L. Cayuela, R. H. Manson, D. A. Coomes, A. Lara, J. M. Rey-Benayas, and A. C. Newton. 2007. Spatial and temporal patterns of forest loss and fragmentation in Mexico and Chile. Pages 14–42 *in* A. C. Newton, editor. Biodiversity loss and conservation in fragmented forest landscapes: the forests of montane Mexico and temperate South America. CAB International, Oxfordshire, United Kingdom.

Ezcurra, E., M. Mazari-Hiriart, I. Pisanty, and A. G. Aguilar. 1999. The Basin of Mexico: environmental issues and sustainability. United Nations University Press, New York, New York, USA.

Feldhamer, G. A., B. C. Thompson, and J. A. Chapman, editors. 2003. Wild mammals of North America: biology, management, and conservation. Second edition. Johns Hopkins University Press, Baltimore, Maryland, USA.

Gallina, S., and L. A. Escobedo-Morales. 2009. Análisis sobre las unidades de manejo (UMAs) de ciervo rojo (*Cervus elaphus* Linnaeus, 1758) y wapiti (*Cervus canadensis* Linnaeus, 1758). Tropical Conservation Science 2:251–265.

García-Marmolejo, G., G. Escalona-Segura, and H. Van Der Wal. 2008. Multicriteria evaluation of wildlife management units in Campeche, Mexico. Journal of Wildlife Management 72:1194–1202.

Gehrt, S. D., S. P. D. Riley, and B. L. Cypher, editors. 2010. Urban carnivores: ecology, conflict and conservation. Johns Hopkins University Press, Baltimore, Maryland, USA.

Geist, H. J., and E. F. Lambin. 2002. Proximate causes and underlying driving forces of tropical deforestation. Bioscience 52:143–150.

Grosse, W. J. 1997. The protection and management of our natural resources, wildlife, and habitat. Oceana, Dobbs Ferry, New York, USA.

Harvey, N. 1996. The reshaping of agrarian policy in Mexico. Pages 103–110 *in* R. Randall, editor. Changing structure in Mexico. M. E. Sharpe, Armonk, New York, New York, USA.

Hodell, D. A., M. Brenner, and J. H. Curtis. 2000. Climate change in the northern tropics and subtropics since the last Ice Age. Pages 13–38 *in* D. L. Lentz, editor. Imperfect balance: landscape transformation in the precolumbian Americas. Columbia University Press, New York, New York, USA.

Hornaday, W. T. 1913. Our vanishing wild life: its extermination and preservation. Charles Scribner's Sons, New York, New York, USA.

Instituto Nacional de Ecología. 2000. Estrategia nacional para la vida silvestre. Instituto Nacional de Ecología, México, D. F., México.

Leopold, A. S. 1959. Wildlife of Mexico: the game birds and mammals. University of California Press, Berkeley, USA.

Llorente-Bousquets, J., and S. Ocegueda. 2008. Estado del conocimiento de la biota. Pages 283–322 *in* J. Soberon, G. Halffter, and J. Llorente-Bousquets, compilers. Capital natural de México. Volume 1. CONABIO, México, D. F., México.

López, C. M., and C. López. 1911. Caza mexicana. Librería de la Vda. de C. Bouret, Paris, France.

Mangun, W. R. 1992. Fish and wildlife policy issues. Pages 3–32 *in* W. R. Mangun, editor. American fish and wildlife policy: the human dimension. Southern Illinois University Press, Carbondale, USA.

Mann, C. C. 2005. 1491: new revelations of the Americas before Columbus. Alfred A. Knopf, New York, New York, USA.

Martínez, A. F., coordinator. 2003. Informe de la situación del medio ambiente en México, compendio de estadísticas ambientales. 2002. SEMARNAT, México, D. F., México.

Martínez, M. L., R. H. Manson, P. Balvanera, R. Dirzo, J. Soberón, L. García-Barrios, M. Martínez-Ramos, P. Moreno-Casasola, L. Rosenzweig, and J. Sarukhán. 2006. The evolution of ecology in Mexico: facing challenges and preparing for the future. Frontiers in Ecology and the Environment 41:259–267.

McCleery, R. A., C. E. Moorman, M. C. Wallace, and D. Drake. 2012. Managing urban environments. Pages 169–191 *in* N. J. Silvy, editor. The wildlife techniques manual: management. Seventh edition. Volume 2. Johns Hopkins University Press, Baltimore, Maryland, USA.

McNeely, J. A., K. R. Miller, W. V. Reid, R. A. Mittermeier, and T. B. Werner. 1990. Conserving the world's biodiversity. IUCN, World Resources Institute, Conservation International, World Wildlife Fund–US, and World Bank, Gland, Switzerland.

Medellín, R. A., C. Manterola, M. Valdéz, D. G. Hewitt, D. Doan-Crider, and T. E. Fulbright. 2005. History, ecology, and conservation of the pronghorn antelope, bighorn sheep, and black bear in Mexico. Pages 387–404 in J.-L. E. Cartron, G. Ceballos, and R. S. Felger, editors. Biodiversity, ecosystems, and conservation in northern Mexico. Oxford University Press, New York, USA.

Melville, E. G. K. 1994. A plague of sheep: environmental consequences of the conquest of Mexico. Cambridge University Press, Cambridge, United Kingdom.

Melville, E. G. K. 2000. Disease, ecology, and the environment. Pages 203–234 *in* W. H. Beezley and M. C. Meyer, editors. The Oxford history of Mexico. Oxford University Press, Oxford, United Kingdom.

Miller, R. R., with the collaboration of W. L. Minckley and S. M. Norris. 2005. Freshwater fishes of Mexico. University of Chicago Press, Illinois, USA.

Morrison, M. L. 2013. Wildlife restoration. Pages 246–261 *in* P. R. Krausman and J. W. Cain III, editors. Wildlife management and conservation: contemporary principles and practices. Johns Hopkins University Press, Baltimore, Maryland.

Mungall, E. C. 2000. Exotics. Pages 736–764 *in* S. Demarais and P. R. Krausman, editors. Ecology and management of large mammals in North America. Prentice Hall, Upper Saddle River, New Jersey, USA.

Ocampo Ruiz, M. A. 2007. Propuesta para establecer unidades de producción de fauna silvestre en ejidos y comunidades indígenas. Pages 157–172 *in* P. Álvarez-Icaza, F. Chapela, and B. Ortiz, compilers. Perspectivas para el uso sostenible de los recursos biológicos de México en siglo XXI. Universidad Iberoamericana Puebla, Puebla, México.

Organ, J. F. 2013. The wildlife professional. Pages 24–33 *in* P. R. Krausman and J. W. Cain III, editors. Wildlife management and conservation: contemporary principles and practices. Johns Hopkins University Press, Baltimore, Maryland, USA.

Pagden, A. R., editor. 1971. Hernan Cortes: letters from Mexico. Grossman, New York, New York, USA.

Peters, C. M. 2000. Precolumbian silviculture and indigenous management of neotropical forests. Pages 203–223 *in* D. L. Lentz, editor. Imperfect balance: landscape transformations in the precolumbian Americas. Columbia University Press, New York, New York, USA.

Peterson, M. J., and P. J. Ferro. 2012. Wildlife health and disease: surveillance, investigation, and management. Pages 181–206 *in* N. J. Silvy, editor. The wildlife techniques manual: research. Seventh edition. Volume 1. Johns Hopkins University Press, Baltimore, Maryland, USA.

Peterson, M. N., and S. L. Rodriguez. 2012. Human dimensions of wildlife management. Pages 1–20 *in* N. J. Silvy, editor. The wildlife techniques manual: management. Seventh edition. Volume 2. Johns Hopkins University Press, Baltimore, Maryland, USA.

Petit, D., J. F. Lynch, R. L. Hutto, J. G. Blake, and R. B. Waide. 1995. Habitat use and conservation in the Neotropics. Pages 145–197 *in* T. E. Martin and D. M. Finch, editors. Ecology and management of Neotropical migratory birds. Oxford University Press, Oxford, United Kingdom.

Primack, R. 2014. Essentials of conservation biology. Sixth edition. Sinauer Associates, Sunderland, Massachusetts, USA.

Primack, R., R. Rozzi, R. Dirzo, and F. Masardo. 2001a. Extinciones. Pages 133–153 *in* R. Primack et al. Fundamentos de conservación biológica: perspectivas latinoamericanas. Fondo de Cultura Económica, México, D. F., México.

Primack, R., R. Rozzi, F. Massardo, and P. Feinsinger. 2001b. Destrucción y degradación del habitat. Pages 183–221 *in* R. Primack et al. Fundamentos de conservación biológica: perspectivas lationoamericanas. Fondo de Cultura Económica, México, D. F., México.

Randel, C. J., III, H. O. Clark Jr., D. P. Newman, and T. P. Dixon. 2012. Habitat conservation planning. Pages 1–20 *in* N. J. Silvy, editor. The wildlife techniques manual: management. Seventh edition. Volume 2. Johns Hopkins University Press, Baltimore, Maryland, USA.

Redman, C. L. 1999. Human impact on ancient environments. University of Arizona Press, Tucson, USA.

Santos-Fita, D., E. J. Naranjo, and J. L. Rangel-Salazar. 2012. Wildlife uses and hunting patterns in rural communities of the Yucatán Peninsula, Mexico. Journal of Ethnobiology and Ethnomedicine 8:38–55.

Seton, E. T. 1898. Wild animals I have known.

Charles Scribner's Sons, New York, New York, USA.

Simonian, L. 1995. Defending the land of the jaguar. University of Texas Press, Austin, USA.

Sisk, T. D., A. E. Castellanos V., and G. W. Koch. 2008. Ecological impacts of wildlife conservation units policy in Mexico. Frontiers in Ecology and the Environment 5:209–212.

Snyder, N. F. R., D. E. Brown, and K. B. Clark. 2009. The travails of two woodpeckers: ivory-bills and imperials. University of New Mexico Press, Albuquerque, USA.

Solís-Cámara, A. B., G. Arnaud-Franco, S. Álvarez-Cárdenas, P. Galina-Tessaro, and J. J. Montes-Sánchez. 2009. Evaluación de la población de cerdos asilvestrados (*Sus scrofa*) y su impacto en la Reserva de la Biósfera Sierra la Laguna, Baja California Sur, México. Tropical Conservation Science 2:173–188.

Stedman-Edwards, P. 2000. Main findings and conclusions of the root causes project. Pages 59–79 *in* A. Wood, P. Stedman-Edwards, and J. Mang, editors. The root causes of biodiversity loss. Earthscan, London, United Kingdom.

Stoleson, S. H., R. S. Felger, G. Ceballos, C. Raish, M. F. Wilson, and A. Búrquez. 2005. Recent history of natural resource use and population growth in northern Mexico. Pages 52–86 *in* J.-L. E. Cartron, G. Ceballos, and R. S. Felger, editors. Biodiversity, ecosystems, and conservation in northern Mexico. Oxford University Press, New York, New York, USA.

Teer, J. G. 2003. Nonnative large mammals in North America. Pages 1180–1187 *in* G. A. Feldhamer, B. C. Thompson, and J. A. Chapman, editors. Wild mammals of North America: biology: management, and conservation. Second edition. Johns Hopkins University Press, Baltimore, Maryland, USA.

Thomas, J. W. 2007. From the management of single species to ecosystem management. Pages 225–240 *in* T. E. Fulbright and D. G. Hewitt, editors. Wildlife science, linking ecological theory and management applications. CRC Press, Boca Raton, Florida, USA.

Thompson, R., and I. Cassaigne. 2017. The empowerment of livestock owners and the education of future generations to reduce human-feline carnivore conflicts. Pages 413–422 *in* C. Castaño-Uribe, C. A. Lasso, R. Hoogesteijn, and E. Payán-Garrido, editors. Conflicto entre felinos y humanos en América Latina. Instituto de Investigación de Recursos Biológicos Alexander von Humboldt, Bogotá, D. C., Colombia.

Toledo, V. M., B. Ortiz-Espejel, L. Cortés, P. Moguel, and M. J. Ordoñez. 2003. The multiple use of tropical forests by indigenous peoples in Mexico: a case of adaptive management. Conservation Ecology 7:9–19.

Valdez, R., J. C. Guzmán-Aranda, F. J. Abarca, L. A. Tarango-Arámbula, and F. C. Sánchez. 2006. Wildlife conservation and management in Mexico. Wildlife Society Bulletin 34:270–282.

Velasco Molina, H. A. 2000. Sobrevivencia en los semidesiertos mexicanos. AGT, Mexico, D. F., México.

Walters, J. R., S. R. Derrickson, D. M. Fry, S. M. Haig, J. M. Marzluff, and J. M. Wunderle Jr. 2010. Status of the California condor (*Gymnogyps californianus*) and efforts to achieve its recovery. Auk 127:969–1001.

Weber, M., G. García-Marmolejo, and R. Reyna-Hurtado. 2006. The tragedy of the commons: wildlife management units in southeastern Mexico. Wildlife Society Bulletin 34:1480–1488.

Wilson, C. 2017. Growing together: economic ties between the United States and Mexico. Woodrow Wilson International Center for Scholars, Washington, DC, USA.

Wilson, C. E., and E. Lee. 2013. Introduction. Pages 4–19 *in* C. E. Wilson and E. Lee, editors. The state of the border report: a comprehensive analysis of the US-Mexico border. Woodrow Wilson International Center for Scholars, Washington, DC, USA.

Wright, A. 2012. Downslope and north: how soil degradation and synthetic pesticides drove the trajectory of Mexican agriculture through the twentieth century. Pages 22–49 *in* C. R. Boyer, editor. A land between waters: environmental histories of modern Mexico. University of Arizona Press, Tucson, USA.

Natural and Human Landscapes in Mexico

2

Antony Challenger

Humans as an Integral Component of Mexico's Natural Ecosystems

This chapter will explore the historical legacy of biogeography, ecology, and land use on the physical landscape of modern Mexico in terms of how these have shaped Mexico's natural ecosystems and how they have been modified by human activities. Natural ecosystems in this context are ecological communities in dynamic nonequilibrium, a shifting steady-state mosaic dominated by vegetation that (for our purposes) may be called primary. However, they are certainly not pristine in the sense that people are or always have been absent (Bormann and Likens 1994, Lentz 2000a). North America, including what is now Mexico, was settled by hunter-gatherers at the end of the Pleistocene, 13,500–11,000 BP, when ecosystems were consolidating their present distributions, species compositions, and trophic structures following the Late Glacial Maximum. The extinction in Mexico of 78 species of herbivorous megafauna, as the probable consequence of overkill by these hunters (perhaps in tandem with climate change), occurred around this time. These extinctions had profound ecological consequences, particularly for the desert scrub formations, semidesert grasslands, and temperate forests of central and northern regions, from which most of these species were lost, such that human agency as a source of change in Mexico's ecosystems is far from novel (Janzen 1986, Martin 1989, Alroy 2001, Koch and Barnosky 2006, Grayson 2007, Goudie 2013).

Beginning around 9,000–7,000 years ago, crop domestication and the development of agriculture allowed advanced cultures with urban ceremonial centers to flourish in central and southern regions while nomadic hunter-gatherers predominated in the arid north. Humans had been an integral and influential part of Mexico's ecosystems and landscapes since the Holocene began 11,700 years ago (Matsuoka et al. 2002, Kraft et al. 2014, Walker et al. 2009), as evidenced by the discovery of pre-Columbian ruins, even entire cities, under purportedly pristine rainforest (Gómez Pompa and Kaus 1992, Denevan 2000, Lentz 2000a, 2000b).

Human modification of Mexico's ecosystems has generated a wide spectrum of change. These changes include (1) alterations in the relative abundance of wild food plants, game, and predatory animal species in ecosystems used for hunting and gathering by traditional (mostly indigenous) cultures; (2) modifications in the cover, complexity, and/or trophic structure of ecosystems due to harvesting, hunting, and farming activities, with large areas converted temporarily or permanently to cropland and pasture; and (3) ecosystems that have been entirely replaced by anthropogenic landscapes through farming and forestry, urban development, and infrastructure (Challenger 1998) (fig. 2.1).

Figure 2.1. Monoculture cropping that still employs semi-terracing and contour plowing, in a region of pine-oak forest in the vicinity of La Malinche volcano, near the border between Puebla and Tlaxcala. Photo by Antony Challenger.

Climate, Geography, and Major Biomes

Mexico is bisected west to east by the Tropic of Cancer, so prevailing climatic conditions to the south are broadly tropical, while the north is situated in the zone of descending dry air typical of the subtropical anticyclonic belt in which the major arid zones of the world occur (Tamayo 1990). This two-dimensional climatic map is overlain topographically by mountain ranges occurring in a broadly north-south disposition, with an average elevation of 2,000–3,000 m (6,562–9,843 ft), although the tallest volcanoes reach 5,000–6,000 m (16,404–19,685 ft) (Tamayo 1990). The cooler air at these elevations provides temperate to cold conditions, while the rain shadows the mountains cast over adjacent plateaus generate or accentuate regional aridity. Meanwhile, precipitation from moist air masses coming into contact with their windward slopes increases soil moisture availability for vegetation and forms rivers, lakes, wetlands, and lagoons (Tamayo 1990). Together, these factors permit the occurrence in Mexico of nine major biomes (climatically similar vegetation), six of which predominate: desert, temperate forest, tropical forest, grassland, wetland, and Mediterranean vegetation. The remaining three—savanna, boreal forest, and tundra—occur in more restricted areas.

The broad distribution of these biomes in Mexico is such that (1) tropical forests cover coastal plains, interior lowlands, and foothills, mostly south of the Tropic of Cancer; (2) temperate forest is confined to mountain ranges; (3) boreal forest occurs in cool, humid enclaves; (4) tundra occurs on the highest peaks; (5) desert scrub predominates on the Baja California Peninsula, extratropical coastal plains, and most of the Central Mexican Plateau except the western fringe; (6) semidesert grassland occurs in the lee of the Sierra Madre Occidental; (7) Mediterranean vegetation is confined to the northwestern Baja Peninsula; (8) wetlands are confined to low-lying coastal regions and endorheic (closed) basins; and (9) savanna often occurs at the juncture between biomes. This distribution coincides with the notional division between the Nearctic and Neotropical realms (fig. 2.2).

These biomes each comprise several major ecosystems, and these in turn vary from region to region in terms of structure, species composition, and relative species abundance. The beta diversity between and within ecological communities is a major facet of Mexico's biological megadiversity (Whittaker 1960, Espinosa et al. 2008, Challenger and Soberón 2008, Sarukhán et al. 2009). This hierarchical approach is simply a construct to allow an appreciation of the diversity of Mexico's natural landscapes, given that the

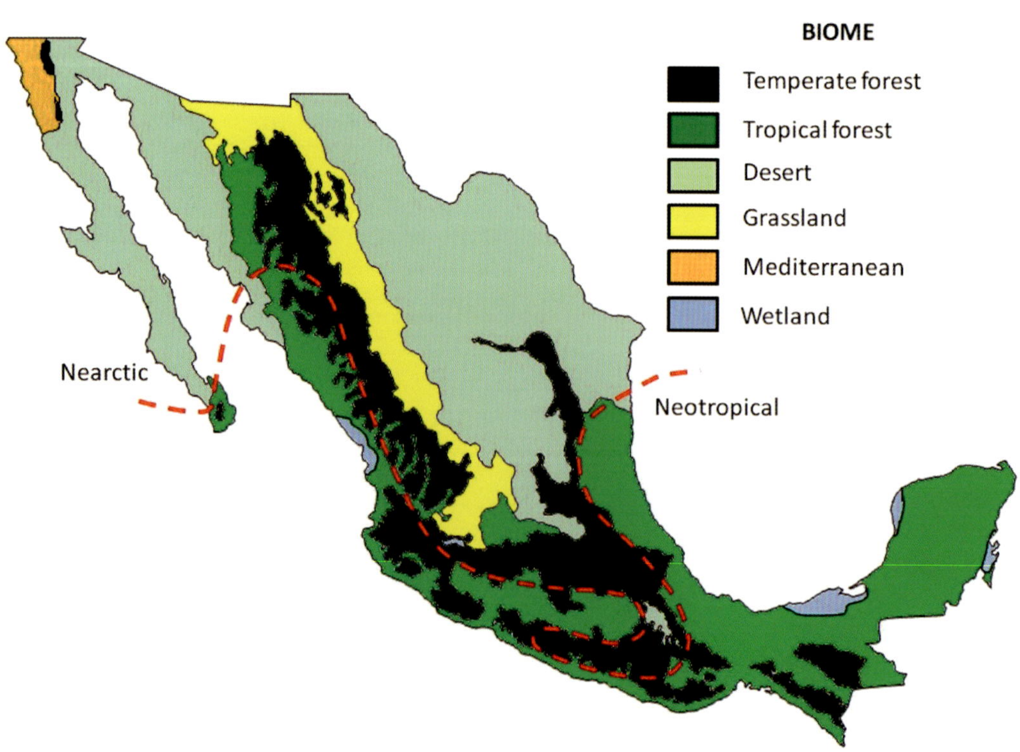

Figure 2.2. Biomes in the Nearctic and Neotropical regions of Mexico. Modified from Challenger (1998, fig. 6, p. 284), and INEGI (2005b).

term “ecosystem” is not constrained by scale: biomes and communities, the biosphere or a drop of pond water are all ecosystems, defined as such only by the boundaries used in defining them for the purpose of study (Tansley 1935, Gignoux et al. 2011). Table 2.1 illustrates these subdivisions and lists 123 vegetation types, based on an analytical review of the literature. Note that many of the English and Spanish terms employed in this table often vary in practice, both between authors and in their usage as applied to different biotic communities. All of these vegetation types, however, have been described in the literature, and most of the names ascribed to them are either in current use or have been used in the past (albeit unsystematically or without consensus). Some of the terms in the table have, however, been modified or coined specifically to provide a technically appropriate term (and/or one that reflects prevailing terminology currently in use in Mexico), which differentiates between vegetation types while conferring conceptual cohesion. In addition to the biomes previously mentioned, table 2.1 includes a subterranean biome because of the importance of Mexico’s cave and other subterranean habitats and their unique flora and fauna.

Mexico’s Natural and Modified Landscapes

In this section, Mexico’s principal natural landscapes are described in outline (in broad hierarchy by biome, ecosystem, and vegetation type), in terms of their structure and species composition, including typical human modifications and uses (Challenger 1998, Challenger and Soberón 2008) (fig. 2.3). Because of space restrictions, not every ecosystem and vegetation type mentioned in table 2.1 is covered.

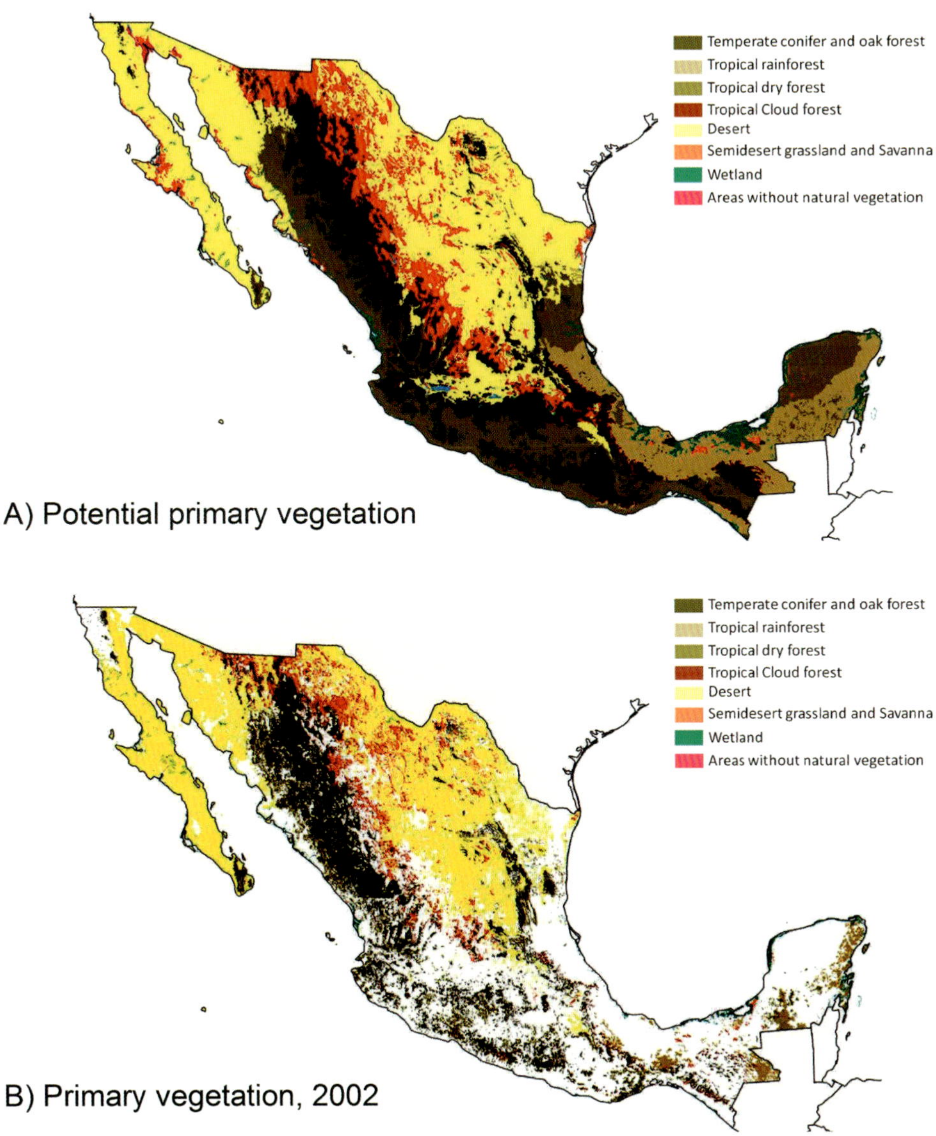

Figure 2.3. Approximate original distribution of Mexico’s principal vegetation types, and the relative reduction in primary (conserved) vegetation resulting from human modification by 2002. From Challenger and Soberón (2008, fig. 3.1, p. 91).

Table 2.1. Mexico's terrestrial biomes, ecosystems, and vegetation types.

Biome	Ecosystem	Vegetation type (number = source)	
		English name	**Spanish name**
Tundra	Alpine tundra	Alpine tundra[1]	Tundra alpina
Grassland	Alpine grassland	Alpine meadow[2, 3]	Pradera de alta montaña
		Alpine bunchgrass land[1]	Zacatonal alpino
	Semidesert grassland	Semidesert grassland[4]	Pastizal semidesértico[5]
		Yucca grassland	Pastizal natural con izotal[7, 8]
		Halophilous grassland	Pastizal halófilo desértico[6, 9]
		Gypsophilous grassland	Pastizal gipsófilo[6, 9]
Boreal forest	Pine-fir forest	Pine-fir forest	Bosque de pino-oyamel[9, 11]
		Pine–Douglas fir forest	Bosque de pino-pseudotsuga[9]
	Fir forest	Fir forest	Bosque de oyamel (o de abeto)[6, 9]
		Douglas fir forest	Bosque de pinabete[5, 9]
		Mixed fir forest	Bosque mixto de pinabete[9]
	Spruce forest	Spruce forest	Bosque de ayarín[6, 9]
Temperate forest	Oak forest	Oak scrub	Chaparral (o encinar arbustivo)[9, 12]
		Oak-palm woodland	Bosque de encino con palma[13]
		Dry oak woodland	Bosque seco de encino[9, 14]
		Oak woodland on lava flow	Encinar de pedregal[15]
		Deciduous oak forest	Bosque de encino caducifolio[16]
		Semideciduous oak forest	Bosque de encino subcaducifolio[9, 17]
		Evergreen oak forest	Bosque de encino perennifolio[9]
		Tropical oak forest	Encinar tropical[18, 19]
		Humid oak forest	Bosque de encino húmedo[9]
	Alder forest	Alder woodland	Bosque de aile[9]
	Oak-pine forest	Oak–pinyon pine forest	Bosque de encino y pino piñonero[9]
		Dry oak-pine woodland	Bosque seco de encino y pino[20]
		Oak-pine forest	Bosque de encino-pino[6, 9]
		Humid oak-pine forest	Bosque de encino-pino húmedo[20]
	Oak-juniper forest	Dry oak-juniper woodland	Bosque seco de encino y enebros[9]
	Juniper forest	Juniper scrub	Matorral de enebros[9]
		Juniper woodland	Bosquete de sabino[9]
		Juniper forest	Bosque de táscate[5, 9]
	Cypress forest	Cypress forest	Bosque de cedro[6, 9]
	Pine-oak forest	Dry pine-oak woodland	Bosque seco de pino y encino[9]
		Pine-oak-palm woodland	Bosque de pino-encino con palma[13]
		Pine-oak forest	Bosque de pino-encino[6, 9]
		Tropical pine-oak forest	Bosque tropical de pino-encino[21]
		Humid pine-oak forest	Bosque de pino-encino húmedo[9]
	Pine forest	Pine scrub	Matorral de pino[6, 9]
		Pinyon pine forest	Piñonar[9]
		Dry pine woodland	Bosque seco de pino[9]
		Pine forest	Bosque de pino[6, 9]
		Pine forest on lava flow	Pinar de pedregal[15]
		High-elevation pine forest	Pinar superior[9]
		Tropical pine forest	Pinar tropical[9, 19]
		Humid pine forest	Bosque de pino húmedo[9]

Biome	Ecosystem	Vegetation type (number = source) English name	Spanish name
Tropical forest	Cloud forest	Pine-oak-*Liquidambar* forest[22]	Bosque de pino, encino y liquidámbar
		Deciduous cloud forest	Bosque deciduo (o de niebla)[23, 24]
		Mexican beech forest	Bosque de haya[25]
		Cloud forest	Bosque mesófilo de montaña[5, 9]
		Oreomunnea forest	Bosque de *Oreomunnea*[21]
		Elfin forest[9]	Bosque enano
		Montane rainforest[26]	Selva alta de montaña
	Tropical rainforest	Lower montane rainforest[26]	Selva alta de montaña baja
		Tall evergreen rainforest	Selva alta perennifolia[6, 23]
		Tall semievergreen rainforest	Selva alta subperennifolia[6, 23]
		Medium evergreen rainforest	Selva mediana perennifolia[6]
		Medium semievergreen rainforest	Selva mediana subperennifolia[6, 23]
		Flooded rainforest	Selva baja perennifolia[6]
		Seasonally flooded rainforest	Canacoital[9, 19]
	Tropical dry forest	Semideciduous tropical forest	Selva mediana subcaducifolia[6, 23]
		Deciduous tropical forest	Selva mediana caducifolia[6]
		Low semideciduous tropical forest	Selva baja subcaducifolia[6]
		Low deciduous tropical forest	Selva baja caducifolia[6, 23]
	Thorn forest	Thorn forest[27]	Selva baja espinosa[6, 9]
		Mesquite woodland	Mezquital[6, 12]
		Tropical mesquite forest	Mezquital tropical[12]
		Seasonally flooded thorn forest	Selva baja subperennifolia (tintal)[6, 19, 28]
	Palm forest	Fan palm forest	Palmar de hoja de abanico[9, 23]
		Feather palm forest	Palmar de hoja pinnada[9, 23]
Wetland	Desert oasis	Nonpalm oasis	Vegetación de oasis sin palmas[37, 38]
		Palm oasis	Vegetación de oasis con palmas[37]
	Sinkhole	Sinkhole vegetation	Vegetación de cenote[36]
	Wet grassland	Wet meadow	Pradera inundable[9, 10]
		Coastal salt grassland	Pastizal halófilo costero[6, 9]
	Riparian vegetation	Tropical gallery forest	Selva de galería[6]
		Temperate gallery forest	Bosque de galería[6, 9]
		Arid-zone gallery vegetation	Vegetación de galería[6]
	Aquatic vegetation	Anchored freshwater vegetation	Popal[6, 9]
		Floating freshwater vegetation	Vegetación flotante[9]
		Submerged freshwater vegetation	Vegetación sumergida[9]
	Reed bed	Sedge or rush bed	Tular[6, 9]
		Reed bed	Carrizal[6, 9]
		Sawgrass bed	Seibadal (o ceibadal o saibadal)[6]
	Swamp forest	Swamp forest	Petén[27]
	Mangrove	Mangrove	Manglar[6, 9]
	Seagrass bed	Seagrass bed	Pradera de pasto marino (ceibadal)[39]
	Coastal sands	Coastal strand	Vegetación de playa[9, 36]
		Foredune vegetation	Vegetación de dunas móviles[6, 9]
		Backdune vegetation	Vegetación de dunas estables[6, 9]

Biome	Ecosystem	Vegetation type (number = source) English name	Spanish name
Savanna	Acacia grassland	Acacia grassland	Pastizal-huizachal[6, 7, 9]
	Mesquite grassland	Mesquite grassland	Pastizal-mezquital[9]
	Savanna	Savanna	Sabana[6, 9, 19]
	Madrean savanna	Madrean savanna	Bosque bajo abierto[6, 9]
	Pine savanna	Pine savanna	Sabana de pino[31]
	Oak savanna	Oak savanna	Sabana de encino[32]
Mediterranean	Mediterranean woodland	Californian pine forest[33]	Bosque mediterráneo de pino[9]
		Californian cypress forest[33]	Bosque mediterráneo de cedro[9]
	Mediterranean chaparral	Coastal sage (or soft chaparral)[34, 35]	Chaparral californiano caducifolio[9]
		Chaparral (or hard chaparral) [35]	Chaparral californiano perennifolio[6, 9]
Desert	Xeric scrubland	Piedmont scrub[29]	Matorral submontano[6, 9]
		Subtropical scrub	Matorral subtropical[6]
		Thorn scrub	Matorral espinoso[9]
		Tamaulipan thorn scrub	Matorral espinoso tamaulipeco[6]
		Mesquite scrub	Mezquital desértico[6, 9]
		Acacia scrub	Huizachal[6, 9]
		Evergreen microphyllous desert[30]	Matorral micrófilo perennifolio[6, 9]
		Deciduous microphyllous desert	Matorral micrófilo caducifolio[6, 9]
	Succulent desert	Prickly pear cactus scrub	Matorral crasicaule tipo *nopalera*[6, 9]
		Columnar cactus desert scrub	Matorral crasicaule tipo *organal*[6, 9]
		Candelabra cactus desert scrub	Matorral crasicaule tipo *cardonal*[6, 9]
		Cholla cactus scrub	Matorral crasicaule tipo *chollal*[6]
		Sarcocaulescent desert[30]	Matorral sarcocaule[6]
		Sarco-crassicaulescent desert	Matorral sarco-crasicaule[6]
		Sarcophyllous desert[30]	Matorral desértico rosetófilo[6, 9]
		Yucca scrub	Izotal[6, 9, 23]
	Azonal lava scrub	San Ángel lava scrub	Pedregal de San Ángel[9, 15]
	Fog desert	Sarco-crassicaulescent fog desert	Matorral sarco-crasicaule de neblina[6]
		Rosetophyllous coastal fog desert	Matorral rosetófilo costero[6]
	Erg desert	Dune field vegetation	Vegetación de desiertos arenosos[6, 23]
	Salt desert	Salt scrub	Vegetación halófila desértica[6, 9, 12]
		Gypsophilous scrub	Vegetación gipsófila[6]
Subterranean	Cave ecosystem	Dry caves[40]	Cuevas
		Flooded caves[40]	Cuevas inundadas

Sources cited:

1. Beaman 1965 cited in Rzedowski 1978, p.224. 21. Miranda and Sharp 1950.
2. Beaman and Andresen 1966, cited in Mc Donald 1993, p. 686. 22. Zuill and Lathrop 1975.
3. Mc Donald 1993. 23. Miranda and Hernández-Xolocotzi 1963.
4. Brown 1982. 24. Williams-Linera 2007.
5. Challenger, 1998. 25. Williams-Linera et al. 2000.
6. INEGI 2005a. 26. Breedlove 1981.
7. Rzedowski and McVaugh 1966, cited in Rzedowski 1978, p.221. 27. Leopold 1959.
8. García Sánchez and Monroy Ata 2005. 28. Flores-Guido and Espejel-Carbajal 1994.
9. Rzedowski 1978. 29. Muller 1939.
10. Bonilla-Barbosa and Novelo Retana 1995. 30. Shreve 1951.
11. Le Sueur 1945, citado en Rzedowski 1978, p.308. 31. Macario Mendoza et al.2005.
12. INEGI 2010a. 32. White 1949, cited in Rzedowski 1978, p. 267.
13. Quero 1994. 33. Minnich 2007.
14. Klink 1973, cited in Rzedowski 1978, p.278. 34. Epling and Lewis 1947, cited in Rzedowski 1978, p.257.
15. Rzedowski 1954. 35. Rundel 2007.
16. Reyes Jaramillo 2006. 36. Berlanga-Robles et al. 2008.
17. Ibarra Cerdeña, 2001. 37. Arriaga et al. 1997.
18. Sousa 1968. 38. INE-SEMARNAP 1999.
19. Pennington and Sarukhán 1998. 39. Huerta 1978.
20. Martin 1958, cited in Rzedowski 1978, p.275.
21. Miranda and Sharp 1950.
22. Zuill and Lathrop 1975.
23. Miranda and Hernández-Xolocotzi 1963.
24. Williams-Linera 2007.
25. Williams-Linera et al. 2000.
26. Breedlove 1981.
27. Leopold 1959.
28. Flores-Guido and Espejel-Carbajal 1994.
29. Muller 1939.
30. Shreve 1951.
31. Macario Mendoza et al.2005.
32. White 1949, cited in Rzedowski 1978, p. 267.
33. Minnich 2007.
34. Epling and Lewis 1947, cited in Rzedowski 1978, p.257.
35. Rundel 2007.
36. Berlanga-Robles et al. 2008.
37. Arriaga et al. 1997.
38. INE-SEMARNAP 1999.
39. Huerta 1978.
40. Tercafs 2001.

Tundra biome

Alpine tundra is confined to frigid environments on the highest peaks of the Neovolcanic Axis, principally the Nevado de Toluca, Popocatépetl, Iztaccíhuatl, La Malinche, Nevado de Colima, and Pico de Orizaba volcanoes. It comprises sparse, open communities of lichens, mosses, tussock grasses (e.g., *Festuca livida, F. tolucensis*) and dwarf plants like sandwort (*Arenaria bryoides*), barely covering more than a fraction of the loose volcanic ash, gravel, and rock. Most of these dwarf plants are perennial and exhibit a rosette form or "pincushion" habit, and their widely separate distributions on "sky islands" (Howell 1947) have favored the evolution of some as narrow endemics, while retaining clear affinity with the North American alpine flora (Scott 1995). Nevertheless, the relative youth of Mexico's volcanic mountains, and the wide geographic and climatic separation of these habitats and floras from similar ones, have reduced inward migration, so these communities tend to be relatively depauperate (McDonald 1993, Moore 2008). Fauna is not abundant in alpine tundra, but insects (including butterflies, in summer), birds (including predators such as hawks), and mammals (particularly rodents) are active during the day (Moore 2008). While the tundra biome is not appropriated for human use, incidental damage by mountaineers and tourists may be locally significant. Climate change poses the biggest long-term threat as higher temperatures threaten to eliminate this vegetation type from mountain summits (Rzedowski 1978, Villers and Trejo 2004).

Grassland biome

Alpine grassland ecosystems comprise two vegetation types: alpine meadows form short swards dominated by creeping dicots, including many endemics (*Argemone subalpina, Sedum clausenii, Thelesperma graminiformis, Poa mulleri*, and others), and alpine bunchgrass lands are dominated by tussock-forming grasses (e.g., *Muhlenbergia* spp.) with erect, flowering dicots such as *Eryngium proteiflorum*. The former occur mostly on mountaintops in the Sierra Madre Oriental, and the latter on peaks of the Neovolcanic Axis, although they coexist at some localities (Rzedowski 1978, McDonald 1993). Their restricted distributions represent a conservation priority, although this is largely met by existing protected areas (Koleff et al. 2009). Fauna includes diverse rodents and lagomorphs with some endemic genera (*Neotomodon*) and species, notably the Mexican volcano mouse (*N. alstoni*) and the volcano rabbit (*Romerolagus diazi*), both of which are restricted to alpine grasslands in the Neovolcanic Axis and listed as endangered (Fa and Morales 1993, Ceballos et al. 2005, Romero and Cervantes 2005).

Alpine grasslands in Mexico have a potential distribution (calculated in the absence of human intervention in the original vegetation) of over 21,400 ha (52,881 ac), which by 2007 was reduced to less than 17,000 ha (42,000 ac) (INEGI 2005b, 2010b). This reduction is the result of human disturbance, which includes mountaineering and tourism, the construction of microwave towers, and fires. Campers cause some fires, but most are set to burn dead grass and stimulate new growth for grazing livestock, the prevalent activity in these ecosystems. Climate change poses the greatest long-term threat (McDonald 1993, Villers and Trejo 2004).

Semidesert grassland ecosystems include four main vegetation types: gypsophilous grassland, halophilous grassland, yucca grassland, and semidesert grassland Gypsophilous and halophilous grasslands occur in endorheic basins in very arid areas, with tobosa grass (*Hilaria mutica*) and the grama grass *Bouteloua chasei* often the respective dominant species, many of the other species being endemic. Yucca grassland (in which species of *Yucca* dominate visually) occurs principally in the Central Mexican Plateau, in the lee of the Sierra Madre Oriental, and on its foothills further south, while semidesert grassland, which also occurs mostly in the Central Mexican Plateau, is found as a continuous northwest-southeast fringe in the lee of the Sierra Madre Occidental, from Chihuahua to Jalisco, and in northern Sonora and parts of Guanajuato. These grasslands are dominated by grasses in the genus *Bouteloua* (blue grama, *B. gracilis*, being the most prevalent), which reaches its highest diversity here, where it probably evolved; these grasses typically form natural swards 20–70 cm (8–28 in) tall (Rzedowski 1978, INEGI 2005a).

Vertebrate diversity (particularly of endemic species), including that of birds and, to a lesser extent, herpetofauna, is lower in these ecosystems than in others, partly due to their lack of structural complexity and partly because of the extinction of native megafauna during the Pleistocene. Mammal diversity is moderately high among small to medium-sized native grazers, burrowers, and rodents and their respective predators.

Semidesert grassland is used for extensive livestock production (mostly cattle ranching), typically with scant planning of grazing regimes, stocking densities that are frequently too high or too low, and inappropriate (European) livestock breeds. This has led to a reduction in the cover, height, and diversity of the sward, a loss of palatable annual grasses (overgrazed), and an increase in less palatable perennial species (undergrazed) (Rzedowski 1978, Challenger 1998, Savory 1999). Near-continuous grazing by cattle is common and is indicated by mats of dead grass, patches of bare earth, severe soil compaction, increased runoff and soil erosion, and an increasing abundance of invasive woody and thorny shrubs like mesquite (*Prosopis* spp.) and acacia (*Acacia* spp.), and some species of cacti (Rzedowski 1978, INEGI 2005a, Savory 1999, Yerena Yamallel et al. 2014). Fire is regularly used to burn off dead plants and stimulate the growth of palatable, fresh grass, and subsidies from federal and state agricultural ministries encourage the sowing of exotic grasses, purportedly to improve the forage quality and productivity of these grasslands. Some of these exotic grasses, typically African buffelgrass (*Cenchrus ciliaris*), outcompete native species, directly or through more efficient fire-ecology adaptations. These exotics become invasive, changing the composition and ecology of semidesert grasslands over vast areas (Búrquez-Montijo et al. 2002, Marshall et al. 2012). Elsewhere, rain-fed agriculture replaces grassland communities entirely, causing localized species extirpation.

Of a potential cover of 16.2 million ha (40 million ac) (around 8.4% of Mexico's land area), semidesert grassland occupied just under 11.9 million ha (29.4 million ac) in 2007, of which only 7.9 million ha (19.5 million ac) was considered primary vegetation, which is likely to be an overestimate (INEGI 2005b, 2010b).

Boreal forest biome

Analogous to fir-dominated taiga in the northern US, Canada, Scandinavia, and Russia, boreal forests in Mexico grow in warm conditions compared to the dark, frigid climes above the Arctic Circle, allowing them to photosynthesize and grow year-round. Without the risk of tree collapse from accumulated snow, they form dense, closed-canopy forests with often sparse understory vegetation, in contrast to the open communities of true taiga. Unlike taiga, boreal forest in Mexico never forms the tree line on high mountains, an ecological role fulfilled by a hardy, high-elevation pine, *Pinus hartwegii*. Confined to cool, humid, temperate mountain slopes and ravines, or surrounding intermontane lakes, boreal forests occur mainly in the Neovolcanic Axis and Sierra Madre Occidental. Dominant taxa are fir (*Abies* spp.) and Douglas fir (*Pseudotsuga* spp.), and local spruce (*Picea* spp.) (Leopold 1959, Rzedowski 1978, Challenger 2003).

Fir forest ecosystems comprise the predominant form of boreal forest in Mexico, although *spruce forest* ecosystems and *pine-fir forest* ecosystems also occur (fig. 2.4). In central Mexico, fir forests are mostly pure stands of sacred fir (*Abies religiosa*), with a ground flora of bunchgrasses, many flowering herbs

Figure 2.4. Pine-fir forest of the Neovolcanic Axis of central Mexico, near Cerro Pelón in the mountains southeast of Zitácuaro, Michoacán. Photo by Antony Challenger.

and shrubs of boreal affinity (*Lonicera*, *Arbutus*, *Penstemon*, *Cirsium*, and *Potentilla*, among others), and some Neotropical temperate taxa such as *Fuchsia* and *Eupatorium*. Fungi are also diverse (e.g., *Amanita* spp., *Boletus* spp., *Russula* spp.), including mycorrhizal fungi that form symbiotic associations with the growing trees (Andrade Torres et al. 2009).

In the Sierra Madre Occidental, either Douglas fir or spruce replaces fir in the *Abies* genus, forming pure or mixed stands (Rzedowski 1978). Faunistically, the fir forests of central Mexico are famous as the winter habitat for the monarch butterfly (*Danaus plexippus*), a migrant from Canada and the northeastern US, but fir forests are rarely the preferred habitat of large species. Nevertheless, as these forests occur as islands in a matrix of predominantly pine forest, they undoubtedly form part of the wider habitat of the coyote (*Canis latrans*), black bear (*Ursus americanus*), puma (*Puma concolor*), and woodpecker (Picidae), among other fauna.

Human uses of boreal forest are diverse. They include (1) the commercial harvesting of trees for construction, timber, paper pulp, cellulose, and essential oils (terpene, turpentine); (2) recreation (surviving fir forests of the Neovolcanic Axis are all in protected areas); (3) urbanization, particularly in the mountains between Mexico City and Toluca, and in the Valle de Bravo tourist area, both in the state of Mexico; (4) sheep farming, which requires understory grasses and shrubs to be burned periodically to stimulate palatable regrowth, at the risk of igniting forest fires (which are infrequent in these humid environments); and (5) conversion to rain-fed, mostly subsistence agriculture (Rzedowski 1978, Challenger 1998, 2003).

These competing human uses have had a detrimental effect on Mexico's boreal forests. Timber harvesting in particular has reduced the distribution of Douglas fir and spruce forests, while harvesting for paper pulp and cellulose, and deforestation to allow the expansion of farming and urbanization, have affected fir forests. Of a total potential area of 440,000 ha (1.08 million ac) (only 0.22% of Mexico's total land area), Mexico's boreal forests occupied only 194,000 ha (469,500 ac) in 2007, of which only 154,000 ha (380,542 ac) was primary vegetation (INEGI 2005b, 2010b).

While these human uses continue to exert pressure on boreal forests, like other montane ecosystems they are particularly vulnerable to climate change. So too is their fauna, including the iconic monarch butterfly. Recent research indicates a potential reduction in the ecological niche (modeled using climatic data) required by sacred fir forests in central Mexico, where the monarch overwinters, of 69.2% by 2030, and 96.5% by 2090 (Sáenz Romero et al. 2012).

Temperate forest biome

Temperate forests comprise three main ecosystems: *pine forest*, *oak forest*, and *pine-oak forest* (in some of which pines outnumber oaks and in others, the reverse), which covers the largest area. *Alder woodland*, *juniper forest*, and *cypress forest* ecosystems also occur locally.

Most of Mexico's temperate forests develop in montane areas and are typically 10–25 m (33–82 ft) in height, with a single stratum of canopy trees, a shrub layer, and an herb layer. Oak forest usually develops at lower, warmer elevations and pines at higher elevations, up to the tree line. Pine and mixed-conifer forests are evergreen, but when oaks form part of the canopy, forests may have varying degrees of deciduousness, depending on the species, local climate, and overall dominance of oaks. Species composition differs considerably between regions, and taken together, temperate forests in Mexico have a higher plant diversity (around 7,000 species) than rainforests (Challenger 1998, Rzedowski 1978, 1993).

Oaks (*Quercus* spp.) and pines (*Pinus* spp.) are the dominant trees of Mexico's temperate forests. If either genus dominates the canopy, it may comprise several species or only a single species, a situation particularly characteristic of pine forests. However, mixed forests with varying levels of canopy dominance shared between these two genera predominate, making up 40% of Mexico's temperate forests, with pine forests comprising 23.6% and oak forests 34.8% (INEGI, 2010b). Despite the predominantly boreal affinity and distribution of pines and oaks, Mexico is the second most important center of global pine diversity, with 56 of the world's 115 or so species, over half of which are endemic, and the second most important center of global oak diversity, with 161 of the world's approximately 450 (estimates range from 300 to 600) species, including 109 endemics (Rzedowski 1993, Styles 1993, Nixon 1993, Llorente Bousquets and Ocegueda 2008, Aguirre Gutiérrez and Duivenvorden 2010, Valencia 2004). Other genera of the tree canopy include *Arbutus* (madrone), *Fraxinus* (ash), *Crataegus* (hawthorn), *Prunus* (cherry), *Platanus* (plane), *Populus* (aspen), *Alnus* (alder), *Juniperus* (juniper), and *Cupressus* (cypress), these last three occasionally dominating the canopy to form alder woodland, juniper forests, and cypress forests. Neotropical trees such as *Oreopanax* spp. may occur in humid forests, but most Neotropical taxa occur in the shrub and herb layers, alongside boreal species. Shrubs include *Agave*, *Berberis*, *Cestrum*, *Fuchsia*, *Helianthemum*, *Mimosa*, *Pernettya*, *Rubus*, *Salvia*, and *Vaccinium*, among others. Herb layer taxa include many species of Asteraceae (particularly in the genera *Cirsium*, *Dahlia*, *Gnaphalium*, *Tagetes*, and *Senecio*) and Poaceae (grasses in the genera *Bromus*, *Festuca*, *Hilaria*, *Muhlenbergia*, and *Stipa* are frequent); other typical herb layer genera are *Castilleja*, *Penstemon*, and *Potentilla*, among others. Ferns, mosses, and lichens increase in diversity and abundance with humidity and shade, while lianas and epiphytes such as orchids and bromeliads increase in diversity and abundance with humidity, warmth, and light and are more typical in oak forests (Rzedowski 1978, Challenger 1998).

In terms of fauna, some vertebrate groups reach their highest levels of endemism in temperate forests, including amphibians (frogs and salamanders), reptiles (lizards), mammals (rodents), and birds (Flores Villela 1993, Fa and Morales 1993, Escalante Pliego et al. 1993, Koleff et al. 2008).

Human modification and use of Mexico's temperate forests range from subtle changes in species composition as a result of hunting and plant collecting, to eradication over hundreds of square kilometers by urbanization and irrigated crop monocultures, as in most of the valleys and plains of the Neovolcanic Axis (compare maps in fig. 2.3). Between these extremes are landscapes that have been typical of Mexico for millennia. These include milpa fields—polycultures of maize (*Zea mays*), beans (*Phaseolus vulgaris*), and squash (*Cucurbita pepo*)—on slopes with stone terraces, or semi-terraces (*metepantli*) of century plant (*Agave americana*), interspersed with remnants of primary forest on steep slopes and ravines, and herbaceous secondary vegetation (made colorful with flowers of *Tagetes*, *Cosmos*, and other Asteraceae) in fallow fields. In large areas of the state of Hidalgo, *Agave americana* is cultivated in contour rows on hill slopes, so the sugary sap of their single inflorescence (called a *quiote* in central Mexico) can be harvested and fermented to make pulque, an alcoholic drink consumed since pre-Hispanic times. In the Milpa Alta district south of Mexico City, nopal cacti (*Opuntia ficus-indica*) are grown in fields terraced with lava rock, supplying much of the nation's demand for this traditional vegetable, the flat pads being the articulated stems of this plant (Wilken 1987, Challenger 1998).

Cultivated hill slopes that are cleared of temperate forest and are managed as described above to conserve the soil are the exception, however. In areas afflicted by poverty, population growth, and land hunger, such as the indigenous subsistence farming communities of the Chiapas Highlands, maize is sown on hills so steep that farmers must tie a rope around their waist and attach it to a live tree upslope to provide safety while working on their precipitous fields. Soil erosion is inevitable under these circumstances. It also predominates in other regions where, despite gentler slopes, the pressure of grazing cattle, sheep, or goats, or overuse of the plow, reduces plant cover and soil integrity to the extent that erosion reveals the bedrock of compact volcanic ash (*tepetate*), as in some municipalities of the state of Mexico, the Mixteca region of Oaxaca, and the Montaña region of Guerrero (Challenger 1998).

Agriculture has almost eradicated oak forests from the gentler slopes and intermontane valleys of the Neovolcanic Axis, and many pine forests have been similarly replaced. Those at higher elevations are used for livestock production in the same way described for boreal forest, but they are more prone to forest fires (Rzedowski 1978). However, the fire ecology adaptations of many pine species, together with their small, wind-dispersed seeds, confer on them a trait typical of ecological pioneer species: the ability to rapidly recolonize a deforested site. This allows pine forests to quickly regenerate in situ following their complete removal, as when they are clear-felled for timber, cellulose, chipboard, and plywood (Styles 1993). These are the main products of Mexico's pine forests, which contributed 79% of the 5.9 million m^3 (208.4 million ft^3) of wood produced by Mexico's commercial forestry sector in 2013, making up 87% of the total value of Mexico's forestry sector (SEMARNAT 2014, pp. 11–15). Nearly 70% of this production comes from temperate forests in just 5 of Mexico's 32 states: Durango, Chihuahua, Michoacán (where pine resin is harvested to make turpentine), Oaxaca, and Veracruz

(SEMARNAT 2014). While pines regenerate well under commercial forestry, many other species of flora and fauna do not, as they require microclimates and niches (such as dead and rotten wood) that standard forestry practices eradicate (Duffy and Meier 1992, Challenger and Dirzo 2009).

Mexico's oak forests rarely produce tall trees with straight boles, so they are less sought after than conifers. They contribute only 8.7% of the total volume and 4.7% of the total value of commercial forestry in Mexico, in the form of square timber (36.9% of oak products), charcoal (27.7%), fuel wood (21.6%), cellulose (8.42%), and sleepers (5.2%) (SEMARNAT 2014). Most of this production comes from the forests of the Sierra Madre Occidental, in Durango (34%), Sonora (10.1%), and Chihuahua (8.6%) (SEMARNAT 2014). Pines and oaks also provide fuel wood and charcoal for cooking in rural Mexico; an estimated 19 million tons per year are gathered from all biomes for this subsistence use (Masera et al. 2006, cited in Balvanera et al. 2009). This practice causes forest degradation, as does illegal logging, which affects mostly pine forests, from which an estimated 13 million m^3 (459 million ft^3) of timber are illegally harvested each year. This is more than twice the volume harvested commercially, which is itself low, inefficient, and costly, providing an incentive for illegal extraction (Torres Rojo 2004, cited in Balvanera et al. 2009). The preeminence of pines in the forestry sector means that silvicultural practices tend to eliminate oaks in favor of pines in mixed pine-oak forests, but the extent of the changes wrought by this process is undocumented and almost impossible to calculate (Castillo and Snook 1984).

Much of Mexico's intensive livestock production (cattle, pigs, and poultry) takes place in areas once covered by temperate forest. Animals kept in stables are fed a balanced concentrate, or alfalfa or oats produced locally. Such enterprises use little land but pollute soil and water with runoff from poorly managed manure. Better management of these by-products is now integral to policies to reduce greenhouse gas emissions (Challenger 1998, Poder Ejecutivo Federal 2014).

In all, human modifications of Mexico's temperate forests have had a significant impact. Of a potential area of 43.5 million ha (107.5 million ac), 20.3 million ha (50.1 million ac) of primary forest remained in 2007, with a further 12.1 million ha (29.8 million ac) of secondary vegetation (INEGI 2005b, 2010b). Computer-generated scenarios of climate change indicate these ecosystems to be at high risk (Villers and Trejo 2004).

Tropical forest biome

Tropical forests comprise three principal ecosystems: cloud forest, rainforest, and deciduous tropical forest, with thorn forest and palm forest occurring locally.

Cloud forest is frequently classified with temperate forest in Mexico because species with temperate affinity dominate the canopy and because these ecosystems occur principally, but not exclusively, in the humid temperate climate zone (Koeppen 1948). But tropical montane cloud forests, as the name suggests, are confined to tropical latitudes, and tropical taxa comprise the bulk of subcanopy species, and indeed of canopy species, closer to the equator (Hamilton et al. 1995).

Cloud forest can vary in structure depending on the vegetation type, but it typically comprises a canopy with occasional emergent trees, a subcanopy tree layer, a shrub layer and an herb layer, and abundant epiphytes, both on canopy and subcanopy branches and on tree trunks (especially mosses). Canopy trees include oaks and Lauraceae, such as *Nectandra* spp. and wild avocado (*Persea* spp.), from which domesticated varieties derive, and trees typical of the Appalachians: *Liquidambar*, *Fagus*, *Carpinus*, *Ostrya*, *Acer*, *Nyssa*, and *Magnolia*. Other taxa typical of cloud forests but no longer found elsewhere in North America are *Cleyera*, *Microtropis*, *Oreomunnea*, and *Deutzia*, which are disjunct with respect to their main center of distribution in Southeast Asia. Their presence here is relictual, evidence of the role of the Bering Strait land bridge in allowing them to migrate into North America, and the hot, humid global climate of the Early Tertiary, which led to the widespread distribution of species with humid-tropical affinity (Wendt 1993, Rzedowski 1996, Challenger 1998, 2003). Other trees come from a third source, the temperate and subtropical regions of the Andes, including *Clusia*, *Podocarpus*, *Brunellia*, and *Oreopanax*, which crossed into Mexico via the Central American land bridge. Some pantropical genera occur in the shrub layer (e.g., Melastomataceae), but the Andes are the source of most Neotropical cloud forest understory shrub species (e.g., Acanthaceae and Piperaceae), diverse families in the order Zingiberales (e.g., Marantaceae, Heliconiaceae), and epiphytes (e.g., Bromeliaceae, Orchidaceae, and Araceae). Neotropical taxa of Amazonian affinity are more prevalent as understory herbs, which dominate in diversity over boreal taxa (Rzedowski 1996, Challenger 1998, Gual-Díaz and González-Medrano 2014).

Despite their diversity, trees make up only 18% of cloud forest flora; epiphytes make up the lion's share with 30% of all cloud forest species, including orchids, which dominate among flowering plants; while ferns, both as epiphytes and in the ground flora, make up 20%. It is the epiphytes and ferns (fig. 2.5), particularly tree ferns, together with the all-enveloping mist, that give these forests their unique appearance and feel (Rzedowski 1996; Challenger 1998, 2003, 2014; Tejero-Díez et al. 2014; Villaseñor and Gual-Díaz 2014).

Figure 2.5. Tree ferns (*Sphaeropteris horrida*) in the understory of a cloud forest near Xalapa, Veracruz. Photo by Antony Challenger.

Unique, too, are many species of cloud forest fauna, some restricted to particular patches of cloud forest (Ochoa-Ochoa and Mejía-Domínguez 2014). Among birds, the horned guan (*Oreophasis derbianus*) is endemic to the Sierra Madre of Chiapas and Guatemala; the resplendent quetzal (*Pharomachrus mocinno*) nests only in cloud forest in southern Mexico and Central America, and the endangered azure-rumped tanager (*Tangara cabanisi*) is a narrow endemic of the Chiapas Sierra Madre (Heath and Long 1991, Stattersfield et al. 1998). No large mammals are cloud forest specialists (however, two genera of deer mice are endemic), although many species range into them, either from tropical rainforest or deciduous forest at lower latitudes, or from temperate forest at higher elevations (Leopold 1959, Ceballos and Oliva 2005, González-Ruíz et al. 2014). Amphibians and reptiles reach some of their highest levels of abundance in cloud forests, with many endemic species, particularly among amphibians, which are also highly threatened (Koleff et al. 2008, Gual-Díaz and Goyenechea Mayer-Goyenechea 2014, Goyenechea Mayer-Goyenechea and Gual-Díaz 2014).

Human use and modification of cloud forest remained small scale until relatively recently. These deeply shaded, wet forests on steep, misty slopes, distributed as a fragmented archipelago among separate mountain ranges, were relatively inhospitable for direct human settlement and agriculture in pre-Hispanic times, and none of Mexico's great civilizations were established in them. With the Spanish Conquest, cloud forest on less steep terrain in central Veracruz was felled and urbanized to create the state capital, Xalapa, as well as Córdoba and other towns; elsewhere, inroads were carved piecemeal to provide land for livestock and agriculture (Williams Linera 2007). During the dictatorship of Porfirio Díaz (1877–1910), the first coffee plantations were sown under the shade of cloud forest trees, once the understory vegetation was cleared, initiating their large-scale transformation (Challenger 1998, Challenger and Dirzo 2009). Sowing *Coffea arabica* under cloud forest trees in traditional plantations produces the highest quality coffee and conserves around 70% of cloud forest biodiversity (Pineda et al. 2005). While such plantations still exist in Veracruz and Chiapas, twentieth-century modifications involved replacing cloud forest trees for shade trees of *Inga* spp., a legume that fixes nitrogen, increasing soil fertility and production at the cost of biodiversity. Since the 1980s, coffee producers have been encouraged to completely remove shade trees and grow coffee in dense plantations under direct sunlight. This yields poor-quality coffee but is more productive, so profits are often higher, although many areas of traditional plantations and primary cloud forest have been lost in this way, and species of flora and fauna locally extirpated (Challenger 1998, Challenger and Dirzo 2009).

Population growth in the poor indigenous and mestizo communities of cloud forest areas, together with conflicts over land and land hunger, has meant cloud forest has been increasingly converted to livestock pasture and agriculture. Traditional subsistence agriculture is a derivative of slash-and-burn practices, so this at least allows cloud forest to regenerate during periods in which fields are left fallow, and traditional land use techniques can accelerate the process. For example, oak saplings are often cut at waist height and left to provide stakes for bean plants in traditional polycultures, but once the field is no longer cultivated they quickly develop foliage, accelerating the natural regeneration of the forest and preventing soil erosion on slopes (Challenger 1998).

Of a potential total cover of 3 million ha (7.4 million ac) (1.5% of Mexico's total land area), only 850,000 ha (2.1 million ac) of primary cloud forest remained in 2007, plus an additional 500,000 ha (1.2 million ac) of secondary vegetation (INEGI 2005b, 2010b). This huge loss of cover is especially significant given the staggering diversity of cloud forests between sites, because in less than half of 1% of Mexico's territory, these forests provide habitat for 6,000 plant species (10% of Mexico's flora), around one-third of which are endemic (Rzedowski 1996, Challenger 1998, Sánchez-Ramos and Dirzo 2014). This combination of high biodiversity and extensive habitat loss means that many cloud forest species are endangered. In fact, 71% of cloud forest tree species are considered to be in one of the IUCN's categories of threat, with 37.9% of species considered endangered or critically endangered, and three already extinct (González-Espinosa et al. 2011, Challenger 2011). An additional and growing threat to these temperature- and humidity-sensitive ecosystems, as in other montane ecosystems, is climate change (Villers and Trejo 2004, Challenger and Dirzo 2009). Indeed, cloud forest is considered among the most threatened of all of Mexico's ecosystems (Challenger 1998, Williams Linera 2007, Sánchez-Ramos and Dirzo 2014).

Tropical rainforest once covered the entire Gulf Coastal Plain, from northern Veracruz into the southern Yucatán Peninsula and eastern Chiapas, with a finger of rainforest entering southern Chiapas from Guatemala along the coast and foothills of the Chiapas Sierra Madre. Today, primary rainforest has been reduced to a handful of relictual areas within its former range (fig. 2.3) (Challenger and Soberón 2008).

Rainforests have a complex structure, comprising a canopy with many emergent trees, at least two subcanopy tree strata, a shrub stratum, and an herb stratum. Woody lianas span the entire height of the forest, and their tangled, knotted stems are a feature of the understory. Abundant epiphytes cover canopy branches, and less frequently tree trunks. Canopy trees include *Terminalia amazonia* (which is often an emergent), *Dialium guianense*, *Bursera simaruba* (the gumbo-limbo tree), *Brosimum alicastrum* (breadnut), *Vochysia hondurensis*, *Guatteria anomala*, hardwoods such as *Cedrela odorata* (Mexican cedar) and *Swietenia macrophylla* (big-leaf mahogany), a variety of strangler figs (*Ficus* spp.), and many others. Woody lianas include *Paullinia* spp. and *Petrea* spp., while understory trees include shade-tolerant taxa such as *Cymbopetalum*, *Pimenta dioica* (allspice), and *Nectandra*, saplings of canopy species, palms of *Astrocaryum mexicanum*, and a diversity of climbing, perennial vines in the Arecaceae (e.g., *Monstera*, *Syngonium*, *Philodendron*). The shrub layer is usually sparse and has many species of parlor palm (*Chamaedorea*) and shade-tolerant shrubs of Melastomataceae, *Piper* spp., and Rubiaceae, while the herb layer comprises young plants of canopy trees, small palms, and ferns (e.g., *Diplazium* spp., *Thelypteris* spp.). Characteristic of the understory are the often enormous root buttresses of canopy trees, and underfoot a tangle of shallow, interwoven roots allow leaf litter to accumulate in pockets, providing cover for snakes. Canopy epiphytes include bromeliads, ferns, orchids, cacti, and begonias (Rzedowski 1978, Challenger 1998, INEGI 2005a).

Rainforest has the highest alpha diversity of any terrestrial ecosystem, with hundreds of species per hectare. Soil microbes, fungi, and insects recycle the constant rain of detritus from the forest interior back into the vegetation, while stingless bees, butterflies, biting insects, and beetles, together with scorpions, millipedes, tarantulas, and other spiders, comprise the major invertebrate groups. Tree frogs, lizards, geckos, and snakes contribute to the diversity of herpetofauna, but it is birds that reach levels of diversity in Mexico's rainforests that are unmatched in other ecosystems (Koleff et al. 2008). Mammals also reach high levels of diversity, with many charismatic species exclusive to rainforest, and while few are endemic, there are several endemic subspecies (Leopold 1959, Ceballos and Oliva 2005, Koleff et al. 2008).

Human use and modification of rainforest includes hunting and gathering, the planting of domesticated and semidomesticated plants in *huertos* or home gardens, and slash-and-burn agriculture, which has been used for millennia in tropical regions the world over. It allows parcels of land to be cycled between agricultural and forest uses, in an efficient system (where there is enough rainforest to minimize the negative impacts of deforestation and accidental fires) that uses natural forest regeneration to reclaim the fallowed milpa after three or four years of cropping. By this time, crop yields will have fallen greatly as soil fertility is leached away by high rainfall, crop pests become abundant, and vigorous weedy plants compete with crops for space, light, and soil nutrients. Once secondary forest has regenerated and is judged to contain sufficient biomass and nutrients to provide ash to fertilize the field, the cycle recommences. These traditional uses often occur simultaneously over short distances, producing a landscape mosaic of cropped and fallowed fields, secondary vegetation of different ages, and patches of primary forest, together with palm-thatched houses and huertos (Challenger 1998).

Commercial agriculture in a rainforest, as elsewhere, replaces the diversity and regenerative properties of the forest, which are the foundations of subsistence production, with monocultures. Such monocultures include plantations of papaya, sugarcane, or banana on irrigated land, and woody fruit trees of citrus and mango on slopes, although locally, cacao (*Theobroma cacao*), vanilla (*Vanilla planifolia*), or *Chamaedorea* palms may be grown under shade trees, creating more diverse agroecosystems (Challenger 1998, Challenger and Dirzo 2009). Deforestation and conversion to pasture for cattle ranching is the most ubiquitous use of rainforest land in Mexico and constitutes another form of monoculture. Some pastures comprise native grass species, induced by grazing on secondary vegetation after slash-and-burn activities, but increasingly large areas are sown to nonnative species such as Bermuda grass (*Cynodon nlemfuensis*) and Guinea grass (*Megathyrsus maximus*). Such pastures may support only a head or two of cattle per hectare where once grew the most diverse ecosystem on Earth. Some diversity is lent to these vast landscapes of grass by shade trees (protected when the forest was felled) and living fencerows of introduced *Gliricidia sepium* along field boundaries, or more commonly by cloned native gumbo-limbo trees (*Bursera simaruba*), propagated from branches formed into stakes and planted. Strangler figs, epiphytes, and other plants invade these linear habitats, which provide niches for rainforest species as well as limited landscape connectivity (Challenger 1998, Laborde et al. 2011).

Of a potential cover of 17.8 million ha (43.9 million ac) in a hypothetical scenario before human alteration, by 2007 primary rainforest had been reduced to 3 million ha (7.4 million ac) (17% of its potential area), with a further 6.2 million ha (15.3 million ac) of secondary vegetation, produced and maintained by slash-and-burn farming, now a principal driver of rainforest destruction (INEGI 2005b, 2010b; Challenger and Dirzo 2009). A recent surge in palm oil cultivation (including for biofuel production), backed by government subsidies, is also driving rainforest deforestation in Campeche and parts of Chiapas (Castro Soto 2015, Massieu Trigo and Acuña Rodarte 2015).

Tropical deciduous forest is one of Mexico's most widespread and characteristic ecosystem types. The following description also includes the thorn forest vegetation type, as it is ecologically similar to and geographically contiguous with deciduous forest, but not the other three vegetation types of the thorn forest ecosystem listed in table 2.1.

Equivalent to Asian monsoon forests, Mexico's tropical deciduous forests or seasonally dry tropical forests (fig. 2.6) (Mooney et al. 1995) are the most northerly expression of tropical forest in North America, and perhaps the most extensive

Figure 2.6. Tropical dry forest consisting of conserved primary vegetation near Huatulco on the Pacific coast of Oaxaca. Photo by Antony Challenger.

representation of these ecosystems in Latin America (Trejo 2010). Typically developing on the rolling topography of piedmont slopes, they produce a landscape of seasonal contrasts, verdant with foliage in the wet season and a mass of drab, leafless branches in the dry season, save for a scattering of gray-green columnar and candelabra cacti, which are a dominant feature of the canopy in all but the most humid subtypes (Trejo 2010). These forests are shorter in stature than rainforest, with a more level canopy and few emergent trees (save the endemic *Conzattia multiflora*), usually with shrub and herb layers but occasionally without one or both. Tropical deciduous forest dominates the Pacific Coastal Plain from Chiapas to Sinaloa, as well as the Balsas Depression, parts of the northern and central Gulf Coastal Plain, the northern Yucatán Peninsula, and the Central Depression of Chiapas (Challenger and Soberón 2008, Bezaury Creel 2010). Mexico's principal crops of maize, beans, and squash have their wild relatives in these ecosystems, where it is probable that Mesoamerican agriculture had its origins and initial development, and they have a long history of human modification (Challenger 1998, Smith 2001, Matsuoka et al. 2002, Kraft et al. 2014).

Typical canopy trees include a few near-evergreen legumes such as Madras thorn (*Pithecellobium dulce*) and guanacaste (*Enterolobium cyclocarpum*), and palms such as *Attalea cohune*, together with many species of deciduous and semideciduous leguminous trees, which frequently dominate in abundance (Trejo 2010, Sousa 2010). Among these are *Acacia* spp., as well as mesquite (*Prosopis* spp.), leadtree (*Leucaena* spp.), *Lonchocarpus*, *Erythrina*, *Calliandra*, *Mimosa*, *Caesalpinia*, and *Lysiloma*, among many others, most of which flower showily in the dry season (Sousa 2010). Several species of fig (*Ficus* spp.) occur, both hemiepiphytic stranglers and free-growing nonstranglers, as do breadnut (*Brosimum alicastrum*), *Cordia* spp., *Malpighia* spp., many Burseraceae (typically *Protium* spp. and *Bursera* spp.), *Ceiba aesculifolia*, and *Bombax* spp. Species like *Cnidoscolus* spp. with spines, thorns, or urticating hairs may also be abundant, particularly in thorn forest. Other trees include relatives of domesticated fruit trees, such as *Jacaratia mexicana*, a relative of papaya, and the wild parent of the domesticated hog plum, *Spondias purpurea*. Arborescent cacti include the prickly pear (*Opuntia excelsa*) and columnar and candelabra cacti in the genera *Cephalocereus*, *Pachycereus*, and *Stenocereus*. Taller, semideciduous forest also has subcanopy trees, including the endemic guayacan (*Guaiacum coulteri*), a highly valued hardwood, and the yucca-like *Beaucarnea* spp.

In terms of the understory, the very dense and diverse shrub layer includes *Cydista* spp., *Euphorbia* spp., *Jatropha* spp., *Bixa orellana*, evergreen *Jacquinia pungens*, *Agave* spp., and cycads (*Zamia* spp. and *Dioon* spp.). Ground flora includes *Bromelia pinguin*, which can form impenetrable spiny mats, as well as grasses and flowering herbs that grow and become visible in the rainy season (e.g., *Pseuderanthemum* spp., *Begonia* spp., and *Callisia fragrans*) and all but disappear in the dry season (Trejo 2010). Lianas and epiphytes (e.g., *Tillandsia* spp.) are smaller and less abundant than in rainforest, but herbaceous creepers (*Dioscorea* spp., *Convolvulus* spp., *Cucurbita* spp., *Phaseolus* spp.) are at their most diverse in tropical deciduous forests (Rzedowski 1978, Challenger 1998, INEGI 2005a, Trejo 2010).

Typical dry forest fauna includes a high diversity of reptiles, with many species of lizard—among which are some of the world's very few poisonous species, such as the beaded lizard (*Heloderma horridum*)—and many species of snake. Other groups with high levels of diversity include scorpions, spiders, butterflies, and bees, including the stingless bees (*Trigona* spp.). Birds and mammals do not reach the levels of diversity of rainforests, but endemism is higher among these and several other groups of fauna (Ceballos et al. 2010).

Human modifications of tropical dry forests include hunting and gathering as well as home gardens and slash-and-burn agriculture, much as in rainforest areas, but fields cut from tropical dry forest can be cultivated for much longer or even permanently, as the strong dry season can reduce populations of pests and weeds to more manageable levels. Secondary vegetation on fallowed milpas (fig. 2.7) may be felled at the shrub stage, after only four to five years, because the majority of species are legumes that fix nitrogen and restore soil fertility relatively rapidly. Traditional landscapes in tropical deciduous forest areas comprise rolling hills with large areas of rain-fed fields (either as traditional polycultures, or increasingly, monocultures), and patches of primary forest on steep slopes and hilltops. Irrigated plantations predominate on valley floors and coastal plains, producing papaya, sugarcane, and other commercial crops. Shrubby secondary vegetation predominates in abandoned milpas, forming low-diversity stands of even-aged, often thorny leguminous species, which may be browsed by goats or cattle, as small-scale livestock production is common. Most of the tropical dry forest of Yucatán is secondary vegetation, including mature trees, and there are many fire-adapted species, likely the result of centuries of fire use by Maya slash-and-burn farmers (Miranda 1958, Rzedowski 1978, Challenger 1998).

Extensive cattle ranching is undertaken in much the same way as in rainforest areas, but with higher stocking densities. This is because pastures are highly productive in the wet season, which, while not an appropriate indicator of total annual stocking rates because of the much lower productivity during the dry season, is nevertheless taken as such by many farmers who wish to maximize herd size. During the dry season it is common for livestock to be fed straw and other crop residues, or indeed a balanced concentrate feed, to maintain their weight. It is equally common, however, to see livestock lose significant weight in the dry season, even to the point of mortality, as water and forage resources decrease dramatically (Villegas Durán et al. 2001).

Figure 2.7. Mosaic of harvested milpas (maize fields, with maize stalks stacked in *mogotes* for use as forage), pasture, and patches of regenerating tropical dry forest of different ages, at the beginning of the dry season in Coacoyula, an ejido near Iguala, Guerrero. Photo by Antony Challenger.

Images of drought and dead livestock (animals allowed to starve, rather than being humanely slaughtered and their carcasses frozen) may then be used unscrupulously, with the unwitting help of the media and public concern, to obtain government assistance for supposedly poor ranchers, a situation that only compounds the social, economic, and ecological problems associated with unsustainable livestock practices in tropical dry forest regions.

Equally common are extensive livestock regimes in which cattle (usually the zebu breed) or goats browse the vegetation directly. Rarely involved in any planned grazing, animals browse the same areas daily, with no respite to allow new plant growth. Under such conditions, the natural vegetation quickly deteriorates, and within a few years tropical deciduous forest is replaced with an induced savanna-like vegetation. This comprises a sparse carpet of grasses and thorny plants on compacted soil, with a scattering of robust, unpalatable species, mostly mature candelabra cacti (survivors of the original forest) and tall, thick-barked, thorny or poisonous woody species (e.g., *Ipomoea arborescens*), with little of the forest's original biodiversity, although some game fauna find these habitats more attractive than closed forest. Such landscapes dominate the Mixteca region of Oaxaca, the Balsas Depression, and many other areas formerly covered with tropical deciduous forest (Leopold 1959, Challenger 1998).

Where the thorn forest vegetation type prevails on the deep, alluvial soils of flat coastal plains north of the Tropic of Cancer, and in the El Bajío region northwest of the Neovolcanic Axis, capital-intensive, mechanized, irrigated agriculture is the predominant land use. In fact, almost none of the original El Bajío thorn forest is left, and of the once extensive thorn forests of the coastal plains of Tamaulipas and Sinaloa, only scattered remnants can be found. Indeed, of all Mexico's vegetation types affected by irrigated farming, thorn forest has received the biggest impact, with at least 26% of its area converted to this use. Most remaining thorn forest has been converted to cattle ranching or degraded by other means (Challenger and Dirzo 2009).

Of a potential area of nearly 4.3 million ha (10.6 million ac), 95% of tropical thorn forest had been lost by 2007, with only 230,000 ha (568,000 ac) of primary vegetation remaining and an additional 480,000 ha (1.1 million ac) of secondary vegetation. In terms of the other ecosystems comprising the bulk of Mexico's tropical deciduous forest, of a potential distribution of nearly 28 million ha (69.1 million ac), only 6.5 million ha (16 million ac) of primary vegetation remains, with an additional 9.1 million ha (22.4 million ac) of mostly shrubby secondary vegetation (INEGI 2005b, 2010b). Human transformation of these vast extensions of forest is lamentable in itself, because the high beta diversity of Mexico's tropical deciduous forests (which are much more diverse than those of Central America) across climatic gradients and between sites means that many unique ecological communities have been lost (Lott and Atkinson 2010, Trejo 2010). Species endemism, one facet of biological diversity, is estimated at around 40% (possibly as high as 60%) for the vascular plants of these ecosystems, and among vertebrate fauna, levels of endemism are notable for amphibians, mammals, birds, and reptiles. Because so much of their habitat has been lost, it is unsurprising that many of these species are listed as threatened (Rzedowski 1991, Lott and Atkinson 2010, Trejo 2010, Ceballos and Valenzuela 2010).

Desert biome

Deserts predominate among the natural landscapes of Mexico, comprising 40% of the nation's territory. They dominate the Central Mexican Plateau and cover the northernmost Gulf and Pacific Coastal Plains, the Baja California Peninsula, and the Tehuacán Valley (the latter is situated in the rain shadow between the mountains of northern Oaxaca and those of the Sierra Madre Oriental). Low rainfall and extreme temperatures in these regions result in low biomass production, such that desert vegetation types comprise open communities with low ground cover.

Two principal ecosystem types account for the majority of Mexico's deserts: *xeric scrubland*, dominated by small-leaved (microphyllous) woody shrubs and trees; and *succulent desert*, in which diverse forms of cacti and *Agave* predominate, together with semisucculent, arborescent, and shrubby species such as *Yucca*, boojum tree, and others. Both ecosystem types do, in fact, comprise a mix of succulent, semisucculent, and woody plants, and many microphyllous shrubs and trees have semisucculent trunks, some of which can photosynthesize. The relative dominance of these plants in a given biotic community, in terms of species composition, relative abundance, and visual dominance, can vary greatly in response to environmental gradients (particularly in terms of precipitation versus potential evapotranspiration), and communities where woody and succulent plants share dominance often occur, particularly at the ecotone between ecosystems (Rzedowski 1978, Challenger 1998, Chapin et al. 2011). While each desert community has low biological diversity, there are many such communities, and their wide distribution, their relative stability over time, and the adaptive responses of their biota to evolutionary pressures have generated enormous biological diversity overall. Mexico's deserts contribute some 6,000 species to Mexico's flora, 20% of the total, of which at least 44% (and perhaps as many as 60%) are endemic, one of the highest levels among Mexico's ecosystems (Rzedowski 1991, 1993).

Among xeric scrubland ecosystems, evergreen microphyllous desert predominates in the most arid regions of the Chihuahuan Desert, the northern Sonoran Desert, and parts of the Baja California Peninsula and is dominated by creosote bush (*Larrea tridentata*), although other shrubs (typically *Ambrosia* spp. and tarwort, *Flourensia cernua*) are also common. There are often few other visible plants, although small cacti (e.g., *Mammillaria* spp.) may be present, as may grasses (e.g., *Hilaria* spp., *Bouteloua* spp.) and short-lived desert ephemerals, which include flowering herbs such as *Amaranthus* spp., *Plantago* spp., *Zaluzania* spp., *Coldenia* spp., and others. These may go unnoticed for years until rain initiates their germination. Subsequent growth, flowering, and seed production occur in only two or three weeks, their short life cycles conferring their ephemeral presence in the ecosystem. Deciduous microphyllous desert is dominated mostly by ocotillo (*Fouquieria splendens*) in the Chihuahuan and Sonoran Deserts, where it occurs in rocky terrain or piedmont habitats. This species may also be among the dominant plants of thorn scrub vegetation types, which include species of woody legumes, typically *Mimosa*, *Acacia*, and mesquite (*Prosopis*), growing not as trees but as shrubs. Acacias and mesquite may form scrubland communities of their own, with few or no other dominant taxa. Piedmont scrub and subtropical scrub vegetation types develop in zones of transition, the former between desert and either oak forest or low deciduous tropical forest, on lower slopes of the Sierra Madre Oriental; and the latter at lower latitudes, in Sonora and on foothills of the Neovolcanic Axis and the Sierra Madre del Sur (Rzedowski 1978, INEGI 2005a, Búrquez and Martínez-Yrizar 2010).

Succulent deserts are dominated, in terms of relative abundance, structural importance, or visual impact, by succulent plants, and in Mexico, they are frequently classified into separate vegetation types according to the dominant taxa. Prickly pear cactus scrub (*nopalera*) is dominated by species in the genus *Opuntia* and covers large areas of the arid central Chihuahuan Desert. Columnar cactus desert scrub is usually dominated by one species of organ pipe cactus, such as saguaro (*Carnegiea gigantea*), old man cactus (*Cephalocereus senilis*), tetecho (*Neobuxbaumia tetetzo*), or other species (e.g., *Cephalocereus columna-trajani*, *Pachycereus pringlei*, etc.). They tend to form open communities in which smaller cacti (e.g., *Mammillaria* spp., *Astrophytum* spp., etc.), agaves, shrubs, and grasses intervene, but in some regions, as on hill slopes in the Tehuacán Valley, they can form dense cactus forests that are difficult to traverse (Miranda and Hernández Xolocotzi 1963, Rzedowski 1978, Dávila et al. 2002, Búrquez and Martínez-Yrizar 2010). Candelabra cactus desert scrub is similar but is dominated by prolifically branching, very large cacti such as *Myrtillocactus geometrizans*, *Stenocereus* spp., and *Pachycereus*. Cholla cactus scrub is dominated by cacti in the *Cylindropuntia* genus, such as the cane or walking-stick cholla (*Cylindropuntia imbricata*) (Rzedowski 1978, INEGI 2005a).

Sarcocaulescent desert occurs in rocky areas of the Baja California Peninsula and on the coast of central Sonora. It is dominated by shrubs and low trees with semisucculent or swollen trunks, typically yellow paloverde (*Parkinsonia microphylla*), ironwood (*Olneya tesota*), elephant tree (*Bursera microphylla*), and *Jatropha* spp. Other species include cacti such as saguaro (*Carnegiea gigantea*) and *Opuntia* spp., and leguminous acacia shrubs. Sarco-crassicaulescent desert develops on the granite hills of the central Baja California Peninsula, species dominance being shared between columnar cacti and other succulents (e.g., *Opuntia* spp.), and trees and shrubs with semisucculent stems including the Baja elephant tree (*Pachycormus discolor*), slipper flower (*Euphorbia lomelii*), and species of ocotillo (*Fouquieria* spp.). Sarcophyllous desert is dominated by succulent or semisucculent plants that grow in rosette form and are dominated by species of *Agave* (*A. lecheguilla*, *A. shawii*, *A. deserti*, etc.). Other species that occur in this vegetation type include candelilla (*Euphorbia antisyphilitica*), guayule (*Parthenium argentatum*), and *Yucca*. When *Yucca* spp. dominate, visually or in terms of cover, such communities (typically of *Y. decipiens*, *Y. filifera*, and *Y. carnerosana*) may be referred to as yucca scrub (Shreve 1951, Rzedowski 1978, INEGI 2005a).

Fog desert occurs only on the Pacific coast of Baja California, where morning fog forms as cold air brought south by oceanic currents meets the warm air over the peninsula. Comprising typical desert communities of shrubs and cacti, fog desert is remarkable for its abundant cover of moisture-dependent lichens (*Ramalina* spp., *Roccella* spp.). Sandy desert vegetation of moving sand dunes (or erg desert) is found mostly in the Altar Desert in the extreme northwest of Sonora, while halophilous desert or salt scrub forms in endorheic basins because of the concentrated salt formed by evaporation. Similar situations in areas of gypsum sediment produce gypsophilous scrub, and both vegetation types are dominated by specialist plants, including many endemics, able to tolerate high salt concentrations (Rzedowski 1978, Davis et al. 1997, INEGI 2005a, 2010a).

The open nature of desert vegetation imposes different requirements on fauna than forested ecosystems do, although some generalist species are shared. Snakes, lizards, scorpions, spiders, grasshoppers, and bees are common but rarely abundant in deserts, although many endemic species exist. Yucca moths (*Tegeticula* spp.) have coevolved with species of *Yucca* to form highly specialized interdependent, mutualistic relationships, such that a given species of *Yucca* may be pollinated only by its particular species of yucca moth (Perry 1983). Birds and mammals of desert ecosystems are low in diversity and abundance, but there are many endemics, including subspecies of elf owl (*Micrathene whitneyi*), skunks, kangaroo rats, and pocket gophers. Specialist desert species include quail; roadrunners (*Geococcyx* spp.); the Gila woodpecker (*Melanerpes uropygialis*), which excavates nest cavities in saguaro cacti; the coyote (*Canis latrans*); the desert cottontail (*Sylvilagus audubonii*); and the pronghorn (*Antilocapra americana*) (Leopold 1959, Challenger 1998, Ceballos and Oliva 2005).

Human modifications of deserts in Mexico traditionally included hunting and gathering, as practiced by indigenous Seri, Pima, Papago, and Yaqui communities, but this has been all but abandoned. Very low rainfall makes rain-fed agriculture almost impossible, but traditional forms of irrigated agriculture persist along permanent rivers, although the area covered is minimal. Capital-intensive irrigated agriculture is possible, fed either from reservoirs in adjacent sierras or, more usually, from fossil water aquifers, pumped to the surface at wells. Of Mexico's 10 million ha (33 million ac) of irrigated land, around 3.5 million ha (11.5 million ac) are situated in deserts (Challenger 1998, Challenger and Dirzo 2009). The impact of pumping water from these aquifers for agriculture has been blamed for the recent desiccation and loss of unique desert wetland habitats, such as the Cuatro Ciénegas region of the Chihuahuan Desert, and it is threatening many narrowly endemic species of flora and fauna with extinction (Souza et al. 2006, Howeth and Brown 2011).

Regional modifications of desert ecosystems have resulted from the overharvesting of commercially useful species, including guayule (*Parthenium argentatum*) and candelilla (*Euphorbia antisyphilitica*) for their latex, jojoba (*Simmondsia chinensis*) for its oil, and lechuguilla (*Agave lecheguilla*) for its fiber (Challenger 1998). The most important transformation of Mexico's desert ecosystems, however, has been their use as rangeland for cattle ranching, in all but the most arid and stony landscapes. Following the Spanish Conquest of Mexico in 1521, the deserts have been grazed without respite. It is almost impossible to know what the composition, relative abundance, and appearance of species would have been in these ecosystems prior to grazing by cattle. However, documented records over the last century or so show increasing cases of soil erosion and desertification, and local and regional species extinctions. Also, invasive species such as mesquite and buffelgrass displace others and alter local ecology (Challenger 1998, Villegas Durán et al. 2001, Búrquez-Montijo et al. 2002, SEMARNAT 2006, Challenger and Dirzo 2009, Marshall et al. 2012). In this context, official statistics on Mexico's deserts should be interpreted with caution, as they indicate that of a potential total cover of 66 million ha (163 million ac), 51 million ha (126 million ac) of primary vegetation (77% of the original cover) remained in 2007, together with a further 4.8 million ha (11.8 million ac) of secondary vegetation (INEGI 2005b, 2010b).

Savanna biome

Savanna vegetation in Mexico comprises three principal types: (1) that which occurs in the transition zone between semidesert grassland and oak woodland along the piedmont of the interior Sierra Madre Occidental, where pines, oaks, and junipers form Madrean and oak savannas; (2) that which occurs in semiarid regions, often in zones of transition between semidesert grassland and xeric scrubland, in which acacias and mesquites are the dominant trees; and (3) that which occurs in tropical zones, which comprise two different forms. One is dominated by the tropical Honduras pine, *Pinus caribaea* var. *hondurensis*, and occurs in very restricted areas of the southern Yucatán Peninsula. The other is the tropical savanna usually described in the literature, which occurs in transition zones between lowland tropical forests and wetlands, where the dominant trees are species also common in tropical dry forests, typically nanche (*Byrsonima crassifolia*), sandpaper tree (*Curatella americana*), calabash tree (*Crescentia cujete*, *C. alata*), and palms (Rzedowski 1978, INEGI 2005a). The ground cover of all these vegetation types is dominated by grasses, and the fauna varies among them, comprising a subset of the fauna of adjacent vegetation with the addition of specialist open-habitat species (Leopold 1959, Ceballos and Oliva 2005).

Modification and use of savanna ecosystems is mostly for extensive cattle ranching, with periodic burning of grass to stimulate new growth. This can affect adjacent forest, often transforming it into an extension of the savanna itself. Tropical savannas, often the product of a high water table and clayey soils, may also be drained and cultivated (Rzedowski 1978, Pennington et al. 2006). The effects of these practices on the biota of savannas are not well documented, although they are likely to be detrimental to flora and fauna not adapted to them. Because of questions of scale and conceptual criteria, official statistics on savanna are not cited here.

Mediterranean biome

Mediterranean vegetation in Mexico is restricted to the northwestern Baja California Peninsula, the southernmost extension of the large California Floristic Province (CFP) of the US. It comprises two ecosystems, Mediterranean conifer woodland and Mediterranean chaparral, which receive winter rainfall and present clear ecological and floristic differences from conifer forests and chaparral vegetation elsewhere in Mexico (Rzedowski 1978, Minnich and Franco-Vizcaíno 2005, Loarie et al. 2008).

Mediterranean conifer forest includes, as dominant species in the canopy, the Juarez pinyon pine (*Pinus juarezensis*, although this species may be a synonym of *P. quadrifolia*), Parry pinyon (*P. quadrifolia*), Jeffrey pine (*P. jeffreyi*), and California incense cedar (*Calocedrus decurrens*), together with several species of oak (Rzedowski 1978, Minnich and Franco-Vizcaíno 2005, Loarie et al. 2008). This vegetation is unique in Mexico, as are many of its fauna species, including the California condor (*Gymnogyps californianus*), which has been successfully reintroduced after several decades of being extinct in the wild (Walters et al. 2010).

Mediterranean chaparral comprises two vegetation types: soft chaparral typically includes California sagebrush (*Artemisia californica*), Munz's sage (*Salvia munzii*), California buckwheat (*Eriogonum fasciculatum*), California brittlebush (*Encelia californica*), and *Lotus scoparius*, among others. Hard chaparral, on slopes at higher elevations, includes chamise (*Adenostoma fasciculatum*), manzanita (*Arctostaphylos* spp.), California lilac (*Ceanothus* spp.), chaparral yucca (*Hesperoyucca whipplei*), several shrub oaks (*Quercus* spp.), and *Cercocarpus* spp., among others (Rzedowski 1978, Minnich and Franco-Vizcaíno 2005, Barbour et al. 2007). These low, dense communities are dominated by perennial shrubs that have evolved fire-ecology adaptations to lightning strikes, including vigorous crown regrowth from live roots (Espejel et al. 2001, Minnich and Franco-Vizcaíno 2005). Chaparral fauna includes several endemic bird species, although most mammal species have more wide-ranging distributions (Leopold 1959, Stattersfield et al. 1998, Ceballos and Oliva 2005).

Human use and modification of Mediterranean ecosystems includes urbanization along the US-Mexico border (for housing and maquiladora industries) and along the Pacific coast, mostly for tourism and homes for American pensioners. Irrigated agriculture has transformed interior plains and valleys; extensive livestock farming and the encroachment of invasive exotic species favored by fire and cattle grazing have destroyed large areas of chaparral; and upland forests are affected mostly by tourism (Minnich and Franco-Vizcaíno 2005, Espejel et al. 2001). Official statistics do not recognize Mediterranean vegetation, but figure 2.3 clearly illustrates that it is the least well conserved on the Baja California Peninsula. Recent studies modeling the effects of climate change indicate probable reductions in distributional range of up to 80% for most of the flora of these communities by the end of this century (Loarie et al. 2008).

Wetlands biome

Wetland ecosystems exist where terrestrial and aquatic environments merge. Coastal wetlands include mangrove and swamp forest often associated with brackish-water lagoons, and coastal strands and dunes on sandy beaches. Estuarine, lake, and marsh wetlands include reed beds, sedge beds, and wet meadows, and other inland wetlands form where the water table is exposed, forming oases in desert regions and sinkholes on the Yucatán Peninsula. Mexico's wetlands are enormously varied, ranging across the full spectra of continental to coastal, lowland to montane, open water to dense vegetation, and cold to warm climates, with all intervening variants (table 2.1) (SEMARNAT et al. 2008).

The dominant fauna of wetlands characteristically includes waterfowl, typically shorebirds, ducks, geese, herons, waders, and others, many of which migrate from Canada and the US and overwinter in Mexico (Carrera and de la Fuente 2003, Lepage 2003). Among typical wetland mammals are two species of river otter and two subspecies of beaver (*Castor canadensis*) in the north of the country (Ceballos and Oliva 2005), while species of wetland herpetofauna include crocodiles, freshwater turtles, and several endemic species of salamander (e.g., the axlotl, *Ambystoma mexicanum*) and frog (Robles Gil et al. 1993).

Principal human uses and modification of wetlands include draining for agricultural use, tourism development, urbanization, and infrastructure, particularly oil refineries and pipelines, which are responsible for polluting thousands of hectares of Mexico's Gulf Coast wetlands (Challenger 1998). Agricultural runoff has also polluted many wetlands, while irrigated agriculture affects wetlands via water removal, causing their desiccation and even disappearance (e.g., all the water bodies of the Comarca Lagunera region of the Chihuahuan Desert are now dry, and more recently the water bodies of the Cuatro Ciénegas Biosphere Reserve, also in the Chihuahuan Desert, are increasingly being affected, driving endemic species of plants, fish, and invertebrates toward extinction) (Challenger 1998, Souza et al. 2006). Over the last 15–20 years, large areas of mangroves have been converted to shrimp farms, particularly on the coast of Sinaloa and Sonora, and these have in turn polluted adjacent wetlands (Carabias and Landa 2006, CONABIO 2008). In all, of a potential cover of around 3.4 million ha (8.4 million ac), primary wetland vegetation had been reduced to around 2.2 million ha (5.4 million ac) by 2007 (INEGI 2005b, 2010b).

Concluding Remarks

Mexico's biological megadiversity, including its game and nongame bird and mammal fauna, is distributed throughout the country, albeit unevenly. Some ecosystems are more important for their species richness, some for their endemic taxa, and others for their game fauna. In terms of landscapes, most of Mexico's original vegetation types remain, although many are in diverse stages of serious deterioration. Many traditional landscapes, inherited from and forming a direct link with pre-Hispanic indigenous cultures, are fast disappearing, replaced by the simplified, ubiquitous landscapes of specialized commercial crop and livestock production. All of Mexico's ecosystems have been largely modified and transformed by human use, and by 2007, primary vegetation, the optimal habitat for most wild species, covered 98.2 million ha (242.6 million ac) of a potential 194 million ha (479.3 million ac), while secondary vegetation contributed 14 million ha (34.5 million ac) (INEGI 2005b, 2010b).

As biodiverse natural and modified landscapes are lost and the monocultural carpet of conventional agriculture and livestock production is rolled out over rural Mexico, the wild flora and fauna that make up Mexico's unique megadiversity are at increasing risk of local and regional extinctions, and ultimately of irrevocable species loss. Slowing, halting, and reversing this process must be the concern of government, wildlife managers, landowners, and citizens.

Literature Cited

Aguirre Gutiérrez, J., and J. F. Duivenvorden. 2010. Can we expect to protect threatened species in protected areas? A case study of the genus *Pinus* in Mexico. Revista Mexicana de Biodiversidad 81:875–882.

Alroy, J. 2001. A multispecies overkill simulation of the End-Pleistocene megafaunal mass extinction. Science 292:1893–1896.

Andrade Torres, A., et al. 2009. La interacción entre abetos y hongos. La Ciencia y el Hombre 22:39–44.

Arriaga, L., S. Díaz, R. Domínguez, and J. L. León. 1997. Pages 69–106 *in* L. Arriaga and R. Rodríguez Estrella, editors. Composición florística y vegetación. Los oasis de la Península de Baja California, México. Centro de Investigaciones Biológicas del Noroeste, La Paz, México.

Balvanera, P., et al. 2009. Estado y tendencias de los servicios ecosistémicos. Pages 185–245 *in* R. Dirzo, R. González, and I. J. March, compilers. Capital natural de México. Volume 2. Estado de conservación y tendencias de cambio. CONABIO, México, D. F., México.

Barbour, M. G., T. Keeler Wolf, and A. A. Schoenherr, editors. 2007. Terrestrial vegetation of California. Third edition. University of California Press, Berkeley, USA.

Berlanga-Robles, C. A., A. Ruíz Luna, and G. de la Lanza Espino. 2008. Esquema de clasificación de los humedales de México. Investigaciones Geográficas. Boletín del Instituto de Geografía 66:25–46.

Bezaury Creel, J. 2010. Las selvas secas del Pacífico mexicano en el contexto mundial. Pages 21–40 *in*

G. Ceballos, L. Martínez, A. García, E. Espinoza, J. Bezaury Creel, and R. Dirzo, editors. Diversidad, amenazas y áreas prioritarias para la conservación de las selvas secas del Pacífico de México. CONABIO and Fondo de Cultura Económica, México, D. F., México.

Bonilla-Barbosa, J. R., and A. Novelo Retana. 1995. Manual de identificación de plantas acuáticas del Parque Nacional Lagunas de Zempoala, México. Cuaderno 26. Instituto de Biología, UNAM, México, D. F., México.

Bormann, F. H., and G. Likens. 1994. Pattern and process in a forested ecosystem: disturbance, development and the steady state based on the Hubbard Brook ecosystem study. Springer, New York, New York, USA.

Breedlove, D. 1981. Introduction to the flora of Chiapas. California Academy of Sciences, San Francisco, USA.

Brown, D. E. 1982. Semidesert grassland. Pages 123–131 *in* D. E. Brown, editor. Biotic communities of the American Southwest: United States and Mexico. Desert Plants 4:1–342.

Búrquez, A., and A. Martínez-Yrizar. 2010. Límites geográficos entre selvas secas y matorrales espinosos y xerófilos: ¿qué conservar? Pages 53–62 *in* G. Ceballos, L. Martínez, A. García, E. Espinoza, J. Bezaury Creel, and R. Dirzo, editors. 2010. Diversidad, amenazas y áreas prioritarias para la conservación de las selvas secas del Pacífico de México. CONABIO and Fondo de Cultura Económica, México, D. F., México.

Búrquez Montijo, A., M. E. Miller, and A. Martínez Yrízar. 2002. Mexican grasslands, thornscrub, and the transformation of the Sonoran Desert by invasive exotic buffelgrass (*Pennisetum ciliare*). Pages 126–146 *in* B. Tellman, editor. Invasive exotic species in the Sonoran region. University of Arizona Press and Arizona-Sonora Desert Museum, Tucson, USA.

Carabias, J., and R. Landa. 2006. Agua, medio ambiente y sociedad: hacia la gestión integral de los recursos hídricos en México. UNAM, COLMEX, and Fundación Gonzalo Río Arronte, México, D. F., México.

Carrera, E., and G. de la Fuente. 2003. Inventario y clasificación de humedales en México. Volume 1. Ducks Unlimited de México, México, D. F., México.

Castillo, P. N., and L. Snook. 1984. Análisis del efecto de la intensidad de la corta sobre la regeneración natural de pinos en un bosque de pino-encino. Ciencia Forestal 47:48–61.

Castro Soto, G. 2015. México: los efectos de la palma africana. http://www.gloobal.net/iepala/gloobal/fichas/ficha.php?entidad=Textos&id=11551&opcion=documento#s5. Accessed 30 Oct 2015.

Ceballos, G., and G. Oliva, coordinators. 2005. Los mamíferos silvestres de México. CONABIO and Fondo de Cultura Económica, México, D. F., México.

Ceballos, G., et al. 2005. Diversidad y conservación de los mamíferos de México. Pages 21–66 *in* G. Ceballos and G. Oliva, coordinators. Los mamíferos silvestres de México. CONABIO and Fondo de Cultura Económica, México, D. F., México.

Ceballos, G., L. Martínez, A. García, E. Espinoza, J. Bezaury Creel, and R. Dirzo, editors. 2010. Diversidad, amenazas y áreas prioritarias para la conservación de las selvas secas del Pacífico de México. CONABIO and Fondo de Cultura Económica, México, D. F., México.

Ceballos, G., and D. Valenzuela. 2010. Diversidad, ecología y conservación de vertebrados. Pages 93–118 *in* G. Ceballos, L. Martínez, A. García, E. Espinoza, J. Bezaury Creel, and R. Dirzo, editors. 2010. Diversidad, amenazas y áreas prioritarias para la conservación de las selvas secas del Pacífico de México. CONABIO and Fondo de Cultura Económica, México, D. F., México.

Challenger, A. 1998. Utilización y conservación de los ecosistemas terrestres de México: pasado, presente y futuro. CONABIO, Instituto de Biología UNAM, and Agrupación Sierra Madre, México, D. F., México.

Challenger, A. 2003. Conceptos generales acerca de los ecosistemas templados de montaña de México y su estado de conservación. Pages 17–44 *in* O. Sánchez, E. Vega, E. Peters, and O. Monroy Vilchis, compilers. Conservación de ecosistemas templados de montaña en México. Diplomado de capacitación y actualización sobre conservación de ecosistemas templados de montaña en México. SEMARNAT-INE and USFWS, México, D. F., México.

Challenger, A. 2011. La neblina se levanta: los árboles del bosque mesófilo de montaña de México, claramente amenazados. Boletín de la Sociedad Botánica de México 89:135–137.

Challenger, A. 2014. Introducción. Pages 21–25 *in*

M. Gual-Díaz and A. Rendón-Correa, compilers. Bosques mesófilos de montaña en México. CONABIO, México, D. F., México.
Challenger, A., and R. Dirzo. 2009. Factores de cambio y estado de la biodiversidad. Pages 37–73 *in* R. Dirzo, R. González, and I. J. March, compilers. Capital natural de México. Volume 2. Estado de conservación y tendencias de cambio. CONABIO, México, D. F., México.
Challenger, A., and J. Soberón. 2008. Los ecosistemas terrestres. Pages 87–108 *in* J. Soberón, R. González, and I. J. March, compilers. Capital natural de México. Volume 1. Conocimiento actual de la biodiversidad. CONABIO, México, D. F., México.
Chapin, F. S., III, P. A. Matson, and P. M. Vitousek. 2011. Principles of terrestrial ecosystem ecology. Second edition. Springer Science and Business Media, New York, New York, USA.
CONABIO. 2008. Manglares de México. CONABIO, México, D. F., México.
Dávila, P., M. del C. Arizmendi, A. Valiente-Banuet, J. L. Villaseñor, A. Casas, and R. Lira. 2002. Biological diversity in the Tehuacán-Cuicatlán Valley, México. Biodiversity and Conservation 11:421–442.
Davis, S. D., V. H. Heywood, O. Herrera-Macbryde, J. Villa-Lobos, and A. C. Hamilton. 1997. Pages 167–171 *in* V. H. Heywood and S. D. Davis, editors. Centres of plant diversity: a guide and strategy for their conservation. Volume 3. The Americas: Cuatro Ciénegas region, México. WWF and IUCN, Gland, Switzerland.
Denevan, W. M. 2000. Foreword. Pages 17–20 *in* D. L. Lentz, editor. Imperfect balance: landscape transformations in the precolumbian Americas. Columbia University Press, New York, New York, USA.
Duffy, D. C., and A. J. Meier. 1992. Do Appalachian herbivorous understories ever recover from clearcutting? Conservation Biology 6:196–201.
Escalante Pliego, P., A. G. Navarro Siguenza, and A. Townsend Peterson. 1993. A geographic, ecological and historical analysis of land bird diversity in Mexico. Pages 281–307 *in* T. P. Ramamoorthy, R. Bye, A. Lot, and J. Fa, editors. Biological diversity of Mexico: origins and distribution. Oxford University Press, New York, New York, USA.
Espejel, I., et al. 2001. La vegetación costera del noroeste de Baja California: sus posibilidades de conservación. Fondo Mexicano para la Conservación de la Naturaleza and UABC, Mexicali, México.
Espinosa, D., et al. 2008. El conocimiento biogeográfico de las especies y su regionalización natural. Pages 33–65 *in* J. Soberón, G. Halffter, and J. Llorente Bousquets, compilers. Capital natural de México. Volume 1. Conocimiento actual de la biodiversidad. CONABIO, México, D. F., México.
Fa, J. E., and L. M. Morales. 1993. Patterns of mammalian diversity in Mexico. Pages 319–361 *in* T. P. Ramamoorthy, R. Bye, A. Lot, and J. Fa, editors. Biological diversity of Mexico: origins and distribution. Oxford University Press, New York, New York, USA.
Flores Guido, J. S., and I. Espejel Carbajal. 1994. Tipos de vegetación de la península de Yucatán. Etnoflora Yucatense. Fascículo 3. Universidad Autónoma de Yucatán, Merida, México.
Flores Villela, O. 1993. Herpetofauna of Mexico: distribution and endemism. Pages 253–280 *in* T. P. Ramamoorthy, R. Bye, A. Lot, and J. Fa, editors. Biological diversity of Mexico: origins and distribution. Oxford University Press, New York, New York, USA.
García Sánchez, R., and A. Monroy Ata. 2005. Micrositios del pasto navajita (*Bouteloua gracilis*) en comunidades de pastizal y de matorral del altiplano mexicano. TIP Revista Especializada en Ciencias Químico-Biológicas 8:61–70.
Gignoux, J., I. D. Davies, S. R. Flynt, and J.-D. Zucker. 2011. The ecosystem in practice: interest and problems of an old definition for constructing ecological models. Ecosystems 14: 1039–1054.
Gómez Pompa, A., and A. Kaus. 1992. Taming the wilderness myth. Bioscience 42:271–279.
González-Espinosa, M., J. A. Meave, F. G. Lorea-Hernández, G. Ibarra-Manriquez, and A. C. Newton. 2011. The Red List of Mexican cloud forest trees. Fauna and Flora International, Cambridge, United Kingdom.
González-Ruíz, N., J. Ramírez-Pulido, and M. Gual-Díaz. 2014. Mamíferos del bosque mesófilo de montaña en México. Pages 305–326 *in* M. Gual-Díaz and A. Rendón-Correa, compilers. Bosques mesófilos de montaña en México. CONABIO, México, D. F., México.
Goudie, A. 2013. The human impact on the natural

environment. Seventh edition. Blackwell Publishing, Oxford, United Kingdom.

Goyenechea Mayer-Goyenechea, I., and M. Gual-Díaz. 2014. Reptiles en el bosque mesófilo de montaña en México. Pages 263–278 *in* M. Gual-Díaz and A. Rendón-Correa, compilers. Bosques mesófilos de montaña en México. CONABIO, México, D. F., México.

Grayson, D. K. 2007. Deciphering North American Pleistocene extinctions. Journal of Anthropological Research 63:185–213.

Gual-Díaz, M., and F. González-Medrano. 2014. Los bosques mesófilos de montaña en México. Pages 27–68 *in* M. Gual-Díaz and A. Rendón-Correa, compilers. Bosques mesófilos de montaña en México. CONABIO, México, D. F., México.

Gual-Díaz, M., and I. Goyenechea Mayer-Goyenechea. 2014. Anfibios en el bosque mesófilo de montaña en México. Pages 249–262 *in* M. Gual-Díaz and A. Rendón-Correa, compilers. Bosques mesófilos de montaña en México. CONABIO, México, D. F., México.

Hamilton, L. S., J. O. Juvik, and F. N. Scatena. 1995. Tropical montane cloud forests. Ecological Studies Series, no. 110. Springer, New York, New York, USA.

Heath, M., and A. Long. 1991. Habitat, distribution and status of the azure-rumped tanager *Tangara cabanisi* in Mexico. Bird Conservation International 1:223–254.

Howell, J. T. 1947. Mono Mesa–Sierra Sky Island. Sierra Club Bulletin 32:15–18.

Howeth, J. G., and W. S. Brown. 2011. *Terrapene coahuila* Schmidt and Owens 1944—Coahuilan box turtle. Pages 049.1–049.13 *in* A. G. J. Rhodin, P. C. H. Pritchard, P. P. van Dijk, R. A. Saumure, K. A. Buhlmann, J. B. Iverson, and R. A. Mittermeier, editors. Conservation biology of freshwater turtles and tortoises. Chelonian Research Monographs. Chelonian Research Foundation, Lawrence, Kansas, USA.

Huerta, L. 1978. Vegetación marina litoral. Pages 328–340 *in* J. Rzedowski. Vegetación de México. Limusa, México, D. F., México.

Ibarra Cerdeña, C. N. 2001. Análisis del crecimiento arbóreo en bosques de encino subcaducifolios de la Sierra de Manantlán, México. Thesis, Centro Universitario de Ciencias Biológicas y Agropecuarias, Universidad de Guadalajara, México.

INEGI. 2005a. Guía para la interpretación de cartografía: uso de suelo y vegetación. INEGI, Aguascalientes, México.

INEGI. 2005b. Carta de vegetación primaria 1:1,000,000. INEGI, Aguascalientes, México.

INEGI. 2010a. Uso de suelo y vegetación. Glosario. INEGI, Aguascalientes, México.

INEGI. 2010b. Carta de uso del suelo y vegetación, escala 1:250,000: Serie IV. INEGI, Aguascalientes, México.

INE-SEMARNAP. 1999. Programa de manejo del Área de Protección de Flora y Fauna Cuatrociénegas. Instituto Nacional de Ecología, SEMARNAP, México, D. F., México.

Janzen, D. H. 1986. Chihuahuan Desert nopaleras: defaunated big mammal vegetation. Annual Review of Ecology and Systematics 17:595–636.

Koch, P. L., and A. D. Barnosky. 2006. Late Quaternary extinctions: state of the debate. Annual Review of Ecology, Evolution, and Systematics 37:215–250.

Koeppen, W. 1948. Climatología. Fondo de Cultura Económica, México, D. F., México.

Koleff, P., et al. 2008. Patrones de diversidad espacial en grupos selectos de especies. Pages 323–364 *in* J. Soberón, G. Halffter, and J. Llorente Bousquets, compilers. Capital natural de México. Volume 1. Conocimiento actual de la biodiversidad. CONABIO, México, D. F., México.

Koleff, P., et al. 2009. Identificación de prioridades y análisis de vacíos y omisiones en la conservación de la biodiversidad de México. Pages 651–718 *in* R. Dirzo, R. González, and I. J. March, compilers. Capital natural de México. Volume 2. Estado de conservación y tendencias de cambio. CONABIO, México, D. F., México.

Kraft, K., et al. 2014. Multiple lines of evidence for the origin of domesticated chili pepper, *Capsicum annuum*, in Mexico. Proceedings of the National Academy of Sciences 111:6165–6170.

Laborde, F. J., S. A. Guevara Sada, and G. Sánchez Ríos. 2011. Diversidad florística en potreros de Los Tuxtlas. Pages 247–259 *in* La biodiversidad en Veracruz: estudio de estado. CONABIO, Gobierno del Estado de Veracruz, Universidad Veracruzana, and Instituto de Ecología, Xalapa, México.

Lentz, D. L. 2000a. Introduction: definitions and

conceptual underpinnings. Pages 1–12 *in* D. L. Lentz, editor. Imperfect balance: landscape transformations in the Precolumbian Americas. Columbia University Press, New York, New York, USA.

Lentz, D. L., editor. 2000b. Imperfect balance: landscape transformations in the Precolumbian Americas. Columbia University Press, New York, New York, USA.

Leopold, A. S. 1959. Wildlife of Mexico: the game birds and mammals. University of California Press, Berkeley, USA.

Lepage, D. 2003. Checklist of birds of Mexico. Bird checklists of the world. Birdlife International Avibase. http://avibase.bsc-eoc.org/checklist.jsp?lang=EN&region=mx&list=clements. Accessed 30 Oct 2015.

Llorente Bousquets, J., and S. Ocegueda. 2008. Estado del conocimiento de la biota. Pages 283–322 *in* J. Soberón, G. Halffter, and J. Llorente Bousquets, compilers. Capital natural de México. Volume 1. Conocimiento actual de la biodiversidad. CONABIO, México, D. F., México.

Loarie, S. R., B. E. Carter, K. Hayhoe, S. McMahon, R. Moe, C. A. Knight, and D. D. Ackerly. 2008. Climate change and the future of California's endemic flora. PloS ONE:e2502. http://journals.plos.org/plosone/article?id=10.1371/journal.pone.0002502. Accessed 20 Aug 2015.

Lott, E. J., and T. H. Atkinson. 2010. Diversidad florística. Pages 63–76 *in* G. Ceballos, L. Martínez, A. García, E. Espinoza, J. Bezaury Creel, and R. Dirzo, editors. 2010. Diversidad, amenazas y áreas prioritarias para la conservación de las selvas secas del Pacífico de México. CONABIO and Fondo de Cultura Económica, México, D. F., México.

Macario Mendoza, P. A., S. A. Torres Pech, and E. F. Cabrera Cano. 1998. Estructura y composición de una comunidad con *Pinus caribaea* var. *hondurensis* (Sénecl.) Barr. y Golf., en el estado de Quintana Roo, México. Caribbean Journal of Science 34:50–57.

Marshall, V. M., M. M. Lewis, and B. Ostendorf. 2012. Buffel grass (*Cenchrus ciliaris*) as an invader and threat to biodiversity in arid environments: a review. Journal of Arid Environments 78:1–12.

Martin, P. S. 1989. Prehistoric overkill: a global model. Pages 354–404 *in* P. S. Martin and R. G. Klein, editors. Quaternary extinctions: a prehistoric revolution. University of Arizona Press, Tucson, USA.

Massieu Trigo, Y. C., and B. O. Acuña Rodarte. 2015. Socio-environmental violence in rural México. Latin American Perspectives 42:67–82.

Matsuoka, Y., Y. Vigouroux, M. M. Goodman, J. Sanchez G., E. Bukler, and J. Doebley. 2002. A single domestication for maize shown by multilocus microsatellite genotyping. Proceedings of the National Academy of Sciences 99:6080–6084.

McDonald, J. A. 1993. Phytogeography and history of the alpine-subalpine flora of northeast Mexico. Pages 681–703 *in* T. P. Ramamoorthy, R. Bye, A. Lot, and J. Fa, editors. Biological diversity of Mexico: origins and distribution. Oxford University Press, New York, New York, USA.

Minnich, R. A. 2007. Southern California forests. Pages 506–538 *in* M. G. Barbour, T. Keeler Wolf, and A. A. Schoenherr, editors. Terrestrial vegetation of California. Third edition. University of California Press, Berkeley, USA.

Minnich, R. A., and E. Franco-Vizcaíno. 2005. Baja California's enduring Mediterranean vegetation: early accounts, human impacts, and conservation status. Pages 370–386 *in* J.-L. E. Cartron, G. Ceballos, and R. S Felger, editors. Biodiversity, ecosystems and conservation in northern Mexico. Oxford University Press, New York, New York, USA.

Miranda, F. 1958. Estudios acerca de la vegetación. Pages 215–271 *in* E. Baltrán, editor. Los recursos naturales del sureste y su aprovechamiento. Volume 2. Estudios particulares. Instituto Mexicano de Recursos Naturales Renovables, México, D. F., México.

Miranda, F., and E. Hernández Xolocotzi. 1963. Los tipos de vegetación de México y su clasificación. Boletín de la Sociedad Botánica de México 28:29–179.

Miranda, F., and A. J. Sharp. 1950. Characteristics of the vegetation in certain temperate regions of eastern Mexico. Ecology 31:313–333.

Mooney, H. A., S. H. Bullock, and E. Medina. 1995. Introduction. *In* S. H. Bullock, H. A. Mooney, and E. Medina, editors. Seasonally dry tropical forests. Cambridge University Press, Cambridge, United Kingdom.

Moore, P. D. 2008. Tundra. Facts on File, New York, New York, USA.

Muller, C. H. 1939. Relations of the vegetation and climatic types in Nuevo Leon, Mexico. American Midland Naturalist 21:687–729.

Nixon, K. C. 1993. The genus *Quercus* in Mexico. Pages 447–458 *in* T. P. Ramamoorthy, R. Bye, A. Lot, and J. Fa, editors. Biological diversity of

Mexico: origins and distribution. Oxford University Prcss, New York, New York, USA.

Ochoa Ochoa, L. M., and N. R. Mejía-Domínguez. 2014. Fauna de los bosques mesófilos de montaña. Pages 237–247 *in* M. Gual-Díaz and A. Rendón-Correa, compilers. Bosques mesófilos de montaña en México. CONABIO, México, D. F., México.

Pennington, R. T., G. P. Lewis, and J. A. Ratter, editors. 2006. Neotropical savannas and seasonally dry forests: plant diversity, biogeography and conservation. Taylor and Francis, Boca Raton, Florida, USA.

Pennington, T. D., and J. Sarukhán. 1998. Árboles tropicales de México: manual para la identificación de las principales especies. Second edition. UNAM and FCE, México, D. F., México.

Perry, N. 1983. Symbiosis: nature in partnership. Blandford, London, United Kingdom.

Pineda, E., G. Halffter, C. E. Moreno, and F. Escobar. 2005. Transformación del bosque de niebla en agroecosistemas cafetaleros: cambios en las diversidades alfa y beta de tres grupos faunísticos. Pages 177–190 *in* G. Halffter, J. Soberón, P. Koleff, and A. Melic, editors. Sobre diversidad biológica: el significado de las diversidades alfa, beta y gamma. Monografías Tercer Milenio. Volume 4. Sociedad Entomológica Aragonesa, Zaragoza, Spain.

Poder Ejecutivo Federal. 2014. Pages 1–95 *in* Programa Especial de Cambio Climático, 2014–2018. Diario Oficial de la Federación (Segunda Sección), 28 April 2014, México, D. F., México.

Quero, H. J. 1994. Flora del Valle de Tehuacán-Cuicatlán. Fascículo 7: Arecaceae. Instituto de Biología, UNAM, México, D. F., México.

Reyes Jaramillo, I. 2006. *Quercus hintonii* Warb.: especie endémica del encinar del SW del estado de México. Contactos 60:64–72.

Robles Gil, P., G. Ceballos, and F. Eccardi. 1993. Mexican diversity of fauna. CEMEX, México, D. F., México.

Romero, F. J., and F. A. Cervantes. 2005. Conejo zacatuche. Pages 832–837 *in* G. Ceballos and G. Oliva, coordinators. Los mamíferos silvestres de México. CONABIO and FCE, México, D. F., México.

Rundel, P. W. 2007. Sage scrub. Pages 208–228 *in* M. G. Barbour, T. Keeler Wolf, and A. A. Schoenherr, editors. Terrestrial vegetation of California. Third edition. University of California Press, Berkeley, USA.

Rzedowski, J. 1954. Vegetación del Pedregal de San Ángel. Thesis, Escuela Nacional de Ciencias Biológicas, Instituto Politécnico Nacional, México, D. F., México.

Rzedowski, J. 1978. Vegetación de México. Limusa, México, D. F., México.

Rzedowski, J. 1991. El endemismo en la flora fanerogámica mexicana: una apreciación analítica preliminar. Acta Botánica Mexicana 15:47–64.

Rzedowski, J. 1993. Diversity and origins of the phanerogamic flora of Mexico. Pages 129–144 *in* T. P. Ramamoorthy, R. Bye, A. Lot, and J. Fa, editors. Biological diversity of Mexico: origins and distribution. Oxford University Press, New York, New York, USA.

Rzedowski, J. 1996. Análisis preliminar de la flora vascular de los bosques mesófilos de montaña de México. Acta Botánica Mexicana 35:25–44.

Sáenz Romero, C., G. E. Rehfeldt, P. Duval, and R. A. Lindig-Cisneros, 2012. *Abies religiosa* habitat prediction in climatic change scenarios and implications for monarch butterfly conservation in Mexico. Forest Ecology and Management 275:98–106.

Sánchez-Ramos, G., and R. Dirzo. 2014. El bosque mesófilo de montaña: un ecosistema prioritaria amenazado. Pages 109–139 *in* M. Gual-Díaz and A. Rendón-Correa, compilers. Bosques mesófilos de montaña en México. CONABIO, México, D. F., México.

Sarukhán, J., et al. 2009. Capital natural de México. Síntesis: conocimiento actual, evaluación y perspectivas de sustentabilidad. CONABIO, México, D. F., México.

Savory, A. 1999. Holistic management: a new framework for decision making. Island Press, Washington, DC, USA.

Scott, G. A. J. 1995. Canada's vegetation: a world perspective. McGill-Queen's University Press, Montreal, Canada.

SEMARNAT. 2006. La gestión ambiental en México. SEMARNAT, México, D. F., México.

SEMARNAT. 2014. Anuario estadístico de la producción forestal, 2013. SEMARNAT, México, D. F., México.

SEMARNAT, CONABIO, and INEGI. 2008. Documento estratégico rector del inventario nacional de humedales. SEMARNAT, México, D. F., México. http://www.inecc.gob.mx/descargas/emc/der.pdf.

Accessed 30 Oct 2015.
Shreve, F. 1951. Vegetation of the Sonoran Desert. Publication 591. Carnegie Institute, Washington, DC, USA.
Smith, B. D. 2001. Documenting plant domestication: the consilience of biological and archaeological approaches. Proceedings of the National Academy of Sciences 98:1324–1326.
Sousa S., M. 1968. Ecología de las leguminosas de Los Tuxtlas, Veracruz. Anales del Instituto de Biología, Universidad Nacional Autónoma de México, Serie Botánica No. 39:121–161.
Sousa S., M. 2010. Centros de endemismo: las leguminosas. Pages 77–91 *in* G. Ceballos, L. Martínez, A. García, E. Espinoza, J. Bezaury Creel, and R. Dirzo, editors. Diversidad, amenazas y áreas prioritarias para la conservación de las selvas secas del Pacífico de México. CONABIO and Fondo de Cultura Económica, México, D. F., México.
Souza, V., et al. 2006. An endangered oasis of aquatic microbial biodiversity in the Chihuahuan Desert. Proceedings of the National Academy of Sciences 103:6565–6570.
Stattersfield, A. J., M. J. Crosby, A. J. Long, and D. C. Wege. 1998. Endemic bird areas of the world: priorities for biodiversity conservation. Birdlife International, Cambridge, United Kingdom.
Styles, B. T. 1993. Genus *Pinus*: a Mexican purview. Pages 397–420 *in* T. P. Ramamoorthy, R. Bye, A. Lot, and J. Fa, editors. Biological diversity of Mexico: origins and distribution. Oxford University Press, New York, New York, USA.
Tamayo, J. L. 1990. Geografía moderna de México. Trillas, México, D. F., México.
Tansley, A. G. 1935. The use and abuse of vegetational concepts and terms. Ecology 16:284–307.
Tejero-Díez, J. D., A. N. Torres-Díaz, and M. Gual-Díaz. 2014. Licopodios y helechos en el bosque mesófilo de montaña de México. Pages 197–220 *in* M. Gual-Díaz and A. Rendón-Correa, compilers. Bosques mesófilos de montaña en México. CONABIO, México, D. F., México.
Tercafs, R. 2001. The protection of the subterranean environment: conservation principles and management tools. Production Services, Luxembourg, Luxembourg.
Trejo, I. 2010. Las selvas secas del Pacífico mexicano. Pages 41–51 *in* G. Ceballos, L. Martínez, A. García, E. Espinoza, J. Bezaury Creel, and R. Dirzo, editors. Diversidad, amenazas y áreas prioritarias para la conservación de las selvas secas del Pacífico de México. CONABIO and Fondo de Cultura Económica, México, D. F., México.
Valencia, S. A. 2004. Diversidad del genero *Quercus* (Fagaceae) en Mexico. Boletín de la Sociedad Botanica de Mexico 75:33–53.
Villaseñor, J. L., and M. Gual-Díaz. 2014. El bosque mesófilo de montaña en México y sus plantas con flores. Pages 221–236 *in* M. Gual-Díaz and A. Rendón-Correa, compilers. Bosques mesófilos de montaña en México. CONABIO, México, D. F., México.
Villegas Durán, G., A. Bolaños Medina, and L. Olguín Prado. 2001. La ganadería en México. Temas selectos de geografía en México, I.5.1. Instituto de Geografía, UNAM, and Plaza y Valdés, México, D. F., México.
Villers, L., and I. Trejo. 2004. Evaluación de la vulnerabilidad en los ecosistemas forestales. Pages 239–254 *in* J. Martínez and A. Fernández Bremauntz, compilers. Cambio climático: una visión desde México. Instituto Nacional de Ecología-SEMARNAT, México, D. F., México.
Walker, M., et al. 2009. Formal definition and dating of the GSSP (Global Stratotype Section and Point) for the base of the Holocene using the Greenland NGRIP ice core, and selected auxiliary records. Journal of Quaternary Science 24:3–17.
Walters, J. R., S. R. Derrickson, D. M. Fry, S. M. Haig, J. M. Marzluff, and J. M. Wunderle Jr. 2010. Status of the California condor (*Gymnogyps californianus*) and efforts to achieve its recovery. Auk 127:969–1001.
Wendt, T. 1993. Composition, floristic affinities, and origins of the canopy tree flora of the Mexican Atlantic slope rain forest. Pages 595–680 *in* T. P. Ramamoorthy, R. Bye, A. Lot, and J. Fa, editors. Biological diversity of Mexico: origins and distribution. Oxford University Press, New York, New York, USA.
Whittaker, R. H. 1960. Vegetation of the Siskiyou Mountains, Oregon and California. Ecological Monographs 30:280–338.
Wilken, G. C. 1987. Good farmers: traditional agricultural resource management in Mexico and Central America. University of California Press, Berkeley, USA.
Williams Linera, G. 2007. El bosque de niebla del

centro de Veracruz: ecología, historia y destino en tiempos de fragmentación y cambio climático. Instituto de Ecología and CONABIO, México, D. F., México.

Williams Linera, G., M. Duvell, and C. Álvarez Aquino. 2000. A relict population of *Fagus grandiflora* var. *mexicana* in a Mexican volcanic crater: structure, phenology, litterfall and dendroecology. Journal of Biogeography 27:1297–1309.

Yerena Yamallel, J. I., et al. 2014. Dynamics of carbon sequestration in abandoned grasslands of northeastern Mexico. Tropical and Subtropical Agroecosystems 17:113–121.

Zuill, H. A., and E. W. Lathrop. 1975. The structure and climate of a tropical montane rain forest and an associated temperate pine-oak-Liquidambar forest in the northern highlands of Chiapas. Anales del Instituto de Biología de la UNAM, Serie Botánica 46:73–118.

3 Wildlife as a Public and Economic Resource in Mexico

Carlos González-Rebeles Islas, Melisa Meztli Méndez Méndez, and Raul Valdez

Wildlife is an economically and culturally important resource in Mexico, especially for rural communities (Pérez-Gil et al. 1988; Ojasti 1993, 2000; Redford and Robinson 1991). Wildlife played an important role in ancient Mexican culture, appearing in art, rituals, sacrifices, and myths. Animals had material and spiritual godlike qualities, and humans believed they needed to care for and respect them (Retana 2006). This mysticism is still relevant today in rural communities, particularly in indigenous cultures. Wildlife is also important as a commodity (e.g., for food and clothing products, and raw material for tools, industry, and medicine) and for ecological services (e.g., seed dispersal, pollination, and pest control).

There are two categories of wildlife use: commercial and noncommercial. Both can be further divided into consumptive (extractive) and nonconsumptive (nonextractive) uses (Pérez-Gil et al. 1996). Commercial use involves an economic gradient, ranging from highly profitable major markets that contribute to the gross domestic product (GDP), to traditional uses whose economic relevance is local (Rozzi et al. 2001). Other aspects of wildlife are generally beyond economic considerations: teaching and research, control of nuisance species, and captive breeding of species at risk under recovery programs. These regulated activities are widely practiced in Mexico.

Wildlife-associated tourism is one of the most popular outdoor attractions (fig. 3.1). Wildlife is a renewable natural resource for which markets already exist, and there is a growing demand for it. Nonconsumptive uses, such as bird-watching, have been neglected but have great potential. The limited amount of recreational public land in Mexico presents an economic opportunity for the ejido and private sectors, which control access to Mexican landscapes with esthetic appeal and ecotourism potential, to provide outdoor recreational opportunities. The existing public lands close to urban areas are overcrowded and inadequately maintained and provide limited outdoor activities, especially relative to wildlife. There are economic incentives for land managers to become involved in outdoor recreation, which can be highly profitable and yield widespread economic benefits. The expenditures of hunters, for example, include trip-related costs (food and lodging, transportation), equipment purchases (hunting equipment, special gear), and outfitter and guide fees. However, wildlife-associated recreation must be sustainably managed to obviate any irreversible damage to wildlife populations and their habitats. Monetary gain from wildlife enterprises should ultimately be aimed at creating a conservation market—that is, stakeholders should be provided with economic incentives to sustainably manage wildlife populations. In this chapter, we describe the several facets of wildlife utilization in Mexico for cultural, economic, conservation, and ecological purposes (table 3.1).

Figure 3.1. Herd of mule deer in a UMA on Valle Colombia Ranch in Coahuila. Wildlife is an important tourist attraction. Photo by David García Solórzano.

Table 3.1. Commercial and noncommercial wildlife use in Mexico with examples of specific consumptive and nonconsumptive activities.

Form of Use	Form of Wildlife Appropriation	Wildlife Use	Examples
Commercial	Consumptive (extractive)	Hunting or harvesting wildlife populations	Sport hunting, commercial hunting, game or wildlife ranching
		Rearing and breeding wildlife in captivity	Wildlife commercial production, game or wildlife farming
	Nonconsumptive (nonextractive)	Observing and appreciating wildlife and habitats	Outdoor recreation, bird and/or wildlife watching, wildlife photography, hiking, camping, ecotourism
Noncommercial	Consumptive (extractive)	Hunting or harvesting wildlife populations	Subsistence hunting, traditional ceremonies, religious or magical rituals, traditional medicine Research, education Culling, predator and pest control
		Rearing and breeding wildlife in captivity	Assisted reproduction for the recovery of endangered species
	Nonconsumptive (nonextractive)	Observing and appreciating wildlife and its habitats	Research, education Religious or magical reverence and appreciation of nature

Hunting

Hunting is the pursuit of wild or feral animals for the purpose of killing or trapping and is one of the oldest consumptive activities practiced by humans. The incentives for hunting include sport (recreation), subsistence (personal food source), and economic gain. There are many hunted species as well as procurement methods; hunting has evolved from subsistence to a commercial and recreational pursuit (Chardonnet et al. 2002). Hunting may be the best economic incentive to initiate wildlife conservation programs because it affords an immediate income with minimal investment. It can have positive economic, sociocultural, and wildlife conservation effects if conducted in a sustainable manner (see chapter 1).

Subsistence hunting

Subsistence hunting is a traditional activity that involves harvesting wild vertebrates and invertebrates mainly for direct consumption to complement human diets. It is not regarded as a commercial activity and persists in indigenous communities; it can be an important local aspect of food security. In rural Latin America, at least 20% of the animal protein consumed by humans comes from wildlife (Redford and Robinson 1991) (fig. 3.2).

As rural communities become more connected and more susceptible to the influences of external markets, some hunters choose to sell their prey to buy domestic animal meat and industrialized products. However, in Mexico, between 30% and 50% of total meat consumption in some rural communities originates from subsistence hunting (Pérez-Gil et al. 1996, Chardonnet et al. 2002, Guerra et al. 2004). Among the most consumed animals in arid and semiarid areas are reptiles, pigeons, quail, wild turkey (*Meleagris gallopavo*), lagomorphs, rodents, white-tailed deer (*Odocoileus virginianus*), collared peccary (*Pecari tajacu*), armadillo (*Dasypus novemcinctus*), and insects (Coleoptera and Hymenoptera). In addition,

Figure 3.2. Subsistence hunting of native and exotic species such as feral pigs are an important source of meat protein for rural communities. Photo by Jorge G. Villarreal-González.

people in temperate regions capture a variety of frogs and salamanders (Mellink et al. 1986, Pérez-Gil et al. 1996, Monroy Vilchis et al. 2008).

In tropical regions, animals consumed include reptiles, cracids, columbids, meleagrids, psittacids, passerines, rodents, opossums, armadillos, skunks, lagomorphs, peccaries, deer, primates, tapir (*Tapirus bairdii*), kinkajou (*Potos flavus*), white-nosed coati (*Nasua narica*), and wild felids: jaguar (*Panthera onca*), ocelot (*Leopardus pardalis*), and puma (*Puma concolor*) (fig. 3.3) (March 1987, Pérez-Gil et al. 1996, Escamilla et al. 2000, Jorgenson 2000, Cárdenas Cervantes and Götz 2015). In central Mexico, people consume great quantities of insects. In Oaxaca, 78 insect species are part of the daily diet in some regions. People grill them, fry them, or incorporate them into traditional recipes. The insects most commonly eaten are in the orders Anoplura, Diptera, Orthoptera, Hemiptera, Homoptera, Lepidoptera, Coleoptera, and Hymenoptera. Grasshoppers (*Oxya* spp.) are a favorite and are commonly sold (Ramos Elourdy et al. 1997).

Figure 3.3. Green iguanas reared in a captive breeding facility in a UMA in Chiapas, Mexico. Several native species are sold commercially for meat and by-products. Photo by Atenas Miranda.

We know little about the biomass extracted annually by subsistence hunters in Mexico. Nor do we know how many permits are issued by government agencies. However, various studies report the existence of local community regulations regarding the use of wild species. These controls are always associated with the beliefs and customs of local inhabitants, their organizational systems, their contact with government and nongovernment institutions, and the relationship between hunting and other income-generating activities (Guerra et al. 2004). However, the lack of quantitative data documenting the numbers of each species collected at the national level is a legitimate concern to conservationists. The need for these data must be balanced with the needs of rural communities for which subsistence hunting is important to food security (Rozzi et al. 2001).

Perhaps because of the economic importance of subsistence hunting to rural families, subsistence hunting has a broader legal definition in the General Wildlife Act (LGVD) (see chapter 21), which legitimizes even some aspects of commercial hunting. The act stipulates that those engaged in subsistence hunting must be inhabitants of rural communities, and the income from the sale of animals, their parts, or derivatives must be used to meet the partial or entire basic needs of hunters and their dependents. The law also mandates that, in order to ensure the sustainable use of wildlife, government agencies at all levels must provide technical assistance, support, and training to those engaged in subsistence hunting (SEMARNAP 2000).

Sport hunting

Sport hunting is widespread and increasing in Mexico. It has been commercialized, and the economic benefits of recreational hunting can be used as an incentive to protect and restore the habitats of game species and, therefore, biodiversity (Chardonnet et al. 2002). Sport hunting consists of searching for, chasing, or stalking wild animals in order to kill them by permissible means to obtain by-products and trophy specimens (SEMARNAP 2000). It is generally practiced by the urban middle and upper classes (Ojasti 1993) and is linked to major ecotourism markets.

In Mexico, estimates of the economic impact of sport hunting include revenues from taxes collected by the federal government, the direct income generated by sport hunting, and the income earned by the services sector. The income generated from the sale of over 52,000 hunting permits during five seasons (2000 to 2005), amounted to more than 35 million Mexican pesos. The number of hunting licenses issued to national hunters increased from fewer than 10,000 during the 2000–2001 season to almost 28,000 in 2007–2008 (SEMARNAT 2009b).

In 2002, the six northern border states reported a gross income of about 2,900 million Mexican pesos, a net total revenue of 666 million pesos, and the creation of more than 43,500 jobs in various service sectors. Sport hunting is practiced principally in northern Mexico, where private hunting enterprises are widespread. Northern Mexico has adopted the American model of sport hunting because most hunters are American, but also because of climatic, socioeconomic, and cultural similarities (UANL et al. 2004).

The species hunted vary; Uribe and Arita (1998) reported 75 game mammals. Among the most economically important big game mammals in northern Mexico is the white-tailed deer (*Odocoileus virginianus texanus*) in Coahuila, Nuevo León, and Tamaulipas. The bighorn sheep (*Ovis canadensis*) is the most economically important big game species in Sonora and Baja California Sur because of the income generated for state and federal agencies and its economic impact at local and regional levels (Pérez-Gil et al. 1996). White-winged doves (*Zenaida asiatica*) were once highly sought by native and foreign hunters in Tamaulipas and are still popular in Baja California, Sinaloa, and Sonora (Uribe and Arita 1998). Criminal activity in northeastern Mexico has significantly curtailed sport hunting in that region (see chapter 20).

In Tuxtlas, Veracruz, sport hunting focuses on mammals and occasionally on birds such as the great curassow (*Crax rubra*). The mammals most in demand are the brocket deer (*Mazama* spp.), paca (*Agouti paca*), collared peccary, white-tailed deer, armadillo, and rabbits (*Sylvilagus* spp). Smaller species are hunted for food; these include the kinkajou, spider monkey (*Ateles geoffroyi*), common agouti (*Dasyprocta punctata*), coati, and cacomistle (*Bassariscus sumichrasti*) (Morales Mavil and Villa Cañedo 1998).

Migratory waterfowl hunting in Mexico is also important, considering the extensive aquatic habitats that attract waterbirds. The states with the largest numbers of ducks and geese collected, in order of importance relative to numbers shot, are Sinaloa, Chihuahua, Yucatán, the state of Mexico, and Tamaulipas (see chapter 7). Raptors, another avian group, include some migratory species that are used in falconry, though this is not a popular pastime. However, it should be strictly regulated because it may contribute to illegal trafficking and exploitation of wildlife; 10 species are used. In most cases, it is a pretext to keep birds of prey as pets (Pérez-Gil et al. 1996).

Management considerations

Management of wild populations for commercial and sport hunting in Mexico varies in intensity. It spans from direct harvest (wildlife cropping or culling) to the planned manipulation of a population and its habitat in order to maintain, increase, or control its abundance; this involves wildlife ranching. This model contrasts with that of exotic game farms, which incorporate wild and domesticated species; their effects on native habitats and wildlife can be detrimental (Weber 1993, Skinner 1996, González-Rebeles 2012).

Diversified livestock is a new management model in Mexico. This scheme combines livestock grazing with game species such as white-tailed deer, wild turkey, white-winged dove, several species of quail, and collared peccary, among others. This form of diversified wild animal husbandry requires integrated rangeland management (FIRA 1998, SEMARNAP 2000, Villareal 2000), and in some cases, ecological restoration (Arguelles 2008). This type of integrated land use, combining livestock production with native biological resources, has been the main focus of private and public wildlife management in Mexico, along with other types of UMAs (see chapter 21).

Commercial Trade in Wildlife

Wildlife commercialization is the selling and buying of free-roaming and captive wild animals and their parts. It involves diverse activities ranging from informal local trade to the organized national and international wildlife market focused on the sale of bushmeat, teeth, antlers, skins, or live specimens. Legal international trade in live wildlife, not including edible fish, amounts to US$15 billion. The illegal market is even more profitable; it includes, for example, the sale of 15 million reptiles or their skins, and half a million parrots annually (Bambaradeniya et al. 2009). The approach varies, ranging from the harvest of specimens in their natural habitats to the direct or indirect management of wild populations of particular interest in order to gain economic benefit. Economic incentives may generate legal but also illegal trade in wildlife; the latter is not sustainable. Emphasis on economic values must be approached cautiously because it may impede long-term habitat enhancement in favor of market values and unsustainable exploitation.

Bushmeat or wild meat is derived from wild vertebrates, excluding fish, obtained from hunting activities. It is usually sold in open markets and is common in rural areas as a complement to income. However, as in most Latin American countries, the economic benefits favor commercial distributors, intermediaries, and collectors, while hunters receive the lowest share (Ojasti 1993). Although the impact of these activities has not been studied, it is assumed that bushmeat sales are often conducted illegally. Bushmeat in Mexico is also sold in urban centers—for example, in restaurants in Catemaco, Veracruz, where up to 19 species are offered; these include mostly mammals, but also several species of reptiles and birds. Several wildlife species are hunted in the south-central region of the central Mexican plateau for the bushmeat market (Mellink et al. 1986).

Street sales of several illegal species and their byproducts have been documented in Veracruz (Morales Mavil and Villa Cañedo 1998). Similar commercial activities occur in San Cristóbal de las Casas, Chiapas (Aranda 1991). On a smaller scale in central Mexico, people sell some species for their parts for several purposes: as ingredients for medicinal recipes, as amulets, and as source material for artisanal leather and fur products. Hunters capture more than 30 wild species of songbirds and ornamental birds for this market. It is common to find permanent vendor stands selling wildlife and animal derivatives in regional public markets (Mellink et al. 1986).

Until the middle of the twentieth century, the commercial use of spotted cats, mustelids, and reptiles for the fur and skin trade was of international importance. However, such activity decreased following the ratification of the Convention on International Trade in Endangered Species of Wild Fauna and Flora (CITES) in 1973 (Chardonnet et al. 2002). Despite these international restrictions, Mexico continues to play a key role in this market. The national and international trade in hides and skins in Mexico has focused primarily on leather and other crafts, fine furs for clothing and accessories, and the footwear industry. These activities use several native species of mammals and reptiles; however, hides from other countries are also imported, making it difficult to determine local harvest estimates. In addition, the market fluctuates frequently in both volume and price, depending on fashions or the abundance of the product. The scarcer the product, the higher the price, and this can lead to overuse and illegal trade (Pérez-Gil et al. 1996).

Between 1980 and 2001, Mexico was one of the main centers for the manufacture and trade of reptile skins. Most of the raw material is imported, mainly from Argentina and Indonesia. Its legality is questionable; in the indicated period, over 50% of the skins of native species were of unknown origin. The quantity of exports was much smaller compared to the volume of imports and reexports. The predominant species included sea turtles (*Chelonia* spp.), though activity diminished after Mexico signed the CITES agreement in 1991; common caimans (*Caiman crocodilus*); and rattlesnakes (*Crotalus* spp.); there were also small quantities of crocodiles (*Crocodylus* spp.) and green iguanas (*Iguana iguana*). Most of these were probably wild trapped. Therefore, the possibility of captive breeding to exploit the potential of this trade should be considered (Arroyo Quiroz et al. 2007).

The local skin market in Mexico is problematic, requiring more attention and regulation. San Cristóbal and Comitán are collection and distribution centers in the highlands and lowlands of Chiapas and the border with Guatemala. The native species marketed include protected and endangered species, which are openly displayed and traded. The skins of these species are used to produce dress accessories, toys, and taxidermy mounts. Twenty-five species belonging to six orders have been identified: Primates, Cingulata, Lagomorpha, Rodentia, Carnivora, and Artiodactyla (Aranda 1991).

Trade in live specimens has been practiced since the pre-Hispanic period. The best-known form of wildlife live trapping that is currently allowed is the commercialization of songbirds and ornamental birds. It provides employment for 3,400 families who specialize in the capture, transport, and sale of about 50 species of birds, mostly songbirds; it was formerly legal to capture pigeons and parrots. During the 1996–1997 season, an estimated 608 permits were issued, most of them to peasants (Contreras et al. 2001). In order to ensure their sustainable use, organized associations of wildlife trappers have been allowed since 2004, under the legal mantle of UMAs (see chapter 21) (SEMARNAT 2009a).

The caged songbird trade generates significant demand nationally and internationally. It has a complex market chain structure. The bird catcher, or *pajarero*, sells directly to the consumer or to an intermediary. The latter can sell to street vendors, to commercial establishments, or to large wholesalers involved in the international trade. In this process, the price of the specimens increases exponentially. The trade in live specimens has serious conservation disadvantages and is not sustainable, since the communities involved earn low profits and thus lack an incentive to manage habitats and populations rationally. For example, a pajarero may earn several US dollars for a rare species, but the price may increase to over a thousand dollars in international markets. The market is regulated, but enforcement is poor. This, coupled with the amplitude and dispersion of the market chain, and continued demand, promotes a significant illegal trade. Furthermore, the lack of population studies, along with the deterioration of habitats, increases the risk of overexploitation (Pérez-Gil et al. 1996). According to regulations in place since 2000, trapping is regulated largely as a subsistence activity.

Psittacids are favored pets in Mexico. Mexico is one of the 10 largest exporters of parrots in the world. Of the 22 species in Mexico, 6 are endemic. Current harvest restrictions have affected foreign trade, but not local trade (Iñigo Elías and Ramos 1991, Cantú et al. 2007). Between 65,000 and 78,500 parrots are captured each year, with a 75% mortality rate associated with the marketing chain. In addition, the monitoring and forfeiture of captured parrots in the illegal market is inefficient; the rate of forfeitures to inspectors from PROFEPA, the federal agency responsible for enforcing wildlife laws, amounts to only 2% of the annual number of smuggled birds (Cantú et al. 2007).

During the 1970s and 1980s, the US was the main destination of parrot exports; currently only a small percentage (4–14%) of the annual illegal catch is smuggled out of the country. Thus, the greatest demand is from internal trade. In recent years, exotic parrot species have become common in the legal and illegal trade (Cantú et al. 2007). Some conservationists favor the sale of exotic species because it may result in a lower demand for native species; however, the accidental release of exotics poses a threat to Mexican species and their ecosystems.

The crafts market is another important focus for commercial hunting by indigenous communities and mestizos. It involves 125 species encompassing 57 families and 91 genera, including marine invertebrates and vertebrates, insects, amphibians, reptiles, birds, and mammals. Their use varies according to regional artisanal traditions. In the tropical zone, the emphasis is on the collection of feathers and the extraction of skins from amphibians such as toads (*Bufo marinus* and *B. valliceps*) and frogs (*Rana palmipes* and *R. berlandieri*); reptiles such as the boa (*Boa constrictor*), rattlesnakes, green iguana, and nauyaca (*Bothrops asper*); and several mammals. The white-tailed deer is the most exploited large mammal in the temperate zone, where, in addition to skins, the antlers and bones are used to make everyday objects and toys (Bravo and López 1999). Morelos and Chiapas are known for their handcrafted clothing, shoes, and various accessories made from the hides and skins of mammals and reptiles. Also, some artisans specialize in taxidermy and feather art; the latter has been practiced since the pre-Hispanic period (Pérez-Gil et al. 1996). In arid and semiarid regions, the Seri Indians of Sonora make dolls from the bones of hares (*Lepus* spp.); they also sew instruments from the skin and bones of deer (*Odocoileus hemionus* and *O. virginianus*) and dance headdresses from the tails, hooves, and tendons. The skins of Gila monsters (*Heloderma suspectum*) and rattlesnakes are used to make belts and necklaces, the latter with vertebrae. In coastal areas, necklaces and other accessories are crafted from marine invertebrates: shells, snails, scallops, and octopuses (Bravo and López 1999).

A close association exists between the areas that provide the largest amount of raw materials, the areas inhabited by indigenous communities, and the areas with high levels of marginalization. This highlights the importance of traditional knowledge and the economic importance of the use and management of wildlife resources (Bravo and López 1999). Some communities, until two decades ago, survived entirely from the handicrafts trade. For example, the Nahua potters of the Alto Balsas region in Guerrero created art characterized by lyrical drawings of birds, wildlife, and plants on bark paper (*amate*) and animal skins. They also made masks and wooden animal figures adorned with the hair, teeth, and skins of wildlife (Good 2009).

The international commercialization of swim bladders from the totoaba (*Totoaba macdonaldi*) is an urgent concern. Totoaba are fish endemic to the Gulf of California and are critically endangered. The demand for their dried swim bladders, which are an ingredient in Asian soup, has imperiled the already declining totoaba populations; bladders sell for thousands of US dollars in China. In the process of catching fish, fishers also capture vaquita porpoises (*Phocoena sinus*), the world's smallest cetacean; the vaquita is also endemic, and the most endangered marine mammal. Although it is illegal to use gill nets over large areas of the vaquita's range, gill nets continue to be used in restricted areas to ensnare the imperiled totoaba. At present, exploitation of the totoaba is uncontrolled (Cisneros-Montemayor and Vincent 2016).

Noncommercial Harvest from Wildlife Populations

Mexico is a country of great cultural richness; at least 26 indigenous regions are distributed across a wide variety of ecosystems (Toledo et al. 2002). Many cultures still practice rites and ceremonies associated with reverence or simply respect for wildlife and their habitats; this influences their actions to protect wildlife resources and to develop rational forms of use. Reverence for wildlife by ethnic groups is evident in their various artistic manifestations and symbolic concepts of nature (e.g., the worship of water, stars, and the fertility of the land), as well as in their various magical-religious manifestations directly related to wildlife. Their activities are both consumptive and noncomsumptive: they harvest wildlife for meat protein and its by-products and derivatives, but they also simply express their appreciation and reverence for wildlife (Pérez-Gil et al. 1996).

Unfortunately, these traditions have been disappearing because of increasing contact with Western cultures. Several ethnobiological studies have recorded the perceptions, attitudes, and knowledge of various Mexican indigenous groups, as well as how they interact with and influence their environment (Caballero 1982, March 1987, Atran et al. 1999,

Haenn 1999, Hellier et al. 1999, Bello et al. 2002, Toledo et al. 2002, Toledo and Boege 2010). The use of wildlife is evident in cultural traditions such as costumes for ceremonial dances; the Seri of Sonora use the skin, hooves, head, and antlers of deer for their famous deer dance (Bravo and López 1999). The concheros dance of the Nahua region makes use of feathers and animal skins for clothes and musical instruments. The Lacandon groups in Chiapas, such as the Najá, use rawhides of agouti and brocket deer (*Mazama americana*) to make ceremonial drums now sold as handicrafts (March 1987).

Traditional medicine

There are few specific studies on the medicinal properties of wildlife compared to those of plants. The literature on Mexican flora and fauna includes studies by Barajas (1951), Hernández (1959), De la Cruz (1991), Ventura (2000), and Torres (2003). Traditional medicine is common among ethnic groups and mestizos and is based on traditional knowledge. Initially, it was a consumptive activity practiced at a subsistence level, but it has recently turned into a locally oriented commercial activity. A study of Tzotzil and Tzeltal native groups in the highlands of Chiapas (Enríquez et al. 2006) identified 74 species (1 species each of mollusk, annelid, arachnid, and fish; 3 crustaceans; 9 birds; 11 amphibians; 12 insects; 15 reptiles; and 20 mammals) used to treat more than 128 diseases recognized by Western medicine, as well as cultural diseases based on pre-Hispanic beliefs. In the state of Mexico, 18 species are used for medicinal purposes (3 reptiles, 5 birds, and 10 mammals) (Monroy Vilchis et al. 2008).

In some cases these remedies, most of which are homemade, are mass-produced and marketed as concoctions or ointments in specialty stores in urban centers. For several decades there have been reports of the commodification of various wild animals and their by-products in Mexico City and surrounding states. The most common are echinoderms and other marine invertebrates, including mollusks, helminths, and arthropods, and fish and terrestrial vertebrates. The use of birds is extensive; in addition to their supposed medicinal properties, they are prescribed as an amulet to find a lover and are purported to have aphrodisiacal properties (Barajas 1951).

In rural communities, a single animal species can serve different purposes: food, medicine, amulets, pets, arts and crafts, and ritual use, among others (Mellink et al. 1986, Juárez et al. 2015). The traditional ecological knowledge and techniques involved in these practices are part of an intangible cultural heritage. Their sociocultural value is difficult to quantify, but it represents an enormous wealth that should be safeguarded (UNESCO 2003).

Education and research

Wildlife also has important educational and research value (fig. 3.4). Scientific collecting is the temporary or permanent capture, removal, or extraction of biological material from the wild in order to obtain scientific knowledge, establish taxonomic inventories, or increase the collections of scientific institutions or museums for purposes of research, education, or dissemination. In Mexico, this activity does not include access to genetic resources for use in biotechnology and bioprospecting (DOF 2006). Another use is in observational studies of individuals, or of the ecological processes of which they are part; this does not involve the collection of organisms. In both education and research, the benefits are public and social, because the knowledge obtained can be used in planning and managing natural resources (Cañas et al. 2008).

Wildlife Control

This is a specialized extractive activity whose objective is to control the population growth of a species to

Figure 3.4. UMA's in rural communities may provide additional economic benefits and educational and conservation opportunities not only from sport hunting but from other activities such as nature photography. Echinocereus stramineus in the state of Nuevo Leon. Photo by Jorge G. Villarreal-González.

address damage concerns; it has no direct commercial objective. The control of harmful species involves relocating specific animals, controlling population growth, and driving away, dispersing, or hindering wild or domestic animals from entering certain areas in order to reduce damage to the environment, to other species, and to humans. This damage can result from changes in the animals' natural habitat or biology, or it can occur because they are outside their natural distribution (DOF 2000, 2006). This may also involve coordinated research in conjunction with educational objectives and information dissemination programs.

This modern approach contrasts with predator control objectives from 1954 to the early 1970s, especially in northern Mexico, in which carnivorous mammals and rodents were eradicated for the benefit of livestock enterprises, sometimes with chemicals that poisoned any animals that consumed the bait or drank contaminated water (e.g., 1080) (Retana 2006). This practice led to the death of nontarget species (domestic animals and carnivorous birds), and to an increase in predator prey. Strychnine and cyanide were also used to eradicate predators; strychnine was even used to promote the population growth of exotic species such as the chukar partridge (*Alectoris chukar*) and the ring-necked pheasant (*Phasianus colchicus*) in Tlaxcala, as well as the native Montezuma quail (*Cyrtonyx montezumae*) (Morales 1970).

The Mexican government promotes programs to prevent the involuntary killing of wildlife, such as at airports, where its presence represents a potential risk for airplanes during landing and takeoff. The Mexico City airport is located in a migratory bird corridor and is adjacent to the Alameda Oriente Ecological Park, the former Texcoco Lake, the Bordo Poniente landfill, and Nabor Carrillo Lake. Since 2003, animal control has been performed using birds of prey and hunting dogs, as well as pyrotechnics, to drive away about 40 species of wildlife, especially local and migratory birds (herons, grackles, pigeons, sparrows, owls, and ducks, among others). Rodents are usually controlled by trapping, netting, and baiting (Román 2009). At the Campeche City airport, a population of white-tailed deer living nearby causes hazards. Capturing them has been encouraged as an alternative to lethal control; they are donated to scientific institutions and zoos (SEMARNAT 2004).

Another instance of wildlife control is the management of exotic and invasive species on islands, such as mice, cats, and goats, as well as mink (*Mustela vison*), wild rabbits, and white-tailed deer. Between 2003 and 2006, at least 16 species of exotic mammals were confirmed on 27 islands in northwestern Mexico. Introduced invasive species must be controlled or eradicated, as they can potentially cause the extinction of insular species (Aguirre et al. 2009).

Captive Breeding

Captive breeding refers to the confined production of wild species through intensive management and husbandry techniques (fig. 3.5). Wildlife husbandry commenced thousands of years ago and has continued until the present; many different species of wild and exotic vertebrates are raised, as well as mollusks and insects. A classic example of the varieties that gave rise to current domestic species is the Mexican turkey (Chardonnet et al. 2002). Domestication requires the consideration of a number of biological parameters for each wild species, such as social behavior, diet, reproductive and growth rates, space requirements, husbandry system design, and medical attention. It is a commercial use of wildlife that focuses on the production of companion animals or pets and breeding stock, including products derived from animals, such as meat and hides.

Captive breeding is also used in the recovery of species whose wild populations are at risk or declining. The goal in this case is not to benefit from commercial gain but to work toward eventual

Figure 3.5. Captive-breeding facilities that produce mule deer for sport hunting are common in Sonora, Mexico. Photo by J. Alfonso Ortega-S.

reintroduction. This model of captive breeding involves rearing both native and exotic wildlife, as well as those species in the process of domestication—for example, red deer (*Cervus elaphus*) and ostrich (*Struthio camelus*), and European and Asiatic pheasants and partridges (Weber 1993, Mockrin et al. 2005, González-Rebeles 2012). An example of a native species that has been successfully managed using captive breeding to augment and establish new wild populations is the desert bighorn sheep (see chapter 20).

In recent years, Mexican crocodiles (*Crocodylus moreletii*), native to the Gulf of Mexico, have been bred in captivity for the leather industry, for meat, and for the taxidermy trade. Although these farms have been commercially successful, they have been criticized because they are concentrated on the Pacific coast, where Mexican crocodiles are not native. Escaped individuals are considered exotic predators that disrupt local food chains and are potential competitors with the native American crocodile (*C. acutus*) of the Pacific coast. They are also a potential disease vector, and the two crocodile species have hybridized (Álvarez Romero et al. 2008).

Traditional captive use of the black-bellied whistling duck (*Dendrocygna autumnalis*) occurs in villages of Coatzacoalcos, Veracruz; the duck is bred for its eggs and meat. It is used for both hunting and captive breeding (Feekes 1991). Also, March and Cuarón (1987) described a program being developed for the management and captive breeding of the white-lipped peccary (*Tayassu pecari*) in Chiapas.

Several exotic wild ungulates are bred intensively in Mexico for sport hunting; however, there is scant information on their management and numbers. These species include Barbary sheep (*Ammotragus lervia*), chital deer (*Axis axis*), domestic sheep (*Ovis aries*), blackbuck antelope (*Antilope cervicapra*), red deer, fallow deer (*Dama dama*), scimitar oryx (*Oryx dammah*), common eland (*Taurotragus oryx*), sika deer (*Cervus nippon*), alpine ibex (*Capra ibex*), and European wild boar. Native species are also captive bred for sport-hunting trophies because they command high market prices; in some cases they have been introduced outside their original distributional range. These include bighorn sheep (fig. 3.6), some subspecies of white-tailed deer (*texanus* and *carmines*), and mule deer (*Odocoileus hemionus*) (Álvarez Romero et al. 2008; C. González-Rebeles, personal

Figure 3.6. This captive-breeding facility produces bighorn sheep for sport hunting in Sonora, Mexico. Photo by J. Alfonso Ortega-S.

observation). These are examples of captive-breeding programs that fall under the category of intensive UMAs. There are many captive-breeding farms in central and southern Mexico that produce meat and meat products, especially of white-tailed deer and collared peccary; however, these are poorly documented (Rojo et al. 2007, 2008).

Nonconsumptive Wildlife Use

Various forms of wildlife use involve no direct appropriation of the resource (nonconsumptive) but may have commercial value. These include recreational activities such as camping; viewing and photographing wildlife (fig. 3.7); environmental studies involving nonconsumptive research; educational programs; and activities that fulfill psychological and spiritual needs (e.g., traditional native rites).

Ecotourism

This environmentally responsible form of tourism consists of outdoor activities that are increasingly standardized and regulated. These activities aim to establish contact with nature and related cultural manifestations in order to promote appreciation and encourage protective actions. Ecotourism can become an efficient and sustainable alternative for income generation and rural development and can also support the conservation of biodiversity, if revenue is reinvested in the protection of wildlife and its habitats. However, intense tourist activities can lead to the rapid erosion of biological and cultural diversity, especially when the tourism infrastructure introduces foreign technologies and attitudes (Massardo et al.

Figure 3.7. The cheje woodpecker (*Melanerpes aurifrons*) in the San Joaquin UMA in Acala, Chiapas, provides an opportunity for nonconsumptive use of wildlife, such as bird watching and photography. Photo by Amado Rios Valdez.

2001, Secretariat CBD et al. 2009). Thus, tourism is a complex undertaking—it requires considering the external and internal political environment, including corruption; establishing management strategies and administrative structures; confronting sustainability issues, along with ecological and economic aspects; respecting local social and cultural values and traditions; and allowing stakeholder and community participation in the project, policy development, and management (Prideaux 2014).

In Mexico, this activity has been increasing significantly in several rural regions. For example, the community of La Ventanilla, Oaxaca, successfully organizes and manages a community-based ecotourism program based on mangroves and observing wild and captive crocodiles. The human population is small and there is consensus regarding the importance of the relationship between nature conservation and human welfare. The community established an appropriate sociopolitical structure, formed a local cooperative, and works in coordination with governmental and nongovernmental organizations. The economic benefits remain in the community and are distributed equitably among the participants. Those who do not form part of the cooperative also support conservation efforts because they indirectly benefit economically from ecotourism (Foucat 2002).

However, some areas are especially sensitive to disturbance, particularly islands. Tourism should be strictly regulated and even prohibited during certain periods in order to reduce damaging effects (e.g., disturbance of populations during the breeding season, trampling of vegetation, and introduction of diseases, invasive species, and pollutants). For example, the more than 230 islands, islets, rock outcroppings, and coastal communities in northwestern Mexico have attracted an increasing number of tourists, resulting in environmental problems. Unfortunately, regulations have not kept pace (Tershy et al. 1999). An example is the boom in whale watching in Laguna San Ignacio and Magdalena Bay, Baja California Sur, which has been commercialized to an extreme degree. The growing migrant community involved in providing ecotourism lacks the organizational and planning skills to sustainably manage the coastal environment; also, corruption is rife. The economic benefits do not remain in the area because there are external service providers. In addition, the numerous tourists and associated boats during the whale-watching season may be detrimentally affecting whales and creating pollution, among other problems. The intense competition for the tourist trade reduces profits and forces local workers to continue extractive activities such as fishing (Young 1999a, 1999b).

Final Considerations

Wildlife remains a fundamental resource that can contribute to national economic development by providing food security for indigenous communities; complementing the economic development of rural communities; conserving ancient cultural traditions; complementing local productive activities; contributing to the GDP through its commercialization in international markets; and providing environmental services. The basic tools for the public management of wildlife populations and their habitats are the extensive and intensive UMAs. However, UMAs should not be used as an isolated resource management model; they should be linked with other types of land use in order to develop a comprehensive strategy for the integration of wildlife resource and production methods (see chapter 21).

The integral and diversified management of natural resources could benefit from the worldview that many indigenous communities still maintain in relation with nature. Several indigenous communities in

Mexico are regionally organized to employ traditional strategies that allow for a diversified and complementary use of their natural resources, combining them with modern forms of conventional production. They work in aquatic and terrestrial ecosystems and obtain a variety of products from plant collecting, fishing, and hunting, in combination with activities such as tree cultivation, animal husbandry, horticulture, and agriculture (Caballero 1982, Challenger 1998, Bocco et al. 2000, Primack et al. 2001, Toledo et al. 2008, Toledo and Boege 2010).

The major challenge for the management of wildlife resources in Mexico is to promote existing market opportunities at the regional level and to promote access to national and international markets, while recognizing the importance of local economies in providing for the welfare and economic development of thousands of communities. The forms of extensive and intensive wildlife management that function within the UMA concept in Mexico offer the opportunity to integrate the original vision of diversifying production through the rational use of wildlife and flora in rural communities, which are major stakeholders, with conventional and unconventional agricultural and livestock production practices.

Literature Cited

Aguirre, A., et al. 2009. Especies exóticas invasoras: impactos sobre las poblaciones de flora y fauna, los procesos ecológicos y la economía. Pages 277–318 *in* Capital natural de México. Volume 2. Estado de conservación y tendencias de cambio. CONABIO, México, D. F., México.

Álvarez Romero, J. G., R. A. Medellín, A. Oliveras, H. Gómez, and O. Sánchez. 2008. Animales exóticos en México: una amenaza para la biodiversidad. CONABIO, Instituto de Ecología-UNAM, and SEMARNAT, México, D. F., México.

Aranda, M. 1991. Wild mammal skin trade in Chiapas, Mexico. Pages 174–180 *in* J. G. Robinson and K. H. Redford, editors. Neotropical wildlife use and conservation. University of Chicago Press, Illinois, USA.

Arguelles, L. 2008. UMAs extensivas de venado cola blanca como instrumento económico de restauración ecológica en el noreste de Morelos. Thesis, UNAM, México, D. F., México.

Arroyo Quiroz, I., R. Pérez-Gil, and N. Leader Williams. 2007. Mexico in the international reptile skin trade: a case study. Biodiversity and Conservation 16:931–952.

Atran, S., D. Medin, N. Ross, E. Lynch, J. Coley, E. Ucan, and V. Vapnarsky. 1999. Folk ecology and commons management in the Maya Lowlands. Proceedings of the National Academy of Sciences 96:7598–7603.

Bambaradeniya, C., et al. 2009. The illustrated atlas of wildlife. University of California Press, Berkeley, USA.

Barajas, E. 1951. Los animales usados en la medicina popular mexicana. Imprenta Universitaria, México, D. F., México.

Bello, E., E. Estrada, L. Mendoza, A. Segundo, and L. Sánchez. 2002. Transdisciplina y sustentabilidad Maya. Ciencia Ergo Sum 9:260–272.

Bocco, G., A. Velázquez, and A. Torres. 2000. Ciencia, comunidades indígenas y manejo de recursos naturales. Un caso de investigación participativa en México. Interciencia 25:64–70.

Bravo, C., and A. López. 1999. Inventario de especies vegetales y animales de uso artesanal. Biodiversitas 5:9–14.

Caballero, J. 1982. Notas sobre el uso de los recursos naturales entre los antiguos Purépecha. Biotica 7:31–42.

Cañas, R., R. Ahuatzi, M. España, and J. Soberón. 2008. Situación legal de la recolecta científica. Pages 215–225 *in* Capital natural de México. Volume 1. Conocimiento actual de la biodiversidad. CONABIO, México, D. F., México.

Cantú, J. C., M. E. Sánchez, M. Grosselet, and J. Silva. 2007. Tráfico ilegal de pericos en México: una evaluación detallada. Defenders of Wildlife and Teyeliz, México, D. F., México.

Cárdenas Cervantes, J. M., and C. M. Götz. 2015. Tafonomía de desechos: modificaciones de restos faunísticos en un solar maya actual y su aplicación a los estudios zooarqueológicos. Revista Archaeobios 1:137–157.

Challenger, A. 1998. Utilización y conservación de los ecosistemas terrestres de México: pasado, presente y futuro. CONABIO, Instituto de Biología-UNAM, and Agrupación Sierra Madre, México, D. F., México.

Chardonnet, P., B. des Clers, J. Fisher, R. Gerhold, F. Fori, and F. Lamarque. 2002. The value of wildlife. Revue Scientifique et Technique 21:15–51.

Cisneros-Montemayor, A. M., and A. Vincent. 2016.

Science, society, and flagship species: social and political history as keys to conservation outcomes in the Gulf of California. Ecology and Society 21:9.

Contreras, A., J. García, A. Guzmán, and J. González. 2001. Aprovechamiento de las aves cinegéticas, de ornato y canoras de Nuevo León, México. Ciencia 4:462–469.

De la Cruz, M. 1991. Libellus de medicinalibus indorum herbis. Manuscrito Azteca de 1552 según traducción latina de Juan Badiano. Fondo de Cultura Económica and Instituto Mexicano del Seguro Social, México, D. F., México.

DOF. 2000. Ley General de Vida Silvestre. DOF. http://www.cddhcu.gob.mx/LeyesBiblio/ref/lgvs/LGVS_orig_03jul00.pdf. Accessed 5 Nov 2009.

DOF. 2006. Reglamento de la Ley General de Vida Silvestre. DOF. http://www.cddhcu.gob.mx/LeyesBiblio/regley/Reg_LGVS.pdf. Accessed 5 Nov 2009.

Enríquez, P., R. M. Mariaca, O. G. Retana, and E. J. Naranjo. 2006. Uso medicinal de la fauna silvestre en los altos de Chiapas, México. Interciencia 31:491–499.

Escamilla, A., M. Sanvicente, M. Sosa, and C. Galindo Leal. 2000. Habitat mosaic, wildlife availability, and hunting in the tropical forest of Calakmul, Mexico. Conservation Biology 14:1592–1601.

Feekes, F. 1991. The black-bellied whistling duck in Mexico: from traditional use to sustainable management? Biological Conservation 56:123–131.

FIRA. 1998. Ranchos cinegéticos. Oportunidad de diversificación, ganadería sustentable. Boletín Informativo FIRA 30:1–100.

Foucat, V. S. A. 2002. Community-based ecotourism management: moving towards sustainability in Ventanilla, Oaxaca, Mexico. Ocean and Coastal Management 45:511–529.

González-Rebeles, C. 2012. Zootecnia aplicada a la fauna silvestre. Pages 295–315 *in* M. E. Trujillo, editor. Introducción a la zootecnia. Second edition. UNAM, México, D. F., México.

Good, C. 2009. El comercio de artesanías: los nahuas del Alto Balsas. El estado del desarrollo económico y social de los pueblos indígenas de Guerrero. UNAM and Secretaría de Asuntos Indígenas del Gobierno del Estado de Guerrero. http://www.nacionmulticultural.unam.mx/Edespig/diagnostico_y_perspectivas/RECUADROS/CAPITULO%205/4%20EL%20COMERCIO%20DE%20ARTESANIAS.pdf. Accessed 10 May 2010.

Guerra, M., E. Naranjo, F. Limón, and R. Mariaca. 2004. Factores que intervienen en la regulación local de la cacería de subsistencia en dos comunidades de la selva Lacandona, Chiapas. Pages 540–550 *in* R. Bodmer, editor. Memorias del VI Congreso Internacional sobre Manejo de Fauna Silvestre en la Amazonía y Latinoamérica. Universidad Nacional de la Amazonia Peruana, Durrell Institute of Conservation and Ecology, University of Kent-Canterbury, and Wildlife Conservation Society, Iquitos, Peru.

Haenn, N. 1999. The power of environmental knowledge, ethnoecology and environmental conflicts in Mexican conservation. Human Ecology 27:477–491.

Hellier, A., A. C. Newton, and S. Ochoa. 1999. Use of indigenous knowledge for rapidly assessing trends in biodiversity: a case study from Chiapas, Mexico. Biodiversity and Conservation 8:869–889.

Hernández, F. 1959. Historia natural de Nueva España. UNAM, México, D. F., México.

Iñigo Elías, E., and M. A. Ramos. 1991. The psittacine trade in Mexico. Pages 380–392 *in* J. G. Robinson and K. H. Redford, editors. Neotropical wildlife use and conservation. University of Chicago Press, Illinois, USA.

Jorgenson, J. P. 2000. Wildlife conservation and game harvest by Maya hunters in Quintana Roo, Mexico. Pages 251–266 *in* J. G. Robinson and E. L. Bennett, editors. Hunting for sustainability in tropical forests. Columbia University Press, New York, New York, USA.

Juárez M., A., C. González-Rebeles I., A. Castillo, E. García F., and M. de J. Ordóñez D. 2015. La vida silvestre manejada como recursos de uso común: estudio de caso en México. Tropical and Subtropical Agroecosystems 18:313–331.

March, I. J. 1987. Los lacandones de México y su relación con los mamíferos silvestres: un estudio etnozoológico. Biótica 12:43–56.

March, I. J., and A. D. Cuarón. 1987. Estado del conocimiento actual sobre *Tayassu pecari* (Artiodactyla, Tayassuidae) en México y actividades de investigación en desarrollo. Boletín Informativo de la Facultad de Ciencias de la Tierra y el Mar, Universidad Nacional de Costa Rica, Heredia, Costa Rica 3:24–40.

Massardo, F., O. Dollenz, and R. Rozzi. 2001. Ecoturismo en el cono austral de América. Pages 303–310 *in* R. Primack, R. Rozzi, P. Feinsinger, R. Dirzo, and F. Massardo, coordinators. Fundamentos de conservación biológica: perspectivas latinoamericanas. Fondo de Cultura Económica, México, D. F., México.

Mellink, E., J. R. Aguirre, and E. García. 1986. Utilización de la fauna silvestre en el Altiplano Potosino-Zacatecano. Colegio de Postgraduados, Montecillo, México.

Mockrin, M. H., E. L. Bennett, and D. T. La Bruna. 2005. Wildlife farming: a viable alternative to hunting in tropical forests? Working Paper No. 23. Wildlife Conservation Society, New York, New York, USA.

Monroy Vilchis, O., L. Cabrera, P. Suárez, M. Mariela Zarco González, C. Rodríguez Soto, and V. Urios. 2008. Uso tradicional de vertebrados silvestres en la Sierra Nanchititla, México. Interciencia 33:308–313.

Morales, C. 1970. Control de la fauna perjudicial. Pages 218–241 *in* R. Hernández, editor. Fauna silvestre: expresiones y planteamientos de un recurso, 1964–1970. Subsecretaría Forestal y de la Fauna, México, D. F., México.

Morales Mavil, J. E., and J. T. Villa Cañedo. 1998. Notas sobre el uso de la fauna silvestre en Catemaco, Veracruz, México. Acta Zoológica Mexicana 73:127–143.

Ojasti, J. 1993. Utilización de la fauna silvestre en América Latina: situación y perspectiva para un manejo sostenible. Guía FAO Conservación No. 25. FAO, Rome, Italy.

Ojasti, J. 2000. Manejo de fauna silvestre neotropical. Series No. 5. Smithsonian Institution/Monitoring and Assessment of Biodiversity Program, Washington, DC, USA.

Pérez-Gil, R., et al. 1988. Fauna silvestre y áreas naturales protegidas. Universo Veintiuno, México, D. F., México.

Pérez-Gil, R., F. Jaramillo, A. M. Muñiz, and M. G. Torres. 1996. Importancia económica de los vertebrados silvestres de México. CONABIO, México, D. F., México.

Prideaux, B. 2014. Findings and research issues. Pages 304–310 *in* B. Prideaux, editor. Rainforest tourism, conservation and management. Routledge, London, United Kingdom.

Primack, R., R. Rozzi, P. Feinsinger, R. Dirzo, and F. Massardo, coordinators. 2001. Fundamentos de conservación biológica: perspectivas latinoamericanas. Fondo de Cultura Económica, México, D. F., México.

Ramos Elourdy, J., J. M. Pino, E. E. Prado, M. A. Pérez, J. L. Otero, and O. L. de Guevara. 1997. Nutritional value of edible insects from the state of Oaxaca, Mexico. Journal of Food Composition and Analysis 10:147–157.

Redford, K. H., and J. G. Robinson. 1991. Subsistence and comercial uses of wildlife in Latin America. Pages 6–23 *in* J. G. Robinson and K. H. Redford, editors. Neotropical wildlife use and conservation. University of Chicago Press, Illinois, USA.

Retana, O. G. 2006. Fauna silvestre de México: aspectos históricos de su gestión y conservación. Universidad Autónoma de Campeche and Fondo de Cultura Económica, México, D. F., México.

Rojo C., A., J. L. Cruz R., and C. Cortés. 2008. Plan de manejo tipo de pecarí de collar en climas aridos y semiáridos del norte de México. SEMARNAT and DGVS, México, D. F., México.

Rojo C., A., J. L. Cruz R., G. Solano C., and R. Hernández L. 2007. Plan de manejo tipo de venado cola blanca en zonas templadas y tropicales de México. SEMARNAT and DGVS, México, D. F., México.

Román, J. 2009. Aves silvestres y migratorias, un peligro para el aeropuerto del D. F. La Jornada. 26 de octubre de 2009. http://www.jornada.unam.mx/2009/10/26/index.php?section=sociedad&article=039nlsoc. Accessed 10 May 2009.

Rozzi, R., R. Primack, and F. Massardo. 2001. Valoración de la biodiversidad. Pages 255–290 *in* R. Primack, R. Rozzi, P. Feinsinger, R. Dirzo, and F. Massardo, coordinators. Fundamentos de conservación biológica: perspectivas latinoamericanas. Fondo de Cultura Económica, México, D. F., México.

Secretariat CBD (Convention on Biological Diversity), World Tourism Organization, and the United Nations Environment Programme. 2009. Tourism for nature and development: a good practice guide. Montreal, Canada. http://www.cbd.int/development/doc/cbd-good-practice-guide-tourism-booklet-web-en.pdf. Accessed 5 Feb 2010.

SEMARNAP. 2000. Estrategia nacional para la vida silvestre. Logros y retos para el desarrollo sustentable 1995–2000. DGFS and INE, México, D. F., México.

SEMARNAT. 2004. Revoca SEMARNAT permiso para matanza de venados cola blanca. La Crónica de Hoy. 8 de mayo de 2004. http://www.cronica.com.mx/nota.php?id_nota=123688. Accessed 10 May 2010.

SEMARNAT. 2009a. Plan de manejo tipo para aves canoras y de ornato. DGVS and SEMARNAT, México, D. F., México.

SEMARNAT. 2009b. Compendio de estadísticas ambientales 2009. Licencias de caza deportiva y cintillos de cobro cinegético expedidos en la temporada cinegética. http://app1.semarnat.gob.mx/dgeia/compendio_2009/compendio_2009/10.100.8.236_8080/ibi_apps/WFServletd101.html. Accessed 23 Jan 2010.

Skinner, J. D. 1996. Game ranching. Pages 65–74 *in* V. J. Taylor and N. Dunstone, editors. The exploitation of mammal populations. Chapman and Hall, London, United Kingdom.

Tershy, B. R., L. Bourillón, L. Metzler, and J. Barnes. 1999. A survey of ecotourism on islands in northwestern Mexico. Environmental Conservation 26:212–217.

Toledo, V. M., P. Alarcón Chaires, P. Moguel, M. Olivo, A. Cabrera, E. Leyequien, and A. Rodríguez Aldabe. 2002. Biodiversidad y pueblos indios en México y centroamérica. Biodiversitas 7:1–8.

Toledo, V. M., N. Barrera Bassols, E. Garcia Frapolli, and P. Alarcón Chaires. 2008. Uso múltiple y biodiversidad entre los mayas yucatecos (México). Interciencia 33:345–352.

Toledo, V. M., and E. Boege. 2010. La biodiversidad, las culturas y los pueblos indígenas. Pages 160–192 *in* V. M. Toledo, coordinator. La biodiversidad de México. Inventarios, manejos, usos, informática, conservación e importancia cultural. Fondo de Cultura Económica and CONACULTA, México, D. F., México.

Torres, A. 2003. Los insectos utilizados en la medicina tradicional de San Francisco Totimehuacan y alrededores del Estado de Puebla. Thesis, UNAM, D. F., México.

UANL, et al. 2004. Beneficios del aprovechamiento sustentable de vida silvestre en el norte de México. UANL, Consejo Estatal de Flora y Fauna de Nuevo León, ANGADI, Instituto Coahuilense de Ecología, and SEMARNAT, Monterrey, Nuevo León, México.

UNESCO. 2003. Convención para la salvaguardia del patrimonio cultural inmaterial. UNESCO, Paris, France. http://unesdoc.unesco.org/Images/0013/001325/132540s.pdf. Accessed 5 May 2010.

Uribe, J., and H. T. Arita. 1998. Distribución, diversidad y conservación de los mamíferos de importancia cinegética en México. Acta Zoológica Mexicana (n.s.) 75:47–71.

Ventura, M. 2000. Evaluación del uso de flora y fauna silvestre en tres comunidades de la Reserva de la Biósfera El Triunfo, Chiapas, México. Thesis, Universidad de Ciencias y Artes de Chiapas, Tuxtla Gutiérrez, México.

Villareal, J. 2000. Manejo del venado cola blanca: manejo y aprovechamiento cinegético. Unión Ganadera Regional de Nuevo León, Monterrey, México.

Weber, M. 1993. Ganadería de ciervos: ¿alternativa de producción animal o amenaza a la conservación de la fauna nativa? Agrociencia 3:99–113.

Young, E. 1999a. Balancing conservation with development in small-scale fisheries: is ecotourism an empty promise? Human Ecology 27:581–620.

Young, E. 1999b. Local people and conservation in Mexico's El Vizcaino Biosphere Reserve. Geographical Review 89:364–393.

Tinamous in Mexico

4

Erick H. Baur, Patricia L. R. Brennan, and Daniel M. Brooks

Mexico is the northern distributional limit of tinamous (order Tinamiformes, family Tinamidae), which are terrestrial Neotropical birds related to flightless ratites, although tinamous are capable of flight. Tinamous reach their greatest diversity in South America and may be subdivided based on ecological and phylogenetic affiliations between open-habitat (Nothurinae or Eudromiinae) and forest (Tinaminae) tinamous (Bertelli and Porzecanski 2004, Bertelli and Chiappe 2005, Remsen et al. 2011). Four species from two genera are native to Mexico, all of which belong to the forest clade: great tinamou (*Tinamus major*), thicket tinamou (*Crypturellus cinnamomeus*), slaty-breasted tinamou (*C. boucardi*), and little tinamou (*C. soui*). The distributions of all four species overlap in southern Mexico, where multiple species are often sympatric (Zimmerman 1957, Estrada et al. 1997), and in some areas, all four tinamous occur sympatrically (Andrle 1967, Puebla-Olivares et al. 2002).

In Mexico as elsewhere, people hunt tinamous for subsistence and sport (Leopold 1959, Cabot 1992, Jorgenson 1995). Many tinamous are highly suitable for management and sustainable use. The reproductive features of most species contribute to high fecundity, which is likely to improve resilience to properly managed harvest pressure. The most commonly hunted Mexican species, the thicket tinamou, thrives in natural and second-growth scrub and can likely be managed compatibly with a broad spectrum of land uses. Several tinamou species can be attracted to imitated or playback calls, which could facilitate selective harvest efforts and offers an appealing element to sport hunting (Sutton 1951, Lancaster 1964a). The great tinamou in particular is renowned for its palatability (Van Tyne 1935, Leopold 1959). Leopold (1959) promoted greater utilization of tinamous, which could be an appropriate means of increasing public interest in and financial support for research and conservation.

Evolution and Taxonomy

Tinamous and flightless ratites belong to the superorder Palaeognathae, a primitive group that is phylogenetically distinct from all other living birds (Prager et al. 1976, Olsen 1985, Cabot 1992, Gill 2007, Harshman et al. 2008). Genome research found that extinct moas (order Dinornithiformes) and tinamous are more closely related to one another than to other palaeognaths, implying that tinamous are embedded among the ratites rather than being a sister group (Phillips et al. 2010, Baker et al. 2014). The oldest palaeognath fossils from the Paleocene and Early Eocene of the western US and England are of volant birds that were morphologically more similar to tinamous than to other modern ratites (Olsen 1985). The earliest known tinamou fossils are from the Middle Miocene in Argentina (Bertelli and Chiappe 2005). Currently, 47 tinamou species are recognized among nine genera (AOU 1998, Remsen et al. 2011).

Multiple subspecies have been described from the entire geographical ranges of all four native Mexican tinamous. Two races of the great tinamou occur in Mexico: *robustus* throughout most of the range of the species, and *percautus* on the Yucatán Peninsula (Cabot et al. 2014). Because of the allopatric distribution of the thicket tinamou, as well as its calling and morphological variability, some researchers have suggested that multiple species are present in Mexico. Six subspecies have been described: *mexicanus* in the north, *sallaei* from Veracruz to Chiapas, *soconuscensis* in the western lowlands of Oaxaca and Chiapas, *vicinior* in the Atlantic lowlands of Chiapas, *goldmani* on the Yucatán Peninsula, and *occidentalis* in the separate Pacific coastal range (Conover 1933, del

Table 4.1. Tinamou species with English and Spanish common names, and subspecies (in parentheses) occurring in Mexico.

Species	English name	Spanish name*
Tinamus major (robustus, percautus)	Great tinamou	Tinamú mayor
Crypturellus cinnamomeus (mexicanus, sallaei, soconuscensis, vicinior, goldmani, occidentalis)	Thicket tinamou	Tinamú canelo
C. boucardi (boucardi)	Slaty-breasted tinamou	Tinamú jamuey
C. soui (meserythrus)	Little tinamou	Tinamú menor

*Spanish common names are based on Escalante et al. (2014).

Hoyo et al. 1992, AOU 1998, Cabot et al. 2014). Only one subspecies of the slaty-breasted tinamou (*boucardi*) and of the little tinamou (*meserythrus*) occur in Mexico (table 4.1) (Cabot et al. 2014).

Description

Mexican tinamous share general anatomical features such as cryptic plumage; a stout, tailless appearance; a long, thin neck; and a dove-like head with a slender bill. Although females are slightly larger than males in many species, this difference is not noticeable under most field conditions (Leopold 1959, Lancaster 1964a, Davies 2002). One is more likely to hear tinamous than to see them, and they can be identified by their distinctive whistle-like calls, examples of which are available on various Internet websites.

Great tinamou

This is the largest Mexican tinamou; adults are at least twice the size of other species, ranging from 795 g to 1,400 g (1.75 to 3 lb) (Leopold 1959, Davies 2002, Brennan 2009). The legs are gray and the plumage on both sexes is grayish brown with slight barring and spots over most of the body, a pale abdomen and throat, a dark crown, and white eye rings. Great tinamou calls have a haunting, trilled quality and are often heard in single or paired phrases that rise and fall in pitch.

Thicket tinamou

This is also referred to as the rufescent tinamou; adults weigh 350 g to 500 g (0.77 to 1.1 lb) (Leopold 1959). This species has red legs and more noticeably barred plumage than other Mexican tinamous. Body plumage is more heavily barred on females. Both sexes have gray plumage on the head and neck, a paler throat, and cinnamon-brown plumage on the underparts (Leopold 1959). There are plumage variations attributed to each subspecies, but the most divergent is the allopatric Pacific race (*occidentalis*), which lacks rufous coloration and is paler than other subspecies (see Conover [1933] and Brodkorb [1939] for subspecies descriptions). Leopold (1959) reported that specimens from northern Mexico are larger than those from the south. The mournful, slightly quavering reproductive calls of this species typically have three syllables in the north but may have only one or two syllables in the south (Sutton 1951, Leopold 1959).

Slaty-breasted tinamou

Also known as the Boucard tinamou, this species is similar in size to the thicket tinamou, with adults weighing 432 g to 485 g (0.95 to 1.06 lb) (Leopold 1959). The legs are red. Males have dark gray plumage on the head, neck, and breast; body plumage is grayish brown with faint barring that is more noticeable on females. Plumage is paler on the throat and undersides in both sexes. Calls are often extended single or paired phrases with a haunting, airy, whistle-like quality.

Little tinamou

Formerly referred to as the pileated tinamou, this is the smallest Mexican tinamou, with adults ranging from 165 g to 268 g (0.36 to 0.59 lb) (Davies 2002). The legs are olive gray and the plumage on the head, neck, and back is dark gray, with a pale throat and rufous undersides in both sexes. Females have a greater amount of rufous plumage than males. Common calls include an alternating duet contact call and an ascending series of five to ten high-pitched, tremulous two-syllable whistles (D. M. Brooks, Houston Museum of Natural Science, personal observation).

Distribution

Great tinamou

The distribution of this species includes humid and semihumid lowlands (less than 1,000 m [3,280 ft] in elevation) from southern Mexico to northern South America (AOU 1998). In Mexico (fig. 4.1), it occurs in parts of Veracruz, Oaxaca, Chiapas, Tabasco, Campeche, and Quintana Roo (Howell and Webb 1995). It requires broadleaf forest and has been locally extirpated where forest cover has been lost (Leopold 1959, Cabot 1992).

Thicket tinamou

This species has the northernmost range of any tinamou and the broadest distribution in Mexico. Its allopatric distribution includes most of the Gulf-Caribbean slope of Mexico, northern Guatemala and Belize, across to the Pacific slope of the Isthmus of Tehuantepec southward through Central America to northwestern Costa Rica, and separately along Mexico's central Pacific coast (AOU 1998). Its distribution in Mexico (fig. 4.2) includes parts of Tamaulipas, Nuevo León, San Luis Potosí, Hidalgo, Puebla, Veracruz, Tabasco, Campeche, Yucatán, Quintana Roo, Oaxaca, and Chiapas, and coastal areas of Sinaloa, Nayarit, Jalisco, Colima, Michoacán, and northern Guerrero (Leopold 1959, Howell and Webb 1995). According to Leopold (1959), the species distribution was formerly continuous along the remaining Pacific coast between Guerrero and Oaxaca. It occurs from sea level to 1,850 m (6,000 ft) (AOU 1998).

Slaty-breasted tinamou

The distribution of this species includes humid and semihumid areas from southern Mexico to northern Costa Rica (AOU 1998). In Mexico (fig. 4.2) it occurs in parts of Veracruz, Oaxaca, Chiapas, Tabasco, Campeche, and Quintana Roo (Howell and Webb 1995). It requires broadleaf forest and has been extirpated from deforested areas. It is generally considered a lowland species (AOU 1998) but may occur at elevations up to 1,800 m (5,900 ft) (Cabot 1992).

Little tinamou

This species is distributed throughout humid and semihumid lowlands from southern Mexico through much of South America (AOU 1998). In Mexico (fig. 4.1) it occupies parts of Veracruz, Oaxaca, Chiapas, Tabasco, Campeche, and Quintana Roo (Howell and Webb 1995). It occurs from sea level to 1,500 m (4,900 ft) (AOU 1998).

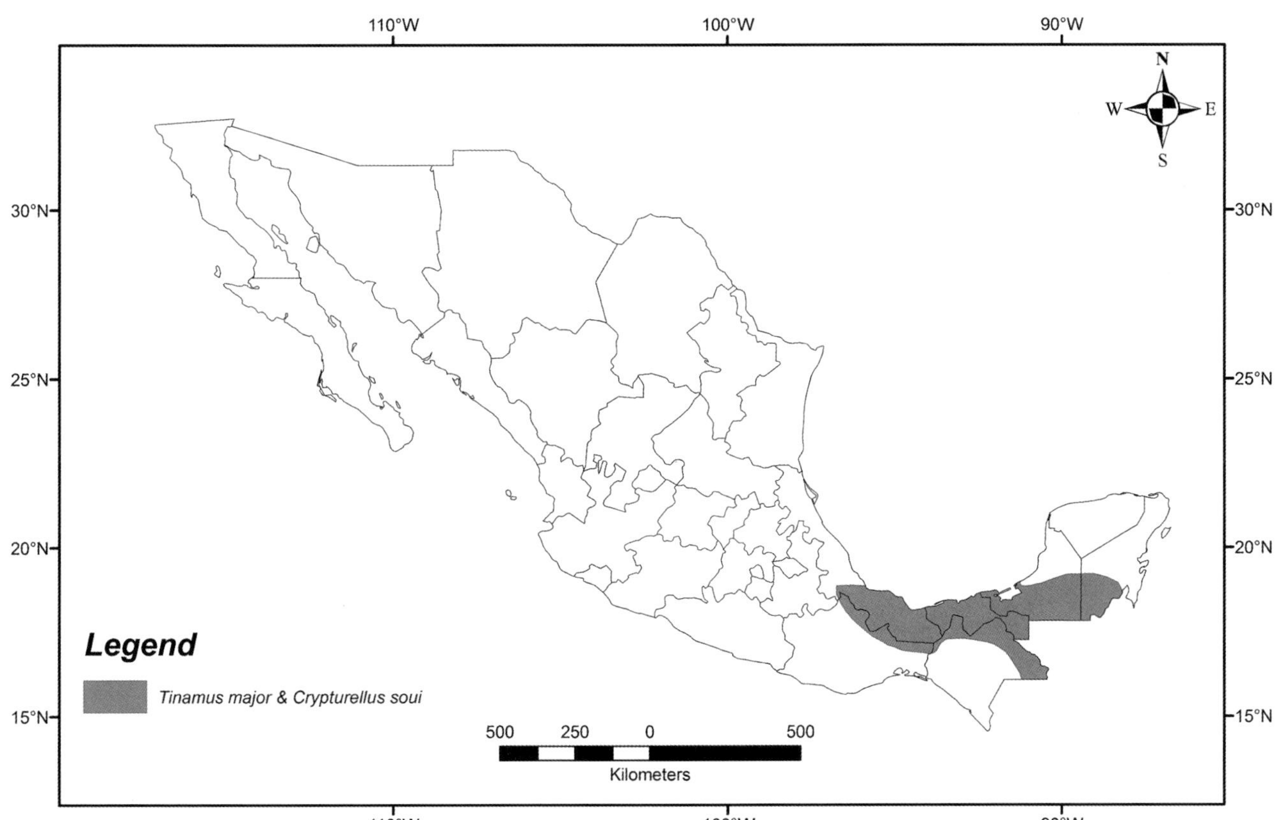

Figure 4.1. Approximate distributions of the great tinamou (*T. major*) and little tinamou (*C. soui*) in Mexico, which are sympatric.

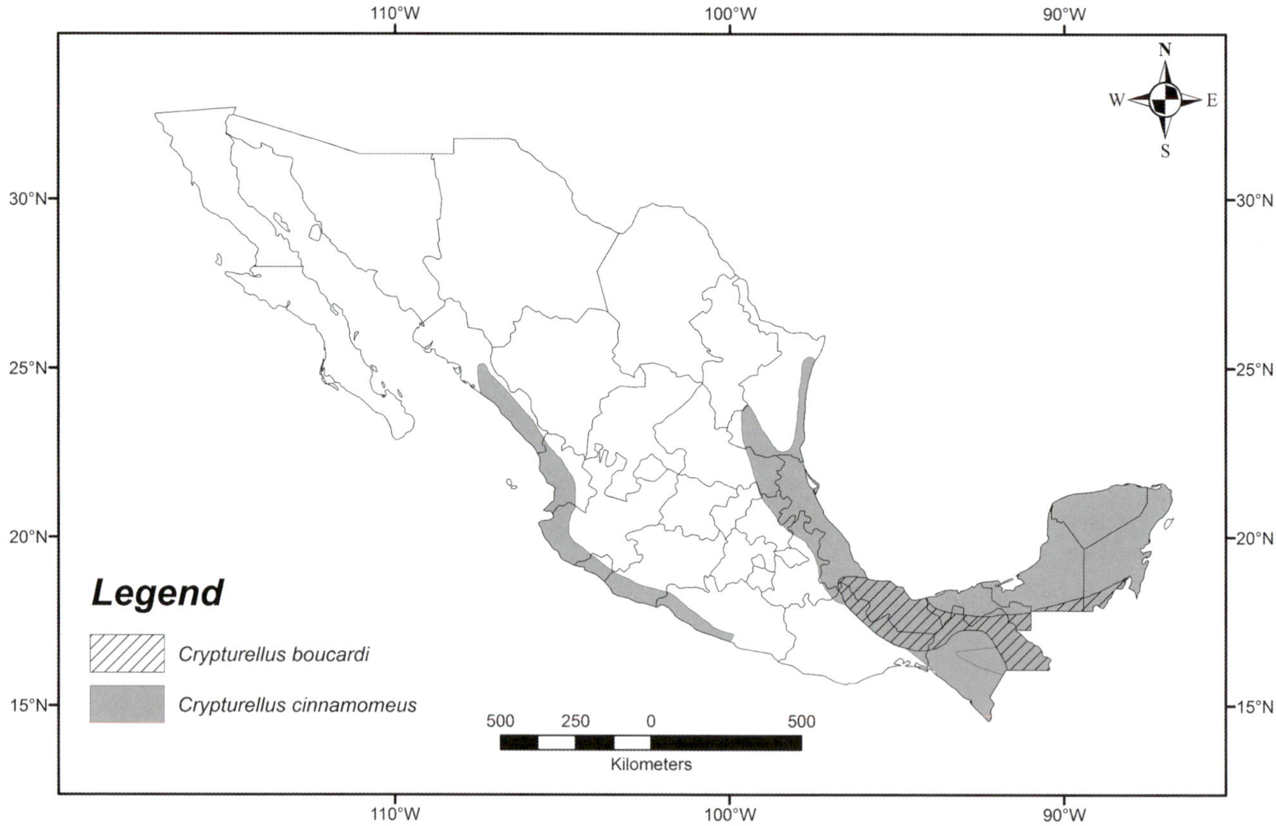

Figure 4.2. Approximate distributions of the slaty-breasted tinamou (*C. boucardi*) and thicket tinamou (*C. cinnamomeus*) in Mexico.

Habitat requirements

Great tinamou

This species is associated with tall broadleaf forests with open understories (Leopold 1959, Skutch 1963, Davies 2002). It occupies primary and secondary forests, fallow plantations and woodlots, riparian forests, and wooded savannas (Zimmerman 1957, Lancaster 1964a, AOU 1998). In northern Guatemala, it was as abundant in low-canopy scrub forests as it was in taller forest types (Baur 2008).

Thicket tinamou

Thick ground cover is a key habitat requirement for this species (fig. 4.3) (Dickey and Van Rossem 1938, Leopold 1959), which occurs from humid to semiarid areas and uses primary and secondary forest or scrub (Leopold 1959, Lancaster 1964a, Cabot 1992). In the north it occupies habitats ranging from thorny scrub dominated by terrestrial bromeliads (*Bromelia pinguin*) to mixed oak-sweetgum (*Quercus-Liquidambar*) forests (Sutton 1951, Leopold 1959). In the south it is often associated with fallow agricultural scrub (Leopold 1959, Lancaster 1964a, Davies 2002), natural scrub forests, and tall forests affected by forest fire (E. H. Baur, personal observation).

Figure 4.3. The thicket tinamou (*C. cinnamomeus*) requires habitat with thick ground cover. Photo by Thibaud Aronson/ Macaulay Library at the Cornell Laboratory of Ornithology.

Slaty-breasted tinamou

This species is associated with tall forests with open understories (Leopold 1959, Lancaster 1964a). In adjacent areas, it may also use scrub forests, fallow plantations, and secondary forests with moderate undergrowth (Lancaster 1964a). Paynter (1957) reported it was common in low, dense, transitional vegetation between pine and broadleaf forests in Chiapas.

Little tinamou

This species inhabits numerous habitat types with thick ground cover. It occupies forest edges, natural scrub, secondary forests and scrub, and various agricultural habitats (Leopold 1959, Skutch 1963, Davies 2002). Skutch (1963) found it to be common in sugarcane, coffee, and banana plantations; maize and cassava fields; and overgrown pastures (fig. 4.4).

Natural History

Diet

Mexican tinamous have similar generalist diets that are dominated by seeds and fruits but also include invertebrates and occasional small vertebrates (Leopold 1959, Lancaster 1964a, Baur 2008). Tinamous do not scratch when foraging, relying on their bills to search through leaf litter. Some species have been observed foraging among mixed flocks at ant swarms (Leopold 1959, Lancaster 1964a).

Great tinamou

Van Tyne (1935) reported *Brosimum alicastrum* and *Pseudolmedia spuria* fruit as dominant food items in northern Guatemala. Leopold (1959) reported *Guatteria* sp. fruit as a dominant food item in Chiapas. Diet composition in northern Guatemala, based on year-round samples ($n = 55$), was dominated by seeds (88%) from 67 plant species, followed by fruit pulp (7%), invertebrates (2%), mineral grit (2%), and minor amounts (1%) of leaves and other plant material (Baur 2008). Seeds and fruits of *B. alicastrum*, *Pseudolmedia spuria*, and *Protium copal* dominated the diet in mass and frequency of occurrence in that study (E. H. Baur, unpublished data). Other important diet items included the fruits and seeds of trees such as *Calophyllum brasiliense*, *Dendropanax arboreus*, *Drypetes* sp., *Manilkara zapota*, and *Mastichodendron foetidissimum*, as well as *Chrysophyllum mexicanum* and *Sabal mauritiiformis*

Figure 4.4. The little tinamou (*Crypturellus soui*) inhabits a variety of habitat types. Photo by Jorge Dangel/Macaulay Library at the Cornell Laboratory of Ornithology.

palms. Invertebrates in the diet included beetles, ants, snails, wasps, spiders, grasshoppers, flies, and moths (E. H. Baur, unpublished data).

Thicket tinamou

Sutton and Pettingill (1942) reported *Bromelia* sp. seeds in all birds collected in Tamaulipas. Dickey and Van Rossem (1938) described a diet of seeds, mineral grit, dung beetles, and other insects in El Salvador. Leopold (1959) reported a variety of seeds and fruits in the diet in Campeche, including a prominent small red fig.

Slaty-breasted tinamou

The spring diet of this species in northern Belize was dominated by fruits and seeds of *Brosimum alicastrum*, *Protium copal*, and *Pseudolmedia spuria*. The diet also included fruits or seeds of tree species such as *Acacia* spp., *Aspidosperma megalocarpon*, *Drypetes brownii*, *Forchhammeria trifoliata*, *Pouteria durlandii*, *Spondias* sp., *Swietenia macrophylla*, and *Trophis racemosa*; shrubs (*Rinorea guatemalensis*); palms (*Chrysophyllum* sp. and *Sabal* sp.); and an agave (*Dracaena americana*). Other diet items included leafcutter (*Atta* sp.) and army ants, termites, moths, beetles, frogs, and lizards (Lancaster 1964a).

Little tinamou

Specimens from Panama consumed insects, a small frog, and vegetative matter and seeds from species of *Amaranthus*, *Oxalis*, *Panicum*, *Paspalum*, *Solanum*, *Scleria*, *Styrax*, *Malvaceae*, and *Euphorbiaceae* (Wetmore 1965). The diet also includes tubers, seeds of *Cyperaceae* spp., ants, termites, millipedes, grasshoppers, weevils, and larval butterflies and moths (Cabot 1992). Lancaster (1964a) observed a pair consuming an unidentified frog.

Reproduction

Mating systems among tinamous vary widely, often including traits of polyandry or promiscuity, and exclusive male incubation and parenting (Handford and Mares 1985, Brennan 2004). Tinamous are ground nesters with precocial young that hatch synchronously (Leopold 1959, Lancaster 1964b, Brennan 2004, Brooks 2015). They lay relatively large, often colorful eggs that have a glossy appearance because of the extreme uniformity of the cuticle on the shell surface (Igic et al. 2015).

Incubating tinamous are extremely reluctant to flush from the nest and will often tolerate being approached closely or even being touched or prodded with a stick or instrument before flushing (Skutch 1963, Cabot 1992, Brennan 2004, 2009). Some species, including the slaty-breasted (Lancaster 1964b) and little tinamous (D. M. Brooks, unpublished data), cover clutches with debris during incubation recesses (Cabot 1992). Incubating thicket tinamous (Sutton 1951) and little tinamous (Skutch 1963) have been observed performing a distraction display after being flushed from the nest that involves walking near the intruder with trembling, lowered wings. Incubating great tinamous (P. L. R. Brennan, personal observation) and slaty-breasted tinamous (Lancaster 1964b) have been observed tossing leaves erratically in the immediate vicinity of the nest after rising for recesses, which may deter nest predators by disturbing chemical cues in the leaf litter.

Great tinamou

This species is promiscuous, pair bonds are temporary, and there are no apparent associations among females (Brennan 2005). In Costa Rica, birds reproduce at approximately one year of age (P. R. L. Brennan, unpublished data). Leopold (1959) reported most nesting in Mexico from April to June, with limited activity later in the year. In northern Guatemala, active nests were found from February through September, with a peak from mid-May to mid-July (Baur 2008). In Costa Rica, most nesting occurs between February and May but continues as late as October (Brennan 2005, 2010). Large (63 × 56 mm, 2.4 × 2.2 in), spheroid, turquoise-blue eggs are laid in scrapes between buttress tree roots (Leopold 1959, Davies 2002, Brennan 2004, 2009). Females lay eggs every other day in the early morning or late afternoon. Clutches produced by a single female have as few as three eggs, whereas clutches formed by multiple females have up to 12 eggs (Cabot 1992, Brennan 2005). In Guatemala (Baur 2008), clutches ranged from three to seven eggs (mode = 4). A 17-day incubation period begins five to nine days after the first egg is laid, and nest attendance is high throughout incubation. Males do not cover clutches during incubation recesses, will continue to incubate partially predated clutches, and aggressively defend clutches and chicks (fig. 4.5) (Brennan 2009, 2010).

Figure 4.5. A male great tinamou (*Tinamus major*) nesting in the roots of a tree and reacting to the presence of an intruder. Photo by Raúl Fournier.

Thicket tinamou

In Mexico, males are territorial during the breeding season from late March until early August (Sutton and Pettingill 1942, Sutton 1951, Paynter 1955, Leopold 1959). In Tamaulipas, Sutton (1951) reported reproductive calling beginning in early March and nesting in April and early May. In Costa Rica, breeding occurs from March to August (Davies 2002). Vegetation usually shelters nests (Sutton 1951, Leopold 1959, Cabot 1992). Eggs are described as various shades of purple (Dickey and Van Rossem 1938, Sutton and Burleigh 1940, Sutton 1951, Cabot 1992). Clutches range from three to seven eggs (mean = 5) that measure approximately 45 × 35 mm (1.77 × 1.37 in) (Leopold 1959). Thicket tinamou young become independent and solitary when they reach half the size of adults (Dickey and Van Rossem 1938).

Slaty-breasted tinamou

The mating system of this species has been described as harem polyandry and alternatively as harem polygyny in conjunction with serial polyandry (Emlen and Oring 1977, Handford and Mares 1985). Van Tyne (1935) and Lancaster (1964b) collected specimens with immature plumage that were in breeding condition, implying that maturity is reached at an early age. In northern Belize, males begin calling and establishing breeding territories in late January, and females begin calling later in the season (Lancaster 1964b). Males typically defend territories from other males via calling duels or occasionally through direct physical aggression. Territoriality declines after mating and ceases entirely once nesting begins (Lancaster 1964b). Breeding males accumulate females one at a time into temporary harems consisting of two to four females. Harems roost with the nesting male, separate during foraging in the morning, and assemble again in the afternoon in response to calls from the nesting male. Following courtship, both sexes cease calling (presumably while laying) until incubation commences, after which females disperse to repeat the process with other males (Lancaster 1964b). Lancaster (1964b) identified a pair of females that transferred together to consecutive harems and speculated that social bonds among females may facilitate communal laying. In that study, some males attending harems exhibited a defensive display in response to intruders that involved repeated vigorous flapping a few feet into the air until the females had left the scene.

Nest placement is not related to territories or home ranges (Lancaster 1964b). Males that lose nests to predators during incubation resume courtship and renest. Clutches of up to 10 purplish-pink eggs averaging 46 × 40 mm (1.81 × 1.57 in) are formed within a 4-day period by multiple females. Nests are located primarily among buttress tree roots and less frequently under fallen trees or vines. Nest attendance is high during a 16-day incubation period, when males do not call. Males use the same routes for approaching and leaving nests (Lancaster 1964b). Brooding males in the presence of chicks are aggressive toward other males. Males attend chicks for approximately two weeks (Lancaster 1964b). Young are nearly adult size at 15 days of age. Some males repeat the entire process of territorial calling, harem acquisition, and cessation of calling up to three times during a single season, leading Lancaster (1964b) to suggest that successful males could raise up to three broods annually.

Little tinamou

This species is considered monogamous; however, no systematic study has been done on its mating behavior in the wild (Skutch 1963). In captivity, pairs breed successfully, and pair mates engage in contact calling and pace restlessly when separated (Brooks 2015). Although breeding pairs (Brooks 2015) and multiple birds (D. Richardson, Dallas World Aquarium, personal communication) may be maintained compatibly in captivity, when unfamiliar birds are introduced into occupied aviaries, residents attack them aggressively, as if in territorial defense (Brooks 2015). Captive females are capable of laying at one year of age. Females perform a precopulatory courtship display that involves walking slowly in circles 0.3 to 0.5 m (0.98 to 1.64 ft) in diameter with the tail oriented upward, and slowly swaying back and forth while the anterior portion of the body is held stationary (Brooks 2015).

Skutch (1963) reported breeding during most of the year in Costa Rica, with nesting peaks in February and September. Most nests were in agricultural habitats (Skutch 1963). Nests can occur under bushes or at the base of trees (Cabot 1992). Captive birds laid eggs more frequently during periods of abundant rainfall (D. M. Brooks, unpublished data) and repeatedly used the same nest sites, which consisted of slight depressions 10–11 cm (3.9–4.3 in) in diameter that were lined with leaf mulch and surrounded by low (10–18 cm, 3.9–7.0 in) vegetation (Brooks 2015). Nest abandonment and false abandonment occurred in response to extreme weather events (floods and heat waves), flushing by humans, or the placement of nests in vulnerable locations.

Eggs are a drab purplish color (Skutch 1963). More than 80% of nests in Costa Rica had two eggs, averaging 43 × 32 mm (1.7 × 1.3 in); the remainder had only one egg (Skutch 1963), although clutches of up to four eggs have been reported (Cabot 1992). Brooks (2015) reported a single clutch of three eggs in captivity that was later abandoned. The nesting male was unable to cover all three eggs simultaneously, and consequently only one developed normally, suggesting a physical constraint favoring the typical clutch size of two eggs. Males exhibit high nest attendance during incubation, taking only 60- to 90-minute recesses each evening during the first 14 days and remaining on the nest for the duration of a 17-day incubation period in captivity (Brooks 2015). In Costa Rica, incubating birds were observed taking recesses in the morning, and incubation lasted 16 days (Skutch 1963). Brooding males exhibit a defensive display that involves ruffling the feathers and extending wings to the ground. The postnatal molt begins on day 5, and by day 25, chicks are fully feathered except for the midsagittal tracts on the head and ventrum, at which stage plumage on both sexes resembles the rich rufous plumage of adult females. Chicks are able to flutter approximately 25 cm (10 in) over distances of 2–3 m (6.5 to 9.8 ft) at 15 days of age and attain full flight at 25 days (Brooks 2015).

Behavior

Where several tinamou species co-occur, they may separate ecologically via different periods of activity or habitat use (Brooks et al. 2001, 2004, Schelsky 2004), although Mexican tinamous exhibit some habitat overlap (Leopold 1959, Lancaster 1964a). Some species, including the great tinamou, appear to be active at night on occasion; however, nocturnal behavior among tinamous is poorly understood (Brooks et al. 2004). Tinamous are generally solitary except when breeding (Leopold 1959, Lancaster 1964a). The great tinamous roost in trees; the other Mexican species roost on the ground (Skutch 1960, 1963; Brennan 2004).

Mexican tinamous share similar evasive and concealment behavior (Leopold 1959, Skutch 1963, Lancaster 1964a, Brennan 2004). When encountered at a distance, individuals walk away quietly and conceal themselves. When approached, they may cease movement and often alter their profile by lowering their head and raising their tail slightly. Tinamous are reluctant to fly and flush only when approached closely, usually over short distances (Leopold 1959, Lancaster 1964a, Brennan 2004).

Communication

Tinamou calls are an iconic element of the Neotropical experience. All four Mexican species call primarily during crepuscular periods, to a limited extent during the day, and occasionally at night (Zimmerman 1957; Leopold 1959; Skutch 1960, 1963; Lancaster 1964a). The great tinamous (Brennan 2005), slaty-breasted tinamou (Lancaster 1964a), and little tinamou (D. M. Brooks, unpublished data) have sexually distinct calls. Slaty-breasted tinamou males call for prolonged periods from prominent landscape features or fallen trees during the breeding season, but they do not call consistently from the same sites (Lancaster 1964a). Skutch (1963) observed incubating little tinamou males calling from the nest until dark in response to calls of nearby cohorts. In captivity the little tinamou calls more frequently following rainfall (D. M. Brooks, unpublished data). Duet calls were more frequently initiated by females than by higher-pitched males, were elicited by separation of paired birds, and often stimulated calling by other pairs (Brooks 2015). The little tinamou begins to make contact calls at six months, and song repertoires are more developed in older birds, suggesting that calling ability develops with age (D. M. Brooks, personal observation).

Movement and territoriality

All Mexican tinamous are resident species (AOU 1998), although some make seasonal movements in response to flooding (Remsen and Parker 1983). Repeated observations of the great tinamou and other species at the same sites over time (P. L. R. Brennan, unpublished data; Brooks 2015) imply either strong site fidelity or strong preferences for particular site characteristics. In Amazonian Peru, the great tinamou was active in upland forest during the day and in edge habitat only at night (Brooks et al. 2001). In South America, *Crypturellus* spp. have been observed moving up to 1 km (0.62 mi) to reach preferred habitats (Schelsky 2004) and flying up to 500 m (1,600 ft) over open water to occupy islands exposed by receding seasonal floods (Remsen and Parker 1983).

Female great tinamou home ranges (3.2–15.0 ha, 7.9–37 ac) were at least twice as large as those of males (0.65–4.0 ha, 1.6–9.8 ac) during the breeding season in Costa Rica (Brennan 2005). Females stayed in small areas for up to eight days before moving to new sites, which is consistent with consecutive mating and laying behaviors. Home ranges overlapped in both sexes without apparent aggression (Brennan 2005). During the breeding season in Belize, male slaty-breasted tinamous males establish exclusive breeding territories in overlapping home ranges of 11–19 ha (27–46.9 ac) (Lancaster 1964b). Most movements occur in the morning (over areas of 2–6 ha [4.9–14.8 ac] or linear distances of 365–550 m [1,200–1,800 ft]) and late afternoon, reaching daily totals of up to 1.6 km (1 mi), and are accompanied by territorial calling as males move erratically in response to other males. One slaty-breasted tinamou brood moved 155 m (500 ft) from the nest site and stayed within a 0.4-ha (0.98-ac) area for over two weeks (Lancaster 1964b). Schäfer (1954) reported little tinamou home ranges of 20 × 50 m (65 × 165 ft) in Venezuela.

Population Dynamics

Little is known about tinamou population structure, but there are reports that populations of all four species have equal sex ratios (Schäfer 1954, Leopold 1959, Lancaster 1964b). Numerous traits of tinamous appear to be adaptive to high natural mortality levels, implying high rates of population turnover (Gill 2007). In the tropics, predation risk for individuals and nests is relatively high in the terrestrial stratum (Martin 1988, Soderstrom 1999). Brennan (2010) found high rates of predation (75%) on great tinamou nests in Costa Rica. Nest predation occurred primarily during incubation when clutches were covered by the male, rather than during laying when eggs were exposed, suggesting that common nest predators do not rely on visual cues to locate nests. Nest predation experiments in Costa Rica did not find a significant difference in predation rates between artificial ground nests with glossy, brightly colored eggs and those with cryptically colored eggs (Janzen 1978).

Some potential nest predators are a threat primarily to clutches, including mammals such as the nine-banded armadillo (*Dasypus novemcinctus*), feral swine (*Sus scrofa domesticus*), collared peccary (*Pecari tajacu*), white-lipped peccary (*Tayassu pecari*), and large snakes (*Pseustes poecilonotus, Spilotes pullatus*, and *Drymarchon* spp.). Potential predators of both clutches and adults in Mexico include the coati (*Nasua narica*), raccoon (*Procyon lotor*), didelphid opossums (*Didelphis* and *Philander* spp.,), skunks (*Conepatus, Mephitis*, and *Spilogale* spp.), tayra (*Eira barbara*), and badger (*Taxidea taxus*). Mammalian predators include coyote (*Canis latrans*), gray fox (*Urocyon cinereoargenteus*), bobcat (*Lynx rufus*), jaguarundi (*Puma yagouaroundi*), ocelot (*Leopardus pardalis*), and margay (*L. wiedii*). Documented avian predators include forest falcons (*Micrastur ruficollis*, *M. semitorquatus*), white hawk (*Leucopternis albicollis*), and ornate hawk-eagle (*Spizaetus ornatus*); a number of other medium- and large-sized tropical raptors are also potential tinamou predators (Lamm 1974, Lyon and Kuhnigk 1985, Thorstrom 2000, Röhe and Antunes 2008).

Status

None of the Mexican tinamou species are globally threatened, although unsustainable hunting practices, deforestation, and other forms of habitat loss have resulted in fragmented distributions of the great and slaty-breasted tinamous, and to a lesser degree the little tinamou (Cabot 1992, IUCN 2011). In 2010, the official status of these three species in Mexico was changed from "Subject to Special Protection" to "Threatened," and the classification of the thicket tinamou was changed to "Subject to Special Protection" (DOF 2002, 2010). These changes were based on a standardized nationwide evaluation of the extinction risk for native species (DOF 2010) rather than on specific systematic efforts to evaluate the status of populations across their respective distributions in the country.

Detailed descriptions of the current distribution and status of Mexican tinamous are unavailable. Some reported density estimates appear robust, but it is unclear whether these represent reproductive aggregations or overall population densities. The great tinamou can be locally common in protected forests (Cabot 1992). In northern Guatemala, great tinamou densities ranged from 5/km^2 (13.1/mi^2) in areas with regular hunting pressure and limited agricultural disturbance, to 15/km^2 (39.2/mi^2) in primary forest without hunting (Baur 2008). The thicket tinamou is common in many areas. Leopold (1959) reported estimates of six to eight breeding pairs of thicket tinamou per 100 ac (30–40/km^2, 78.3–104.4/mi^2) in Tamaulipas. Mapping of male territories, with the assumption of equal sex ratios, indicated that slaty-breasted tinamou densities in northern Belize approached 20/km^2 (52.2/mi^2) in areas where it was most common (Lancaster 1964b). In Peru, little tinamou densities ranged from 15–19/km^2 (39.2–49.6/mi^2) in preferred habitats to 4–6/km^2 (10.4–15.7/mi^2) in other habitats (Schelsky 2004).

Management

Harvest management

We do not fully understand the response of tinamou populations to harvest pressure, and research is needed to develop evidence-based management recommendations for these species in Mexico. However, existing management principles for galliform species can guide related efforts. Subsistence hunting, which has been shown to have a negative relationship with great tinamou population density (Baur 2008), is widespread, especially in southern Mexico where most remaining forests are community managed (Weber et al. 2006). Efforts to control subsistence hunting or mitigate its impacts should be prioritized where management objectives include the conservation or sustainable use of tinamou populations.

Currently, the thicket tinamou is the only species people are permitted to hunt in Mexico; the slaty-breasted and great tinamous are officially threatened, and the little tinamou is not considered a game species because of its diminutive size. With adequate control of timing and intensity, populations of the three largest Mexican tinamous could likely be harvested sustainably and integrated into game management efforts to improve local food security in community-managed forests or to diversify sport-hunting options in UMAs. Until more information is available, restricting hunting from the fall through January should minimize interference with reproductive activities.

Habitat management

Little is known about how populations of Mexican tinamous respond to specific land-use practices. For the time being, only general management recommendations can be made based on habitat affiliations. Habitat management for great and slaty-breasted tinamou populations should emphasize forest conservation and restoration. In Guatemala, the great tinamou was less likely to occupy forest fragments than smaller *Crypturellus* species, and occupancy by the great tinamou exhibited a negative relationship with forest fragment density at the landscape scale, implying an adverse edge effect (Thornton et al. 2012). Habitat management efforts could target slopes and other areas less productive for agriculture on private or communal property, and the protection of existing forest and secondary growth from fire or other disturbance. In areas adjacent to existing forest tracts, reforestation and silvicultural treatments that expedite the maturation of secondary growth would likely increase available habitat and connectivity between existing forest fragments.

Habitat management for thicket and little tinamou populations should emphasize conservation of natural scrub communities and retention of the maximum possible area of secondary scrub in deforested areas, preferably in contiguous tracts and areas that improve connectivity among habitat patches. In dry and semiarid areas, habitat management for the thicket tinamou could be as simple as timing brush control efforts to avoid interfering with reproductive activities and allowing scrub to remain or develop along linear landscape features such as ridges, drainages, arroyo channels, and riparian areas.

Coordination with other land uses

Little information is available to inform land-use decisions with respect to tinamou population management in Mexico. Anecdotal evidence suggests that the management of thicket and little tinamou populations may be consistent with the production of cattle and other livestock, some plantation agriculture, and traditional swidden agriculture that maintains large areas in fallow growth. Forest fragment occupancy by the little tinamou in Guatemala exhibited a negative relationship with distance from human settlements, implying that management for this species could be appropriate in relatively disturbed habitats (Thornton et al. 2012). Rotational grazing; the use of trees that are important in tinamou diets for shade, forage, or as living fences or pasture borders; and mitigation of livestock damage to vegetation at stock tanks, ponds, and riparian areas by restricting livestock access should be explored as means of conserving or improving habitat availability and quality. Avoiding or intensively controlling burning and other brush control efforts immediately prior to and during breeding activity peaks also deserves consideration, along with retaining as much fallow growth as possible.

Areas subject to low-density, selective, certified timber extraction on the southern Yucatán Peninsula did not exhibit negative impacts on the abundance of great, thicket, or slaty-breasted tinamous (Radachowsky 2004). Reforestation and timber enrichment efforts or plantations should be explored as potential forestry practices for tinamou population management. Valuable timber species such as *Aspidosperma megalocarpon*, *Calophyllum brasiliense*, *Dendropanax arboreus*, *Mastichodendron foetidissimum*, *Pouteria amygdalina*, and *Swietenia macrophylla* provide important food resources for tinamous and should be considered in such efforts (Lancaster 1964a).

Management Challenges in Mexico

The relative obscurity of Mexican tinamous poses a challenge to management. Whereas the popularity of sport hunting for waterfowl, doves, and wild turkeys has led to research efforts and increased international cooperation and policy development in Mexico, tinamous have failed to attract similar attention. Because of the high degree to which some UMAs depend on foreign sport hunters (Valdez et al. 2006), it may be helpful to raise awareness of tinamous internationally, such as among residents of the border areas of the US. Such efforts could be as simple as featuring a thicket tinamou hunt in popular press magazines or related cable television programs. Alternatively, efforts could focus on promoting awareness of tinamous among the international and domestic memberships of user-based conservation groups already working in Mexico (e.g., Ducks Unlimited, National Wild Turkey Federation, Safari Club International). Efforts to promote domestic interest in tinamous could focus on Mexican sport-hunting organizations and UMA administrators.

It is remarkable that half a century after Leopold (1959) proclaimed the virtues of the great and thicket tinamous, we have made little advancement in our understanding of these or the other tinamou species native to Mexico. Tinamous merit increased research attention to learn more about their intriguing ecological and biological aspects (Brennan 2004), as well as to develop effective policies for their conservation and management. We need a greater understanding of their population dynamics, habitat suitability, dispersal, and population responses to different land uses and controlled harvest scenarios.

General challenges to tinamou management in Mexico include many issues that affect wildlife management nationwide. High rates of deforestation, rural poverty, limited law enforcement and rural governance capacity, and the challenges inherent in developing and administering regionally appropriate policies are wildlife management challenges that also apply to tinamou conservation (Valdez et al. 2006, Weber et al. 2006). Unfortunately, many of these problems appear to be most severe in southern Mexico, where tinamou diversity is greatest (Bray and Wexler 1996, Weber et al. 2006).

Literature Cited

Andrle, R. F. 1967. Birds of the Sierra de Tuxtla in Veracruz, Mexico. Wilson Bulletin 79:163–187.

AOU. 1998. Check-list of North American birds. Seventh edition. American Ornithologists' Union, Washington, DC, USA.

Baker, A. J., O. Haddrath, J. D. McPherson, and A. Cloutier. 2014. Genomic support for a moa-tinamou clade and adaptive morphological convergence in flightless ratites. Molecular Biology and Evolution 31:1686–1696.

Baur, E. H. 2008. Structure of a lowland Neotropical galliform bird guild. Thesis, University of Florida, Gainesville, USA.

Bertelli, S., and L. M. Chiappe. 2005. Earliest tinamous (Aves: Palaeognathae) from the Miocene of Argentina and their phylogenetic position. Natural History Museum of Los Angeles County, Contributions in Science 502:1–20.

Bertelli, S., and A. L. Porzecanski. 2004. Tinamou (Tinamidae) systematics: a preliminary combined analysis of morphology and molecules. Ornitología Neotropical 15:1–7.

Bray, D. B., and M. B. Wexler. 1996. Forest policies in Mexico. Pages 217–228 *in* R. Randall, editor. Changing structure in Mexico. M. E. Sharpe, Armonk, New York, USA.

Brennan, P. L. R. 2004. Techniques for studying the behavioral ecology of forest-dwelling tinamous (Tinamidae). Ornitología Neotropical 15:329–337.

Brennan, P. L. R. 2005. The breeding biology and ecology of great tinamous (*Tinamus major*): female joint-nesting, extra-pair paternity and natural history. Dissertation, Cornell University, Ithaca, New York, USA.

Brennan, P. L. R. 2009. Incubation in great tinamou (*Tinamus major*). Wilson Journal of Ornithology 121:506–511.

Brennan, P. L. R. 2010. Clutch predation in great tinamous (*Tinamus major*) and implications for the evolution of egg color. Journal of Avian Biology 41:419–426.

Brodkorb, P. 1939. New subspecies of birds from the district of Soconusco, Chiapas. Occasional Papers of the Museum of Zoology Number 401. University of Michigan, Ann Arbor, USA.

Brooks, D. M. 2015. Behavior, reproduction, and development in little tinamou (*Crypturellus soui*). Wilson Journal of Ornithology 127:761–765.

Brooks, D. M., L. Pando-Vasquez, A. Ocmin-Petit, and J. Tejada-Renjifo. 2001. Resource separation in a Napo-Amazonian gamebird community. Pages 213–225 *in* D. M. Brooks and F. Gonzalez-Garcia, editors. Biology and conservation of cracids in the new millennium. Miscellaneous Publications Number 2. Houston Museum of Natural Science, Texas, USA.

Brooks, D. M., L. Pando-Vasquez, A. Ocmin-Petit, and J. Tejada-Renjifo. 2004. Resource separation in a Napo-Amazonian tinamou community. Ornitología Neotropical 15:323–328.

Cabot, J. 1992. Tinamiformes. Pages 112–144 *in* J. del Hoyo, A. Elliot, and J. Sargatal, editors. Handbook of birds of the world. Volume 1. Lynx Edicions, Barcelona, Spain.

Cabot, J., F. Jutglar, E. F. J. Garcia, P. Boesman, and C. J. Sharpe. 2014. Eastern thicket tinamou (*Crypturellus cinnamomeus*). *In* J. del Hoyo, A. Elliot, J. Sargatal, D. A. Christie, and E. de Juana, editors. Handbook of the birds of the world alive. Lynx Edicions, Barcelona, Spain. http://www.hbw.com/node/52429. Accessed 24 Jun 2015.

Conover, H. B. 1933. The races of the tinamou *Crypturellus cinnamomeus.* Proceedings of the Biological Society of Washington 46:113–118.

Davies, S. J. J. F. 2002. Ratites and tinamous. Oxford University Press, New York, New York, USA.

del Hoyo, J., N. Collar, D. A. Christie, E. F. J. Garcia, and C. J. Sharpe. 1992. Western thicket tinamou (*Crypturellus occidentalis*). *In* J. del Hoyo, A. Elliot, J. Sargatal, D. A. Christie, and E. Juana, editors. Handbook of the birds of the world alive. Lynx Edicions, Barcelona, Spain. http://www.hbw.com/node/467081. Accessed 24 Jun 2015.

Dickey, D. R., and A. J. Van Rossem. 1938. The birds of El Salvador. Field Museum of Natural History, Zoology Series 23:1–609.

DOF. 2002. Norma Oficial Mexicana NOM-059-ECOL-2001. Protección ambiental—Especies nativas de México de flora y fauna silvestres—Categorías de riesgo y especificaciones para su inclusión, exclusión o cambio—Lista de especies en riesgo. Miércoles 6 de marzo de 2002 (Segunda Sección). SEMARNAT, México, D. F., México. www.ine.gob.mx/ueajei/norma59a.html. Accessed 18 Jun 2011.

DOF. 2010. Norma Oficial Mexicana NOM-059-ECOL-2010. Protección ambiental—Especies nativas de México de flora y fauna silvestres—Categorías de riesgo y especificaciones para su inclusión, exclusión o cambio—Lista de especies en riesgo. Jueves 30 de diciembre de 2010 (Segunda Sección). SEMARNAT, México, D. F., México. www.ine.gob.mx/ueajei/norma59a. html. Accessed 18 Jun 2011.

Emlen, S. T., and L. W. Oring. 1977. Ecology, sexual selection, and the evolution of mating systems. Science 197:215–223.

Escalante, P., A. M. Sada, and J. R. Gil. 2014. Lista de nombres comunes de las aves de México. UNAM and CIPAMEX, México, D. F., México.

Estrada, A., R. Coates-Estrada, and D. A. Meritt Jr. 1997. Anthropogenic landscape changes and avian diversity at Los Tuxtlas, Mexico. Biodiversity and Conservation 6:19–43.

Gill, F. B. 2007. Ornithology. Third edition. W. H. Freeman, New York, New York, USA.

Handford, P., and M. A. Mares. 1985. The mating systems of ratites and tinamous: an evolutionary perspective. Biological Journal of the Linnaean Society 25:77–104.

Harshman, J., E. L. Braun, M. J. Braun, C. J. Huddleston, R. C. K. Bowie, J. L. Chojnowski, S. J. Hackett, K. L. Han, R. T. Kimball, B. D. Marks, K. J. Miglia, W. S. Moore, S. Reddy, F. H. Sheldon, D. W. Steadman, S. J. Steppan, C. C. Witt, and T. Yuri. 2008. Phylogenomic evidence for multiple losses of flight in ratite birds. Proceedings of the National Academy of Sciences 105:13462–13467.

Howell, S. N. G., and S. Webb. 1995. A guide to the birds of Mexico and northern Central America. Oxford University Press, New York, New York, USA.

Igic, B., D. Fecheyr-Lippens, M. Xiao, A. Chan, D. Hanley, P. R. L. Brennan, T. Grim, G. I. N. Waterhouse, M. E. Hauber, and M. D. Shawkey. 2015. A nanostructural basis for gloss of avian eggshells. Journal of the Royal Society Interface 12:20141210.

IUCN. 2011. IUCN Red List of Threatened Species. www.iucnredlist.org. Accessed 30 Jun 2011.

Janzen, D. H. 1978. Predation intensity on eggs on the ground in two Costa Rican forests. American Midland Naturalist 100:467–470.

Jorgenson, J. P. 1995. Maya subsistence hunters in Quintana Roo, Mexico. Oryx 29:49–57.

Lamm, D. W. 1974. White hawk preying on the great tinamou. Auk 91:845–846.

Lancaster, D. A. 1964a. Life history of the Boucard tinamou in British Honduras. Part I: distribution and general behavior. Condor 66:165–181.

Lancaster, D. A. 1964b. Life history of the Boucard tinamou in British Honduras. Part II: breeding biology. Condor 66:253–276.

Leopold, A. S. 1959. Wildlife of Mexico: the game birds and mammals. University of California Press, Berkeley, USA.

Lyon, B., and A. Kuhnigk. 1985. Observations on nesting ornate hawk eagles in Guatemala. Wilson Bulletin 97:141–147.

Martin, T. E. 1988. On the advantages of being different: nest predation and the coexistence of bird species. Proceedings of the National Academy of Sciences 85:2196–2199.

Olsen, S. L. 1985. The fossil record of birds. Pages 79–252 *in* D. S. Farner, J. R. King, and K. C. Parkes, editors. Avian biology. Volume 8. Academic Press, New York, New York, USA.

Paynter, R. A., Jr. 1955. The ornithogeography of the Yucatán Peninsula. Peabody Museum of Natural History Bulletin 9:1–397.

Paynter, R. A., Jr. 1957. Birds of Laguna Ocotal. Pages 249–285 *in* R. A. Paynter Jr., editor. Biological investigations in the Selva Lacandona, Chiapas, Mexico. Bulletin of the Museum of Comparative Zoology 116:249–285.

Phillips, M. J., G. C. Gibb, E. A. Crimp, and D. Penney. 2010. Tinamous and moa flock together: mitochondrial genome sequence analysis reveals independent losses of flight among ratites. Systematic Biology 59:90–107.

Prager, E. M., A. C. Wilson, D. T. Osuga, and R. E. Feeney. 1976. Evolution of flightless land birds on southern continents: transferrin comparison shows monophyletic origin of ratites. Journal of Molecular Evolution 8:283–294.

Puebla-Olivares, F., E. Rodríguez-Ayala, B. E. Hernández-Baños, and A. G. Navarro. 2002. Status and conservation of the avifauna of the

Yaxhilan Natural Monument, Chiapas, Mexico. Ornitología Neotropical 13:381–396.

Radachowsky, J. 2004. Effects of certified logging on wildlife in community and industrial forest concessions of northern Guatemala. Wildlife Conservation Society, New York, New York, USA.

Remsen, J. V., Jr., C. D. Cadena, A. Jaramillo, M. Nores, J. F. Pacheco, J. Pérez-Emán, M. B. Robbins, F. G. Stiles, D. F. Stotz, and K. J. Zimmer. 2011. A classification of the bird species of South America. AOU, Washington, DC. http://www.museum.lsu.edu/~Remsen/SACCBaseline.html. Accessed 30 Jun 2011.

Remsen, J. V., Jr., and T. A. Parker III. 1983. Contribution of river-created habitats to bird species richness in Amazonia. Biotropica 15:223–231.

Röhe, F., and A. P. Antunes. 2008. Barred forest falcon (*Micrastur ruficollis*) predation on relatively large prey. Wilson Journal of Ornithology 120:228–230.

Schäfer, E. 1954. Zur biologie des steisshuhness *Nothocercus bonapartei*. Journal für Ornithologie 95:219–223.

Schelsky, W. M. 2004. Research and conservation of forest dependent tinamou species in Amazonia, Peru. Ornitología Neotropical 15:317–321.

Skutch, A. F. 1960. The great tinamou of the tropical forest. Animal Kingdom 62:179–183.

Skutch, A. F. 1963. Life history of the little tinamou. Condor 65:224–231.

Soderstrom, B. 1999. Artificial nest predation rates in tropical and temperate forests: a review of the effects of edge and nest site. Ecography 22:455–463.

Sutton, G. M. 1951. The rufescent tinamou. Wilson Bulletin 63:67–68.

Sutton, G. M., and T. D. Burleigh. 1940. Birds of Valles, San Louis Potosí, Mexico. Condor 42:259–262.

Sutton, G. M., and O. S. Pettingill Jr. 1942. Birds of the Gomez Farias region, southwestern Tamaulipas. Auk 59:1–34.

Thornton, D. H., L. C. Branch, and M. E. Sunquist. 2012. Response of large galliforms and tinamous (Cracidae, Phasianidae, Tinamidae) to habitat loss and fragmentation in northern Guatemala. Oryx 46:567–576.

Thorstrom, R. 2000. The food habits of sympatric forest-falcons during the breeding season in northeastern Guatemala. Journal of Raptor Research 34:196–202.

Valdez, R., J. C. Guzmán-Aranda, F. J. Abarca, L. A. Tarango-Arámbula, and F. C. Sánchez. 2006. Wildlife conservation and management in Mexico. Wildlife Society Bulletin 34: 270–282.

Van Tyne, J. 1935. The birds of northern Peten, Guatemala. Miscellaneous Publications Number 27. Museum of Zoology, University of Michigan, Ann Arbor, USA.

Weber, M., G. García-Marmolejo, and R. Reyna-Hurtado. 2006. The tragedy of the commons: wildlife management units in southeastern Mexico. Wildlife Society Bulletin 34:1480–1488.

Wetmore, A. 1965. Birds of the Republic of Panama. Part I. Tinamidae (tinamous) to Rhynchopidae (skimmers). Smithsonian Miscellaneous Collections 150:1–483.

Zimmerman, D. A. 1957. Spotted-tailed nightjar nesting in Veracruz, Mexico. Condor 59:124–127.

Curassows, Guans, and Chachalacas in Mexico

5

Roel R. Lopez, Nova J. Silvy, Markus J. Peterson, Daniel M. Brooks, Erick H. Baur, and Fernando González-García

With the exception of one species, the family Cracidae comprises forest and brushland bird species (i.e., chachalacas, curassows, and guans) endemic to the Neotropics of Mexico and Central and South America. Mexico is the northernmost distributional range for many of these species groups, except for the plain chachalaca (*Ortalis vetula*), which extends into the Rio Grande Valley of South Texas (Peterson 2000). Cracids, especially chachalacas and guans, are primarily arboreal, whereas other galliforms (e.g., turkeys, pheasants, and quails) are ground dwellers. This chapter focuses on the biology and management of three cracid groups that occur in Mexico: chachalacas, curassows, and guans.

Eight species and five genera of cracids occur in Mexico: the plain chachalaca, rufous-bellied chachalaca (*O. wagleri*) (fig 5.1), west Mexican chachalaca (*O. poliocephala*) (fig 5.2), white-bellied chachalaca (*O. leucogastra*) (fig. 5.3); highland guan (*Penelopina nigra*), crested guan (*Penelope purpurascens*), horned guan (*Oreophasis derbianus*), and great curassow (*Crax rubra*) (figs. 5.4 and 5.5) (Leopold 1959). Only the horned guan is typically found in the temperate uplands of Chiapas, Mexico (González-García 2001).

Figure 5.1. Rufous-bellied chachalaca (*O. wagleri*). Mathew Garvin/Macaulay Library at the Cornell Lab of Ornithology.

Species range in size from chachalacas (the size of a female pheasant) to guans (medium-sized turkey) to curassows (large turkey) (Brooks and Strahl 2000, Brooks and Fuller 2006).

Roughly one-half of the 50 cracid species are threatened at some level (Brooks et al. 2006). Cracids are of great social and economic value to indigenous populations and campesinos (Jorgenson 1995, 1997, González-García et al. 2001). As frugivores,

Figure 5.2. West Mexican chachalaca (*O. poliocephala*). Louis Imbeau/Macaulay Library at the Cornell Lab of Ornithology.

Figure 5.3. White-bellied chachalaca (*O. leucogastra*). Erick Hernandez/Macaulay Library at the Cornell Lab of Ornithology.

Figure 5.4. Great curassow female (*Crax rubra*). Don Coons/Macaulay Library at the Cornell Lab of Ornithology.

Figure 5.5. Great curassow male (*Crax rubra*). James Bozeman/Macaulay Library at the Cornell Lab of Ornithology

cracids serve an important ecological function by maintaining forested ecosystems and regenerating plant communities through their seed scarification and dispersal (Érard and Théry 1994, Santamaria-G. and Franco 1994, Peres and van Roosmalen 1996, Sedaghatkish 1996). More recently, their increased popularity among ecotourists, principally bird-watchers, has provided an economic incentive to conserve cracids. In addition to their intrinsic and biological value, cracids also serve as an important source of meat protein (Delacour and Amadon 1973). Several studies (Silva and Strahl 1991, Begazo 1997, Begazo and Bodmer 1998) have indicated that the largest avian biomass removed in Neotropical forests by subsistence hunters consisted of cracids. Cracid populations most threatened with overharvest were negatively and positively correlated, respectively, with distance to a city and city population size (Brooks 1999). These studies demonstrate the significance of cracids to the livelihood of many cultures in Mexico and other regions of Latin America.

Habitat destruction and unsustainable hunting are primary factors in cracid declines throughout their distribution (Brooks and Strahl 2000, Brooks and Fuller 2006). From a wildlife management perspective, it is important to note the sharp differences in population response among these species to human activities. In general, curassows and guans depend on climax tropical and montane forests and respond negatively to significant timber harvesting (Brooks 2006). Further, these species cannot withstand significant hunting pressure, whether subsistence or recreational (Brooks 1999, Peres 2000, 2001). It may not be practical to separate the effects of habitat deterioration and hunting pressure on curassows and guans because both practices are detrimental to cracid populations. Throughout Mexico and Latin America, the status of several of these species ranges from near threatened (e.g., greater curassow) to vulnerable (e.g., highland guan) to critically endangered (e.g., horned guan; Brooks et al. 2006). In contrast, chachalacas fare well in second-growth forests and brushland, and most of the species can sustain significant hunting pressure (Brooks 1997a, Brooks and Fuller 2006). This is not to suggest that chachalacas do not prefer climax tropical and montane forests; rather, they readily adapt to dense brushlands (i.e., second-growth forests) resulting from timber harvest (Brooks 1997b). Many chachalacas are heavily hunted throughout their range, with few observed impacts on populations (Arriaga-U. and Baquero 1997). In our discussion of wildlife management practices, we organize our recommendations based on these important differences in response to human activities between curassows and guans versus chachalacas. We also discuss the life histories of cracids in three broad categories (i.e., chachalacas, guans, and curassows).

Taxonomy and Evolution

Cracids are a primitive family of game birds that originated nearly 50 million years ago. During the Oligocene era, much of North America was tropical, including the northern plains (Wetmore 1956). The earliest fossil records are those of a primitive bird, considered to be arboreal, found in the state of Wyoming (del Hoyo 1994). Other fossil forms (approximately 30 million years old) related to chachalacas have been found in South Dakota (Tordoff and MacDonald 1957). Cracids probably originated in North and Central America and expanded to South America coincident with climate and habitat changes (Brodkorb 1964). Recent fossils of contemporary cracids (e.g., *Crax*, *Penelope*), approximately 20,000 years old, have been found in their current distributional range (del Hoyo 1994).

Distribution

Chachalacas

In Mexico, chachalaca species are widely distributed along the Gulf-Caribbean slope from the lower Rio Grande Valley of Texas and the Pacific coast of western Mexico (fig. 5.6), southward to northern Honduras and north-central Nicaragua. Although their historical distribution was probably similar to their current distribution, many populations are disjunct because of agricultural and urban development, rendering isolated populations more vulnerable (Peterson 2000, Brooks 2006, Brooks et al. 2006). Other chachalaca

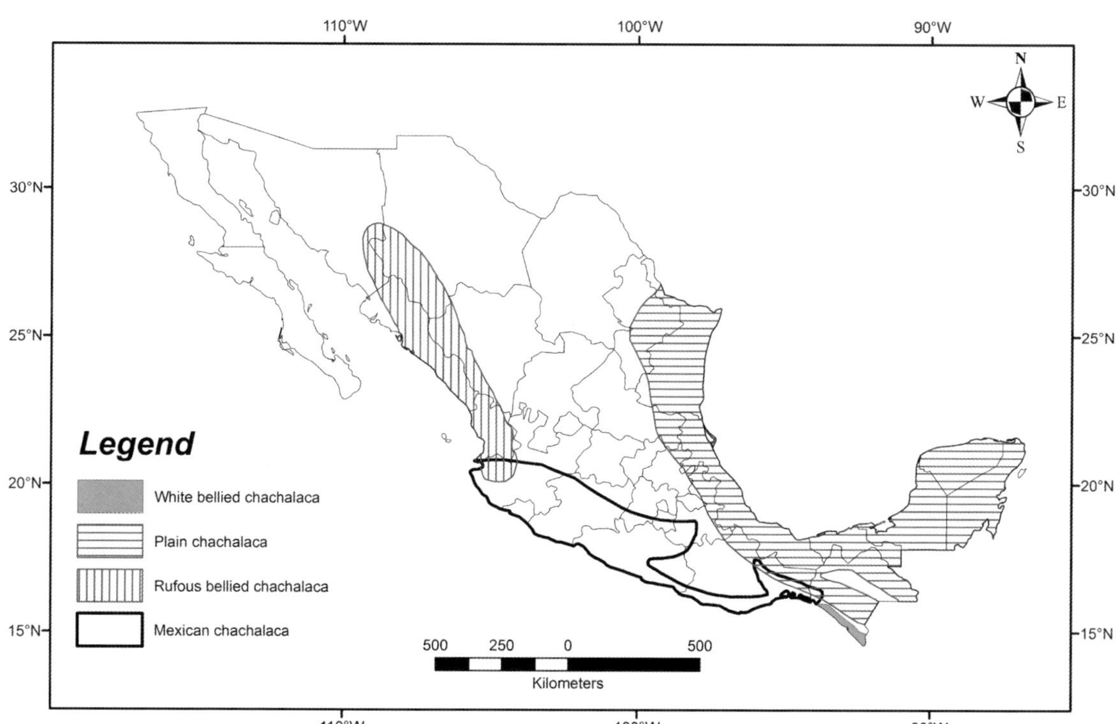

Figure 5.6. Distribution of the white-bellied chachalaca (*Ortalis leucogastra*), plain chachalaca (*O. vetula*), rufuous-bellied chachalaca (*O. wagleri*), and west Mexican chachalaca (*O. poliocephala*) in Mexico.

species occurring within or adjacent to the range of the plain chachalaca (fig. 5.7) include the west Mexican chachalaca and white-bellied chachalaca (Wagner 1953, Vaurie 1965, 1968, Delacour and Amadon 1973). The ranges of the plain and west Mexican chachalaca overlap on the Isthmus of Tehuantepec in Oaxaca (Peterson 2000). These two species occur in parapatry with the white-bellied chachalaca in coastal western Chiapas. The rufous-bellied chachalaca occurs in northwestern Mexico (Sibley and Monroe 1990).

Figure 5.7. Plain chachalaca (*Ortalis vetula*) is the only species with a distributional range that extends into the Rio Grande Valley of South Texas. Jordan Broadhead/Macaulay Library at the Cornell Lab of Ornithology.

Hybridization among these species probably does not occur in the wild (Wagner 1953, Leopold 1959, Vaurie 1965, 1968, Vuilleumier 1965).

Curassows

The great curassow is the only species in the genus *Crax* in Mexico and includes two subspecies—the mainland form (*rubra*), distributed along the coastal region of eastern Mexico; and the Cozumel curassow (*griscomi*), occurring on the island of Cozumel, Mexico (fig. 5.8) (Delacour and Amadon 1973, del Hoyo 1994, Martínez-Morales 1999). Curassows inhabit mature tropical forests between sea level and 1,200 m (0 and 3,937 ft) (Arguedas et al. 1997, Delgado 1997, McCoy 1997, Midence 1997, Sermeño 1997, Vannini and Rockstroh 1997), and occasionally up to 2,000 m (6,562 ft) (Vaughan 1983). Mainland populations have been extirpated in portions of their range (Howell and Webb 1995, González-García et al. 2001), with remaining populations highly fragmented. In some areas, populations are locally abundant (Martínez-Morales 1996), and a favorite game species (Jorgenson 1997).

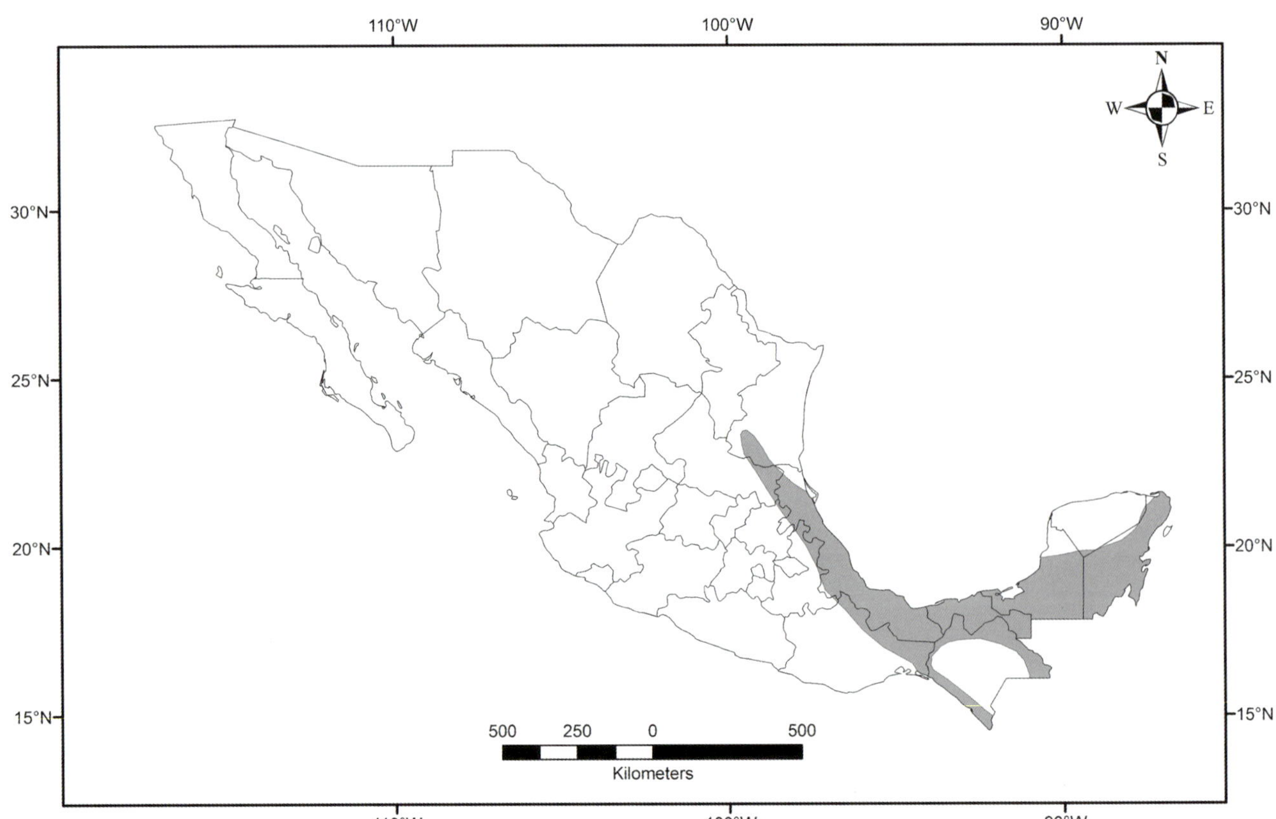

Figure 5.8. Distribution of the great curassow (*Crax rubra*) in Mexico.

Guans

Guans in Mexico inhabit cloud forests at elevations of 1,000–3,350 m (3,280–11,483 ft) (fig. 5.9). Horned guans are found on the slopes of the Sierra Madre de Chiapas, primarily in humid evergreen forests (fig. 5.10) (González-García et al. 2001). Highland guans occupy a variety of forest types in Chiapas and Oaxaca, including deciduous cloud, pine-oak, and coniferous forests at 900–3,300 m (2,953–10,827 ft) (Delacour and Amadon 1973, Eisermann et al. 2006), but they may also occupy adjacent low-canopy secondary forests (Eisermann 2005). In Mexico, the crested guan occurs in tropical forests of northern Sinaloa and central Tamaulipas, southward toward the forests of Chiapas and the Yucatán Peninsula. Seasonally, guans appear to move altitudinally, likely following fruit blooms.

Description

Chachalacas

Chachalacas are similar in size to female ring-necked pheasants (*Phasianus colchicus*) (Peterson 2000), except chachalacas have a broader, longer tail and a different plumage pattern. The general body color is olive drab; west Mexican and white-bellied chachalacas have much whiter underparts and/or white-tipped rectrices (Delacour and Amadon 1973). As the common name suggests, the abdomen and undertail coverts of the rufous-bellied chachalaca are rufous (reddish brown). All species are characterized by long legs, necks, and tails and lack wattles and caruncles, which are common in other cracid species (Delacour

Figure 5.10. A horned guan (*Oreophasis derbianus*) in natural habitat in Mexico. Photo by Fernando González García.

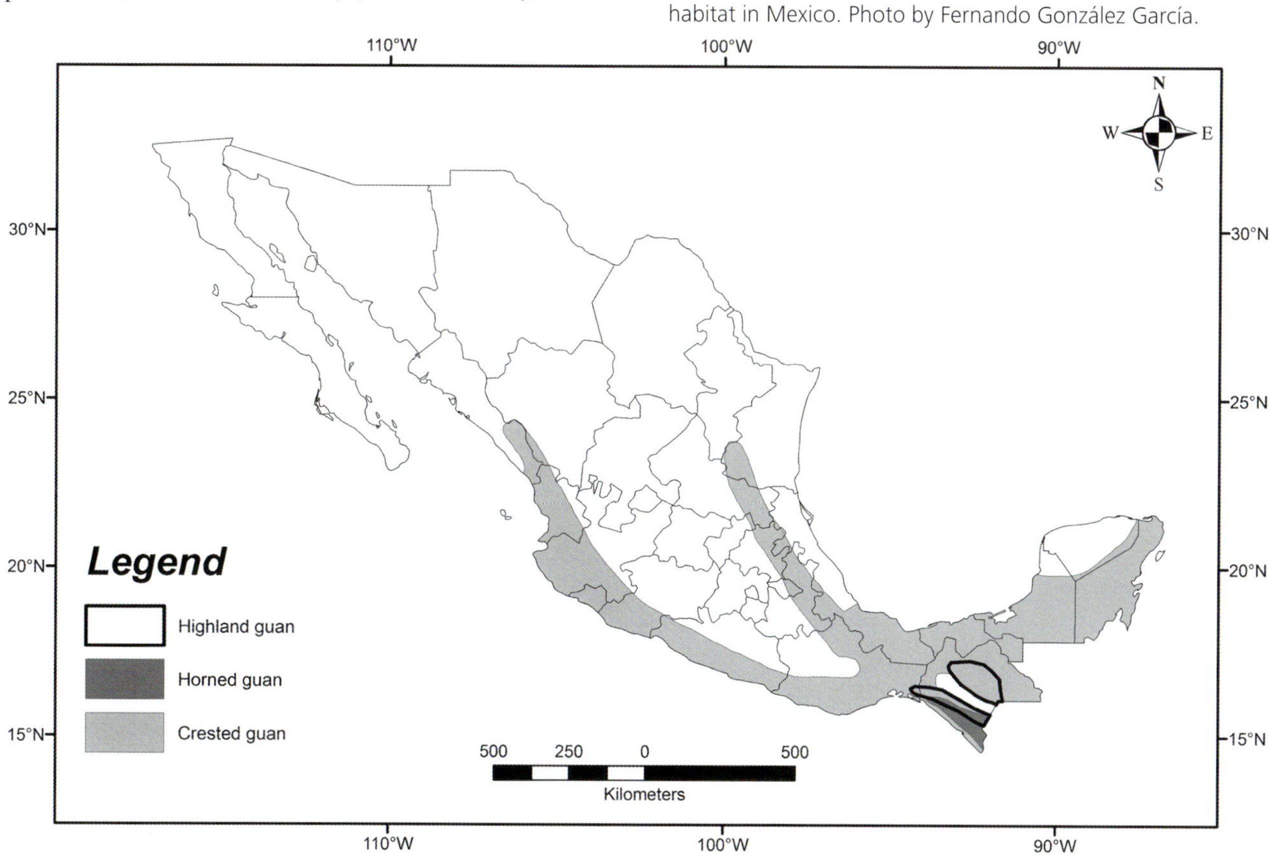

Figure 5.9. Distribution of the highland guan (*Penelopina nigra*), horned guan (*Oreophasis derbianus*), and crested guan (*Penelope purpurascens*) in Mexico.

and Amadon 1973). There is relatively little variation in body size, except for the smaller bodies of white-bellied chachalaca and certain populations of plain chachalaca (Vaurie 1965, 1968). Males are only slightly larger than females, and sexes can be differentiated by the presence of a tracheal loop found only in adult males (Peterson 2000).

Curassows

The great curassow is a magnificent bird approximately the size and shape of a large turkey with a crest of curly feathers. The general body color varies between sexes, with males being all black, except for a white abdomen and crissum, and females having cinnamon brown on the underside, shading to black on the neck and tail (Delacour and Amadon 1973). Rufous (reddish-brown) and barred (yellow, black, and white) morphs are possible in females (Delacour and Amadon 1973). Although it has been suggested that these morphs vary along a latitudinal gradient (del Hoyo 1994), different morphs have been observed in the same flock (barred and plain morphs in Belize) contemporaneously (Zimmer 1999). Males are slightly larger (3,600–4,600 g, 7.9–10 lb) than females (3,100–4,200 g, 6.8–9.3 lb).

Guans

Guans have the form of a pheasant or turkey, and the long tail typical of cracids. The crested guan is about the size of a small turkey (1,600–2,400 g, 3.5–5.3 lb), with a dusky olive-brown color. Breast feathers are edged with white, with the throat sparsely covered with bristles and a small red wattle. The crested guan is named after its normally erect crown feathers and is sexually monomorphic. The highland guan is the smallest guan in Mexico, approximately the size of a chicken (800–1,000 g, 1.8–2.2 lb). Common anatomical characteristics include a long tail, bare throat wattle, and low crest. Males are black, while females are barred brown—both have orange legs and feet, and the male has a red bill (Delacour and Amadon 1973). Horned guans are approximately the size of a turkey (2,000–2,300 g, 4.4–5 lb). The horned guan has a pronounced bright red casque that is in the form of a "horn" (hence its name). Both sexes are predominantly black with whiter underparts, red feet and legs, and a yellow bill (Delacour and Amadon 1973). Females are smaller than males and have a less pronounced "horn."

Life History

Chachalacas

This section will focus primarily on studies of plain chachalacas because the four species found in Mexico are similar behaviorally and ecologically. The plain chachalaca has been studied the most, but we do include information on the other species. While there are no known studies of rufous-bellied chachalaca, the two remaining species (west Mexican and white-bellied chachalaca) have been studied in detail. In El Salvador, diets included the fruits of 11 plant genera (at least 12 species) (Simas 1979). Studies on behavior and reproduction of west Mexican chachalaca were carried out by Rowley (1984) and Gurrola (1985). This species consumed the fruits of 24 plant genera (at least 27 species), as well as beetles, ants, crickets, snails (*Triodopsis* sp.), and soil (Gurrola 1985, Berlanga 1991), and was found to be an important disperser of *Spondias purpurea* and *Ficus cotinifolia* (Berlanga 1991, Mandujano et al. 1994). The forest falcon (*Micrastur semitorquatus*) is a known predator (Delacour and Amadon 1973) of plain chachalacas.

Diet

Chachalacas are primarily folivorous and frugivorous but also eat invertebrates and other plant material such as flowers and seeds (fig. 5.11) (Leopold 1959, Oberholser 1974, Marion 1976, Christensen et al. 1978). They are readily domesticated or semidomesticated and consume feed supplied by humans (e.g., cracked corn, milo, sunflower seeds) (Oberholser 1974, Marion 1976). In a year-round food-habit study of plain chachalacas, conducted in the Lower

Figure 5.11. Although primarily folivorous and frugivorous, chachalacas also eat other plant parts like the flowers of *Ehretia anacua*. Photo by Carlos Abrego.

Rio Grande Valley of Texas (Marion 1976), plant species provided more than 90% of the diet by volume and included coyotillo (*Karwinskia humboldtiana*), Mexican ash (*Fraxinus berlandieriana*), bloodberry rougeplant (*Rivina humilis*), and anaqua (*Ehretia anacua*). Food consumption was related to food availability, likely as with other species in the genus. Based on 205 diet samples of specimens from tropical lowland forest, Baur (2008) found that diets included the seeds of 86 plant species, and overall diets comprised seeds (82%), 3–7% each for pulp, leaf, and grit and snail parts, and less than 2% each for other animals, flowers, and stems. Approximately one-half (47%) of the diet was shared with sympatric galliforms (Baur 2008). Baur (2008) found diet overlap of 65% with crested guan, 49% with great tinamou (*Tinamus major*), 39% with great curassow, and 33% with ocellated turkey (*Meleagris ocellata*).

Reproduction and behavior

The only reproductive study of any of the four chachalaca species included herein was conducted on plain chachalacas in South Texas (Marion 1978, Marion and Fleetwood 1978). Little information for other species is available; however, a few generalities can be extrapolated for chachalacas regarding latitude, altitude, and habitat conditions relative to the timing of reproduction (Peterson 2000). Chachalacas are probably monogamous, with pair formation beginning when individuals are still in winter feeding flocks (Wagner 1953, Leopold 1959, Delacour and Amadon 1973, Marion and Fleetwood 1978). Typically, in early spring (February–March in their northern range), an increase in activity and loud calling signals the beginning of the breeding season (Marion and Fleetwood 1978). Once mating occurs, chachalaca pairs begin nest construction and egg laying. Most nests are in trees, tall shrubs, or vines supported by trees or shrubs (Wagner 1953, Leopold 1959, Peterson 2000); however, some studies report that chachalacas occasionally nest on the ground between shrubs (Alvarez del Toro 1952). In Chiapas, nests were reported to be 6–10 m (19.7–32.8 ft) above the ground (Wagner 1953), whereas in the geographically proximal lowlands of Guatemala, nests ranged from 1 to 18 m (3.3 to 59 ft) above the ground (mean = 5.7 m, or 18.7 ft) in trees averaging 16.9 cm (6.7 in) DBH; 73% of nests were in trees (Baur 2008). Peak nesting season is late spring but varies considerably by latitude (March–July) (Alvarez del Toro 1952, Wagner 1953, Paynter 1955, Leopold 1959, Marion and Fleetwood 1978, Baur 2008); it is triggered by seasonal temperature changes in northern regions, or the beginning of the wet season in southern areas (Wagner 1953, Delacour and Amadon 1973). Nesting pairs are territorial, and males aggressively defend nests during the incubation period (Delacour and Amadon 1973, Marion 1974a). Incubation is approximately 25–28 days (Peterson 2000), with a clutch size of two to three eggs (Bent 1932, Oberholser 1974, Peterson 2000, Baur 2008). Chicks are brooded by both parents, parental care is provided for approximately six to eight months (Marion and Fleetwood 1978), and offspring are present from March to October (peaking from April to June) (Baur 2008).

Chachalacas are not strong fliers and instead make short flights from tree to tree or glide through dense vegetation with skill and accuracy (Bent 1932, Wagner 1953, Leopold 1959, Oberholser 1974). One of the most striking characteristics of chachalacas is the loud, raucous calls a flock emits when in full chorus (Peterson 2000). The resultant sound has been previously described as making "the valleys ring" (Baird 1857); "loud and simply indescribable" and "deafening" (Sennett 1878); "a pandemonium" (Sutton 1951); "ear-splitting" (Delacour and Amadon 1973); and comparable to the sound of howler monkeys (*Alouatta* spp.) (Oberholser 1974). To humans, this overlapping chorus sounds much like *cha-cha-lac-a*—hence the onomatopoeic common name for this genus (Delacour and Amadon 1973, Oberholser 1974, Peterson 2000). Pairs and family groups (3–5) are the basic social unit (Leopold 1959, Delacour and Amadon 1973, Marion 1974a, Peterson 2000), but flocks of 10–20 form during the fall and winter (Smith 1910, Teale 1965, Stiles and Skutch 1989) and probably represent transitory family group aggregations resulting from factors such as clumped food availability (Sutton 1951, Teale 1965, Marion 1976, Brooks 1997a).

Population dynamics

There are no reliable survey methods for chachalacas because of their secretive nature and the dense habitats they typically occupy, although road transects can provide a relative population index (Brooks 1997b). Marion et al. (1981) and Marion (1982) used loud prerecorded calls to estimate density in South Texas for plain chachalacas; they reported a density of 2.5 birds/ha (1 bird/ac), or approximately 15,000–18,000 birds for the three-county region (Marion 1974b). Though absolute density estimates for chachalaca species are not available, overall populations appear to be doing well throughout their range, despite considerable hunting pressure wherever sufficient suitable habitat remains (Lowery and Dalquest 1951, Leopold 1959, Delacour and Amadon 1973). Chachalacas are permanent residents in the areas they occupy (Peterson 2000).

Few studies have been conducted on chachalacas to determine survival and mortality factors. Incidental recapture of 10 banded plain chachalacas revealed that 1 male was 8 years of age or more, 3 females and 1 male were 7 years old or more, and the 5 remaining individuals were 5 years of age or more (Marion and Fleetwood 1974); thus, adults probably commonly live 5–8 years or more in the wild. Balda and Schemnitz (1997) translocated 17 birds to other parts of South Texas: 13 had lost or failed transmitters, 2 were killed by predators, and 2 were successfully tracked. Translocation to other sites in South Texas was not successful (Balda and Schemnitz 1997). Sex ratios of wild-trapped plain chachalacas were not significantly different ($p > 0.05$) from a 1:1 ratio (Marion and Fleetwood 1978). Similarly, of 26 birds collected in Chiapas, Mexico, the sex ratio was 12:14 males to females (Wagner 1953). The primary mortality factor for chachalacas is nest predation. In South Texas, mammalian predators such as raccoons (*Procyon lotor*) and Virginia opossum (*Didelphis virginiana*), and snakes such as Texas indigo (*Drymarchon corais erebennus*), accounted for 44% and 25% of 47 incubated-nest losses, respectively (Marion and Fleetwood 1978). Of 77 nests, only 3 (4%) were predated in Guatemalan lowlands (Baur 2008). The population-level significance of nonhuman predation, if any, is unknown. The black spider monkey (*Ateles paniscus*) and kinkajou (*Potos flavus*) might prey on nests in Chiapas, Mexico (Wagner 1953). Coyotes (*Canis latrans*), domestic dogs, bobcats (*Lynx rufus*), hawk-eagles (*Spizaetus* spp.), and Harris's hawks (*Parabuteo unicinctus*) are known or probable predators of juveniles and adults in Chiapas (Wagner 1953). Baur (2008) found annual harvest by humans to be 16% of all game birds harvested during the first year of study, and 25% the second year; respective first- and second-year total game bird biomass was 3% and 6%. Chachalacas represented the least-harvested galliform in total galliform biomass but were the third most harvested game birds in number of individuals, most of which were usually taken 0–4 km (0–2.5 mi) from camp (Baur 2008). Other recorded mortality factors for chachalacas included exposure to cold weather, which likely limits their northern distributional range (Balda 1989, Balda and Schemnitz 1997). Cold temperatures (–11.7°C, or 10.9°F) killed thousands of plain chachalacas in South Texas in February 1898 (Smith 1910).

Population management

Although market hunters pursued chachalacas historically (Peterson 2000), most direct exploitation today is by subsistence hunters; the taste and texture of chachalaca flesh are not desirable. These species have been domesticated by capturing chicks and using domestic hens to incubate eggs. Habitat destruction through clear-cutting continues to be the primary factor likely restricting their distribution (Peterson 2000, González-García et al. 2001). In some regions, native habitat has been replaced with vast industrialized, agricultural landscapes, while urban sprawl and other habitat conversions have destroyed native habitats. Unregulated exploitation, particularly in sites near towns and villages, can depress population numbers even where habitat is otherwise suitable. Despite these challenges, chachalacas fare well compared to other species of cracids under these environmental pressures. Population management strategies should include providing suitable habitat—both climax and second-growth forests and brushlands—and monitoring hunting pressure on local populations. Basic research on topics such as survival and reproduction is needed to establish a database to better manage populations.

Great curassow

Diet

The primary diet of the great curassow is fruit, but it also eats invertebrates and other plant materials (Leopold 1959). Using 267 samples, Baur (2008) found 74 species of seeds in the diet, with overall diet consisting of seeds (70%), grit and snail parts (20%), pulp (7%), 2% each of leaves and other animals, and less than 1% of flowers and stems. Sermeño (1997) reported that curassows consumed fruits of 15 plant species, leaves of 4 plant species, and 4 species of invertebrates. Some of the more common species of fruits consumed were represented by the following plant families: Moraceae (at least two species of *Ficus*, and *Brosimum alicastrum* and *B. terrabanum*), Sapotaceae (*Manilkara chicle* and *Mastichodendron capiri*), and Urticaceae (*Cecropia peltata* and *C. obtusifolia*) (Sermeño 1997). Rivas (1995) reported that curassow consumption included fruits and seeds of approximately 44 plant species, and over 22 species of arthropods; thus, curassows perform an important ecosystem service by dispersing and scarifying seeds in tropical montane forests (Rivas 2004). Dietary overlap is extensive with sympatric galliforms. Baur (2008) found 86% overlap with great tinamou, 68% with crested guan, 58% with ocellated turkey, and 39% with plain chachalaca, with a mean overlap of 63% for all four species.

Reproduction and behavior

Although curassows mature at two years of age, they often will not successfully reproduce until their third year (Alvarez del Toro 1952, Guido 1985). Pair formation begins following courtship, and males actively participate in nest construction and chick rearing, but only the females incubate (Guido 1985, Sermeño 1986). Baur (2008) provided the following breeding chronology: male booming calls from January to August (peaking March–June), nesting April–June (peaking in April), and offspring present April–October (peaking June–July).

The male has an elaborate courtship display that includes "strutting" with raised tail feathers and emitting a loud, booming "hum" or "oomp" from a tall treetop (Sutton and Pettingill 1942, Alvarez del Toro 1952, Garcia and Brooks 1997). When calling, the male sits on a branch or stands on the ground with lowered head and tail and "booms" to lure the female (Delacour and Amadon 1973; C. Garcia, personal observation). The male will offer the hen pieces of fruits or nuts, green leaf fragments, or even pebbles (D. Brooks, personal observation). The nest is made of branches and leaves (both dried and green), typically placed in the fork of a tree and often covered with vines, and measures 26–46 cm (9.4–18 in) in diameter, 16–27 cm (6.3–10.6 in) high at the sides, and 8–13 cm (3.1–5.1 in) deep (Guido 1985, Sermeño 1997). Nests are typically placed 4–10 m (13–33 ft) high in the tree but may be as high as 30 m (66 ft). Nest trees average 20 cm (7.9 in) DBH; at least 31 species of trees have been recorded to contain nests, with the most frequent being *Brosimum alicastrum*, *Faramea occidentalis*, *Hirtella racemosa*, *Inga punctata*, *Licania retifolia*, *Lonchocarpus salvadorensis*, *Rheedia edulis*, and *Swartzia simplex* (Sutton and Pettingill 1942, Guido 1985, Sermeño 1986, 1997, Baur 2008). After nest construction, the female lays two (rarely three) large white eggs, each averaging 250 g (0.6 lb), with an incubation period of 31–33 days. The female will take recesses off the nest ranging from 20 to 120 minutes. Chicks are brooded with biparental care for approximately eight months (Guido 1985, Quinto 1981, Sermeño 1997; J. Estudillo, personal communication).

In addition to the booming "oomp" call described above, alarm calls to warn of a potential predator are also made through a series of short "piit witt" calls (D. Brooks, personal observation). The basic social unit for the curassow is primarily individuals or small groups of up to 6 birds (Sermeño 1997, Zimmer 1999). Occasionally curassows form flocks of 20 or more birds of both sexes, but flocks typically do not exceed 10 to 15 birds (Guido 1985, Sermeño 1997, Zimmer 1997).

Population dynamics

Curassows are considered threatened (mainland population) to critically endangered (Cozumel population), and they have been extirpated from many areas, resulting in highly fragmented, localized populations (Howell and Webb 1995, Brooks and Strahl 2000, Rios and Muñoz 2006). Densities were estimated at 0.4–1.2 birds/km^2 (1–3.1 birds/mi^2) (minimum population = 500 individuals) in Costa Rican forest fragments (Vaughan 1983, McCoy 1997); 0.9 birds/km^2 (2.3 birds/mi^2) (97–304 total individuals) on Cozumel Island (Martinez-Morales 1999); 1.4 birds/km^2 (3.6 birds/mi^2) within Punta Manabique, a protected area along the Guatemalan Atlantic coast (Eisermann 2004); and 6–8 individuals along about 45 km (28 mi) of transect in Tikal, Guatemala (Donegan 2001).

There are few studies on longevity, mortality, or recruitment for great curassow populations (Arguedas et al. 1997). A captive female lived 24 years, laying two eggs per clutch for 23 of those years (Taibel 1940). Martinez-Morales (1999) reported a sex ratio of 1:1 for the Cozumel population, with approximately 80% of the estimated population being adults. Baur (2008) reported that of 24 nests, only 1 (4%) was predated. Predators include Neotropical felids, some canids and mustelids, large raptors that specialize on large vertebrates (e.g., *Spizaetus*), and humans (González-García et al. 2001). Parasites include ticks, cestodes, *Capillaria* sp., and *Eimeria* sp. (Delacour and Amadon 1973, Palomares 1980, García 1998). Baur (2008) found annual harvest by humans to be 278 individuals (47% of all game bird harvest), or 946 kg (2,086 lb) of biomass (58% of total game bird biomass). This represented by far the most-hunted game bird in the region, and individuals were usually harvested 4–8 km (2.5–5 mi) from camp (Baur 2008).

Population management

Principal threats to the great curassow are habitat loss or degradation and hunting (Brooks and Strahl 2000, Rios and Muñoz 2006). The size of the curassow makes it an easy target for human hunters (i.e., for human consumption, sport, and illegal pet trafficking) (Brooks and Strahl 2000, Rios and Muñoz 2006). Even though the species can withstand some level of habitat alteration (McCoy 1997), the great curassow is sensitive to changes in habitat structure. In some regions, native habitat has been replaced with vast industrialized, agricultural landscapes; in addition, urban sprawl, along with other habitat conversions, has also displaced this species (Vannini and Rockstroh 1997). Population management strategies include providing suitable habitat—both climax and pristine montane tropical forests—and monitoring and regulating hunting pressure on local populations (Rios and Muñoz 2006). More detailed research on life history and survivorship is needed to better manage these populations.

Guans

Diet

The primary diet for guans, as for other cracids, is fruits, but they also eat other plant materials (e.g., tender leaves, orchid flowers) and some invertebrates (Leopold 1959, González-García 2005). González-García (2005) reported that horned guans consumed at least 57 plant species. In Guatemala, horned guans consume fruits of *Dendropanax arboreus*, *Symplocos hartwegii*, *Phoebe* sp., and nectar of *Chiranthodendron pentadactylon* (Gonzalez-García et al. 2006a; O. Mendez, personal communication). Alvarez del Toro (1952) observed highland guans "on the ground scratching in the leaf litter for fruits, insects, and tender green plants," although their diet may also include crabs, small rodents, and salamanders (Rowley 1984, González-García et al. 2001). Highland guans have been recorded eating fruits of 24 different genera in El Triunfo, Mexico (F. González-García, personal observation), and 10 genera in Montecristo, El Salvador (Pullen 1978), 6 of which were different genera from those in Mexico, with a total of at least 33 different species of fruit. Based on 142 samples from Guatemalan lowlands, Baur (2008) found 72 species of seeds in the diet of the crested guan, with overall diet consisting of seeds (79%), 9% each for pulp and grit/snail parts, and less than 1% each for leaves, flowers, stems, and other animals. In Costa Rican dry forest, crested guans consumed 38 species of plants from 27 families, with the diet constituting 76% fruit, 21% flowers, leaves, and tender shoots, and 3% ants, but only when fruit was scarce (Pacheco 1994). Leopold (1959) examined bird crops and found that the crested guans consumed primarily fruit, including that of the capulin selvaje (*Bumelia peninsularis*); additionally, he reported the consumption of xocotl plums (*Spondias mombin*) as a dominant spring food (fig. 5.12). McCoy (1997) noted crested guans consuming *Bumelia peninsularis*, *Chione mexicana*, *Guatteria* sp., and *Symplocos* spp. Seeds of *Tetragastris panamensis* and *Virola surinamen-*

Figure 5.12. *Spondias mombin* plums provide spring food for the crested guan. Photo by Marco Schmidt.

sis are among the seeds dispersed by the crested guan (fig. 5.13) (Howe and Vande Kerckhove 1980, 1981; Howe 1984, 1993). Dietary overlap between crested guans and sympatric galliforms is extensive. Baur (2008) found 73% overlap with great tinamou, 68% with great curassow, 65% with plain chachalaca, and 57% with ocellated turkey, with a mean overlap of 66% for all four species.

Reproduction and behavior

Reproduction may start as early as two years of age in crested and highland guans (Wagner 1953, Leopold 1977). The breeding season for horned and crested guans begins in the spring and continues through the summer (March–August) (Alvarez del Toro 1952, Wagner 1953, Baur 2008). The reproductive season of the highland guan is from January to June, peaking between March and April, and is initiated by calls of the male, the first of which are heard sporadically during November and December (Jiménez 2002); in El Salvador, O. Komar (personal observation) reported a female incubating on 21 July.

During courtship, male crested guans raise their crest feathers, display the bare part of their throat, which is an intense red color, and emit vocalizations that can be heard as far as 0.5 km (0.3 mi), primarily during crepuscular periods (Schäefer 1953; F. González-García, personal observation). The whistle of the male highland guan during courtship is its best-known vocalization; the male also flies from its perch, producing a sound like a stampede or fabric tearing (Dickey and Van Rossem 1938, Salvin and Godman 1902, Andrle 1967), and makes short hops among branches close to the female, raising and lowering its tail with great agility (Jiménez 2002). Pair formation of highland guans begins in January, and nesting behavior begins in February (Jiménez 2002). Horned guans emit courtship calls primarily from November through May, when males make short flights from tree to tree, constantly emitting bellowing vocalizations to lure females; males will also offer fruit and leaf fragments to females, either regurgitated or directly (González-García 1995, Gómez de Silva et al. 1999, González-García et al. 2001). The social system of the horned guan is serial polygyny; the male has access to three or four females (González-García 1995). Most guans are probably polygamous, with limited to no assistance from the males during the nesting season (e.g., in nest construction, brooding, and rearing) (Wagner 1953, González-García 1994, 1995; González-García et al. 2001). The courtship call is elaborate and includes "drumming" (i.e., rapid beating of wings) (Salvin and Godman 1897–1904, Sutton and Pettingill 1942, Alvarez del Toro 1952), similar to the sound of the ruffed grouse (*Bonasa umbellus*).

Figure 5.13. The seeds of *Virola surinamensis* are consumed and dispersed by the crested guan. Photo by Feroze Omardeen

In Mexico, horned guans build their nest of bromeliad roots, orchids, and dead leaves; it measures 32 × 30 cm (12.6 in × 11.8 in) and is located in upper, isolated canopy trees with an average height of 19 m (62 ft), in species including *Matudaea trinervia*, *Ternstroemia lineata*, *Quercus* sp., and *Clethra* sp. (González-García 1994, 1995, 1997). In contrast, Méndez (2000) found a Guatemalan horned guan nest at a height of 7.8 m (25.6 ft) in *Chiranthodendron pentadactylon*. The female horned guan will take one to four 30-minute breaks per day to eat, drink, or dust bathe (González-García 1995, 1997). All but 1 of 16 highland guan nests were in trees, averaging 5 m (16 ft) above the ground (range = 2.4–12.1 m, 7.9–39.7 ft) (Rowley 1984); similarly, F. González-García (personal observation) found nests 1–15 m (3.3–49 ft) above the ground. Highland guans generally construct their nest in forks of medium-sized trees; external measurements are 13 cm (5.1 in) deep × 30 cm (11.8 in) in diameter, and cup measurements are 4.5 cm (1.8 in) deep × 20 cm (7.9 in) in diameter; the nest is made of interwoven leaves and green twigs and is covered with dry leaves, moss, and sometimes pine needles (Pullen 1983, González-García et al. 2001, Jiménez 2002). Baur (2008) found crested guan nests invariably in trees, at a height ranging from 4 to 15 m (13 to 49 ft) (mean = 9.75 m, 32 ft), in trees averaging 26.3 cm (10.4 in) DBH. The nest of the crested guan is a large, bulky structure covered with branches and both green and dry leaves (Rowley 1984). Nest construction and egg laying (clutch size = 2) begin immediately following the mating season, with an incubation period of approximately 35–36 days in horned guans (Wagner 1953, Rowley 1984, González-García 1995, Baur 2008), 25–28 days estimated in highland guans (González-García et al. 2001), and 26–28 days in captive crested guans (data not available for wild birds) (Taibel 1957). Horned guan chicks hatch within an hour or two of one another, stay in the nest for three to seven days, and then leave the nest, where they "peep" for the female to reunite the group (González-García 1995). The chicks of the highland guan take approximately 18 hours to hatch and dry. They leave the nest the day after hatching, after which the female diminishes her arboreal activity and stays with the chicks (Andrle 1967, Pullen 1978, Rowley 1984, González-García et al. 2001; F. González-García, personal observation). Crested guan offspring have been encountered from May to September, with a peak in July (Baur 2008). The chicks leave the nest only minutes after hatching, where they initially live on the ground but quickly shift to trees (Leopold 1977).

When disturbed on the ground, crested guans immediately fly high and land on tree branches or fly strongly for short distances; the main reaction to a potential threat, however, is to remain silently perched high in the branches, where their location can be revealed by displaced fruit that falls to the ground (Pacheco 1994, González-García et al. 2001). The "cauk, cauk, cauk" call of crested guans is emitted up to 144 times/minute (Kilham 1978); other vocalizations can be compared to pig "grunts," but the usual alarm call is a "honk" (Lowery and Dalquest 1951, Slud 1964, McCoy 1997). The whistle of the male highland guan is used for territoriality, and courtship that involves two phases: first, the male perches on a branch with an open bill and head thrown backward, emitting a loud, guttural whistle (like that of a human) that lasts 2.5–3 seconds and is phonetically similar to a "caahahahuiuiui." This is generally followed by the other vocalization, which is structured in three phrases; the first two phrases are two whistles each, and the third phrase has a variable number of whistles (Pullen 1978, Jiménez 2002). Male and female horned guans make four and eight types of vocalizations, respectively (González-García 1995). The estimated area of activity for a nesting female horned guan is 9 ha (22 ac), and 8 ha (19.8 ac) for a male (González-García 1991). They are typically solitary, in pairs, or in small groups of up to five (Andrle 1967, Parker et al. 1976, Pullen 1978, 1983). The small groups appear to be females with young subadults (González-García et al. 2001). Apart from this, little is known about guan social structure and spacing.

Population dynamics

The density of horned guans at El Triunfo was estimated to be 4.5–6.7 birds/km^2 (11.8–17.6 birds/mi^2) (González-García 1995, Gómez de Silva et al. 1999). The entire population in Mexico was estimated at 4,500–6,700 individuals (González-García 2005). Highland guan density estimates are similar in Guatemala (1–10 birds/km^2, or 2.6–26 birds/mi^2) (Eisermann 2004, Renner 2003), though some estimates exceed 25 birds/km^2 (65 birds/mi^2) (Morales 1991, 2004). Brooks and Strahl (2000) considered the latter density estimates unrealistic. In El Salvador, abundance (per four hours of observation) averaged 1 bird in pine-oak forest, 0.7 in cloud forest, and 0.3 in cypress plantations (Komar 2002), whereas Pullen (1983) recorded 121 sightings during 126 days in Montecristo Reserve. Reliable density estimates for crested guans are lacking. The status of these species ranges from lower risk (crested guan) to vulnerable (highland guan) to critically endangered (horned guan) (Brooks et al. 2006, Eisermann et al. 2006, González-García et al. 2006a). Outside of natural history studies of guans (González-García 1995, Delacour and Amadon 1973), there are few studies on longevity, mortality, and recruitment.

Possible predators on eggs and chicks of horned and highland guans include green toucanet (*Aulacorhynchus prasinus*), red-tailed hawk (*Buteo jamaicensis*), and fulvous owl (*Strix fulvescens*). Various carnivores that prey on adults include tayra (*Eira barbara*), kinkajou (*Potos flavus*), coati (*Nasua narica*), jaguar (*Panthera onca*), and puma (*Puma concolor*) (González-García et al. 2001). Primary predators of crested guan include ocelot (*Leopardus pardalis*), margay (*Leopardus wiedii*), and puma (*Puma concolor*) (Schäefer 1953). Raptors such as the ornate hawk-eagle (*Spizaetus ornatus*) attack adult crested guans (Kilham 1978), and birds such as toucans probably prey on eggs in the nest (González-García et al. 2001). Of 19 crested guan nests in Guatemalan lowlands, none were preyed on (Baur 2008). Baur (2008) found annual harvest of crested guans by humans to be 7% of the total game bird harvest during the first year of study, and 17% the second year; respective first- and second-year biomass quantities were 6% and 16% of total game bird biomass.

Population management

Principal threats to guans are habitat loss or degradation and unregulated hunting (Brooks and Strahl 2000, González-García 2005, González-García et al. 2001, 2006a). In recent years, uncontrolled forest fires have threatened prime guan habitat. Asbjornsen et al. (2005), for example, reported a 60% reduction in the Chimalapas National Forest Reserve. Population management strategies include providing suitable habitat and continued protection from hunting pressure (e.g., the horned guan is officially protected in Mexico) (Brooks and Strahl 2000, González-García et al. 2006a). Little is known about guan life histories; basic research addressing survival and reproduction is needed to better manage these populations.

Habitat and Management

Habitat requirements

Chachalacas occupy a variety of subtropical and tropical habitat types ranging from altered landscapes (e.g., plantations, agriculture-brushland mosaics, urban areas, degraded forests) to lowland moist forests, tropical mangroves, and lowland deciduous thorn shrublands (del Hoyo 1994, Stattersfield et al. 1998, Peterson 2000). In Mexico, chachalacas are rarely found at elevations above 1,300 m (4,265 ft) (del Hoyo 1994, Peterson 2000) and are classified as species of "least concern" (Brooks et al. 2006). In general, and unlike other cracids, chachalacas thrive in the thickets and brushlands that often arise after the clearing of tropical forests (Alvarez del Toro 1952, Leopold 1959, Delacour and Amadon 1973, Martínez-Sánchez 1997, Peterson 2000).

In contrast, guans and curassows occupy various subtropical and tropical habitats ranging from wet premontane and montane broad-leaved forests to lowland moist forests and tropical mangroves (Eisermann et al. 2006, González-García et al. 2006a, 2006b, Rios and Muñoz 2006). The general habitats guans and curassows prefer, however, are typically areas with less human disturbance (e.g., climax forest) (Eisermann et al. 2006; González-García et al. 2006a, 2006b; Rios and Muñoz 2006). Highland and horned guans typically occupy habitats at higher elevations (700–3,330 m, 2,297–10,827 ft), whereas the crested guan and great curassow are found at lower elevations (0–2,000 m, 0–6,562 ft), with the latter group in many cases overlapping with chachalacas (Escobar-Ortíz 1997). In Mexico, the status of guans and curassows ranges from near threatened (e.g., great curassow), to vulnerable (e.g., highland guan), to critically endangered (e.g., horned guan) (Eisermann et al. 2006, González-García et al. 2006a, Rios and Muñoz 2006).

Management

Habitat destruction and unsustainable hunting practices are principal threats to cracids throughout Mexico (Eisermann et al. 2006, González-García et al. 2006a, 2006b, Rios and Muñoz 2006). The sharp differences in the population responses of chachalacas versus those of guans and the great curassow related to these primary threats have already been noted. Cracid management recommendations to address threats in Mexico fall into three broad categories: (1) forest management (i.e., protection, restoration, and coordination of timber management practices); (2) management of hunting pressure (i.e., numbers taken, access to species); and (3) ecological research related to population dynamics and species response to habitat management practices. Each of these will be discussed individually.

Forest management

Management of cracids in Mexico requires continued protection and preservation of remaining habitats for these species (González-García et al. 2006b). This is particularly important for species that depend on climax forests, such as montane or cloud forest habitats (e.g., horned guan, highland guan). Deforestation is prevalent throughout Mexico and much of Central America. For example, despite having over 55 million ha (135 million ac) of forest cover in 2000 (the second-highest area of any Latin American country next to Peru), Mexico lost over

630,000 ha (1.6 million ac) per year from 1990 to 2000, a greater yearly loss than in any other Latin American country (Brooks 2006). This has resulted in habitat fragmentation and increased human access, which collectively relegate cracid populations to isolated habitat patches where they are more vulnerable to extirpation by subsistence hunters or other human activities (Martínez-Sánchez 1997). National parks and other protected areas can serve to protect cracids in Mexico (McCoy 1997, Midence 1997). However, legal and physical protection of habitat and protection of populations are not necessarily synonymous. Although designating protected areas is a useful conservation strategy, such designation does not necessarily prevent fuel-wood collection, grazing, illegal hunting, or even extraction of valuable timber resources, particularly near the boundaries of protected areas (Bruner et al. 2001).

Hunting

A second management recommendation for cracid conservation in Mexico is gaining a better understanding of the effects of unrestricted hunting and controlling it where appropriate. Illegal and unregulated subsistence hunting can limit cracid populations and can cause local extirpations, particularly in habitat "islands" near villages or local communities (McCoy 1997, Vannini and Rockstroh 1997, González-García et al. 2006b). Impacts from hunting obviously vary among the different cracid species in Mexico. In such cases, active regulation of hunting of the more accessible populations is recommended (González-García et al. 2006b).

Historically, hunting statutes and regulations have been instituted but not often heeded by subsistence hunters (Leopold 1959). Ambiguously worded statutes also serve to render legal protection uncertain (Vannini and Rockstroh 1997). Even if laws were clarified, however, insufficient agents and funding would inhibit their meaningful enforcement. Strictly regulated hunting practices should be initiated for cracid populations that occupy sufficiently large areas of suitable habitat and that can tolerate some level of hunting (e.g., chachalacas). Thus, chachalaca hunting, for example, would continue to provide human food and recreation as well as badly needed funds for species conservation (Waggerman 1979, McCoy 1997). This option deserves further exploration. An increase in basic knowledge regarding the ecology of these species—and their anticipated population responses to external variables such as land-use changes (e.g., patch size and condition) and other human-related influences (e.g., subsistence hunting and other mortality factors)—is required to establish sound management of these species in Mexico.

Research

The biology of most cracid species is poorly understood, despite great improvements in the last two decades (Brooks et al. 2006). These species need more basic research for informed wildlife management. High-priority research needs include (1) surveys of local population densities and total population sizes; (2) estimates of rates of cloud-forest destruction and degradation within the range of each species; (3) studies of population responses to land-use changes, particularly estimates of population dynamics (e.g., survival, mortality factors, reproduction); and (4) assessment of the economic and biological impacts of nature tourism and ecotourism related to cracids (González-García et al. 2006b). Other important areas of research include the impact of climate change on cracid populations, the impact of wildfires in primary cloud forests, and the benefits of forest restoration practices to cracid populations.

Management Challenges in Mexico

The greatest management challenges for cracid conservation in Mexico are protecting remaining habitat for cracid species, particularly those requiring mature forest, and improving hunting impacts where appropriate. The former will require incentives for forest conservation and restoration. Currently, the rate of deforestation in Mexico is driven by traditional uses (e.g., timber exploitation, agriculture, mining). A shift in management approaches will require fundamental changes in the value of these areas beyond traditional uses (e.g., timber, minerals) to include alternative or emerging values (e.g., ecotourism, regulated sport hunting). Conservation of these areas through these alternative approaches can ultimately serve to protect habitats critical to cracids throughout Mexico. To improve hunting impacts, basic knowledge of these impacts on cracid population dynamics, followed by implementation of pragmatic management strategies based on these data, is required. Additionally, increased enforcement coupled with improved reserve design and hunter access will require greater coordination in the management of remaining habitats for cracid populations.

Literature Cited

Alvarez del Toro, M. 1952. Los animales silvestres de Chiapas. Ediciones del Gobierno del Estado, Tuxtla Gutiérrez, Chiapas, México.

Andrle, R. F. 1967. Notes on the black chachalaca (*Penelopina nigra*). Auk 84:169–172.

Arguedas, N., P. Mayer, D. Milligan, and L. Nordstrom. 1997. Plan de recuperación para *Crax rubra* en Costa Rica. Pages 249–262 *in* S. D. Strahl, S. Beaujon, D. M. Brooks, A. J. Begazo, G. Sedaghatkish, and F. Olmos, editors. The Cracidae: their biology and conservation. Hancock House, Blaine, Washington, USA.

Arriaga-U., I., and B. Baquero. 1997. Criterios para la clasificación de grupos por edad y aspectos reproductivos en la guacharaca (*Ortalis ruficauda*) en el estado Cojedes. Pages 48–60 *in* S. D. Strahl, S. Beaujon, D. M. Brooks, A. J. Begazo, G. Sedaghatkish, and F. Olmos, editors. The Cracidae: their biology and conservation. Hancock House, Blaine, Washington, USA.

Asbjornsen, H., N. Velázquez-Rosas, R. García-Soriano, and C. Gallardo-Hernández. 2005. Deep ground fires cause massive above- and below-ground biomass losses in tropical montane cloud forests in Oaxaca, Mexico. Journal of Tropical Ecology 21:427–434.

Baird, S. F. 1857. Birds of the boundary, with notes by the naturalists of the survey. United States and Mexican boundary survey. Government Printing Office, Washington, DC, USA.

Balda, W. E. 1989. Evaluation of success of plain chachalaca (*Ortalis vetula*) transplants in South Texas. Thesis, New Mexico State University, Las Cruces, USA.

Balda, W. E., and S. D. Schemnitz. 1997. Evaluation of success of plain chachalaca (*Ortalis vetula*) transplants in South Texas. Pages 201–210 *in* S. D. Strahl, S. Beaujon, D. M. Brooks, A. J. Begazo, G. Sedaghatkish, and F. Olmos, editors. The Cracidae: their biology and conservation. Hancock House, Blaine, Washington, USA.

Baur, E. H. 2008. Structure of a lowland Neotropical galliform bird guild. Thesis, University of Florida, Gainesville, USA.

Begazo, A. J. 1997. Use and conservation of the Cracidae in the Peruvian Amazon. Pages 449–459 *in* S. D. Strahl, S. Beaujon, D. M. Brooks, A. J. Begazo, G. Sedaghatkish, and F. Olmos, editors. The Cracidae: their biology and conservation. Hancock House, Blaine, Washington, USA.

Begazo, A. J., and R. E. Bodmer. 1998. Use and conservation of Cracidae (Aves: Galliformes) in the Peruvian Amazon. Oryx 32:301–309.

Bent, A. C. 1932. Life histories of North American gallinaceous birds. Bulletin of the United States National Museum 162:1–490.

Berlanga, G. H. A. 1991. Las aves frugívoras de Chamela, Jalisco. Su recurso vegetal y su papel en la dispersión de semillas. Thesis, UNAM, México, D. F., México.

Brodkorb, P. 1964. Catalogue of fossil birds. Part 2. Bulletin of the Florida State Museum 8:195–335.

Brooks, D. M. 1997a. The influence of habitat structure upon species evenness and diversity. Texas Journal of Science 49:247–254.

Brooks, D. M. 1997b. Population and ecological parameters of the Chaco chachalaca (*Ortalis canicollis*). Pages 412–417 *in* S. D. Strahl, S. Beaujon, D. M. Brooks, A. Begazo, G. Sedaghatkish, and F. Olmos, editors. The Cracidae: their biology and conservation. Hancock House, Blaine, Washington, USA.

Brooks, D. M. 1999. *Pipile* as a protein source to rural hunters and Amerindians. Pages 42–50 *in* D. M. Brooks, A. J. Begazo, and F. Olmos, editors. Biology and conservation of the piping guans (*Pipile*). Special Monograph 1. Cracid Specialist Group, Houston, Texas, USA.

Brooks, D. M. 2006. The utility of hotspot identification for forest management: cracids as bioindicators. Acta Zoologica Sinica 52:199–201.

Brooks, D. M., L. Cancino, and S. L. Pereira, editors. 2006. Conserving cracids: the most threatened family of birds in the Americas. Miscellaneous Publication Number 6. Houston Museum of Natural Science, Texas, USA.

Brooks, D. M., and R. A. Fuller. 2006. Biology and conservation of cracids. Pages 9–21 *in* D. M. Brooks, L. Cancino, and S. L. Pereira, editors. Conserving cracids: the most threatened family of birds in the Americas. Miscellaneous Publication Number 6. Houston Museum of Natural Science, Texas, USA.

Brooks, D. M., and S. D. Strahl. 2000. Cracids: status survey and conservation action plan. IUCN, Gland, Switzerland.

Bruner, A. G., R. E. Gullison, R. E. Rice, and G. A. B. da Fonseca. 2001. Effectiveness of parks in protecting tropical biodiversity. Science 291:125–128.

Christensen, Z. D., D. B. Pence, and G. Scott. 1978. Notes on food habits of the plain chachalaca from the Lower Rio Grande Valley. Wilson Bulletin 90:647–648.

Delacour, J., and D. Amadon. 1973. Curassows and related birds. American Museum of Natural History, New York, New York, USA.

Delgado, F. S. 1997. Situación de los cracidos de Panamá. Pages 345–349 *in* S. D. Strahl, S. Beaujon, D. M. Brooks, A. J. Begazo, G. Sedaghatkish, and F. Olmos, editors. The Cracidae: their biology and conservation. Hancock House, Blaine, Washington, USA.

del Hoyo, J. 1994. Family Cracidae. Pages 310–363 *in* J. del Hoyo, A. Elliot, and J. Sargatal, editors. Handbook of the birds of the world. Volume 2. Lynx Edicions, Barcelona, Spain.

Dickey, D. R., and A. J. Van Rossem. 1938. The birds of El Salvador. Field Museum of Natural History, Zoological Series 23:1–609.

Donegan, T. M. 2001. Recent records of cracids from Tikal, Petén, Guatemala, with notes on their conservation. Boletín de Grupo de Especialistas en Crácidos 12:4–7.

Eisermann, K. 2004. Estatus de *Crax rubra* en punta de Manabique, Guatemala: hábitat, tamaño de la población e impacto humano. Boletín de Grupo de Especialistas en Crácidos 18:4–8.

Eisermann, K. 2005. Noteworthy bird observations in Alta Verapaz, Guatemala. Bulletin of the British Ornithologists Club 125:3–11.

Eisermann, K., N. Herrera, and O. Komar. 2006. Highland guan (*Penelopina nigra*). Pages 85–89 *in* D. M. Brooks, L. Cancino, and S. L. Pereira, editors. Conserving cracids: the most threatened family of birds in the Americas. Miscellaneous Publication Number 6. Houston Museum of Natural Science, Texas, USA.

Érard, C., and M. Théry. 1994. Frugivorie et ornithocorie en fôret Guyanaise: l'exemple des grans oiseaux terrestres et de la *Penelope marail*. Alauda 62:27–31.

Escobar-Ortíz, E. R. 1997. Situación actual del pavo de cacho *Oreophasis derbianus* en Guatemala. Page 217 *in* S. D. Strahl, S. Beaujon, D. M. Brooks, A. J. Begazo, G. Sedaghatkish, and F. Olmos, editors. The Cracidae: their biology and conservation. Hancock House, Blaine, Washington, USA.

Garcia, C., and D. M. Brooks. 1997. Evolution of *Crax*: sociobiology and phylogeny using behavioral and ecological characters. Pages 401–410 *in* S. D. Strahl, S. Beaujon, D. M. Brooks, A. Begazo, G. Sedaghatkish, and F. Olmos, editors. The Cracidae: their biology and conservation. Hancock House, Blaine, Washington, USA.

García, F. L. M. 1998. Determinación de parásitos gastrointestinales de crácidos silvestres libres y en cautiverio. Thesis, UNAM, México, D. F., México.

Gómez de Silva, G. H., F. González-García, and M. P. Casillas-Trejo. 1999. Birds of the upper cloud forest of El Triunfo, Chiapas, Mexico. Ornitología Neotropical 10:1–26.

González-García, F. 1991. Observaciones sobre la ecología y biología reproductiva del pavón *Oreophasis derbianus* en la Reserva de la Biosfera El Triunfo, Chiapas, México. IV Congreso de Ornitología Neotropical, Quito, Ecuador.

González-García, F. 1994. Behavior of horned guans in Chiapas, Mexico. Wilson Bulletin 106:357–365.

González-García, F. 1995. Reproductive biology and vocalizations of the horned guan *Oreophasis derbianus* in Mexico. Condor 97:415–426.

González-García, F. 1997. Conducta de anidación del pavón *Oreophasis derbianus* (Aves, Cracidae) en la Reserva de la Biosfera El Triunfo, Chiapas, México. Pages 147–166 *in* S. D. Strahl, S. Beaujon, D. M. Brooks, A. J. Begazo, G. Sedaghatkish, and F. Olmos, editors. The Cracidae: their biology and conservation. Hancock House, Blaine, Washington, USA.

González-García, F. 2001. Estado de conservación del pavón *Oreophasis derbianus* en el campo y en cautiverio en México. Pages 147–166 *in* D. M. Brooks and F. González-García, editors. Cracid ecology and conservation in the new millennium. Miscellaneous Publication Number 2. Houston Museum of Natural Science, Texas, USA.

González-García, F. 2005. Dieta y comportamiento de forrajeo del pavón *Oreophasis derbianus* en la Reserva de la Biósfera El Triunfo, Chiapas, México. Thesis, UNAM, México, D. F., México.

González-García, F., D. M. Brooks, and S. D. Strahl. 2001. Estado de conservación de los crácidos en México y Centroamérica. Pages 1–50 *in* D. M.

Brooks and F. González-García, editors. Cracid ecology and conservation in the new millennium. Miscellaneous Publication Number 2. Houston Museum of Natural Science, Texas, USA.

González-García, F., J. A. Rivas R., and A. J. Cóbar-C. 2006a. Horned guan (*Oreophasis derbianus*). Pages 35–40 *in* D. M. Brooks, L. Cancino, and S. L. Pereira, editors. Conserving cracids: the most threatened family of birds in the Americas. Miscellaneous Publication Number 6. Houston Museum of Natural Science, Texas, USA.

González-García, F., J. A. Rivas-R., A. J. Cóbar-C., K. Eisermann, N. Herrera, O. Komar, M. M. Rios, M. C. Muñoz, and D. M. Brooks. 2006b. Northern Central America: Mexico, Guatemala, Belize, El Salvador, Honduras and Nicaragua. Pages 114–116 *in* D. M. Brooks, L. Cancino, and S. L. Pereira, editors. Conserving cracids: the most threatened family of birds in the Americas. Miscellaneous Publication Number 6. Houston Museum of Natural Science, Texas, USA.

Guido, M. M. Y. 1985. Proyecto de reproducción de pavos silvestres (*Crax rubra*) en El Imposible, Zona II, San Benito. Informe interno. Servicio de Parques Nacionales y Vida Silvestre, Ministerio de Agricultura y Ganadería, San Salvador, El Salvador.

Gurrola, H. M. A. 1985. Hábitos de alimentación, reproducción y comportamiento de la chachalaca (*Ortalis p. poliocephala*, Aves: Cracidae) en la región costera de Chamela, Jalisco. Thesis, UNAM, México, D. F., Mexico.

Howe, H. F. 1984. Implications of seed dispersal by animals for tropical reserve management. Biological Conservation 30:261–281.

Howe, H. F. 1993. Aspects of variation in a Neotropical seed dispersal system. Vegetation 107/108:149–162.

Howe, H. F., and G. A. Vande Kerckhove. 1980. Nutmeg dispersal by tropical birds. Science 210:925–927.

Howe, H. F., and G. A. Vande Kerckhove. 1981. Removal of wild nutmeg (*Virola surinamensis*) crops by birds. Ecology 62:1093–1106.

Howell, S. N. G., and S. Webb. 1995. A guide to the birds of Mexico and northern Central America. Oxford University Press, New York, New York, USA.

Jiménez, G. J. 2002. Contribución al conocimiento de la biología, distribución y abundancia del pajuil *Penelopina nigra* (Aves: Cracidae, Fraser 1850) en la Reserva de la Biosfera, El Triunfo, Chiapas, México. Thesis, UNAM, México, D. F., Mexico.

Jorgenson, J. P. 1995. Maya subsistence hunters in Quintana Roo, Mexico. Oryx 29:49–57.

Jorgenson, J. P. 1997. Cambios en los patrones de la cacería de subsistencia a través de mejoramientos socio-económicos: el ejemplo de los cazadores Mayas en México. Pages 31–40 *in* T. G. Fang, R. E. Bodner, R. Aquino, and M. H. Valqui, editors. Manejo de fauna silvestre en la Amazonia. Universidad Arturo Prat, University of Florida, and United Nations Development Program/Global Environment Facility, La Paz, Bolivia.

Kilham, L. 1978. Alarm call of crested guan when attacked by ornate hawkeagle. Condor 80:347–348.

Komar, O. 2002. Birds of Montecristo National Park, El Salvador. Ornitología Neotropical 13:167–193.

Leopold, A. S. 1959. Wildlife of Mexico: the game birds and mammals. University of California Press, Berkeley, USA.

Leopold, S. A. 1977. Fauna silvestre de México. IMERNAR, México, D. F., México.

Lowery, G. H., and W. W. Dalquest. 1951. Birds from the state of Veracruz, Mexico. University of Kansas Publications, Museum of Natural History 3:531–649.

Mandujano, S., S. Gallina, and S. H. Bullock. 1994. Frugivory and dispersal of *Spondias purpurea* (Anacardiaceae) in a tropical deciduous forest in Mexico. Revista de Biología Tropical 42:107–114.

Marion, W. R. 1974a. Ecology of the plain chachalaca in the Lower Rio Grande Valley of Texas. Dissertation, Texas A&M University, College Station, USA.

Marion, W. R. 1974b. Status of plain chachalaca in South Texas. Wilson Bulletin 86:200–205.

Marion, W. R. 1976. Plain chachalaca food habits in South Texas. Auk 93:376–379.

Marion, W. R. 1978. Nesting ecology of the plain chachalaca in South Texas. Wilson Bulletin 90:386–395.

Marion, W. R. 1982. Plain chachalaca. Pages 57–58 *in* D. E. Davis, editor. CRC handbook of census methods for terrestrial vertebrates. CRC Press, Boca Raton, Florida, USA.

Marion, W. R., and R. J. Fleetwood. 1974. Longevity of chachalacas. Bird-Banding 45:178.

Marion, W. R., and R. J. Fleetwood. 1978. Nesting ecology of plain chachalaca in South Texas. Wilson Bulletin 90:386–395.

Marion, W. R., T. E. O'Meara, and D. S. Maehr. 1981. Use of playback recordings in sampling elusive or secretive birds. Studies in Avian Biology 6:81–85.

Martínez-Morales, M. A. 1996. The Cozumel curassow: abundance, habitat preference and conservation. Thesis, University of Cambridge, United Kingdom.

Martínez-Morales, M. A. 1999. Conservation status and habitat preferences of the Cozumel curassow. Condor 101:14–20.

Martínez-Sánchez, J. C. 1997. Deforestación y conservación de cracidos en Nicaragua: un informe preliminar. Pages 340–344 *in* S. D. Strahl, S. Beaujon, D. M. Brooks, A. J. Begazo, G. Sedaghatkish, and F. Olmos, editors. The Cracidae: their biology and conservation. Hancock House, Blaine, Washington, USA.

McCoy, M. 1997. Country report on cracids of Costa Rica. Pages 298–313 *in* S. D. Strahl, S. Beaujon, D. M. Brooks, A. J. Begazo, G. Sedaghatkish, and F. Olmos, editors. The Cracidae: their biology and conservation. Hancock House, Blaine, Washington, USA.

Méndez, O. 2000. Conducta de anidación del pavo de cacho *Oreophasis derbianus* en el Volcán Tolimán, Guatemala. IV Congreso de la Sociedad Mesoamericana para la Biología y la Conservación. Fundación Natura, Ciudad de Panamá, Panamá.

Midence, S. F. 1997. Informe de la familia Cracidae en Honduras. Pages 335–339 *in* S. D. Strahl, S. Beaujon, D. M. Brooks, A. J. Begazo, G. Sedaghatkish, and F. Olmos, editors. The Cracidae: their biology and conservation. Hancock House, Blaine, Washington, USA.

Morales, C. L. 1991. Estimados de población de la chachalaca negra (*Penelopina nigra*) en el sur-oeste de Guatemala por el método de censo de crácidos de Strahl y Silva. Informe final. Fundación Interamericana de Investigación Tropical, Universidad de San Carlos, Guatemala, Guatemala.

Morales, J. R. 2004. Programa piloto de conservación y manejo de especies cinegéticas en dos comunidades de la Reserva de Biosfera Sierra de las Minas. Segundo informe técnico parcial. Organización Nacional para la Conservación y el Ambiente, Guatemala, Guatemala.

Oberholser, H. C. 1974. The bird life of Texas. University of Texas Press, Austin, USA.

Pacheco, S. C. C. 1994. Hábitos alimentarios y uso estacional de hábitat de la pava crestada (*Penelope purpurascens*) en el bosque seco tropical, Parque Nacional Santa Rosa, Costa Rica. Thesis, Universidad Nacional de Costa Rica, Heredia, Costa Rica.

Palomares, H. E. 1980. Estudio coproparasitoscopico de algunas aves exóticas criadas en cautiverio. Thesis, UNAM, México, D. F., México.

Parker, T. A., III., S. Hilty, and M. Robbins. 1976. Birds of El Triunfo cloud forest, Mexico, with notes on the horned guan and other species. American Birds 30:779–782.

Paynter, R. A., Jr. 1955. The ornithogeography of the Yucatán Peninsula. Bulletin of the Peabody Museum of Natural History 9:1–347.

Peres, C. A. 2000. Effects of subsistence hunting on vertebrate community structure in Amazonian forests. Conservation Biology 14:240–253.

Peres, C. A. 2001. Perils in parks or parks in peril? Reconciling conservation in Amazonian reserves with and without use. Conservation Biology 15:793–797.

Peres, C. A., and M. G. M. van Roosmalen. 1996. Avian dispersal of "mimetic seeds" of *Ormosia lignivalvis* by terrestrial granivores: deception or mutualism? Oikos 75:249–258.

Peterson, M. J. 2000. Plain chachalaca (*Ortalis vetula*). *In* The birds of North America. Philadelphia Academy of Natural Sciences, Pennsylvania, and American Ornithologists' Union, Washington, DC, USA.

Pullen, T. M., Jr. 1978. A field study of the black chachalaca (*Penlopina nigra*, Fraser 1850) in El Salvador. Unidad de Parques Nacionales, Vida Silvestre, and Ministerio de Agricultura y Ganadería, San Salvador, El Salvador.

Pullen, T., Jr. 1983. Sound production and reproductive biology of the highland guan in El Salvador's Montecristo cloud forest. American Birds 37:948–950.

Quinto, A. J. F. 1981. Observaciones y reproducción de *Crax rubra* en condiciones seminaturales en San Felipe Bacalar, Quintana Roo, México. Pages 249–259 *in* J. Estudillo, editor. Primer simposio internacional de la familia Cracidae. UNAM, Morelos, México.

Renner, S. 2003. Structure and diversity of cloud forest bird communities in Alta Verapaz, Guatemala, and implications for conservation. Dissertation, Georg-August-Universität, Göttingen, Germany.

Rios, M. M., and M. C. Muñoz. 2006. Great curassow (*Crax rubra*). Pages 104–108 *in* D. M. Brooks, L. Cancino, and S. L. Pereira, editors. Conserving cracids: the most threatened family of birds in the Americas. Miscellaneous Publication Number 6. Houston Museum of Natural Science, Texas, USA.

Rivas, J. A. 2004. Endozoocoria en los crácidos (Aves: Galliformes) de Petén-Guatemala. Mesoamerica 8:59–60.

Rivas, R. J. 1995. Preferencias alimenticias del faisán o pajuil (*Crax rubra rubra* L.) en condiciones naturales. Thesis, Universidad de San Carlos, Guatemala, Guatemala.

Rowley, J. S. 1984. Breeding records of land birds in Oaxaca, Mexico. Proceedings of the Western Foundation of Vertebrate Zoology 2:74–224.

Salvin, O., and F. D. Godman. 1897–1904. Biologia Centrali-Americana: Aves. Volume 3. Taylor and Francis, London, United Kingdom.

Santamaría-G., M., and A. M. Franco. 1994. Historia natural del pajuil (*Mitu salvini*) y densidades poblacionales de cracidos en un bosque de la Amazonia Colombiana. Informe final. Wildlife Conservation Society, New York, New York, USA.

Schäefer, E. 1953. Estudio bioecológico comparativo sobre algunos Cracidae del norte y centro de Venezuela. Boletín de la Sociedad Venezolana de Ciencias Naturales 15:30–63.

Sedaghatkish, G. 1996. The importance of seed dispersers in the conservation of useful wild plant species: a case study of the avian family Cracidae. Thesis, University of Maryland, College Park, USA.

Sennett, G. B. 1878. Notes on the ornithology of the Lower Rio Grande of Texas, from observations made during the season of 1877. Bulletin of the US Geological and Geographical Survey of the Territories 4:1–66.

Sermeño, M. A. 1986. Alimentación y reproducción del pajuil *Crax rubra* en El Salvador. Thesis, Universidad de El Salvador, San Salvador, El Salvador.

Sermeño, M. A. 1997. Situación de la familia Cracidae en El Salvador. Pages 320–324 *in* S. D. Strahl, S. Beaujon, D. M. Brooks, A. J. Begazo, G. Sedaghatkish, and F. Olmos, editors. The Cracidae: their biology and conservation. Hancock House, Blaine, Washington, USA.

Sibley, C. G., and B. L. Monroe. 1990. Distribution and taxonomy of birds of the world. Yale University Press, New Haven, Connecticut, USA.

Silva, J. L., and S. Strahl. 1991. Human impact on populations of chachalacas, guans and curassows (Galliformes: Cracidae) in Venezuela. Pages 37–52 *in* J. G. Robinson and K. H. Redford, editors. Neotropical wildlife use and conservation. University of Chicago Press, Illinois, USA.

Simas, K. J. 1979. Observations on the white-bellied chachalaca *Ortalis leucogastra*. Informe no publicado. Unidad de Parques Nacionales, Vida Silvestre, and Ministerio de Agricultura y Ganadería, El Salvador, San Salvador.

Slud, P. 1964. The birds of Costa Rica: distribution and ecology. Bulletin of the American Museum of Natural History 128:1–430.

Smith, A. P. 1910. Miscellaneous bird notes from the Lower Rio Grande. Condor 12:93–103.

Stattersfield, A. J., M. J. Crosby, A. J. Long, and D. C. Wege. 1998. Endemic bird areas of the world: priorities for biodiversity conservation. BirdLife International, Cambridge, United Kingdom.

Stiles, F. G., and A. R. Skutch. 1989. A guide to the birds of Costa Rica. Christopher Helm, London, United Kingdom.

Sutton, G. M. 1951. Mexican birds: first impressions. University of Oklahoma Press, Norman, USA.

Sutton, G. M., and O. S. Pettingill Jr. 1942. Birds of the Gomez Farias region, southwestern Tamaulipas. Auk 59:1–34.

Taibel, A. M. 1940. Osservazioni sulla riproduzione in cattivitá di *Crax globicera globicera* Linneos. Rivista Italiana di Ornitologia 10:93–126.

Taibel, A. M. 1957. Riproduzione in cattivitá di *Penelope purpurascens* e *Penelope pileata*. Rivista Italiana di Ornitologia 3:3–28.

Teale, E. W. 1965. Wandering through winter. Dodd, Mead, New York, New York, USA.

Tordoff, H. B., and J. R. MacDonald. 1957. A new bird (family Cracidae) from the early Oligocene of South Dakota. Auk 74:174–184.

Vannini, J. P., and P. M. Rockstroh. 1997. The status of cracids in Guatemala. Pages 326–334 *in* S. D. Strahl, S. Beaujon, D. M. Brooks, A. J. Begazo, G. Sedaghatkish, and F. Olmos, editors. The Cracidae: their biology and conservation. Hancock House, Blaine, Washington, USA.

Vaughan, C. 1983. A report on dense forest habitat for endangered wildlife species in Costa Rica. Editorial Universidad Nacional, Heredia, Costa Rica.

Vaurie, C. 1965. Systematic notes on the bird family

Cracidae. No. 2. Relationship and geographical variation of *Ortalis vetula*, *Ortalis poliocephala*, and *Ortalis leucogaster*. American Museum Novitates 2222:1–36.

Vaurie, C. 1968. Taxonomy of the Cracidae (Aves). Bulletin of the American Museum of Natural History 138:131–260.

Vuilleumier, F. 1965. Relationships and evolution within the Cracidae. Bulletin of the Museum of Comparative Zoology 134:1–27.

Waggerman, G. L. 1979. Brush country pheasants. Texas Parks and Wildlife 37:20–22.

Wagner, H. O. 1953. Die Hockohühner der Sierra Madre de Chiapas, Mexiko. Veröffentlichungen aus dem Überseemuseum Bremen 2:105–128.

Wetmore, A. 1956. A fossil guan from the Oligocene of South Dakota. Condor 58:234–235.

Zimmer, B. 1997. Observations on a large flock of greater curassow (*Crax rubra*). Bulletin of the Cracid Specialist Group 5:10–12.

Zimmer, B. 1999. Observations on a barred morph of the greater curassow (*Crax rubra*) in Belize. Bulletin of the Cracid Specialist Group 8:18–20.

Cranes in Mexico

6

Felipe Chávez-Ramírez

Two species of cranes occur in North America: sandhill (*Grus canadensis*) and whooping (*G. americana*). The sandhill crane is the most abundant of the world's cranes, while the whooping crane is one of the rarest bird species in the world, with 308 in the wild as of 2014. The whooping crane, which was a common winter migrant to the Mexican Altiplano or Mexican Plateau (Allen 1952), has been extirpated in Mexico and currently winters in only a small section of the Texas coast (Urbanek and Lewis 2015), although there are unconfirmed sightings in Chihuahua. Sandhill cranes winter in the Mexican Altiplano, primarily in the Chihuahuan Desert ecoregion (Chavez-Ramirez 2005, Lopez-Saut et al. 2011), and to a lesser extent along the two coasts. There are historical reports of both species in central Mexico, particularly at Lake Chapala, Jalisco (Allen 1952), although there are no recent reports from that far south. A former major wintering area for whooping and sandhill cranes was the now nonexistent Lake Mayran in Coahuila and Durango, previously formed by the internally draining Nazas River, which has been dry since dams were constructed between 1936 and 1946 for crop irrigation.

Sandhill cranes are hunted in Mexico for sport and food, although in low numbers (tens to low hundreds per year) compared to other migratory game species. Rural inhabitants are familiar with wintering sandhill cranes, and 43% of those interviewed believe cranes have a negative economic effect because cranes feed on drying or stored crops (Barcelo Llanes et al. 2012). Many recognize cranes by their calls and associate them with the beginning of winter. Several Mexican sayings refer to the crane's seasonal migration and behavior. For example, "está aleando la grulla" (the crane is flapping its wings) refers to decreasing temperatures in fall and winter; and "a tu tierra grulla, que esta no es la tuya" (to your native land, crane, this is not yours) alludes to the crane's migratory behavior. The whooping crane is listed as endangered, and the sandhill as subject to special protection in Mexico (SEMARNAT 2010).

Taxonomy

The earliest-known gruid fossils are from the Eocene, 54–37 million years ago (mya) (Feduccia 1999). The Gruidae family is classified into two subfamilies: Balearicinae (ancient cranes) with two species in one genus (*Balearica*), and Gruinae (modern or true cranes) with 13 species in three genera. There have been 11 species in the Balearicinae over the last 50 million years, but only two are extant, and these are crowned cranes endemic to Africa. Fossil records from North America, including a whole skeleton from Nebraska dated to the Oligocene, are believed to be crowned cranes (Voorhies 1981, Feduccia and Voorhies 1992). Modern fossil cranes appear in the Miocene, 23.0–5.3 mya. There are 13 extant crane species, but 20 are known to have existed during the Holocene (Meine and Archibald 1996).

There are three crane genera: *Anthropoides*, *Bugeranus*, and *Grus*. Within the genus *Grus*, there are four distinct groupings based on morphological, ethological, and genetic differences (Archibald 1976, Krajewski 1989). The sandhill crane is in its own group, while the whooping crane is in a group of five species, along with the Eurasian (*G. grus*), hooded (*G. monachus*), black-necked (*G. nigricollis*), and red-crowned (*G. japonensis*) cranes. The two other groups consist of the lone Siberian crane (*G. leucogeranus*), and a group of three: sarus (*G. antigone*), brolga (*G. rubicundus*), and white-naped cranes (*G. vipio*). The sandhill crane shares characteristics with the group of five and the group of three, which suggests it is ancestral to both groups. The oldest-known sandhill crane fossils are from the Miocene (5.3 mya).

There are six sandhill crane subspecies, based on anatomical and migratory behavioral differences

(Tacha et al. 1985, Johnson et al. 2005, Krapu and Brandt 2008). Subspecies include three that are migratory: lesser sandhill (*canadensis*); greater sandhill (*tabida*), the most abundant; and Canadian sandhill (*rowani*); and three that are nonmigratory: Florida (*pratensis*), Mississippi (*pulla*), and Cuban (*nesiotes*). The three migratory subspecies occur in Mexico during the winter and are differentiated primarily by size (greater is largest, Canadian is intermediate, and lesser is smallest) and breeding range (most lessers breed in Alaska, northern Canada, and northeastern Siberia; Canadians in subarctic and boreal Canada; and greaters in the US and southern Canada). Of the three migratory subspecies, only the lesser is genetically distinct. The other five are genetically indistinct. The nonmigratory populations are genetically isolated (Jones et al. 2005).

Distribution

Most information available for sandhill cranes in Mexico concerns range distribution (Tacha et al. 1992, 1994, Howell and Webb 1995, Drewien et al. 1996). The distribution previously reported was inaccurate (Chavez-Ramirez 2005, Lopez Saut et al. 2011). Sandhill cranes occupy a large area in north-central Mexico (Chihuahuan Desert ecoregion) from November to February. The present distribution extends from Chihuahua and Coahuila in the north to Zacatecas and San Luis Potosí in the south (fig. 6.1) (Chavez-Ramirez 2005, Lopez Saut et al. 2011). Sandhill cranes are also present in smaller numbers along the coastal Gulf of Mexico in Tamaulipas and along the Gulf of California in Sonora and Sinaloa. Although sandhill cranes occupy a large portion of Mexico during the winter months, they are present in low densities and in clustered distributions in and near roosting wetlands. In the Central Highlands, their distribution appears to be limited by open areas such as grasslands and desert scrub vegetation in the western portion of the range in Chihuahua and Durango.

Description

Cranes have long necks and legs and are the tallest living birds in North America; the whooping crane is the tallest. The sandhill crane is up to 1.2 m (3.9 ft) tall (fig. 6.2). Lesser sandhill females weigh an average of 3.34 kg (7.36 lb), and males weigh 3.75 kg

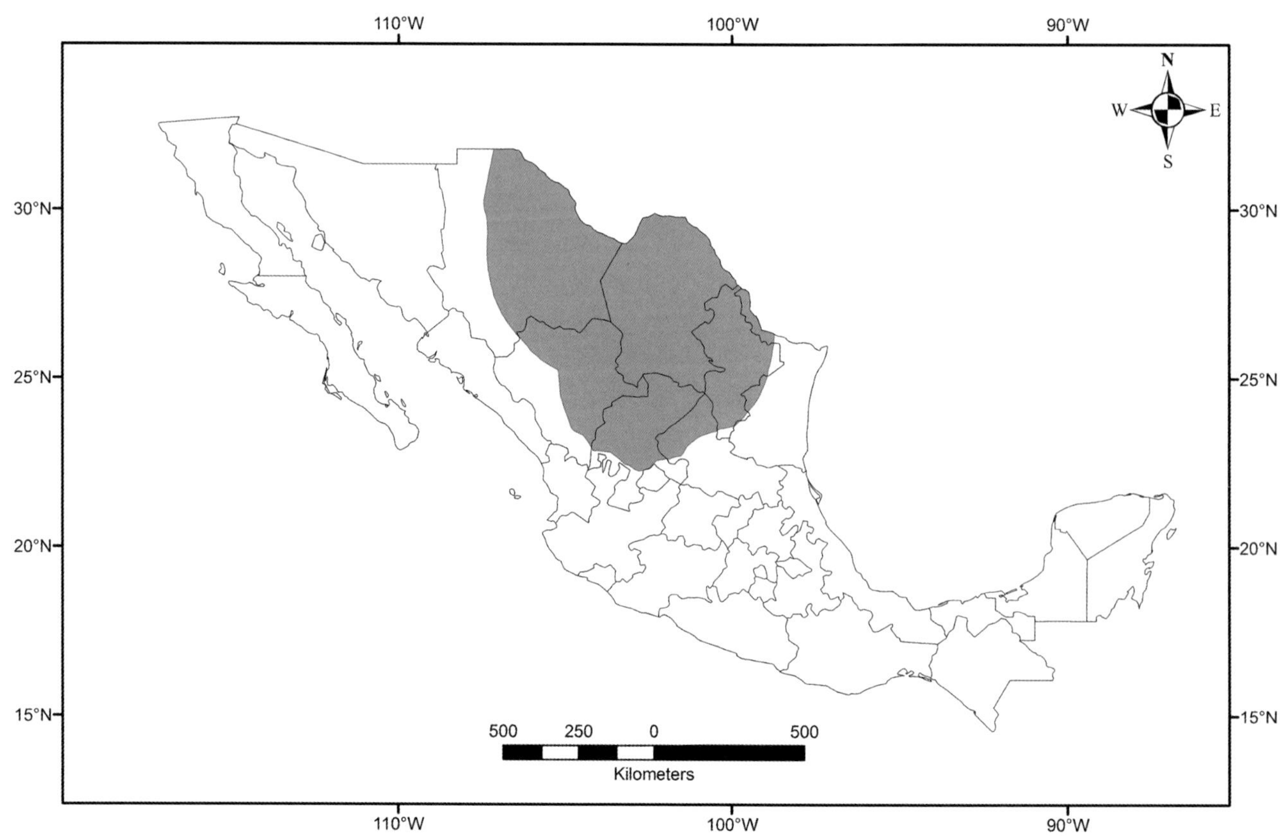

Figure 6.1. Distribution of sandhill cranes in Mexico.

(8.25 lb) (Tacha et al. 1992). Sandhill cranes usually have 10 primaries, 16 secondaries, and 12 short rectrices. The obvious and distinctive bustle, which is confused with a tail, is formed by elongated tertials, not the actual rectrices. Both males and females have monochromatic plumage and cannot be easily differentiated, except that males average slightly taller and heavier than females. Color is overall gray in adults, with red skin on the crown that extends to below the eye and the back of head, and they usually have a white cheek patch. The color of cranes varies from lighter to darker gray, occasionally appearing brownish. The rusty and brownish color, particularly on the back and wings, is usually due to soil substances that are rubbed onto the feathers via preening. In winter, subadults are distinguished by a feathered crown and an overall rusty-brownish color on the head, neck, and upper back.

Figure 6.2. The sandhill crane is the most abundant crane in the world. Photo by Felipe Chávez-Ramírez.

Life History

Sandhill cranes are one of the most-studied bird species in North America; however, winter ecology information for Mexico is limited and has long needed study (Tacha et al. 1994, Urbanek and Lewis 2015, Meine and Archibald 1996).

Diet

Sandhill cranes are omnivorous, feeding mostly on plant material and to a lesser extent on animal matter (Urbanek and Lewis 2015). Cultivated grains are the primary food when present (Tacha et al. 1992). In wintering grounds, sandhill cranes leave their roost every morning in search of food. They are common in crop fields in northern Mexico. Primary food items consumed in Mexico, as determined by fecal analysis, are corn, followed by oats and sorghum. Noncultivated foods include grass seeds (Barcelo Llanes, 2012). The same diet items were consumed from Chihuahua to Zacatecas (Barcelo Llanes 2012).

Reproduction

Sandhill cranes are perennially monogamous, forming a pair bond as early as three years old, with more than 90% mated by age seven. Mate switches are rare, particularly after successful breeding, but individuals will readily remate when an original mate dies (Littlefield 1981, Tacha 1988). Pair bonding occurs primarily during spring migration (Tacha 1988), but cranes are commonly observed performing courtship displays during winter. Successful reproduction can occur at age five but is most common at age eight (Tacha et al. 1989). Cranes lay two eggs one to two days apart, but they raise a single chick. Incubation begins with the first egg and lasts for 30 days, with both parents sharing incubation. Most commonly a single brood is raised per breeding season; however, sandhill cranes are known to renest if eggs or chicks are lost. Sex ratios appear to be slightly biased toward females (54–60% of the population).

Average annual adult reproduction (cranes older than three years of age) in *G. c. tabida* was 0.35 young per year; 64% of adults reproduced successfully. Estimated recruitment rates for postfledgling to adult ratios for different populations were between 6% and 18% (Tacha et al. 1994).

Longevity

The oldest-known wild sandhill crane was a 30-year-old bird in the Rocky Mountain population (R. Drewien, personal communication). For the midcontinental population, the oldest-known bird was 19.4 years old (Klimkiewicz and Futcher 1989). Among nonmigratory populations, the oldest included a 26-year-old from Mississippi (S. Hereford, personal communication), a 21.6-year-old from Florida (Tacha et al. 1992), and several cranes that were banded as adults in Cuba that were at least 15 years old.

Physiology and growth

Chicks are nidifugous and able to leave the nest and feed within 24 hours of hatching. Chicks weigh 114–115 g (4.02–4.05 oz) (Walkinshaw 1973, Watanabe 1996) and reach adult weights around day 55 (Watanabe 1996). Tarsi (Watanabe 1996) and legs (Baldwin 1977) are near adult size between day 40 and 60. Wing chord and bill length reach near adult size by day 55–60 (Baldwin 1977, Watanabe 1996). Full growth is attained at 10 months (Tacha et al. 1985).

Energy requirements

Cranes maintain a constant level of body protein throughout the year, but lipid levels vary. Lipid levels are constant from fall through late winter, but they increase while the birds are staging in Nebraska in late winter and early spring. During staging, sandhill cranes can increase body mass by 25% because of lipid accumulation (Iverson and Vohs 1982, Krapu et al. 1985) from consuming corn, a high-energy food. Loss of body lipids occurs during prenesting and nesting periods because of increased energy expenditures associated with egg production and territorial defense, and reduced food availability and consumption.

Water requirements

Sandhill cranes drink water daily, and this water may have a salinity of up to 20 ppt (Haley 1987). Cranes do not seek shade during the day, preferring to rest in open, exposed grasslands or wetlands, although subtropical sandhills do seek shade during midday in the summer.

Thermal relationships

Sandhill cranes are homeothermic, but some subspecies (e.g., *pratensis*) are poikilothermic at hatching and do not become homeothermic until 24 hours posthatching. *G. c. tabida* is partially homeothermic at hatching and becomes fully homeothermic 9 hours posthatching; however, *rowani* and *canadensis* are fully homeothermic at hatching.

Behavior and Social Structure

Sandhill cranes are highly social during the nonbreeding season. Large roosting flocks, numbering in the tens of thousands, are not uncommon in wintering grounds. Large flocks are observed primarily when birds are roosting, because they tend to form smaller flocks when feeding. Size of roosting flocks is influenced by wetland size. The basic social unit is a mated pair or family group (pair with chick) (Tacha 1988), which congregates with other family units and unmated subadults to form large flocks. Sandhill cranes have nesting territories and aggressively defend their boundaries against conspecifics.

Movements

Most sandhill crane populations are migratory, flying south in the fall and north during late winter and spring. They are overwintering migrants in Mexico, some traveling from as far away as Siberia, Alaska, Canada, and the northern continental US (Krapu et al. 2011). Sandhill cranes use wetlands for overnight roosts. Their winter presence in Mexico is based on the availability of wetlands, which in northern Mexico can be highly variable depending on annual precipitation patterns (fig. 6.3). The number of sandhill cranes in a specific wetland can vary tenfold from one year to the next, most likely as a result of the availability of a roosting area. It is not known whether sandhill cranes show fidelity to specific wintering areas or are facultative migrants. Because of the large differences in abundance from year to year in a given site, they do not appear to have high fidelity to specific wintering wetlands. It is not clear whether sandhill cranes roost the entire winter in the same wetland or move to different wetlands throughout the winter.

Daily movement patterns in winter involve flying from roosting to feeding and resting sites, which may or may not be the same as roosting sites. The first

Figure 6.3. A flock of several hundred sandhill cranes leaving their night roosting site to feed in Lake Babicora, Chihuahua, Mexico. Photo by Antonio Chávez-Ramírez.

movement occurs early in the morning, when birds leave to feed in crop fields up to 20 km (12.4 mi) away, although some cranes in Zacatecas probably travel up to 35 km (21.7 mi). After foraging for a few hours, cranes move to a resting site, which may be in wetlands, bare areas near wetlands, crop fields, or short grasslands. After resting, cranes return to a feeding site and eventually fly back to a roosting site at dusk.

Communication

Sandhill cranes are well known for their loud bugling or trumpeting rattle calls. These include the unison call given during pair bonding activities and courtship, during which both birds call in a duet while holding their beaks straight up in the air. The unison call may also be used in aggressive interactions, as when cranes intrude into neighboring territories. A second loud call is the guard call, which is given in alarm or in an aggressive display, such as after spotting a predator or other potential threat or during territorial defense. Adult cranes have several loud nonrattle calls of less intensity, including several short *purrt* sounds commonly heard as cranes fly overhead or when they roost and forage in fields. Shorter, lower calls may serve to announce their presence when they arrive in or depart from an area or to maintain contact with other birds in flight and on the ground. Juvenile cranes make a distinctive trilling whistle, heard commonly as they fly, which apparently helps maintain contact with adults. The juvenile whistle may also be heard during food begging.

Spacing

Sandhill cranes are territorial during breeding. Nesting territories range from 10 ha (24.7 ac) in Idaho (Drewien 1973) to 85 ha (210 ac) in the Upper Peninsula of Michigan (Walkinshaw 1973). In Siberia, sandhill crane nests were spaced an average of 6.5 km (4 mi) apart (Watanabe 2003). Annual home ranges for *pratensis* average 1.4 ha (3.4 ac) in central Florida (Nesbitt and Williams 1990) and 93 ha (230 ac) in Okefenokee Swamp in Georgia (Bennett 1989).

Population Dynamics

Sandhill cranes wintering in Mexico occur in low densities throughout their winter range. It is possible to see up to 30,000 cranes in a single wetland roost; however, a more common number for wetlands is in the low hundreds or teens. The largest roosting flocks of thousands are in Chihuahua, and flock size decreases southward, with the smallest groups in Zacatecas and San Luis Potosí. Recent estimates suggest that up to 100,000 cranes may winter in wetlands of the Mexican Altiplano.

The migratory sandhill crane population in North America is probably stable (Kruse et al. 2009) or slightly increasing, and it is estimated to consist of 640,000 to 800,000 birds. Their breeding range seems to be expanding to the south of their current reproductive range. New locality records for wintering grounds in Mexico (Lopez Saut et al. 2011) may simply reflect previously unrecorded winter range based on extensive recent fieldwork.

Predation is not a significant adult mortality factor, although bald eagles (*Haliaeetus leucocephalus*), golden eagles (*Aquila chrysaetos*), and coyotes (*Canis latrans*) can prey on adult cranes. Bobcats (*Lynx rufus*) have been significant predators of reintroduced whooping cranes in Florida and of sandhill cranes in Mississippi. Eggs, hatchlings, and prefledged chicks can be prey for mammals and avian predators, including coyotes, bobcats, foxes, raccoons (*Procyon lotor*), wolves (*Canis lupus*), bears (*Ursus* spp.), red-tailed hawks (*Buteo jamaicensis*), crows (*Corvus* spp.), and horned owls (*Bubo virginianus*). Overwinter mortality of age classes is unknown. Known causes of mortality in Mexican wintering grounds include collision with power lines and hunting; predation has not been documented.

Population Management

The size of the midcontinental population, which is the primary group wintering in Mexico, is estimated by conducting aerial survey transects and analyzing photo-corrected images during spring migration in the Platte and North Platte River valleys in Nebraska in the US. With the addition of cranes counted outside survey areas, a total population estimate can be made (Kruse et al. 2009). Similar techniques are used for the Rocky Mountain population in the San Luis Valley of New Mexico, where most of this population concentrates in spring. There are no formal or official counts of wintering cranes in Mexico. Accurate winter counts are lacking because the complete distribution of subspecies wintering areas in Mexico is unknown. However, major roosting sites could be surveyed annually to determine abundance and population trends. Any changes in abundance could reflect changes in annual dispersion patterns rather than actual changes in population numbers. Counts would have to be conducted simultaneously in a large portion of the winter distribution to obtain a reliable abundance estimate.

Habitat Requirements

Habitat needs for wintering sandhill cranes include open, shallow wetlands for roosting and resting and agricultural fields for feeding. Wetlands are also important stopover sites when cranes migrate to and from their final wintering locations. Wetlands used by cranes are shallow and open, with low, gradual slopes (less than 20°), little or no emergent vegetation, and visual obstruction more than 100 m (328 ft) away. Roosting wetlands vary greatly in size, but cranes use smaller bodies of water more often than large, permanent bodies such as lakes. However, the largest crane flocks are found in large but shallow water bodies such as Lakes Babicora and Mexicanos in Chihuahua. Sandhill cranes appear to be found most commonly in low-salinity wetlands. Most wetlands used by cranes in the Chihuahuan Desert are ephemeral, dependent on precipitation, or permanent if fed by springs. During spring staging in Nebraska, most cranes roost in the shallow Platte River, feed on waste corn in agricultural fields, and rest in managed grazed or burned grasslands interspersed with wet meadows. Other stopover areas include ponds, lakes, and marshes; feeding sites are primarily agricultural fields.

Habitat Management

Many wetlands in northern Mexico are shallow and have limited to no vegetation, characteristics preferred by sandhill cranes. More than from direct management, most wetlands would benefit from protection of the surrounding landscape. Excluding or limiting potential disturbances in roosting wetlands is recommended. Cranes use grazed or burned grasslands for resting and occasionally roosting. Livestock grazing in areas surrounding roosting wetlands may benefit cranes if it does not negatively impact the wetland itself. Most mixed and tall grasslands that have not been grazed are not used by sandhill cranes; thus, moderate cattle grazing is compatible with crane use. In some areas such as mixed and tallgrass prairie, grazing (or burning) is essential to create habitat conditions favorable to cranes. During winter in Mexico, most sandhill crane foods consist of agricultural crops. Agricultural fields may be necessary in the landscape surrounding potential roosts.

Management Challenges in Mexico

Management challenges in northern Mexico include maintaining existing wetlands. Most wetlands used by sandhill cranes are not formally protected and are therefore potentially at risk from conversion to agricultural development. Hunting records need to be maintained and evaluated to determine the effect of harvesting on wintering cranes. Evaluating the sources of mortality for overwintering sandhill cranes is also important. Monitoring long-term abundance of cranes in Mexican wetlands and surrounding habitats would enable better understanding of site fidelity and inter- and intra-annual population dynamics.

Literature Cited

Allen, R. P. 1952. The whooping crane. Research Report Number 3. National Audubon Society, New York, New York, USA.

Archibald, G. W. 1976. The unison call of cranes as a useful taxonomic tool. Dissertation, Cornell University, Ithaca, New York, USA.

Baldwin, J. H. 1977. A comparative study of sandhill crane subspecies. Dissertation, University of Wisconsin, Madison, USA.

Barcelo Llanes, I. 2012. Winter ecology of sandhill cranes (*Grus canadensis*) in northern Mexico. Dissertation, University of Nebraska, Lincoln, USA.

Barcelo Llanes, I., J. C. Guzmán-Aranda, F. Chávez-Ramírez, and L. A. Powell. 2012. Rural inhabitant perceptions of sandhill cranes in wintering areas of northern Mexico. Human Dimensions of Wildlife 17:301–307.

Bennett, A. J. 1989. Movements and home range of Florida sandhill cranes. Journal of Wildlife Management 53:830–836.

Chavez-Ramirez, F. 2005. New locations and range extension of wintering sandhill cranes in central northern Mexico. Proceedings of the North American Crane Workshop 9:173–178.

Drewien, R. C. 1973. Ecology of Rocky Mountain greater sandhill cranes. Dissertation, University of Idaho, Moscow, USA.

Drewien, R. C., W. M. Brown, and D. S. Benning. 1996. Distribution and abundance of sandhill cranes in Mexico. Journal of Wildlife Management 60:270–285.

Feduccia, A. 1999. The origin and evolution of birds. Second edition. Yale University Press, New Haven, Connecticut, USA.

Feduccia, A., and M. R. Voorhies. 1992. Crowned cranes (Gruidae: *Balearica*) in the Miocene of Nebraska. Los Angeles County Museum of Natural History, Sciences Series 36:239–248.

Haley, D. E. 1987. Drinking sites and requirements of wintering sandhill cranes. Proceedings of the 1985 International Crane Workshop 4:172–179.

Howell, S., and S. Webb. 1995. A guide to the birds of Mexico and northern Central America. Oxford University Press, Oxford, United Kingdom.

Iverson, G. C., and P. A. Vohs Jr. 1982. Estimating lipid content of sandhill cranes from anatomical measurements. Journal of Wildlife Management 46:478–483.

Johnson, D. H., J. E. Austin, and J. A. Shaffer. 2005. A fresh look at the taxonomy of midcontinental sandhill cranes. Proceedings of the North American Crane Workshop 9:37–45.

Jones, K. L., G. L. Krapu, D. A. Brandt, and M. V. Ashley. 2005. Population genetic structure in migratory sandhill cranes and the role of Pleistocene glaciations. Molecular Ecology 14:2645–2657.

Klimkiewicz, M. K., and A. G. Futcher. 1989. Longevity records of North American birds. Supplement 1. Journal of Field Ornithology 60:469–494.

Krajewski, C.1989. Phylogenetic relationships among cranes (Gruiformes: Gruidae) based on DNA hybridization. Auk 106:603–618.

Krapu, G, L., and D. A. Brandt. 2008. Spring migratory habits and breeding distribution of lesser sandhill cranes that winter in west-central New Mexico and Arizona. Proceedings of the North American Crane Workshop 10:43–49.

Krapu, G. L., D. A. Brandt, K. L. Jones, and D. H. Johnson. 2011. Geographic distribution of the mid-continent population of sandhill cranes and related management implications. Wildlife Monographs 175:1–38.

Krapu, G. L., G. C. Iverson, K. J. Reineke, and C. M. Boise. 1985. Fat deposition and usage by Arctic nesting sandhill cranes during spring. Auk 102:362–368.

Kruse, K. L., D. E. Sharp, and J. A. Dubovsky. 2009. Status and harvest of sandhill cranes: Mid-Continent, Rocky Mountain and Lower Colorado River Valley populations. Administrative report. US Fish and Wildlife Service, Denver, Colorado, USA.

Littlefield, C. D. 1981. A probable record of intraspecific egg dumping for sandhill cranes. Auk 98:631.

Lopez-Saut, E., F. Chavez-Ramirez, and R. Rodriguez-Estrella. 2011. New records of wintering grounds for sandhill cranes in Mexico. Waterbirds 34:239–246.

Meine, C. D., and G. W. Archibald, editors. 1996. The cranes: status survey and conservation action plan. IUCN, Gland, Switzerland, and Cambridge, United Kingdom.

Nesbitt, S., and K. S. Williams. 1990. Home range and habitat use of Florida sandhill cranes. Journal of Wildlife Management 54:92–96.

SEMARNAT. 2010. Norma Oficial Mexicana NOM-059-ECOL-2001. Protección ambiental—Especies nativas de México de flora y fauna silvestres—Categorías de riesgo y especificaciones para su inclusión, exclusión o cambio—Lista de especies en riesgo. DOF, México, D. F., México.

Tacha, T. C. 1988. Social organization of sandhill cranes from mid-continental North America. Wildlife Monographs 99:1–37.

Tacha, T. C., D. E. Haley, and P. A. Vohs. 1989. Age of sexual maturity in sandhill cranes from mid-continental North America. Journal of Wildlife Management 53:43–46.

Tacha, T. C., S. A. Nesbitt, and P. A. Vohs. 1992. Sandhill crane. *In* The birds of North America. Academy of Natural Sciences, Philadelphia, and American Ornithologists' Union, Washington, DC, USA.

Tacha, T. C., S. A. Nesbitt, and P. A. Vohs. 1994. Sandhill crane. Pages 76–94 *in* T. C. Tacha and C. E. Braun, editors. Migratory shore and upland game bird management in North America. International Association of Fish and Wildlife Agencies, Washington, DC, USA.

Tacha, T. C., P. A. Vohs, and W. D. Warde. 1985. Morphometric variation of sandhill cranes from mid-continental North America. Journal of Wildlife Management 49:246–250.

Urbanek, R. P., and J. C. Lewis. 2015. Whooping crane (*Grus americana*). *In* The birds of North America online. Cornell Laboratory of Ornithology, Ithaca, New York. http://bna.birds.cornell.edu/bna/species/153. doi:10.2173/bna.153. Accessed 5 Aug 2015.

Voorhies, M. R. 1981. Ancient ashfall creates a Pompeii of prehistoric animals. National Geographic 159:66–75.

Walkinshaw, L. H. 1973. Cranes of the world. Winchester Press, New York, New York, USA.

Watanabe, T. 1996. Comparison of behavioral ontogeny and ethology of whooping crane and greater sandhill crane chicks. Thesis, Southern Connecticut State University, New Haven, USA.

Watanabe, T. 2003. A comparative breeding ecology of lesser sandhill cranes (*Grus canadensis canadensis*) and Siberian cranes (*G. leucogeranus*) in eastern Siberia. Dissertation, Texas A&M University, College Station, USA.

Ducks and Geese in Mexico

7

Raul Valdez, Fernando Clemente-Sánchez,
and Jack C. Eitniear

Ducks, geese, and swans, or collectively waterfowl or wild fowl, are classified in the order Anseriformes and family Anatidae. They are cosmopolitan in distribution, excluding the Antarctic. Anatids are subdivided into 4 subfamilies, 12 tribes, 49 genera, and 158 species. Six species have become extinct, including the Labrador duck (*Camptorhynchus labradorius*) of the northeastern American Atlantic coast, 6 are critically endangered, 9 are endangered, and 12 are vulnerable. North American anatids are classified in 4 subfamilies, 6 tribes, and 43 species. They are some of the most abundant aquatic birds in the Nearctic, with an estimated breeding population of 49.5 million in 2015 (Dooley et al. 2015). In North America, they vary in size from a swan with a wingspan of 240 cm (7.9 ft) and weighing up to 15 kg (32.8 lb), to the diminutive green-winged teal, which weighs 350 g (less than 1 lb). Gregarious and vocal, they are associated with aquatic habitats, spending most of their time in marshes, swamps, streams, lakes, ponds, reservoirs, marine estuaries, and lagoons. However, shoreline and upland habitats are essential for nesting and feeding for some species. Most line their nests with down, and the downy chicks are nidifugous. Waterfowl have readily adapted to human landscapes, such as stock tanks and larger impoundments, and agricultural areas such as rice, corn, and wheat fields are important food sources. In North America, most species breed in northerly latitudes, some as far north as the Arctic fringes and tundra, and they migrate south in the winter to the warmer climates of the southern US, Mexico, and the Neotropics. They are vulnerable to hunting because they congregate in limited aquatic habitats (Alderfer 2014, Baldassarre 2014, Carboneras 2014).

Historically, ducks and geese have been important in Mexico both economically and culturally. Wild ducks were a significant source of meat protein for

Figure 7.1. A redhead, one of the many duck species that winter in Mexico. Photo by Greg Lasley.

Mexican cultures (fig. 7.1). Montezuma, king of the Aztec Empire when the Spaniards invaded Mexico in 1519, had an aviary that included numerous species of ducks valued for their esthetic qualities as well as for their meat and feathers. In lakes and wetlands surrounding the empire's capital, Tenochtitlan, (incorporated into modern Mexico City), there were 23 duck species, some in vast populations. Hunted regionally, these ducks were sold in the bustling city market. Waterfowl hunting tactics included spearing and netting (see chapter 1). The introduction of firearms greatly increased the hunting harvest, and it was not until the 1940s that federal authorities ended large-scale commercial duck hunting in central Mexico (see chapter 21). Presently, waterfowl in Mexico are among the favorite game birds of sport hunters and continue to be a source of protein for rural populations who practice subsistence hunting (Arellano and Rojas 1956). Sport hunting in Mexico generates millions of dollars; the potential recreational value of waterfowl has not been realized. However, waterfowl harvest in Mexico is minimal; it amounts to less than 10% of the annual harvest in California, and less than 1% of the harvest estimate in the US (Kramer et al. 1995). Waterfowl can offer excellent recreational opportunities for national and foreign tourists (see chapter 3).

Studies on anatid biology in Mexico are few and of short duration. These have focused mainly on the distribution and abundance of migratory species. Current waterfowl population statuses and trends, except those of the black brant (*Branta bernicla*), are unknown, creating a serious gap in our knowledge of North American waterfowl. Mexico is of major importance in providing essential stopover and overwintering habitats for millions of North American waterbirds and waterfowl. Resident waterfowl (occurring year-round) include five duck species and one subspecies of mallard. Comprehensive seminal book references on the waterfowl of North America, including Mexico, are Leopold (1959), Johnsgard (1978, 2010), Carboneras (1992, 2014), Howell and Webb (1995), Baldassarre and Bolen (2006), Alderfer (2014), and Baldassarre (2014). In this chapter, we focus on published information on anatids and their habitats in Mexico.

Description

Anatids have a long neck, small head, and thick tongue. Except in mergansers, the bill is long and somewhat flattened, with combed serrations at the edges, which typically function to strain small organisms. In general, the legs are set back on the body, and the feet in most are webbed to assist in swimming. The wings are relatively short and pointed, and the tail is short and square or slightly rounded. Outer feathers cover a layer of down for insulation, and they have an oil gland for feather maintenance. The powerful, muscular wings help in maintaining sustained, rapid, migratory flight over long distances, and in taking off from water surfaces. Most species are monogamous or seasonally monogamous, and most nest on the ground, although some species nest in tree cavities. There is a high degree of sexual dimorphism in most species in the subfamily Anatinae, with females inconspicuous in varying shades of brown, probably a cryptic adaptation for concealment during the incubation period, and most males with contrasting bright colors.

Male ducks, unlike most other birds, have a vascularized penis in the cloaca, which emerges and penetrates the female's cloaca to achieve fertilization. Mating occurs on the water in most species. Many species practice brood parasitism, in which two or more females contribute eggs to a single clutch. Waterfowl are notorious for their ability to hybridize. For example, the mallard is known to interbreed with 23 species in the genus *Anas*, resulting in fertile offspring, and with species in 7 other genera. In the wild in North America, mallards hybridize with 7 species. Species in northern Eurasia and North America migrate long distances to warmer southern climates during the winter. Waterfowl residing in tropical regions are less likely to make annual migrations.

Taxonomy and Evolution

Anatid classification above the genus level remains unresolved because of the unclear relationships among ancestral and living species. Anatids are probably most closely related ancestrally to galliforms (quail and related forms), and to two other families of anseriforms: Anhimidae (screamers, with three extant species in South America), and Anseranatidae (Australian magpie goose). An anseriform fossil from the Upper Paleocene (61 million years ago [mya]) and anatid fossils from the Upper Eocene (50–40 mya) have been found, but it was probably during the Miocene (23 mya) that major speciation occurred, including the emergence of some modern genera. Several species assignable to modern taxa were present by the Pleistocene, about 2 mya (Howard 1964, Baldassarre and Bolen 2006). Anatids formed two principal evolutionary lineages: geese and swans (Anserinae), and most ducks (Anatinae).

The 33 species of waterfowl in Mexico are classified into four subfamilies (Dendrocygninae, Oxyurinae, Anserinae [divided into two tribes], and Anatinae [divided into four tribes]) (table 7.1). The dendrocygnines include the whistling ducks, of which two species occur in Mexico; and two oxyurine species, the masked and ruddy duck. Geese are classified in the subfamily Anserinae and tribe Anserini: five goose species occur in Mexico. The other anatids in Mexico are in the subfamily Anatinae, which is divided into four tribes; these include the Anatini or dabbling ducks (ten species), and Aythyini or diving ducks (five species). Dabbling ducks feed by dipping their body to feed and take flight in an initial vertical ascent, whereas pochards feed by diving and require a running start on the water surface before taking flight. The rest are classified into two tribes: Cairinini (two species: Muscovy and wood duck), and Mergini (eight species), which includes the mergansers (three

Table 7.1. Thirty-three species of ducks and geese (family Anatidae) occurring in Mexico, including subfamilies (Sf.) and tribes (Tr.), with English and Spanish common names.

Subfamilies/tribes	Species[a]	English name	Spanish name[b]
Sf. Dendrocygninae (whistling ducks)	*Dendrocygna autumnalis*	black-bellied whistling duck	pijije ala blanca
	Dendrocygna bicolor	fulvous whistling duck	pijije canelo
Sf. Oxyurinae (stiff-tailed ducks)	*Nomonyx dominicus*	masked duck	pato enmascarado
	Oxyura jamaicensis	ruddy duck	pato tepalcate
Sf. Anserinae Tr. Anserini (geese)	*Chen caerulescens*	snow goose	ganso blanco
	Chen rossii	Ross's goose	ganso de Ross
	Branta bernicla	black brant	ganso de collar
	Branta canadensis	Canada goose	ganso canadiense
	Anser albifrons	greater white-fronted goose	ganso careto mayor
Sf. Anatinae Tr. Cairinini (perching ducks)	*Cairina moschata*	Muscovy duck	pato real
	Aix sponsa	wood duck	pato arcoiris
Tr. Anatini (dabbling ducks)	*Anas americana*	American wigeon	pato chalcuán
	Anas strepera	gadwall	pato friso
	Anas fulvigula	mottled duck	pato tejano
	Anas platyrhynchos	mallard	pato de collar
	Anas discors	blue-winged teal	cerceta ala azul
	Anas cyanoptera	cinnamon teal	cerceta canela
	Anas clypeata	northern shoveler	pato cucharón norteño
	Anas acuta	northern pintail	pato golondrino
	Anas crecca	green-winged teal	cerceta ala verde
Tr. Athyini (pochards)	*Aythya valisineria*	canvasback	pato coacoxtle
	Aythya americana	redhead	pato cabeza roja
	Aythya collaris	ring-necked duck	pato pico anillado
	Aythya marila	greater scaup	pato-boludo mayor
	Aythya affinis	lesser scaup	pato-boludo menor
Tr. Mergini (scoters and mergansers)	*Melanitta perspicillata*	surf scoter	negreta nuca blanca
	Melanitta fusca	white-winged scoter	negreta ala blanca
	Melanitta americana	black scoter	negreta negra
	Bucephala albeola	bufflehead	pato monja
	Bucephala clangula	common goldeneye	pato chillón
	Lophodytes cucullatus	hooded merganser	mergo cresta blanca
	Mergus merganser	common merganser	mergo mayor
	Mergus serrator	red-breasted merganser	mergo copetón

[a]Classification based on AOU (1998).
[b]Spanish names based on Escalante et al. (2014).

species), scoters (three species), and two related species: bufflehead and common goldeneye (Carboneras 2014, Baldassarre 2014, Alderfer 2014).

Resident waterfowl in Mexico consist of black-bellied whistling duck, fulvous whistling duck, Muscovy duck, Mexican duck (a subspecies of mallard), mottled duck, and masked duck. Blue-winged teal, redhead, ruddy ducks, and common mergansers have been recorded breeding in Mexico, but only rarely; the mallard breeds locally but irregularly in northern Baja California. Also, nonbreeding individuals of some species, usually immatures, may oversummer, principally in northern Mexico (Howell and Webb 1995). The garganey, long-tailed duck, Eurasian wigeon, and harlequin duck are vagrants (occurring outside their normal range) and have occasionally been sighted; we do not include these species. The cackling goose is also not included because it is rare in Mexico. Two species of swans have been recorded in Mexico, but these have been extirpated, so we do not include them in the extant bird fauna. Refer to Howell and Webb (1995), Baldassarre (2014), and Alderfer (2014) for range maps and color illustrations of anatids recorded in Mexico. For detailed natural history accounts of individual North American species, refer to Baldassarre (2014).

Waterfowl of Special Significance in Mexico

Five species of waterfowl in Mexico merit special mention. These are Muscovy duck, because of its cultural value—it is one of only two native game birds domesticated in Mexico; Mexican duck, a subspecies of the mallard, because of its almost endemic distribution in Mexico; two species of whistling duck, because their center of distribution in North America is Mexico; and the black brant, a goose, because most of its North American population winters in Mexico. These species are also representative of Mexican waterfowl diversity. Here we describe their distributions and life histories.

Muscovy duck

Muscovy duck (*pato real*) is a very large, heavy-bodied perching duck (fig. 7.2). The feathers on the crown and on the rear of the neck are somewhat elongated, forming crests. The extensive bare skin on the face is mostly black, with red or pink caruncles. The plumage is dark in wild Muscovies; it is primarily black with a greenish gloss, although the wing coverts are white (on both the upper and lower surfaces of the wing). The sexes are generally similar, but females, with a body length of 71–76 cm (28–30 in) and a weight of 1.10–1.47 kg (2.4–3.2 lb), are significantly smaller than males, which measure 76–84 cm (30–33 in) and weigh 1.99–4.00 kg (4.3–8.8 lb) (Leopold 1959). The plumage of females is duller (less glossy) and has less white in the wings; females also lack a crest and facial caruncles. Juveniles are even duller (dark brown rather than black), with very little white in the wing. In Mexico, Muscovies are nonmigratory, inhabiting wetlands on Pacific and Atlantic slopes. They are most abundant along the Gulf Coast from northern Tamaulipas (fragmented populations along the Rio Grande) and central Nuevo León, south into the Yucatán Peninsula (rare), but they are frequently encountered in Belize southward through the remainder of Central and South America (Eitniear et al. 2015). Although the breeding season is variable, Muscovy ducks typically breed during the local wet season. They nest singly, often in tree cavities between 2 and 18 m (6.5 and 59 ft) above the ground. Like most waterfowl, they add a down lining. They sometimes nest in caves, or in dense vegetation on the banks of streams. The clutch size is 8–15 (average 10) eggs, which are subelliptical to oval and white to greenish white; they measure 63 × 47 mm (2.5 × 1.8 in). Incubation lasts about 35 days. Although relatively silent, the male produces a low hissing sound, and the female has a short, weak "quack" (Woodyard and Bolen 1984).

Muscovy ducks are opportunistic omnivores, feeding on stems, seeds, grasses, aquatic plants, small vertebrates, and invertebrates. Food items from 15 Muscovy ducks collected in Veracruz included

Figure 7.2. Muscovy duck hen with her ducklings. Note the bare skin on her head. Photo by Michael Patrikeev.

58% water lilies (*Nymphaea* spp.), 26% soldier flies (Diptera, Stratiomyidae), 14% black mangroves (*Avicennia nitida*), 8% hoverflies (Syrphidae), and trace amounts of additional aquatic plants and invertebrates (Woodyard 1982).

The Muscovy duck is a protected resident species under the category "In Danger of Extinction" (SEMARNAT 2002). The global population trend appears to be decreasing, but the decline is not sufficiently rapid to approach the threshold for "Vulnerable" (> 30% decline over 10 years or three generations) (Wetlands International 2016). The estimated total world population is 50,000–499,999 birds (BirdLife International 2012). With numbers declining in Mexico during the 1970s, Ducks Unlimited (DUMAC) installed artificial nest boxes under the assumption that the lack of predator-resistant nesting sites was limiting population growth. From 1981 to 1994, DUMAC installed 7,400 nest boxes, of which 23% were destroyed by various factors. Boxes were installed in Nuevo León, Tamaulipas, Veracruz, Campeche, and Yucatán. Unfortunately, because these ducks live in forested areas in small numbers, population determination was problematic; anecdotal evidence suggests the nest box program was successful (Alonzo-Parra et al. 2006, Markum and Baldassarre 1989).

Bones of a Muscovy duck have been dated to the Pleistocene in Panama, documenting its presence in Mesoamerica for more than two million years. In pre-Columbian times, the Muscovy duck was prized for its feathers, which were used to make the rich feather cloaks that Aztec rulers were so fond of. The bird portrayed in the mythical "bird-serpent" motif of the famed Aztec god-hero Quetzalcoatl is the Muscovy duck. Spaniards brought Muscovies, previously captured in Brazil and domesticated in Europe, to Mexico, which then mixed with birds already domesticated by indigenous Indians. A significant source of protein as well as a prominent feature of Mexico's cultural history, wild Muscovies continue to inhabit tropical wetlands, while domesticated varieties populate ejidos in rural areas (Whitley 1973).

Mexican duck

The Mexican duck, known as the *pato mexicano* or *pato triguero* in Mexico, is listed as threatened (SEMARNAT 2002). It is considered a subspecies of the mallard, but its classification is contentious. The male is strikingly different from the northern male mallard; both the female and male Mexican duck resemble a female mallard. Males weigh an average of 1 kg (2.2 lb) (range 849–1,243 g, 1.9–2.7 lb), and females 1 kg (2.2 lb) (range 647–1,264 g, 1.4–2.8 lb). Because the Mexican duck hybridizes with the mallard at the northern edge of its distribution, it has been classified as a subspecies. However, hybridization decreases progressively southward from the northern hybrid zone, which indicates it could merit full species status. A study of Mexican ducks from five subgroups found that ducks in the northern Altiplano differed genetically from those in southern areas, which indicates minimal intergroup gene flow, as well as a nonmigratory species (Colón Quezada 2012). Clinal variation in plumage colors also suggests population subdivisions.

Leopold (1959) described the Mexican duck as a bird of the temperate uplands of Mexico; it occurs in arid and semiarid habitats at elevations of 900–2,500 m (2,950–8,200 ft) on the central plateau. Its range extends from the southwestern US (Arizona, New Mexico, and West Texas), to Mexico, where it occurs south from Chihuahua and Durango to the central volcanic belt and eastern and western Tamaulipas, southeastern San Luis Potosí, and Morelos. It is a resident species, perhaps with nomadic movements related to aquatic habitat availability, but it is adapted to disturbed habitats in agriculturally developed areas.

Most studies of Mexican ducks focus on taxonomic issues; the only detailed ecological study was conducted in the 1970s. That study indicated that Mexican ducks were continuously distributed in 16 states, and 85–90% were in the western Central Highlands (Altiplano). In the dry season, flocks consisted of 100–5,000 birds and were composed of mated pairs, which fed in crop fields. When the rainy season began, flocks disbanded into pairs to initiate nesting, mostly in seasonal wetlands. They nested in crop fields, pastures, and woodlands, at distances of 50 m to 0.5 km (164 ft to 0.3 mi) from water, although nesting over water was also documented. Peak nesting occurred during June and July, after the onset of summer rains. Breeding extended from mid-April until November, when the last young fledged. Clutches averaged 7.7 eggs (range 2–10). Broods averaged 4.9 ducklings (range 3–9). Broods of highly protective females were reared on a variety of human-induced landscapes. An aerial survey during the dry season (May to early June) led to an estimate of 55,500 birds in the study area, of which 86% occurred in the Central Highlands, and an estimated 78,000 in a potential postbreeding population (Williams 1980).

In the Lerma Marsh in the state of Mexico, the fall diet was dominated by aquatic plant seeds, forbs, corn, and *Sagittaria* tubers. In the El Llano region in Aguascalientes, agricultural and livestock enterprises created hundreds of water bodies used by several duck species, including the Mexican duck. These water storage ponds created favorable aquatic habitats readily available for waterfowl (Medina Torres 1996, Medina Torres et al. 2007).

Midwinter surveys by the US Fish and Wildlife Service (USFWS) during 1960–2000 indicated that 16% of the Mexican duck population occurred in the Northern Highlands, and 84% in the Central Highlands. During 1991–2000, populations in the same areas were 31% and 61%, respectively. Populations fluctuated widely during the long term. Fifteen priority sites, which accounted for 70% of the midwinter Mexican duck population, have been identified. Of these 15 sites, 10 sites qualify as wetlands of international importance; 4 are in the Northern Highlands and 11 in the Central Highlands, of which 8 are distributed along the highly degraded Río Lerma drainage (Pérez-Arteaga et al. 2002a).

Black brant

Of the four black brant subspecies, only the Pacific subspecies (*B. b. nigricans*) occurs in Mexico; 75% of the population winters there (fig. 7.3). The Mexican government lists this subspecies as threatened (SEMARNAT 2002). The brant is the smallest goose in North America; male winter weights average 1.5 kg (3.2 lb) and are usually 10% heavier than females. Like other geese species, they are monogamous and mate for life. They nest on the ground and average 3.6 eggs per pair. Brant are vegetarian and highly dependent on eelgrass (*Zostera marina*) during migration. They winter on the Pacific coast from Alaska south to Mexico. They migrate in early November, and their spring migration to Alaska peaks in mid-March. Beginning in the 1950s, they shifted their primary wintering areas from the northwestern coast of the US to Mexico. In Mexico, they winter on the west coast of the Baja California Peninsula and on the mainland coasts of western Mexico in Sonora and northern Sinaloa. They make a 54-hr direct nonstop flight of more than 5,000 km (3,100 mi) from their staging area on the Izembek Lagoon on the Alaska Peninsula to the Baja California Peninsula, at a ground speed of 75–100 km/hr (46–62 mi/hr); they lose about 32% of their body weight during migration. During the midwinter counts of 1998–2001, 86% were located on the western coastal San Quintín, Scammon's, San Ignacio, and Magdalena Lagoons of Baja California. Midwinter counts since 1970 have ranged from a high of 186,000 in 1981 to a low of 68,432 in 2015. They averaged 126,700 from 1970 to 2008, and 89,915 from 2005 to 2015 (excluding 2009, when a survey was not conducted) (table 7.2). During the latter 10-year period, 80% oc-

Figure 7.3. Black brant. Photo by Mike Morel.

Table 7.2. Total numbers of brant by survey unit for Baja California and the west coast of Mexico during 2005–2015.

Year	2005	2006	2007	2008	2010	2011	2012	2013	2014	2015	Mean
Baja	59,960	87,483	65,250	83,856	71,688	61,153	101,571	71,607	68,290	44,533	71,539
West	14,068	14,254	13,932	19,443	23,389	12,464	9,873	23,566	28,869	23,899	18,376
Total	74,028	101,737	79,182	103,299	95,077	73,617	111,444	95,173	97,159	68,432	89,915

Source: Olson (2015).
Note: Surveys were not conducted in 2009.

curred on the western Baja California Peninsula, and 20% on the west coast; the highest number recorded (111,444) was in 2012. Relative to hunting, in the past most of the harvest occurred in Bahía de San Quintín; 1,741 were harvested during the winter of 1974–1975, and 6,532 in 1975–1976. Brant harvest surveys from 1974–1975 to 2000–2001 reported an average annual harvest of 1,408, an insignificant decrease compared to the total wintering population (Lewis et al. 2013, Baldassarre 2014, Olson 2015).

Whistling ducks

In Mexico, there are two species of whistling ducks (fulvous and black-bellied), so named because of their high-pitched whistling calls. They are primarily tropical and have long legs and necks. Both sexes are similar in size, plumage, and voice. In North America, they are most numerous in Mexico and are resident: populations do not migrate but are better described as nomadic. In Mexico, they occur principally along both coasts, occasionally on the Baja California Peninsula, and south to South America. In the US, they occur as breeding populations in coastal Texas, southern Louisiana, and Florida. Fulvous whistling ducks also occur in southern California, and black-bellied whistling ducks extend into southern Arizona. They do not line their nests with down. Unlike in almost all other species of ducks, in this species both sexes assist in incubation and brood rearing. Their main food consists of vegetable matter (Carmona et al. 2011, Baldassarre 2014).

Fulvous whistling duck

The common name “fulvous” refers to the tawny (brownish-orange) color over most of this duck’s body. Fulvous whistling ducks have an almost pantropical distribution, occurring on four continents. They average about 700 g (1.5 lb). In Mexico, they occur along the Pacific coast from southern Sonora to Chiapas, and on the Gulf Coast from the Texas border south to Campeche. Populations that breed in Texas and Louisiana winter primarily in Mexico. In the US, their habitats are characterized by shallow freshwater and brackish wetlands, and they make extensive use of flooded pastures, grasslands, and rice cultivation. In Mexico, they occur in open marshes, flooded fields, and lagoons, roosting by day in dense vegetation. They nest on the ground and rarely if ever perch in trees. An east coast winter survey in Mexico in 2000 recorded 17,691, of which 98% occurred in the Tabasco lagoons in southeastern Mexico; during the 2006 survey, 9,124 were recorded, of which 91% occurred in Tabasco. Most fulvous whistling ducks on the west coast of Mexico were recorded near Los Mochis, Sinaloa. The west coast population averaged 9,500 during the 1981–1984 surveys but decreased to an average of 210 during 1997–2006. Those in the interior highlands averaged 597 during surveys from 1947 to 2006 (Howell and Webb 1995, Baldassarre 2014).

Black-bellied whistling duck

The distinct black belly, pink feet, and white eye ring differentiate this species (fig. 7.4). It averages about 750 g (1.6 lb). Restricted to North and South America, it occurs in Mexico primarily along both coasts, throughout the Yucatán Peninsula, and in tropical areas. Preferred habitats include wooded marshes, swampy forests, lagoons, mangroves, and flooded fields. It usually nests in tree cavities, sometimes distant from water, and is adapted to perching. It will readily use nest boxes. Intraspecific brood parasitism is common. During midwinter surveys in Mexico from 1978 to 1994, an average of 58,000 were recorded, with 44,000 on the east coast, 14,000 on the west coast, and fewer than 100 in the interior. The largest concentrations were on the east coast between the city of Veracruz and Tamiahua Lagoon, Veracruz (Howell and Webb 1995, Baldassarre 2014).

Wintering Areas

Most of the 33 species of anatids in Mexico are there only in the winter, arriving between September

Figure 7.4. Black-bellied whistling ducks are very abundant in the wetlands adjacent to the Mexican coast. Photo by Alejandro Martínez-Palacio.

and December and departing between March and May. Waterfowl are one of the four bird groups that account for 60% of the regular northern winter visitors in Mexico. Most of the waterfowl wintering in Mexico are at the southern edge of their winter range. Only five species migrate as far south as South America (Howell and Webb 1995). The migratory travel routes of waterfowl from their breeding grounds in northern North America to wintering habitats in Mexico include three principal North American flyways: Mississippi, Central, and Pacific. The mainland west coast of Mexico receives birds primarily from the Pacific Flyway, with smaller numbers from the Central Flyway. The east coast wintering sites are occupied by waterfowl from the Central and Mississippi Flyways. We list 34 sites (14 on the Pacific side, 10 on the east coast, and 10 in the Central Highlands) (fig. 7.5) considered important based on midwinter surveys conducted by the USFWS. Some population estimates are based on an early survey but are important for establishing trends, and the site descriptions provide a basis for comparing present conditions. Also, they highlight the need for resurveying these sites.

Pacific coast

The Pacific coast wintering grounds consist of two distinct physiographic regions: the Baja California Peninsula, and the mainland west coast in Sonora, Sinaloa, and Nayarit (fig. 7.5). The Pacific coast of Baja California consists of large, shallow, open-water lagoons with eelgrass beds (*Zostera marina*), an important food source, especially for black brant. The Bahía de San Quintín (43 km², 26.7 mi²) has extensive beds of eelgrass in its open waters. The Scammon's area complex (305 km², 189.5 mi²) consists of a series of coastal lagoons extending more than 200 km (124 mi) and comprising Laguna Manuela (north), Laguna Guerrero Negro (middle), and Laguna Ojo de Liebre (south), which has extensive areas converted to salt evaporation ponds. The Colorado River Delta (100 km², 62 mi²) has freshwater habitat and encompasses more than 80 km (49.7 mi) of the lower Colorado River, which terminates in the Gulf of California; much of the delta is degraded because of agricultural development, urbanization, and reduced river flows (Glenn et al. 2001). The more southern wintering sites are

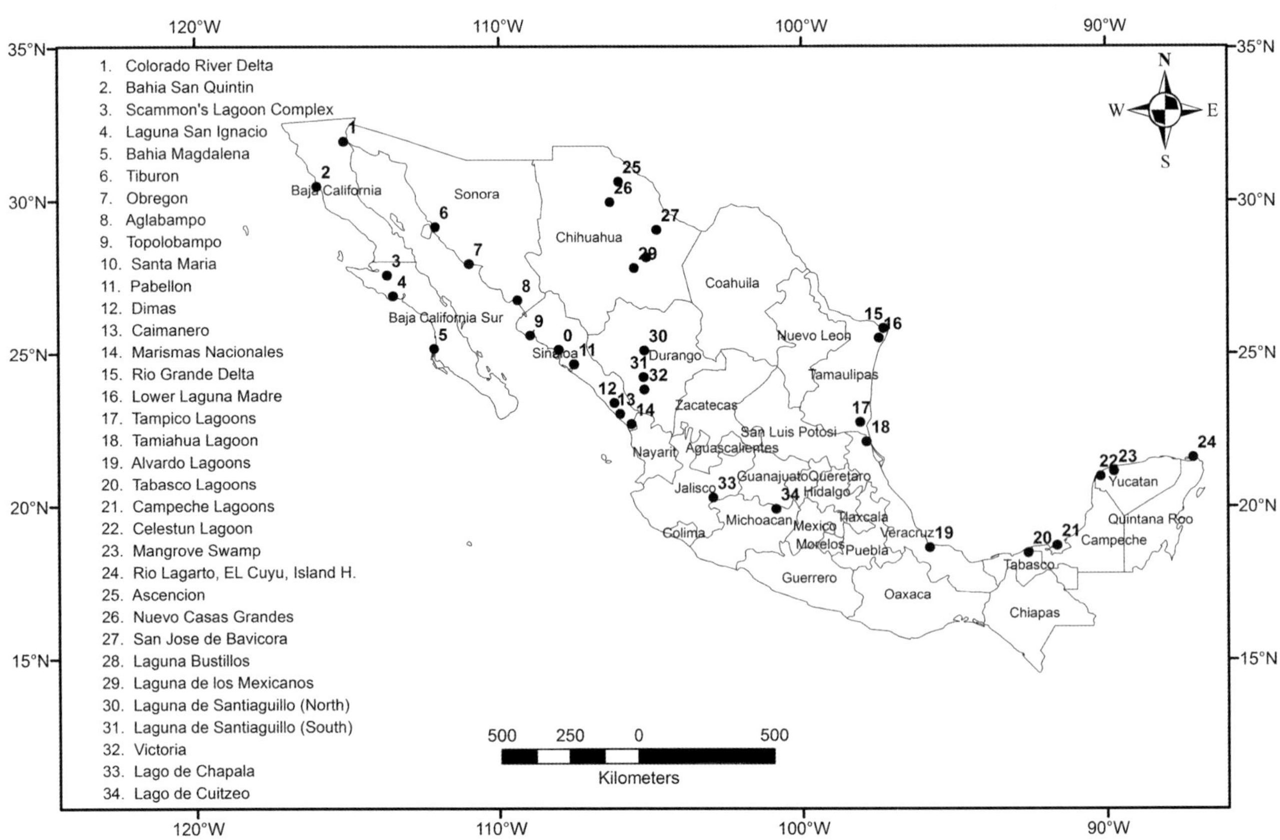

Figure 7.5. Major wintering areas for waterfowl in Mexico.

the Laguna San Ignacio Complex (202 km², 125.5 mi²), and the Bahía Magdalena (240 km², 149 mi²).

The mainland west coast (fig. 7.5), the most important waterfowl area, has nine major wetlands characterized by shallow coastal lagoons with freshwater from rivers and agricultural irrigation canals; the four northernmost units have eelgrass. In some areas, cultivated grains such as rice and wheat provide a food source. The Obregon unit includes an extensive agriculturally developed area, as does the Topolobampo area, but to a lesser degree. The Ensenada de Pabellones has abundant small mollusks, an important food of diving ducks. The Marismas Nacionales habitats consist of broad, shallow, saline lagoons, mangrove-bordered pools, and extensive marshes and salt flats.

Twenty-four species of waterfowl regularly winter on Mexico's Pacific coast. An average of 45,000 ducks, 2,000 geese, and 99,000 black brant were recorded between 1980 and 1985 in the Baja California segment, representing 4% of ducks, less than 1% of geese, and 76% of black brant. Most of the diving ducks occur in saltwater bays; the most common species are lesser scaup and redheads. During 1976–1985, an average of 1.1 million ducks, 5,500 geese, and 30,100 brant wintered along the mainland west coast. The largest duck numbers were of pintail, green-winged teal, shoveler, lesser scaup, ruddy duck, redhead, and cinnamon teal. The peak period with the highest numbers of all species was in December. Extensive areas of the mainland west coast and adjacent desert areas have been cleared for agricultural development, and numerous water storage reservoirs have been constructed. Consequently, these areas attract ducks. For example, duck populations in the Topolobampo region increased from 21,000 in 1951 to 743,000 in 1963. Pintails and wigeons were the most numerous in agricultural areas. In the coastal lagoons, wigeon grass (*Ruppia maritima*), snails, and moist-soil plants in adjacent areas were the most important foods along the mainland west coast (Kramer and Migoya 1989).

East coast

In this region there are at least 17 wintering areas, but only 7 have been surveyed, covering the coastal states of Tamaulipas, Veracruz, Tabasco, Campeche, Yucatán, and Quintana Roo along 2,970 km (1,857 mi) of shoreline (fig. 7.6). From 1970 to 1988, 19 duck and 3 goose species were recorded, and the total number of ducks averaged 1,135,691 (Baldassarre et al. 1989). Here, we review the three most important wintering sites on the east coast: Laguna Madre, Tabasco lagoons, and Alvarado lagoons (fig. 7.5).

The most important site along the Tamaulipas coast is the Laguna Madre (2,030 km², 1,261 mi²). This is the wintering site for 39% of ducks surveyed on the east coast from 1970 to 1988, with an average duck population of 439,765 (39% dabbling and 61% diving ducks). The most numerous species were redhead (49%), northern pintail (23%), scaup (mostly lesser scaup) (10%), and wigeon (10%). The Laguna Madre in South Texas and Mexico is the wintering site for about 80% of the world's population of redheads; about 35% of the continental population winters in Mexico. The Tabasco lagoons (20,000 km², 12,420 mi²), which extend into Campeche and are among the most important wetlands in the Neotropics, receive floodwaters from several rivers to form freshwater delta wetlands. The wetlands have nine habitat types, including beach communities, mangrove forest, marshes, lakes and ponds, and gallery forest along streams and channels; they also have the most extensive mangrove forest in Mexico (1,100 km², 683 mi²). The Los Pantanos de Centla Biosphere Reserve (3,027 km², 1,880 mi²) is within these wetlands (ParksWatch 2003). Winter duck populations averaged 225,194, of which dabbling ducks made up 94%; the blue-winged teal was the most numerous (70%) and made up 48% of the population of this species on the east coast. The Alvarado Lagoons (5,380 km², 3,340 mi²) consist of

Figure 7.6. Coastal wetland in Yucatán, a major waterfowl wintering area. Photo by Eduardo Carrera.

a delta and coastal lowland system, principally south of Alvarado, Veracruz, and contain about 2,500 km^2 (1,550 mi^2) of wetlands. Waterfowl populations declined in the area because of an increase in undesirable aquatic plants such as water hyacinth (*Eichhornia crassipes*). Waterfowl populations averaged 109,523 ducks during 1970–1986. Dabbling ducks were the most numerous, of which blue-winged teal was most abundant, followed by whistling ducks and American wigeon (Saunders and Saunders 1981, Baldassarre et al. 1989).

Central Highlands

This region provides habitats for numerous waterfowl but has declined in importance because of the degradation, loss, and pollution of formerly productive wetlands (fig. 7.5). This decline has occurred because of wetland siltation, resulting from soil erosion caused by overgrazing and deforestation. In addition, reduced flows because of dams, water diversions to irrigate extensive agricultural lands, and dense human populations have contributed to declines in habitat quantity and quality. This region is also predicted to be subjected to more serious and lasting droughts because of climate change. In some years, drought has deleteriously affected habitats and caused significant decreases in waterfowl numbers. Also, because of variation in annual rainfall, the sizes of water bodies vary greatly from year to year. Nonetheless, important wintering habitats remain for migratory and resident waterfowl. In agricultural areas, artificial impoundments have created important habitats for waterfowl and have enhanced food resources, in some cases compensating for the loss of natural water systems. The most important wetlands (lakes, lagoons, and reservoirs) in terms of waterfowl numbers are in the Southern Highlands in Michoacán, Jalisco, and Guanajuato. Wetlands in the northern states of Chihuahua and Durango host lower numbers of wintering waterfowl. Here we describe three representative wetlands (Saunders and Saunders 1981, Arreguín Cortés et al. 2011).

Lake Babicora, in northern Chihuahua, is 19 km × 8 km (11.8 mi × 5 mi), its size varying depending on rainfall. It consists of shallow lakes and channels and was formerly renowned for wintering one of the highest concentrations of sandhill cranes in North America. When it was first surveyed in 1947, there were nearly 50,000 snow geese, 120 mallards, 450 gadwalls, 275 green-winged teal, and 28,000 sandhill cranes, but within a greatly reduced surface area of 1.6 km × 0.8 km (1 mi × 0.5 mi). When it was surveyed again during 1961–1965, there were about 46,000 ducks on average (range: 9,000–80,000). White-fronted geese averaged about 1,600 (range: 40–2,638); snow geese averaged more than 13,000 (range: 700–34,000). The duck population averaged about 8.2% of the total Highlands wintering population. The most common ducks were green-winged teal, shoveler, and pintail (Saunders and Saunders 1981).

Lake Chapala, between Jalisco and Michoacán, is 80 km (50 mi) long and 12.5 km (7.8 mi) wide (1,100 km^2, 424 mi^2). It is Mexico's largest lake and formerly wintered the most ducks. It is fed chiefly by the Río Lerma. This lake, the principal source of water for the city of Guadalajara, has been devastated by siltation, upstream damming, the diking and draining of delta lagoons and marshes, and the widespread presence of water hyacinth. Also, it has been highly polluted by municipal, industrial, and agricultural wastes, but construction of water purification plants along the Río Lerma has lessened the pollution. Numbers of ducks and geese have dropped dramatically: the total number of ducks observed declined from 128,210 in 1948 to 17,402 in 1965. Numbers of geese have also dropped precipitously. The duck species with the highest numbers observed were shoveler, pintail, green-winged teal, gadwall, and lesser scaup. During 1951–1965, the average number of ducks constituted 10.2% of the total average of ducks observed in the Highlands.

Lake Pátzcuaro is one of Mexico's most important waterfowl wintering sites. It is approximately 15 km × 17.5 km (9.3 mi × 10.8 mi) and has been much affected by siltation and invasion of water hyacinth. It has become a major tourism center, which has probably reduced its importance as a waterfowl site. In the first survey in 1948, a total of 17,530 ducks were observed, but the number increased to 63,930 in January 1950. Estimates in some of the following years were 16,890 in 1951, 118,930 in 1955, and 22,705 in 1960, with an annual average of 46,500 (Saunders and Saunders 1981).

Waterfowl Abundance

Waterfowl counts

Since 1948, the USFWS and Mexico's federal government have conducted midwinter aerial counts in Mexico, including the inland wetlands (north and central) and the Pacific and Atlantic coasts. From 1976 through 1985, 4.7 million waterfowl (8% of the total waterfowl population of North America) wintered in Mexico, of which 1.5 million (48%) wintered in Mexico's Pacific coast bays and estuaries in 1985. Waterfowl were most abundant in Mexico in the early 1980s. Ducks represented an average of 90% and geese 10% of waterfowl counted during 1961–2000, and 84% and 16%, respectively, during 1991–2000 (Saunders and Saunders 1981, Pérez-Arteaga and Gaston 2004).

From 1961 through 2000, the annual duck population averaged 2,118,405, while geese averaged 241,744. However, the short-term trend from 1981 to 2000 showed a slight increase in annual duck and geese averages to 2,211,419 and 260,481, respectively. The annual average number for duck and geese populations in Mexico from 1961 through 2000 was 2,360,149 waterfowl, with a maximum of 3,825,500 ducks in 1982 and a minimum of 379,500 geese in 1973. During this 40-year period, the total numbers of waterfowl did not change significantly, but there has been a decline of almost 4% per year since 1981. The mean numbers of 10 waterfowl species (American wigeon, gadwall, green-winged teal, mallard, northern pintail, blue-winged teal, cinnamon teal, ring-necked duck, lesser scaup, and Canada goose) during 1991–2000 were significantly below the long-term mean (1961–2000). Only the 1991–2000 mean of combined snow and Ross's geese populations was significantly higher than the long-term mean (Pérez-Arteaga and Gaston 2004).

The five most abundant species during 1961–2000 were northern pintail, blue-winged teal, northern shoveler, green-winged teal, and lesser scaup. However, the five most abundant species during 1991–2000 were blue-winged teal, northern shoveler, redhead, northern pintail, and green-winged teal. The most abundant species over the long term has been the northern pintail; at least 12% of the continental population winters in Mexico. However, numbers have decreased in Mexico since 1978, probably because of their decline in the breeding grounds. Lesser scaup have also declined in North America; Mexican populations have declined at a rate of 8% per year since 1981. Redheads have declined during the last 20 years in the Laguna Madre region of Mexico, probably because of habitat changes (Pérez-Arteaga and Gaston 2004).

Density

Lake Cuitzeo, in Michoacán, is one of the most important wetlands for waterfowl in Mexico's lacustrine region. During the winter season of 1996–1997, 11 species were recorded; the Mexican and ruddy duck were resident species, while the remaining 9 were migratory and arrived at different times. Other species recorded were the American wigeon, northern shoveler, redhead, ring-necked duck, and canvasback, and the three teals. Population densities showed a higher abundance for the canvasback, with 2.5 birds/ha (1.04 birds/ac) of water surface, followed by the ring-necked duck with 1.0 birds/ha (0.41 birds/ac); American wigeon with 0.9 birds/ha (0.38 birds/ac); Mexican duck with 0.7 birds/ha (0.29 birds/ac); blue-winged teal with 0.5 birds/ha (0.2 birds/ac); ruddy duck with 0.4 birds/ha (0.16 birds/ac); redhead and green-winged teal with 0.3 birds/ha (0.12 birds/ac); northern shoveler with 0.2 birds/ha (0.08 birds/ac); and cinnamon teal with 0.1 birds/ha (0.04 birds/ac) (Aguilar Valdéz 1998).

At Pantanos de Centla in Tabasco from December 1999 to December 2000, only 5 anatid species were recorded of the potential 22 that could have occurred during winter. This was possibly caused by heavy rains that created abundant aquatic habitats, which enabled waterfowl to disperse to adjacent wetlands. The most abundant species was the blue-winged teal, followed by the black-bellied whistling duck, a resident species. The black-bellied whistling duck had the highest density, with 13.3 birds/ha (5.5 birds/ac), followed by the blue-winged teal with 10.3 birds/ha (4.2 birds/ac). The Muscovy duck, also resident, had a low abundance, probably because of illegal subsistence hunting; it was restricted to mangroves, where it perched on trees. The green-winged teal was present only for short periods, and the lesser scaup was present only occasionally (Córdova Ávalos 2002).

Between August 1994 and April 1995 in the El Llano region of Aguascalientes, 14 duck and 2 geese species were recorded. The most abundant species were the green-winged teal and northern shoveler. The period of residence varied from almost seven months during winter for the green-winged teal, to 15 days for the lesser scaup. The density of species (birds/ha) varied according to period and wetland type. From September to October 1994, waterfowl density varied from 0.8 to 20.9 birds/ha (0.3 to 8.7

birds/ac); from December 1994 to January 1995, density varied from 5.7 to 100.6 birds/ha (2.3 to 41.9 birds/ac); and from February to March 1995, it varied from 4.2 to 52.4 birds/ha (1.75 to 21.8 birds/ac) (Medina Torres 1996).

Management and Research Needs

Mexico's wetlands are of vital importance to North American waterfowl management. They provide critical habitats for breeding, migratory, and wintering waterfowl as well as many other aquatic birds. Waterfowl conservation efforts in Canada and the US will be ineffective without Mexico's participation in protecting wintering waterfowl and habitats. It is in these wintering wetlands that millions of waterfowl must replenish their nutritional reserves to enable them to migrate to their nesting and rearing areas. Mexico's participation in international agreements is an integral element in the conservation and management of wetlands. Mexico is a member of the Ramsar Convention (1986) and the North American Waterfowl Management Plan (NAWMP), which commit participating nations to cooperate in wetlands conservation programs. The mission of Ramsar includes the designation of suitable wetlands of international importance, to ensure their effective management and to cooperate internationally on transboundary wetlands and shared species. The goals of NAWMP include ensuring abundant waterfowl populations and habitats and their sustainable use, and supporting waterfowl and wetland conservation through cooperative ventures by the US, Canada, and Mexico. The North American Wetlands Conservation Act (NAWCA), enacted by the US Congress in 1989, provides funding to encourage wetland conservation through public-private partnerships. NAWCA provided US$2.8 million for conservation projects in Mexico in fiscal year 2012 (NAWMP 2002, Carrera and de la Fuente 2003, RCS 2010, Morzaria-Luna et al. 2014).

Harvest limits in Mexico are determined by the federal government under the advisement of a waterfowl subcommittee. Most legal hunting in Mexico is conducted in Wildlife Management Units for Conservation and Sustainable Use (UMAs) with a hunting guide. Each UMA employs a technician who monitors game population numbers and requests waterfowl licenses from the federal government. As of 2015, there were 12,524 UMAs in Mexico, with 25,551 hunting licenses (2013–2014) sold (see chapter 3). Of the total number of hunting licenses sold, approximately 4,000 were strictly for waterfowl. Of the 521 UMAs approved in 2012 for harvesting waterfowl, 73% were in states in northern Mexico. According to a hunter activity survey conducted some years ago, 75% of waterfowl hunters were from the US, and 80% of those were from Texas and California. The USFWS estimated (2014–2015 season) that Texas has 86,300 active duck hunters and 30,300 goose hunters, but the percentage of Texas waterfowl hunters that travel to Mexico is small. While Mexico could potentially attract additional hunters from Texas, the continued violence in the border states, stemming largely from drug cartels, is having a significant negative impact. Data from Ducks Unlimited (DUMAC) indicate that in the past six years, UMAs in northern Mexico have experienced a 60–80% hunter cancellation rate (E. Carrera, DUMAC, personal communication).

Unfortunately, Mexico has not prioritized wetland conservation, even though water is a diminishing resource (Alcocer et al. 2000). Public awareness of worsening water shortages in Mexico is forcing the government to confront what is already a crisis (see chapter 1). Hence, it is in the best interest of the Mexican public that Mexican natural resource agencies prioritize aquatic ecosystems. Also, the Migratory Bird Treaty Act commits Mexico to protect, maintain, and monitor habitats for migratory bird populations. This requires wildlife biologists to conduct basic field research and accumulate the knowledge necessary to efficiently manage habitats. However, unless basic ecological information and methodologies are generated, management goals and objectives cannot be met. A primary need is to conduct annual terrestrial and aerial surveys to determine the numbers, trends, distribution, and composition of wintering waterfowl populations, and to quantify waterfowl harvest. Mexico has failed to participate in winter waterfowl surveys since 2000. Inadequately staffed government agencies lack the funding and trained personnel to conduct surveys and studies. Life history research needs include nutritional and physiological studies of waterfowl in Mexico. Wetland inventories are also needed, including agriculturally developed areas important to waterfowl, as well as water quality monitoring and mediation of pollution sources. The existence of wetlands needs to be justified to the public; this will require public awareness programs and will also provide economic incentives for rural populations to participate in and champion wetland conservation (see chapter 1).

The current system of protected wetlands needs to be expanded, and new habitats should be created to compensate for those that have deteriorated or have been destroyed. There are seven designated Ramsar sites in Mexico, but additional undesignated sites meet the Ramsar convention criteria, indicating a need to critically evaluate national wetlands for potential inclusion in the Mexican and international protected areas systems. In conjunction with designating additional waterfowl use areas, degraded areas should be restored and habitat improvement programs initiated. Some protected areas are ineffective because of unresolved social, political, and economic issues. An example of threatened wetlands in Mexico is the Pantanos de Centla Biosphere Reserve, one of the most important wetlands in the world. It contains 12% of all of Mexico's aquatic and subaquatic vegetation and 255 bird species, many of which are migratory waterbirds. However, this biosphere, which is inadequately staffed, is threatened with habitat destruction because of mismanaged cattle ranching, unregulated fishing and hunting, illegal fires to expand cattle ranching and farming, oil extraction, contamination, an expanding human population, and lack of law enforcement (ParksWatch 2003). Concerted effort should be made to ensure that protected areas function sustainably (Pérez-Arteaga et al. 2002b).

Many coastal wetlands still provide healthy ecosystems that can support numerous wintering waterfowl, and some of these include protected areas. Economic incentives for the rural sector will be essential to a successful waterfowl program in Mexico (see chapters 1 and 22). Of primary importance is that Mexico, Canada, and the US cooperate in joint ventures to conserve and manage North American waterfowl. Additionally, cooperation will be required from NGOs such as DUMAC, private initiatives, academic institutions, state and federal agencies, and rural communities to accomplish national and international waterfowl conservation goals. It is essential to incorporate a comprehensive national legal and policy framework in Mexico with an integrated cooperative joint waterfowl conservation initiative (Kramer and Migoya 1989, Morzaria-Luna et al. 2014).

Acknowledgments

We wish to thank the Graduate School (Colegio de Postgraduados) at Campus San Luis Potosí for permitting F. Clemente Sánchez to dedicate the needed time to coauthor this chapter, and Eduardo Carrera for providing useful information on hunting in Mexico.

Literature Cited

Aguilar Valdéz, B. C. 1998. Hábitat y abundancia de anátidos en el Lago de Cuitzeo, Michoacán. Thesis, Colegio de Postgraduados, Montecillo, Estado de México, México.

Alcocer, J., E. Escobar, and L. E. Marin. 2000. Epicontinental aquatic systems of Mexico in the context of hydrology, climate, geography and geology. Pages 1–13 *in* M. Munawar, S. G. Lawrence, I. F. Munawar, and D. F. Malley, editors. Aquatic ecosystems of Mexico: status and scope. Backhuys, Leiden, the Netherlands.

Alderfer, J., editor. 2014. National Geographic complete birds of North America. 2014. Second edition. National Geographic, Washington, DC, USA.

Alonzo-Parra, D., E. Bestard-Barrera, and A. Zaldivar-Jimenez. 2006. Manual para la conservación del pato real mexicano y su hábitat. Final report. DUMAC, Garza Garcia, Nuevo León, México.

AOU. 1998. AOU checklist of North and Middle American birds. Seventh edition. American Ornithologists' Union, Philadelphia, Pennsylvania, USA.

Arellano, M., and P. Rojas. 1956. Aves acuáticas migratorias en México I. Instituto Mexicano de Recursos Naturales Renovables, México, D. F., México.

Arreguín Cortés, F. I., M. López Pérez, and H. Marengo Mogollón. 2011. Mexico's water challenges for the 21st century. Pages 21–38 *in* U. O. Spring, editor. Water resources in Mexico. Springer, Berlin, Germany.

Baldassarre, G. 2014. Ducks, geese, and swans of North America. Revised edition. Johns Hopkins University Press, Baltimore, Maryland, USA.

Baldassarre, G. A., and E. G. Bolen. 2006. Waterfowl ecology and management. Second edition. Krieger Publishing, Malabar, Florida, USA.

Baldassarre, G. A., A. R. Brazda, and E. R. Woodyard. 1989. The east coast of Mexico. Pages 407–425 *in* L. M. Smith, R. L. Pederson, and R. M. Kaminski, editors. Habitat management for migrating and

wintering waterfowl in North America. Texas Tech University Press, Lubbock, USA.
BirdLife International. 2012. *Cairina moschata*. IUCN Red List of Threatened Species 2012:e.T22680061A40109661. http://dx.doi.org/10.2305/IUCN.UK.2012–1.RLTS.T22680061A40109661.en. Accessed 20 June 2016.
Carboneras, C. 1992. Order Anseriformes. Pages 536–628 *in* J. del Hoyo and J. Gargatal, editors. Handbook of the birds of the world. Volume 1. Ostrich to ducks. Lynx Edicions, Barcelona, Spain.
Carboneras, C. 2014. Order Anseriformes. Pages 124–146 *in* J. del Hoyo and N. J. Collar, editors. Illustrated checklist of the birds of the world. Volume 1. Non-passerines. Lynx Edicions, Barcelona, Spain.
Carmona, R., V. Ayala-Pérez, and A. Gutiérrez Aguilar. 2011. New and noteworthy waterfowl records at artificial wetlands from Baja California Sur, Mexico. Revista Mexicana de Biodiversidad 82:721–726.
Carrera, E., and G. de la Fuente. 2003. Inventario y clasificación de humedales en México. Parte I. Ducks Unlimited de México, Monterrey, México.
Colón Quezada, D. 2012. Estructura genética del pato mexicano (*Anas diazi*) en México. Thesis, UNAM, Instituto de Biología, México, D. F., México.
Córdova Ávalos, A. 2002. Situación actual de la población de anátidos en la Reserva de la Biósfera Pantanos de Centla, Frontera, Tabasco. Thesis, Colegio de Postgraduados, Montecillo, Estado de México, México.
Dooley, J., K. Fleming, P. Garretson, W. Rhodes, and N. Zimpfer. 2015. Waterfowl population status, 2015. USFWS, Laurel, Maryland, USA.
Eitniear, J., R. Bribiesca-Formisano, C. Rodríguez-Flores, C. Soberanes-González, and M. C. Arizmendi. 2015. Muscovy duck (*Cairina moschata*), Neotropical birds online. Cornell Laboratory of Ornithology, Ithaca, New York, USA. http://neotropical.birds.cornell.edu. Accessed 21 Jun 2016.
Escalante, P., A. M. Sada, and J. R. Gil. 2014. Lista de nombres comunes de las aves de México. UNAM and CIPAMEX, México, D. F., México.
Glenn, E. P., F. Zamora-Arroyo, P. L. Nagler, M. Briggs, W. Shaw, and K. Flessa. 2001. Ecology and conservation biology of the Colorado River delta, Mexico. Journal of Arid Environments 49:5–15.
Howard, H. 1964. Fossil Anseriformes. Pages 233–248 *in* J. Delacour, editor. Waterfowl of the world. Country Life, London, United Kingdom.
Howell, S. N. G., and S. Webb. 1995. A guide to the birds of Mexico and northern Central America. Oxford University Press, Oxford, United Kingdom.
Johnsgard, P. A. 1978. Ducks, geese and swans of the world. University of Nebraska Press, Lincoln, USA.
Johnsgard, P. A. 2010. Waterfowl of North America. Second edition. University of Nebraska Press, Lincoln, USA.
Kramer, G. W., E. Carrera, and D. Zaveleta. 1995. Waterfowl harvest and hunter activity in Mexico. Transactions of the North American Wildlife and Natural Resources Conference 60:243–250.
Kramer, G., and R. Migoya. 1989. The Pacific coast of Mexico. Pages 507–528 *in* L. M. Smith, R. L. Pederson, and R. M. Kaminsky, editors. Habitat management for migrating and wintering waterfowl in North America. Texas Tech University Press, Lubbock, USA.
Leopold, A. S. 1959. The wildlife of Mexico: the game birds and mammals. University of California Press, Berkeley, USA.
Lewis, T. L., D. H. Ward, J. S. Sedinger, A. Reed, and D. V. Derksen. 2013. Brant (*Branta bernicla*). The birds of North America online. Cornell Laboratory of Ornithology, Ithaca, New York, USA. http://bna.birds.cornell.edu/bna/species/337. Accessed 12 Jan 2016.
Markum, D. E., and G. A. Baldassarre. 1989. Breeding biology of Muscovy ducks using nest boxes in Mexico. Wilson Bulletin 101:621–626.
Medina Torres, S. M. 1996. Distribución, abundancia y uso de hábitat de patos y gansos en la región de El Llano, Aguascalientes. Thesis, Colegio de Postgraduados, Montecillo, Estado de México, México.
Medina Torres, S. M., M. M. Olivas, and E. García Moya. 2007. Uso y selección por el pato mexicano (*Anas diazi*) en la región del Llano, Aguascalientes-Jalisco, México. Acta Zoológica Mexicana (n.s.) 23:163–181.
Morzaria-Luna, H. N., A. Castillo-López, G. D. Anemann, and P. Turk-Boyher. 2014. Conservation strategies for coastal wetlands in the Gulf of California, Mexico. Wetlands Ecology and Management 22:267–288.
NAWMP. 2002. Strengthening the biological foundations: 2003 North American Waterfowl Management Plan update. http://birdhabitat.fws.gov/

NAWMP/images/FirstDraftforReviewbyPlanStakeholdersS.pdf. Accessed 20 Mar 2016.

Olson, S. M., compiler. 2015. Pacific Flyway data book, 2015. USFWS, Division of Migratory Bird Management, Vancouver, Washington, USA.

ParksWatch. 2003. Pantanos de Centla Biosphere Reserve. http://www.parkswatch.org/parksprofiles/pdf/pcbr_eng.pdf. Accessed 24 Apr 2016.

Pérez-Arteaga, A., and K. J. Gaston. 2004. Wildfowl population trends in Mexico, 1961–2000: a basis for conservation planning. Biological Conservation 115:343–355.

Pérez-Arteaga, A., K. J. Gastón, and M. Kershaw. 2002a. Population trends and priority conservation sites for Mexican duck *Anas diazi*. Bird Conservation International 12:35–52.

Pérez-Arteaga, A., K. J. Gastón, and M. Kershaw. 2002b. Undesignated sites in Mexico qualifying as wetlands of international importance. Biological Conservation 107:47–57.

RCS. 2010. International cooperation: guidelines and other support for international cooperation under the Ramsar Convention on Wetlands. Fourth edition. Ramsar Convention Secretariat, Gland, Switzerland.

Saunders, G. B., and D. C. Saunders. 1981. Waterfowl and their wintering grounds in Mexico, 1937–1964. Resource Publication 138. USFWS, Washington, DC, USA.

SEMARNAT. 2002. NOM-059-ECOL-2001. Protección ambiental, que determina las especies y subespecies de flora y fauna silvestre terrestre y acuática en peligro de extinción, amenazadas y las sujetas a protección especial, y que establece especificaciones para su protección. SEMARNAT, Diario Oficial de la Federación del 16 de marzo de 2002, México, D. F., México.

Wetlands International. 2016. Waterbird population estimates. wpe.wetlands.org. Accessed 20 Jun 2016.

Whitley, G. R. 1973. The Muscovy duck in Mexico. Anthropological Journal of Canada 11:2–6.

Williams, S. O., III. 1980. The Mexican duck in Mexico: natural history, distribution, and population status. Dissertation, Colorado State University, Fort Collins, USA.

Woodyard, E. R. 1982. Some ecological aspects of the ecology of Muscovy ducks in Mexico. Thesis, Texas Tech University, Lubbock, USA.

Woodyard, E. R., and E. G. Bolen. 1984. Ecological studies of Muscovy ducks in Mexico. Southwestern Naturalist 29:453–461.

8 Quail in Mexico

Fidel Hernández, E. Alejandro Lozano-Cavazos, Gilberto Chávez-León, David García-Solórzano, and Leonard A. Brennan

Quail are inextricably linked to the culture, history, and people of Mexico. The significant role that quail have played in the daily life of historical Mexico can be appreciated by a simple visit to the Flora and Fauna room of the Templo Mayor Museum in Mexico City. Here one can find vestiges of the diverse uses of quail for medicine, food, and religious purposes involving species such as scaled quail, northern bobwhite, and Montezuma quail. Early cultures throughout Mesoamerica had quail, particularly tree-quail (*Dendrortyx* spp.), in captivity for long periods for domestication (Valadéz 2003). It is likely that quail will continue to play an important role in Mexico's culture. Quail hunting is becoming increasingly popular among both Mexican and US sportsmen and women. Thousands of quail are harvested annually throughout the country, particularly in its northern region. In addition, hundreds of birders visit Mexico each year to view rare or endemic bird species, some of which include quail. Both quail hunting and ecotourism represent emerging but largely undeveloped enterprises in Mexico. Therefore, a conservation-based framework is needed to guide the management of Mexico's wildlife resources. It is our goal to propose such a framework for quail in this chapter. We begin with a brief summary of the taxonomy and evolution of Mexican quail; discuss their population ecology, habitat needs, and management; and conclude with a discussion of the challenges facing quail conservation in Mexico.

Taxonomy and Evolution

Ornithologists recognize 15 species of New World quail (family Odontophoridae) in Mexico (Carroll 1995). These 15 species are represented by eight genera. Mexico is considered to be near the center of the evolutionary history and geographic origin of these birds (table 8.1). Johnsgard (1973) hypothesized that the earliest forms of New World quail were derived from ancestral cracids during the mid-Tertiary era, about 30 million years ago (mya). The earliest-known fossils of New World quail date to the Early Miocene, about 20 mya (Brodkorb 1964). Gutiérrez et al. (1983) hypothesized that the genera *Dendrortyx* and *Odontophorus* were two primitive and distinct lineages that emerged about 16 mya. Genetic data presented by Gutiérrez et al. (1983) suggested that the mountain quail (*Oreortyx pictus*) originated about 13 mya in western North America and was separate from a preancestral form of the *Colinus-Callipepla* complex at this time. During the past several million years, the *Colinus-Callipepla* complex developed into at least eight species with relatively distinct habitat affinities. During the same period, the endemic species in Mexico evolved into the forms and distributions present today. While the biogeographic data are sketchy, the basic hypothesis is that New World quail originated in Central America and intermittently radiated northward and retreated southward in response to Pliocene and Pleistocene glacial cycles, resulting in the present-day assortment of species and diverse geographic distributional patterns. Southward radiation into South America was also a key aspect of the evolution of New World quail.

Ecologically, the 15 species of quail in Mexico can be divided into two broad groups based on their habitat affinities. Five of these species are typically found in humid, mesic forested habitats, and 10 are found in more xeric brushy, grassy, or rangeland habitats (table 8.1). Species accounts for Mexican quail are provided at the end of this chapter. The material for the species accounts was drawn liberally from publications by Leopold (1959, 1977), Johnsgard (1973), Madge and McGowan (2002), and Brennan (2007).

Table 8.1. Common names (English and Spanish) and conservation status of quail species in Mexico.

Habitat affinity / Common name	Spanish name	Scientific name	Conservation status: SEMARNAT	IUCN
Forest				
Semiarboreal (humid forest)				
Bearded wood-partridge	Gallina-de-monte veracruzana	*Dendrortyx barbatus*	Endangered	Vulnerable
Long-tailed wood partridge	Gallina-de-monte coluda	*Dendrortyx macroura*	Special Protection	Least Concern
Buffy-crowned wood partridge	Gallina-de-monte centroamericana	*Dendrortyx leucophrys*	Threatened	Least Concern
Singing quail	Codorniz silbadora	*Dactylortyx thoracicus*	Special Protection	Least Concern
Spotted wood-quail	Codorniz bolonchaco	*Odontophorus guttatus*	Special Protection	Least Concern
Rangeland				
Terrestrial-digging (woodland)				
Montezuma quail	Codorniz de Moctezuma	*Cyrtonyx montezumae*	Low Risk[a]	Least Concern
Ocellated quail	Codorniz ocelada	*Cyrtonyx ocellatus*	Threatened	Near Threatened
Terrestrial-scratching (grassland and scrub, chaparral)				
Elegant quail	Codorniz elegante	*Callipepla douglasii*	Low Risk	Least Concern
Scaled quail	Codorniz escamosa	*Callipepla squamata*	Low Risk	Least Concern
Gambel's quail	Codorniz de Gambel	*Callipepla gambelii*	Low Risk	Least Concern
California quail	Codorniz de California	*Callipepla californica*	Low Risk	Least Concern
Mountain quail	Codorniz de montaña	*Oreortyx pictus*	Low Risk	Least Concern
Banded quail (Barred quail)	Codorniz barrada; Codorniz listada	*Philortyx fasciatus*	Low Risk	Least Concern
Northern bobwhite	Codorniz norteña	*Colinus virginianus*	Low Risk[b]	Near Threatened
Black-throated bobwhite (Yucatán bobwhite)	Codorniz Yucateca	*Colinus nigrogularis*	Low Risk	Least Concern

Sources: SEMARNAT (2002), International Union for Conservation of Nature (IUCN).

[a]The subspecies Pacific Montezuma quail (*C. m. sallei*) is considered threatened, and the subspecies Merriam's Montezuma quail (*C. m. merriami*) is considered endangered.

[b]The masked bobwhite (*C. v. ridgwayi*), a subspecies, is considered endangered.

Population Ecology

Population status

Mexico is important for quail conservation in Latin America because of its great species diversity (McGowan et al. 1995). Unfortunately, the population status of most Mexican quail is unknown because of a lack of basic ecological knowledge. Even popular Mexican game species such as northern bobwhite, scaled quail, and California quail, which traditionally generate research interest among sportsmen, have remained largely unstudied in Mexico (fig. 8.1). Despite this lack of information, a broad generalization of conservation status can be made of the two groups of quail in Mexico: species inhabiting grassland or scrub

Figure 8.1. The population status of most Mexican quail is unknown because basic ecological knowledge is lacking. This is true even for popular Mexican game species such as the Montezuma quail. Photo by Alejandro Lozano.

rangeland are more common and widespread, whereas forest-dwelling species are less abundant, have more restricted distributions, and are of greater conservation concern. For the latter group, deforestation resulting from factors such as agriculture and coffee plantations has reduced the amount of existing habitat.

Several publications have addressed the conservation status of Mexican quail (Carroll 1995, McGowan et al. 1995, Eitniear et al. 2004). These conservation assessments consider the bearded tree-quail to be critically endangered and of the highest conservation priority of all Mexican quail (Carroll and Eitniear 2000). This conservation designation coincides with the Official Mexican Regulation for Protection of Native Wildlife NOM-059-SEMARNAT-2001 (SEMARNAT 2002), which formally identified the bearded tree-quail as an endangered species in Mexico. This federal document also identified an additional five species and two subspecies as quail of conservation concern. Specifically, it identified two species as threatened (buffy-crowned tree-quail and ocellated quail), and three species as warranting special protection (long-tailed tree-quail, spotted wood-quail, and singing quail). In addition, the report identified one endangered subspecies (masked bobwhite, *Colinus virginianus ridgwayi*) and one threatened subspecies (Pacific Montezuma quail, *Cyrtonyx montezumae sallei*). Collectively, 40% of Mexican quail (6 of 15 species) are considered of conservation concern by the federal government (SEMARNAT 2002).

Ecology and Life History

Little is known about the ecology and life history of Mexican quail (Leopold 1959, Johnsgard 1973, 1988). Several recent studies address this general lack of knowledge. Species studied have included the bearded tree-quail (Baccus 2004), long-tailed tree-quail (Chávez-León 2004a, 2004b, Chávez-León and Velázquez 2004, Chávez-León et al. 2004), singing quail (Eitniear and Aragon 2004), banded quail (Urbina-Torres and Zaiñhana-Ortiz 2004, MacGregor-Fors 2006), elegant quail (Brooks 2004), and Montezuma quail (Robles et al. 2002, García-Solórzano 2005, Téllez et al. 2005, Palafox et al. 2007). One of the most studied quail species in Mexico is the endangered masked bobwhite, which has been researched intensively for more than 50 years (reviewed by Hernández et al. 2006). Although these studies have contributed important information, much remains unknown regarding their ecology.

Available information suggests that the general life histories and ecologies of Mexican quail are similar to those of the better-studied North American species. We rely on this more extensive literature base to provide a broad overview of population ecology, and we refer to studies specific to Mexican species when such information is available. Like most Galliformes, quail experience relatively high annual mortality. Annual survival rates typically range from 20% to 30%, although much variability exists. For example, annual survival has been estimated at 2–21% for northern bobwhite (Cox et al. 2004), 10–17% for scaled quail (Campbell et al. 1973, Schemnitz 1994), 14–59% for California quail (Emlen 1940), and 18–59% for Montezuma quail (Brown 1989). It is worth noting that some of these survival estimates were obtained using radiotelemetry, which has been debated as a valid technique to estimate survival because of a potential negative bias (Guthery and Lusk 2004, Folk et al. 2007). No annual survival estimates are available for Mexican quail; however, it is thought that the forest-dwelling species in central Mexico experience relatively higher survival rates than the rangeland species found in the north (Carroll 1995). This supposition is plausible because forest quail possess life-history traits more characteristic of *K*-selected species (e.g., larger body size, smaller clutches), which generally experience higher survival than *r*-selected species (e.g., rangeland quail) (Pianka 1970).

Predation is the primary cause of quail mortality and nest failure (Hurst et al. 1996). In general, raptors are thought to pose a greater threat to adult quail, whereas mammalian predators represent a greater risk to nests (Rollins and Carroll 2001). Raptors considered to be important quail predators include Cooper's hawk (*Accipiter cooperii*), sharp-shinned hawk (*A. striatus*), northern harrier (*Circus cyaneus*), Harris's hawk (*Parabuteo unicinctus*), and great-horned owl (*Bubo virginianus*). Common nest predators include raccoon (*Procyon lotor*), opossum (*Didelphis virginiana*), skunk (*Mephitis* spp.), bobcat (*Lynx rufus*), coyote (*Canis latrans*), several species of foxes and snakes, and numerous other species.

Quail are able to sustain such levels of mortality because of their high reproductive potential, which results from their early sexual maturity, large clutch size, high propensity to renest, and flexible breeding systems. In *Colinus* (Baldini et al. 1952) and *Callipepla* (Johnsgard 1988) species, sexual maturity is achieved at approximately 150–180 days. Clutch size is also relatively large (10–15 eggs), although forest quail in central Mexico have smaller clutches (4–6 eggs) (Carroll 1995). Egg hatchability, or the percentage of eggs from a clutch that hatch, tends to be quite high (> 90%).

Nest success is low (30–50%), but quail compensate by renesting and even producing multiple broods. Renesting following nest failure has been reported with great frequency in virtually all New World quail studied (Johnsgard 1988). Multiple brooding, or the successful hatching of two or more broods in a single season, is less well documented but has been reported in northern bobwhite (Stanford 1972), scaled quail (Wallmo 1956b), California quail (Francis 1965), and Gambel's quail (Gullion 1956). An additional characteristic that contributes to the reproductive potential of quail is their flexible breeding system. Quail have traditionally been considered monogamous, but research suggests that they possess a more complex mating system (Burger et al. 1995, Faircloth 2008). For example, rapid multiclutching has been reported in mountain quail (Gutiérrez and Delehanty 1999), during which a male incubates one nest while the female simultaneously incubates a second. Serial polyandry, or the sequential pairing of females with different males to lay multiple clutches, has been reported in Montezuma (Stromberg 2000), Gambel's (Brown et al. 1998), and California quail (Francis 1965). A third type of breeding system, ambisexual polygamy, has been documented in northern bobwhite (Curtis et al. 1993). This breeding system involves both males and females forming sequential pairs with different individuals (i.e., serial polygyny and serial polyandry, respectively) to produce multiple nests. Collectively, all these reproductive strategies contribute to the high reproductive capacity of quail.

Population Structure

The population structure of quail has been studied primarily from the perspective of sex (male:female) and age (juvenile:adult) ratios. In quail populations, sex ratios tend to be slightly skewed toward males, which generally represent about 55–60% of the adult population. The general trend is that the sex ratio is equal (50:50) at hatching but gradually becomes skewed toward males as the cohort progresses from hatchlings to juveniles to adults. For example, the percentages of males in the juvenile and adult segments of a population have been reported as 58% versus 63% in Montezuma quail (Brown and Gutiérrez 1980), 49% versus 58% in California quail (Emlen 1940), and 51% versus 62% in northern bobwhite (Leopold 1945). These ratios suggest that females suffer higher mortality than males. It is generally assumed that higher female mortality results from the stress and risks associated with nesting, because the highest disparity in sex ratio occurs in late spring during breeding (Wallmo 1956a). However, alternative explanations involving sexual selection have been proposed (Brown and Gutiérrez 1980).

Age ratios (reported here as the number of juveniles per adult) in quail typically range from about 1.5 to 3.5 for most species but can vary greatly between years, depending on environmental conditions. We use the term "juvenile" for birds that have not entered their first breeding season. Age ratios have been reported as 2.1–6.5 in Gambel's quail (Raitt and Ohmart 1968), 1.5–8.3 in scaled quail (Campbell et al. 1973), 0.0–13.5 in mountain quail (Delehanty 1997), and 0.6–7.0 in northern bobwhite (Kiel 1976). As would be expected, age ratios are the most skewed toward juveniles during fall (i.e., immediately after reproduction has occurred in summer) and tend to decrease with time. In Gambel's quail, age ratios decreased from 3.3 (fall) to 2.3 (winter), to 2.5 (spring), to 1.5 (summer) (Raitt and Ohmart 1968). Such a pattern suggests that juveniles sustain higher mortality than older groups. However, it should be noted that the primary coverts of juvenile plumage, used to determine age in quail (Leopold 1939), are replaced by adult plumage when birds are from 11 to 17 months of age (Emlen 1940). Thus, the decrease in age ratio with time is also due, in part, to juvenile birds molting into adult plumage.

Population Dynamics

Quail populations, particularly those inhabiting xeric environments such as northern bobwhite, scaled quail, Gambel's quail, and California quail, are strongly influenced by weather. Populations in these dry landscapes are strongly correlated with the amount and timing of rainfall, increasing during years of abundant precipitation and decreasing drastically during drought (fig. 8.2). During unfavorable conditions, a large percentage—in some

Figure 8.2. Quail populations inhabiting xeric environments are strongly influenced by precipitation, increasing during years of abundant precipitation and decreasing drastically during drought. Photo by Aaron Verinder.

circumstances up to 50%—of females in a population may not nest (Johnsgard 1988, Hernández et al. 2005). In addition, the population may remain in coveys until the arrival of favorable breeding conditions, as has been documented in mountain quail and Gambel's quail (Leopold 1959), and northern bobwhite (Lehmann 1984). Consequently, the population dynamics of these temperate species have been described as "boom-and-bust." The mesic environment of humid forests, as well as life-history traits more characteristic of *K*-selected species, makes it unlikely that forest quail in Mexico undergo such drastic population fluctuations.

The demographic process that exerts the greatest influence on population change likely depends on the environment the species inhabits. Johnsgard (1988) suggested that annual population change is more closely related to variable reproduction in xeric landscapes, whereas it may be more influenced by survival in more mesic landscapes. This generalization is supported by population models. In northern bobwhite, population models suggest that population change may be more sensitive to survival in the southeastern forests of the US (Sandercock et al. 2008), whereas in southwestern semiarid rangeland, it appears more sensitive to reproduction (DeMaso 2008).

Harvest

Quail hunting in Mexico currently occurs within the framework of Management Units for Wildlife Conservation (UMA), as established by the Ministry of the Environment and Natural Resources (SEMARNAT). UMAs represent properties registered with the federal government that operate under an approved management plan for wildlife conservation. Properties can be private ranches, communal lands (ejidos), or even wildlife cooperatives comprising several landowners. Landowners are given the right to use UMAs and share responsibility with the federal government for the conservation of wildlife and its habitat. There are two basic UMA models: extensive and intensive (SEMARNAT 2009a). The extensive UMA model is based on the conservation of free-ranging native wildlife, whereas the intensive UMA model is focused on captive wildlife. There also are two types of wildlife usage within UMAs: extractive and nonextractive. Management units with extractive usage have the right to capture, collect, raise, or harvest wildlife. Management units with nonextractive usage involve activities associated with wildlife in their natural habitat that do not involve their removal, such as ecotourism and wildlife photography (see chapter 21).

Legal hunting of quail in Mexico is practiced on what are called "hunting ranches," which represent extensive UMAs with permission for extractive usage. To obtain authorization for hunting, each UMA annually requests the number of harvest permits, or *cintillos*, needed for the property. The level of harvest is based on a population assessment conducted by a wildlife biologist. Each harvest permit authorizes the harvest of 15 individual quail. Thus, if the estimated harvest for a UMA is 300 quail, then 20 harvest permits are solicited. Presently, 9 quail species in 28 of the 31 Mexican states are authorized for harvest by SEMARNAT through the General Office of Wildlife (DGVS) (table 8.2). Harvest restrictions have been established for the bearded tree-quail and masked bobwhite because of their endangered status, as well as the threatened buffy-crowned tree-quail, ocellated quail, and Montezuma quail subspecies *sallei* (SEMARNAT 2002). The long-tailed tree-quail, spotted wood-quail, and singing quail also receive special protection.

Quail harvest information in Mexico is scarce and difficult to obtain because the regulation (i.e., issuance, housing, and record keeping) of harvest permits is now decentralized. From 1997 to 2005, hunting permits were assigned exclusively by the national headquarters of SEMARNAT and stored in this central location. In 2006, however, when the granting of hunting permits became decentralized, state offices of SEMARNAT began issuing permits, and individual states became responsible for regulating their respective hunting permits (see chapter 21). Consequently, a centralized database of quail harvest permits in Mexico no longer exists.

The most recent hunting season with harvest data for all states is 2005–2006. It is important to note that the harvest data reported here are based on the number of hunting permits issued, which may not represent the actual number of quail harvested. For example, if 1,000 hunting permits were issued for a given year, then harvest rates for that year would be estimated at 15,000 quail (1,000 permits × 15 quail/permit). Nevertheless, harvest rates based on hunting permits provide some insight into quail hunting intensity in Mexico. A summary of the 2005–2006 data indicates that northern bobwhite had the greatest harvest pressure (77,575 specimens), followed by California quail (42,286), scaled quail (39,051), Gambel's quail (23,090), and banded quail (11,712) (table 8.3). Species with considerably lower harvest rates included mountain quail (1,470), Montezuma quail (765), elegant quail (656), and

Table 8.2. Quail species and states authorized for quail harvest in Mexico for the 2008–2009 hunting season.

Common name	Scientific name	States
Black-throated bobwhite	*Colinus nigrogularis*	Campeche, Yucatán
Northern bobwhite	*Colinus virginianus*	Aguascalientes, Chiapas, Coahuila, Guanajuato, Guerrero, Jalisco, Michoacán, Morelos, Nuevo León, Oaxaca, Puebla, Querétaro, Sinaloa, Tabasco, Tamaulipas, Tlaxcala, Veracruz
California quail	*Callipepla californica*	Baja California, Baja California Sur
Elegant quail	*Callipepla douglasii*	Durango, Nayarit, Sinaloa, Sonora
Gambel's quail	*Callipepla gambelii*	Baja California, Sonora
Scaled quail	*Callipepla squamata*	Aguascalientes, Chihuahua, Coahuila, Durango, Guanajuato, Guerrero, Jalisco, Michoacán, Nuevo León, Oaxaca, Querétaro, San Luis Potosí, Sonora, Tamaulipas, Zacatecas
Montezuma quail	*Cyrtonyx montezumae*	Durango, Hidalgo, Jalisco, Oaxaca, Puebla, Tlaxcala
Mountain quail	*Oreortyx pictus*	Baja California
Banded quail	*Philortyx fasciatus*	Colima, Morelos, Puebla

Source: SEMARNAT (2008a).

Table 8.3. Authorized quail harvest in Mexico for the 2005–2006 hunting season.

Species	State	Number of specimens
Scaled quail *(Callipepla squamata)*	Coahuila	16,569
	Nuevo León	12,500
	Tamaulipas	8,670
	Zacatecas	540
	Durango	525
	Estado de México	120
	San Luis Potosí	100
Black-throated bobwhite *(Colinus nigrogularis)*	Campeche	495
Northern bobwhite *(Colinus virginianus)*	Tamaulipas	46,881
	Coahuila	10,140
	Guerrero	7,350
	Morelos	4,362
	Jalisco	3,962
	Querétaro	2,000
	Nuevo León	1,110
	Zacatecas	1,000
	Durango	525
	Guanajuato	125
	Estado de México	120
California quail *(Callipepla californica)*	Baja California	39,250
	Baja California Sur	3,036
Elegant quail *(Callipepla douglasii)*	Sonora	480
	Jalisco	176
Gambel's quail *(Callipepla gambelii)*	Baja California	16,580
	Sonora	6,510
Mountain quail *(Oreortyx picta)*	Baja California	1,470
Banded quail *(Philortyx fasciatus)*	Guerrero	7,350
	Morelos	4,362

Source: SEMARNAT (2006).

Table 8.4. Authorized quail harvest in Mexico for the 2008–2009 hunting season (incomplete data).

Species	State	Specimens
Scaled quail *(Callipepla squamata)*	Coahuila Nuevo León	990 180
California quail *(Callipepla californica)*	Baja California	29,285
Gambel's quail *(Callipepla gambelii)*	Baja California	30,400
Mountain quail *(Oreortyx pictus)*	Baja California	845

Source: SEMARNAT (2008b).

Yucatán bobwhite (495) (table 8.3). Harvest rates for the 2008–2009 hunting season are the most recently published; however, they include only partial information regarding four quail species (table 8.4).

Habitat Management

Wildlife species are adapted to and require specific vegetation communities and structure to thrive as dictated by their life history and ecology. Rangeland quail, for example, require perennial bunchgrasses for nesting, bare ground to facilitate movement and foraging, and woody cover for protection from predators and the elements. Forest quail, on the other hand, require intact forests and dense canopy cover for protection and deep, thick forest litter for foraging. Thus, habitat management and conservation allow the persistence and livelihood of quail by providing for their habitat needs.

Habitat management for quail should occur within the context of usable space in order to be most effective (Guthery 1997). Usable space is defined as an area that meets the physical, physiological, and behavioral needs of quail (Guthery 1997). The abundance of quail in an area depends on the amount of usable space and the prevailing environmental conditions. Because we do not have control over the weather, the primary means through which we can influence quail abundance is by increasing usable space. Quail populations are maximized when all of an area is usable all of the time.

It would be impractical to discuss specific habitat management strategies for each quail species in

Mexico or to provide an in-depth discussion of all habitat management practices, given the diversity of both quail and these practices. Thus, we discuss habitat management only in very general terms. For ease of discussion, we have categorized management practices into those commonly occurring on rangelands (i.e., prescribed fire, mechanical treatments, and grazing) and those occurring on forests (e.g., silvicultural systems).

Rangeland management

Prescribed fire

Controlled or prescribed fires in Mexico have been conducted primarily to restore ecosystems that evolved with this natural process. We use the term "controlled fire" to describe fire applied in a confined and planned manner, whereas a "prescribed fire" is a controlled fire that is conducted under prescribed conditions to achieve a specific management objective. The use of prescribed fire as a management tool to promote wildlife habitat or biodiversity has been limited to nonexistent in Mexico. The ability to use prescribed fire as a management tool is important because of its relatively low cost and positive ecological effects (e.g., nutrient cycling) (fig. 8.3). Naturally, trained and experienced personnel should conduct prescribed fires and employ the necessary safety precautions to ensure a safe burn. Technical and legal specifications for conducting prescribed burns on forest and grazing lands are given by the Official Mexican Regulation NOM-015-SEMARNAT/SAGARPA-2007 (SEMARNAT 2009b).

The environmental conditions (i.e., temperature, humidity, and wind speed) under which prescribed burning should occur depend on various factors such as the management objective, the amount and type of fine fuel (i.e., volatile vs. nonvolatile), and the ecoregion, among others (Scifres 1980). We refer the reader to Scifres (1980) and Wright and Bailey (1982) for an in-depth discussion regarding the application of prescribed fire, and we recommend that landowners interested in using prescribed fire as a management tool seek the assistance of technical

Figure 8.3. Prescribed fire is an important management tool because of its relatively low cost and positive ecological effects, such as litter removal and nutrient cycling. Photo by Marc Bartoskewitz.

personnel. Generally speaking, a plan for a prescribed fire to manage quail habitat should consider what habitat feature is limiting the amount of usable space for quail. This in turn will influence the conditions (e.g., season of burning, environmental conditions) under which prescribed fire is best applied. For example, if excess grass cover is reducing the amount of usable space in an area, then conducting a prescribed fire during the winter may be advisable, because prescribed fires during this season are generally able to remove excess grass cover without damaging vegetation. In addition, prescribed fires during winter are considerably safer than those at other times of year, because temperatures are lower and prevailing winds tend to be more calm and constant. If excess brush cover limits usable space, then conducting a prescribed fire during the summer may be advisable because hot summer fires result in greater brush mortality. However, it should be noted that hot summer fires involve considerably more risk. Temperatures are highest and humidity lowest during this time of year, which can make controlling a fire difficult. In addition to safety concerns, quail may be nesting during the summer. Because of these two considerations, prescribed fires should be conducted during summer only when specifically required by management objectives.

Prescribed fire plans for quail may consider factors that may not directly create usable space but may be important for rangeland health or other wildlife. For example, managers may consider the vulnerability of warm-season plants or cool-season forbs to fire when planning a prescribed burn for quail. Warm-season plants make up much of Mexico's native grasslands, and cool-season forbs have high dietary importance for many wildlife species. Conducting a prescribed fire during spring, therefore, may or may not be desirable, depending on whether the vegetative growth of warm-season plants has begun. During this stage of growth (elongation), the growing points of warm-season plants are more exposed and therefore more vulnerable to the heat of a prescribed fired. Similarly, plans for prescribed fires during the fall may consider the growth cycle of cool-season forbs.

Mechanical treatments

Several mechanical treatments are available to manage rangelands, such as roller chopping, root plowing, and chaining (Hamilton et al. 2004). Briefly, a roller chopper is a large cylindrical drum 2–5 m (8–15 ft) wide, with longitudinal blades, that is pulled by a bulldozer to chop brush as it rolls over woody plants. A root plow is a large plow 3–5 m (10–16 ft) wide that is pulled at a depth of 30–35 cm (12–14 in) by a bulldozer to uproot brush. Chaining involves two bulldozers that pull a long, heavy chain in a U shape to uproot brush. The chain is about 60–120 m (200–400 ft) long and weighs about 60–112 kg/m (40–75 lb/ft).

Each method is appropriate under certain conditions. For example, on rangeland dominated by relatively young plants of resprouting woody species, root plowing may be preferred. In these situations, roller chopping would only make the problem worse by stimulating more resprouting, whereas chaining would simply "glide" over the flexible young plants without uprooting them. It should be noted that root plowing may not be advisable as a general practice because it can reduce woody plant diversity (Fulbright 1996) and cause high mortality of perennial grasses (Wiedemann 1997). On rangeland that has sandy soils and mature plants with sturdy trunks, chaining would probably be best because trees would be too large for root plowing or roller chopping, and the characteristics of the soil and plants would facilitate uprooting via chaining. We focus on roller chopping because of its widespread and recent use in Mexico.

The use of the roller chopper is relatively new in Mexico. It is approved by the Official Mexican Regulation NOM-020-RECNAT-2001 (SEMARNAT 2001) for the rehabilitation, enhancement, and conservation of grazing lands. It has been applied extensively throughout northeastern Mexico, primarily to enhance habitat for white-tailed deer (*Odocoileus virginianus*). However, since 2003, nonprofit organizations such as Pronatura Noreste have begun using roller choppers to restore rangeland degraded by overgrazing and soil compaction. These restoration programs have treated approximately 25,000 ha (61,700 ac) to date across northeastern Mexico. Relative to quail, the roller chopper has not been used extensively in Mexico as a habitat management tool. This is in sharp contrast to quail management in South Texas, where roller chopping has a long history of use in managing quail habitat and has been used to reduce brush canopy cover, stimulate forb growth, and promote perennial grasses.

Studies evaluating roller chopping in Mexico are

few, but available data suggest that roller chopping may be a viable tool to create usable habitat for quail. A recent study documented greater species richness and evenness on roller-chopped areas compared to control sites (Casas-Domínguez 2008). The generally degraded conditions of most Mexican rangelands suggest that such a vegetation response could possibly create usable quail habitat. However, we need to generate more knowledge regarding the use of roller chopping on different vegetation communities in Mexico, and its efficacy in promoting usable space for quail and other wildlife species.

Grazing

Grazing lands in Mexico have traditionally been subjected to unsustainable stocking rates for many generations (Casas-Domínguez 2008). Overutilization has resulted in rangeland degradation involving erosion of soils, reduction of primary production, and loss of biodiversity. This historical misuse of rangelands resulted in the creation of the Technical and Consulting Commission for the Regional Determination of Stocking Rates (COTECOCA 2002) during the mid-1960s. Its goal was to determine sustainable stocking rates for livestock on Mexico's grazing lands. This commission delineated homogeneous regions of forage capacity throughout the country based on climate, topography, soil type, and vegetation. Although the formation of COTECOCA was a step toward proper grazing management, it has not prevented overutilization. Many other variables not considered by the commission can influence stocking rates, such as variable weather patterns, prior management history, and variability in forage production among grazing units. These additional variables seem to have precluded sustainable management in some situations. For example, using simulation models, Díaz-Solis et al. (2003) determined that stocking rates recommended by COTECOCA were too high to maintain sustainability in the semiarid rangelands of Coahuila. This general finding is supported by the current degraded condition of grazing lands throughout much of that state.

Proper stocking rates are important for creating usable habitat for quail. The level of grazing that quail can tolerate depends on the species. For example, northern bobwhite and Montezuma quail are generally more sensitive to grazing pressure than species inhabiting more xeric landscapes, such as scaled quail and Gambel's quail. In addition, how grazing can be used to create usable space for quail will depend on the habitat needs of the species and on the prevailing environmental conditions. In general, quail need more herbaceous cover on arid and semiarid rangeland because plant growth is naturally slow in these ecosystems, and less herbaceous cover in more mesic environments, where plant growth is more rapid (Spears et al. 1993). Consequently, lighter grazing pressure is necessary in xeric regions to create the proper seral stages for quail, whereas heavier grazing pressure is needed in mesic regions. This latter point is important, because although lack of grass cover is often the most common management challenge on grazed rangelands, too much cover may occur in mesic environments, especially if livestock grazing has been removed for an extended period (fig. 8.4). Lozano-Cavazos (2009) documented that the abundance of northern bobwhite and other grassland bird species in the coastal prairie of Texas decreased and became virtually absent in areas with excessive grass cover and little to no bare ground. Using grazing as a tool to create usable habitat for quail therefore involves a balance between too little and too much grass cover.

Forest management

Historical overview

The forests of Mexico are characterized by their diversity and expansiveness. A third of Mexico (32.8%) is covered by forest. Of 63.6 million ha (157.1 million ac) of forestlands, 52% are temperate forests and 48% are tropical, dry tropical, and humid forests (Palacio-Prieto et al. 2000). Mexican forests contain the greatest diversity of pine (*Pinus* spp., 52) and oak (*Quercus* spp., 130) species in the world (Perry 1991, Castilleja 1996), of which 70% are endemic to Mexico. Most (59%) of the forests are owned by ejidos and indigenous communities (FAO 2006). However, only about a third (28%) of these landowners conduct management activities and commercial harvest (OIMT 2005). The remaining forestland is under private or state ownership, the latter representing a relatively smaller portion (Barton and Merino 2005).

The commercial harvest of forests in Mexico began at the end of the nineteenth century. During this time, large concessions of land were allocated to

Figure 8.4. Proper stocking rates are important for creating usable habitat for quail. Lack of grazing has decreased the suitability of this area for quail. Photo by Fidel Hernández.

private companies to harvest wood, construct roads, and open mines. No regulations existed at the time for the extraction of wood or changes in land use. Therefore, the richest forests of the country were exploited. Large quantities of timber were extracted from tropical rainforests in southeastern Mexico. In addition to unregulated commercial harvest, urban policies during the 1960s and 1970s encouraged the migration of people from crowded urban areas in the center of the country to tropical zones, resulting in further degradation. Farmers also received subsidies from the National Program of Livestock Production and the National Commission of Clearing to finance the removal of forests. Collectively, these policies eliminated about 80% of the original forests that existed in Mexico. It was not until after the Mexican Revolution (1910–1920) that the regulation of forestlands and harvest began. Forest management in Mexico has since progressed through several historical periods. Currently, the Law for Sustainable Forest Development and the General Law of Ecological Equilibrium and Environmental Protection require that (1) all timber harvest be executed under authorized management plans, and (2) environmental impact studies be conducted for all forest extractions (see chapter 21).

Management for quail

The current composition and structure of Mexican forests are the result of past exploitation and harvest systems. Other disturbances such as overgrazing, unplanned or illegal timber harvest, intentional fires, and clearing for agriculture and cattle production have also left their footprint. Because 6 of the 15 quail species in Mexico are found in forests and are protected by law because of their conservation risk (SEMARNAT 2002), proper forest management is imperative. In general, healthy forest ecosystems equal usable habitat for forest quail (Leopold 1959). Forest management programs should be directed toward the protection and conservation of forest ecosystems, including soil, flora, fauna, and water. The Official Mexican Regulations have established specifications to guide sound forestry practices, as well as protocols to mitigate adverse effects of timber harvest on the ecosystem. For example, the regulations prescribe that clear-cutting should be avoided on rugged terrain and highly erodible soils. The regulations also have specifications regarding the design and construction of roads, bridges, and campsites in forests. We recommend that areas inhabited by plant or animal species listed as rare, threatened, or subject to special protection be excluded from timber harvest.

Conservation and Management Challenges

The challenges facing quail conservation in Mexico can be categorized into four broad areas: (1) limitations of the UMA concept, (2) incongruence between harvest and conservation legislation, (3) illegal hunting, and (4) habitat loss.

The UMA concept represents an advancement over prior methods of game harvesting, because wildlife is given an economic value and management occurs within a conservation framework. However, the concept has several limitations. One is that no formal evaluation exists to determine whether UMAs are actually achieving the goal of sustainable use. An analysis of the 2005–2006 hunting-season quotas for nine species of game birds and mammals in Baja California (including Gambel's quail, mountain quail, and California quail) concluded that no regional studies of wildlife populations existed (Contreras-Gil et al. 2007). Consequently, in some cases, hunting quotas may have not been aligned with the natural reproduction of populations in the region (Contreras-Gil et al. 2007), thereby illustrating the need for demographic research to establish appropriate harvest regulations (e.g., timing, length, level). The study also highlighted an array of deficiencies associated with the UMA process, including the lack of an entity and formal procedure to collect, synthesize, and interpret information; the lack of a standardized population-monitoring program; and the lack of an official map georeferencing all UMAs. Another limitation of the UMA concept is that expenses associated with their functioning increase the cost of legal harvest, thereby causing poaching to appear more lucrative. A solution would be to conduct official inspections and enforce legal sanctions and fines for all violations (Huerta 2005). However, the agency responsible for such enforcement (Federal Attorney for Environmental Protection, PROFEPA) does not have the personnel or infrastructure to implement these recommendations. Finally, UMA personnel often lack the technical training to properly execute the required objectives of UMAs. Technical workshops on species biology and population monitoring techniques would strengthen UMA performance. These actions are particularly important if the goals of conservation and socioeconomic benefits for rural communities are to be achieved.

A second conservation challenge involves the incongruence between harvest and conservation legislation. Although the modernization of wildlife laws represents an advancement in terms of wildlife conservation, the current regulations represent, in the best of cases, an ideal or theoretical standard that most often is not fully followed. One example involves the hunting-season schedule and the species-at-risk list (SEMARNAT 2009a). The 2009–2010 hunting-season regulations permit the hunting of nine quail species in 29 Mexican states. Among them is Montezuma quail, which also happens to be listed by the Official Mexican Regulation NOM-059-SEMARNAT-2001 (SEMARNAT 2002) as a species with special protection, and its Pacific subspecies (*sallei*) is threatened. Thus, it is incongruent that this species is considered both at risk by the Official Mexican Regulation, and a game bird subject to legal harvest. Other species may also be in similar circumstances, as permitted by Articles 85 and 87 of the General Wildlife Law. These articles specify that priority is given to the collection and capture of species for restoration, repopulation, and reintroduction efforts, but they do not explicitly prohibit hunting.

Illegal hunting of quail is a third conservation challenge. It continues today because of either disregard for regulations or ignorance of them. Subsistence hunting is practiced primarily in southeastern Mexico and in remote regions of the country. It may persist more because of tradition or social custom rather than necessity. In addition, in areas where agricultural grains are grown, some quail species, such as the banded quail and the Yucatán bobwhite, are considered pest species and are killed in large quantities during the crop harvest season. The effects of the regulation of subsistence hunting by the General Law of Wildlife, and the impacts of poaching on wildlife populations, need to be evaluated.

Finally, habitat loss, primarily in forests, is a fourth conservation challenge. Habitat loss has placed at risk almost half of Mexican quail, most of which are Neotropical in distribution and inhabit forests and jungles. Addressing this conservation challenge will be difficult because of the rapid rate at which forests are being lost. Communal forest management that conserves native vegetation cover has been suggested to be as effective as protected areas in preventing deforestation (Durán-Medina et al. 2007). However, practices to mitigate the effects of timber harvest on wildlife are generally not followed in ejidos and communal forests. A study comparing the abundance of long-tailed tree-quail in a protected area and a forest where an indigenous community practiced selective harvest, resin extraction, and grazing revealed densities of 42.9 birds/km² (112.8 birds/mi²), and 7.8 birds/km² (20.5 birds/mi²), respectively (Chávez-León 2004a). These findings indicate that even though the indigenous community had succeeded in preventing deforestation, it had not implemented the required mitigation practices to protect wildlife or habitat.

Ultimately, quail conservation in Mexico hinges on habitat conservation. This will be achieved through protecting natural communities (primarily forests), reducing the deforestation rate, and conserving native habitat. Accomplishing such a task will require the organized participation of society, and solutions to the socioeconomic issues that are causing the overexploitation of natural resources. We contend that the following initiatives are critical if quail conservation and management are to advance in Mexico: (1) the creation of an entity and formal procedure to collect, synthesize, and interpret UMA information, a standardized population-monitoring program, and an official map of UMAs (Contreras-Gil et al. 2007); (2) a formal method for evaluating UMAs to determine whether conservation and socioeconomic goals are being achieved; (3) a regional-scale monitoring program to determine mean, maximum, and atypical hunting quotas (this will require visits to sites soliciting atypical quotas prior to authorization, assuming they are justified; technical training for UMA personnel in quail ecology, population-monitoring techniques, and sustainable harvest); (4) demographic research to determine appropriate hunting regulations (e.g., timing, length, level); (5) amendments to current laws to explicitly prohibit harvest of species considered at risk; (6) enforcement of legal sanctions and fines for illegal hunting; and (7) conservation of native habitat through appropriate means such as protected areas, management, and sustainable use.

Acknowledgments

We thank the Caesar Kleberg Wildlife Research Institute, the Richard M. Kleberg Jr. Center for Quail Research, and Texas A&M University–Kingsville for providing financial and logistical support during the writing of this chapter. We thank J. P. Carroll and F. S. Guthery for providing helpful comments on an earlier version of this chapter.

Species Accounts

The following species accounts of quail species in Mexico are arranged in the taxonomic order suggested by Madge and McGowan (2002).

Bearded wood-partridge (*Dendrortyx barbatus*)

Distribution. East-central Mexico to southeastern San Luis Potosí, Hidalgo, and Puebla, northeastern Querétaro, and northern Oaxaca. *Mass.* Males 459 g (1 lb); females 405 g (0.9 lb). *Plumage and sexual dimorphism.* Reddish-brown tree-quail with rufous chest, blue-gray face and neck. Stout red bill; legs and feet orange to bright red. Sexes similar. Long tail (130–160 mm) (5–6.3 in) with 12 rectrices. *Diet.* Seeds and small fruits, black beans, corn in captivity. *Reproduction.* Perhaps April–June, but known only in captivity, where observed constructing shallow scrape nest lined with palm leaves. *General habitat.* Humid, montane cloud forests with dense understory, also extends to pine-oak zone. *Behavior.* Largely unknown. Forms small coveys of three to five birds, roosts in trees and small shrubs. Whistles loudly, similar to other congeners. Considered sedentary. *Population dynamics.* Unknown.

Long-tailed wood-partridge *(Dendrortyx macroura)*

Distribution. West-central and southern Mexico, Jalisco to Oaxaca, central mountains from Colima to Veracruz. *Mass.* Males 433 g (0.95 lb); females 390 g (0.86 lb). *Plumage and sexual dimorphism.* Long, rounded tail 122–175 mm (4.8–6.9 in). Head and throat black with white streaks above and below eyes. The bill, legs, and skin around the eyes are geranium pink. Body is chestnut gray, with spots and streaks of rufous on chest and flanks. Sexes alike. *Diet.* Legumes, flower buds, small fruits, and acorns. *Reproduction.* Courtship behavior poorly documented, scant information on incubation or fledging periods. Thought to be mid-April through July. *General habitat.* Montane forest 1,200–3,300 m (3,937–10,827 ft) above sea level, with dense understory. *Behavior.* Considered furtive and shy, escapes by running or flying short distances (50 m or less) with a noisy beating of wings, and a short and sharp alarm call. Roosts in trees and shrubs. Scratches for food in forest leaf litter. *Population dynamics.* Not known. Population abundance documented as ranging from 7.8 birds/km^2 (managed forests) to 42.9 birds/km^2 (107 birds/mi^2) in unmanaged forests (Chávez-León 2004a, 2004b).

Buffy-crowned wood-partridge (*Dendrorytx leucophrys*)

Distribution. Central America, southeastern Chiapas (Mexico) south through Guatemala, Honduras, and eastern El Salvador and northern Costa Rica. *Mass.* Males 397 g (0.89 lb); females 340 g (0.75 lb). *Plumage and sexual dimorphism.* Long tail, similar to other tree-quail (138–160 mm) (5.4–6.3 in). Forecrown and throat pale buff to white with fine, pale-streaked black near ears; rest of crown and neck chestnut streaked and spotted. Underparts bluish gray with brown streaks on flanks. Sexes similar. *Diet.* Seeds, flower buds, invertebrates. *Reproduction.* Apparently tied to rains, but largely undocumented. *General habitat.* Humid and semihumid montane pine-oak (*Pinus* spp.–*Quercus* spp.) forests, 300–3,000 m (984–9,843 ft) above sea level. *Behavior.* Sedentary, coveys of 4–12 birds. Roosts in shrubs and small trees. *Population dynamics.* Unknown.

Mountain quail (*Oreortyx pictus*)

Distribution. Extreme western North America from southern Washington and southwestern Nevada south to Baja California. *Mass.* Males 235 g (0.52 lb); females 230 g (0.51 lb). *Plumage and sexual dimorphism.* Long, erect, black crown plume; head and throat blue gray; mantle olive brown gray; side flanks streaked white against a chestnut-brown background. Sexes alike. *Diet.* Fruits and seeds from *Lithophragma, Quercus, Stellaria, Rhus, and Trifolium* extremely important. *Reproduction.* March–July, clutch 6–10 eggs, males guard nests and incubate. *General habitat.* Mixed evergreen forest, chaparral, and riparian woodlands. *Behavior.* Secretive and shy. Seems to have affinity for areas of open water in arid regions. Often forages by digging. *Population dynamics.* Compared to *Callipepla* and *Colinus*, which often exhibit population booms and busts, mountain quail populations are relatively stable from year to year. Considered by Gutiérrez (1980) to be more of a *K*-selected than an *r*-selected species, where habitat remains stable and available.

Scaled quail (*Callipepla squamata*)

Distribution. Ranges from southeastern Arizona, northern New Mexico, eastern Colorado, and southwestern Kansas to west-central and South Texas, through northern and central Mexico. *Mass.* Males 179 g (0.4 lb); females 173 g (0.38 lb). *Plumage and sexual dimorphism.* Head gray brown, with fluffy white crest. Mantle and chest radiate from blue-gray to tawny brown belly, with "fish-scale" margins on contour feathers. Wings gray brown and flanks streaked with white. Sexes very similar, but females have a gray throat with dark streaking whereas males have a buff throat with no streaking. *Diet.* Range of weed and grass seeds such as *Croton* and *Panicum*; also prefers legumes and leafy green vegetation where available. *Reproduction.* Late June to September or October, apparently timed to take advantage of summer rains. Clutch 9–16 eggs, incubates 23 days, young reach adult size in 77–105 days. Brood size 7.8–11.5. Males known to brood young while females produce second clutch. *General habitat.* Desert grassland, brush, and thornscrub, sea level to 2,450 m (8,038 ft). Will use sites with human structures and junk such as abandoned farming equipment. *Behavior.* Somewhat sedentary, although movements of up to 100 km (62 mi) known. Prefers to run rather than flush. Can form large winter coveys of up to 200 birds. *Population dynamics.* Can be irruptive in response to favorable weather conditions during breeding seasons with above-average rainfall in good habitat.

Elegant quail (*Callipepla douglasii*)

Distribution. Sonora to southwestern Chihuahua, Sinaloa, Durango, Nayarit, and northwestern Jalisco. *Mass.* Males 175 g (0.38 lb); females 169 g (0.37 lb). *Plumage and sexual dimorphism.* Overall grayish brown with white spots on underparts. Males more gray than females. Crest on males is tawny brown, on females dusky darker brown. *Diet.* Legumes apparently most favored, also takes other seeds and insects.

Reproduction. Begins each year in April–May, but little else known except clutch size (8–12 eggs) and incubation (22–23 days). *General habitat.* Arid and semiarid brushy woodland, thorn forest, and desert scrub, sea level to 1,000 m (3,300 ft). *Behavior.* Considered sedentary. Coveys of 6–20 individuals. Roosts in thick brush and vines. *Population dynamics.* Unknown.

California quail (*Callipepla californica*)

Distribution. Northern Oregon, east to Idaho and Nevada, south to southern Baja California. *Mass.* Males 176 g (0.38 lb); females 162 g (0.36 lb). *Plumage and sexual dimorphism.* Both sexes show characteristic topknot crest, that of males larger. Males have black face and top of head, golden forecrown, with face and top of head rimmed in white border. Chest solid blue gray, back of neck spotted with whitish flecks. Mantle and wings brown gray, sides blue gray flecked with white streaks, belly with white grading to gold contour feathers with fish-scale pattern. Female generally drab brown gray, with light flecks on back and wings, grayish-black flecks on flanks and belly. *Diet.* Forb seeds and legumes. Leafy greens form bulk of diet in winter and spring. *Reproduction.* April–August, depending on location. Nest a grass-lined depression, clutch 6–17 eggs, incubation 22–23 days. Males documented to assist with chick rearing. Double brooding documented. Chicks fully grown in 70 days, and adultlike plumage developed in 21 weeks. *General habitat.* Somewhat of a habitat generalist, found in a variety of areas such as scrub, grassland, croplands, riparian zones, and chaparral to 2,800 m (9,186 ft). *Behavior.* Considered sedentary. Documented to form huge coveys (up to 1,000 birds), but most range from 25 to 40 in size. Home ranges vary in size from 6.8 to 18.0 ha (16.8 to 44.5 ac). *Population dynamics.* Classic *r*-selected quail species; populations strongly influenced by precipitation, increasing during favorable breeding and habitat conditions.

Gambel's quail (*Callipepla gambelii*)

Distribution. East-central California to southern Nevada, Arizona, New Mexico, southwestern Texas and south into Mexico and northeastern Baja California. *Mass.* Males 161 g (0.35 lb); females 156 g (0.34 lb). *Plumage and sexual dimorphism.* Similar to California quail, but slightly larger and grayer. Both sexes have crest feathers that hang forward over face, those of females shorter than those of males. Males have black face and throat with white border and chestnut crown, chestnut flanks streaked with white, gray-brown back and wings, and tawny belly with black spot. Females duller than males with brown-gray face and whitish streaks, and contour feathers grading into fish-scale pattern on edges of belly. *Diet.* Legumes, leafy greens when available, other seeds, and arthropods. *Reproduction.* Nests April–August, built in scratched depression, clutch 9–16 eggs. Incubates 23 days. Male known to both guard nests and incubate. Will produce second broods in favorable conditions. *General habitat.* Desert scrub, riparian washes, irrigated agricultural areas from sea level to 1,800 m. *Behavior.* Covey sizes range to 40 or more. Maximum recorded movement relatively small, 16 km (9.9 mi). Home ranges documented from 7.6 to 38 ha in size. *Population dynamics.* Considered relatively stable, compared to many other species of declining Odontophoridae. Will respond favorably and dramatically to summer rains with high annual productivity.

Banded quail (*Philortyx fasciatus*)

Distribution. Endemic in Mexico; Jalisco and Colima to Puebla and Guerrero. *Mass.* Males 130 g (0.29 lb); females 126 g (0. 28 lb). *Plumage and sexual dimorphism.* Drab brownish black above with black stripe behind eye. Small brown crest feathers; flanks and belly background nearly white with broken bar pattern of brownish stripes. Light gray-white throat. Sexes alike. *Diet.* Poorly documented, but presumed to be seeds, leafy greens, and arthropods, as with most quail. *Reproduction.* Undocumented in wild, but thought to be from July to September, which is considerably later than bobwhites in same area near Acapulco. Nests in lined depression with partial grass roof, clutch 3–7, incubated by female only for 21–23 days. *General habitat.* Dry grassland scrub, open thorn scrub, sea level to 1,600 m (5,249 ft). *Behavior.* Prefers to hide in grassy cover. Forms coveys of up to 30 birds, but average size is about 12. *Population dynamics.* Unknown but considered locally common and stable.

Northern bobwhite (*Colinus virginianus*)

Distribution. Largest distribution of any New World quail. Southern New England states, west through southern Great Lake states to southeastern Wyoming, eastern Colorado, south into the eastern third of Mexico, and south to Guatemala. Endangered subspecies *ridgwayi* still exists in Sonora and extreme southern Arizona, but status unclear. *Mass.* Highly variable, with larger birds (up to 220 g, or 0.49 lb) in northern parts of range and smaller birds (< 150 g, or < 0.33 lb) in southern parts. *Plumage and sexual dimorphism.* Males typically have white throat and forecrown stripe that runs over each eye and down middle side of neck. Back tawny brown gray (depending on region) with black flecks, flanks chestnut with whitish stripes, and belly gold brown with black flecks. Females more drab throughout, lacking white on face and much more muted on flanks. *Diet.* Highly variable but dominated by seeds, leafy greens, and arthropods. Female diet during breeding documented to be 25% arthropods by biomass. *Reproduction.* Not monogamous, as asserted by Madge and McGowan (2002). Rather, breeding system considered ambisexually polygamous, with both males and females known to raise more than one brood with more than one mate during a single breeding season. Evidence of nesting found during all 12 months in South Texas, but typical nesting season is April–August. Clutch size averages about 12–15 eggs, incubation is 23 days. Will double brood if conditions favorable, and renest if nest fails. *General habitat.* Highly variable, ranging from open deciduous woodlands and parklike pine forests to prairie shrubland and thornscrub. *Behavior.* Complex repertoire of vocalizations documented by Stokes (1967). Forages on ground, scratching lightly through scattered litter for food. Forms coveys of 4–20 birds in winter. Roosts on ground in circle, facing outward. *Population dynamics.* Highly variable depending on region and weather. In mesic portions of range, where habitat is maintained through management, populations characteristically stable from year to year. Irruptive in semiarid, subtropical areas in relation to rainfall. Will decrease quickly in relation to drought or overgrazing.

Black-throated bobwhite (*Colinus nigrogularis*)

Distribution. Yucatán Peninsula. *Mass.* Males 137 g (0.30 lb); females 139 g (0.31 lb). *Plumage and sexual dimorphism.* Also known as the Yucatán bobwhite. Species has a large black throat patch. Breast and back of neck differ from northern bobwhite by having bolder whitish feathers with chestnut margins on shoulders, grading to black margins on flank and belly. Females similar to northern bobwhite, except for more white feathers on belly. *Diet.* Undocumented. *Reproduction.* Undocumented, but presumed similar to that of northern bobwhite. *General habitat.* Savanna with brush and trees. Apparently associated with *Agave* clumps. *Behavior.* Shy, wary, prefers to remain well hidden. Presumably similar to behavior of northern bobwhite. *Population dynamics.* Considered locally abundant on Yucatán Peninsula, with estimated abundance of up to 2.5 birds/ha, but method of enumeration not documented.

Spotted wood-quail (*Odontophorus guttatus*)

Distribution. Central Veracruz, northern Oaxaca, to Chiapas and the Yucatán. *Mass.* Males 360 g (0.89 lb); females 288 g (0.6 lb). *Plumage and sexual dimorphism.* Crown brown blackish, with broad crest that is black in females and orange in males. Black throat streaked with white. Underparts olive to reddish brown. *Diet.* Small bulbs, tubers, Diptera and Coleoptera larvae and pupae. *Reproduction.* Nests in May–June, no information on nests or eggs. *General habitat.* Deep rainforest, under dense shade of tree canopy. *Behavior.* Reluctant to fly. Typically runs to evade danger or crouches to hide in leaf litter. *Population dynamics.* Not known.

Singing quail (*Dactylortyx thoracius*)

Distribution. Jalisco to central Guerrero, southern Tamaulipas, central Veracruz, Chiapas, and lowland forests of Yucatán. *Mass.* 170–260 g (0.37–0.57 lb). Males slightly larger than females. *Plumage and sexual dimorphism.* Sexes similar from dorsal view; chestnut brown with black, olive, and buff streaks. Male has orange on throat and cheeks, female dull gray on face and throat. *Diet.* Onions, bulbs, seeds, larvae, and pupae dug from ground. *Reproduction.* Breeds from March to August in all parts of range. *General habitat.* Cloud forest, oak–sweet gum forest, and humid canyons. On Yucatán Peninsula found in lowland deciduous forest. Associated with deep leaf litter in closed-canopy forests. *Behavior.* Noted for having the most complex and melodious vocal repertoire of any galliform. Males and females thought to remain paired through breeding season, but this is not known conclusively. During nonbreeding season, remains grouped in coveys of 5–10 birds. Large feet and claws clearly adapted for digging. *Population dynamics.* Density estimated at 3.5 pairs/40 ha (0.04 pairs/100 ac), but method of enumeration not documented (Leopold 1959, Warner and Harrell 1957). Study by Eitniear and Aragon (2004) indicated a possible increasing trend on El Cielo Ranch during 15 years of Christmas Bird Counts (1972–1996).

Montezuma quail *(Cyrtonyx montezumae)*

Distribution. Extreme southwestern US, south into central hills and mountains of Mexico. *Mass.* Males 194 g (0.43 lb); females 176 g (0.39 lb). *Plumage and sexual dimorphism.* Males have striking harlequin or clown-like head pattern of interswirled white and black arches on each side around eye. Head crest somewhat flattened and tawny brown. Back tawny brown with white streaks on back of neck and dark brown-black flecks. Flanks have nearly solid black background with numerous white spots. Belly chestnut. Females much drabber, lacking most white and black present on males. Mantle and contour feathers tawny with whitish streaks. Face grayish with tawny crest and tawny spot behind and below each eye. *Diet.* Specializes on tubers and bulbs, mostly those of *Oxalis*. *Reproduction.* Late March–September, depending on location and summer rainfall patterns. Scrape nest lined with grass and roof dome, sometimes with side entrance. Clutch 6–28 eggs, incubates 25–26 days. *General habitat.* Dry, brushy grassland and pine-oak woodland with grassy understory, usually between 1,000 and 3,000 m. *Behavior.* Extremely wary and reluctant to flush or run; seems to prefer to "freeze" and hide. Roosts on ground in circle facing outward, like bobwhites. Aggressive scratcher and digger for preferred foods. *Population dynamics.* Populations highly localized because of specific habitat affinity. Year-to-year variation fluctuates greatly in relation to annual production, which is closely tied to summer monsoons.

Ocellated quail *(Cyrtonyx ocellatus)*

Distribution. South of Montezuma quail range in Mexico, to Honduras. *Mass.* Males 281 g (0.62 lb); females 182 g (0.40 lb). *Plumage and sexual dimorphism.* Very similar to Montezuma quail. Male lacks black flanks with white polka dots but is otherwise similar. Females also extremely similar to Montezuma quail except face is more distinctly patterned with black and bordered with gray white. *Diet.* Thought to be similar to that of Montezuma quail. *Reproduction.* Undocumented but assumed to be analogous to that of Montezuma quail. *General habitat.* Pine-oak woodland and brush, 750–3,050 m (2,460–10,000 ft). *Behavior.* Presumed similar to that of Montezuma quail. *Population dynamics.* Undocumented but considered uncommon to rare in Mexico.

Literature Cited

Baccus, J. T. 2004. Status of the bearded wood-partridge: with comments on its conservation in Querétaro. Pages 9–12 *in* J. C. Eitniear, J. T. Baccus, S. L. Dingle, and J. P. Carroll, editors. Conservation of quail in the Neotropics. Center for the Study of Tropical Birds, San Antonio, Texas, USA.

Barton, B. D., and L. Merino P. 2005. La experiencia de las comunidades forestales en México. INE-SEMARNAT, México, D. F., México.

Brennan, L. A., editor. 2007. Texas quails: ecology and management. Texas A&M University Press, College Station, USA.

Brodkorb, P. 1964. Catalog of fossil birds: Part 2. (Anseriformes through Galliformes). Bulletin of the Florida State Museum of Biological Sciences 8:195–335.

Brooks, D. M. 2004. Observations of elegant quail (*Callipepla douglasii*) in southern Nayarit, México. Pages 83–84 *in* J. C. Eitniear, J. T. Baccus, S. L. Dingle, and J. P. Carroll, editors. Conservation of quail in the Neotropics. Center for the Study of Tropical Birds, San Antonio, Texas, USA.

Brown, D. E. 1989. Arizona game birds. University of Arizona Press, Tucson, USA.

Brown, D. E., and R. J. Gutiérrez. 1980. Sex ratios, sexual selection, and sexual dimorphism in quails. Journal of Wildlife Management 44:198–202.

Brown, D. E., J. C. Hagelin, M. Taylor, and J. Galloway. 1998. Gambel's quail (*Callipepla gambelii*). *In* The birds of North America. Philadelphia Academy of Sciences, Pennsylvania, and American Ornithologists' Union, Washington, DC, USA.

Burger, L. W., M. R. Ryan., T. V. Dailey, and E. W. Kurzejeski. 1995. Reproductive strategies, success, and mating systems of northern bobwhite in Missouri. Journal of Wildlife Management 59:417–426.

Campbell, H., D. K. Martin, P. E. Ferkovich, and B. K. Harris. 1973. Effects of hunting and some other environmental factors on scaled quail in New Mexico. Wildlife Monographs 34:1–49.

Carroll, J. P. 1995. Family Odontophoridae. Pages 412–433 *in* J. del Hoyo and A. Elliot, editors.

Handbook of the birds of the world. Volume 2. Lynx Edicions, Barcelona, Spain.

Carroll, J. P., and J. C. Eitniear. 2000. Quails in Mexico: needs and opportunities. Proceedings of the National Quail Symposium 4:191–193.

Casas-Domínguez, N. Y. 2008. Evaluación del uso del rodillo aireador para la rehabilitación de tierras de pastoreo y el contenido de carbono en zonas semiáridas del noreste de México. Thesis, Instituto Tecnológico y de Estudios Superiores de Monterrey, México.

Castilleja, G. 1996. Mexico. Pages 193–205 *in* C. S. Harcourt and J. A. Sayer, editors. The conservation atlas of tropical forests: the Americas. Simon and Schuster, New York, New York, USA.

Chávez-León, G. 2004a. Long-tailed tree-quail (*Dendrortyx macroura*): current knowledge and research needs. Pages 15–26 *in* J. C. Eitniear, J. T. Baccus, S. L. Dingle, and J. P. Carroll, editors. Conservation of quail in the Neotropics. Center for the Study of Tropical Birds, San Antonio, Texas, USA.

Chávez-León, G. 2004b. Abundance of the long-tailed tree-quail in managed and unmanaged pine-oak forests. Pages 27–32 *in* J. C. Eitniear, J. T. Baccus, S. L. Dingle, and J. P. Carroll, editors. Conservation of quail in the Neotropics. Center for the Study of Tropical Birds, San Antonio, Texas, USA.

Chávez-León, G., and A. Velázquez. 2004. Abundance of the long-tailed wood-partridge (*Dendrortyx macroura*) in a temperate coniferous forest. Journal of Field Ornithology 75:345–352.

Chávez-León, G., A. Velázquez, A. Fregoso, and G. Bocco. 2004. Habitat associations of the long-tailed wood-partridge (*Dendrortyx macroura*) in a managed coniferous forest in Michoacán, Mexico. Biodiversity and Conservation 13:1943–1960.

Contreras-Gil, J., R. Martínez-Gallardo, and E. Mellink. 2007. Tasas de aprovechamiento cinegético autorizadas en Baja California, México. Revista Latinoamericana de Recursos Naturales 3:129–137.

COTECOCA. 2002. Manual de la organización de la Comisión Técnico Consultiva de Coeficientes de Agostadero. COTECOCA, México, D. F., México.

Cox, S. A., A. D. Peoples, S. J. DeMaso, J. J. Lusk, and F. S. Guthery. 2004. Survival and cause-specific mortality of northern bobwhites in western Oklahoma. Journal of Wildlife Management 68:663–671.

Curtis, P. D., B. S. Mueller, P. D. Doerr, C. F. Robinette, and T. DeVos. 1993. Potential polygamous breeding behavior in northern bobwhite. Proceedings of the National Quail Symposium 3:55–63.

Delehanty, D. J. 1997. Mountain quail reproduction and restoration in the Great Basin. Dissertation, University of Nevada, Reno, USA.

DeMaso, S. J. 2008. Population dynamics of northern bobwhites in southern Texas. Dissertation, Texas A&M University–Kingsville and Texas A&M University, College Station, USA.

Díaz-Solis, H., M. M. Kothmann, W. T. Hamilton, and W. E. Grant. 2003. A simple ecological sustainability simulator (SESS) for stocking rate management on semiarid grazing lands. Agricultural Systems 76:655–680.

Durán-Medina, E., J.-F. Mas, and A. Velázquez. 2007. Cambios en las coberturas de vegetación y usos del suelo en regiones con manejo forestal comunitario y áreas naturales protegidas de México. Pages 267–299 *in* D. B. Bray, L. Merino Pérez, and D. Barry, editors. Los bosques comunitarios de México. Manejo sustentable de paisajes forestales. INE-SEMARNAT, México, D. F, México.

Eitniear, J. C., and A. T. Aragon. 2004. Changing densities of singing quail on the El Cielo Biosphere, Tamaulipas, Mexico: 1949–1996. Pages 5–7 *in* J. C. Eitniear, J. T. Baccus, S. L. Dingle, and J. P. Carroll, editors. Conservation of quail in the Neotropics. Center for the Study of Tropical Birds, San Antonio, Texas, USA.

Eitniear, J. C., J. T. Baccus, S. L. Dingle, and J. P. Carroll, editors. 2004. Conservation of quail in the Neotropics. Center for the Study of Tropical Birds, San Antonio, Texas, USA.

Emlen, J. T., Jr. 1940. Sex and age ratios in survival of the California quail. Journal of Wildlife Management 4:92–99.

Faircloth, B. C. 2008. An integrative study of social and reproductive systems in northern bobwhite (*Colinus virginianus*): a non-migratory, avian species bearing precocial young. Dissertation, University of Georgia, Athens, USA.

FAO. 2006. Evaluación de los recursos forestales mundiales 2005—Hacia la ordenación forestal sostenible. FAO, Rome, Italy.

Folk, T. H., et al. 2007. Estimates of survival from radiotelemetry: a response to Guthery and Lusk. Journal of Wildlife Management 71:1027–1033.

Francis, W. J. 1965. Double broods in California quail. Condor 67:541–542.

Fulbright, T. E. 1996. A theoretical basis for planning woody plant control to maintain species diversity. Journal of Range Management 49:554–559.

García-Solórzano, D. 2005. Anidación *in-situ*, incubación y crianza *ex-situ* de la codorniz Moctezuma (*Cyrtonyx montezumae montezumae*). Thesis, UNAM, D. F., México.

Gullion, G. W. 1956. Let's go quail hunting. Biological Bulletin 2. Nevada Fish and Game Commission, Las Vegas, USA.

Guthery, F. S. 1997. A philosophy of habitat management for northern bobwhites. Journal of Wildlife Management 61:291–301.

Guthery, F. S., and J. J. Lusk. 2004. Radiotelemetry studies: are we radiohandicapping northern bobwhites? Wildlife Society Bulletin 32:194–201.

Gutiérrez, R. J. 1980. Comparative ecology of the mountain and California quails in the Carmel Valley, California. Living Bird 18:71–93.

Gutiérrez, R. J., and D. J. Delehanty. 1999. Mountain quail (*Oreortyx pictus*). *In* The birds of North America. Philadelphia Academy of Sciences, Pennsylvania, and American Ornithologists' Union, Washington, DC, USA.

Gutiérrez, R. J., R. M. Zink, and S. Y. Yang. 1983. Genic variation, systematic, and biogeographic relationships of some galliform birds. Auk 100:33–47.

Hamilton, W. T., A. McGinty, D. N. Ueckert, C. W. Hanselka, and M. R. Lee. 2004. Brush management: past, present, future. Texas A&M University Press, College Station, USA.

Hernández, F., F. Hernández, J. A. Arredondo, F. C. Bryant, L. A. Brennan, and R. L. Bingham. 2005. Influence of precipitation on demographics of northern bobwhites in southern Texas. Wildlife Society Bulletin 33:1071–1079.

Hernández, F., W. P. Kuvlesky Jr., R. W. DeYoung, L. A. Brennan, and S. A. Gall. 2006. Recovery of rare species: case study of the masked bobwhite. Journal of Wildlife Management 70:617–631.

Huerta, H. M. C. 2005. Unidades de manejo para la conservación de la vida silvestre. Pages 151–152 *in* L. Villaseñor, editor. 2005. La biodiversidad en Michoacán: estudio de estado. CONABIO, SUMA-Michoacán, and UMSNH, Morelia, México.

Hurst, G. A., L. W. Burger, and B. D. Leopold. 1996. Predation and galliform recruitment: an old issue revisited. Transactions of the North American Wildlife and Natural Resources Conference 61:62–76.

Johnsgard, P. A. 1973. Grouse and quails of North America. University of Nebraska Press, Lincoln, USA.

Johnsgard, P. A. 1988. The quails, partridges, and francolins of the world. University of Nebraska Press, Lincoln, USA.

Kiel, W. H. 1976. Bobwhite quail population characteristics and management implications in South Texas. Transactions of the North American Wildlife Conference 41:407–420.

Lehmann, V. W. 1984. Bobwhites in the Rio Grande Plain of Texas. Texas A&M University Press, College Station, USA.

Leopold, A. S. 1939. Age determination in quail. Journal of Wildlife Management 3:261–265.

Leopold, A. S. 1945. Sex and age ratios among bobwhite quail in southern Missouri. Journal of Wildlife Management 9:30–34.

Leopold, A. S. 1959. Wildlife of Mexico: the game birds and mammals. University of California Press, Berkeley, USA.

Leopold, A. S. 1977. The California quail. University of California Press, Berkeley, USA.

Lozano-Cavazos, E. A. 2009. Avian community response to brush reduction in Texas coastal prairie habitats. Dissertation, Texas A&M University–Kingsville, USA.

MacGregor-Fors, I. 2006. Nota sobre la distribución de *Philortyx fasciatus*, una especie endémica de México. Huitzil 7:30–31.

Madge, S., and P. McGowan. 2002. Pheasants, partridges and grouse. Princeton University Press, New Jersey, USA.

McGowan, P. J. K., S. D. Dowell, J. P. Carroll, and N. J. Aebischer. 1995. Partridges, quails, francolins, snowcocks, and guineafowl: status, survey, and conservation action plan 1995–1999. IUCN and World Pheasant Association, Gland, Switzerland.

OIMT. 2005. Consecución del Objetivo 2000 y la ordenación forestal sostenible en México: resumen analítico. Consejo Internacional de Maderas Tropicales por la Misión de Diagnóstico, 39th session. OIMT, Yokohama, Japan.

Palacio-Prieto, J. L., et al. 2000. La condición actual de los recursos forestales en México: resultados del inventario forestal nacional. Investigaciones Geográficas 43:183–203.

Palafox, L. H., G. D. M. Martínez, J. I. V. Hernández, J. L. A. Carbajal, F. C. Sánchez, and M. H. Téllez. 2007. Densidad poblacional y uso del hábitat de la codorniz Moctezuma (*Cyrtonyx montezumae*) en Hualtepec, Hidalgo. Veterinaria México 33:255–263.

Perry, J. P. 1991. The pines of Mexico and Central America. Timber Press, Portland, Oregon, USA.

Pianka, E. 1970. On *r* and *K* selection. American Naturalist 104:592–597.

Raitt, R. J., and R. D. Ohmart. 1968. Sex and age ratios in Gambel quail of the Rio Grande Valley, southern New Mexico. Southwestern Naturalist 13:27–34.

Robles, J. T., S. I. Zimbrón, G. D. M. Martínez, C. Z. Hernández, F. C. Sánchez, and L. A. T. Arámbula. 2002. Densidad de la codorniz Moctezuma (*Cyrtonyx montezumae*) en la región noroeste del estado de México, México. Veterinaria México 33:255–263.

Rollins, D., and J. P. Carroll. 2001. Impacts of predation on northern bobwhite and scaled quail. Wildlife Society Bulletin 29:39–51.

Sandercock, B. K., W. E. Jensen, C. K. Williams, and R. D. Applegate. 2008. Demographic sensitivity of population change in northern bobwhite. Journal of Wildlife Management 72:970–982.

Schemnitz, S. D. 1994. Scaled quail (*Callipepla squamata*). *In* The birds of North America. Academy of Natural Sciences, Philadelphia, Pennsylvania, and American Ornithologists' Union, Washington, DC, USA.

Scifres, C. J. 1980. Brush management: principles and practices for Texas and the southwest. Texas A&M University Press, College Station, USA.

SEMARNAT. 2001. Norma Oficial Mexicana NOM-020-SEMARNAT-2001, Que establece los procedimientos y lineamientos que se deberán observar para la rehabilitación, mejoramiento y conservación de los terrenos forestales de pastoreo. DOF (Primera Sección), 10 de diciembre de 2001.

SEMARNAT. 2002. Norma Oficial Mexicana NOM-059-SEMARNAT-ECOL-2001, Protección ambiental—Especies nativas de México de flora y fauna silvestres—Categorías de riesgo y especificaciones para su inclusión, exclusión o cambio—Lista de especies en riesgo. SEMARNAT. DOF (Segunda Sección), 6 de marzo de 2002, pp. 95–190, México, D. F., México.

SEMARNAT. 2006. UMAs con tasa de aprovechamiento autorizada 2005–2006. DGVS, México, D. F., México.

SEMARNAT. 2008a. Épocas hábiles 2008–2009. DGVS, México, D. F., México.

SEMARNAT. 2008b. UMAs con tasa de aprovechamiento autorizada 2008–2009. DGVS, México, D. F., México.

SEMARNAT. 2009a. SNIARN, Dirección General de Estadística e Información Ambiental. http://www.semarnat.gob.mx/informacionambiental//. Accessed 1 July 2009.

SEMARNAT. 2009b. Norma Oficial Mexicana NOM-015-SEMARNAT/SAGARPA-2007, Que establece las especificaciones técnicas de métodos de uso del fuego en los terrenos forestales y en los terrenos de uso agropecuario. DOF (Primera Sección), 16 de enero de 2009, pp. 23–91, México, D. F., México.

Spears, G. S., F. S. Guthery, S. M. Rice, S. J. Demaso, and B. Zaiglin. 1993. Optimum seral stage for northern bobwhites as influenced by site productivity. Journal of Wildlife Management 57:805–811.

Stanford, J. A. 1972. Second broods in bobwhite quail. Proceedings of the National Quail Symposium 1:22–27.

Stokes, A. W. 1967. Behavior of the bobwhite. Auk 84:1–33.

Stromberg, M. R. 2000. Montezuma quail (*Cyrtonyx montezumae*). *In* The birds of North America. Philadelphia Academy of Sciences, Pennsylvania, and American Ornithologists' Union, Washington, DC, USA.

Téllez, M. H., G. D. M. Martínez, C. Z. Hernández, F. C. Sánchez, L. A. T. Arámbula, and R. Valdez. 2005. Estimación de la capacidad nutricional de carga para *Cyrtonyx montezumae*. Revista Científica 15:27–32.

Urbina-Torres, F., and L. Zaiñhana-Ortiz. 2004. Aspectos de la dieta y anidación de la codorniz rayata (banded quail) *Philortyx fasciatus*. Pages 13–14 *in* J. C. Eitniear, J. T. Baccus, S. L. Dingle, and J. P. Carroll, editors. Conservation of quail in the Neotropics. Center for the Study of Tropical Birds, San Antonio, Texas, USA.

Valadéz, A. R. 2003. La domesticación animal. Plaza y Valdés, México, D. F., México.

Wallmo, O. C. 1956a. Determination of sex and age of scaled quail. Journal of Wildlife Management 20:154–158.

Wallmo, O. C. 1956b. Ecology of scaled quail in West Texas. Texas Game and Fish Commission, Austin, USA.

Warner, D. W., and B. E. Harrell. 1957. The systematics and biology of the singing quail, *Dactylortyx thoracicus*. Wilson Bulletin 69:123–148.

Wiedemann, H. T. 1997. Factors to consider when sculpting brush: mechanical treatment options. Pages 88–95 *in* D. Rollins, D. N. Ueckert, and C. G. Brown, editors. Brush Sculptors Symposium Proceedings. Texas Agricultural Extension Service, Abilene, USA.

Wright, H. A., and A. W. Bailey. 1982. Fire ecology: United States and southern Canada. John Wiley, New York, New York, USA.

Turkeys in Mexico

9

Erick H. Baur, Sanford D. Schemnitz,
Alberto Lafón-Terrazas, and Lovett E. Williams Jr.

Turkeys are gallinaceous birds (order Galliformes, family Phasianidae) that originated in North America. There are two extant turkey species, the wild turkey (*Meleagris gallopavo*) and the ocellated turkey (*M. ocellata*), and both are native to Mexico. The wild turkey is broadly distributed in North America and classified by regional subspecies, including two subspecies in Mexico: Gould's (*mexicana*) and Rio Grande (*intermedia*). The ocellated turkey is not divided into subspecies.

Historical Perspective

Turkeys have profound historical and cultural significance in Mexico. The Maya, Nahua, and Pueblo cultures associated turkeys with rain and related deities. Images of turkeys appear consistently on pottery and pictorial codices (Tozzer and Allen 1910, Aldrich 1967, Lafón-Terrazas 1997). Archaeological evidence indicates that both turkey species were kept in captivity at pre-Columbian sites (Schorger 1966, Steadman 1980, Götz 2008). In addition to their culinary value, captive turkeys provided feathers and bones valued for ornamental and utilitarian purposes (Schorger 1966, Aldrich 1967).

The domestic turkey is probably derived from a third subspecies, the Mexican wild turkey (*gallopavo*), which occurred in central Mexico but is now extirpated (Aldrich 1967, Speller et al. 2010). Domestic turkeys were common in the Tehuacán Valley of central Mexico over 2,000 years ago (Flannery 1967) and their remains have been found well outside the wild turkey's historical range at preclassic Mayan sites in Guatemala from as early as 300 BC to 100 AD (Thornton et al. 2012), and the Gulf and Caribbean coasts, where their remains occur in numerous classic and post-classic Mayan sites (Hamblin 1984, Götz 2008). By the 1500s, they were an important component of indigenous cultures throughout Mexico and surrounding regions (Schorger 1966, Aldrich 1967). Chronicles of the conquest of Mexico describe local economies based on the production of maize (*Zea mays*) and turkeys (Leopold 1948), and the use of turkeys as tribute to indigenous leaders (Prescott 1843).

Evolution and Taxonomy

Turkeys are classified in the subfamily Meleagridinae but were formerly in the family Meleagrididae (AOU 1998). The oldest fossil turkeys are from the Miocene in North America (Steadman 1980). Several extinct species in the genus *Meleagris* have been described from the Late Pliocene and Pleistocene (Aldrich 1967, Steadman 1980). The immediate ancestor of both modern turkey species was likely a Pleistocene turkey that occupied the entire Gulf of Mexico region during a glacial peak when lower sea levels and a drier climate would have favored savanna-like habitat along the coast (Steadman 1980). Subsequent rising sea levels and the development of humid forests to the west and south would have isolated populations on the Yucatán Peninsula that evolved into the ocellated turkey. Gould's and Rio Grande wild turkeys are genetically distinct from the other subspecies, and Gould's is the most genetically homogeneous race, as well as the most genetically divergent from the other subspecies (Mock et al. 2002).

The ocellated turkey was reassigned to *Meleagris* based on osteological similarities to other members of the genus (Steadman 1980) nearly a century after it was separated into the genus *Agriocharis* by Chapman (1896). The close relationship between the two species is further supported by similarities in their morphology, behavior, and life histories (Steadman et al. 1979, Gonzalez et al. 1998, Williams et al. 2010). Hybridization of ocellated and domestic turkeys via artificial insemination has been documented, but not crosses with wild turkeys (Schorger 1966). Williams et al. (2010) reported observations of flocks of ocellated and domestic turkey hybrids in Guatemala that had reproduced readily in captivity over multiple generations.

Distribution

Wild turkey

Historically, the wild turkey occurred in the Great Lakes region of southern Canada; east of the Great Plains, along the Gulf Coast, and in parts of the southern Rocky Mountains in the US; and in the Sierra Madre Occidental and Sierra Madre Oriental mountain ranges and adjacent areas of northern and central Mexico (Schorger 1966). Leopold (1948) found no credible evidence of wild turkeys south of the Río Balsas valley (50 km [31 mi] south of Mexico City). Gould's wild turkey formerly occupied both slopes of the Sierra Madre Occidental in northwestern Mexico, from the Mexico-US border south to Jalisco (Aldrich 1967). By the 1990s, the Gould's race remained on approximately 14 million ha (34.5 million ac) in Mexico (fig. 9.1), mostly in Chihuahua (69%), followed by Durango (12%), Sonora (9%), Zacatecas (6%), and parts of Sinaloa, Aguascalientes, Nayarit, and Jalisco (Lafón-Terrazas 1997). More recently, Gould's turkey has been introduced in Guanajuato and Hidalgo (SEMARNAT 2011). The historical distribution of the Rio Grande wild turkey in Mexico included both slopes of the Sierra Madre Oriental and adjacent areas from the Mexico-US border south to central Veracruz (Aldrich 1967). Information on its current distribution in Mexico is limited; however, it occurs in parts of Coahuila, Nuevo León, Tamaulipas, and San Luis Potosí and has been introduced in Hidalgo (Howell and Webb 1995, SEMARNAT 2011).

Ocellated turkey

This species is endemic to the Yucatán Peninsula and adjacent areas of northern Guatemala and Belize. Its historical distribution in Mexico (fig. 9.1) also included parts of eastern Tabasco and Chiapas (AOU 1998). Currently, it is essentially restricted to parts of Campeche, Yucatán, and Quintana Roo (Calmé and Sanvicente 2000, Kampichler et al. 2010).

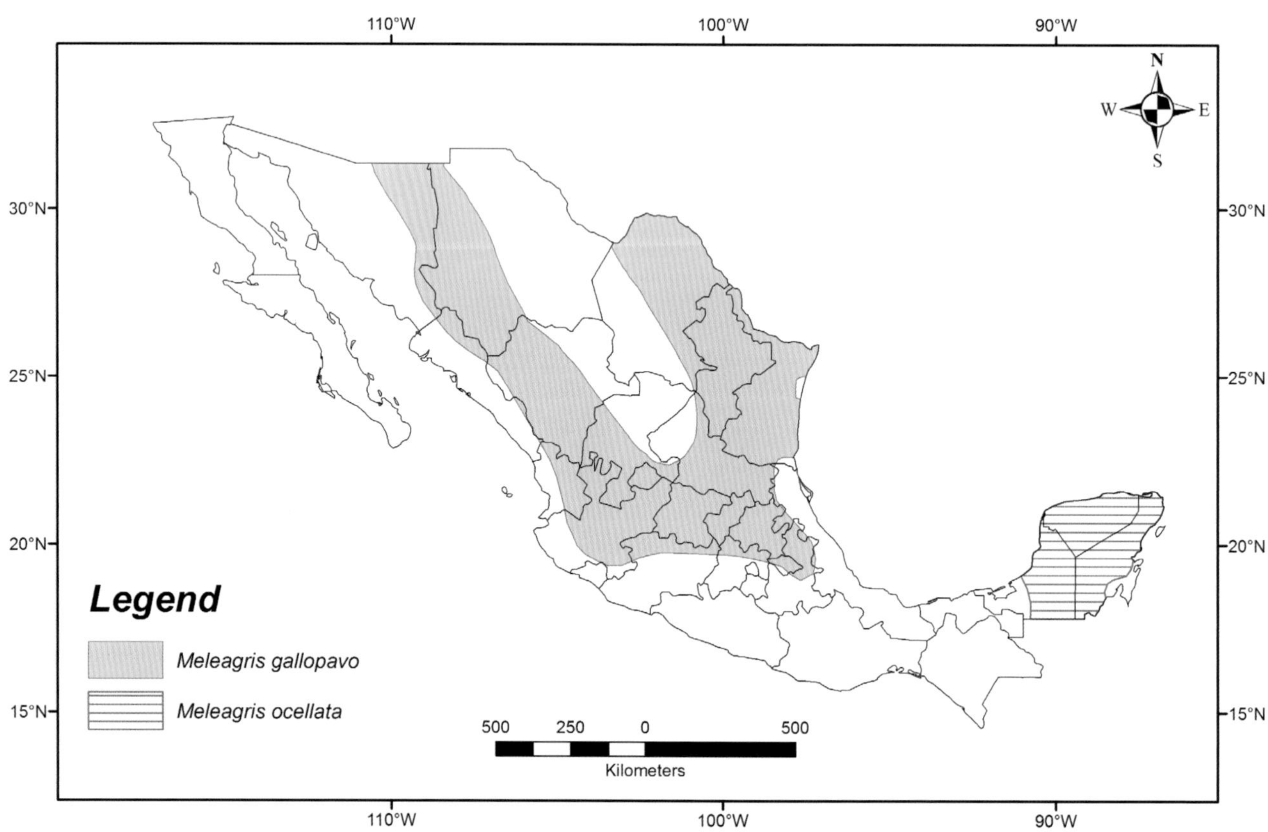

Figure 9.1. Approximate historical distributions of the wild turkey (*M. gallopavo*) and ocellated turkey (*M. ocellata*) in Mexico.

Description

Both turkey species share several morphological traits (Schorger 1966, Lewis 1967, Williams et al. 2010), including significant sexual dimorphism of certain features. In both sexes, the skin is exposed on the head and upper neck, having only a light coating of fine, bristlelike feathers, but female turkeys (hens) have relatively less exposed skin than males. On males, parts of the exposed skin become swollen during the breeding season. Small permanent protuberances (caruncles) form on the exposed skin, which are more developed on males (Lewis 1967, Williams et al. 2010). The skin on the throat of both genders forms a dewlap. Turkeys have a soft frontal process (snood) on the rostrum that is more developed on males than on females, and on the wild turkey relative to the ocellated turkey (Schorger 1966, Williams et al. 2010).

Turkeys are heavy bodied with long tails and strong legs that are adapted to terrestrial habits. Adults have reddish-pink legs, and those of younger birds are paler. Male turkeys become larger than females at an early age and by adulthood they are one and a half to nearly twice the size of hens of similar age (Lewis 1967, Williams 2006, Williams et al. 2010). Male turkeys normally develop a single tarsal spur on the back of each leg above the foot that consists of a bony core with a keratinized epidermal sheath. Extra spurs and the absence of spurs on males, and small spurs or spur-like scutes on hens are unusual but occur in both species (Williams 1966, Steadman 1980, Williams et al. 2010).

The distal ends of the largest contour feathers on the body are square shaped on male turkeys and more rounded on hens, with distinct, gender-dependent coloration (Lewis 1967, Williams 2006, L. E. Williams, personal communication). Rectrices and tail coverts have dark vermiculated barring along most of the feather length and distinct coloration on the distal end. Body plumage is primarily dark brown, with an iridescent surface that reflects bright metallic hues. The primary and secondary wing feathers have a dark base color with variable degrees of white barring. The greater secondary wing coverts form a conspicuous, metallic-copper patch on the shoulder that is useful for distinguishing between adult and yearling birds in the spring, when this patch is noticeably larger and more uniform in shape on adults (Williams 1961, Steadman et al. 1979). Adult males of both species develop a fat deposit (breast sponge) on the breast (Korschgen 1967, Williams et al. 2010).

Wild turkey

In Mexico, this species is commonly known as *guajolote silvestre* or *cócono silvestre*. The exposed skin on the head and neck is generally a pale bluish color; however, this color changes during aggressive intraspecific interactions and notably in males during courtship, ranging to variable intensities of white or red, including the caruncles, dewlap, and snood, which often assume a pink or reddish color that contrasts with the surrounding skin (Lewis 1967). Caruncles vary in size and shape and are most developed on the lower part of the neck. The bristlelike feathers on the head and neck are more developed in the Gould's turkey than in other races (Williams 2002). Adult body mass fluctuates by approximately 20% during the course of the year, being heaviest in early spring at the onset of breeding and the least heavy in early summer (Bailey and Rinell 1967). The breast sponge on adult males attains its greatest size immediately prior to the breeding season (Korschgen 1967). In spring (March–April), spurs of yearling males are typically less than 1.3 cm (0.5 in) long and have a rounded tip; on two-year-old males, spurs range from 1.9 to 2.5 cm (0.75 to 1 in) and are blunt; and on three-year-old and older birds, the spurs reach or exceed 3.2 cm (1.25 in) and are slightly curved with a sharp point (Williams 1981).

Male wild turkeys have a tuft of specialized hairlike feathers (beard) on the breast that grows continuously, often reaching lengths of 20 to 30 cm (8–12 in) on adults (Schorger 1966). Males with multiple beards and bearded hens occur infrequently (Lewis 1967). Male plumage is more iridescent and distinctive than that of hens, which are more cryptically colored. The distal ends of the larger contour feathers on the body are black on male wild turkeys and lighter colored on hens (Lewis 1967). The rectrices and tail coverts have a brown base on the larger vermiculated section, followed distally by a black subterminal bar, and a lighter-colored tip that varies by subspecies (fig. 9.2).

Gould's wild turkey is the largest subspecies in all dimensions (Aldrich 1967). Mean body masses of specimens from Chihuahua were 10.2 ± 0.51 kg (22.5 ± 1.1 lb) (n = 34) for adult males; 6.7 ± 0.66 kg (14.8 ± 1.5 lb) (n = 20) for yearling males; 6.2 ±

Figure 9.2. Adult male Gould's wild turkey. Photo by Lovett E. Williams Jr.

Figure 9.3. Adult male Rio Grande wild turkeys strutting. Photo by Poncho Ortega Jr.

0.6 kg (13.7 ± 1.3 lb) (n = 25) for adult hens; and 4.7 ± 0.48 kg (10.4 ± 1.1 lb) (n = 38) for yearling hens (Lafón-Terrazas 1997). Mean body masses of Rio Grande turkeys in the US during fall and winter were 7.9 kg (17.4 lb) (n = 226) for adult males; 5.9 kg (13 lb) (n = 197) for yearling males; 3.9 kg (8.6 lb) (n = 427) for adult hens; and 3.4 kg (7.5 lb) (n = 310) for yearling hens (Lewis 1967). The tips of the rectrices and tail coverts are nearly white on Gould's turkey (fig. 9.2) and buff or rusty beige on the Rio Grande (fig. 9.3). The white barring on the primary and secondary wing feathers is more prominent on the Gould's turkey than on the Rio Grande race. The legs and feet of Gould's turkeys are darker, and the spurs are typically shorter and fail to develop more frequently than in other subspecies (Williams 2006).

Ocellated turkey

In Mexico, this bird is commonly known as *pavo ocelado* or simply *pavo*. The head and neck are a light blue that does not vary, and the bare skin is finely wrinkled on both sexes (Williams et al. 2010). There is a fleshy appendage (crown) on top of the head that becomes enlarged during breeding on males and is rudimentary on females. Caruncles are much more defined than those of the wild turkey and are more consistently sized (≤ 0.5 cm, ≤ 0.2 in) and shaped (spherical). Most caruncles are a conspicuous golden orange, but they often fade to reddish or bluish on the lower neck (fig. 9.4). The number and distribution of caruncles vary, but they tend to be clustered over the eyes and on the rostrum and crown, and sparsely distributed on the remainder of the head and neck. There is often a single caruncle at the tip of the snood on males. Eyelids are reddish orange in both genders.

Ocellated turkeys are smaller than wild turkeys. The body masses of birds collected during the spring in Guatemala were 4.2–5.7 kg (9.3–12.5 lb) with a mean of 5.0 kg (11 lb) (n = 233) for adult males; 3.6–4.4 kg (7.9–9.7 lb) with a mean of 3.9 kg (8.6 lb) (n = 8) for yearling males; and 3.3–3.9 kg (7.3–8.6 lb) with a mean of 3.5 kg (7.7 lb) (n = 4) for

Figure 9.4. Adult male ocellated turkey. Photo by Lovett E. Williams Jr.

adult hens (Williams et al. 2010). The ocellated turkey lacks a beard and has relatively longer and more pointed spurs than the wild turkeys. In the spring (March–April), yearling males have blunt spurs that are typically shorter than 2.5 cm (1 in); two-year-old males have sharp spurs approximately 3.8 cm (1.5 in) in length; and three-year-old and older birds have sharp, curved spurs that reach or exceed 4.4 cm (1.7 in) (Williams et al. 2010).

Body plumage on the ocellated turkeys reflects a similar bright iridescent green in both sexes, but the iridescence is less vivid on hens (fig. 9.5). The distal edges of contour feathers on the body are golden bronze on males and a duller golden green on hens. Secondary feathers have more white barring than on the wild turkey, forming a distinctive white wing patch in flight. Rectrices and tail coverts have a light gray base on the vermiculated section, a metallic-blue subterminal patch outlined in black, and a coppery-bronze terminal bar. Tail feathers are more rounded and separate at the tips when fanned relative to the sharp edges and more uniform shape of the tail on the wild turkey (compare figs. 9.3 and 9.5).

Habitat Use

Wild turkey

Gould's turkey habitat in Chihuahua and Sonora is characterized by rough terrain dissected by steep rocky canyons above 1,600 m (5,250 ft) (Lafón-Terrazas 1997). In Chihuahua, most radio-tagged birds were located in mixed pine-oak and pine-juniper forests (42%), followed by oak forest (30%), open habitats and rangeland (17%), pine forest (10%), and agricultural areas (1%) (Lafón-Terrazas 1997). In New Mexico, Gould's turkeys made greater use of pinyon pine (*Pinus edulis*) and juniper (*Juniperus deppeana*) woodlands with pinyon ricegrass (*Piptochaetium fimbriatum*) ground cover in the winter, and they spent more time in riparian habitats during spring and summer (York and Schemnitz 2003). In that study, birds also used savannas with Toumey oak (*Quercus toumeyi*) overstory and bull muhly grass (*Muhlenbergia emersleyi*) ground cover; mixed Toumey oak and manzanita (*Arctostaphylos pungens*) scrub; riparian areas in canyons with Chihuahua pine (*P. leiophylla*) and silverleaf oak (*Q. hypoleucoides*) understory; mixed Chihuahua pine and silverleaf oak forests; and open Emory oak (*Q. emoryi*) woodlands with bear grass (*Nolina microcarpa*) and sideoats grama (*Bouteloua curtipendula*) ground cover.

Figure 9.5. Adult male ocellated turkey strutting. Photo by Lovett E. Williams Jr.

Phillips (1911) reported that Rio Grande wild turkeys occurred in pine-oak forests up to 1,800 m (5,900 ft), and in lowlands dominated by mesquite grassland in Mexico. Rio Grande turkeys in Nuevo León exhibited preferential use of *Acacia rigidula–A. berlandieri* scrub associated with *Guaiacum angustifolium*, *Eysenhardtia texana*, *Koeberlinia spinosa*, *Lippia graveolens*, *Opuntia engelmannii*, *Leucophyllum frutescens*, *Ziziphus obtusifolia*, *Aloysia wrightii*, and *Aloysia gratissima* (Carrillo-Reyes 2003). In Texas, riparian areas in canyons and draws with thick juniper stands are used by Rio Grande turkeys during winter. In summer, adult males occupy open mixed hackberry (*Celtis* spp.), live oak (*Q. virginiana*), and honey mesquite (*Prosopis glandulosa*) woodlands with grassy ground cover, and hens with broods occupy open areas with bunchgrasses, hay meadows, and mixed grassland-scrub (Beasom and Wilson 1992).

Ocellated turkey

In Guatemala, habitats include all forest types, marshes, savannas, and agricultural areas (Gonzalez et al. 1998, Baur 2008). Tall upland forests used year-round are dominated by *Brosimum alicastrum*, *Pouteria reticulata*, and *Manilkara zapota*, with relatively open understories that often include Piper sp. and subcanopy palms such as *Chrysophyllum mexicanum* and *Chamaedorea* spp. Dense, low-canopy scrub forests in poorly drained areas dominated by

Haematoxylon campechianum, *Bucida buceras*, and *Vitex gaumeri*, often with understories dominated by *Bravaisia tubiflora* scrub or sedge ground cover, are also used year-round, particularly during the dry season, when turkeys make extensive use of relatively open areas exposed as seasonal flooding recedes. Ocellated turkeys frequent open areas for foraging and breeding, including agricultural fields, scrubby pastures, vehicular trails, abandoned forest camps, and timber-staging areas (Williams et al. 2010).

Life History

Diet

Both turkey species have similar generalist dietary habits and foraging behaviors. Turkey diets include mast and other seeds, fruits, leaves, invertebrates, and occasionally small vertebrates (Gaumer 1883, Schorger 1966, Korschgen 1967, Steadman et al. 1979, Baur 2008). They have muscular gizzards and regularly consume grit. Diets vary according to food availability, reflecting seasonal changes and interannual variability of mast and forage production (Korschgen 1967; E. H. Baur, unpublished data). Both species consume a variety of agricultural crops when available (Gaumer 1883, Leopold 1948, Korschgen 1967). Turkeys generally forage on the ground by scratching through leaf litter and the soil surface to expose food items, alternating the foot used in a repeated four-stroke movement (Korschgen 1967, Steadman et al. 1979). Turkeys also pluck food items from low vegetation and often strip grass seeds from the stalk with a quick head movement (Dalke et al. 1946). Foraging tends to be concentrated during morning and afternoon periods (Davis 1949, Steadman et al. 1979). Breeding males concentrate on reproductive activities during the spring and feed less often, relying heavily on fat reserves of the breast sponge (Korschgen 1967, Steadman et al. 1979).

Wild turkey

Important year-round food items of the Gould's turkey in Chihuahua were fruits of alligator juniper (*Juniperus deppeana*), manzanita (*Arctostaphylos pungens*), and fragrant sumac (*Rhus aromatica*); insects, which dominated the summer diet; and grasses such as bulbed panic grass (*Panicum bulbosum*), wild oats (*Avena sativa*), and blue grama (*Bouteloua gracilis*) (Lafón-Terrazas 1997). In New Mexico, the diet also included juniper, manzanita, and sumac fruits year-round (York and Schemnitz 2003). Other important diet items in New Mexico included pinyon pine (*Pinus edulis*) seeds, acorns, and leaves of mustard (Cruciferae) and barnyard grass (*Echinochloa crusgalli*). Additional important grasses in the diet included pinyon ricegrass (*Piptochaetium fimbriatum*) and sideoats grama (*Bouteloua curtipendula*). Both studies were conducted during periods of extended drought (York and Schemnitz 2003); mast-producing species known to be important for turkeys (Leopold 1948), such as oak, were poorly represented. Important Gould's turkey diet items during the spring in Aguascalientes included Aztec clover (*Trifolium amabile*), tenleaf woodsorrel (*Oxalis decaphylla*), and *Eryngium* sp. (Márquez-Olivas et al. 2005). Diets included maize in Chihuahua and Aguascalientes (Lafón-Terrazas 1997, Márquez-Olivas et al. 2005).

Rio Grande turkey diets vary regionally across a relatively larger distribution than Gould's. Most available diet information is from Texas, where important mast-producing species are oaks, pecan (*Carya illinoinensis*), and walnuts (*Juglans* spp.) (Schorger 1966, Cathey et al. 2004). Important fruit sources include hackberries (*Celtis* spp.), *Bumelia* spp., sumacs (*Rhus* spp.), western soapberry (*Sapindus drummondii*), Chickasaw plum (*Prunus angustifolia*), ground cherry (*Physalis* sp.), red hawthorn (*Crataegus* sp.), and cacti such as prickly pear (*Opuntia atrispina*) and trujillo (*O. leptocaulis*) (Beck and Beck 1955, Schorger 1966, Petersen 2007). Important seed sources include *Croton* spp., honey mesquite (*Prosopis glandulosa*), groundsel (*Senecio ampullaceus*), crownbeard (*Verbesina encelioides*), and southern prickly ash (*Zanthoxylum clava-herculis*) (Beck and Beck 1955, Schorger 1966, Cathey et al. 2004). Important grasses in the diet include bristlegrass (*Setaria* spp.), western wheatgrass (*Agropyron smithii*), and *Panicum* spp. (Cathey et al. 2004, Petersen 2007). Dense-flower bladderpod (*Lesquerella densiflora*) and western ragweed (*Ambrosia psilostachya*) are important diet items in western Texas, where grasshoppers and beetles are important in spring, and summer diets also include peanuts and maize (Petersen 2007).

Ocellated turkey

Leopold (1948) reported that birds foraged in agricultural areas during the fall in Yucatán and

Campeche, where they ate *Solanum hirtum* blossoms, wild cucumber (Cucurbitaceae sp.), and crops such as maize and cassava (*Manihot esculenta*) (Gaumer 1883). Steadman et al. (1979) described the spring diet at Tikal National Park in Guatemala, an archaeological site with extensive areas of open and early-successional vegetation. Half (51%) of the observed foraging involved stripping grass seeds, followed by feeding on leaves (31%) of at least 14 species (*Dicliptera assurgens*, *Justicia* sp., *Ruellia* sp., *Begonia* sp., *Parthenium hysterophorus*, *Rivinia* sp., maidenhair fern [*Adiantum tenerum*], *Piper* sp., and *Psychotria* sp.); fruits (8%) of *Brosimum alicastrum*, *Bursera simaruba*, *Ficus* sp., and palms (*Chamaedorea* spp.); flowers (7%); and insects and grit (3%). Winter diet at Tikal included leaves of annual ragweed (*Ambrosia artemisiifolia*), spiderworts (*Zebrina* spp.), wild grapes (*Vitis* spp.), and *Paspalum* spp.; seed heads of *Paspalum conjugatum*; palm nuts; fruit (*Brosimum alicastrum*); and invertebrates such as leafcutter ants (*Atta cephalotes*), beetles, and moths (Sugihara and Heston 1981).

A year-round study in Guatemala found over 100 plant species and a broad spectrum of invertebrates in the diet (Baur 2008, Williams et al. 2010). Diet composition (excluding grit) was dominated by seeds (32%), followed by mast (19%), invertebrates (17%), leaves (11%), fruit pulp (11%), small fruits (9%), and flowers and stems (1%) (Williams et al. 2010). The most important fruits were from trees such as *Brosimum alicastrum*, *Bursera simaruba*, *Pseudolmedia spuria*, *Calophyllum brasiliense*, *Manilkara zapota*, and the canopy palm (*Sabal mauritiiformis*) (E. H. Baur, unpublished data). The most prominent invertebrate in the diet was a snail that proliferates in flooded forests during the wet season. Larval and adult beetles, millipedes, grasshoppers, and leafcutter ants were also important. In a landscape dominated by primary forest, the diet often included items from isolated agricultural fields such as black bean leaves, maize grains, and early-successional forbs and grasses (Williams et al. 2010).

Soils at that study site are karstic, with very limited availability of hard minerals (Baur 2008). Most mineral grit found in turkeys consisted of small pieces (≤ 1 cm, ≤ 0.4 in) of chert. Most gizzard specimens contained shell fragments of hard-coated seeds of *Bursera simaruba* and to a lesser degree *Protium copal*, which may have compensated for mineral grit scarcity (Williams et al. 2010). Gizzard contents also included lead shotgun pellets, a 22-caliber brass, and an obsidian arrowhead shard.

Water requirements

Turkeys have minimal water requirements. Metabolic water derived from the digestion of carbohydrates can supply much of a turkey's water needs (Williams 2002). Turkeys also derive water from green leaves, succulent plant material, and invertebrates in the diet. In Guatemala, the consumption of leaf material by ocellated turkeys increased from a frequency of about 50% during the wet season to 70% during the dry season (Williams et al. 2010).

Turkeys may meet water requirements using artificial or natural water sources, including dew. In arid and semiarid zones occupied by Gould's and Rio Grande wild turkeys, natural water sources are scarce but important habitat components (Korschgen 1967). Riparian habitats in these areas are also often important roosting and foraging areas. Greater soil humidity in riparian areas increases vegetation diversity and productivity as well as soil invertebrate availability, and in some landscapes only riparian areas support trees suitable for roosting (Figert 1989, York 1991, Lafón-Terrazas 1997, York and Schemnitz 2003). Artificial water sources such as livestock tanks and rain catchments can be important wild turkey management tools in arid and semiarid environments (Glazener 1967, Korschgen 1967).

Reproduction

The two turkey species share a number of breeding and reproductive traits. Adult males compete for access to multiple hens, which prevents or limits mating opportunities for most males (Bailey 1967, Steadman et al. 1979). Yearling hens can breed (Van Tyne 1935, Williams and Austin 1988), and although yearling males may exhibit reproductive behaviors, they usually do not breed until their second year (Schorger 1966, Bailey and Rinell 1967, Williams et al. 2010).

Annual reproductive cycles follow a prescribed sequence, beginning with the initiation of breeding calls by adult males that continue throughout the breeding season, followed by courtship, mating, nesting, and brooding young (Bailey and Rinell 1967, Baur 2008). Male turkeys engage in courtship behavior that includes strutting and plumage displays

(Bailey 1967, Steadman et al. 1979). Mated hens abandon spring flocks and become solitary while nesting. Hens nest and brood young. Vegetative cover often conceals ground nests. Hens lay large clutches of buff-colored, speckled eggs over approximately two weeks, and incubation lasts 26–28 days (Bailey and Rinell 1967, Williams and Austin 1988, Gonzalez et al. 1998). Hens that lose nests to predators often attempt to renest (Williams and Austin 1988, Gonzalez et al. 1998).

Wild turkey

This species is promiscuous, with adult males defending access to multiple hens that may mate with more than one male (Krakauer 2008). Reproductive calling by males is called gobbling, and adult males are often referred to as gobblers in the US. Intensity and duration of gobbling vary according to the local influences of weather, hunting pressure, and hen behavior (Schorger 1966, Kienzler et al. 1996, Williams 2006). Gobbling usually begins in February or early March in the US and increases steadily until mating begins (Bailey and Rinell 1967). Male wild turkeys extend their head forward when gobbling.

Strutting males (figs. 9.6 and 9.7) ventilate their plumage, which gives them an enlarged appearance, spread their tail feathers in a fanlike display, and produce an airy sound described as a "pulmonic puff" (Schorger 1966) or "chump" (Williams 1984). When dominant adult males strut, they are intolerant of yearling males. They do not gobble while strutting, which may contribute to the perception of two seasonal calling peaks. Gobbling steadily increases until hen receptivity reaches its seasonal maximum, at which point males concentrate on strutting and mating, resulting in a lull in gobbling behavior until most hens abandon spring flocks to nest, after which gobbling behavior increases again (Williams 2006).

Nesting generally begins in late April or early May in the US (Everett et al. 1980, Vangilder and Kurzejeski 1995). During the laying process, hens cover the nest with dry vegetation during recesses (fig. 9.8) (Williams and Austin 1988). During the incubation period, hens take recesses every few days and leave clutches uncovered (Williams and Austin 1988). Reported clutch sizes range from 7 to 18 eggs, with a mean of 10 to 14 eggs (Schorger 1966).

The percentage of hens that attempt to nest each year is generally high (> 85%) (Vangilder 1992) and

Figure 9.6. Adult male Rio Grande wild turkey displaying near a hen. Photo by Jacob White.

Figure 9.7. Adult male and female ocellated turkeys during the reproductive season. Photo by Lovett E. Williams Jr.

Figure 9.8. Adult female Gould's wild turkey with poults. Photo by Nathan Fyffe.

the percentage of hens that attempt to renest after losing their first nest is lower than for initial nesting attempts (Williams and Austin 1988, Vangilder and Kurzejeski 1995). Yearling hens lay smaller clutches, are more likely to abandon nests after being disturbed, and are less likely to renest than older hens (Williams and Austin 1988). Hens that lose nests earlier in the nesting process are more likely to renest than hens that lose nests at later stages.

Nesting success, or the percentage of nests that hatch at least one egg, averages 35% for wild turkeys (Schorger 1966). Hen success is the percentage of hens that successfully incubate a clutch during the entire season, including renesting attempts. Among multiple-year studies, mean annual success rates for adult hens ranged from 22% to 68%, with considerable interannual differences (Glidden and Austin 1975, Everett et al. 1980, Vangilder and Kurzejeski 1995).

Most gobbling by the Gould's turkey in Chihuahua occurs from mid-March through mid-May and may continue until June (Lafón-Terrazas 1997). The mean date of initiation of incubation in a Gould's population in New Mexico was 24 May (Schemnitz et al. 1990). Most nests (85%, n = 13) in Chihuahua were in oak forest and on eastern exposures (62%) (Lafón-Terrazas 1997). In that study, seven nests (54%) were located at the base of slopes, and the rest were found equally in sloped terrain and above slopes. Mean distance to the nearest watering source was 111.9 ± 72.4 m (367 ± 237.5 ft). Mean clutch size was 9.4 eggs in 1994, and 6.0 eggs in 1995. In 1994, hatching peaked on 12 July, with a mean of 7.6 eggs (81%) hatched per clutch; in 1995 the mean was 3.9 eggs (65%) hatched per clutch.

Gobbling by the Rio Grande turkey in Texas begins as early as the first week of February (Schorger 1966). Nesting in Texas occurs from February to August, with a peak from April to July (Cathey et al. 2004). In areas of Texas with stable Rio Grande turkey populations, 78–94% of the hens attempted to nest, nesting success was 17%, and hen success ranged from 21% to 35%, with a mean of 28% (Melton 2007). Renesting rates in that study were 47–80%, with a mean of 60%, and first nesting dates consistently ranged from early April to the end of May over three years.

Ocellated turkey

The breeding call of the male ocellated turkey is distinct from that of the wild turkey and difficult to describe (Leopold 1948). A complete breeding call begins with a series of deep, muffled sounds or drum tones that increase in volume and tempo before culminating in a loud percussive sound similar to a woodpecker drumming (Steadman et al. 1979, Gonzalez et al. 1998, Williams et al. 2010). Males pull their head toward their body at the culmination of the call, in contrast to the posture of the wild turkey when gobbling (Steadman et al. 1979).

There is considerable variation in the frequency and temporal pattern of reproductive calling by the ocellated turkey (Steadman et al. 1979, Williams et al. 2010). In Guatemala, calling usually begins in late February or early March, peaks in mid to late April, declines in May, and ends in June, although in some years, calling may last until early July (Gonzalez et al. 1998, Baur 2008). Daily calling behavior usually begins before sunrise from the roost; continues regularly during early morning; appears to decrease during the day, although this may be a function of calling being muffled by dense vegetation while birds are on the ground; increases again in late afternoon until after roosting; and may continue until after dark (Steadman et al. 1979, Williams et al. 2010).

In forested habitat, breeding male ocellated turkeys often stay in restricted areas, which they patrol alone daily, calling occasionally and often producing drum tones in short series, without culminating in a complete breeding call. These areas usually include natural or artificial clearings with adequate space for strutting and display, and they appear to be breeding territories. Breeding males will often investigate playback recordings of breeding calls in these areas (Williams et al. 2010).

Williams et al. (2010) reported that some adult males do not exhibit courtship behaviors during the breeding season. These birds do not make breeding calls, drum, strut, assume courtship postures, or express any reproductive intentions in the presence of hens. They do not exhibit any enlargement of the crown or exposed skin. These adult males do not appear to patrol or defend breeding territories, and they sometimes accompany flocks with hens (Williams et al. 2010). The cause of this condition, and whether it is temporary or permanent, are not known.

The percentage of hens monitored by Gonzalez et al. (1998) that attempted to nest ranged from 75% to 100% per year, hen success was 62%, and 40% of hens that lost nests attempted to renest. Baur (2008) found active nests in April, May, and June. Nests occurred in all natural forest types, and in agricultural habitats ranging from recently cut primary forest to fallow field woodlots (Williams et al. 2010). Half of the nests (49%) were found in stands of bracken fern (*Pteridium* sp.) that sometimes form in fallow fields. This may be maladaptive because these fern stands burn quickly and thoroughly and are often ignited unintentionally by farmers preparing adjacent areas for planting after turkeys have begun nesting. Most forest nests were placed among the buttress roots of large trees (Baur 2008). Gaumer (1883) reported clutch sizes of 8–16 eggs in Yucatán, and Leopold (1948) reported clutches of 8–15 eggs based on field observations in Yucatán and Campeche. Gonzalez et al. (1998) reported an average clutch size of 8.8 ± 2.5 (n = 5). Williams et al. (2010) reported clutch sizes of 3–12 eggs (n = 36) with a mean of 8.4 ± 2.0 eggs, including clutches that were incomplete when observed. Gonzalez et al. (1998) reported hatching from the end of May until the beginning of July. Brood flocks have been observed from March until October, with a peak in May (Baur 2008).

Survival and mortality factors

Little is known about the population dynamics of either species in Mexico, where predator communities, hunting characteristics, and environmental conditions represent important differences with respect to survival and mortality in areas where related research has been conducted. Both turkey species possess traits that contribute to high fecundity (e.g., large clutches, nesting by yearling hens, renesting) that evolved in response to high rates of natural mortality. Annual survival rates of eastern wild turkey (*M. g. silvestris*) hens in the US range from 45% to 80% (Little et al. 1990, Palmer et al. 1993, Vangilder and Kurzejeski 1995). Annual survival rates of male eastern wild turkeys in hunted populations are generally lower, often ranging from 30% to 40% (Diefenbach and Vreeland 2010). By virtue of its tropical environment, the ocellated turkey is likely to face a broader array of natural enemies than the wild turkey (Hughes et al. 2005, Baur 2008). McRoberts (2014) reported a survival rate of 50.5% for adult ocellated turkeys in Campeche.

Coyotes (*Canis latrans*) and bobcats (*Lynx rufus*) were the main predators (56.7%) of radio-tagged Gould's turkeys in Chihuahua (Lafón-Terrazas 1997). Other common predators of adult wild turkeys are gray foxes (*Urocyon cinereoargenteus*), golden eagles (*Aquila chrysaetos*), and great horned owls (*Bubo virginianus*). Mammalian predators in tropical Mexico include jaguar (*Panthera onca*), puma (*Puma concolor*), jaguarundi (*P. yagouaroundi*), ocelot (*Leopardus pardalis*), and probably margay (*L. wiedii*). Tropical avian predators include not only large raptors (*Harpyhaliaetus* sp., *Harpia* sp., *Morphnus* sp., and *Spizaetus* spp.), but also much smaller species such as the collared forest falcon (*Micrastur semitorquatus*) (Thorstrom 2000).

Wild turkey poult survival is generally low (20–40%), and most predation on poults occurs during the first two weeks after hatching before they develop the ability to fly (Glidden and Austin 1975, Everett et al. 1980, Speake et al. 1985, Vangilder and Kurzejeski 1995, Hughes et al. 2005). Rio Grande turkey poults in Texas had an average life span of six days, and 94% of documented predation occurred within 14 days of hatching (Melton 2007). In Guatemala, only 15% of ocellated turkey poults survived to the age of four months (Gonzalez et al. 1996).

Predation is the leading cause of nest failure in wild turkeys (Cook 1972, Vangilder and Kurzejeski 1995, Hughes et al. 2005). Predation of Rio Grande turkey nests in Texas ranged from 59% to 88% (Melton 2007). Gonzalez et al. (1998) reported ocellated turkey hen survival rates during nesting and brooding of 0.75 (n = 4) in 1993, and 0.60 (n = 6) in 1994.

Raccoons (*Procyon lotor*) are the most common turkey nest predators in the US, although the coati (*Nasua narica*) may assume that role in Mexico (Hughes et al. 2005, Williams et al. 2010). Other documented predators of Rio Grande turkey nests in Texas include bobcat, skunks (*Mephitis* and *Spilogale* spp.), gray fox, feral swine (*Sus scrofa domesticus*), nine-banded armadillo (*Dasypus novemcinctus*), rat snakes (*Pantherophis* spp.), and raven (*Corvus corax*) (Melton 2007, Dreibelbis et al. 2008). In Mexico, additional potential nest predators include didelphid opossums (*Didelphis* and *Philander* spp.), collared peccary (*Pecari tajacu*), white-lipped peccary (*Tayassu pecari*), tayra (*Eira barbara*), badger (*Taxidea taxus*), and large snakes (*Pseustes poecilonotus*, *Spilotes pullatus*, and *Drymarchon* spp.).

Subsistence hunting can be highly destructive to turkey populations because it is often relentless and unselective for age or gender. Lafón-Terrazas (1997) reported the loss of 10 radio-tagged turkeys (33% of the study sample) to poachers in Chihuahua. Subsistence hunting is a widespread threat to turkey populations in rural Mexico, particularly in the south (Weber et al. 2006, McRoberts 2014), where most remaining forest is community managed (Bray and Wexler 1996). Many of the traditional subsistence hunting methods used in Mexico, such as wooden pens, fishhooks baited with maize, snares, and nocturnal roost shooting, are illegal.

Behavior

Social organization

Throughout most of the year, wild turkeys associate with one of four types of flocks: hens with broods, juvenile or yearling males, broodless hens, or adult males (Williams 2006). Although ocellated turkeys also tend to associate with single-gender or age-class flocks, they do not segregate as strictly by these criteria (Williams et al. 2010). Both species begin life in brood flocks composed of a hen and her poults, although two and sometimes three hens with broods may coalesce into larger brood flocks (Bailey 1967, Gonzalez et al. 1998). Juvenile wild turkey males separate from brood flocks late in the year (Bailey 1967), whereas juvenile or yearling male ocellated turkeys may associate compatibly with broodless hens or cohorts of either sex throughout the year (Williams et al. 2010).

Flocks of adult male wild turkeys remain associated through the winter (Bailey 1967). In spring, adult male flocks separate either as single individuals or as small flocks composed of a dominant bird and a variable number of allied males. Flocks of allied adult males fight as a unit against similar flocks (Bailey 1967).

During the winter, adult male ocellated turkeys may associate with both family and adult male-only flocks (Sugihara and Heston 1981, Williams et al. 2010). In open habitats, breeding adult male ocellated turkeys may associate with one another during the spring, but in such cases dominant males perform most reproductive behaviors, defend small display areas, and tolerate only limited breeding behaviors by subordinate males (Steadman et al. 1979). Williams et al. (2010) observed a few flocks of adult male ocellated turkeys in the spring composed entirely of birds that did not appear to be in breeding condition.

Roosting

Individual members of wild turkey flocks often roost in separate trees within a relatively small area. In Chihuahua, Gould's turkeys often roost in tall Chihuahua or Apache pines (Lafón-Terrazas 1997) but also use oak, sycamore, and madrone trees (Márquez-Oliveras et al. 2007). Tree height at 37 roost sites averaged 23.5 ± 6.6 m (77 ± 21.6 ft) and ranged from 14 to 43 m (46 to 141 ft) (Lafón-Terrazas 1997). Roost trees are often adjacent to clearings or open slopes, which facilitate access, and they are typically tall, with a large diameter and a relatively open crown with horizontal branches (Lafón-Terrazas 1997). Roost proximity to water sources averaged 1.5 km (1 mi) in New Mexico (York 1991) and 147 m (482 ft) in Chihuahua (Lafón-Terrazas 1997).

In many areas occupied by the Rio Grande turkey in the US, large trees are uncommon. Turkeys in these areas often roost on artificial structures (fig. 9.9) such as fences, power lines, electrical towers, or similar structures (Beasom and Wilson 1992). During winter, turkeys in these areas exhibit high fidelity to roost sites, which are usually located along streams or in protected canyons (Glazener 1967). Rio Grande turkeys sometimes congregate at winter roost sites in large flocks of 200–500 birds (Walker 1948, Taylor 1951). Adult males may use winter roost sites year-round, but hens often move considerable distances. Brood flocks sometimes roost in mesquite (*Prosopis* spp.) and hackberry (*Celtis* spp.) trees (Beasom and Wilson 1992).

Ocellated turkey flocks typically roost together in the same tree. Roost trees can be a variety of species but typically have open canopies with horizontal limbs, such as canopy emergents, or are located near sharp habitat edges with lower vegetation or clearings, such as savannas, marshes, pastures, agricultural fields, or roads. Birds roost on horizontal limbs 10–20 cm (4–8 in) in diameter, at heights between 6 and 22 m (20–72 ft) (Steadman et al. 1979). Mixed flocks of juvenile or yearling males and hens of both age classes may roost together, but breeding males roost separately (Williams et al. 2010).

Figure 9.9. Rio Grande wild turkeys using man-made roosts. Photo by Brandon Mitchell.

Movements and spacing

Wild turkey populations make seasonal movements in some areas (Ligon 1946, Thomas et al. 1966). Typically, turkeys have smaller home ranges in areas with higher-quality habitat than they do in areas with poorer habitat (Wigley et al. 1986). Daily movements of Gould's turkeys in Chihuahua did not vary among seasons, but turkeys moved from higher-altitude pine forests to lower oak forests in spring. Average daily movements were 3.7 ± 1.8 km (2.3 ± 1.1 mi) for hens (n = 18), and 2.9 km ± 1.8 km (1.8 ± 1.1 mi) (n = 8) for adult males. Annual home ranges of adult hens in Chihuahua ranged from 3,012 to 5,843 ha (7,465 to 14,438 ac) and averaged 3,199 ha (7,905 ac) for adult males (Lafón-Terrazas 1997). Mean annual Gould's turkey hen home ranges were 4,385 ha (10,836 ac) (York and Schemnitz 2003), 2,367 ha (5,849 ac) (Zornes 1993), and 4,384 ha (10,833 ac) (York 1991) in New Mexico. Mean annual home range of adult males in New Mexico was 6,742 ha (16,666 ac), and movements across the Mexico-US border were documented (Zornes 1993).

In Texas, Rio Grande wild turkeys exhibit seasonal movements between winter roosting areas and spring and summer ranges (Beasom and Wilson 1992, Phillips 2004). Annual home ranges of yearling hens in Texas were consistently larger than those of adult hens and males of both age classes (Phillips 2004, Hall et al. 2006). Mean annual home ranges were 1,908 ± 112 ha (4,714 ± 277 ac) for hens, and 1,578 ± 127 ha (3,889 ± 314 ac) for males (Hall et al. 2006). Adult hen home ranges were 1,527 ha (3,773 ac) in Texas and 4,003 ha (9,892 ac) in Kansas; and ranges of adult males were 1,941 ha (4,796 ac) in Texas and 3,398 ha (8,397 ac) in Kansas (Hall et al. 2006). Most birds used the same winter roosting areas each year; however, some turkeys, particularly yearling hens, dispersed to other winter ranges (Phillips 2004).

Home ranges of radio-collared ocellated turkeys in Campeche were much smaller than those of the wild turkeys mentioned above, and annual patterns also suggest seasonal movements (McRoberts 2014). In Guatemala, breeding male turkeys often patrol and display in areas of scrub forest that are completely submerged by seasonal floods during the wet season but are relatively devoid of vegetation during the dry season (Williams et al. 2010). Radio-collared ocellated turkey hens moved between 0.2 and 8.0 km

(0.12 and 5 mi) from trapping sites to nest sites, with a mean distance of 2.4 km (1.5 mi) (Gonzalez et al. 1998). Hens with broods spent most of their time in tall upland forest. Home range sizes of hens with broods ranged from 27.6 to 1,247 ha (68 to 3,081 ac) (Gonzalez et al. 1998).

Population Management

Survey methods

Estimation of turkey abundance is complicated by the difficulty of reconciling fluid population dynamics with meaningful temporal and spatial scales. Turkey populations fluctuate considerably over the course of each year because of mortality and reproduction. Populations are at their annual minimum entering the breeding season, increase by 200–300% or more at the peak of hatching, and are steadily reduced by predation and other mortality factors until the reproductive cycle is repeated. Variability in mortality, immigration, emigration, and reproductive success results in annual fluctuations of up to 50% of long-term mean population size in wild turkeys (Mosby 1967).

Turkey behavior further complicates estimates of their abundance. Flock movements result in significant changes in the number of turkeys in each area. Turkeys may be much more detectible at some periods of the year because of seasonal changes in behavior and habitat use. Turkeys in well-protected areas may become habituated to human activity and be easy to observe, whereas individuals in populations subject to hunting often avoid human activity by shifting habitat use and daily activity patterns.

In Durango, Garza and Servín (1993) noted that temporary aggregations of Gould's turkeys in sampled areas could result in robust density estimates. Lafón-Terrazas (1997) evaluated several survey methods for Gould's turkey populations in Chihuahua, including gobbling counts, roost surveys, line transects, bait site counts, winter flock counts, camera stations, radiotelemetry, landowner interviews, and hunter success surveys. Seasonal or annual roost surveys provided adequate indices to monitor population trends, but the use of combined methods was recommended. In Texas and Kansas, morning roost counts were more effective for monitoring Rio Grande turkey population trends than night-vision devices, infrared cameras, or video surveillance, and hen-poult counts correlated well with reproductive parameters at the local level, but not at the regional level (Butler 2006). In that study, aerial survey methods were less biased but more expensive than ground-based transects and roadside counts.

Dense tropical vegetation often complicates making meaningful estimates of ocellated turkey abundance. Steadman et al. (1979) estimated ocellated turkey densities using line transects based on observations of calling males and local flock composition at Tikal National Park in Guatemala. Williams et al. (2010) mentioned similar monitoring efforts for harvested populations in adjacent management units. Baur (2008) applied a systematic line-transect effort over annual periods and generated density estimates for dominant forest types using the DISTANCE software (Research Unit for Wildlife Population Assessment, University of St Andrews, Scotland), which yielded more robust density estimates than those described by Steadman et al. (1979) and Williams et al. (2010).

Population trends and current status

Wild turkey

Leopold (1948, 1959) expressed concern for the future of the wild turkey in Mexico, citing deforestation and agricultural development as principal agents of the species' decline. Despite Leopold's concerns, the wild turkey has not only persisted in Mexico but has benefited from reintroduction efforts and appears to be rebounding. The wild turkey was removed from the list of threatened and endangered species in recent years, having formerly been classified as subject to special protection (DOF 2002, 2010).

Healthy Gould's turkey populations in pine-oak-juniper woodland near Colonia Pacheco, Chihuahua, had densities of approximately 4.1 birds/km^2 (10.7 birds/mi^2) in 1937, and 12.4 birds/km^2 (32.6 birds/mi^2) in 1948 (Leopold 1948, 1959). Leopold (1948) reported Rio Grande turkey densities of approximately 16.5 birds/km^2 (43.4 birds/mi^2) in the Sierra de Tamaulipas and eastern San Luis Potosí. Lafón and Schemnitz (1996) reported a Gould's turkey population density of 1.2 birds/km^2 (3.2 birds/mi^2) in west-central Chihuahua during a drought, which increased in subsequent years to 1.7–2.8 birds/km^2 (4.4–7.3 birds/mi^2) (Lafón-Terrazas 1997). Garza and Servín (1993) estimated Gould's turkey densities in the spring by habitat type in La Michilía Biosphere Reserve in Durango that ranged from 1.6 to 6.0 birds/km^2 (4.2 to 15.7 birds/mi^2) based on direct observations, and from 1.4 to 10.7 birds/km^2 (4.2 to 28.2 birds/mi^2) based on calls. Some current management guidelines in Mexico consider density ranges of 3–20 birds/km^2 (8–53 birds/mi^2) as normal, and densities greater than 20 birds/km^2 (53 birds/mi^2) as high (Garza-Herrera and Aragón-Piña 2011).

Ocellated turkey

Leopold (1948, 1959) was more optimistic about the ocellated turkey's future prospects because of its evasive behavior and ability to use secondary and agricultural habitats. Despite that confidence, researchers were voicing concern by the 1990s after the species had become dramatically reduced or eliminated from many areas (Gonzalez et al. 1996, Calmé and Sanvicente 2000). The ocellated turkey has been listed as threatened in Mexico for the last two decades (DOF 2002, 2010). Kampichler et al. (2010) expressed concerns that current conservation efforts may not be sufficient to reverse this trend.

Steadman et al. (1979) estimated conservative densities of 1 bird/km^2 (3 birds/mi^2) in Tikal National Park in Guatemala. Density estimates reported by Baur (2008) in areas north of the park ranged from 3 to 25 birds/km^2 (8 to 66 birds/mi^2) in protected areas, and from 5 to 8 birds/km^2 (13 to 21 birds/mi^2) in areas subjected to hunting. Densities were similar in both dominant forest types, and there was considerable interannual variation regardless of protection level. Large aggregations of ocellated turkeys have been reported in areas of mechanized agriculture in Campeche, where they are protected from subsistence hunting (J. T. McRoberts, University of Missouri, personal communication).

Hunting management

Although much of what is known about the management of hunted wild turkey populations in the US is relevant to Mexico, there are such significant contextual differences between the two countries, and even within Mexico itself, that a great deal of research and innovation is still needed to address hunting sustainability issues. One important difference between the two nations is the relative prevalence of poorly regulated subsistence hunting in many areas of Mexico, which has been cited as a serious threat to the survival of populations of both species (Gonzalez et al. 1996, Lafón-Terrazas 1997). Strategies to curtail and mitigate subsistence hunting in the region include control of vehicular access to areas that support turkey populations, improvement of local governance and law enforcement capacity, participation by local communities in conservation initiatives, public education programs, and generation of local employment opportunities. However, the success or failure of any strategy will depend on the ability to find appropriate context-specific approaches. Mexico has made significant advances in the regulation of sport hunting and wildlife management in recent decades, and with continued progress, hunting can contribute effectively to the conservation of both turkey species in Mexico, just as it has in the US.

Wild turkey

Sport hunting for this species is often offered by UMAs that also offer mule deer (*Odocoileus hemionus*) and white-tailed deer (*O. virginianus*) hunting. Currently only spring hunting is allowed, typically from March or April until May (SEMARNAT 2011). SEMARNAT (2008) requires abundance monitoring of harvested wild turkey populations based on line-transect sampling, point counts, or bait station counts, and it provides explicit design, sampling, and analysis guidelines for developing requisite turkey management plans. Refer to Garza-Herrera and Aragón-Piña (2011) for additional information on wild turkey management in the context of UMAs in Mexico.

The use of artificial feeders for wild turkeys and the release of pen-raised turkeys have been reported in UMAs in northern Mexico (Lafón-Terrazas 1997). Mechanized supplemental feeding concentrates turkeys, which may increase the risk of predation, disease, and poaching, and in some cases it is simply not effective (Carrillo-Reyes 2003). Releasing captive-raised turkeys poses the threat of introducing diseases to wild populations.

Ocellated turkey

This species is often offered by commercial sport hunting operations in combination with opportunities to hunt great curassow (*Crax rubra*), crested guan (*Penelope purpurascens*), coati, or brocket deer (*Mazama* spp.). Ocellated turkey hunting is limited to the spring, typically from February through May (SEMARNAT 2011). For ocellated turkeys, SEMAR-

NAT (2009) requires abundance monitoring of harvested populations using line transects sampled in the fall and postharvest, and it recommends using the DISTANCE software for calculating density estimates.

Traditional commercial sport hunting in southern Mexico has been criticized for failing to adequately provide economic incentives for conservation at the local level, raising questions about whether policies that are successful in the north will achieve similar conservation outcomes for the ocellated turkey (Weber et al. 2006). Under proper circumstances, sport hunting can offer sufficient financial incentives for local residents to mitigate their own impacts on ocellated turkey populations, and it can even provide a lucrative substitute for subsistence hunting (Williams et al. 2010, Baur et al. 2012). The community-based ocellated turkey program featured in Williams et al. (2010) prioritized the generation of local employment and other means of distributing benefits locally, including the support of community-based forest management. There is evidence that selective spring sport hunting can accompany population growth, especially in areas where subsistence hunting is controlled (Baur et al. 2012, McRoberts 2014).

Based on extensive experience with both subsistence and sport hunting, and observations of ocellated turkey populations at various sites throughout the species' range, Williams et al. (2010) recommended promoting hunting of males only, discouraging roost shooting among subsistence hunters, and delaying the opening of hunting seasons until the beginning of April to avoid conflicts with reproductive activities. They emphasized the need for better understanding of the factors that influence the reproductive condition of adult males to help determine the impacts of hunting on local populations. Additional research is needed on reproductive calling, particularly calling by yearling males, and on how site conditions affect the ability to hear calls, in order to better relate observed calling to turkey abundance. Although ocellated turkeys can be attracted to bait, supplemental feeding may not be beneficial or cost effective because of the year-round availability of food resources (Williams et al. 2010).

Livestock management

Wild turkey management experiences in the US, and observations of ocellated turkey populations at a cattle ranch in Belize over multiple years (L. E. Williams Jr., personal communication), make it reasonable to expect that populations of both turkey species can be managed compatibly with cattle production in Mexico. Unfortunately, there is a dearth of relevant information specific to Mexico, especially considering the vast range of conditions under which cattle and turkeys coexist in the country. Zornes (1993) reported that 56% of all Gould's turkey locations in southwestern New Mexico were in areas grazed by livestock, and that turkeys utilized areas with high livestock densities if escape cover was nearby. Stocking levels that allow a maximum utilization of 35% of the herbaceous vegetation are recommended for improving habitat quality for Gould's turkeys in New Mexico (Schemnitz and Zornes 1995). In Nuevo León, Rio Grande turkey hens with broods exhibited preferential use of cattle pastures (Carrillo-Reyes 2003). In south-central Texas, a grazing rotation system was shown to improve habitat used by Rio Grande turkeys for nesting and feeding (Merrill 1975), and similar grazing management practices would likely benefit turkey populations in cattle-producing areas of Mexico.

Riparian areas are critical components of wild turkey habitat in semiarid and arid areas, and mitigation of livestock degradation in similar areas of Mexico is likely to benefit wild turkeys and other wildlife populations (Carrillo-Reyes 2003). Fencing at water sources may restrict livestock access, improve herbaceous growth, and encourage the development of escape cover. Zeedyk and Clothier (2009) described methods of enhancing degraded riparian habitats by restoring a meandering pattern to incised stream channels. Other management practices for improving riparian habitat for wild turkey populations in New Mexico, such as the use of permeable dams and site treatments to encourage the regeneration of native vegetation such as willow and cottonwood, are also likely to be effective in similar areas of northern Mexico.

Vegetation management

There is considerable need for research on the impacts of agricultural practices and other land uses on populations of both turkey species in Mexico. As with turkey hunting management, practices that are known to be effective management tools under similar environmental conditions in the US can serve as a blueprint to guide relevant research in Mexico.

Wild turkey

Many of the vegetation management guidelines for Merriam's wild turkey (*M. g. merriami*) are applicable to Gould's turkey populations in the southwestern US (Hoffman et al. 1993) and are also likely applicable to similar areas that support Gould's and even some Rio Grande turkey populations in Mexico. Likewise, many management guidelines for areas that support Rio Grande turkey populations in coastal and southern areas of Texas (Cathey et al. 2004) are also likely to be relevant in similar areas of Mexico. Leopold (1959) recommended the management of mountain pine forests for watershed protection and timber production to protect wildlife, and particularly wild turkey habitat in Mexico. Bear grass (*Nolina microcarpa*) provides excellent cover for nesting and brooding by Gould's turkeys in Chihuahua and is also harvested locally for making traditional brooms; therefore Lafón-Terrazas (1997) recommended managing grass harvests in areas that support wild turkey populations to avoid conflicts with reproductive activities.

Ocellated turkey

In forested areas, clearings are very important to ocellated turkeys for feeding and courtship display (Steadman et al. 1979, Baur 2008, Williams et al. 2010). Because openings are scarce in the relatively dense forest vegetation characteristic of much of the range of the ocellated turkey in Mexico, assessment of the availability and quality of clearings as brood habitat, and the effects of site treatments for the maintenance of artificial and natural clearings, would help determine locally appropriate management practices (Williams et al. 2010). Maintenance of clearings and roadside rights-of-way by mowing or controlled burning may conflict with nesting if performed from April to June; research is therefore needed to determine the appropriate timing and intensity of these and other vegetation control practices in Mexico. Ocellated turkey populations could likely benefit from clearings associated with selective timber extraction operations provided that related vehicular access did not increase exposure to hunting pressure (Baur et al. 2012). Careful research attention should be given to the effects of forestry practices on ocellated turkey populations, given the significant geographical overlap of the species' distribution with areas where commercial and community-based forestry are common in Mexico.

Because ocellated turkeys make extensive use of agricultural habitats, substantial research is needed regarding the effects of a vast spectrum of crops, agricultural systems, and related practices on habitat quality for turkeys. The widespread use of fire to prepare agricultural areas for planting in southern Mexico is another practice that deserves research attention. Appropriate control of agricultural burning would not only mitigate conflicts with nesting, which would likely convey the additional benefit of reducing forest fire risk during the height of the dry season, and might also offer a means of improving foraging and brood habitat.

Management challenges in Mexico

The recovery of the wild turkey in much of North America is one of the most remarkable wildlife management success stories, and although a range of distinct governmental, cultural, and socioeconomic factors confronts the prospect of extending this recovery over much of Mexico, there are many reasons to believe that such an achievement is not only possible but likely (Valdez et al. 2006). The turning point for the wild turkey in the US came after centuries of decline, when the impending extinction of the species was widely expected. Despite similar historical declines in turkey abundance and distribution in Mexico, there are many indications that the inflection point has already passed, and that these trends are being reversed.

Another challenge to the successful recovery of the two turkey species in Mexico is access to existing information. An immense quantity of information relevant to the management and conservation of turkeys in Mexico is sequestered in reports submitted to and produced by SEMARNAT and its predecessors, as well as other federal and state agencies, domestic and international conservation organizations, and especially theses, dissertations, and other research documents from academic institutions across the country. Because of the incredible scope of geographic, cultural, and environmental differences under which turkey populations exist in Mexico, the compilation of such information, its conversion into a readily transmissible digital format, and the development of an accessible online interface to deliver it would provide an immeasurable benefit to turkey research and conservation efforts throughout the country and should be given high priority.

Many turkey populations in Mexico coexist with some of the country's most marginalized citizens. In many rural areas, residents that rely on local wildlife resources are the de facto determinants of the fate of turkey populations. Every effort should be made to co-opt rural communities into the conservation of local turkey populations through sustainable use, mitigation of the impacts of land management practices and natural resource use via practical mechanisms, and meaningful incentives. The Wildlife Conservation and Production Diversification in the Rural Sector program, and the participation of communal landholdings in the UMA system, are promising steps in that direction (see chapter 21). The continued evolution of the UMA system is necessary to help ensure that regional and species-specific considerations are integrated in the regulatory, administrative, and technical elements of this system (Weber et al. 2006, García-Marmolejo et al. 2008).

The need for research on the wild turkey and ocellated turkey, their utilization, and many factors that impact their populations in Mexico cannot be overstated. The continued development of appropriate conservation and management solutions will benefit from collaborative research and innovative conservation approaches to specific conditions in Mexico. The protection of habitat and Rio Grande wild turkey populations in Coahuila and Nuevo León, and support for wild turkey research and management initiatives by CEMEX, provide an excellent model that may prove replicable in other parts of the country, and with other corporate sponsors. There are many examples of international cooperation among universities and both international and domestic organizations on wild turkey and ocellated turkey research in Mexico. For example, the National Wild Turkey Federation (NWTF) formalized agreements with SEMARNAT in 2005 and has since been engaged in collaborative research and restoration efforts with a number of Mexican partner organizations. We hope these efforts will lead to increased interest and support for research and conservation among the public, academic institutions, private sector businesses and organizations, and public institutions that will ensure the momentum needed for the continued recovery of the wild turkey and ocellated turkey in Mexico.

Literature Cited

Aldrich, J. W. 1967. Taxonomy, distribution, and present status. Pages 17–44 *in* O. H. Hewitt, editor. The wild turkey and its management. The Wildlife Society, Washington, DC, USA.

AOU. 1998. Check-list of North American birds. Seventh edition. American Ornithologists' Union, Washington, DC, USA.

Bailey, R. W. 1967. Behavior. Pages 93–112 *in* O. H. Hewitt, editor. The wild turkey and its management. The Wildlife Society, Washington, DC, USA.

Bailey, R. W., and K. T. Rinell. 1967. Events in the turkey year. Pages 73–91 *in* O. H. Hewitt, editor. The wild turkey and its management. The Wildlife Society, Washington, DC, USA.

Baur, E. H. 2008. Structure of a lowland Neotropical galliform bird guild. Thesis, University of Florida, Gainesville, USA.

Baur, E. H., L. E. Williams Jr., R. B. McNab, V. H. Ramos, J. Radachowsky, and M. Guariguata. 2012. Multiple forest use through commercial sport hunting: lessons from a community-based model from the Petén, Guatemala. Forest Ecology and Management 268:112–120.

Beasom, S. L., and D. Wilson. 1992. Rio Grande turkey. Pages 306–330 *in* J. G. Dickson, editor. The wild turkey: biology and management. Stackpole, Mechanicsburg, Pennsylvania, USA.

Beck, J. R., and D. O. Beck. 1955. A method for nutritional evaluation of wildlife foods. Journal of Wildlife Management 19:198–205.

Bray, D. B., and M. B. Wexler. 1996. Forest policies in Mexico. Pages 217–228 *in* R. Randall, editor. Changing structure in Mexico. M. E. Sharpe, Armonk, New York, USA.

Butler, M. J. 2006. Evaluation of survey techniques for Rio Grande wild turkey populations in the southern Great Plains. Dissertation, Texas Tech University, Lubbock, USA.

Calmé, S., and M. Sanvicente. 2000. Distribución actual, estado poblacional y evaluación del estado de protección del pavo ocelado (*Agriocharis ocellata*). El Colegio de la Frontera Sur, Chetumal, Mexico. http://www.conabio.gob.mx/institucion/proyectos/resultados/InfR114.pdf. Accessed 20 Oct 2010.

Carrillo-Reyes, A. 2003. Uso de hábitat del guajalote

silvestre (*Meleagris gallopavo intermedia*) en Lampazos de Naranjo, Nuevo León. Thesis, UANL, San Nicolás de los Garza, México.

Cathey, J. C., K. Melton, J. Dreibelbis, B. Cavney, S. L. Locke, S. J. DeMaso, T. W. Schwertner, and B. Collier. 2004. Rio Grande wild turkey in Texas: biology and management. Texas Cooperative Extension, Texas A&M University, College Station, USA.

Chapman, F. M. 1896. Notes on birds observed in Yucatan. Bulletin of the American Museum of Natural History 8:271–290.

Cook, R. L. 1972. A study of nesting turkeys in the Edwards Plateau of Texas. Proceedings of the Annual Conference of the Southeastern Association of Game and Fish Commissioners 26:236–244.

Dalke, P. D., A. S. Leopold, and D. L. Spencer. 1946. The ecology and management of the wild turkey in Missouri. Technical Bulletin 1. Missouri Conservation Commission, Jefferson City, USA.

Davis, H. E. 1949. The American wild turkey. Small-Arms Technical, Georgetown, South Carolina, USA.

Diefenbach, D. R., and W. C. Vreeland. 2010. Annual survival and spring harvest rates of male wild turkeys in New York, Ohio, and Pennsylvania. Pennsylvania Cooperative Fish and Wildlife Research Unit, Pennsylvania State University, University Park, USA.

DOF. 2002. Norma Oficial Mexicana NOM-059-ECOL-2001. Protección ambiental—Especies nativas de México de flora y fauna silvestres—Categorías de riesgo y especificaciones para su inclusión, exclusión o cambio—Lista de especies en riesgo. Miércoles 6 de marzo de 2002 (Segunda Sección), SEMARNAT, México, D. F. www.ine.gob.mx/ueajei/norma59a.html. Accessed 15 Jun 2011.

DOF. 2010. Norma Oficial Mexicana NOM-059-ECOL-2010. Protección ambiental—Especies nativas de México de flora y fauna silvestres—Categorías de riesgo y especificaciones para su inclusión, exclusión o cambio—Lista de especies en riesgo. Jueves 30 de diciembre de 2010 (Segunda Sección), SEMARNAT, México, D. F. www.ine.gob.mx/ueajei/norma59a.html. Accessed 15 Jun 2011.

Dreibelbis, J. Z., K. B. Melton, R. Aguirre, B. A. Collier, J. Hardin, N. J. Silvy, and M. J. Peterson. 2008. Predation of Rio Grande wild turkey nests on the Edwards Plateau, Texas. Wilson Bulletin 120:906–910.

Everett, D. D., D. W. Speake, and A. K. Maddox. 1980. Natality and mortality of a north Alabama wild turkey population. Proceedings of the National Wild Turkey Symposium 4:117–126.

Figert, D. E. 1989. Status, reproduction, and habitat use of Gould's turkeys in the Peloncillo Mountains of New Mexico. Thesis, New Mexico State University, Las Cruces, USA.

Flannery, K. V. 1967. Vertebrate fauna and hunting practices. Pages 132–177 *in* D. S. Byers, editor. Prehistory of the Tehuacan Valley. Volume 1. Environment and subsistence. University of Texas Press, Austin, USA.

García-Marmolejo, G., G. Escalona-Segura, and H. Van der Wal. 2008. Multicriteria evaluation of wildlife management units in Campeche, Mexico. Journal of Wildlife Management 72:1194–1202.

Garza, A., and J. Servín. 1993. Estimación de la población y utilización del hábitat del cócono silvestre (*Meleagris gallopavo*, Aves: Phasianidae) en Durango, México. Ecología Austral 3:15–23.

Garza-Herrera, A., and E. E. Aragón-Piña. 2011. Conceptos ecológicos, métodos y técnicas para la conservación, manejo y aprovechamiento del cócono o guajolote silvestre. Pages 193–227 *in* O. Sánchez, P. Zamorano, E. Peters, and H. Moya, editors. Temas sobre conservación de vertebrados silvestres en México. SEMARNAT, México, D. F., México.

Gaumer, G. F. 1883. Notes on *Meleagris ocellata*, Cuvier. Transactions of the Annual Meetings of the Kansas Academy of Sciences 8:60–62.

Glazener, W. C. 1967. Management of the Rio Grande turkey. Pages 453–492 *in* O. H. Hewitt, editor. The wild turkey and its management. The Wildlife Society, Washington, DC, USA.

Glidden, J. W., and D. E. Austin. 1975. Natality and mortality of wild turkey poults in southwestern New York. Proceedings of the National Wild Turkey Symposium 3:48–54.

Gonzalez, M. J., H. B. Quigley, and C. I. Taylor. 1996. Habitat use, reproductive behavior, and survival of the ocellated turkey in Tikal National Park, Guatemala. Proceedings of the National Wild Turkey Symposium 7:193–199.

Gonzalez, M. J., H. B. Quigley, and C. I. Taylor. 1998. Habitat use and reproductive ecology of the ocellated turkey in Tikal National Park, Guatemala. Wilson Bulletin 110:505–510.

Götz, C. M. 2008. Coastal and inland patterns of faunal exploitation in the prehispanic northern Maya lowlands. Quaternary International 191:154–169.

Hall, G. I., M. J. Butler, M. C. Wallace, W. B. Ballard, D. C. Ruthven, R. L. Houchin, R. T. Huffman, R. S. Phillips, and R. D. Applegate. 2006. Rio Grande wild turkey home ranges in the southern Great Plains. Proceedings of the Annual Conference of the Southeastern Association of Fish and Wildlife Agencies 60:36–42.

Hamblin, N. 1984. Animal use by the Cozumel Maya. University of Arizona Press, Tucson, USA.

Hoffman, R. W., H. G. Shaw, M. A. Rumble, B. F. Wakeling, C. M. Mollohan, S. D. Schemnitz, R. Engel-Wilson, and D. S. Hengel. 1993. Management guidelines for Merriam's wild turkeys. Colorado Division of Wildlife, Denver, USA.

Howell, S. N. G., and S. Webb. 1995. A guide to the birds of Mexico and northern Central America. Oxford University, New York, New York, USA.

Hughes, T. W., J. L. Tapley, J. E. Kennamer, and C. P. Lehman. 2005. The impacts of predation on wild turkeys. Proceedings of the National Wild Turkey Symposium 9:117–126.

Kampichler, C., S. Calmé, H. Weissenberger, and S. L. Arriaga-Weiss. 2010. Indication of a species in an extinction vortex: the ocellated turkey on the Yucatán Peninsula, Mexico. Acta Oecologica 36:561–568.

Kienzler, J. M., T. W. Little, and W. A. Fuller. 1996. Effects of weather, incubation, and hunting on gobbling activity in wild turkeys. Proceedings of the National Wild Turkey Symposium 7:61–68.

Korschgen, L. J. 1967. Feeding habits and food. Pages 137–198 *in* O. H. Hewitt, editor. The wild turkey and its management. The Wildlife Society, Washington, DC, USA.

Krakauer, A. H. 2008. Sexual selection and the genetic mating system of wild turkeys. Condor 110:1–12.

Lafón, A., and S. D. Schemnitz. 1996. Distribution, habitat use, and limiting factors of Gould's turkey in Chihuahua, Mexico. Proceedings of the National Wild Turkey Symposium 7:185–191.

Lafón-Terrazas, A. 1997. Distribution, habitat use, and ecology of Gould's turkey in Chihuahua, Mexico. Dissertation, New Mexico State University, Las Cruces, USA.

Leopold, A. S. 1948. The wild turkeys of Mexico. Transactions North American Wildlife Conference 13:393–400.

Leopold, A. S. 1959. Wildlife of Mexico: the game birds and mammals. University of California Press, Berkeley, USA.

Lewis, J. C. 1967. Physical characteristics and physiology. Pages 45–72 *in* O. H. Hewitt, editor. The wild turkey and its management. The Wildlife Society, Washington, DC, USA.

Ligon, J. S. 1946. History and management of Merriam's wild turkey. New Mexico Game and Fish Commission, Santa Fe, USA.

Little, T. W., J. M. Kienzler, and G. A. Hanson. 1990. Effects of fall either-sex hunting on survival in an Iowa wild turkey population. Proceedings of the National Wild Turkey Symposium 6:119–125.

Márquez-Olivas, M., E. Garcia-Moya, C. Gonzáles-Rebeles Islas, and L. A. Tarango-Arámbula. 2005. Diet composition of reintroduced wild turkey (*Meleagris gallopavo mexicana*, Gould, 1856) in Sierra Fria, Aguascalientes, Mexico. Veterinaria México 63:395–409.

Márquez-Olivas, M., E. Garcia-Moya, C. Gonzáles-Rebeles Islas, and H. Vaquera-Huerta. 2007. Roost site characteristics of wild turkey (*Meleagris gallopavo mexicana*) in Sierra Fria, Aguascalientes, Mexico. Revista Mexicana de Biodiversidad 78:163–173.

McRoberts, J. T. 2014. Investigations into the ecology and management of ocellated turkeys in Campeche, Mexico. Dissertation, Texas Tech University, Lubbock, USA.

Melton, K. B. 2007. Reproductive ecology of Rio Grande wild turkey in the Edwards Plateau of Texas. Thesis, Texas A&M University, College Station, USA.

Merrill, L. B. 1975. Effect of grazing management practices on wild turkey habitat. Proceedings National Wild Turkey Symposium 3:108–112.

Mock, K. E, T. C. Theimer, O. E. Rhodes Jr., D. L. Greenberg, and P. Keim. 2002. Genetic variation across the historical range of the wild turkey (*Meleagris gallopavo*). Molecular Ecology 11:643–657.

Mosby, H. S. 1967. Population dynamics. Pages 113–198 *in* O. H. Hewitt, editor. The wild turkey and its management. The Wildlife Society, Washington, DC, USA.

Palmer, W. A., J. E. Stys, D. R. Smith, and J. D. Burk. 1993. Survival rates of wild turkey hens in loblolly pine plantations in Mississippi. Journal of Wildlife Management 57:783–789.

Petersen, B. E. 2007. Rio Grande wild turkey diets in the Texas panhandle. Thesis, Texas Tech University, Lubbock, USA.

Phillips, J. C. 1911. A year's collecting in the state of Tamaulipas, Mexico. Auk 4:219.

Phillips, R. S. 2004. Movements, survival, and reproduction of Rio Grande wild turkeys in the Texas Panhandle. Thesis, Texas Tech University, Lubbock, USA.

Prescott, W. H. 1843. History of the conquest of Mexico and history of the conquest of Peru. Modern Library, New York, New York, USA. (Reprint 1979)

Schemnitz, S. D., D. E. Figert, and R. C. Willging. 1990. Ecology and management of Gould's turkeys in southwestern New Mexico. Proceedings of the National Wild Turkey Symposium 6:72–83.

Schemnitz, S. D., and M. L. Zornes. 1995. Management practices to benefit Gould's turkeys in the Peloncillo Mountains, New Mexico. Pages 461–464 *in* L. F. DeBano, editor. Biodiversity and management of the Madrean Archipelago: the sky islands of southwestern United States and northwestern Mexico. Diane Publishing, Darby, Pennsylvania, USA.

Schorger, A. W. 1966. The wild turkey: its history and domestication. University of Oklahoma Press, Norman, USA.

SEMARNAT. 2008. Plan de manejo tipo para guajalote silvestre. Dirección General de Vida Silvestre, México, D. F. http://www.semarnat.gob.mx. Accessed 18 Jul 2011.

SEMARNAT. 2009. Plan de manejo tipo de pavo ocelado (*Meleagris ocellata*) en vida libre. Dirección General de Vida Silvestre, México, D. F. http://www.semarnat.gob.mx. Accessed 18 Jul 2011.

SEMARNAT. 2011. Épocas hábiles de aprovechamiento de aves y mamíferos silvestres para la temporada *2011–2012*. Dirección General de Vida Silvestre, México, D. F. http://www.semarnat. gob.mx. Accessed 15 Jun 2011.

Speake, D. W., R. Metzler, and J. McGlincy. 1985. Mortality of wild turkey poults in northern Alabama. Journal of Wildlife Management 49:472–474.

Speller, C. F., B. M. Kemp, S. D. Wyatt, C. Monroe, W. D. Lipe, U. M. Arndt, and D. Y. Yang. 2010. Ancient mitochondrial DNA analysis reveals complexity of indigenous North American turkey domestication. Proceedings of the National Academy of Sciences 107:2807–2812.

Steadman, D. W. 1980. A review of the osteology and paleontology of turkeys (Aves: Meleagridinae). Natural History Museum of Los Angeles County, Contributions in Science 330:131–207.

Steadman, D. W., J. Stull, and S. W. Eaton. 1979. Natural history of the ocellated turkey. World Pheasant Association 4:15–37.

Sugihara, G., and K. Heston. 1981. Field notes on winter flocks of the ocellated turkey (*Agriocharis ocellata*). Auk 98:396–398.

Taylor, W. P. 1951. Notes on the Rio Grande wild turkey in central Texas. Proceedings of the Oklahoma Academy of Sciences 1949:110–114.

Thomas, J. W., C. V. Hoozer, and R. G. Marburger. 1966. Wintering concentrations and seasonal shifts in range in the Rio Grande turkey. Journal of Wildlife Management 30:34–49.

Thornton, E. K., K. F. Emery, D. W. Steadman, C. Speller, R. Matheny, and D. Yang. 2012. Earliest Mexican turkeys (*Meleagris gallopavo*) in the Maya Region: implications for pre-Hispanic animal trade and the timing of turkey domestication. PLoS ONE 7:e42630. http://www.plosone.org/article/info%3Adoi%2F10.1371%2Fjournal.pone.0042630. Accessed 20 Aug 2012.

Thorstrom, R. 2000. The food habits of sympatric forest-falcons during the breeding season in northeastern Guatemala. Journal of Raptor Research 34:196–202.

Tozzer, A. M., and G. M. Allen. 1910. Animal figures in the Maya codices. Peabody Museum Papers 4:283–372.

Valdez, R., J. C. Guzmán-Aranda, F. J. Abarca, L. A. Tarango-Arámbula, and F. C. Sánchez. 2006. Wildlife conservation and management in Mexico. Wildlife Society Bulletin 34: 270–282.

Vangilder, L. D. 1992. Population dynamics. Pages 144–164 *in* J. G. Dickson, editor. The wild turkey: biology and management. Stackpole, Harrisburg, Pennsylvania, USA.

Vangilder, L. D., and E. W. Kurzejeski. 1995. Population ecology of the eastern wild turkey in northern Missouri. Wildlife Monographs 130:3–50.

Van Tyne, J. 1935. The birds of northern Peten, Guatemala. Miscellaneous Publications Number

27. Museum of Zoology, University of Michigan, Ann Arbor, USA.

Walker, E. A. 1948. Rio Grande turkey in the Edwards Plateau of Texas. Federal Aid Project 22–R, 1946–1948. Texas Game, Fish, and Oyster Commission, Austin, USA.

Weber, M., G. García-Marmolejo, and R. Reyna-Hurtado. 2006. The tragedy of the commons: wildlife management units in southeastern Mexico. Wildlife Society Bulletin 34:1480–1488.

Wigley, B. T., J. M. Sweeney, and M. E. Garner. 1986. Wild turkey home ranges in the Ouachita Mountains. Journal of Wildlife Management 50:540–544.

Williams, L. E., Jr. 1961. Notes on wing molt in the yearling wild turkey. Journal of Wildlife Management 25:439–440.

Williams, L. E., Jr. 1966. Wild turkeys with supernumerary leg spurs. Auk 84:113–114.

Williams, L. E., Jr. 1981. The book of the wild turkey. Winchester Press, Tulsa, Oklahoma, USA.

Williams, L. E., Jr. 1984. The voice and vocabulary of the wild turkey. Real Turkeys, Gainesville, Florida, USA.

Williams, L. E., Jr. 2002. Hunting the Gould's wild turkey in Mexico. Real Turkeys, Cedar Key, Florida, USA.

Williams, L. E., Jr. 2006. Wild turkey hunting and management. Real Turkeys, Cedar Key, Florida, USA.

Williams, L. E., Jr., and D. H. Austin. 1988. Studies of the wild turkey in Florida. Florida Game and Fresh Water Commission, Technical Bulletin Number 10. University Press of Florida, Gainesville, USA.

Williams, L. E., Jr., E. H. Baur, and N. F. Eichholz. 2010. The ocellated turkey in the land of the Maya. Real Turkeys, Cedar Key, Florida, USA.

York, D. L. 1991. Habitat use, diet, movements and home range of Gould's turkey in the Peloncillo Mountains of New Mexico. Thesis, New Mexico State University, Las Cruces, USA.

York, D. L., and S. D. Schemnitz. 2003. Home range, habitat use, and diet of Gould's turkey, Peloncillo Mountains, New Mexico. Southwestern Naturalist 48:231–240.

Zeedyk, B., and V. Clothier. 2009. Let the water do the work: induced meandering, an evolving method for restoring incised channels. Quivira Coalition, Santa Fe, New Mexico, USA.

Zornes, M. L. 1993. Ecology and habitat evaluation of Gould's wild turkeys in the Peloncillo Mountains, New Mexico. Thesis, New Mexico State University, Las Cruces, USA.

10 Pigeons and Doves in Mexico

Nova J. Silvy and Roel R. Lopez

Pigeons and doves are classified in the family Columbidae and are distributed worldwide, except in polar regions; they are most abundant in temperate and tropical zones. There are more than 300 species, depending on the authority consulted (Peters 1961, Grzimek 1975, Goodwin 1983, Gibbs et al. 2001). Pigeons and doves are stout-bodied birds with short necks, fleshy legs, and short, slender bills with a fleshy cere covering the base of the bill, which contains the nares (Baskett et al. 1993, Atherton et al. 1982). The feral rock pigeon (*Columba livia*), common in American cities, is referred to as a pigeon in the US.

Columbids use sticks to build relatively flimsy nests, which may be placed in trees, on ledges, on the ground, or on human structures in urban areas, depending on species preferences (fig. 10.1). They lay one or two eggs, and both parents care for the young, which leave the nest after 7–28 days, depending on the species. Most feed predominantly on seeds and fruit. Unlike most other birds, doves and pigeons produce "crop milk," which is secreted by a sloughing of fluid-filled cells from the lining of the crop, a muscular pouch that is an extension of a bird's esophagus used to store excess food prior to digestion. Both sexes produce this highly nutritious substance to feed the young.

Figure 10.1. White-tipped dove nestling in nest. Photo by Jordan Giese.

Overall, columbids are highly maneuverable in flight. Pigeons and doves are either arboreal (feeding in trees) or terrestrial (feeding on the ground). Most species in Mexico are terrestrial. Worldwide, columbids vary greatly in size: the largest (up to 74 cm; 29 in) are the ground-dwelling crowned pigeons (*Goura* spp.) of New Guinea, and the smallest is the plain-breasted ground dove (*Columbina minuta*) of the New World (14–16 cm; 5.1–6.3 in) (Baskett et al. 1993, Aldrich 1993, Gibbs et al. 2001).

Plumage coloration is variable. Granivorous species tend to have dull plumage, with a few exceptions, whereas frugivorous species have brightly colored plumage. *Ptilinopus* fruit doves are some of the brightest-colored pigeons, but most pigeons and doves have subdued colors (Aldrich 1993). Males and females may have similarly colored plumage or be sexually dimorphic. In addition to bright colors, pigeons may sport crests or other ornamentation (Grzimek 1975, Goodwin 1983, Gibbs et al. 2001, Wells and Wells 2001).

In Mexico, columbids are some of the most common birds. They are widespread, and several species occur in large numbers, making them readily visible in and around cities and rural countryside. People admire their esthetic qualities, pleasing cooing vocalizations, and the excellent sport-hunting opportunities they provide. Hunters from Mexico and the US are particularly interested in the mourning dove and white-winged dove. Hunters also can take red-billed and band-tailed pigeons. These birds are important economically since numerous foreign and native hunters pay to hunt these challenging game birds, and they also are an important source of meat protein for rural communities.

Taxonomy and Evolution

The terms "pigeon" and "dove" are usually used to represent the larger and smaller species, respectively,

Figure 10.2. Eurasian collared dove. Photo by Michael Vaughn.

but there is no taxonomic basis for distinguishing between them. Columbids are classified in the order Columbiformes, which includes the sandgrouse (family Pteroclididae) and the extinct dodos (family Raphidae). Dodos are considered distant relatives of pigeons and doves (Goodwin 1983, Gibbs et al. 2001). The first pigeons probably evolved in Australasia; however, columbids are not well represented in the fossil record, and no truly primitive forms have been found. Fossils of an Early Miocene pigeon were found in New Zealand, but most if not all fossils belong to extant genera (Grzimek 1975).

In Mexico, there are 22 native extant, 3 nonnative, and 2 extinct columbid species. The 3 nonnative species are the (1) rock pigeon, (2) Eurasian collared dove (*Streptopelia decaocto*) (fig. 10.2), and (3) spotted dove (*S. chinensis*). The rock pigeon (advertised by sport-hunting outfitters as the blue rock pigeon) and Eurasian collared dove are large enough and locally abundant enough to be hunted; however, they are more common in urban areas. The two extinct species are the passenger pigeon (*Ectopistes migratorius*) and Socorro dove (*Zenaida graysoni*), which is extinct in the wild. The passenger pigeon, which became extinct a century ago, was the largest native pigeon of Mexico. This species was recorded from Mexico City, Puebla, Veracruz, and Tabasco; two mounted specimens in the National Museum of Natural History in Mexico City are labeled "Mexico." The Socorro dove was endemic to small Pacific islands off Mexico's western coast; there are about 200 in captivity, and

Figure 10.3. White-crowned pigeon. Photo by Michael Vaughn.

efforts are underway to reintroduce it to its native habitat (Leopold 1959, Howell and Webb 1995, Butchart et al. 2006).

The 22 extant native columbids in Mexico include six pigeons, six ground doves, seven doves, and three quail doves (table 10.1). Pigeons (genus *Columba*) include six large species (28–38 cm, 11–15 in) that are in demand for sport and subsistence hunting. The white-crowned pigeon (fig. 10.3) is the rarest of the six species and is found mostly in the Yucatán. The six ground doves (genera *Columbina* [four species] and *Claravis* [two species]) are small and have reddish eyes and pinkish legs. The seven doves are classified in two genera (*Zenaida* and *Leptotila*), and the three quail doves in one genus (*Geotrygon*). The three dove species in the genus *Zenaida* are medium sized and

Table 10.1. Species of native pigeons and doves occurring in Mexico, with English and Spanish common names.

Species[a]	English name	Spanish name[b]
Pigeons		
Columba fasciata	Band-tailed pigeon	Paloma de collar
C. cayennensis	Pale-vented pigeon	Paloma colorada
C. leucocephala	White-crowned pigeon	Paloma corona blanca
C. flavirostris	Red-billed pigeon	Paloma morada
C. nigrirostris	Short-billed pigeon	Paloma triste
C. speciosa	Scaled pigeon	Paloma escamosa
Ground doves		
Columbina inca	Inca dove	Tórtola cola larga
C. passerina	Common ground dove (fig. 10.4)	Tórtola coquita
C. talpacoti	Ruddy ground dove	Tórtola rojiza
C. minuta	Plain-breasted ground dove	Tórtola pecho liso
Claravis mondetoura	Maroon-chested ground dove	Tórtola pecho morado
C. pretiosa	Blue ground dove	Tórtola azul
Doves and quail doves		
Zenaida asiatica	White-winged dove	Paloma ala blanca
Z. macroura	Mourning dove	Paloma huilota
Z. aurita	Zenaida dove	Paloma aurita
Leptotila verreauxi	White-tipped dove	Paloma arroyera
L. jamaicensis	Caribbean dove	Paloma caribeña
L. plumbeiceps	Gray-headed dove	Paloma cabeza ploma
L. cassinii	Gray-chested dove	Paloma pecho gris
Geotrygon montana	Ruddy quail dove	Paloma-perdiz rojiza
G. albifacies	White-faced quail dove	Paloma-perdiz cara blanca
G. lawrencii	Purple-backed quail dove	Paloma-perdiz tuxtleña

[a] Taxonomy based on Howell and Webb (1995).
[b] Spanish names based on Escalante et al. (2014).

Figure 10.4. Common ground dove. Photo by Oscar Perez.

have dark gray bills, pale blue orbital rings, and pinkish legs. They include two of the most widespread and numerous species in Mexico: white-winged dove (*Z. asiatica*) and mourning dove (*Z. macroura*). The four dove species in the genus *Leptotila* have yellowish eyes, dark reddish orbital rings, blackish bills, and red to pinkish-red legs. The three quail doves are wary and favor shady forest understory (Howell and Webb 1995).

Leopold (1959), based on his field studies in the late 1940s, described 12 species large enough and abundant enough to be hunted, some of them heavily. These included the (1) red-billed pigeon, (2) band-tailed pigeon, (3) scaled pigeon, (4) short-billed pigeon, (5) mourning dove, (6) white-winged dove, (7) white-tipped dove, (8) gray-headed dove, (9) Caribbean dove, (10) gray-chested dove, (11) ruddy quail dove, and (12) white-faced quail dove.

Mourning and white-winged doves have been intensively studied, and red-billed and band-tailed pigeons to a lesser degree, but there is substantial published information on the natural history and management of these four species. There is limited information on all the other species. Therefore, we discuss in detail only the mourning dove, white-winged dove, red-billed pigeon, and band-tailed pigeon. For specific information on the life histories and distributions of all native columbids in Mexico, including color illustrations, refer to Howell and Webb (1995) and Gibbs et al. (2001).

Red-Billed Pigeon

Description

The red-billed pigeon (fig. 10.5) is similar in size to the rock dove and is a dark grayish or bluish gray, with dull brownish gray on the mantle and scapulars and light gray under the wing. The head, neck, and breast are drab purple or dull reddish purple, and the tail is a dark slate gray to grayish black. Eyes are orange to red and eyelids are red; the bill is pale with a red base; and the legs and feet are dark red. Sexes are similar, but females are slightly duller than males; males have more gray on top of the head, and reddish brown on the lesser wing coverts. They weigh 241–336 g (8.1–15.0 oz) (Dunning 1992).

Figure 10.5. Red-billed pigeon. Photo by Paul van Giersbergen.

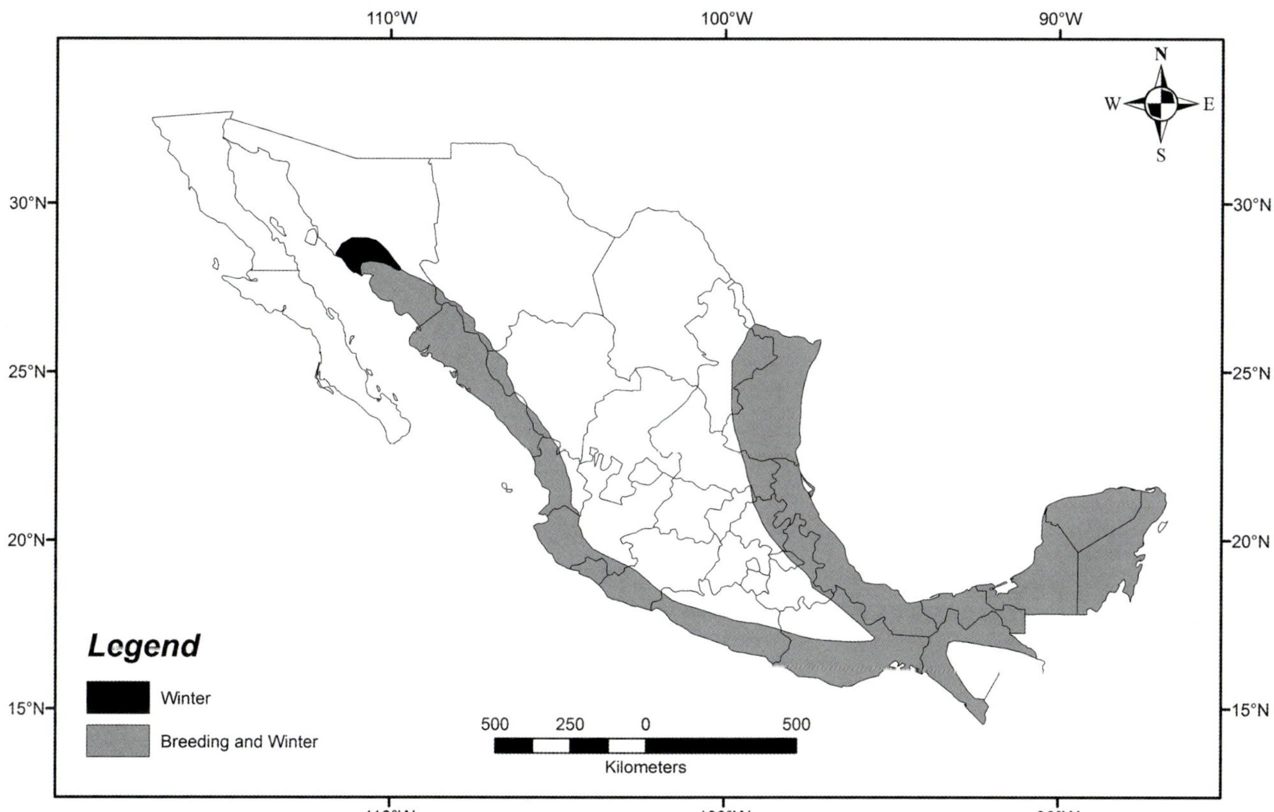

Figure 10.6. Distribution of red-billed pigeon in Mexico. After Lowther (2002).

Range in Mexico

The red-billed pigeon occurs along both coasts of Mexico and in central Sonora, southern Chihuahua, Durango, San Luis Potosí, and Nuevo León, and north to southernmost Texas (fig. 10.6). It also occurs in coastal plains and foothills from central Sonora south to Chiapas and from northern Tamaulipas south to the Yucatán Peninsula (Leopold 1959, Howell and Webb 1995, Russell and Monson 1998). Of the several tropical pigeons, the red-billed pigeon is the most widely distributed in Mexico and is locally abundant (Leopold 1959).

Population ecology

Between 1,000 and 3,000 red-billed pigeons have been declared annually by US hunters returning from Mexico (del Hoyo et al. 1997). In Sonora, vast fields of sorghum, corn, and sesame seed attract thousands of mourning and white-winged doves, and it is possible to collect red-billed pigeons in this region. There is little information on this species in Texas and no information on populations in Mexico (Lowther 2002). It has declined dramatically in the lower Rio Grande region, where it is uncommon and in danger of extirpation, with only 30 birds documented (Brush 1998). Local extirpations have occurred in Texas (Brush and Cantu 1998). In the evergreen forest floodplain of the Río Corona in central Tamaulipas, censuses during 1973–1974 of 8 ha (19.7 ac) plots totaled five pairs in January; three pairs in April (breeding noted); and two pairs in June (Gehlbach et al. 1976, Gehlbach 1987).

By August, birds in Mexico form winter flocks to search for food, but there are no long seasonal migrations (Leopold 1959). Martin et al. (1954) noted red-billed pigeons at all elevations in the pine-oak forests of the Sierra de Tamaulipas in April and June 1949. Birds are resident throughout their range except in Sonora, where they move northward along the coast and farther inland during fall (Russell and Monson 1998). The northernmost populations in Texas and northeastern Mexico are migratory (AOU 1998). In Mexico, the red-billed pigeon inhabits forest patches, but in more humid regions it forages in open pastureland, agricultural fields, and areas with scattered trees, tall brush, and bamboo (Sutton and Pettingill 1942, Peterson and Chalif 1973, Terres 1980, AOU 1983, DeGraaf et al. 1991, Brush 1998).

Diet

Red-billed pigeons forage in trees and feed mostly on fleshy fruits such as guava (*Psidium* spp.) and possum grape (*Cissus incisa*), berries, acorns (*Quercus* spp.), tree buds, and large quantities of seeds in winter (Bent 1932, Leopold 1959, Oberholser 1974, Terres 1980, Goodwin 1983, Stiles and Skutch 1989, DeGraaf et al. 1991, Brush 1998), but they will drink and consume salt on the ground (Dickey and van Rossem 1938) and forage in cultivated fields (Skutch 1964, Stiles and Skutch 1989, Russell and Monson 1998). They also are known to aggregate around salt sources (Dickey and van Rossem 1938). Like other pigeons and doves, they drink water by sucking it from rivers at sandbars (Sennett 1879, Lowery and Dalquest 1951, Oberholser 1974) or from pools in intermittent streams (Stager 1957).

In northeastern Mexico, the summer diet was dominated by the purplish berry of the logwood tree (*Condalia obovata*) and wild grapes (*Vitis cinerea*), and in December by shriveled berries from a grove of corongorro trees (*Zizyphus sonorensis*) (Leopold 1959). Pigeons ate palm seeds and figs (*Ficus* spp.) (Wetmore 1943), and ripe fruits from groves of nanchi trees (*Byrsonima crassifolia*) (Lowery and Dalquest 1951) in southern Veracruz. Leopold (1959), in Nuevo León, found seeds of corn and of Johnson grass (*Sorghum halepense*) in specimens killed in winter; in eastern San Luis Potosí, he noted hundreds of pigeons feeding in cornfields. They are known to forage in grain stubble, corn, and sorghum fields.

Reproduction and behavior

Little research has been done on the breeding ecology of the red-billed pigeon. It breeds in the lower and middle Rio Grande Valley in South Texas and adjacent Tamaulipas, Mexico, and from Mexico's Río Corona south into Central America (Brush 1998, DeGraaf et al. 1991). Breeding populations extend to both the east and west coasts of Mexico (del Hoyo et al. 1997). Lowther (2002) noted that nests have been found in Mexico from March to August, indicating multiple broods per year. During the nuptial flight, the male makes a circle with deep, slow wing strokes, slapping its wings together above and below as it flies, and returns to alight near the female (Leopold 1959). Nests are flimsy platforms of sticks, built on horizontal branches or in crotches of trees or bushes. The male brings nest materials to the female, who constructs the nest. Leopold (1959), Eitniear and Breeden (2003), and Eitniear and Aragon-Tapia (2005) noted that a single white egg was a normal clutch; however, clutches of two eggs have been reported. The breeding season may be prolonged, with several broods per year (Bent 1932). Both sexes incubate eggs and tend young (Harrison 1978, Terres 1980, del Hoyo et al. 1997, Brush 1998). Leopold (1959) noted that fledged young closely resembled adults but were smaller and grayer and not as richly colored. Eitniear and Aragon-Tapia (2005) observed a red-billed pigeon nest being predated by a groove-billed ani (*Crotophaga sulcirostris*).

Population dynamics

There is no information on annual and lifetime reproductive success, number of broods reared per season, lifespan and survivorship, disease, and causes of mortality, other than hunting (Lowther 2002).

Deforestation and damage to riparian communities and agricultural irrigation projects have led to its decline in South Texas and adjoining Mexico (del Hoyo et al. 1997). Habitat loss and deterioration, as well as the effects of droughts and freezes on large forest trees, are probably responsible for the decline. Dewatering and the succession of riparian woodland and subtropical evergreen forest to thorn-scrub communities have deteriorated habitat quality (Brush and Cantu 1998). High nest predation rates, as inferred from improved breeding success on islands without predators, also may have caused some local population declines (Brush 1998).

Population management

Remnant populations of red-billed pigeons in the Lower Rio Grande Valley are vulnerable to extirpation from widespread degradation of riparian woodlands and subtropical evergreen forests, while some Mexican populations appear secure. Restoration of native habitats for the northern population will be most promising where water and land management can restore mature riparian forests (Brush 1998). Special habitat requirements include tall, dense brush with small patches of tall trees (DeGraaf et al. 1991), and in arid environments, mature woodlands near water (Terres 1980). In the north, wooded islands are favored nesting sites (Brush 1998).

Research is needed to determine site fidelity, home range, utilization of agricultural grains, nesting success, survivorship, population trends, effects of forest disturbances, and relationships between red-billed pigeon populations in Texas, Mexico, and elsewhere (Brush 1998). Foraging ecology in northern riparian forests where little fruit is available also needs further research (Dickey and Van Rossem 1938).

Band-Tailed Pigeon

Description

The head, throat, chest, and underparts of the band-tailed pigeon (fig. 10.7) are purplish drab, paling to white on the lower tail coverts, with a white crescent across the back of the neck. The foreback and hind neck are iridescent bronze. The back, wing coverts, rump, and tail are gray, the tail fading to a pale gray band at the tip. The wings are pale gray, with dark tips noticeable in flight. The yellow bill is tipped with black. The feet and legs are yellow. Both sexes are similar in size and appearance. Young birds are more grayish below, without a white crescent or iridescence on the back of the neck. Birds weigh 342–364 g (12–12.8 oz), with males averaging 7% larger than females (Keppie and Braun 2000).

Figure 10.7. Band-tailed pigeon. Photo by C. J. Randel.

Distribution in Mexico

The band-tailed pigeon occurs from eastern Sonora and western Chihuahua south through the broad Sierra Madre Occidental of western Mexico, and from southeastern Coahuila south through the Sierra Madre Oriental of eastern Mexico, as well as the Sierra Madre del Sur in the south, with all ranges converging in Oaxaca; its range continues south from the Sierra Madre of Chiapas (fig. 10.8) (Howell and Webb 1995). Leopold (1959) noted the band-tailed pigeon was associated with pine-oak highlands throughout Mexico, and flocks could be seen throughout the higher mountains from Baja California to Chiapas. Birds bred in most or all of this range in local colonies but were not uniformly distributed. During the fall, there was a southward influx of birds from the western US into northern Baja California and probably into Sonora and Chihuahua.

Population ecology

Diet

There are no known differences in diet between sexes or for immatures once they leave the nest. Band-tailed pigeons normally eat grain, fruit, acorns, pine nuts, and inflorescences of trees and shrubs (Neff 1947). Leopold (1959) noted that band-tailed pigeons eat mostly fruit, including pokeberries (*Phytolacca rugosa*), wild cherries (*Prunus serotina*),

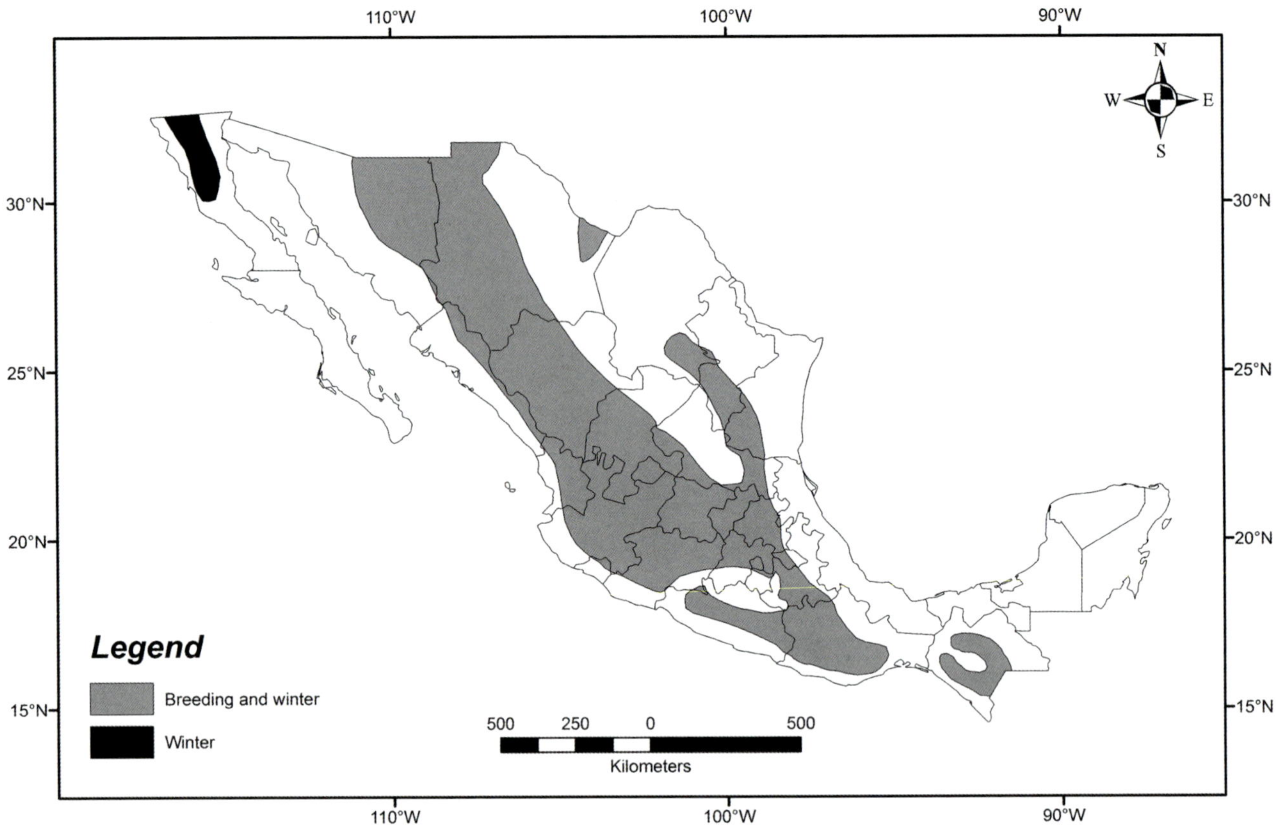

Figure 10.8. Distribution of band-tailed pigeon in Mexico. After Keppie and Braun (2000).

manzanita berries (*Arctostaphylos* spp.), and elderberries (*Sambucus* spp.) in the summer, and mast (acorns especially) in the winter. Grains may be most widely consumed in the US, and fruit and acorns are probably greatest in Mexico (Keppie and Braun 2000); however, Leopold (1959) observed birds eating wheat during summer months in Mexico.

Band-tailed pigeons are known to frequent mineral springs in autumn in Sonora to obtain sodium (Russell and Monson 1998). Leopold (1959) noted that in winter, during the dry season, band-tailed pigeons came to water daily, usually in midmorning. In the US, the greatest use of mineral sites was during summer to early autumn (March and Sadleir 1972, Sanders 2000). Radio-marked band-tailed pigeons in Oregon used mineral sites about once a week (Leonard 1998). Visits to mineral sites in other areas of the US were less common, but birds may have a more variable summer diet and may make greater use of mineral-rich grit (Keppie and Braun 2000).

Reproduction and behavior

Many adults are paired when they first arrive on nesting range; hence, birds may pair on winter range or during migration (Peeters 1962). The courtship display of a male includes loud and persistent cooing and a nuptial flight consisting of a high circle, usually 183 m (600 ft) or more in diameter, made with the wings extended horizontally, the tips making short, fast, fluttering strokes, after which the male returns to alight near the female (Leopold 1959). The nesting season extends from May to August. Nests are active most months of the year; an early nest with an egg was observed on 26 December in Baja California Sur (Thayer 1909), and the latest a nest was found was on 18 August near Pacheco, Chihuahua (Leopold 1959).

Nests are constructed over three to six days (Willard 1916, Peeters 1962, Silovsky 1969). Each pair of pigeons may complete several nesting cycles during the summer, and in western Oregon, the mean number of nests initiated was 1.69 per adult (Leonard 1998). Circumstantial evidence indicates that autumn nesting in the southern US varies with the abundance of acorns (Gutiérrez et al. 1975). Peeters (1962) noted the male may lead the female to a particular location in a tree, but the female brings most of the twigs to the nest. The nest is a saucer-shaped platform with loosely intertwined twigs, similar to the nest of the mourning dove. Eggs are white (40.7 × 28.5 mm, 1.6 × 1.1 in), and the normal clutch size is one egg (Keppie and Braun 2000). Although two eggs are common (Neff 1947), there is no evidence of three-egg clutches, even though some females contain three enlarged follicles (Gutiérrez et al. 1975). Incubation lasts about 18 days (16–22 days) (Campbell 1912, MacGregor and Smith 1955, Silovsky 1969). Both sexes share incubation, with the female incubating at night and the male during the day (Neff 1947, Peeters 1962, Skutch 1964, 1983). Young pigeons fledge after three to four weeks in the nest; adults may renest (Leopold 1959).

Nesting parents produced crop milk for about 30 days, from 2–3 days before hatching to 2–3 days after immatures left the nest. Experimental removal of captive adults showed that a single parent successfully fed a single squab 6–9 days posthatch (Zeigler 1971). Squabs fed three times daily during the first 5–7 days posthatch; thereafter, they fed one or two times daily. They may move from the nest onto a limb at 15 days posthatch, and fledge at 22–29 days (Neff and Niedrach 1946, MacGregor and Smith 1955, Peeters 1962). Young may remain near the nest in an adjacent tree for two days (Peeters 1962). Immatures at bait sites may beg in front of adults (Houston 1963), and adults frequently feed free-flying young (Keppie and Braun 2000).

Population dynamics

Leopold (1959) observed that band-tailed pigeons were not heavily hunted in Mexico, and most of the hunting was for subsistence. Locally, there was evidence of substantial killing at watering holes, as local hunters would wait at the water for birds that alighted to drink, killing several birds with each shot. There was little sport hunting for band-tailed pigeons in Mexico.

No data are available for daily survival rates for band-tailed pigeons in Mexico. In Oregon, daily survival rate during the 21-day incubation period was 0.98 (79% for 21 days), and 0.99 (89% over 24 days) for the nestling period (Leonard 1998, Keppie and Braun 2000). Estimated annual reproductive success was 1.26 immatures per female, with a nest survival rate of 69% (Leonard 1998). Nest success was similar in California (78% of 18 nests with eggs, 71% of 17 nests with eggs, 88% of 16 hatched eggs) (MacGregor and Smith 1955). In Oregon, Leonard (1998) observed that 62% females reared at least one young to the fledging stage. Eggs and squabs may fall from nests (MacGregor and Smith 1955, Silovsky 1969) because of strong winds.

The oldest band-tailed pigeon from Washington was at least 18 years old (Klimkiewicz and Futcher 1989), and a specimen from Oregon was close to 22 years old (Jarvis and Passmore 1992). Estimated mean daily survival of adults from May to September in Oregon was 0.99 (Leonard 1998). Mean annual adult survival from banded bird recoveries was 73% in Colorado (554 recoveries, 1969–1974) (Kautz and Braun 1981), and 64% in Oregon (1,008 recoveries, 1965–1973) (Jarvis and Passmore 1992). Estimated mean annual survival for immatures was 66% in Colorado (113 recoveries, 1969–1974), with estimates varying considerably among years (Kautz and Braun 1981).

The only disease affecting population numbers is trichomoniasis, caused by the protozoan *Trichomonas gallinae*. Of 265 birds tested in Colorado and Arizona, 10.9% were positive for the protozoan, but none showed evidence of disease (Stabler 1951, Sileo and Fitzhugh 1969). When Colorado band-tailed pigeons were inoculated with a virulent Pacific coast strain, 22 of 23 died an average of 15.4 days later (Stabler and Braun 1979). This virulent strain caused substantive losses in California: an estimated 15,000 individuals in 1988 (Braun 1994).

Although population changes in Mexico are unknown, studies in the US indicate declines. The band-tailed pigeon was more abundant on the US Pacific coast than in interior states: 2.9–7.1 million versus fewer than 250,000 in the early 1970s, estimated from harvest reports and band-recovery rates (Braun 1994, Keppie and Braun 2000). Populations decreased by 3% per year in nearly all breeding bird survey areas from 1966 to 1998 (Keppie and Braun 2000). Pigeons visiting mineral springs in Oregon have decreased 70% since 1950 (Jarvis and Passmore 1992, Keppie and Braun 2000).

Population management

This species is subjected to unregulated hunting in Mexico, Central America, and South America. Within the US, hunting is allowed in all states within the species' range except Texas (Case and Hughes 2011). Migrant birds from the US may experience heavier hunting mortality when wintering in Mexico and Central America. Glover (1953) was concerned that harvests in coastal forests of California would cause declines in band-tailed pigeons; however, Marcot (1984) noted high densities of pigeons in young shrub stands resulting from timber harvest. It is listed as a species of concern in Canada (Dunn et al. 1999), but not in the US. Its numbers apparently are greatly decreasing and may surpass a 50% reduction (Keppie and Braun 2000). Research and management recommendations have been proposed by Braun (1994), but lack of funds prevents implementation (Keppie and Braun 2000). Keppie and Braun (2000) noted that implementation of reliable methods to survey abundance was of greatest importance.

Mourning Dove

Description

The mourning dove is medium sized and has a long, graduated, pointed tail with a white tip. Tail and wing feathers are gray except for the black-bordered white tips on the tail. There are black spots on the wing coverts and near the ears (fig. 10.9). Eyes are brown, bordered by light blue bare skin. Bills are delicate and black. Feet and legs are dull red. Males are larger than females and are slightly more brightly colored, with a pinkish wash on the breast and a brighter blue-gray color on the top of the head and hind neck. Juveniles can be distinguished from adults by the light buff tips of their feathers, which give them a scaly appearance, and their age as determined by primary feather replacement up to 124 days of age (Swank 1955, Bivings and Silvy 1980, Morrow et al. 1985, 1992). Weight is 96–130 g (3.3–4.5 oz).

Distribution in Mexico

The mourning dove is the most abundant and widespread North American game bird (Grue 1977, Grue et al. 1982), and one of the most important game species in Mexico (Leopold 1959). Mourning doves offer quality sport hunting in the US and Mexico (Keeler 1977). They breed abundantly in northern and central Mexico, in Baja California to 26° N latitude, and in the interior to Jalisco and Hidalgo, sparingly to Oaxaca (Tamazulapan and Tlacolula). Their breeding range also includes the Revillagigedo and Tres Marias Islands of western Mexico (Aldrich 1993). Leopold (1959) noted they were much more numerous in winter along the coasts as well as inland, and south to the arid central valley of Chiapas. Their winter range covers all of Mexico (Aldrich 1993). Mourning doves started migrating into northern Mexico in September, but it was November before they reached Guerrero in any large numbers; the northward migration occurred mostly in March (Leopold 1959).

Population ecology

Leopold (1959) noted that a substantial breeding population of mourning doves in northern and central Mexico was augmented in winter by considerable migration from the US. Blankenship and Reeves (1970) suggested that about 7% of mourning doves breeding in the US wintered in Mexico. Seven percent of doves banded near Turlock, California, were recovered in Mexico (Channing 1979). Breeding mourning doves banded in South Dakota

Figure 10.9. Male (top) and female (bottom) mourning doves. Photos by Michael Vaughn.

migrated south through Texas to west-central Mexico (Dunks et al. 1982), and 5% of immature mourning doves banded in central Texas were recovered in Mexico (Bivings et al. 1984). About 13.2% of mourning doves banded in the Central Management Unit (CMU) of the US, 0.2% banded in the Eastern Management Unit (EMU), and 6.2% banded in the Western Management Unit (WMU) were recovered in Mexico. Of 902 direct recoveries in Mexico of doves banded in the US during 1967–1975, 76% originated from CMU states, 23% from WMU states, and fewer than 2% from EMU states (Tomlinson 1993). The most important areas for US doves harvested in Mexico were the Western Highlands (Jalisco, Michoacán, Guanajuato, Colima, and Guerrero), with 78% of recoveries from that region. The extensive agricultural development and water sources of the Western Highlands provided excellent habitat for wintering doves. The Northwest Coast region was the next most important harvest area (10%); however, most of the doves reported from this region were probably migrating farther south into the Western Highlands when harvested. Southern Sinaloa, which had irrigated farmland, was probably the only location where doves wintered in the region. The Central Highlands, which surround Mexico City, provided a moderate proportion of the Mexican harvest (6%), with few doves being recovered in the Northern Highlands and the Northeast Coast regions. No birds were recovered on the Yucatán Peninsula (Tomlinson 1993). Leopold (1959) noted the greatest concentrations of mourning doves in the arid, temperate uplands, and diminishing numbers toward the tropical coasts.

Diet

Mourning doves are primarily granivorous ground feeders (Lewis 1993). Seeds and other adult foods are added to the squab diet as they mature (fig. 10.10). By the time fledglings are 16 days old, feeding is done mainly by the male and often continues for 27 days posthatch (Hitchcock and Mirarchi 1984).

Figure 10.10. Two young mourning dove squabs in a nest waiting to be fed. Photo by Larry Marcy.

Fledglings begin feeding themselves by 17 days and can be independent by about 21 days. By the time the young begin flying, their diet is the same as that of adults (Lewis 1993). Fledglings tend to gather in loose groups apart from the breeding adults.

Mourning doves feed almost entirely on plant seeds, which they consume from the ground surface. About 1% of their diet is of animal origin and consists mostly of snails, which are ingested most frequently during the nesting season (Lewis 1993) and which probably fulfill the calcium requirement of the rapidly growing young. They prefer small seeds but often take corn kernels and, rarely, acorns. Leopold (1959) found a wide variety of seeds in dove crops; some of the dominant ones were pigweed, sunflower (*Helianthus annuus*), turkey mullein (*Croton setigerus*), doveweed (*Croton* spp.), rattlebox (*Crotalaria pumila*), beans (*Desmodium* spp.), various large-seeded grasses, and cultivated crops such as corn, wheat, beans, and sesame. The legs and feet of doves are small and weak, and they cannot scramble through thick weeds or grass but must move about on essentially bare ground. Perhaps for this reason they favor arid regions where ground cover is sparse and seeds that fall to the surface are easily gathered.

In Texas, doves fed mainly in or near agriculture fields (Dillon 1961). They obtained most of their foods from areas disturbed by cultivation, grazing, ditch and roadside maintenance, or similar practices that encouraged annuals and cultivated plants (Browning 1962).

Mourning doves require surface freshwater for drinking on a regular basis and fly to drink in the morning and evening after feeding (McClure 1943, Elder 1956, Slade 1969). They occupy desert habitats, but only where they can fly to surface water every few days. They can function for four to five days without water or succulent foods and can drink brackish water with up to 0.15 molar sodium chloride, which enables them to use saline springs in desert areas, but they will lose weight (Lewis 1993). Grit also is an essential component of the diet but functions primarily to grind seeds in the gizzard; salt may be necessary for proper egg viability (McClure 1943). Doves also eat hard seeds, which may supplement grit in the diet (Lewis 1993).

Reproduction and behavior

Mourning doves are monogamous and form strong bonds that persist for at least one nesting season (Sayre and Silvy 1993). Doves pair early and have a long nesting season, raising several broods a year. Their breeding season is among the longest of all North American birds, and dove nests are active every month of the year in the southernmost breeding range (Peters 1961). However, most nesting in the northern US begins in mid-March and ends in September, while nesting in southern states starts in February and continues into October (Sayre and Silvy 1993). Even though some adults may remain paired during winter, most doves are probably unpaired at the onset of the breeding season. Unpaired males during the breeding season devote considerable time to perch cooing and other displays to attract a mate, whereas paired males engage in nest construction, incubation during the day, and feeding young posthatch (Sayre and Silvy 1993).

Leopold (1959) noted that breeding doves were easily located, not only by their conspicuous movements to and from the nesting area but also by the persistent cooing of the male: a long, accented "coo" followed by two or three shorter notes at a lower pitch. The courtship flight of the male also was distinctive: a high circle with deep wing strokes that slapped the primaries together above and below, followed by a stately glide on fixed wings back to the female.

The male takes the initiative in nest site selection and may begin searching for a new site before the young of the previous nesting have fledged. During nest building, the male selects small twigs and delivers them individually to the female, who arranges them in the nesting platform (Nice 1922). Nest construction usually takes seven to ten hours but may take two to four days (Blockstein 1986). Time required for courtship, nest site selection, and nest construction varies from three to seven days (Cowan 1952, Mackey 1954). Mourning doves are determinate layers with a genetically controlled clutch size of two eggs; however, reports of dove nests containing more than two eggs are not uncommon (Moore and Pearson 1941, Quay 1951). Doves are unusual because they lay white eggs in open nests. White egg color may have been a selective factor for development of constant incubation (Westmoreland and Best 1986).

On average, mourning dove pairs require about 32 days (± 5) to complete a nesting cycle, including 3–7 days for courtship and nest building, 14–15 days for incubation, and 11–15 days for brooding (Swank

1955, Sayre and Silvy 1993). Unlike in many other bird species, the male and female share in incubation and brooding. Male doves typically incubate eggs from midmorning until late afternoon (fig. 10.11). Adults, particularly the male, feed the fledged young for 10–12 days but normally begin renesting within 2–5 days of the fledging date. Adult females may nest as often as five to six times in a single season in warm southern climates, but they average two to three attempts per season in northern states (George 1988).

An average of 3.6 young are produced annually per breeding pair in the US. Nest success rates varied from 0 to 82% and averaged 48% (Sayre and Silvy 1993). In treeless areas, most mourning doves build nests on the ground. Ground-nesting doves in Oklahoma had lower nest success (29%) than did tree-nesting doves (49%) in the same area (Downing 1959). However, ground-nesting doves were more successful (23%) than doves nesting in trees (16%) in Texas (Soutiere and Bolen 1976). Bivings and Silvy (1981) observed a 29% nesting success for doves nesting in urban areas of central Texas. In another study, nesting success in the same area was 26.6% (Munoz et al. 2008). Most seasonal nesting studies indicate increases in nest success as the season progresses (Sayre and Silvy 1993). Lowered nest success in the early months of the breeding season has been attributed to weather conditions, predation, and a greater likelihood of nest desertion. Bivings and Silvy (1994) also observed that most eggs in September were lost to weather. Based on 1,556 nests studied from February through October, 22% of eggs laid survived to fledging, with the greatest mortality occurring during the first week after laying (Morrow and Silvy 1982). Higher mortality was observed for eggs (56%) than for nestlings (45%), but there were seasonal differences. Early in the nesting season, egg mortality was higher than nestling mortality; this was reversed during July and September, when nestling mortality was higher. Total mortality differed among months. Except for February, when nesting cycle mortality was 68%, total mortality was high (82–85%) from July to September, declined to 64–70% from July to September, and increased to 82% in September and October.

When selecting nesting sites, mourning doves tend to scatter and use whatever sites are available. They are closely associated with farmlands but will nest in many vegetation types. Nests occur at the edges of

Figure 10.11. Female mourning dove on a nest in late afternoon. Photo by Michael Vaughn.

fields, pastures, or clearings but are seldom found in densely wooded areas. Some areas with high levels of habitat interspersion are selected for nesting. Nests are built in coniferous and deciduous trees, shrubs, and vines. Tree nesting is most common, but mourning doves also will nest on human structures, and on the ground in the absence of trees or shrubs. Nests are often several miles from water, and the parents have to fly that distance at least once a day to drink, and long distances to find feeding areas. Nesting habitat is not usually a limiting factor. Like nesting cover, roosting cover is provided by various vegetation types. These include coniferous and deciduous trees, brushy thickets, dead snags, and power-line rights-of-way. Hedgerows and shelterbelts also provide excellent roosting sites (Leopold 1959, Grue et al. 1982).

Population dynamics

Estimates of the nationwide population in the US in autumn ranged from 350 million to 600 million birds (Reeves 1993). Approximately 13.8 million mourning doves were harvested in the US during 2014 (Seamans 2015). The number of mourning doves harvested has steadily declined from an estimated 40–50 million birds taken during the 1980s (Baskett and Sayre 1993). The mourning dove is the primary North American game bird, with numbers harvested exceeding those of all other migratory game birds combined. The average lifespan of mourning doves banded in the US is about 1.0 year for immatures and 1.3 years for adults, but there is a longevity record of 19.3 years (Clapp et al. 1983). Despite being one of the 10 most abundant birds in the continental US (Robbins et al. 1986), mourning doves have been declining since 1966 (USFWS 2005). This also is probably true in Mexico.

Mourning dove populations experience high annual mortality rates, depending on geographical region: 50–75% of the population dies each year (Otis et al. 2008). Significant causes of mortality include predation, hunting, weather, disease, and chemical toxicosis, but the nature of mortality interactions is not understood (Otis et al. 2008). Mortality rates in the eastern US reached 58% per year for adults and 69% for immatures (Baskett and Sayre 1993). In central Texas, estimated daily survival for 36,735 mourning dove eggs ranged from 98.6% in February to 84.0% in October (Morrow and Silvy 1983). Estimated daily nestling survival for 24,356 nestlings ranged from 98.1% in February to 95.4% in October. Also in central Texas, based on 1,170 banded adults, 1,223 free-flying juveniles, and 1,338 nestlings, adult survival was 46.2%, juvenile survival was 18.0%, and nestling survival was 19.5% (Bivings et al. 1984).

Predators of adult mourning doves include falcons and hawks and, during nesting, corvids, grackles, blue jays (*Cyanocitta cristata*) (Morrow and Silvy 1982), fox squirrels (*Sciurus niger*), and house cats. Snakes prey on eggs and young. Several diseases and parasites affect mourning doves; however, only infections of trichomoniasis and avian pox can result in death by starvation (Conti 1993). The trichomoniasis flagellate resides in the mouth, esophagus, and crop and causes a cheesy, yellowish growth in the mouth and esophagus, making swallowing impossible (Schulz et al. 2005). Avian pox virus causes lesions on the skin and membranes of the mouth and upper respiratory tract, resulting in difficulty with seeing and eating. Avian pox is highly contagious and is transmitted by blood-sucking insects (Conti 1993).

Ingestion of lead shot by mourning doves can be a major problem on managed annual hunting fields in Mexico. In the US, ingestion rates for hunter-killed mourning doves range from 2% to 6.5% (Otis et al. 2008). Some wild doves ingested from 24 to 43 lead pellets (Schulz et al. 2002, Franson et al. 2009), and up to 92% of birds that ingested lead shot died; birds not dying were more prone to predation (Schulz et al. 2006). Schulz et al. (2006) estimated that 8.8–15 million doves may die each year from ingesting lead pellets.

Leopold (1959) noted the number of mourning doves killed for sport and subsistence hunting in Mexico was minor in view of their abundance. Mourning doves were taken not only by shotguns but also by clap nets and baited walk-in traps. Sadler (1993) estimated the annual mourning dove harvest in Mexico during the late 1970s and early 1980s to be about 3.2 million doves. There was weak evidence that hunting mortality was additive (Otis 2002).

Population management

Approximately 1 million US hunters annually harvest about 20 million birds (Dolton and Rau 2006), which represent about 7% of the fall population (Otis et al. 2008). Additional harvest occurs in Mexico and Central America (Sadler 1993), but its impacts on overall harvest are unknown.

In general, the primary management tool is hunting regulations, including season length, bag limits, and shooting hours. Hunting regulations in Mexico are set annually. Mourning dove hunting seasons, season lengths, and daily bag and possession limits vary from year to year. Since 1980, individual state hunting regulations have often been set by zones within states. Hunting is allowed in 28 states but prohibited in Quintana Roo and Yucatán, where mourning doves are scarce. Mexico is not bound to the US prohibition of hunting from 10 March to 1 September or to the 3.5-month maximum hunting period, because these restrictions are not specified in the Migratory Bird Treaty Act (Reeves 1993). For example, in Tamaulipas, the 2012–2013 mourning dove season was from 24 August 2012 to 17 February 2013, concurrent with the US white-winged dove season. Possession limits were 190 birds for each species, a liberal number compared to US standards.

Early and late harvest seasons may affect mourning dove annual production, especially in the southern US and Mexico. August-initiated nests in central Texas contributed 20% of the total year's fledglings in 1978 and 26% in 1979; September-initiated nests contributed only 5% of total nests and 6% of fledged young. When one parent was removed during nesting, nests with young less than six days of age had a significant reduction in success (79.5%) (Bivings and Silvy (1981).

Primary research and management needs in the US that also apply to Mexico include assessment of (1) breeding population status, (2) mortality, and (3) productivity; (4) relationships with habitat, (5) biology, (6) habitat management, and (7) actions to accomplish these goals (Braun et al. 1993). Maintenance of mourning doves in a healthy, productive state is of primary importance. Basic management goals should be to provide adequate food, water, cover, nest sites, and song posts within the daily flight range (Grue at al. 1982). Ultimately, the goal should be to provide diverse vegetation types with these habitat requisites and preferred foods throughout the year, not just during the hunting season.

White-Winged Dove

Description

The white-winged dove is medium sized, somewhat larger than the mourning dove, and has a square tail similar to that of a pigeon (fig. 10.12). It is the only dove within its distribution with a conspicuous white wing patch. The top of the head and foreback are purplish gray, blending to drab brown on the back, middle wing coverts, and two central tail feathers. Flight feathers are dark gray and finely edged with white, and distal wing coverts are white, forming an elongated white patch on the spread wing. The rump is gray, and the tail (except the central feathers) is dark gray at the base, has a subterminal band of black, and is broadly white at the tip. The sides of the neck are an iridescent bronze, with a black spot on each side. Underparts are tawny brown, blending posteriorly to pale gray on the abdomen and undertail coverts. The bill is black, and the legs and feet are red. The bare skin surrounding the eyes is bright blue, and the eyes have a bright orange iris. Juveniles resemble adults but lack a black cheek patch; also, the feet are a dull reddish brown, the head is smaller, and the iris is brown to light orange. Weight is 130–196 g (4.5–6.9 oz). Males are slightly larger than females.

Figure 10.12. White-winged dove. Photo by Michael Vaughn.

Range in Mexico

White-winged doves occur from the US-Mexico border through Baja California and south to central Veracruz, Michoacán, and Guerrero; the northern Yucatán Peninsula; Pacific slopes of Oaxaca; and Chiapas. They also occur on several islands in the Gulf of California, the Tres Marias Islands off the west coast of Mexico, and Cozumel Island east of the Yucatán Peninsula (Howell and Webb 1995, Schwertner et al. 2002). Leopold (1959) noted the white-winged dove was abundant along both coasts, especially in thorn forest and tropical deciduous forests, and occurred sparingly in the arid interior of Mexico. The resident population was augmented in winter by a small migrating population from the southwestern US. The highest summer populations were at the tip of the Baja California Peninsula and in coastal Sinaloa, Nayarit, Guerrero, Tamaulipas, and the lower Río Balsas Valley. Lesser numbers occurred in the Tehuantepec region of Oaxaca, along the dry tip of the Yucatán Peninsula, and in the upper Río Balsas Basin.

Population ecology

Whitewings are noticeably more gregarious than mourning doves at all times of year. In winter they fly in flocks, and during the breeding season they are colonial nesters, unlike mourning doves, which nest in isolation. Arnold (1943), who studied whitewings nesting in Arizona, concluded that formation of nesting colonies was a function of limited habitat, which forced birds to congregate in favored areas. However, Leopold (1959) found that white-winged doves along the Mexican coasts nested in colonies even where suitable habitat seemed almost limitless. Nesting white-winged doves were conspicuous because of the loud and persistent cooing of males in and around a large colony of thousands of pairs. The male frequently performed nuptial flights much like those of the mourning dove.

Diet

In Mexico, whitewings ate fruits of palms, figs (*Ficus* spp.), mesquite (*Prosopis* spp.), hackberry (*Celtis* spp.), *Condalia* spp., and larger cacti (Leopold 1959). In South Texas, they ate fruits of granjeno (*Celtis pallida*), anacua (*Ehretia anacua*), coma (*Sideroxylon celastrinum*), lime prickly-ash (*Zanthoxylum fagara*), brasil (*Condalia hookeri)*, privet (*Forestiera angustifolia*), and pigeonberry (*Rivina humilis*) (Cottam and Trefethen 1968, Schacht et al. 1995). In open fields and pastures, they ate seeds of sunflower (*Helianthus annuus*), dove weed (*Croton* spp.), spurge (*Euphorbia spathulata*), panic grass (*Panicum* spp.), bristlegrass (*Setaria* spp.), and other grasses and legumes. In central Texas, doves consumed large white seeds of exotic Chinese tallow (*Sapium sebiferum*), berries of gum bumelia (*Sideroxylon lanuginosum*), and pecan fragments crushed by vehicle traffic. In West Texas and New Mexico, seeds of leatherweed (*Jatropha dioica*) were an important food item (Schwertner et al. 2002).

Where available, domestic grains, including sunflower, milo, corn, beans, sesame, melon seeds, safflower, and wheat, are important foods (Leopold 1959, Cottam and Trefethen 1968, Dolton 1975, Waggerman and Sorola 1977, West 1993, Schacht et al. 1995, Cunningham et al. 1997). Consumption of specific food items typically correlated closely with availability (Jenks 1983). Pruitt et al. (2008) noted the importance of native seeds in white-winged dove diets dominated by agricultural grains. In urban areas of Texas (e.g., San Antonio), nesting whitewings foraged at bird feeders near urban nesting sites early in the breeding season, shifting late in the season to ripening grain in rural areas (West 1993, West et al. 1993). Juveniles did not participate in feeding flights to rural areas to the same extent as adults, instead remaining in the city to feed. Whitewings occasionally ingest small stones, snails, and bone fragments (Schwertner et al. 2002). Stones aid in digestion, and shells and bones provide calcium for production of crop milk and eggshells (Cottam and Trefethen 1968, Haughey 1986, Schacht et al. 1995).

In the Sonoran Desert, nectar, pollen, and fruit of saguaro (*Carnegiea gigantea*) dominated the diet of nesting white-winged doves (Haughey 1986, Wolf and del Rio 2000). Other desert foods include seeds and fruit of ocotillo (*Fouquieria splendens*), elephant tree (*Bursera microphylla*), organ-pipe cactus (*Stenocereus thurberi*) (Brown 1989), Mexican jumping bean (*Sapium biloculare*), sangre de drago (*Jatropha cuneata*) (Haughey 1986), and various other legumes (Neff 1940).

White-winged doves require water on a regular basis in dry, hot environments but can satisfy water requirements by eating moist fruit (MacMillan and Trost 1966). They prefer to drink during the morning and late afternoon in open areas, such as bare ditch or stream banks, sand or gravel bars, and stock ponds or troughs (Schwertner et al. 2002).

Reproduction and behavior

In North America, migrants arrive on breeding grounds in late March, with males arriving first (Schwertner et al. 2002). In the southwestern US and northern Mexico, whitewings occupy humid to semiarid dense, thorny woodlands, cactus–palo verde communities, and riparian woodlands. They prefer to nest in the interior of woodlands (Hayslette 1996). Highest nest densities occurred historically in dense mixed woodlands dominated by Texas ebony (*Chloroleucon ebano*) in Tamaulipas, Coahuila, Nuevo León, and Veracruz. They also nest in oak-juniper (*Quercus* spp.–*Juniperus* spp.) forests, in salt cedar (*Tamarix* spp.), and in citrus groves in northern and central Mexico (Cottam and Trefethen 1968, Brown 1989). Leopold (1959) noted that nests were usually in groves of low, dense trees such as mesquite or acacia, but some were in clumps of cactus or tangles of vine. Along the coasts of Mexico, whitewings readily nest in black mangrove (*Avicennia germinans*) forests. In urban areas, they prefer older residential shade and ornamental trees (Cottam and Trefethen 1968, Oberholser 1974, Jenks 1983, Small et al. 1989, George 1991, West 1993, West et al. 1993, George et al. 1994, Breeden et al. 2004, 2008). Winter habitat does not differ significantly from breeding habitat (Waggerman and Sorola 1977, George et al. 2000).

Pairs may be formed before arrival on breeding grounds and are monogamous through at least one breeding season (Cottam and Trefethen 1968). They defend nest sites and cooing perches; territory size may depend on nest density (Saunders 1940, Cottam and Trefethen 1968, Alamia 1970, Viers 1970, Brown 1989, Swanson 1989, Swanson and Rappole 1993). In Texas and northern Mexico, they often congregate in colonies to nest in dense woodland and riparian areas (Oberholser 1974). Colonies are traditionally defined as at least 25 pairs/ha (≥

10.4 pairs/ac), and colonial pairs may renest more often than dispersed nesters (Cottam and Trefethen 1968). White-winged doves will nest as scattered pairs in other habitats throughout their range. In northeastern Mexico, colonies are usually located in areas of dense vegetation near freshwater (Blankinship 1970). Habitat conditions usually dictate the number of doves that nest in a given locale. When adequate amounts of food, water, and nesting cover are present, doves occur in high nesting densities. The highest density recorded was in Tamaulipas, Mexico, in June 1973, when 141 active nests were recorded on a 0.1 ha (0.24 ac) belt transect (1,392 nests/ha, 580 nests/ac) in native-brush habitat 30 km (18.6 mi) east of San Fernando (Waggerman 1973). Large colonies are declining as birds expand into urban areas with dispersed nesting trees. In nesting colonies in South Texas and northeastern Mexico, territories may encompass only part of a nesting tree (Cottam and Trefethen 1968, Alamia 1970, Swanson and Rappole 1993). In colonies, nests are as close as 25 cm (9.8 in) apart (Blankinship 1970).

Nest building by migrants usually begins in late April to mid-May in northeastern Mexico, usually immediately following arrival in breeding areas (Blankinship 1970). Nesting by resident urban populations in Texas begins earlier, by early to mid-March (Passmore 1981, West et al. 1993, Hayslette and Hayslette 1999). Usually there are two nesting peaks or seasons, the timing of which varies annually (Cunningham et al. 1997). Some remain on the wintering grounds to breed (Blankinship et al. 1972). Eggs are usually laid 24 hours apart. They also hatch 24 hours apart, after 15–20 days of incubation. Young fledge at 13–18 days. Parents continue to feed the young for up to 30 days. They usually renest immediately following predation of a previous nest. The number of broods varies with environmental conditions but averages two per season, and usually one per season in dispersed desert-nesters. Subsequent nesting was initiated 4–15 days following the fledging of the previous brood (Neff 1940, Saunders 1940, Alamia 1970, Blankinship 1970).

Males select and defend the territory and nesting site, and females select the nest location within a territory (Cottam and Trefethen 1968). Nest construction can require more than five days, depending on disruptions, the availability of materials, and other factors. It most often takes one to two days (Cottam and Trefethen 1968, Williams 1971). Males gather small twigs and bring them to the female, who builds the nest (Alamia 1970). Eggs are creamy white to creamy buff and have a dull texture (Wetmore 1920, Neff 1940); there are generally two eggs (Neff 1940, Saunders 1940, Leopold 1959, Cottam and Trefethen 1968). The first egg is laid late in the afternoon after completion of the nest, and the second egg is laid the following morning (Cottam and Trefethen 1968).

Both sexes incubate, the male during the day and the female at night. Male territorial defense leaves the nest vulnerable to nest predators, which are common among colonial nests, in which 90% of eggs laid may be predated (Schwertner et al. 2002). If one mate disappears, the other will continue to care for the nest (Neff 1940, Cottam and Trefethen 1968). Length of incubation varies individually and geographically. In South Texas, incubation is usually 14 days, and in Arizona, usually 15–17 days, although occasionally up to 20 days (Neff 1940). Longer incubation periods may be required in colder, mountainous regions (Cottam and Trefethen 1968).

Both adults produce crop milk, but first-day feedings also may include fruit pulp and soft seeds (Schwertner et al. 2002). By 3–4 days, feedings contain foods more similar to adult diets. Fledging usually occurs at 13–18 days of age but can occur at 8 days (Neff 1940, Cottam and Trefethen 1968). At 13 days, young generally have the appearance of adults and participate in activities similar to those of adults (Cottam and Trefethen 1968). Fledglings remain together and in their territory for around 14 days after fledging (Saunders 1940). They are fed by their parents for more than one month, although they are capable of feeding independently at around three weeks (Cottam and Trefethen 1968). Fewer than 29% of whitewings attempt second broods (Alamia 1970, Williams 1971, Schacht et al. 1995).

Population ecology

There are no estimates of numbers of white-winged doves in Mexico, except for some local nesting colonies, but in some places they are enormously abundant. In areas where they are concentrated, it is not unusual to see thousands milling over croplands or around watering sources. Because of their abundance, they are a highly popular game bird and offer excellent hunting opportunities. They are hunted for food locally in Mexico (Leopold 1959). Between 1966 and

1982, the population in Tamaulipas increased from approximately 5 million to 12–16 million, despite the destruction of a number of nesting colonies. This increase was probably the result of a shift to dispersed nesting habitats, combined with increased food supplies because of agricultural development (Purdy and Tomlinson 1991). Their distributional and breeding range on the Pacific coast of southern Mexico includes a large percentage of farmland. These populations are probably subject to the same effects of agricultural development, both positive and negative, as are populations in the US and northern Mexico (Waggerman and Sorola 1977).

Habitat loss and fragmentation are among the most critical threats to white-winged doves, particularly within their historical range in South Texas and Mexico. With the introduction of mechanized farming practices and a burgeoning human population, agricultural development and concomitant destruction of nesting habitat in Tamaulipas are expected to increase (Purdy and Tomlinson 1991, George et al. 2000). In Tamaulipas, Sanchez-Johnson et al. (2009) located 31 previously recorded colonies, of which 13 were active and 18 inactive. Brush clearing was the cause for 78% of the 18 inactive colonies, followed by weather catastrophes (56%) and overharvest (39%). Habitat loss may not have significantly reduced the US population for two reasons: habitat loss in exurban areas may have been negated by use of urban environments, and northward expansion may have increased the size of the US population, despite declines within its historical range (Schwertner et al. 2002).

Increased food availability from agricultural development in some areas in Tamaulipas since the mid-1970s resulted in a rapidly expanding population estimated at 16–19 million birds (Ortega Melendez and Zamora Treviño 1984). These attracted increasing numbers of American hunters, who harvested an estimated 1–1.4 million birds in northeastern Mexico (George 1993). The number of hunters in northeastern Mexico has greatly decreased because of security reasons.

In South Texas, fledging success was 28.4% in citrus orchards (Cottam and Trefethen 1968), ranging from 18% in one colony (Marsh and Saunders 1942) to 35.9% in native brush between 1951 and 1966 (Cottam and Trefethen 1968). In South Texas, 36.0% of eggs hatched, and of those that hatched, 93.5% produced fledglings (Hayslette et al. 2000). Fledging success was 30.5–34.7% in colonial nesting areas, while in dispersed nesting sites (< 25 pairs/ha, 10.4 pairs/ac), it was 49.6% (Cottam and Trefethen 1968). Another study in the same region (Swanson 1989), however, failed to detect a significant difference in fledging success between high- and low-density areas (46.4% vs. 55.2%, respectively). Nest survival for urban-nesting doves in San Antonio, Texas, averaged 46% over two years (West et al. 1993). Nest survival was 58% in an urban colony in Kingsville, Texas, with all losses occurring prior to hatching (Hayslette and Hayslette 1999). In Arizona, nestling success was 67% (Neff 1940), and in birds nesting in salt cedar habitats, fledging success rates ranged from 74% to 86% (Cunningham et al. 1997). Fledging success was 39% for doves nesting in salt cedar habitats along the Rio Grande in the Trans-Pecos region of Texas (Gallucci 1978). Seasonal production was 1.9 young/breeding pair in South Texas, and in Arizona there were 1.44 juveniles per adult, based on the ratio of juveniles to adults in the harvest (Cottam and Trefethen 1968, Stair 1970).

The longest reported life span of a wild white-winged dove was 21 years 9 months (Clapp et al. 1983). Survival rates of white-winged doves varied from 55% (1968–1976) to 71% (1960–1965) for banded birds (George et al. 2000). Birds banded in northern Mexico during the late 1960s exhibited higher survival rates than birds banded in South Texas during the same period, probably owing to heavier hunting pressure in Texas (Schwertner et al. 2002). Band recovery data from South Texas and northern Mexico indicate a disproportionately high mortality among young birds. Nestlings banded in 1961–1965 had an 18% annual survival rate, versus 60% and 65.7% for juvenile and adult-banded birds, respectively (George et al. 2000). Hatch-year survival varied from 21% to 36%, and after-hatch-year bird survival varied from 46% to 54%, depending on their banding location in Texas (Collier et al. 2012). In Arizona, the combined juvenile and adult mortality rate was 23.1%. Mortality rates of adults from hunted areas were between 43.8% and 72.2%; juvenile mortality rates in the same areas were 60% to 80% (Brown 1989).

Predation and disease

Major nest predators in Mexico include Tamaulipas crow (*Corvus imparatus*), white-throated magpie-jay

(*Calocitta formosa*), and brown jay (*Cyanocorax morio*) (Cottam and Trefethen 1968, Blankinship 1970). In Texas, the great-tailed grackle (*Quiscalus mexicanus*) is the primary nest predator (Schwertner et al. 2002). Great-tailed grackles often nest in high densities in habitats with white-winged dove colonies, taking eggs or nestlings when adult doves are off the nest (Cottam and Trefethen 1968). Eggs and nestlings are most susceptible; there is typically higher predation on eggs than nestlings (Blankinship 1966, Swanson and Rappole 1993).

Mammalian nest predators include wood rats (*Neotoma* spp.), deer mice (*Peromyscus* spp.), gray fox (*Urocyon cinereoargenteus*), Norway rat (*Rattus norvegicus*), and black rat (*R. rattus*) in areas of dense ground cover (Blankinship 1966, Cunningham et al. 1997). House cats (*Felis domesticus*) take nestlings around houses and in urban sites. Other species include Virginia opossum (*Didelphis virginianus*), raccoon (*Procyon lotor*), and ringtail (*Bassariscus astutus*). Neff (1947) mentions dogs and coyotes (*Canis latrans*) as predators of young that have fallen to the ground. Coyotes, bobcats (*Lynx rufus*), and house cats are opportunistic predators of adults (Schwertner et al. 2002).

Snakes and lizards prey on eggs and nestlings (Saunders 1940, Cottam and Trefethen 1968, Schwertner et al. 2002), and snakes occasionally consume dead or injured adult doves (Saunders 1940, Cottam and Trefethen 1968, Blankinship et al. 1972, Mannan and Boal 2000). Peregrine falcons (*Falco peregrinus*), Harris's hawks (*Parabuteo unicinctus*), Cooper's hawks (*Accipiter cooperii*), and other raptors occasionally take adults (Schwertner et al. 2002).

Trichomoniasis, considered a serious disease in columbids, appears to be nonpathogenic to white-winged doves, although it has been identified frequently in the species. In a sample of 171 white-winged doves from southern and central Texas, all but one were infected with *trichomoniasis*, but none exhibited clinical symptoms (Glass et al. 2001), and there was a prevalence rate of 100% in a sample of 51 in Texas (Stabler 1961).

Population management

The white-winged dove occurs throughout its historical breeding range. Its viability seems secure across its range, despite population declines that are primarily a result of habitat loss and excessive harvest in the early twentieth century (Schwertner et al. 2002). Bag limits in Mexico are generous by US standards, and hunting regulations often are poorly enforced (George 1993). Although the current population in Mexico appears to be able to sustain a harvest of this magnitude, hunting pressure and harvest should be carefully monitored (Tomlinson 1988). Brown et al. (1977) suggested that harvest should not exceed 25% of the breeding population.

Schwertner et al. (2002) recommended priorities for future white-winged dove research, of which the following are pertinent to Mexico: (1) develop reliable estimators of population abundance, (2) obtain reliable estimates of demographic variables such as survival and recruitment, (3) address disease transmission and competition for resources, (4) determine the effect of hunting on breeding populations, (5) develop a better understanding of nesting-habitat requirements, (6) determine nutritional requirements, especially as they relate to foraging strategy, and (7) investigate the ecology of wintering birds in Mexico.

Efforts have been made in Mexico regarding some of these research priorities. Nichols et al. (1986) estimated nest detection probabilities for white-winged dove nest transects in Tamaulipas. Much of the research in the US has application in Mexico. Silvy and Rollins (2004) highlighted the biology, research, and management of white-winged and mourning doves in Texas, while Silvy (2004) discussed ongoing research projects on these two species. Mathewson (2002) studied nest-site selection and partitioning among sympatric white-winged, mourning, and Inca doves in central Texas. Similar research efforts in Mexico are lacking. Most of the dove species in Mexico have not been studied. Also, the effects of the invasive Eurasian collared dove on American doves remain unknown.

Research and Management Challenges

Because of the lack of information on the status, natural history, and management of doves and pigeons, research is needed to better manage rare and hunted populations in Mexico. There is a need to develop a reliable, standardized nationwide census method for columbids and to determine estimates of population abundance. Because several species are migratory, abundance information is needed for both wintering and breeding populations. This information needs to be compared with similar information obtained for species that cross natural boundaries during migration. Coordination and cooperation between the US and

Mexico are essential to managing migratory species. In addition, research is needed to obtain reliable estimates of survival and recruitment, especially for rare and hunted dove and pigeon populations in Mexico.

Literature Cited

Alamia, L. A. 1970. Renesting activity and breeding biology of the white-winged dove (*Zenaida asiatica*) in the Lower Rio Grande Valley of Texas. Thesis, Texas A&M University, College Station, USA.

Aldrich, J. W. 1993. Classification and distribution. Pages 47–54 *in* T. S. Baskett, M. W. Sayre, R. E. Tomlinson, and R. E. Mirarchi, editors. Ecology and management of the mourning dove. Stackpole, Harrisburg, Pennsylvania, USA.

AOU. 1983. Check-list of North American birds. Sixth edition. American Ornithologists' Union, Washington, DC, USA.

AOU. 1998. Check-list of North American birds. Seventh edition. American Ornithologists' Union, Washington, DC, USA.

Arnold, L. W. 1943. The western white-winged dove in Arizona. Arizona Game and Fish Commission, Phoenix, USA.

Atherton, N. W., M. E. Morrow, A. E. Bivings IV, and N. J. Silvy. 1982. Shrinkage of spiral plastic leg bands with resulting leg damage to mourning doves. Proceedings of the Annual Conference of the Southeastern Association of Fish and Wildlife Agencies 36:666–670.

Baskett, T. S., and M. W. Sayre. 1993. Characteristics and importance. Pages 1–6 *in* T. S. Baskett, M. W. Sayre, R. E. Tomlinson, and R. E. Mirarchi, editors. Ecology and management of the mourning dove. Stackpole, Harrisburg, Pennsylvania, USA.

Baskett, T. S., M. W. Sayre, R. E. Tomlinson, and R. E. Mirarchi, editors. 1993. Ecology and management of the mourning dove. Stackpole Books, Harrisburg, Pennsylvania, USA.

Bent, A. C. 1932. Life histories of North American gallinaceous birds. Bulletin 162. US National Museum, Washington, DC, USA.

Bivings, A. E., IV, M. E. Morrow, and N. J. Silvy. 1984. Movement and survival of mourning doves banded preseason in Texas. Proceedings of the Annual Conference of the Southeastern Association of Fish and Wildlife Agencies 38:153–157.

Bivings, A. E., IV, and N. J. Silvy. 1980. Primary feather molt of mourning doves in central Texas. Proceedings of the Annual Conference of the Southeastern Association of Fish and Wildlife Agencies 34:410–413.

Bivings, A. E., IV, and N. J. Silvy. 1981. The contribution of September mourning dove nesting to total production in central Texas. Proceedings of the Annual Conference of the Southeastern Association of Fish and Wildlife Agencies 35:32–36.

Bivings, A. E., IV, and N. J. Silvy. 1994. Success of single-parent mourning dove nests in September. Proceedings of the Annual Conference of the Southeastern Association of Fish and Wildlife Agencies 48:236–239.

Blankenship, L. H., and H. M. Reeves. 1970. Mourning dove recoveries from Mexico. Special Scientific Report, Wildlife 135. USFWS, Washington, DC, USA.

Blankinship, D. R. 1966. The relationship of white-winged dove production to control of great-tailed grackles in the Lower Rio Grande Valley of Texas. Transactions of the North American Wildlife and Natural Resource Conference 34:45–58.

Blankinship, D. R. 1970. White-winged dove nesting colonies in northeastern Mexico. Transactions of the North American Wildlife and Natural Resource Conference 35:171–182.

Blankinship, D. R., J. G. Teer, and W. H. Kiel Jr. 1972. Movements and mortality of white-winged doves banded in Tamaulipas, Mexico. Transactions of the North American Wildlife and Natural Resource Conference 37:312–325.

Blockstein, D. E. 1986. Reproductive behavior and parental investment of mourning dove. Dissertation, University of Minnesota, Minneapolis, USA.

Braun, C. E. 1994. Band-tailed pigeon. Pages 61–74 *in T. Tacha and C. E. Braun, editors.* Migratory shore and upland game bird management in North America. International Association of Fish and Wildlife Agencies, Washington, DC, USA.

Braun, C. E., N. J. Silvy, T. S. Baskett, and R. E. Tomlinson. 1993. Research and management needs. Pages 507–513 *in* T. S. Baskett, M. W. Sayre, R. E. Tomlinson, and R. E. Mirarchi, editors. Ecology and management of the mourning dove. Stackpole, Harrisburg, Pennsylvania, USA.

Breeden, J. B., F. Hernandez, R. L. Bingham, N. J. Silvy, and G. L. Waggerman. 2004. An evaluation of survey methods for white-winged doves in urban areas of Texas. Proceedings of the Annual

Conference of the Southeastern Association of Fish and Wildlife Agencies 58:274–281.

Breeden, J. B., F. Hernández, R. L. Bingham, N. J. Silvy, and G. L. Waggerman. 2008. Effects of traffic noise on auditory surveys of urban white-winged doves. Wilson Journal of Ornithology 120:384–389.

Brown, D. E. 1989. Arizona game birds. University of Arizona Press, Tucson, USA.

Brown, D. E., D. R. Blankinship, P. K. Evans, W. H. Kiel, G. L. Waggerman, and C. K. Winkler. 1977. White-winged dove (*Zenaida asiatica*). Pages 246–272 *in* G. L. Sanderson, editor. Management of migratory shore and upland game birds in North America. University of Nebraska Press, Lincoln, USA.

Browning, B. M. 1962. Food habits of the mourning dove in California. California Fish and Game 48:91–115.

Brush, T. 1998. Recent nesting and current status of red-billed pigeon along the Rio Grande in southern Texas. Bulletin of the Texas Ornithological Society 31:22–26.

Brush, T., and A. Cantu. 1998. Changes in the breeding bird community of subtropical evergreen forest in the Lower Rio Grande Valley of Texas, 1970s–1990s. Texas Journal of Science 50:123–133.

Butchart, S. H. M., A. J. Stattersfield, and N. J. Collar. 2006. How many bird extinctions have we prevented? Oryx 40:266–278.

Campbell, C. J. 1912. The band-tailed pigeon in North Dakota. Auk 29:539–540.

Case, D. J., and S. J. Hughes, editors. 2011. Priority information needs for band-tailed pigeons, Zenaida doves, white-tipped doves and scaly-naped pigeons: a funding strategy. D. J. Case and Associates, Mishawaka, Indiana, USA.

Channing, E. 1979. Movements of banded mourning dove near Turlock, California. California Fish and Game 65:23–35.

Clapp, R. B., M. K. Klimkiewicz, and A. G. Futcher. 1983. Longevity records of North American birds: Columbidae through Paridae. Journal of Field Ornithology 54:123–137.

Collier, B. A., S. R. Kremer, C. D. Mason, M. J. Peterson, and K. W. Calhoun. 2012. Survival, fidelity, and recovery rates of white-winged doves in Texas. Journal of Wildlife Management 76:1129–1134.

Conti, J. A. 1993. Diseases, parasites, and contaminants. Pages 205–224 *in* T. S. Baskett, M. W. Sayre, R. E. Tomlinson, and R. E. Mirarchi, editors. Ecology and management of the mourning dove. Stackpole, Harrisburg, Pennsylvania, USA.

Cottam, C., and J. B. Trefethen. 1968. White-wings: the life history, status, and management of the white-winged dove. Van Nostrand, Princeton, New Jersey, USA.

Cowan, J. B. 1952. Life history and productivity of a population of western mourning doves in California. California Fish and Game 38:505–521.

Cunningham, S. C., R. W. Engel-Wilson, P. M. Smith, and W. B. Ballard. 1997. Food habits and nesting characteristics of sympatric mourning and white-winged doves in Buckeye-Arlington Valley, Arizona. Technical Report 26. Arizona Game and Fish Department, Phoenix, USA.

DeGraaf, R. M., V. E. Scott, R. H. Hamre, L. Ernst, and S. H. Anderson. 1991. Forest and rangeland birds of the United States. USDA, Washington, DC, USA.

del Hoyo, J., A. Elliot, and J. Sargatal. 1997. Handbook of the birds of the world. Volume 4. Sandgrouse to cuckoos. Lynx Edicions, Barcelona, Spain.

Dickey, D. R., and A. J. Van Rossem. 1938. The birds of El Salvador. Field Museum of Natural History, Zoological Series 23:1–609.

Dillon, O. W., Jr. 1961. Mourning dove foods in Texas during September and October. Journal of Wildlife Management 25:334–336.

Dolton, D. D. 1975. Patterns and influencing factors of white-winged dove feeding activity in the Lower Rio Grande Valley of Texas and Mexico. Thesis, Texas A&M University, College Station, USA.

Dolton, D. D., and R. D. Rau. 2006. Mourning dove population status, 2006. USFWS, Laurel, Maryland, USA.

Downing, R. L. 1959. Significance of ground nesting by mourning doves in northwestern Oklahoma. Journal of Wildlife Management 23:117–118.

Dunks, J. H., R. E. Tomlinson, H. M. Reeves, D. D. Dolton, C. E. Braun, and T. P. Zapatka. 1982. Migration, harvest, and population dynamics of mourning doves banded in the Central Management Unit, 1967–77. Special Scientific Report, Wildlife 249. USFWS, Washington, DC, USA.

Dunn, E. H., D. J. T. Hussell, and D. A. Welsh. 1999. Priority-setting tool applied to Canada's land birds based on concern and responsibility for species. Conservation Biology 13:1404–1415.

Dunning, J. B., Jr., editor. 1992. CRC handbook of avian body masses. CRC Press, Boca Raton, Florida, USA.

Eitniear, J. C., and A. Aragon-Tapia. 2005. Nesting characteristics of red-billed pigeons (*Patagioenas flavirostris*) in northeast Mexico. Ornitología Neotropical 16:279–281.

Eitniear, J. C., and J. Breeden. 2003. Variation in red-billed pigeon clutch size. Bulletin of the Texas Ornithological Society 36:6–8.

Elder, J. B. 1956. Watering patterns of some desert game animals. Journal of Wildlife Management 20:368–378.

Escalante, P., A. M. Sada, and J. R. Gil. 2014. Lista de nombres comunes de las aves de México. UNAM and CIPAMEX, México, D. F., México.

Franson, J. C., S. P. Hansen, and J. H. Schulz. 2009. Ingested shot and tissue lead concentrations in mourning doves. Pages 1–12 *in* R. T. Watson, M. Fuller, M. Pokras, and W. G. Hunt, editors. Ingestion of lead from spent ammunition: implications for wildlife and humans. The Peregrine Fund, Boise, Idaho, USA.

Gallucci, T. 1978. The biological and taxonomic status of the white-winged doves of the Big Bend of Texas. Thesis, Sul Ross State University, Alpine, Texas, USA.

Gehlbach, F. R. 1987. Natural history sketches, densities, and biomass of breeding birds in evergreen forests of the Rio Grande, Texas, and Rio Corona, Tamaulipas, Mexico. Texas Journal of Science 39:241–251.

Gehlbach, F. R., D. O. Dillon, H. L. Harrell, S. E. Kennedy, and K. R. Wilson. 1976. Avifauna of the Río Corona, Tamaulipas, Mexico: northeastern limit of the tropics. Auk 93:53–65.

George, R. R. 1988. Mourning doves in Texas. Texas Parks and Wildlife Department, Austin, USA.

George, R. R. 1991. The adaptable whitewing. Texas Parks and Wildlife Magazine 49:10–15.

George, R. R. 1993. White-winged dove banding analysis. Final Report, Federal Aid Project W-128-R, Job 6. Texas Parks and Wildlife Department, Austin, USA.

George, R. R., R. E. Tomlinson, R. W. Engel-Wilson, G. L. Waggerman, and A. G. Spratt. 1994. White-winged dove. Pages 29–50 *in* T. C. Tacha and C. E. Braun, editors. Migratory shore and upland game bird management in North America. Allen Press, Lawrence, Kansas, USA.

George, R. R., G. L. Waggerman, D. M. McCarty, R. E. Tomlinson, D. Blankinship, and J. H. Dunks. 2000. Migration, harvest and population dynamics of white-winged doves banded in Texas and northeastern Mexico, 1950–1978. Texas Parks and Wildlife Department, Austin, USA.

Gibbs, D., E. Barnes, and J. Cox. 2001. Pigeons and doves: a guide to the pigeons and doves of the world. Yale University Press, New Haven, Connecticut, USA.

Glass, J. W., A. M. Fedynich, M. F. Small, and S. J. Benn. 2001. *Trichomonas gallinae* in an expanding population of white-winged doves from Texas. Southwestern Naturalist 46:234–237.

Glover, F. A. 1953. A nesting study of the band-tailed pigeon (*Columba f. fasciata*) in north-western California. California Fish and Game Bulletin 39:397–407.

Goodwin, D. 1983. Pigeons and doves of the world. Cornell University Press, Ithaca, New York, USA.

Grue, C. E. 1977. Classification, inventory, analysis, and evaluation of the breeding habitat of the mourning dove in Texas. Dissertation, Texas A&M University, College Station, USA.

Grue, C. E., R. R. Reid, and N. J. Silvy. 1982. Correlation of habitat parameters with mourning dove call counts in Texas. Journal of Wildlife Management 46:186–195.

Grzimek, B., editor. 1975. Grzimek's animal life encyclopedia. Volume 8. Birds II: Grouse–mousebirds. Van Nostrand Reinhold, New York, New York, USA.

Gutiérrez, R. J., C. E. Braun, and T. P. Zapatka. 1975. Reproductive biology of the band-tailed pigeon in Colorado and New Mexico. Auk 92:665–677.

Harrison, C. J. O. 1978. A field guide to nests, eggs, and nestlings of North American birds. William Collins and Sons, Glasgow, United Kingdom.

Haughey, R. A. 1986. Diet of desert-nesting western white-winged doves *Zenaida asiatica mearnsi*. Thesis, University of Arizona, Tucson, USA.

Hayslette, S. E. 1996. Factors affecting reproduction of white-winged, white-tipped, and mourning doves in the Lower Rio Grande Valley of Texas. Thesis, Texas A&M University–Kingsville, USA.

Hayslette, S. E., and B. A. Hayslette. 1999. Late and early season reproduction of urban white-winged doves in southern Texas. Texas Journal of Science 51:173–180.

Hayslette, S. E., T. C. Tacha, and G. L. Waggerman. 1996. Changes in white-winged dove reproduction in southern Texas, 1954–93. Journal of Wildlife Management 60:298–301.
Hayslette, S. E., T. C. Tacha, and G. L. Waggerman. 2000. Factors affecting white-winged, white-tipped, and mourning dove reproduction in the Lower Rio Grande Valley. Journal of Wildlife Management 64:286–295.
Hitchcock, R. R., and R. E. Mirarchi. 1984. Duration of dependence of wild fledgling mourning doves on parental care. Journal of Wildlife Management 48:99–108.
Houston, D. B. 1963. A contribution to the ecology of the band-tailed pigeon, *Columba fasciata*, Say. Thesis, University of Wyoming, Laramie, USA.
Howell, S. N. G., and S. Webb. 1995. A guide to the birds of Mexico and northern Central America. Oxford University Press, New York, New York, USA.
Jarvis, R. L., and M. F. Passmore. 1992. Ecology of band-tailed pigeons in Oregon. Biological Report 6. US Fish and Wildlife Service, Corvallis, Oregon, USA.
Jenks, R. S. 1983. Distribution, abundance, and habitat of white-winged dove in New Mexico. Thesis, New Mexico State University, Las Cruces, USA.
Kautz, J. E., and C. E. Braun. 1981. Survival and recovery rates of band-tailed pigeons in Colorado. Journal of Wildlife Management 45:214–218.
Keeler, J. E. 1977. Mourning dove (*Zenaida macroura*). Pages 274–298 *in* G. C. Sanderson, editor. Management of migratory shore and upland game birds in North America. International Association of Fish and Wildlife Agencies, Washington, DC, USA.
Keppie, D. M., and C. E. Braun. 2000. Band-tailed pigeon (*Patagioenas fasciata*). *In* The birds of North America. Academy of Natural Sciences, Philadelphia, Pennsylvania, and American Ornithologists' Union, Washington, DC, USA.
Klimkiewicz, M. K., and A. G. Futcher. 1989. Longevity records of North American birds. Supplement 1. Journal of Field Ornithology 60:469–494.
Leonard, J. P. 1998. Nesting and foraging ecology of band-tailed pigeons in western Oregon. Dissertation, Oregon State University, Corvallis, USA.
Leopold, A. S. 1959. Wildlife of Mexico: the game birds and mammals. University of California Press, Berkeley, USA.
Lewis, J. C. 1993. Foods and feeding ecology. Pages 181–204 *in* T. S. Baskett, M. W. Sayre, R. E. Tomlinson, and R. E. Mirarchi, editors. Ecology and management of the mourning dove. Stackpole, Harrisburg, Pennsylvania, USA.
Lowery, G. H., and W. W. Dalquest. 1951. Birds from the state of Veracruz, Mexico. University of Kansas Publications, Museum of Natural History 3:531–649.
Lowther, P. E. 2002. Red-billed pigeon (*Patagioenas flavirostris*). *In* The birds of North America. Academy of Natural Sciences, Philadelphia, Pennsylvania, and American Ornithologists' Union, Washington, DC, USA.
MacGregor, W. G., and W. M. Smith. 1955. Nesting and reproduction of the band-tailed pigeon in California. California Fish and Game Bulletin 41:315–326.
Mackey, J. P. 1954. Some aspects of mourning dove behavior related to reproduction. Thesis, Ohio State University, Columbus, USA.
Macmillan, R. E., and C. H. Trost. 1966. Water economy and salt balance in white-winged and Inca doves. Auk 83:441–456.
Mannan, R. W., and C. W. Boal. 2000. Home range characteristics of male Cooper's hawks in an urban environment. Wilson Bulletin 112:21–27.
March, G. L., and R. M. F. S. Sadleir. 1972. Studies on the band-tailed pigeon (*Columba fasciata*) in British Columbia. II. Food resource and mineral-gravelling activity. Syesis 5:279–284.
Marcot, B. G. 1984. Habitat relationships of birds and young-growth Douglas-fir in northwestern California (silviculture, diversity, community, clearcutting, avian). Dissertation, Oregon State University, Corvallis, USA.
Marsh, E. G., and G. B. Saunders. 1942. The status of the white-winged dove in Texas. Wilson Bulletin 54:145–146.
Martin, P. S., C. R. Robins, and W. B. Heed. 1954. Birds and biogeography of the Sierra de Tamaulipas, an isolated pine-oak habitat. Wilson Bulletin 61:38–57.
Mathewson, H. A. 2002. Nest-site selection and partitioning among sympatric white-winged, mourning, and Inca doves in Mason, Texas. Thesis, Texas A&M University, College Station, USA.
McClure, H. E. 1943. Ecology and management of

the mourning dove, *Zenaidura macroura* (Linn.) in Cass County, Iowa. Research Bulletin 310. Iowa Agricultural Experiment Station, Ames, USA.

Moore, G. C., and A. M. Pearson. 1941. The mourning dove in Alabama. Alabama Department of Conservation, Montgomery, USA.

Morrow, M. E., A. E. Bivings IV, and N. J. Silvy. 1985. Feather replacement for predicting hatching phenologies of mourning doves. Proceedings of the Annual Conference of the Southeastern Association of Fish and Wildlife Agencies 39:499–505.

Morrow, M. E., and N. J. Silvy. 1982. Nesting mortality of mourning doves in central Texas. Proceedings of the Annual Conference of the Southeastern Association of Fish and Wildlife Agencies 37:537–542.

Morrow, M. E., and N. J. Silvy. 1983. A comparison of individual and nest survival of mourning dove and implications to nesting studies. Proceedings of the Annual Conference of the Southeastern Association of Fish and Wildlife Agencies 37:19–22.

Morrow, M. E., N. J. Silvy, and W. G. Swank. 1992. Post-juvenal primary feather molt of wild mourning doves in Texas. Proceedings of the Annual Conference of the Southeastern Association of Fish and Wildlife Agencies 46:194–198.

Munoz, A. M., R. A. McCleery, R. R. Lopez, and N. J. Silvy. 2008. Nesting ecology of mourning doves in an urban landscape. Urban Ecosystems 11:257–267.

Neff, J. A. 1940. Notes on nesting and other habits of the western white-winged dove in Arizona. Journal of Wildlife Management 4:279–290.

Neff, J. A. 1947. Habits, food, and economic status of the band-tailed pigeon. North American Fauna 58:1–76.

Neff, J. A., and R. J. Niedrach. 1946. Nesting of the band-tailed pigeon in Colorado. Condor 48:72–74.

Nice, M. M. 1922. A study of the nesting of mourning doves. Part 1. Auk 39:457–474.

Nichols, J., R. E. Tomlinson, and G. Waggerman. 1986. Estimating nest detection probabilities for white-winged dove nest transects in Tamaulipas, Mexico. Auk 103:826–828.

Oberholser, H. C. 1974. The bird life of Texas. Volume 1. University of Texas Press, Austin, USA.

Ortega Melendez, H., and H. V. Zamora Treviño. 1984. Study of the white-winged dove in Tamaulipas, 1984. Unpublished. Fifth Latin American Workshop on Function and Management of National Wildlife Refuges. Santa Ana National Wildlife Refuge, Alamo, Texas, USA.

Otis, D. L. 2002. Survival models for harvest management of mourning dove populations. Journal of Wildlife Management 66:1052–1063.

Otis, D. L., J. H. Schulz, D. Miller, R. E. Mirarchi, and T. S. Baskett. 2008. Mourning dove (*Zenaida macroura*). *In* The birds of North America. Academy of Natural Sciences, Philadelphia, Pennsylvania, and American Ornithologists' Union, Washington, DC, USA.

Passmore, M. F. 1981. Population biology of the common ground dove and ecological relationships with mourning and white-winged doves in South Texas. Dissertation, Texas A&M University, College Station, USA.

Peeters, H. J. 1962. Nuptial behavior of the band-tailed pigeon in the San Francisco Bay area. Condor 64:445–470.

Peters, H. S. 1961. The past status and management of the mourning dove. Transactions of the North American Wildlife and Natural Resource Conference 26:371–374.

Peterson, R. T., and E. L. Chalif. 1973. A field guide to Mexican birds: Mexico, Guatemala, Belize, El Salvador. Houghton Mifflin, New York, New York, USA.

Pruitt, K. D., D. G. Hewitt, N. J. Silvy, and S. Benn. 2008. Importance of native seeds in white-winged dove diets dominated by agricultural grains. Journal of Wildlife Management 72:433–439.

Purdy, P. C., and R. E. Tomlinson. 1991. The eastern white-winged dove: factors influencing use and continuity of the resource. Pages 255–265 *in* J. G. Robinson and K. H. Redford, editors. Neotropical wildlife use and conservation. University of Chicago Press, Illinois, USA.

Quay, T. L. 1951. Mourning dove studies in North Carolina. North Carolina Wildlife Resources Commission, Raleigh, USA.

Reeves, H. M. 1993. Mourning dove hunting regulations. Pages 429–448 *in* T. S. Baskett, M. W. Sayre, R. E. Tomlinson, and R. E. Mirarchi, editors. Ecology and management of the mourning dove. Stackpole, Harrisburg, Pennsylvania, USA.

Robbins, C. S., D. Bystrak, and P. H. Geissler. 1986. The breeding bird survey: its first fifteen, 1965–1979. Resource Publication 157. USFWS, Washington, DC, USA.

Russell, S. M., and G. Monson. 1998. The birds of Sonora. University of Arizona Press, Tucson, USA.

Sadler, K. C. 1993. Mourning dove harvest. Pages 449–458 *in* T. S. Baskett, M. W. Sayre, R. E. Tomlinson, and R. E. Mirarchi, editors. Ecology and management of the mourning dove. Stackpole, Harrisburg, Pennsylvania, USA.

Sanchez-Johnson, Y., F. Hernandez, D. G. Hewitt, G. L. Waggerman, E. J. Redeker, H. O. Melendez, H. V. Z. Trevino, and J. A. Roberson. 2009. Status of white-winged dove nesting colonies in Tamaulipas, Mexico. Wilson Journal of Ornithology 121:338–346.

Sanders, T. A. 2000. Habitat availability, dietary mineral supplement, and measuring abundance of band-tailed pigeons in western Oregon. Dissertation, Oregon State University, Corvallis, USA.

Saunders, G. B. 1940. Eastern white-winged dove (*Melopelia asistica*) in southeastern Texas. Unpublished report. USFWS, Washington, DC, USA.

Sayre, M. W., and N. J. Silvy. 1993. Nesting and production. Pages 81–104 *in* T. S. Baskett, M. W. Sayre, R. E. Tomlinson, and R. E. Mirarchi, editors. Ecology and management of the mourning dove. Stackpole, Harrisburg, Pennsylvania, USA.

Schacht, S. J., T. C. Tacha, and G. L. Waggerman. 1995. Bioenergetics of white-winged dove reproduction in the Lower Rio Grande Valley of Texas. Wildlife Monographs 129:1–31.

Schulz, J. H., A. J. Bermudez, and J. J. Millspaugh. 2005. Monitoring presence and annual variation of trichomoniasis in mourning doves. Avian Diseases 49:387–389.

Schulz, J. H., J. J. Millspaugh, A. J. Bermudez, X. Gao, T. W. Bonnot, L. G. Britt, and M. Paine. 2006. Acute lead toxicosis in mourning doves. Journal of Wildlife Management 70:413–421.

Schulz, J. H., J. J. Millspaugh, B. E. Washburn, G. R. Wester, J. T. Lanigan III, and J. C. Franson. 2002. Spent shot availability and ingestion on areas managed for mourning doves. Wildlife Society Bulletin 30:112–120.

Schwertner, T. W., H. A. Mathewson, J. A. Roberson, M. Small, and G. L. Waggerman. 2002. White-winged dove (*Zenaida asiatica*). *In* The birds of North America. Academy of Natural Sciences, Philadelphia, Pennsylvania, and American Ornithologists' Union, Washington, DC, USA.

Seamans, M. E. 2015. Mourning dove population status, 2015. USFWS, Division of Migratory Bird Management, Washington, DC, USA.

Sennett, G. B. 1879. Further notes on the ornithology of the lower Rio Grande of Texas. Bulletin of the US Geological and Geographical Survey of the Territories, Washington, DC, USA.

Sileo, L., Jr., and E. L. Fitzhugh. 1969. Incidence of trichomoniasis in the band-tailed pigeons of southern Arizona. Journal of Wildlife Disease 5:146.

Silovsky, G. D. 1969. Distribution and mortality of the Pacific coast band-tailed pigeon. Thesis, Oregon State University, Corvallis, USA.

Silvy, N. J. 2004. Recent research on mourning and white-winged doves in Texas. Pages 8–14 *in* N. J. Silvy and D. Rollins, editors. Dove biology, research, and management in Texas. Texas A&M University Research and Extension Center, San Angelo, USA.

Silvy, N. J., and D. Rollins, editors. 2004. Dove biology, research, and management in Texas. Texas A&M University Research and Extension Center, San Angelo, USA.

Silvy, N. J., and D. Zimmerman, editors. 1967. Kansas upland game birds. Kansas State University Student Chapter of the Wildlife Society, Manhattan, USA.

Skutch, A. F. 1964. Life histories of Central American pigeons. Wilson Bulletin 76:211–247.

Skutch, A. F. 1983. Birds of tropical America. University of Texas Press, Austin, USA.

Slade, N. A. 1969. Factors affecting mourning dove use of water in artificial catchment basins in a dryland farming area of Utah. Thesis, Utah State University, Logan, USA.

Small, M. F., R. A. Hilsenback, and J. F. Scudday. 1989. Resource utilization and nesting ecology of the white-winged dove (*Zenaida asiatica*) in central Trans-Pecos, Texas. Texas Journal of Agricultural and Natural Resources 3:37–38.

Soutiere, E. C., and E. G. Bolen. 1976. Mourning dove nesting on tobosa grass-mesquite rangeland sprayed with herbicides and burned. Journal of Range Management 29:226–231.

Stabler, R. M. 1951. A survey of Colorado band-tailed pigeons, mourning doves, and wild common pigeons for *Trichomonas gallinae*. Journal of Parasitology 37:471–472.

Stabler, R. M. 1961. A parasitological survey of fifty-one eastern white-winged doves. Journal of Parasitology 47:309–311.

Stabler, R. M., and C. E. Braun. 1979. Effects of a California-derived strain of *Trichomonas gallinae* on Colorado band-tailed pigeons. California Fish and Game Bulletin 65:56–58.

Stager, K. E. 1957. The avifauna of the Tres Marias Islands, Mexico. Auk 74:413–432.

Stair, J. L. 1970. Chronology of the nesting season of white-winged doves *Zenaida asiatica mearnsi* (Ridgway) in Arizona. Thesis, University of Arizona, Tucson, USA.

Stiles, F. G., and A. F. Skutch. 1989. A guide to the birds of Costa Rica. Cornell University Press, Ithaca, New York, USA.

Sutton, G. M., and O. S. Pettingill. 1942. Birds of the Gomez Farias region, southwestern Tamaulipas. Auk 59:1–34.

Swank, W. G. 1955. Feather molt as an ageing technique for mourning doves. Journal of Wildlife Management 19:412–414.

Swanson, D. A. 1989. Breeding biology of the white-winged dove (*Zenaida asiatica*) in South Texas. Thesis, Texas A&I University, Kingsville, USA.

Swanson, D. A., and J. H. Rappole. 1993. Breeding biology of the eastern white-winged dove in southern Texas. *Southwestern Naturalist* 38:68–71.

Terres, J. K. 1980. The Audubon Society encyclopedia of North American birds. Alfred A. Knopf, New York, New York, USA.

Thayer, J. E. 1909. Letter to the editor. Condor 11:142–143.

Tomlinson, R. E. 1988. Hunter declarations at border ports of doves and pigeons bagged in Mexico. Unpublished report. USFWS, Albuquerque, New Mexico, USA.

Tomlinson, R. E. 1993. Migration. Pages 57–80 *in* T. S. Baskett, M. W. Sayre, R. E. Tomlinson, and R. E. Mirarchi, editors. Ecology and management of the mourning dove. Stackpole, Harrisburg, Pennsylvania, USA.

USFWS. 2005. Mourning dove national strategic harvest management plan. USFWS, Washington, DC, USA.

Viers, C. E., Jr. 1970. The relationship of calling behavior of white-winged doves to population and production in southern Arizona. Dissertation, University of Arizona, Tucson, USA.

Waggerman, G. L. 1973. White-winged dove harvest regulations. Federal Aid Project W-30-R-28. Texas Parks and Wildlife Department, Austin, USA.

Waggerman, G. L., and S. Sorola. 1977. Whitewings in the winter. Texas Parks and Wildlife Magazine 35:6–9.

Wells, J. V., and A. C. Wells. 2001. Pigeons and doves. Pages 319–325 *in* C. Elphick, J. B. Dunning Jr., and D. A. Sibley, editors. The Sibley guide to bird life and behavior. Alfred A. Knopf, New York, New York, USA.

West, L. M. 1993. Ecology of breeding white-winged dove in the San Antonio metropolitan area. Thesis, Texas Tech University, Lubbock, USA.

West, L. M., L. M. Smith, R. S. Lutz, and R. R. George. 1993. Ecology of urban white-winged doves. Transactions of the North American Wildlife and Natural Resource Conference 59:70–77.

Westmoreland, D., and L. B. Best. 1986. Incubation continuity and the advantage of cryptic egg coloration in mourning doves. Wilson Bulletin 98:297–300.

Wetmore, A. 1920. Observations on the habits of the white-winged dove. Condor 22:140–146.

Wetmore, A. 1943. The birds of southern Veracruz, Mexico. Proceedings of the United States National Museum 93:215–340.

Willard, F. C. 1916. Nesting of the band-tailed pigeon in southern Arizona. Condor 18:110–112.

Williams, L. 1971. A renesting study of the white-winged dove (*Zenaida asiatica* [Linnaeus]) in the Lower Rio Grande Valley of Texas. Thesis, Texas A&I University, Kingsville, USA.

Wolf, B. O., and C. M. del Rio. 2000. Use of saguaro fruit by white-winged doves: isotopic evidence of a tight ecological association. Oecologia 124:536–543.

Zeigler, D. L. 1971. Crop-milk cycles in band-tailed pigeons and losses of squabs due to hunting pigeons in September. Thesis, Oregon State University, Corvallis, USA.

Coyote in Mexico

11

Lucina Hernández and John W. Laundré

Coyotes (*Canis latrans*) have had a controversial and often difficult-to-define role in the ecology of wildlife in North America. They are not large carnivores, usually weighing less than 20 kg (44 lb), but they are not small, easily outweighing their cousins the foxes (*Vulpes* spp.). They are "medium" predators, or mesopredators, but can also be top predators, preying on large mammals such as adult deer (*Odocoileus* spp.). They are vulnerable to larger predators but can feed on smaller ones. Though not alone in having a diverse diet and ecological role, they are probably the most successful canid. Their diet consists of prey of many sizes. Like many small mammalian predators, they are omnivores, readily eating fruit and vegetable matter. In general, defaunation is size selective, with larger species exhibiting greater risks of decline or extinction than smaller ones (Purvis et al. 2000, Cardillo 2003). However, coyotes have expanded beyond their historical geographic range during the previous 300 years. Their success has created ecological and socioeconomic challenges.

The multiple ecological roles played by coyotes have led to conflicts with humans. Among pre-Columbian cultures, the coyote's status varied from respected fellow predator to vile trickster or spoiler. Today that dual role continues, as the coyote is respected by some and hated by others. Ecologically, coyotes play an important role in limiting small mammal populations. Because they have dietary preferences for medium- to smaller-sized mammals, such as rabbits, hares, and mice (Hernández et al. 1994, Hernández et al. 2002), this is a valid argument. However, they also prey on domestic animals (Lafón-Terrazas 1987, Bueno-Cabrera et al. 2005) and on wild prey consumed by humans, such as deer and wild turkeys (*Meleagris gallopavo*) (Truett 1979, Hamlin et al. 1984, Niedzielski and Bowman 2014).

The coyote is the most ecologically successful mesopredator in Mexico, occurring from northern temperate deserts and forests to humid tropical forests (fig. 11.1). It assumes diverse ecological roles, from top predator to prey. It is the focus of many

Figure 11.1. Male coyote in the *matorrales* of Nuevo León, Mexico. Photo by Jorge G. Villarreal González.

socioeconomic conflicts between humans, especially regarding livestock enterprises and wildlife management programs. The primary management concern is whether coyotes should be indiscriminately controlled and even extirpated, or whether they should be considered an integral component of Mexico's wildlife communities and valued for their ecological services. In this chapter, we explore the coyote's adaptability and its ecological role in wildlife communities.

Historical Perspective

Value to native peoples

When we manage wildlife species, our beliefs, biases, and cultural backgrounds affect our decisions and strategies. In any discussion of the ecology and management of the coyote in Mexico, we should consider how Mexico's rich cultural heritage has shaped past and present views of this species. In native cultures, the coyote is portrayed as a trickster or a clever antihero (Dove 1990, Erdoes and Ortiz 1998). In Aztec legends, the coyote is called Huehuecoyotl, meaning "old coyote," and it is considered the god of the arts, the lord of music and dance, and a guide from youth to adulthood. Mexicans linked coyotes with good luck, but nowadays, the word "coyote" refers to a dishonest person, and *coyotear* means to take advantage of, cheat, or surprise someone. Although early Mexicans believed coyotes were tricksters who had shape-shifting powers, they also admired coyotes for their cunning and strength. This fits our modern perception of coyotes—we admire their ability to survive but also despise their shrewdness and adversarial relationship with us.

Extirpation and recovery

In Mexico and the rest of North America, the coyote is a successful species that has expanded beyond its historical distributional range. Coyotes occur near, and even in, some of the larger cities in northern Mexico (personal observations). Coyotes have also become more abundant than is historically evident in southern tropical Mexico (Hidalgo-Mihart et al. 2004a), and they now extend to the southern border of Mexico.

Taxonomy and Evolution

Coyotes are one of the oldest extant members of the genus *Canis*. They evolved from small vulpine ancestors in the Early to Middle Pliocene (Nowak 1978). Coyotes have remained less specialized than larger canids such as wolves (*C. lupus*). The 19 recognized subspecies (Nelson 1932, Jackson 1951, Hall 1981) in North America, occurring from Alaska to Costa Rica, indicate that coyotes have become regionally differentiated. Ten subspecies occur in Mexico. However, it is improbable that a vagile species such as the coyote could maintain subspecies integrity, considering the short distances and lack of barriers separating recognized subspecies (Hall 1981). Thus, the validity of these subspecies is doubtful, and a taxonomic revision based on a molecular genetic analysis similar to that conducted for pumas (*Puma concolor*) (Culver et al. 2000) is needed.

Distribution

The coyote is currently the most widely distributed native mesopredator in North America (Ceballos et al. 2006). Leopold (1959) opined that coyotes did not occur originally in the tropical forests of southern Mexico in Campeche, Quintana Roo, and Yucatán (Sosa Escalante et al. 1997). However, recent evidence indicates that coyotes were present in southern Mexico and Central America as early as the Pleistocene and Early Holocene and were present during European colonization (Hidalgo-Mihart et al. 2004a), but they were probably less abundant than they are now. Leopold (1959) hypothesized that because wolves preyed on coyotes, wolves kept coyote densities low in the northern pine forests of the Sierra Madre Occidental, as well as in the plains and more arid areas. If we consider the negative impact wolves have had on coyotes in Yellowstone National Park in the US since wolves were reestablished (Murray Berger and Gese 2007), Leopold's contention may have some validity. After the extermination of wolves from northern Mexico in the 1950s, coyotes probably became abundant in the region (fig. 11.2).

Human-caused habitat alterations have also contributed to the increase in coyote populations. The clearing of forests stimulated an increase in small mammal populations, and the introduction of small domestic stock provided an alternate food source. In tropical Mexico, these habitat changes probably resulted in sufficient increases in coyote abundance to allow confirmation of their presence (Hidalgo Mihart et al. 2013). Coyote densities probably differ among habitats, being lower in forested than in desert and semiarid areas.

Figure 11.2. Coyotes in northern Mexico probably became abundant after extermination of wolves. Photo by Jorge G. Villarreal González.

Description

General description

Adult males weigh 8 to 20 kg (14 to 44 lb), and females 7 to 18 kg (15 to 40 lb). Gier (1975) reported that weights of coyotes from Mexican deserts averaged less (11.5 kg, 25.3 lb) than those of northern animals (~18 kg, 40 lb). Weights of male and female coyotes from the northern Chihuahuan Desert in the US averaged 11.6 ± 1.4 kg (25.5 ± 3 lb) (SE, n = 21) and 9.5 ± 1.0 kg (20.9 ± 2.2 lb) (n = 13), respectively (Windberg et al. 1997). Average weights of males and females from the northern Sierra Madre Occidental were 13.6 ± 1.45 kg (30 ± 3.19 lb) (n = 8) and 11.0 ± 2.2 kg (24.2 ± 4.85 lb) (n = 6), respectively (Servín et al. 2003). In the Bolsón de Mapimí in the central Chihuahuan Desert in Mexico, the average weight of females was 8.8 ± 0.31 kg (19.4 ± 0.68 lb) (range: 5–11 kg, 11–24.2 lb, n = 32), and that of males was 10.3 ± 0.38 kg (22.7 ± 0.83 lb) (range: 7–13 kg, 15.4–28.6 lb, n = 27) (Hernández et al., unpublished data). Weights of coyotes from southern Mexico are unknown.

Coyotes vary in color, being grayer and having more blackish tints in northwestern North America. In Mexico, their color varies from a fulvous shade in desert areas (Jackson 1951) to gray with reddish tints in forested tropical areas (personal observation). Their body fur is often longer in northern Mexico but can be short and coarse in tropical areas. The tail usually has a distinctive black tip, but the black tip may be absent in southern tropical areas.

Life History

Diet

Hundreds of diet studies using scat and stomach analyses have shown that coyotes consume nonfood items such as wood and plastic products, a wide variety of plant material including berries, seeds, and grass, and invertebrates and vertebrates (Johnson and Hansen 1979, MacCracken and Uresh 1984, Crimmins et al. 2012). Coyotes often deposit their scat on trails and roads, which facilitates collection of samples for diet studies. Diet is the most studied aspect of coyote life histories.

Coyote food habits in Mexico appear to be as diverse as in other areas (fig. 11.3) (Delibes et al. 1986, Hernández and Delibes 1994, Hernández et al. 1994, Sanabria et al. 1995, Guerrero et al. 2002, Guerrero et al. 2004, Álvarez-Castañeda and González-Quintero 2005, Martínez García et al. 2014). In desert environments, mammals—mostly rodents and lagomorphs—were major food items (Vela 1985, Delibes et al. 1986, Arnaud 1992, Hernández and Delibes 1994, Sanabria et al. 1995,

Grajales-Tam 1998, Grajales-Tam et al. 2003, Martínez Calderas 2005, González-Quintero 2001, Grajales-Tam and González-Romero 2014). In the drier Sonoran Desert, reptiles and arthropods increased in importance (Hernández et al. 1994). In coastal tropical dry forest and mountainous oak-pine forest, small mammals and plant material shared importance (Salas 1987, Delibes et al. 1989, Esparza 1991, Servin and Huxley 1991, González et al. 1992, Chacón de la Cruz 1994, Sandoval-Sandoval 1998, López Soto and Badii 2000, Hidalgo-Mihart et al. 2001, Guerrero et al. 2004, Díaz-Díaz et al. 2005, Gaona et al. 2005). In an island population in the Gulf of California near the Baja California Peninsula, birds and fish dominated diets (Álvarez-Castañeda and González-Quintero 2005). Domestic livestock and large mammalian prey were rarely present in any of the diet studies, constituting less than 5% of coyote diets (Hernández and Delibes1994, Hernández et al. 1994).

Although coyotes are opportunistic foragers (Van Vuren and Thompson 1982, Todd and Keith 1983), they are highly selective feeders, concentrating on specific prey groups (Johnson and Hansen 1979, MacCracken and Hansen 1987, Hernández et al. 2002). Their preference for lagomorphs over most of their range, in particular hares (*Lepus* spp.), represents an optimal foraging strategy (MacCracken and Hansen 1987, Hernández et al. 2002). The capture of one black-tailed jackrabbit (*Lepus californicus*) provides a coyote with energy for approximately two days, the equivalent of 9 to 10 rodents per day (Laundré and Hernández 2003a). Coyotes will select lagomorphs over small mammals, even in habitats where the latter are more abundant (Laundré et al. 2009).

Reproduction

Mating and breeding

The onset of estrus in females is variable and can occur anytime over a two-month period starting in mid-January. Once begun, estrus normally lasts for about four weeks (Kennelly 1978, Carlson and Gese 2008). Estrus in northern areas of the US and Canada occurs as much as two weeks later (Kennelly 1978), probably because of the later onset of spring. In the northern arid regions of Mexico, the commencement of the rainy season, approximately in June, affects the

Figure 11.3. Coyotes consume diverse foods, including insects, small mammals, and fruits such as prickly pear tunas. Photo by Jorge G. Villarreal González.

reproductive patterns of many species. There is not enough information on the timing of coyote reproductive cycles in northern Mexico to know whether they are synchronizing their reproductive efforts with precipitation. The timing of reproduction in tropical southern areas is unknown, but it is probably sensitive to tropical patterns of productivity.

Gestation

Gestation ranges from 60 to 63 days (Kennelly 1978). Pups exhibit a standard growth curve and attain weights of around 180 g (6.34 oz) by birth. Litter size varies from 1 to 8 or more (Kennelly 1978), but it usually ranges from 4.4 to 6.2 pups (Bekoff 1977). Coyote density and prey abundance seem to affect average litter size; lower litter sizes were reported during periods of low prey density or higher coyote density (Knowlton 1972). Sex ratio at birth is 1:1 (Bekoff 1977). Reproductive parameters of Mexican coyote populations are unknown.

Longevity

Coyotes in captivity are known to live up to 18 years (Young 1951). In the wild, few animals live more than 6 to 8 years (Gier 1968, Mathwig 1973). The oldest reported wild coyotes were 13.5 to 14.5 years old (Knowlton 1972, Nellis and Keith 1976). Longevity data are not available from Mexico.

Physiology

Energy and prey requirements

Though little is known of the energy requirements of coyotes in Mexico, there have been several studies of energetics and resulting prey needs of coyotes north of Mexico. In any energetic study, a principal component that determines most of the energy needed by an animal is its body mass. Kleiber (1961) initially established the basic relationship between body mass and energy needs with his formula for basic basal metabolic rate (BMR = W0.75, W = body mass). There have been attempts to estimate the more realistic resting metabolic rate (RMR) for unrestrained animals (Okarma and Koteja 1987). The second most important contribution to metabolic demand is from movement. Various investigators have attempted to estimate the magnitude of this component (Taylor et al. 1970, Corts and Lindzey 1984). Others have combined estimates of RMR and energy demands for movement to estimate the total daily energy demand, or field metabolic rate (FMR), of various species (Powell 1979, Ackerman et al. 1986, Laundré and Hernández 2003a). Ackerman et al. (1986) and Laundré and Hernández (2003a) added reproductive costs to better estimate the annual energy demands of animals.

Although energetic studies of coyotes in Mexico are not available, there are estimates for coyotes in the US. These estimates, with some adjustments, should be representative of the energetic demand of coyotes in Mexico. Golightly and Ohmart (1983) estimated a daily metabolic rate in kilojoules (kJ) of 3,722 kJ (890.4 kcal) for confined western desert coyotes (body weight 10 kg, 22 lb). Perkins (1992) used food consumption rates and estimates of daily travel to calculate winter FMR rates of northeastern coyotes. For coyotes weighing 15–16 kg (33–35.2 lb), estimates ranged from 6,309 to 10,861 kJ/day (1,508 to 2,596 kcal/day). Laundré and Hernández (2003b) used estimates of daily travel based on telemetry data to calculate FMR for western coyotes. Male weights averaged 11.4 ± 0.35 kg (25.1 ± 0.77 lb) (n = 11; range: 9–12.3 kg, 19.8–27.1 lb), and females averaged 9.4 ± 0.53 kg (20.7 ± 1.16 lb) (n = 8; range: 6.3–10.4 kg, 14.5–22.9 lb). Males traveled an average of 26.3 ± 1.4 km/day (16.3 ± 0.86 mi/day), and females traveled 27.6 ± 1.4 km/day (17.1 ± 0.86 mi/day). Estimated FMR of males and females was 1,170.1 ± 29 kcal/day (4,892.0 ± 121.2 kJ/day) and 988.6 ± 44.3 kcal/day (4,132.3 ± 185.2 kJ/day), respectively.

There are no published FMR data from Mexico, but there are estimates derived from GPS-collared coyotes in the Mapimí Biosphere Reserve in the Chihuahuan Desert (J. Laundré, unpublished data). Based on body weight and daily distance traveled, the estimated FMR for females was 864.5 ± 144.1 kcal/day (n = 5), and for males it was 929 ± 138.4 kcal/day (n = 4). These estimates are lower than those for northern populations. The differences in FMR between Mapimí coyotes and their northern cousins could result from differences in RMR, or in activity levels. RMR is determined primarily by animal body mass. Coyotes from the Chihuahuan Desert were similar in weight to those from Idaho (females: 9.9 ± 0.7 kg, 21.8 ± 1.5 lb; males: 12.5 ± 0.3 kg, 27.5 ± 0.6 lb). RMR should be comparable for both areas and should not contribute to differences in FMR. Relative to coyote activity levels in the Chihuahuan Desert, average daily travel for females and males was 9.9 ± 2.2 km/day (6.1 ± 1.3 mi/day) and 10.4 ± 2.1 km/day (6.4 ± 1.3 mi/day), respectively. These distances are substantially lower than those of coyotes in Idaho and are probably the major contributing factor to FMR differences with coyotes from other regions.

The lower daily travel distances for coyotes in Mapimí were like those reported by Servín et al. (2003) for coyotes in forested habitat in the Sierra Madre Occidental of northern Mexico (females: 12.5 ± 3.52 km SD, 7.7 ± 2.1 mi, n = 23; males: 16.5 ± 4.9 km SD, 10.2 ± 3 mi, n = 68). These data indicate that coyotes from southern areas travel less per day than their northern counterparts. The reason for these differences is unknown but might be related to differences in prey density, prey distribution, or temperature patterns.

Estimates of the metabolic demand of coyotes make it possible to estimate how many prey items a coyote would have to consume to meet its metabolic demand. In the US, Litvaitis and Mautz (1980) used feeding trials to estimate minimum requirements of captive coyotes. Their results indicated that a 12.9 kg (28.4 lb) coyote would need to consume 8 deer, 105 hares, or 4,800 mice per year. Their estimates did not include the energy demand of activity. Laundré and Hernández (2003a) used their estimated energy budgets for coyotes, as well as published caloric values of prey types, to calculate prey requirements of free-ranging coyotes in the western US. They estimated that nonreproductive male coyotes would need to consume 14.6 ± 0.8 deer, 192.0 ± 11 lagomorphs, or 3,681 ± 204 small mammals per year. Nonreproductive females would require 12.3 ± 1.3 deer, 162.2 ± 17 lagomorphs, or 3,110 ± 331 small mammals per year. A reproductive female coyote would need to consume 37 more lagomorphs or 700 additional small mammals per year to provide the additional energy for gestation, lactation, and pup dependence after lactation.

If the lower activity levels of coyotes in Mapimí are adjusted to calculate metabolic demand (Laundré and Hernández 2003a), a nonreproductive male coyote would need to consume approximately 11.6 deer, 152 lagomorphs, or 2,923 small mammals per year. Nonreproductive female coyotes would need to consume approximately 10.8 deer, 142 lagomorphs, or 2,720 small mammals per year. These estimates provide an initial approximation of the prey demands of Mexican coyotes.

Water requirements

Little research has been done on the water requirements of coyotes in general. Canids' evaporative loss of body moisture increases with temperature (Golightly and Ohmart 1983). However, how coyotes replace this loss is not well documented. Because prey tissue usually consists of more than 80% water, much of the water demand by coyotes is probably satisfied by prey consumption. Whether coyotes need to drink freestanding water is debatable. In monitoring coyotes for over 80 months, mostly during 24-hour periods in the summertime in the cold desert of the northwestern US, J. W. Laundré never recorded individuals visiting freestanding water sources (unpublished data). Likewise, in our GPS study of coyotes in the Chihuahuan Desert of northern Mexico, most animals monitored continuously over nine weeks did not visit water sources. The few animals that had access to freestanding water sources rarely visited them (Arias-Del Razo, unpublished data). In more humid areas, where dew occurs and where abundant freestanding water is available, coyotes probably occasionally supplement their water needs by drinking. However, it appears that the lack of available water is not a limiting factor for coyotes.

Thermal relationships

Coyotes in North America experience temperature extremes of −30°C in winter to 30°C in summer. Their ability to tolerate extreme temperatures depends on physiological and behavioral adaptations. In northern areas, coyotes grow thick insulating winter fur, which they lose to accommodate higher summertime temperatures. During periods of cold temperatures, coyotes probably seek microhabitats with reduced wind and direct sunlight. In warmer areas, coyotes likely seek shaded areas to rest. Except for the denning season, there is no evidence that coyotes seek dens as a thermal-regulating behavior. They have not been documented seeking freestanding water to remain cool, but it remains a possibility.

Behavior and Ecology

Social structure

Coyotes form annual pair bonds not only for copulation. The pair bond persists through gestation and pup rearing to pup dispersal, and there is evidence that surviving individuals will pair bond again the subsequent year (Bekoff and Gese 2003).

The female cares intensively for pups in the den while the male forages and provides food to the female. Within a few weeks, the male and female both forage to feed the growing pups. At about two to three weeks, pups leave the den (Bekoff 1977) and join the parents, first at mobile resting locations and eventually traveling with the parents within the home range. During these times, group members keep in contact by vocalizations (see section on communication), as they do when they hunt in different areas or cooperate in joint hunting forays. Group hunting continues until dispersal of the young in the fall.

Because the family dissolves toward the end of the year, most coyotes do not exist in established packs as wolves do. However, under certain circumstances, persistent family groups, or packs of coyotes, have been recorded. The most notable cases occur where large ungulates, mainly elk (*Cervus canadensis*), concentrate in the winter (Bowen 1981, Gese et al. 1996). Under these circumstances, coyotes scavenge dead elk, and family groups persist and defend territories from other groups (Bowen 1981). This is an exceptional situation, however, and most coyotes, including those in Mexico, probably exhibit seasonal family group structure (fig. 11.4).

Marking

Coyotes scent mark by depositing feces and urine in specific locations. Both sexes engage in scent marking, and it is probably related primarily to defining their territorial boundaries (Gese and Ruff 1997).

Movements

Some predators use ambush tactics, while others actively seek and pursue their prey (Sanderson 1966). Coyotes spend much time actively searching for food. Using radiotelemetry, Andelt and Gipson (1979) reported that male and female coyotes in Nebraska in the US traveled 10.1 and 11.8 km/day (6.2 and 7.3 mi/day), respectively. Laundré and Keller (1981) reported daily averages of 28.3 km/day (17.5 mi/day) for males and 22.1 km/day (13.7 mi/day) for females in Idaho. In the Sierra Madre Occidental in northern Mexico, Servín et al. (2003) reported average daily distances of 13.8 and 12.5 km/day (8.5 and 7.7 mi/day) for males and females, respectively. Similar distances traveled per day were also found for Chihuahuan Desert coyotes in the Mapimí Biosphere Reserve (Hernández et al. 1993; Arias-Del Razo, unpublished data). Coyotes are primarily crepuscular and are most active during sunrise and sunset (Andelt and Gipson 1979, Laundré and Keller 1981, Smith et al. 1981, Hernández et al. 1993, Servín et al. 2003, Hidalgo-Mihart et al. 2009, Arias-Del Razo et al. 2011). Coyotes periodically travel beyond their home ranges (Harrison et al. 1991). One coyote (personal observation), collared in the Mapimí Biosphere Reserve, traveled approximately 120 km (74.5 mi) and then returned close to where it was trapped. Dispersing juvenile coyotes can travel several hundred kilometers before establishing home ranges (Harrison et al. 1991, Harrison 1992, Rosatte 2002, Kolbe and Squires 2004) (fig. 11.5).

Figure 11.4. Young male coyote in the Mapimí Biosphere Reserve, northern Mexico. Photo by John W. Laundré.

Figure 11.5. Female adult coyote trapped for a study in the Mapimí Biosphere Reserve, northern Mexico. Photo by John W. Laundré.

Communication

Coyotes vocalize extensively. The coyote's reputation as a "song dog" is legendary (Leydet 1988); its scientific name means "barking dog." Coyotes have a well-developed repertoire of howls and yelps (McCarley 1975). Lehner (1978) recognized 11 different vocalizations ranging from growls to the classic howl. The vocalizations vary in complexity and the information they convey. They can be used in an agonistic context (e.g., growling) or for greeting conspecifics (e.g., whining) (Lehner 1978). The most familiar vocalization is the howl, which can be broken down into different categories, such as a lone howl versus a group howl (Lehner 1978). The howl conveys information primarily over long distances within and among groups (Mitchell et al. 2006). Many coyote vocalizations contain individually specific characteristics, enabling coyotes to recognize individuals (Mitchell et al. 2006). The amount of howling appears to increase during the reproductive season (Laundré 1981, Gese and Ruff 1998), but not, as is popularly believed, with the full moon (Laundré, unpublished data; Bender et al. 1996). Seasonal changes are probably related to the different communication needs of coyotes at different times of the year (Laundré 1981, Gese and Ruff 1998).

In addition to their vocal capabilities, coyotes will often use facial expressions to communicate emotions such as aggression, fear, pleasure, and uncertainty (Lehner 1978). Coupled with these facial expressions are body postures that reinforce the signals given (Lehner 1978). Coyotes use these combined forms of communication in intimate circumstances—when courting, reuniting with family members, establishing dominance hierarchies, or engaging in territorial disputes between family groups. Coyotes also exchange information olfactorily. By scent marking with anal gland secretions, urine, and feces, coyotes announce and recognize territorial limits, communicate reproductive status, and convey information in passive-submissive social contexts (Lehner 1978, Gese and Ruff 1997, Allen et al. 1999).

Spacing

Coyote home ranges vary from less than 10 km² (6.2 mi²) to more than 90 km² (55.9 mi²) (Hibler 1977, Bowen 1982, Laundré and Keller 1984). In general, male home ranges are larger than those of females (Laundré and Keller 1984). In the Sierra Madre Occidental in northern Mexico, Servin and Huxley (1995) reported home ranges of 10.6 ± 0.96 km² (6.5 ± 0.59 mi²) for males and 6.1 ± 1.3 km² (3.8 ± 0.8 mi²) for females. In the Chihuahuan Desert, Hernández et al. (1993) estimated home ranges of 15.2 and 16.8 km² (9.4 and 10.4 mi²) for a male and a female coyote, respectively. De Ita Vélez (2007) and Arias-Del Razo (unpublished data) reported home ranges of 10.4 to 23.7 km² (6.4 to 14.7 mi²) for males, and 9.9–20.6 km² (6.1–12.8 mi²) for females in the same area. In a tropical dry forest on the west coast of Mexico, home ranges varied from 10.9 to 43.7 km² (6.7 to 27.1 mi²) (Hidalgo-Mihart et al. 2004b). In southern Mexico, estimated home ranges averaged 3.5 km² (2.1 mi²) (Marín-Sánchez et al. 2015).

Several factors contribute to variation in home range size, including prey density, coyote density, habitat, and methods used to estimate home range size (Laundré and Keller 1984). Coyotes likely determine their home range by a variety of factors important to their innate needs for survival. The size of an area is not as important as how, why, and when individuals use it. Advances in GPS technology are allowing investigators to follow animal movements with striking detail and over extended periods. Recent GPS studies of coyote movements in the Mapimí Biosphere Reserve confirm previous data indicating that coyotes do not use their home range uniformly (Laundré and Keller 1981, Arias-Del Razo et al. 2012). Over a nine-week period, coyotes used some areas of their home ranges intensely, and others were not used (Arias-Del Razo et al. 2012). The use of certain areas also appears to be related to the time of day, as in day versus night. Why and when they use certain areas within home ranges is probably related to foraging strategies (Laundré and Keller 1981). For example, Hidalgo-Mihart et al. (2004b) found that coyotes using an open landfill in coastal Mexico had significantly smaller home ranges than animals outside the landfill.

Coyotes are considered territorial, especially during the reproductive and pup-rearing seasons (Gese 2001, Neale et al. 2007). Although they actively scent mark and vocalize their presence in an area, as well as aggressively defend their boundaries (Gese 2001), most telemetry studies have found considerable overlap in areas of use (Andelt and Gipson 1979, Bowen 1982, Kitchen et al. 2000, Kamler et al. 2005). This is also the case for coyotes in Mexico (De Ita Vélez 2007).

Some of the overlap occurs because of variation in spatial use of adjoining territories over time (Kitchen et al. 2000). Often, the overlap is from transient animals that have not established a territory. However, their use of areas occupied by resident animals puts into question the degree of territoriality in coyotes. Their territorial behavior is flexible seasonally and annually, and it probably evolved to fit their survival needs in the variable environments they occupy.

Population Dynamics

How a population functions, or its population dynamics, is a major consideration in understanding wildlife populations. The population size of a species may be considered a representation of how much energy that species can capture or carve out for itself as the energy flows through the ecosystem. Each individual represents a certain quantity of energy, an energy packet as it were, in the population, and the survival, mortality, and reproduction of these individuals determines the fate of the population. How a species captures additional energy via reproduction and how it loses that energy via mortality can help us better understand the role it plays in the ecosystem, as well as how we may be able to manage that energy flow. Thus, various aspects of energy flow, such as how much is in the population at a given time (density), or how much is being added or lost (productivity and survival), become important in understanding the total process of energy flow into and out of the population. Each factor will be considered individually, but all of them working together produce the results we see regarding the past, current, and just as importantly, future trends in energy flow through the species.

Density

The first factor to consider regarding a species' population dynamics is density. This helps determine how much energy is currently trapped in the population, and the change in that energy over space and time. Strictly defined, density is the number of individuals per unit area. Thus, to have an accurate estimate of density (absolute density), we would need to account for all individuals in an area, a formidable and likely impossible task. Density is usually our best estimate, with an unknown degree of error. However, if we assume the error is similar among studies, we can compare these density estimates on a relative basis (relative density) to make comparisons among different populations. Because estimating the absolute density, or even relative density, of a mobile and secretive predator such as the coyote is extremely difficult, most investigators use relative abundance indices, which they obtain using several techniques. In the case of the coyote, the most commonly used technique is scent stations (Linhart and Knowlton 1975). Coyote abundance is expressed in the number of visits per scent station per night and can be used for relative comparisons over time and among different areas. Here we report results from the few studies that provide absolute density estimates, and studies of relative abundance. From these comparisons, we can get a better picture of whether coyote abundance differs among habitat types and over time.

Of the few coyote density studies, none are from Mexico (Clark 1972, Camenzind 1978, Babb and Kennedy 1989, Gese et al. 1989, Hein and Andelt 1995). Density estimates vary from fewer than 10 coyotes/100 km^2 (16.1 coyotes/100 mi^2) (Clark 1972, Boutin et al. 1995) to more than 250/100 km^2 (403/100 mi^2) (Gese et al. 1989). Much of this variation can be attributed to differences in methodology (Clark 1972, Hein and Andelt 1995). This makes it difficult to extrapolate these data to Mexico. Added to this is the evidence that density, within a given study, varies as much as seven times from the lowest to the highest annual estimates over a period of several years (Clark 1972, Boutin et al. 1995).

These annual density fluctuations are supported by relative abundance data. In a regional analysis of scent station data over the western half of the US, Linhart and Knowlton (1975) reported positive or negative percentage changes from one year to the next in every state, with a maximum of 131%. Over a 23-year study in the northern Chihuahuan Desert, Windberg et al. (1997) reported scent station indices fluctuating annually from a low of fewer than 50 visits/1,000 station nights to more than 140 visits/1,000 station nights. Also in the Chihuahuan Desert, Hernández et al. (unpublished data) reported that coyote abundance varied from 16 to 226 visits/1,000 nights. Similar abundance levels are reported from a one-year study in a more northern part of the Mexican Chihuahuan Desert (Rodríguez Martínez 2011). Changes in abundance in other desert areas (Clark 1972) were related to prey abundance, primarily of black-tailed jackrabbits, which was affected by precipitation and plant productivity

(Hernández et al., unpublished data). Similar biotic and abiotic habitat interactions are evident in other studies (Clark 1972, Todd and Keith 1983, Boutin et al. 1995, O'Donoghue et al. 1997, Ponce Guevara et al. 2005). Coyote long-term abundance data from forested temperate or tropical areas is lacking, but short-term data (one to two years) indicate similar relative abundance (Cruz-Jácome et al. 2015, Castro Cuamatzi 2015). Coyote abundance in forested areas in Mexico is probably also related to annual fluctuations in density of their principal prey.

Population trends and current status

Coyotes have a stable population trend in most of Mexico. In southern states, sightings could indicate increases in coyote abundance, but this has not been verified. If so, it is probably the result of an increase in semiopen agricultural development.

Productivity and recruitment

Female coyotes reproduce once a year. The average in utero litter size is six pups (Kennelly 1978, Sacks 2005). However, because of various preparturition and postparturition factors, litter sizes are around five (Kennelly 1978). Recruitment into a stable population is usually low because of territorial behavior. The lack of space means that most young coyotes disperse from their natal area (Harrison 1992), and at any given time, there seems to be a high number of transient young animals in the population. However, over much of the range of coyotes, including Mexico, animals are hunted indiscriminately and are often the object of control operations. Under these circumstances, dispersing young animals quickly replace losses of resident individuals. Whether these new residents are locally produced animals or dispersers from other areas is unknown, but they are probably a mixture.

Survival and mortality factors

Coyotes can potentially live up to 18 years (Manville 1953) but rarely achieve that in the wild. Andrews and Boggess (1978) reported a first-year mortality rate of 35.8% for coyote pups in Iowa. Annual survival rates of coyotes vary from 87% to 90% in protected populations, 68% to 74% in lightly exploited populations, and 47% to 64% in moderate to heavily exploited ones (Kamler and Gipson 2004). Human hunting and trapping is the major mortality factor for coyotes. Because coyotes prey on game species and domestic livestock, humans kill them over most of their range.

Predators

Besides humans, other predators do not commonly prey on coyotes, although pumas will do so occasionally (Ackerman et al. 1984; personal observations). One female puma in New Mexico was documented to have killed 45 coyotes in less than one year (Ron Thompson, personal observation). In Mexico, coyotes were reported in the diet of pumas from Sonora (Rosas-Rosas et al. 2003, Cassaigne et al. 2016), and we found several incidences of pumas killing coyotes in the Mapimí Biosphere Reserve (personal observation). Because coyotes constitute a small percentage of the puma's diet (Ackerman et al. 1984, Rosas-Rosas et al. 2003), pumas are not considered a major threat. Leopold (1959) reported that wolves were major predators of coyotes, and that they were rarely found together. The negative impact of reintroduced wolves on the abundance and distribution of coyotes in Yellowstone National Park supports this observation (Murray Berger and Gese 2007).

Parasites and diseases

A wide range of parasites affect coyotes (Gier et al. 1978). Ectoparasites include mange mites, ticks, lice, and fleas. Two kinds of mange afflict coyotes: *Demodex canis* and scabies (*Sarcoptes scabiei*). Occasional outbreaks can occur over extensive geographical areas. Ticks commonly infect coyotes, and the severity varies with the habitat, season, and year. Lice infections are rare, which may indicate a natural resistance to these parasites (Gier et al. 1978). Although most if not all coyotes are infected with fleas, they are usually a nuisance rather than a health hazard (Gier et al. 1978).

A variety of flatworms and roundworms infect coyotes internally (Gier et al. 1978). Flatworm infections are not common or widespread; they include the lung fluke (*Paragonimus kellicotti*), intestinal flukes (*Alaria* spp.), and tapeworms (*Taenia* spp.). Tapeworms are often the most common endoparasites reported; 60% to 95% of the population may be parasitized (Gier et al. 1978). The more common roundworm infections include intestinal worms (*Toxascaris* spp.) and hookworms (*Ancylostoma* spp.). Heartworms (*Dirofilaria immitis*) also infect coyotes. The severity of infection varies regionally but does not appear to have detrimental effects on population numbers.

Coyotes are also susceptible to distemper, rabies, and parvovirus (Guo et al. 1986, Gese et al. 1991, Grinder and Krausman 2001a, Allison et al. 2013). However, little is known about how widespread such infections are or whether they have any impact on population numbers. There are indications that at times, deaths from these viruses can be substantial.

Emigration and immigration

Because of the territorial social structure of coyotes, more pups are produced annually than can be absorbed locally. Thus, many transient animals emigrate to other parts of the coyote's historical range and beyond, contributing to its success in colonizing new areas. Most dispersing individuals are young of the year of both genders (Harrison 1992). The propensity of both males and females to disperse probably explains why coyotes have been able to spread into unoccupied areas faster than other species such as pumas, in which males are the principal dispersers (Laundré and Hernández 2003b).

Population Management

Coyotes in Mexico are unprotected. Although they are recognized as providing a beneficial service by controlling rodents (SEMARNAP 2000), they are not afforded adequate legal protection. Some states have established an official hunting season for coyotes, but many UMAs petition SEMARNAT for permission to hunt coyotes without regard for season (E. Martinez, personal communication). In areas of livestock operations, coyotes are listed as predators and can be killed when necessary to protect livestock. In addition, even in established coyote hunting seasons, the number, gender, and age that can be killed are unrestricted.

Habitat requirements and management

Coyotes occur in a variety of habitats, including rural and urban areas (Grinder and Krausman 2001b, Hidalgo-Mihart et al. 2006). Management invariably involves reducing coyote numbers or restricting their presence in an area. Coyotes are highly adaptable to habitat changes, so habitat manipulations are not a likely management tool.

Management challenges in Mexico

Balancing the ecological value of coyotes as a top predator with their negative effects on important wildlife and domestic animal populations is a major challenge to coyote management in Mexico. The effects of coyotes on ecosystem structure and function should not be ignored or treated lightly. Top predators are an integral part of ecosystems and the evolutionary process (Leopold 1966, Henke and Bryant 1999, Berger et al. 2001, Schmitz et al. 2008) and provide ecosystem benefits that cascade through entire landscapes (Ripple and Beschta 2004, Schmitz 2006, Estes et al. 2011, Ripple et al. 2014). Severely reducing or completely removing predators, which has been practiced indiscriminately in the past, leads to cascading negative effects on ecosystems that can affect our reliance on the services these ecosystems provide (Henke and Bryant 1999, Borrvall and Ebenman 2006, Myers et al. 2007, Manning et al. 2009, Estes et al. 2011).

Changing public perception of the negative impacts of coyotes is another major challenge for wildlife agencies. Coyotes provide an important service by preying primarily on lagomorphs and small mammals that damage farm crops. Negative impacts of predators are often the result of human-induced environmental changes that can be addressed by implementing habitat management solutions. Coyote predation of game species is usually not significant and does not justify the time, money, and effort necessary to implement predator control programs; the solution is to provide adequate secure habitat for the desired species. The benefits of multispecies management programs far outweigh single-species management that attempts to sacrifice one species for another at the expense of ecosystem health.

There is no doubt that coyotes kill domestic livestock. What is debatable is the seriousness of the problem (Sheldon 1973, Love et al. 1974). Most diet studies show that coyotes rarely feed on livestock, and when they do, it is often a small portion of their diet (Delibes et al. 1986, Hernández and Delibes 1994, Hernández et al. 1994, Sanabria et al. 1995, Guerrero et al. 2002, Guerrero et al. 2004, Álvarez-Castañeda and González-Quintero 2005). However, in some cases, predation of livestock by coyotes can be significant (Lafón-Terrazas 1987, Stoddart et al. 2001). How can we reconcile these two extreme observations? The answer is that not all coyotes prey on livestock; the problem is often localized, and only a few coyotes are responsible. Also, the severity of the problem varies with the type of livestock. Cattle and horses are less prone to coyote predation than sheep and goats (Bueno-Cabrera et al. 2005). In one area of the Chihuahuan Desert, losses to other factors, including robbery, were higher than losses to predation (Bueno-Cabrera et al. 2005). Husbandry practices have a major impact on the severity of livestock losses to predators (Bueno-Cabrera et al. 2005). Ranchers can often reduce their losses to coyotes through increased vigilance over their flocks, use of guard dogs and protective collars, and other antipredator techniques (Burns et al. 1996, Andelt and Hopper 2000, Mitchell et al. 2004, Shivik 2006, Rigg et al. 2011, Tumenta et al. 2013). However, in rural areas of Mexico, socioeconomic factors may affect the feasibility of these techniques. Wildlife managers need to work in partnership with livestock owners to develop economic and ecological solutions to depredation problems (Knowlton et al. 1999). Managers can build such partnerships through local workshops where people can exchange information and develop mutual trust (Bueno Cabrera 2004).

In contrast to the negative effects of coyote livestock predation, coyotes consume high numbers of lagomorphs and rodents that can compete with livestock for forage (Lafón-Terrazas 1987). The reduction or control of these potential forage competitors can be especially beneficial in arid and semiarid habitats where forage is often limited. The problem of coyote predation on livestock is not simple and can be an emotional one. However, because problems are usually localized, control operations should also be localized. The lessons learned from widespread unrestricted control and elimination of large predators should not be forgotten.

The ultimate solution is to manage coyotes as valuable components of viable ecosystems and agroecosystems. Coyotes should be recognized for their positive economic effects and the ecological services they provide in maintaining healthy ecosystems on which all Mexicans depend. The many ecological roles of coyotes, and their socioeconomic impacts, greatly complicate the implementation of sound management practices, but it is critical that managers develop prudent strategies.

Literature Cited

Ackerman, B. B., F. G. Lindzey, and T. P. Hemker. 1984. Cougar food habits in southern Utah. Journal of Wildlife Management 48:147–155.

Ackerman, B. B, F. G. Lindzey, and T. P. Hemker. 1986. Predictive energetics model for cougars. Pages 333–352 *in* S. D. Miller and D. D. Everett, editors. Cats of the world: biology, conservation, and management. National Wildlife Federation, Washington, DC, USA.

Allen, J. J., M. Bekoff, and R. L. Crabtree. 1999. An observational study of coyote (*Canis latrans*) scent-marking and territoriality in Yellowstone National Park. Ethology 105:289–302.

Allison, A. B., D. J. Kohler, K. A. Fox, J. D. Brown, R. W. Gerhold, V. I. Shearn-Bochsler, E. J. Dubovi, C. R. Parrish, and E. C. Holmes. Frequent cross-species transmission of parvoviruses among diverse wild carnivore hosts. Journal of Virology 87:2342–2347.

Álvarez-Castañeda, S. T., and P. González-Quintero. 2005. Winter-spring food habits of an island population of coyote *Canis latrans* in Baja California, Mexico. Journal of Arid Environments 60:397–404.

Andelt, W. F., and P. S. Gipson. 1979. Home range, activity, and daily movements of coyotes. Journal of Wildlife Management 43:944–951.

Andelt, W. F., and S. N. Hopper. 2000. Livestock guard dogs reduce predation on domestic sheep in Colorado. Journal of Range Management 53:259–267.

Andrews, R. D., and E. K. Boggess. 1978. Ecology of coyotes in Iowa. Pages 249–265 *in* M. Bekoff, editor. Coyotes: biology, behavior, and management. Academic Press, New York, New York, USA.

Arias-Del Razo, I., L. Hernández, J. Laundré, and O. Myers. 2011. Do predator and prey activity patterns match? A study of coyotes (*Canis latrans*), and lagomorphs (*Lepus californicus* and *Sylvilagus audubonii*). Journal of Arid Environments 75:112–118.

Arias-Del Razo, I., L. Hernández, J. Laundré, and L. Velasco-Vázquez. 2012. The landscape of fear: habitat use by a predator (*Canis latrans*) and its main prey (*Lepus californicus* and *Sylvilagus audubonii*). Canadian Journal of Zoology 90:683–693.

Arnaud, G. 1992. Ecología alimenticia del coyote (*Canis latrans*) en un región ganadera del norte del estado de Nuevo León, México. Thesis, UNAM, México, D. F., México.

Babb, J. G., and M. L. Kennedy. 1989. An estimate of

minimum density for coyotes in western Tennessee. Journal of Wildlife Management 53:186–188.
Bekoff, M. 1977. *Canis latrans*. Mammalian Species 79:1–9.
Bekoff, M., and E. M. Gese. 2003. Coyote. Pages 467–481 *in* G. A. Feldhamer, B. C. Thompson, and J. A. Chapman, editors. Wild mammals of North America: biology, management, and conservation. Second edition. Johns Hopkins University Press, Baltimore, Maryland, USA.
Bender, D. J., E. M. Bayne, and R. M. Brigham. 1996. Lunar condition influences coyote (*Canis latrans*) howling. American Midland Naturalist 136:413–417.
Berger, J., P. B. Stacey, L. Bellis, and M. P. Johnson. 2001. A mammalian predator-prey imbalance: grizzly bear and wolf extinction affect avian Neotropical migrants. Ecological Applications 11:947–960.
Borrvall, C., and B. Ebenman. 2006. Early onset of secondary extinctions in ecological communities following the loss of top predators. Ecology Letters 9:435–442.
Boutin, S., et al. 1995. Population changes of the vertebrate community during a snowshoe hare cycle in Canada's boreal forest. Oikos 74:69–80.
Bowen, W. D. 1981. Variation in coyote social organization: the influence of prey size. Canadian Journal of Zoology 59:639–652.
Bowen, W. D. 1982. Home range and spatial organization of coyotes in Jasper National Park, Alberta. Journal of Wildlife Management 46:201–216.
Bueno Cabrera, A. 2004. Impacto del puma (*Puma concolor*) en ranchos ganaderos del Área Natural Protegida "Cañon de Santa Elena," Chihuahua. Thesis, Instituto de Ecología, Xalapa, Veracruz, México.
Bueno-Cabrera, A., L. Hernández-Garcia, J. Laundré, A. Contreras-Hernandez, and H. Shaw. 2005. Cougar impact on livestock ranches in the Santa Elena Canyon, Chihuahua, Mexico. Pages 141–149 *in* R. A. Beausoleil and D. A. Martorello, editors. Proceedings of the Eighth Mountain Lion Workshop, Olympia, Washington, USA.
Burns, R. J., D. E. Zemlicka, and P. J. Savarie. 1996. Effectiveness of large livestock protection collars against depredating coyotes. Wildlife Society Bulletin 24:123–127.
Camenzind, F. J. 1978. Behavioral ecology of coyotes on the National Elk Refuge, Jackson, Wyoming. Pages 267–294 *in* M. Bekoff, editor. Coyotes: biology, behavior, and management. Academic Press, New York, New York, USA.
Cardillo, M. 2003. Biological determinants of extinction risk: why are smaller species less vulnerable? Animal Conservation 6:63–69.
Carlson, D. A., and E. M. Gese. 2008. Reproductive biology of the coyote (*Canis latrans*): integration of mating behavior, reproductive hormones, and vaginal cytology. Journal of Mammalogy 89:654–664.
Cassaigne, I. R., et al. 2016. Diet of pumas (*Puma concolor*) in Sonora, Mexico, as determined by GPS kill sites and molecular identified scat, with comments on jaguar (*Panthera onca*) diet. Southwestern Naturalist 61:125–132.
Castro Cuamatzi, H. M. 2015. Riqueza y abundancia de mamíferos de talla mediana ante la presencia humana y animales domésticos en el Parque Nacional La Malinche, Tlaxcala. Thesis, Universidad Autónoma de Tlaxcala, Tlaxcala, México.
Ceballos, G., S. Blanco, C. González, and E. Martínez. 2006. *Canis latrans* (coyote). Distribución potencial, escala 1:1000000. Instituto de Biología. UNAM, México, D. F., México. http://www.conabio.gob.mx/informacion/gis/layouts/can_latragw.png. Accessed 23 Jun 2016.
Chacón de la Cruz, J. E. 1994. La alimentación del coyote en relación con la disponibilidad de alimento en la Reserva de la Biosfera "La Michilía," Durango. Thesis, Universidad Juárez del Estado de Durango, Gómez Palacio, Durango, México.
Clark, F. W. 1972. Influence of jackrabbit density on coyote population change. Journal of Wildlife Management 36:343–356.
Corts, K. E., and F. G. Lindzey. 1984. Basal metabolism and energetic cost of walking in cougars. Journal of Wildlife Management 48:1456–1458.
Crimmins, S. M., J. W. Edwards, and J. M. Houben. 2012. *Canis latrans* (coyote) habitat use and feeding habits in central West Virginia. Northeastern Naturalist 19:411–420.
Cruz-Jácome, O., E. López-Tello, C. A. Delfín-Alfonso, and S. Mandujano. 2015. Riqueza y abundancia relativa de mamíferos medianos y grandes en una localidad en la Reserva de la Biosfera Tehuacán-Cuicatlán, Oaxaca, México. Therya 6:435–448.
Culver, M., W. E. Johnson, J. Pecon-Slattery, and S. J. O'Brien. 2000. Genomic ancestry of the Ameri-

can puma (*Puma concolor*). Journal of Heredity 91:186–197.

De Ita Vélez, C. I. 2007. Uso del ámbito hogareño de los coyotes (*Canis latrans*) en la Reserva de la Biosfera Mapimí. Thesis, Instituto de Ecología, Xalapa, Veracruz, México.

Delibes, M., L. Hernandez, and F. Hiraldo. 1986. Datos preliminares sobre la ecología del coyote y el gato montés en el sur del desierto de Chihuahua, México. Historia Natural 6:77–80.

Delibes, M., L. Hernandez, and F. Hiraldo. 1989. Comparative food habits of three carnivores in western Sierra Madre, Mexico. Zeitschrift für Säugetierkunde 54:107–110.

Díaz-Díaz, D., J. Martínez-Vázquez, and R. M. González-Monroy. 2005. Dieta del coyote (*Canis latrans caggotis*) en el Parque Nacional Pico de Orizaba, Puebla. *In* 18 Congreso Nacional de Zoología, México, D. F., México.

Dove, M. 1990. Coyote stories. University of Nebraska Press, Lincoln, USA.

Erdoes, R., and A. Ortiz. 1998. American Indian trickster tales. Penguin Books, New York, New York, USA.

Esparza, J. A. 1991. Variaciones estacionales en la dieta de mamíferos carnívoros en la Estación Científica Las Joyas. Thesis, Universidad de Guadalajara, Jalisco, México.

Estes, J. A., et al. 2011. Trophic downgrading of Planet Earth. Science 333:301–306.

Gaona, S., G. López-Ortega, E. Hernández-Pérez, and D. Lozano-Cuenca. 2005. Hábitos alimentarios del coyote (*Canis latrans cagotis*) en el Parque Nacional Iztaccihuatl-Popocatépetl, Estado de México. *In* 18 Congreso Nacional de Zoología, México, D. F., México.

Gese, E. M. 2001. Territorial defense by coyotes (*Canis latrans*) in Yellowstone National Park, Wyoming: who, how, where, when, and why. Canadian Journal of Zoology 79:980–987.

Gese, E. M., O. J. Rongstad, and W. R. Mytton. 1989. Population dynamics of coyotes in southeastern Colorado. Journal of Wildlife Management 53:174–181.

Gese, E. M., and R. L. Ruff. 1997. Scent-marking by coyotes, *Canis latrans*: the influence of social and ecological factors. Animal Behaviour 54:1155–1166.

Gese, E. M., and R. L. Ruff. 1998. Howling by coyotes (*Canis latrans*): variation among social classes, seasons, and pack sizes. Canadian Journal of Zoology 76:1037–1043.

Gese, E. M., R. L. Ruff, and R. L. Crabtree. 1996. Social and nutritional factors influencing the dispersal of resident coyotes. Animal Behaviour 52:1025–1043.

Gese, E. M., R. D. Schultz, O. J. Rongstad, and D. E. Andersen. 1991. Prevalence of antibodies against canine parvovirus and canine distemper virus in wild coyotes in southeastern Colorado. Journal of Wildlife Diseases 27:320–323.

Gier, H. T. 1968. Coyotes in Kansas. Kansas Agricultural Experiment Station, Kansas State University, Manhattan, USA.

Gier, H. T. 1975. Ecology and social behavior of the coyote. Pages 247–262 *in* M. W. Fox, editor. The wild canids. Dogwise Publishing, Wenatchee, Washington, USA.

Gier, H. T., S. M. Kruckenberg, and R. J. Marler. 1978. Parasites and diseases of coyotes. Pages 37–71 *in* M. Bekoff, editor. Coyotes: biology, behavior, and management. Academic Press, New York, New York, USA.

Golightly, R. T., Jr., and R. D. Ohmart. 1983. Metabolism and body temperature of two desert canids: coyotes and kit foxes. Journal of Mammalogy 64:624–635.

González, P. G., V. M. Sánchez, L. I. Iñiguez, and E. Santana. 1992. Patrones de actividad del coyote (*Canis latrans*), la zorra gris (*Urocyon cinereoargenteus*) y el tlacuache (*Didelphis virginiana*) en la Sierra de Manantlán, Jalisco. Anales Instituto de Biología, UNAM 63:293–299.

González-Quintero, E. P. 2001. Dieta del coyote *Canis latrans lepticus* (Carnivora: Canidae) en la Isla San Luis Gonzaga, B. C., México. Thesis, Universidad Autónoma de Tamaulipas, Tampico, México.

Grajales-Tam, K. 1998. Dieta estacional del coyote (*Canis latrans*) en el Desierto del Vizcaíno, B. C. S. y su impacto potencial sobre el berrendo peninsular (*Antilocapra americana peninsularis*). Thesis, UNAM, Iztacala, México.

Grajales-Tam, K. M., and A. G. González-Romero. 2014. Determinación de la dieta estacional del coyote (*Canis latrans*) en la región norte de la Reserva de la Biosfera Mapimí, México. Revista Mexicana de Biodiversidad 85:553–564.

Grajales-Tam, K. M., R. Rodríguez-Estrella, and J.

Cancino Hernández. 2003. Dieta estacional del coyote *Canis latrans* durante el periodo 1996–1997 en el Desierto de Vizcaíno, Baja California Sur, México. Acta Zoológica Mexicana (n.s.) 89:17–28.

Grinder, M., and P. R. Krausman. 2001a. Morbidity-mortality factors and survival of an urban coyote population in Arizona. Journal of Wildlife Diseases 37:312–317.

Grinder, M. I., and P. R. Krausman. 2001b. Home range, habitat use, and nocturnal activity of coyotes in an urban environment. Journal of Wildlife Management 65:887–898.

Guerrero, S., M. H. Badii, S. S. Zalapa, and J. A. Arce. 2004. Variación espacio-temporal en la dieta del coyote en la costa norte de Jalisco, México. Acta Zoológica Mexicana (n.s.) 20:145–157.

Guerrero, S., M. H. Badii, S. S. Zalapa, and A. E. Flores. 2002. Dieta y nicho de alimentación del coyote, zorra gris, mapache, y jaguarondi en un bosque tropical caducifolio de la costa sur del estado de Jalisco, México. Acta Zoológica Mexicana (n.s.) 86:119–137.

Guo, W., J. F. Evermann, W. J. Foreyt, F. F. Knowlton, and L. A. Windberg. 1986. Canine distemper virus in coyotes: a serologic survey. Journal of the American Veterinary Medical Association 189:1099–1100.

Hall, E. R. 1981. The mammals of North America. Volume 2. Second edition. John Wiley and Sons, New York, New York, USA.

Hamlin, K. L., S. J. Riley, D. Pyrah, A. R. Dodd, and R. J. Mackie. 1984. Relationships among mule deer fawn mortality, coyotes, and alternate prey species during the summer. Journal of Wildlife Management 48:489–499.

Harrison, D. J. 1992. Dispersal characteristics of juvenile coyotes in Maine. Journal of Wildlife Management 56:128–138.

Harrison, D. J., J. A. Harrison, and M. O'Donoghue. 1991. Predispersal movements of coyote (*Canis latrans*) pups in eastern Maine. Journal of Mammalogy 72:756–763.

Hein, E. W., and W. F. Andelt. 1995. Estimating coyote density from mark-resight surveys. Journal of Wildlife Management 59:164–169.

Henke, S. E., and F. C. Bryant. 1999. Effects of coyote removal on the faunal community in western Texas. Journal of Wildlife Management 63:1066–1081.

Hernández, L., and M. Delibes. 1994. Seasonal food habits of coyotes, *Canis latrans*, in the Bolsón de Mapimí, southern Chihuahuan Desert, Mexico. Zeitschrift für Säugetierkunde 59:82–86.

Hernández, L., M. Delibes, and E. Ezcurra. 1993. Activity pattern, home range, and habitat preference by coyotes (*Canis latrans*) in the Mapimi Biosphere Reserve of the Chihuahuan Desert, Mexico. Doñana Acta Vertebrata 20:276–283.

Hernández, L., M. Delibes, and F. Hiraldo. 1994. Role of reptiles and arthropods in the diet of coyotes in extreme desert areas of northern Mexico. Journal of Arid Environments 26:165–170.

Hernández, L., R. R. Parmenter, J. W. Dewitt, D. C. Lightfoot, and J. W. Laundré. 2002. Coyote diets in the Chihuahuan Desert: more evidence for optimal foraging. Journal of Arid Environments 51:613–624.

Hibler, S. J. 1977. Coyote movement patterns with emphasis on home range characteristics. Thesis, Utah State University, Logan, USA.

Hidalgo-Mihart, M. G., L. Cantú-Salazar, S. E. Carrillo-Percastegui, and C. A. López-González. 2009. Daily activity patterns of coyotes (*Canis latrans*) in a tropical deciduous forest of western Mexico. Studies on Neotropical Fauna and Environment 44:77–82.

Hidalgo-Mihart, M. G., L. Cantú-Salazar, A. González-Romero, and C. A. López-González. 2004a. Historical and present distribution of coyote (*Canis latrans*) in Mexico and Central America. Journal of Biogeography 31:2025–2038.

Hidalgo-Mihart, M. G., L. Cantú-Salazar, C. A. López-González, E. C. Fernandez, and A. González-Romero. 2004b. Effect of a landfill on the home range and group size of coyotes (*Canis latrans*) in a tropical deciduous forest. Journal of Zoology (London) 263:55–63.

Hidalgo-Mihart, M. G., L. Cantú-Salazar, C. A. López-González, P. G. Martínez-Gutiérrez, E. C. Fernandez, and A. González-Romero. 2006. Coyote habitat use in a tropical deciduous forest of western Mexico. Journal of Wildlife Management 70:216–221.

Hidalgo-Mihart, M. G., L. Cantú-Salazar, C. A. López-González, E. Martínez-Meyer, and A. González-Romero. 2001. Coyote (*Canis latrans*) food habits in a tropical deciduous forest of western Mexico. American Midland Naturalist 146:210–216.

Hidalgo-Mihart, M. G., F. M. Contreras-Moreno, L. A. Péze-Solano, and C. Hernández-Lara. 2013. Primeros registros de coyote (*Canis latrans*) en Campeche, México. Revista Mexicana de Biodiversidad 84:1012–1017.

Jackson, H. H. T. 1951. Part 2. Classification of the races of coyotes. Pages 227–341 *in* S. P. Young and H. H. T. Jackson. The clever coyote. Wildlife Management Institute, Washington, DC, USA.

Johnson, M. K., and R. M. Hansen. 1979. Coyote food habits on the Idaho National Engineering Laboratory. Journal of Wildlife Management 43:951–956.

Kamler, J. F., W. B. Ballard, P. R. Lemons, R. L. Gilliland, and K. Mote. 2005. Home range and habitat use of coyotes in an area of native prairie, farmland, and CRP fields. American Midland Naturalist 153:396–404.

Kamler, J. F., and P. S. Gipson. 2004. Survival and cause-specific mortality among furbearers in a protected area. American Midland Naturalist 151:27–34.

Kennelly, J. J. 1978. Coyote reproduction. Pages 73–93 *in* M. Bekoff, editor. Coyotes: biology, behavior, and management. Academic Press, New York, New York, USA.

Kitchen, A. M., E. M. Gese, and E. R. Schauster. 2000. Long-term spatial stability of coyote (*Canis latrans*) home ranges in southeastern Colorado. Canadian Journal of Zoology 78:458–464.

Kleiber, M. 1961. The fire of life: an introduction to animal energetics. John Wiley and Sons, New York, New York, USA.

Knowlton, F. F. 1972. Preliminary interpretations of coyote population mechanics with some management implications. Journal of Wildlife Management 36:369–382.

Knowlton, F. F., E. M. Gese, and M. M. Jaeger. 1999. Coyote depredation control: an interface between biology and management. Journal of Range Management 52:398–412.

Kolbe, J. A., and J. R. Squires. 2004. Long distance movement by a coyote within the Rocky Mountains. Northwest Science 78:344–345.

Lafón-Terrazas, A. 1987. Análisis de factores que influyen en la pérdida de becerros por depredación del coyote *Canis latrans* en cuatro áreas del estado de Chihuahua. Thesis, Universidad Autónoma de Chihuahua, Chihuahua, México.

Laundré, J. W. 1981. Temporal variation in coyote vocalization rates. Journal of Wildlife Management 45:767–769.

Laundré, J. W., J. M. M. Calderas, and L. Hernández. 2009. Foraging in the landscape of fear, the predator's dilemma: Where should I hunt? The Open Ecology Journal 2:1–6.

Laundré, J. W., and L. Hernández. 2003a. Total energy budget and prey requirements of free-ranging coyotes in the Great Basin Desert of the western United States. Journal of Arid Environments 55:675–689.

Laundré, J. W., and L. Hernández. 2003b. Factors affecting dispersal in young male pumas. Pages 151–160 *in* S. A. Becker, D. D. Bjornlie, F. G. Lindzey, and D. S. Moody, editors. Proceedings of the Seventh Mountain Lion Workshop, Lander, Wyoming, USA.

Laundré, J. W., and B. L. Keller. 1981. Home-range use by coyotes in Idaho. Animal Behaviour 29:449–461.

Laundré, J. W., and B. L. Keller. 1984. Home-range size of coyotes: a critical review. Journal of Wildlife Management 48:127–139.

Lehner, P. N. 1978. Coyote communication. Pages 127–162 *in* M. Bekoff, editor. Coyotes: biology, behavior, and management. Academic Press, New York, New York, USA.

Leopold, A. 1966. A Sand County almanac and essays on conservation from Round River. Oxford University Press, New York, New York, USA.

Leopold, A. S. 1959. Wildlife of Mexico: game birds and mammals. University of California Press, Berkeley, USA.

Leydet, F. 1988. The coyote: defiant song dog of the west. University of Oklahoma Press, Norman, USA.

Linhart, S. B., and F. F. Knowlton. 1975. Determining the relative abundance of coyotes by scent station lines. Wildlife Society Bulletin 3:119–124.

Litvaitis, J. A., and W. W. Mautz. 1980. Food and energy use by captive coyotes. Journal of Wildlife Management 44:56–61.

López Soto, J. H., and M. H. Badii. 2000. Depredación en crías de venado cola blanca (*Odocoileus virginianus texanus*) por coyote (*Canis latrans*) en una unidad de manejo y aprovechamiento del norte de Nuevo León, México. Acta Zoológica Mexicana (n.s.) 81:135–138.

Love, R., T. S. Baskett, R. D. Sparrowe, R. J. Baldauf, and M. Shelton. 1974. The coyote as a livestock predator. Bioscience 24:324.

MacCracken, J. G., and R. M. Hansen. 1987. Coyote feeding strategies in southeastern Idaho: optimal foraging by an opportunistic predator? Journal of Wildlife Management 51:278–285.

MacCracken, J. G., and D. W. Uresh. 1984. Coyote foods in the Black Hills, South Dakota. Journal of Wildlife Management 48:1420–1423.

Manning, A. D., I. J. Gordon, and W. J. Ripple. 2009. Restoring landscapes of fear with wolves in the Scottish Highlands. Biological Conservation 142:2314–2321.

Manville, R. H. 1953. Longevity of the coyote. Journal of Mammalogy 34:390.

Marín-Sánchez, A. I., M. Briones-Salas, R. López-Wilchis, and J. Servín. 2015. Ámbito hogareño de coyote (*Canis latrans*) en un bosque templado de la Sierra Madre de Oaxaca, México. Revista Mexicana de Biodiversidad 86:440–447.

Martínez Calderas, J. M. 2005. Forrajeo óptimo del coyote en la Reserva de la Biosfera de Mapimí. Thesis, Universidad Autónoma de Ciudad Juárez, México.

Martínez García, J. A., G. D. Mendoza Martínez, F. X. Plata P., O. C. Rosas Rosas, L. A. Tarango Arámbula, and L. C. Bender. 2014. Use of prey by sympatric bobcat (*Lynx rufus*) and coyote (*Canis latrans*) in the Izta-Popo National Park, Mexico. Southwestern Naturalist 59:167–172.

Mathwig, H. J. 1973. Food and population characteristics of Iowa coyotes. Iowa State Journal of Research 47:167–189.

McCarley, H. 1975. Long-distance vocalizations of coyotes (*Canis latrans*). Journal of Mammalogy 56:847–856.

Mitchell, B. R., M. M. Jaeger, and R. H. Barrett. 2004. Coyote depredation management: current methods and research needs. Wildlife Society Bulletin 32:1209–1218.

Mitchell, B. R., M. M. Makagon, M. M. Jaeger, and R. H. Barrett. 2006. Information content of coyote barks and howls. Bioacoustics 15:289–314.

Murray Berger, K., and E. M. Gese. 2007. Does interference competition with wolves limit the distribution and abundance of coyotes? Journal of Animal Ecology 76:1075–1085.

Myers, R. A., J. K. Baum, T. D. Shepherd, S. P. Powers, and C. H. Peterson. 2007. Cascading effects of the loss of apex predatory sharks from a coastal ocean. Science 315:1846–1850.

Neale, J. C. C., B. N. Sacks, and K. M. Blejwas. 2007. Coyote flight movements relative to territory boundaries: an experiment in the field. American Midland Naturalist 158:162–167.

Nellis, C. H., and L. B. Keith. 1976. Population dynamics of coyotes in central Alberta, 1964–68. Journal of Wildlife Management 40:389–399.

Nelson, E. W. 1932. Remarks on coyotes with a description of a new subspecies from El Salvador. Proceedings of the Biological Society of Washington 45:223–226.

Niedzielski, B., and J. Bowman. 2014. Survival and cause-specific mortality of the female eastern wild turkey at its northern range edge. Wildlife Research 41:545–551.

Nowak, R. M. 1978. Evolution and taxonomy of coyotes and related *Canis*. Pages 3–16 *in* M. Bekoff, editor. Coyotes: biology, behavior, and management. Academic Press, New York, New York, USA.

O'Donoghue, M., S. Boutin, C. J. Krebs, and E. J. Hofer. 1997. Numerical responses of coyotes and lynx to the snowshoe hare cycle. Oikos 80:150–162.

Okarma, H., and P. Koteja. 1987. Basal metabolic rate in the gray wolf in Poland. Journal of Wildlife Management 51:800–801.

Perkins, P. J. 1992. Winter diet and bioenergetics of eastern coyotes: a review. Pages 87–99 *in* A. H. Boer, editor. Ecology and management of the eastern coyote. Fredericton Wildlife Research Unit, University of New Brunswick, Canada.

Ponce Guevara, E., K. Pelz Serrano, and C. A. Lopez González. 2005. Coyote abundance in relation to habitat characteristics in Sierra San Luis, Sonora, México. USDA Forest Service Proceedings, RMRS-P-36. Rocky Mountain Research Station, Fort Collins, Colorado, USA.

Powell, R. A. 1979. Ecological energetics and foraging strategies of the fisher (*Martes pennanti*). Journal of Animal Ecology 48:195–212.

Purvis, A., P. Agapow, J. Gittleman, and G. Mace. 2000. Nonrandom extinction and the loss of evolutionary history. Nature 288:328–330.

Rigg, R., S. Findo, M. Wechselberger, M. I. Gorman, C. Sillero-Zubiri, and D. W. MacDonald. 2011.

Mitigating carnivore-livestock conflict in Europe: lessons from Slovakia. Oryx 45:272–280.

Ripple, W. J., and R. L. Beschta. 2004. Wolves and the ecology of fear: can predation risk structure ecosystems? BioScience 54:755–766.

Ripple, W. J., et al. 2014. Status and ecological effects of the world's largest carnivores. Science 343:151–163.

Rodríguez Martínez, A. 2011. Distribución y abundancia del coyote (*Canis latrans*) en el centro del Desierto Chihuahuense en México. Thesis, Universidad Autónoma de Querétaro, Querétaro, México.

Rosas-Rosas, O. C., R. Valdez, L. C. Bender, and D. Daniel. 2003. Food habits of pumas in northwestern Sonora, Mexico. Wildlife Society Bulletin 31:528–535.

Rosatte, R. C. 2002. Long distance movement by a coyote, *Canis latrans*, and red fox *Vulpes vulpes*, in Ontario: implications for disease-spread. Canadian Field-Naturalist 116:129–131.

Sacks, B. N. 2005. Reproduction and body condition of California coyotes (*Canis latrans*). Journal of Mammalogy 86:1036–1041.

Salas, A. 1987. Hábitos alimenticios de la zorra (*Urocyon cinereoargenteus*), coyote (*Canis latrans*) y gato montés (*Lynx rufus*) en la Sierra Purépecha, Estado de Michoacán. Pages 136–144 *in* Memorias del IX Congreso de Zoología, Tabasco, México.

Sanabria, B., A. Ortega-Rubio, and C. Arguelles-Méndez. 1995. Food habits of the coyote in the Vizcaíno Desert, México. Ohio Journal of Science 95:289–291.

Sanderson, G. C. 1966. The study of mammal movements: a review. Journal of Wildlife Management 30:215–235.

Sandoval-Sandoval, M. R. 1998. Hábitos de alimentación del coyote *Canis latrans*, Merriam en la región de Tenacatita, Jalisco. Thesis, Universidad de Guadalajara, Jalisco, México.

Schmitz, O. J. 2006. Predators have large effects on ecosystem properties by changing plant diversity, not plant biomass. Ecology 87:1432–1437.

Schmitz, O. J., J. H. Grabowski, B. L. Peckarsky, E. L. Preisser, G. C. Trussell, and J. R. Vonesh. 2008. From individuals to ecosystem function: toward an integration of evolutionary and ecosystem ecology. Ecology 89:2436–2445.

SEMARNAP. 2000. Estrategia nacional para la vida silvestre: logros y retos para el desarrollo sustentable 1995–2000. INE/DGVS, México, D. F., México.

Servin, J., and C. Huxley. 1991. La dieta del coyote en un bosque de encino-pino de la Sierra Madre Occidental de Durango, México. Acta Zoológica Mexicana (n.s.) 44:1–26.

Servin, J., and C. Huxley. 1995. Coyote home range size in Durango, Mexico. Zeitschrift für Säugetierkunde 60:119–120.

Servín, J., V. Sánchez-Cordero, and S. Gallina. 2003. Distance traveled daily by coyotes *Canis latrans* in a pine-oak forest in Durango, Mexico. Journal of Mammalogy 84:547–552.

Sheldon, M. 1973. Some myths concerning the coyote as a livestock predator. BioScience 23:719–720.

Shivik, J. A. 2006. Tools for the edge: what's new for conserving carnivores. BioScience 56:253–259.

Smith, G. J., J. R. Cary, and O. J. Rongstad. 1981. Sampling strategies for radio-tracking coyotes. Wildlife Society Bulletin 9:88–93.

Sosa Escalante, J., S. Hernández, A. Segovia, and V. Sánchez-Cordero. 1997. First record of the coyote, *Canis latrans* (Carnivora: Canidae), in the Yucatan Peninsula, Mexico. Southwestern Naturalist 42:494–495.

Stoddart, L. C., R. E. Griffiths, and F. F. Knowlton. 2001. Coyote responses to changing jackrabbit abundance affect sheep predation. Journal of Range Management 54:15–20.

Taylor, C. R., K. Schmidt-Nielsen, and J. L. Raab. 1970. Scaling of energetic cost of running to body size in mammals. American Journal of Physiology 219:1104–1107.

Todd, A. W., and L. B. Keith. 1983. Coyote demography during a snowshoe hare decline in Alberta. Journal of Wildlife Management 47:394–404.

Truett, J. C. 1979. Observations of coyote predation on mule deer fawns in Arizona. Journal of Wildlife Management 43:956–958.

Tumenta, P. N., H. H. de Iongh, P. J. Funston, and H. A. Udo de Haes. 2013. Livestock depredation and mitigation methods practiced by residents and nomadic pastoralists around Waza National Park, Cameroon. Oryx 47:237–242.

Van Vuren, D., and S. E. Thompson Jr. 1982. Opportunistic feeding by coyotes. Northwest Science 56:131–135.

Vela, C. E. L. 1985. Determinación de la composición de la dieta del coyote, *Canis latrans*

Say, por medio del análisis de heces en tres localidades del estado de Chihuahua. Thesis, UANL, Monterrey, México.

Windberg, L. A., S. M. Ebbert, and B. T. Kelly. 1997. Population characteristics of coyotes (*Canis latrans*) in the northern Chihuahuan Desert of New Mexico. American Midland Naturalist 138:197–207.

Young, S. P. 1951. Its history, life habits, economic status and control. Part 1. Pages 1–226 *in* S. P. Young and H. H. T. Jackson. The clever coyote. Wildlife Management Institute, Washington, DC, USA.

12 Black Bear in Mexico

Jonás A. Delgadillo Villalobos, Santiago Gibert Isern, Mario Alberto García Aranda, Ramiro Velázquez Rincón

Mexico has experienced the extirpation of two large carnivores, the Mexican gray wolf (*Canis lupus baileyi*) and the grizzly bear (*Ursus arctos horribilis*), principally from predator control through the late 1960s (Brown 1985, Treviño and Jonkel 1986). Human intolerance also almost led to the extirpation of the Mexican black bear (*Ursus americanus*) (Leopold 1959). Although the black bear has been protected since 1970, illegal predator control and poaching (e.g., for sport and commercial sale of gallbladders) continue (Baker 1956, Leopold 1959, SEMARNAT-CONANP 2009). Several remnant populations have persisted, and some are recovering with the help of better legal protection and law enforcement, as well as assistance from private landowners. However, inadequate population-specific information still impairs management and conservation efforts. A first step is reviewing and integrating extant knowledge to stimulate field research and monitoring, which is the purpose of this chapter.

Historical Perspective

Although Mexico's native peoples recorded the presence of black bears in stone engravings and cave paintings in Coahuila and Nuevo León, little is known about the ancient significance of bears—for instance, their use as a food resource, or their cultural and spiritual meaning. Since the 1800s, the black bear has been a coveted sport-hunting trophy (Leopold 1959, Villa 1977) and has also been hunted as a livestock predator (Baker 1958). During campaigns to eradicate coyotes and Mexican gray wolves, black bear populations were extirpated over extensive areas. Cattle ranchers were rewarded with predator bounties for black bears, a practice that continued through the 1960s. Although predator control and hunting are now illegal, these practices continue in some areas.

Bears have been negatively impacted in northern Mexico, where many large ranches have remained intact, by (1) habitat fragmentation caused by forest exploitation and mining, (2) urban development, (3) construction of communication infrastructure, and (4) expansion of farmland—although some areas where former farms have been abandoned have temporarily reverted to bear habitat. These factors eventually reduced the black bear population to only 20% of its original distribution in Mexico (INE-SEMARNAP 1999). Small populations persist in remote areas in northern Mexico in the Sierra Madre Oriental and Occidental. By the 1960s, the black bear was scarce, with remnant populations in Chihuahua, Durango, and Coahuila (Baker 1956, 1958, Leopold 1959, Villa 1977). Fortunately, the black bear now occurs in most of its historical range (Leopold 1959, Pelton et al. 1998). In Mexico, bears originally occurred in 12 states and currently occur in 9 (fig. 12.1).

Populations in the US and Mexico have been increasing during the last 20 years (Williamson 2002), and in Mexico, they have been reported even where they were believed to be extirpated. In some areas of Coahuila, black bear populations have recovered, mainly in the Sierra del Carmen, Sierra La Encantada, Sierra de Santa Rosa, and Serranías del Burro, where there are continuous and protected habitats (McKinney and Delgadillo, unpublished data). Recent range extensions have also been reported in the state of Nuevo León and possibly in Tamaulipas, Chihuahua, Sonora, San Luis Potosí, and Zacatecas. Sightings have been reported in the Sierra Gorda in Querétaro and Guanajuato, where they had not been previously reported, even though Leopold (1959) mentioned their possible occurrence farther south of their known range at the time. Black bears in the Serranías del Burro, Coahuila, are listed as a species under special protection, whereas those in other areas are still considered endangered (SEMARNAT 2010). Nevertheless, overall the black bear is listed as a species of least concern by the IUCN (2009).

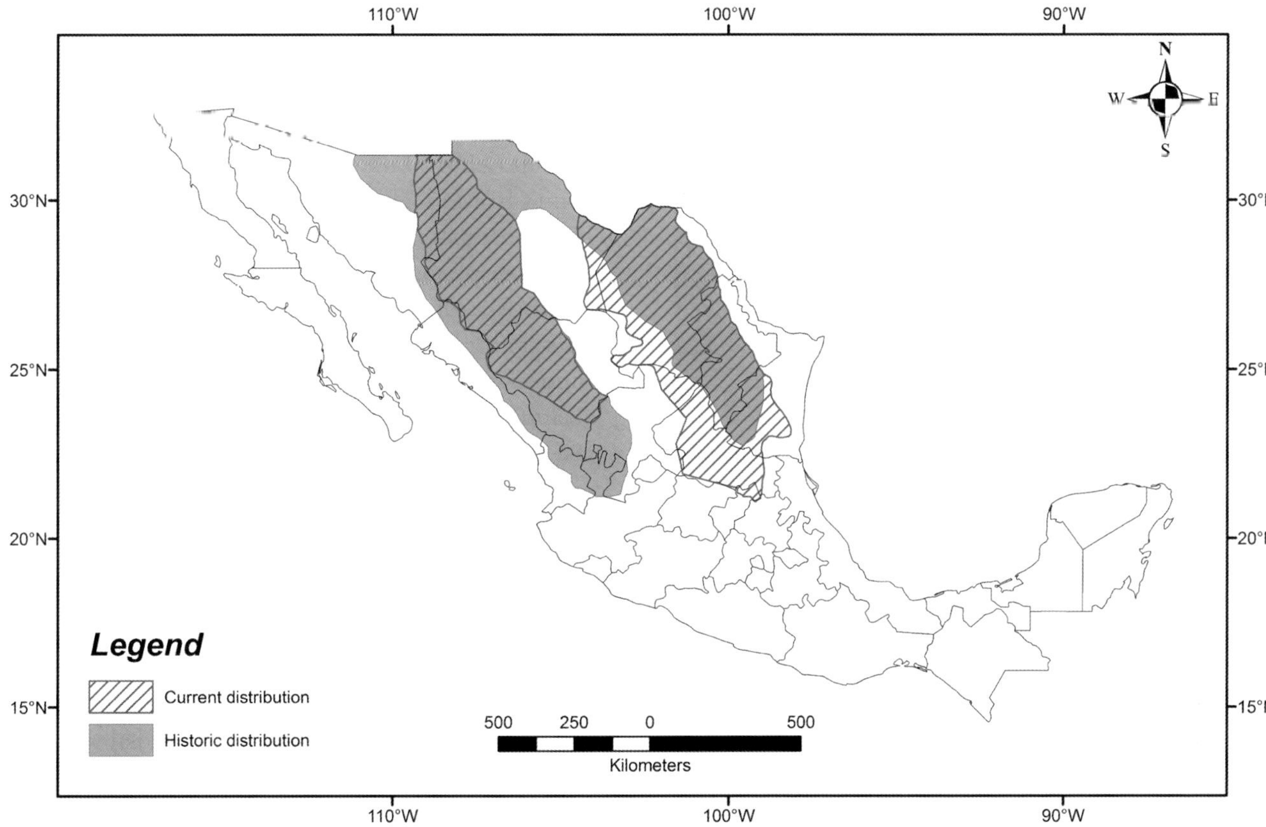

Figure 12.1. Geographical distribution of the black bear in Mexico.

Ranchers who coexist with black bears and even accept economic losses, as well as landowners who have developed wildlife conservation areas, are largely responsible for the success of black bear conservation in Mexico (Medellín et al. 2005). Black bear travel corridors unite mountain ranges in northern Coahuila with Big Bend National Park and Black Gap Wildlife Management Area in Texas (McKinney et al. 2006, Hewitt and Doan-Crider 2008). In addition, several mining companies in northern Coahuila have contributed to black bear population recovery by providing safe havens. As a result, black bears have extended their range to southern Coahuila, crossed the desert, and reached other mountain ranges with suitable habitat (Contreras-Balderas et al. 2007; D. González, personal communication). Likewise, they have expanded into West Texas, where in the 1980s black bear sightings were sporadic (Lobello 1989); they have now reached a breeding population level (Onorato and Hellgren 2001, McKinney and Pitman 2001, Onorato et al. 2004, Van Den Bussche et al. 2009). Additional assistance to black bear recovery has been provided by private landowners, state and federal agencies, and NGOs (Doan-Crider and Moctezuma Orozco 2014).

Black bears were formerly widely distributed in the mountains of Nuevo León, but populations decreased drastically until they occurred only in the most remote regions (Leopold 1959). During the 1980s and 1990s, black bears were still considered rare in Nuevo León (Jiménez 1981, Carrillo 1981, Zepeda-González et al. 1997, Moreno-Valdez 1998). During the past few years, black bears have established a population in the Sierra Plegada, Nuevo León, where they damage cornfields and orchards (N. Sanchez, personal communication). Other distributional records include northern mountains such as the Sierras de Picachos, de Gomas, de Bustamante, and Lampazos (Niño-Ramírez 1989, Salgado-De Los Santos et al. 2005, Juárez-Casillas 2006, Salgado-De Los Santos 2007, Mijangos 2009). These mountain ranges create a biological corridor that connects important forested areas in Querétaro, Guanajato, San Luis Potosí, Tamaulipas, and Nuevo León with mountainous regions in the Chihuahuan Desert in Coahuila. There, the Sierra de Santa Rosa, Serranías del Burro, Sierra La Encantada, and Sierra

Maderas del Carmen have become famous for their abundance of bears.

In Tamaulipas, black bears have never been considered abundant (Álvarez 1963). However, they inhabit El Cielo Biosphere Reserve and the Sierra de San Carlos and Sierra de Tamaulipas (Moreno-Valdez 1991, Vargas and Hernández 2001, Moreno-Valdez and Vázquez-Farías 2005, Lara 2006, Carvajal et al. 2007, Carrera-Treviño et al. 2015). In their southern historical distribution, as in Zacatecas and San Luis Potosí, sporadic records indicate the presence of black bears in remote areas of the Sierra Madre Oriental, as well as in the Sierra Gorda in Querétaro and Guanajuato (INE-SEMARNAP 1999, SEMARNAT-CONANP 2005). A dead bear was recorded in northern Hidalgo, which could be its southernmost distribution (Rojas-Martínez and Juárez-Casillas 2013).

In contrast with black bear recovery in the Sierra Madre Oriental, that in the Sierra Madre Occidental has been slower, but recent records indicate an increase in remote areas. In the 1950s in Durango, black bears were still abundant but were restricted to the western mountains, where populations had been controlled by cattle ranchers (Leopold 1959, Baker and Greer 1962). Bears were considered extirpated in that region until an isolated population was recorded during a state mammal inventory (Servín et al. 1997, Sosa 1998). In Sonora, bears are restricted to the Sierra de San Luis, an area contiguous with Chihuahua (Sierra-Corona et al. 2005, Sáyago and López 2005, Rodríguez-Martínez et al. 2006). They have also been recorded in the Ajos-Bavispe region of Sonora. In Chihuahua, bears occur in most of their historical range, but sightings are still rare, even in areas where they were once common (Vargas 1997, Calderón-Domínguez et al. 2006, Delfín-Alfonso et al. 2011; R. Uranga, personal communication). In some ranches in the Sierra del Nido in central Chihuahua, landowners have developed a tolerant attitude.

The growing black bear population has awakened public interest in the species, which has encouraged federal and state agencies to develop conservation and management strategies. There is political pressure to declare the back bear a game species, especially in areas of northern Coahuila and Nuevo León, where bears are considered abundant and are thought to be livestock predators. Sport hunting would supposedly reduce economic losses from cattle predation and provide revenue from hunters. On many properties with diversified livestock and wildlife programs, the black bear is included in management strategies. In addition, one of the most important conservation decisions, apart from the establishment of a closed season, was the declaration of the Coahuila government in 1999 that the black bear was the state symbol of natural resource conservation (Gobierno del Estado de Coahuila 2009).

Taxonomy and Distribution

Ursids occur in North and South America, Europe, and Asia. Of the four subfamilies, only American black bears are classified in the Ursinae, which is endemic to North America. The American black bear probably evolved from a smaller species, *Ursus minimus*, which occurred in Eurasia and increased in size during its evolution (McLellan and Reiner 1994). Fossil records of black bears in North America were more common by the Late Pliocene and Pleistocene; it is estimated that black bears populated the continent 1.8 million years ago (mya) (Wooding and Ward 1997). Although categorized as carnivores, most ursids are primarily omnivores (McLellan and Reiner 1994). There are 16 recognized black bear subspecies, 3 of which occur in Mexico: *eremicus*, *machetes*, and *amblyceps* (Baker 1956, Hall 1981, IUCN 2009).

A dearth of genetic information makes it difficult to define the geographic limits of the three subspecies in Mexico. Genetic samples from the transborder corridor region, which includes the Serranías del Burro and Sierra Maderas del Carmen in Coahuila, and Big Bend National Park and Black Gap Wildlife Management Area in Texas, indicate that these populations share similar genetic characteristics, which are indicative of *eremicus* (Onorato and Hellgren 2001, Onorato et al. 2004, Onorato et al. 2006). The same studies show that whereas endemic black bears have long since been extirpated from South Texas, immigrants from northern Mexico have established a pioneering population. Although there are few studies of the Sierra Madre Occidental and Sierra Madre Oriental populations, the black bear populations in northeastern Sonora belong to *amblyceps*, in contrast to Chihuahuan Desert bear populations (Varas 2007).

On the basis of the historical distribution of the three recognized subspecies in Mexico, *eremicus* occurred in Coahuila, Nuevo León, Tamaulipas, Zacatecas, and San Luis Potosí; *machetes* in Sinaloa, Sonora, Durango, Chihuahua, Zacatecas, Jalisco, and Nayarit; and *amblyceps* in central and northern Sonora and Chihuahua (Baker 1956, Leopold 1959, IUCN 2009).

Since its appearance in North America, the black bear has been associated with forests (Wooding and Ward 1997). Nevertheless, it easily adapts to habitat changes and human-dominated landscapes. In Mexico, it occurs principally in forested areas of the Sierra Madre Occidental and Oriental and has a strong association with transition areas of the Chihuahuan Desert. Arid zones are not a limiting dispersal factor. Forested areas occupied by bears range from temperate forest to tropical rainforest (Leopold 1959, Baker and Greer 1962, Hewitt and Doan-Crider 2008).

Description

The black bear is one of the largest carnivores in Mexico, reaching a total length of 150 to 210 cm (59 to 82.6 in); shoulder height varies between 50 and 90 cm (19.6 and 35.4 in). The body is robust, with moderately long, rounded ears and a short tail. It has plantigrade feet. The front paws of an adult bear older than three years vary between 8.5 and 13.8 cm (3.4 and 5.4 in) in length and width; the hind paws measure 15.7–22.5 cm (6.1–8.8 in) long and 11.3–12 cm (4.4–4.7 in) wide. There are five digits on each paw, with nonretractable claws approximately 3 cm (1.1 in) long (INE-SEMARNAP 1999, McKinney and Delgadillo 2004). Continent-wide, adult males attain weights of 60–225 kg (132–496 lb); females are about two-thirds as large as males, generally weighing 40–80 kg (88–176 lb) (Garshelis 2009). In Mexico, however, both sexes tend to be smaller, with males seldom exceeding 130 kg (286 lb), and females 70 kg (154 lb) (McKinney and Delgadillo 2004). Adult males can sometimes be distinguished by their relatively large size, robustness, and scarring from fighting; adult females are not readily distinguished from adolescent males unless sex organs or cubs are visible (fig. 12.2) (McKinney and Delgadillo 2004, Stringham 2010).

Contrary to their name, black bears exhibit a wide range of coat colors. Coats can be cinnamon to gray in the Sierra Madre Occidental (*machetes*), while those from the Sierra Madre Oriental (*eremicus*) are usually black, but brown also occurs. Across North America in general, fur tends to be the darkest black where bears inhabit deeply shadowed forests, whereas lighter colors predominate where bears are more exposed to sunlight and overheating. Genetics governs the color of fur when it first grows, as well as how readily that color bleaches in the sun. Bleaching accounts for hair being lightest on the back, and parts of the legs. Bleaching is progressive, and thus most extensive just prior to annual shedding; newly grown fur tends to be darker.

In especially hot areas and in poorly nourished bears, growth of new hair may be delayed until fall, when accumulation of body fat and protein reserves peaks (S. Stringham, unpublished data). Consequently, summer fur tends to be relatively "short, dry and splotchy," compared to new fur grown in fall (Doan-Crider 1995). Poor, bleached fur, along with body thinness, is especially common among juveniles during the first summer after they become independent of their mother and must fend for themselves. Some bears from the Sierra Madre Oriental have a white chest patch, which in extreme cases is V shaped and resembles a collar.

Figure 12.2. Female black bear with cub in northern Mexico. Photo by Santiago Gibert Isern.

Biology

Diet

The black bear is omnivorous (fig. 12.3), and plant tissue makes up 70–90% of its diet (Herrera-González 1999, Delgadillo 2001, McKinney and Delgadillo 2004, Sáyago and López 2005, Rodríguez-Martínez et al. 2006, Juárez-Casillas 2006, Delgadillo et al. 2007, Moreno 2008). In high-elevation mountainous areas, primary foods include acorns (*Quercus* spp.), manzanita (*Arctostaphylos pungens*), Texas madrone (*Arbutus xalapensis*), junipers (*Juniperus* spp.), squawroot (*Conopholis mexicana*), pinyons (*Pinus* spp.), plums (*Prunus* spp.), simple-leaf tick clover (*Desmodium psilophyllum*), wild grapes (*Vitis* spp.), chokecherry (*Prunus virginiana*), and a variety of other fruits. At lower elevations, bears feed on the soft tissue of sotol (*Dasylirion* spp.) and yucca (*Yucca* spp.), as well as on mesquite pods (*Prosopis glandulosa*), prickly pear (*Opuntia* spp.), persimmon (*Diospyros texana*), and agarita (*Mahonia trifoliolata*), among others. Grasses and succulents such as sotol and yucca are especially important during spring before higher-calorie mast is available (Hellgren 1993, Doan-Crider 1995). Bears prey on or scavenge white-tailed deer (*Odocoileus virginianus*), mule deer (*O. hemionus*), javelina (*Pecari tajacu*), domestic cattle, insects, small mammals, and miscellaneous carrion (Niño-Ramírez 1989, Herrera-González 1999, McKinney and Pittman 2001, Delgadillo 2001, McKinney and Delgadillo 2004, Sáyago and López 2005, Sierra-Corona et al. 2005, Delgadillo et al. 2007, Moreno 2008, Martínez et al. 2014).

Hibernation

All age-sex classes may den in high-altitude mountain habitats where there is negligible food during winter. At lower elevations, where the climate is warmer and winter food available, as in the Serranías del Burro, Coahuila, and in northern Mexico, denning is most typical of pregnant females and may be delayed until December or January, shortly before cubs are born.

Figure 12.3. Adult male black bear consuming acorns in northern Mexico. Photo by Jonás A. Delgadillo Villalobos.

Other age-sex classes remain at least semiactive year-round (Leopold 1959, Doan-Crider 1995, 2003), as is typical in subtropical Florida (McDaniel 1974). Irrespective of elevation, females with neonates typically remain denned until April or May, when their cubs are large enough and strong enough to travel with their mother as she forages. Den sites range from soil beds or piles of vegetation hidden from view by rocks or trees, to cavities in or under trees or rock ledges, to deep caves in soil or rocks (Doan-Crider 1995, 2003, McKinney and Delgadillo 2004).

Onset of hibernation is more closely related to food scarcity than to weather (Leopold 1959). Pregnant females tend to den first and longest, allowing cubs to mature to a certain level before emerging in spring. Hibernation by other bears ends when spring foods become available. Adult males enter dens latest and emerge earliest in spring, whereupon they begin replenishing fat reserves in preparation for the mating season. Once adult females commence estrus, adult males focus on finding and courting females.

Energy requirements

Most black bear movements relate to their search for food. Bears tend to lose one-quarter to one-third of their body weight while hibernating. Their daily activity increases in summer and fall, when many of their preferred foods are present. Bears consume large amounts to build up fat reserves to meet their energy needs during winter and early spring, when foods are scarce (Eagle and Pelton 1983, Hellgren and Vaughan 1988). Pregnant females remain in their winter dens for four to five months without consuming water or food. Their own energy needs and the production of milk for cubs depend on their body reserves of fat and protein (McDonald and Fuller 2005).

Spring foods (e.g., grasses and succulents such as sotol and yucca) tend to be relatively high in protein and low in energy (Beeman 1975). Most calories are obtained during summer and fall from fruits and seeds of oak, manzanita, juniper, mesquite, and Texas madrone (Costello et al. 2001, McKinney and Pittman 2001, Doan-Crider 2003, McKinney and Delgadillo 2004, Ryan et al. 2007).

Water requirements

Little is known about the relationship between black bears and water availability in Mexico. Metabolically, they require 7–8 liters (1.8–2.1 gal) daily. Water consumption increases when food consumption increases (Rogers 1987). Submersion in water can also be important for cooling during hot weather. According to Rogers (1993), water availability could limit the distribution of black bears in some areas, but it is probably not a factor in population density. In high mountains with temperate forests in northern Mexico, which normally support sizable populations, the average annual precipitation is greater than in lower regions, and water is readily available in rivers, creeks, and watering holes. In semiarid zones, where management of cattle and wildlife includes artificial watering sources such as stock tanks and dams, bears can exploit these sources to meet their water requirements, gaining access to habitats where they might not otherwise survive (D. González, personal communication).

Social behavior

Adult males and females associate with one another in pairs or small groups for several days at a time during the mating season. One or more courting adult males will follow an estrous female for several days. The males compete for her favor, usually through dominance displays such as sumo strutting, but occasionally with bouts of combat that seldom last more than 10 seconds (Stringham 2010). One or more males may eventually mate with her, which induces ovulation. Whether the ova are fertilized by the dominant male is yet unclear, but in some cases, a litter has at least two sires. Promiscuity may confuse males about which of them sired a litter and might inhibit those males from injuring or killing the cubs (Stringham 1983, 1984).

Outside the mating season, adult black bears spend little time socializing with one another except at concentrations of prime foods such as fruit or nuts, where cost/benefit ratios favor sharing the resource hierarchically over trying to exclude one another entirely. During hot weather, they may also aggregate at watering sources. Young bears are more tolerant of each other and seek opportunities to play, usually by sparring and chasing, which teaches them combat skills and strengthens social bonds. Playful sparring also determines dominance hierarchies without risking injury.

During a serious conflict, opponents tend to approach each other with their heads lowered and jaws pointed toward the ground, signaling both reluctance and readiness for combat. Each bear might also pant-huff, pop its lips, and snap its teeth together several times, perhaps followed by lunging forward or running forward several paces before swatting the ground, a tree, or a boulder. When a bear's head is lowered, the hair on its neck or shoulders might appear to be erected, but this is not piloerection as exhibited by canids. On occasions where an opponent is charged, the charging bear usually veers aside or stops just out of the opponent's reach. Although commonly called "bluff" charges, these are more accurately called "intimidation" charges. Their function is to frighten the opponent into fleeing, without risk of combat (Stringham 2010). Females tend to be most aggressive while protecting cubs but generally prefer to flee after sending their cubs up a tree, where they are out of danger.

Bears mark trails and roadsides by clawing, biting, and rubbing against trees, wooden posts (especially those treated with creosote), and cabins (LeCount 1986, McKinney and Delgadillo 2004). Bears frequently use the same areas, and their tracks can be visible along trails (McKinney and Delgadillo 2004). Generally, bears are diurnal and crepuscular, but some are nocturnal, especially while hunting deer fawns, javelina newborns, and calves, which they usually capture in bedding sites. Stressful encounters with humans can also cause a shift to primarily nocturnal activity.

Home range

The most influential factors in bear habitat selection and movements are food, water, and cover for concealment and thermoregulation. In several areas of Mexico, home ranges are similar in size to those recorded in other areas of North America. Those of adult males vary between 30 and 120 km^2 (11.5 and 46.3 mi^2), whereas those of females are between 10 and 30 km^2 (3.8 and 11.5 mi^2) (Powell et al. 1997; Costello et al. 2001; Doan-Crider 2003; McKinney and Delgadillo, unpublished data). In Sonora, the home range of an adult male was 345 km^2 (133 mi^2), and that of a subadult male 474 km^2 (183 mi^2) (Rodríguez-Martínez et al. 2006). Black bear home ranges vary depending on habitat conditions, being larger where food is scarce during at least part of the year. Home ranges that encompass a wide range of elevations are more likely to have food during all seasons, and thus to be smaller (Jonkel and Cowan 1971).

Females with infants have the smallest home ranges, but these increase in size as cubs mature and need more food. By the time cubs are weaned and become independent, a female's home range is usually larger than it will be before her next litter is born. Once cubs dissociate from the mother, they tend to continue sharing her home range for one or more years, whereupon daughters tend to establish a permanent range near the mother's and sons tend to disperse up to 100 km (62.1 mi) away. During this wandering phase, a subadult male may travel over an area larger than he will ever establish as a permanent home range. Home ranges can overlap; females and adult and subadult males can have overlapping home ranges but avoid each other (Rogers and Allen 1987; Powell et al. 1997; Doan-Crider 2003; Auger 2003; McKinney and Delgadillo, unpublished data).

Density

Depending on management and habitat quality, bear density in North America varies from 0.77 bears/km^2 (2 bears/mi^2) in the forested western US to 0.007–0.009 bears/km^2 (0.018– 0.023 bears/mi^2) in South Texas where desert vegetation predominates and forested areas form pockets in the Chihuahuan Desert (McKinney and Pittman 2001, Onorato and Hellgren 2001). In Mexico, population studies are scarce, and population density has been poorly documented. In the Burro Mountains of Coahuila, the estimated population density was 0.72 bears/km^2 (1.89 bears/mi^2), one of the highest densities reported in North America (Doan-Crider 1995). Subsequent monitoring in the same area revealed a density of 0.62 bears/km^2 (1.6 bears/mi^2) (INE-SEMARNAP 1999). In a UMA in the Sierra Madre Oriental in the Gomas and Milpillas Mountains in northeastern Nuevo León, monitoring with movement sensor cameras indicated a density of 0.11 bears/km^2 (0.28 bears/mi^2) (Salgado et al. 2006). In the Lampazos Range, in an area bordering Nuevo León and Coahuila, another study used noninvasive DNA analysis to determine a density of 1.0 bear/km^2 (2.6 bears/mi^2) (Mijangos 2009). The Sierra Madre Occidental populations have been little studied, but one study conducted in the San Luis Mountains and Ajos Mountains in Sonora estimated a density of 14.0 bears/100 km^2 (36.8 bears/100 mi^2) and 2.0 bears/100 km^2 (5.2 bears/100 mi^2), respectively (Espinosa-Flores et al. 2012).

Productivity and Recruitment

Reproduction

The age at which females reach sexual maturity and produce their first cubs tends to be inversely related to nutritional status. One index of nutritional status is age- and sex-specific body weight. In populations where females average no more than 60 kg (132.2 lb), a female typically first gives birth as a five- to seven-year-old, whereas first whelping commonly occurs during the third year where females are especially well nourished and average around 90 kg (198.4 lb) (Stringham 1990). A regression analysis of seven other North American black bear populations indicated that the relatively small body size of Mexican bears predicts that a female will first give birth at four years old. This has been confirmed in Coahuila (Doan-Crider 2003, Doan-Crider and Hellgren 1996).

Mating occurs during June–August. The embryonic blastocyst floats free in the uterus for up to five months. Then, if the mother is sufficiently well nourished, the blastocyst implants and commences a two-month gestation. Cubs are typically born during the last week of January or first week of February (Doan-Crider 2003). At birth, cubs weigh about 250 g (8.8 oz) (Leopold 1959, INE-SEMARNAP 1999). Cubs remain in the den until their weight has increased nearly eightfold (Eiler et al. 1989).

Once out of the den, the cubs accompany their mother on short walks, which become longer as they grow and require more food. Cubs remain in daily contact with their mother until the spring of their second year, whereupon they become independent and she ceases lactation. At the age of nearly 1.5 years, cubs become fully independent, and their mother again becomes sexually active. Females that succeed in raising a litter to independence typically breed every other year (Leopold 1959, Costello et al. 2001). However, if a litter is lost during its year of birth, the mother is sometimes able to breed again that year, reducing her interbirth interval to one year.

Continent-wide, a female's first litter usually consists of one or two cubs. More-mature females typically give birth to two or three cubs. Litter size is also directly related to nutritional status, and exceptional females can produce four cubs (Jonkel and Cowan 1971, Elowe and Dodge 1989, Stringham 1990, Costello et al. 2001). Variation in litter size appears less a function of natality than of postnatal survival, as a result of better-nourished mothers producing more milk (Hellgren 1988). Following the severe drought of 2000–2001 in northern Mexico, no infants were known to have survived until their mother emerged from her winter den (personal observation).

In northern Coahuila, the number of cubs varies between 2.5 and 3.3 per female every two years (Doan-Crider 1995, 2003; McKinney and Delgadillo, unpublished data). That is more than double the litter size one would expect given the small body size of Mexican bears—judging from a regression analysis of litter size versus body size for seven other black bear populations (Stringham 1990). This discrepancy suggests a genetic adaptation that shunts more nutrition into reproduction than into growth, compared to other populations. Relatively small body size might be advantageous in habitats where famine is common, or where bears need to minimize overheating during hot weather, as is common in much of Mexico (S. Stringham, personal communication).

Survival and mortality factors

Cubs are most vulnerable during two stages. The first is during infancy, when they leave the winter den with their mother and are still so small, weak, and slow that they easily fall victim to predators or accidents (e.g., drowning). The second is at 1.5 years of age, when they dissociate from their mother and have to fend for themselves. The main predators of juvenile bears are other bears and larger canids or felids (Leopold 1959, Stringham 1983, 1984, LeCount 1987, Doan-Crider 2003). In northern Coahuila, it is common to find adult male bear droppings containing cub hair, bones, and claws. Subadult males are at higher risk than subadult females, because young males are targeted by adult males and disperse into areas where they are unfamiliar with food sources and hazards.

In northern Coahuila, the annual survival rate in the Burro Mountains (fig. 12.4) was 0.94 for females, 1.0 for subadult males, and 0.80 for cubs (Doan-Crider 1995, 2003). In the Sierra Maderas del Carmen, the survival rate was 0.98 for adult females and males, and 1.0 for subadults and cubs during the time cubs were with their mothers. Survival of bears in these areas is relatively high compared to that of other bear populations in North America, which demonstrates that the quality of their habitat has possibly favored the fast recovery of their populations and their expansion to adjacent areas. Bears occasionally break their hips, possibly from falling out of trees, causing them to die of starvation. Deaths of females caused by other bears for unknown reasons have been documented (McKinney and Delgadillo, unpublished data).

Population Status

Many factors that led to the extinction of large carnivores in northern Mexico still exist. Humans still kill bears, whether by poaching or by shooting them while they are feeding on livestock. Highways in northern Mexico have recently contributed to bear mortality, especially in Coahuila and Nuevo León, where bears have been run over, possibly during intermountain movements (G. Salgado de los Santos and N. Sánchez, personal communication). Illegal trade in pet cubs is another threat. Only recently have wildlife authorities recognized the magnitude of these threats to bear populations (PROFEPA 2008).

Figure 12.4. Female black bear with two cubs in open habitat in northern Mexico. Photo by Jonás A. Delgadillo Villalobos.

As their populations increase, bears wander into cities and towns, where people tolerate and feed them; this has led to an increase in bear-human interactions and has created a coexistence problem. Many bears are captured and placed in zoos, while others are relocated to other areas. Relocated bears may be at a disadvantage because they must adapt to a new habitat or search for a suitable one.

Habitat reduction, degradation, and fragmentation have negatively affected black bears in Mexico. Forests have been logged and lands converted to other uses such as agriculture, livestock production, and urban expansion, as well as the construction of transportation infrastructure, electrical networks, and dams, and mining exploitation. It is estimated that in a 26-year period (1976–2002), nearly 454,794 ha (1.1 million ac) of temperate forests were lost, of which 298,049 ha (736,495 ac) were oak, oak-pine, and pine-oak forests (SEMARNAT 2008a). Large-scale forest fires have also contributed to black bear habitat reduction in northeastern Mexico; these fires are a natural phenomenon that occurs regularly in Mexico. Nevertheless, in the last few years, they have been so severe that important areas of black bear forest and shrubland habitats have been lost.

The American black bear is not endangered according to the IUCN Red List of Threatened Species (IUCN 2009), but it could reach that point if its exploitation is not properly regulated. However, it is listed in Appendix II of CITES.

Of the three subspecies in Mexico, only *eremicus* is listed as endangered, except for the Serranías del Burro population, which is included in a special protection category according to the Mexican Official Rule NOM-059; it establishes categories of risk for species in Mexico (SEMARNAT 2010). The inclusion of both of the other subspecies in this category is being considered (De la Maza-Benignos 2014).

Population Management

The black bear requires large areas of contiguous habitat to ensure its minimum population size, genetic diversity, population viability, and adequate year-round food supply, even during harsh years. Hence, management, evaluation, and monitoring of populations must be undertaken at a regional level (Rogers 1993, Rudis 1995). A population of between 50 and 100 adult animals is the absolute minimum to ensure genetic diversity in a population.

Specifically, in the case of the black bear, there should be 30–40 adult females within an area of roughly 300–400 km² (115–154 mi²) of high-quality habitat. Male home ranges overlap female territories, so they would be included in this required area (Shaw 1985, Rogers and Allen 1987).

At present, sport hunting of black bears in Mexico is prohibited because of the bears' protected legal status and because of the scarcity of scientific information about the actual distribution and population status of the species. Successful conservation, and eventually harvest management, will require detailed information on each population and the impacts to which it is subjected by adverse weather and human activities.

Only populations in northern Coahuila could provide controlled exploitation, if a permanent monitoring program were established to assess harvestable yield and to ensure population growth and natural dispersion of animals. The Serranías del Burro in northern Coahuila has a potential area of 3,600 km² (1,389 mi²), which is voluntarily protected by cattlemen (Doan-Crider 1995, 2003). This mountain range, which along with the Sierra del Carmen, Sierra La Encantada, and Sierra de Santa Rosa totals more than 10,000 km² (3,861 mi²) of occupied habitat, could provide a sustained yield of black bears without risking their population viability.

One of the principal problems of black bear management in Mexico is the high expense of population monitoring. Innovative techniques include the identification of individual bears using DNA analysis. However, Mexico lacks trained personnel and specialized laboratories to analyze hair samples. Additionally, this method is too expensive to apply extensively. Another method is the use of cameras with movement sensors. Its major disadvantage is the great difficulty of distinguishing genders, and the near impossibility of distinguishing individuals. Data on population distribution, abundance, density, age-sex ratios, and age- and sex-specific rates of reproduction and recruitment are essential for large-scale management, and for determining effects cause by humans and climate change.

Habitat Requirements

Black bears in Mexico occupy diverse habitats (fig. 12.5), even semiarid areas and shrublands. However, bears occur particularly in montane forested areas throughout North America. They prefer habitats with small clearings and a variety of vegetation associations. As with other large mammals, they require extensive contiguous habitats with available water and food. Where seasonal abundances of prime foods are widely separated, habitats need to be especially large (Rogers 1987, Powell et al. 1997, Auger 2003).

Forests that abound in mast-producing oaks, junipers, Texas madrone, and pinyons govern bear movements, reproduction, and survival in Mexico, much as these or other mast producers govern black bears elsewhere in North America (Leopold 1959, Hellgren 1993, Costello et al. 2001). In Mexico, black bear distribution along the Sierra Madre Oriental and Occidental is related mainly to the presence of pine and oak forests. The same is true for populations that live in semiarid areas in Coahuila and the eastern side of Chihuahua, where they inhabit mountain

Figure 12.5. Female black bear and cub feeding on oak acorns. Photo by Jonás A. Delgadillo Villalobos.

slopes and highlands surrounded by submontane shrublands and rosetophyllous scrub. These island-like mountain ranges (sky islands) are surrounded by desert. They are commonly covered with a rich and diverse vegetation composed of forbs, shrubs, and trees—mainly pinyon pines, oaks, and junipers—that provide food most of the year, without which bears could not survive.

Management and Conservation Challenges in Mexico

Habitat management

Mexico lacks specific habitat management programs. Nevertheless, black bears have indirectly benefited from the management of livestock because water availability has increased and illegal hunting has diminished. In some areas, landowners have included ramps inside deep-water units to facilitate the exit of animals that could become trapped. In some areas, the black bear has equally benefited from the establishment of water and food sources for white-tailed deer.

Coordination with cattle management

In northern Mexico, most cattle ranches have diversified and now include wildlife management programs. Sport hunting enterprises have been economically beneficial. Black bears have benefited by the large extensions of habitat and watering sources that private ranches provide, as well as from the restricted access that limits illegal hunting. However, the black bear has become a serious predator of livestock—mainly calves, goats, and sheep—and causes considerable losses, an issue that must be incorporated in conservation management projects. There have been proposals to allow sport hunting on these ranches to reduce depredations and raise income to compensate ranchers for livestock losses.

Coordination with forest management

Historically, forest management has been conducted without consideration of wildlife needs. However, the governmental agency Comisión Nacional Forestal is now developing forest management programs that include ecosystem conservation schemes that benefit bear populations. Like livestock management, sustained forest utilization does not necessarily represent a serious problem for a species as adaptable as the black bear. Properly designed forest clearings in the aftermath of logging allow the growth of forbs and grasses preferred by bears. Forest management with rotating tree harvest, and appropriately limited forest exploitation, could be compatible with black bear populations in Mexico.

Coordination with other land uses

Although the presence of bears in an area is sometimes viewed as a problem, it is more optimistically viewed as an advantage because bears are charismatic animals that can attract tourist dollars. In some parts of North America, bear viewing sites of only a few hectares generate millions of dollars and constitute an economic cornerstone for small rural communities. Some areas in northern Mexico also have an excellent potential for ecotourism activities such as photography, camping, and trekking without posing a problem for bears or humans. Some protected areas in Mexico support black bear populations that could be an economic incentive to promote economically viable, community-based conservation programs.

Final Considerations

Although information on Mexican black bears is scarce, it consistently indicates significant population increases. One consequence is increasing conflict with humans, especially in ranching and farming areas. These growing conflicts, along with the lower vulnerability of some populations, have led to reduced protection for bears. The population conservation status in the Serranías del Burro in Coahuila (subspecies *eremicus*) has been downgraded from endangered to special protection, which legalizes controlled exploitation of black bears in the region and could help compensate for the economic losses they cause. This management change, which is currently stalled because of the lack of an adequate monitoring system, could prevent overexploitation. The black bear is one of 25 species of priority concern in Mexico (SEMARNAP-INE 1997). This resulted in the formation of the Black Bear Conservation and Utilization Project (PREP) (INE-SEMARNAP 1999) within the Action Program for the Conservation of Species (PACE) (SEMARNAT-CONANP 2012). The Directorate of Priority Species for Conservation (PROCER) of the National Commission of Natural Protected Areas (CONANP) has established guidelines whose main objectives are the recovery of priority species. Natural Protected Areas (ANPs) in

Mexico are increasing and now include 176 areas in different categories (CONANP 2013); black bears have been recorded in 22 of these. Such areas can play an important role in black bear conservation, especially those that have extensive forests. ANPs have also been effective in maintaining plant communities (Figueroa and Sánchez-Cordero 2008).

Another conservation success in Mexico, mainly in the northern region, is the establishment of UMAs, whose objective is to create opportunities for legal, sustainable wildlife use in ways that promote alternative production schemes compatible with environmental protection. These management units are engaged mainly with white-tailed deer, mule deer, and collared peccary, but in some units where black bears occur, the bears have been included in conservation management plans (SEMARNAT 2008b).

The black bear in Mexico is an example of a successful recovery program. Although it was extirpated over a large area, it now reoccupies much of its former range and even extends into areas where it has not previously been recorded. There is reason to be optimistic about the future of the black bear in Mexico; current private landowner, national, and state black bear conservation initiatives will continue to ensure its long-term survival.

Acknowledgments

We thank Raul Valdez and J. Alfonso Ortega for the opportunity to contribute a chapter, as well as those individuals committed to the conservation of black bears in Mexico. Thanks primarily to Bonnie R. McKinney and Diana Doan-Crider, who have spent considerable time conducting bear research in Mexico and sharing their experiences with younger generations. I am also grateful to Romelia Medina Guillén, Cynthia Reséndiz, Nancy Armendáriz, Adrian Rosales, and Deneb Galindo for their writing collaboration. Nahum Sánchez, Daniel González, Jorge Luis Guerrero, Alfonso Martínez, Carlos López, Pedro Calderón, Rogelio Carrera, and Gilberto Salgado have generated infinite amounts of information and currently contribute significantly to black bear conservation.

Literature Cited

Álvarez, T. 1963. The recent mammals of Tamaulipas, Mexico. University of Kansas Museum of Natural History 14:365–473.

Auger, J. 2003. Movement of Utah black bear: implications for management and conservation. Western Black Bear Workshop 8:72–80.

Baker, R. H. 1956. Mammals of Coahuila, Mexico. University of Kansas Museum of Natural History 9:125–335.

Baker, R. H. 1958. The future of wildlife in northern Mexico: a problem in conservation education. Transactions of the North American Wildlife Conference 23:567–575.

Baker, R. H., and J. K. Greer. 1962. Mammals of the Mexican state of Durango. Publications of the Museum, Michigan State University, Biological Series 35:25–154.

Beeman, L. E. 1975. Population characteristics, movements and activities of the black bear (*Ursus americanus*) in the Great Smoky Mountains National Park. Thesis, University of Tennessee, Knoxville, USA.

Brown, D. E. 1985. The grizzly in the Southwest: documentation of an extinction. University of Oklahoma Press, Norman, USA.

Calderón-Domínguez, P., E. Carreón-Hernández, A. Lafón-Terrazas, and J. C. Guzmán-Aranda. 2006. Distribución espacio-temporal de las poblaciones de oso negro (*Ursus americanus*) en el estado de Chihuahua. International Conference on Bear Research and Management 18:15.

Carrera-Treviño, R., L. Martínez-García, and I. Lira-Torres. 2015. Primer registro de oso negro *Ursus americanus eremicus* (Carnivora: Ursidae) en un bosque tropical subperennifolio en la Reserva de la Biosfera El Cielo, Tamaulipas, México. Therya 6:653–660.

Carrillo, R. 1981. Notas sobre el oso negro (*Ursus americanus eremicus* Merriam 1904) en la Sierra Madre Oriental del estado de Nuevo León, México. Thesis, UANL, Monterrey, México.

Carvajal, S. V., D. S. Maher, and A. Caso. 2007. The black bear in a tropical cloud forest: local community participation in the El Cielo Biosphere Reserve project. International Conference on Bear Research and Management 18:3.

CONANP. 2013. Áreas Naturales Protegidas de México. http://www.conanp.gob.mx. Accessed 10 Oct 2013.

Contreras-Balderas, A. J., D. J. Hafner, J. H. López-Soto, J. Ma Torres-Ayala, and S. Contreras-Arquieta. 2007. Mammals of the Cuatrocienegas

Basin, Coahuila, Mexico. Southwestern Naturalist 52:400–409.

Costello, C. M., D. E. Jones, K. A. H. Green, R. M. Inman, K. H. Inman, B. C. Thomson, R. A. Deitner, and H. B. Quigley. 2001. A study of black bear ecology in New Mexico with models for population dynamics and habitat suitability. Federal Aid in Wildlife Restoration Project W-131-R. New Mexico Department of Game and Fish, and Hornocker Wildlife Institute, Santa Fe, USA.

De la Maza-Benignos, M., editor. 2014. Plan de acción para la conservación y recuperación de especies de fauna silvestre prioritaria en el estado de Chihuahua. Pronatura Noreste and Gobierno del Estado de Chihuahua, México.

Delfín-Alfonso, C. A., C. A. López-González, and N. E. Lara-Díaz. 2011. El oso negro americano en el noroeste de México: recuperación de registros de ocurrencia. Acta Zoológica Mexicana 27:777–801.

Delgadillo, J. A. 2001. Ecología nutricional del oso negro en la Sierra Maderas del Carmen, Coahuila, México. Thesis, UANL, Linares, México.

Delgadillo, J. A., B. McKinney, D. Onorato, S. Mitchell, M. Pittman, S. Skyles, and L. Harveson. 2007. Diet of three black bear populations in the Chihuahuan Desert region of northern Coahuila, Mexico and Western Texas. International Conference on Bear Research and Management 18:57.

Doan-Crider, D. L. 1995. Population characteristics and home range dynamics of the black bear in northern Coahuila, Mexico. Thesis, Texas A&M University–Kingsville, USA.

Doan-Crider, D. L. 2003. Movements and spatiotemporal variation in relation to food productivity and distribution, and population dynamics of the Mexican black bear in the Serranías del Burro, Coahuila, Mexico. Dissertation, Texas A&M University–Kingsville, USA.

Doan-Crider, D. L., and E. C. Hellgren. 1996. Population characteristics and winter ecology of black bears in Coahuila, Mexico. Journal of Wildlife Management 60:398–407.

Doan-Crider, D., and O. Moctezuma Orozco. 2014. American black bear. Pages 518–521 *in* G. Ceballos, editor. Mammals of Mexico. Johns Hopkins University Press, Baltimore, Maryland, USA.

Eagle, T. C., and M. R. Pelton. 1983. Seasonal nutrition of black bear in the Great Smoky Mountains National Park. International Conference on Bear Research and Management 5:94–101.

Eiler, J. H., W. G. Wathen, and M. R. Pelton. 1989. Reproduction in black bears in the southern Appalachian Mountains. Journal of Wildlife Management 53:353–360.

Elowe, K. D., and W. E. Dodge. 1989. Factors affecting black bear reproductive success and cub survival. Journal of Wildlife Management 53:962–968.

Espinosa-Flores, M. E., N. E. Lara-Díaz, and C. A. López-González. 2012. Tamaño poblacional del oso negro (*Ursus americanus*) en dos islas de El Cielo del noreste de Sonora, México. Therya 3:413–415.

Figueroa, F., and V. Sánchez-Cordero. 2008. Effectiveness of natural protected areas to prevent land use and land cover change in Mexico. Biodiversity and Conservation 17:3223–3240.

Garshelis, D. L. 2009. Family Ursidae (bears). Pages 448–497 *in* D. E. Wilson and R. A. Mittermeier, editors. Handbook of the mammals of the world. Volume 1. Carnivores. Lynx Edicions, Barcelona, Spain.

Gobierno del Estado de Coahuila. 2009. Acuerdo por el que se declara al oso negro como símbolo de la conservación de los recursos naturales de Coahuila. http://www.semac.gob.gob. Accessed 10 Sep 2009.

Hall, E. R. 1981. The mammals of North America. Second edition. John Wiley and Sons, New York, New York, USA.

Hellgren, E. C. 1988. Great Dismal Swamp National Wildlife Refuge black bear study status report, East. Workshop Black Bear Research and Management 8:242–243.

Hellgren, E. C. 1993. Status, distribution, and summer food habits of black bear in Big Bend National Park. Southwestern Naturalist 38:77–80.

Hellgren, E. C., and M. R. Vaughan. 1988. Seasonal food habits of black bears in Great Dismal Swamp, Virginia–North Carolina. Proceedings of the Annual Conference of the Southeastern Association of Fish and Wildlife Agencies 42:295–305.

Herrera-González, D. E. 1999. Ecología de la dieta del oso negro (*Ursus americanus*) en el área natural protegida "Maderas del Carmen," en el estado de Coahuila. Thesis, UANL, Monterrey, México.

Hewitt, D., and D. Doan-Crider. 2008. Metapopulations, food and people: bear management in northern Mexico. Pages 165–181 *in* T. E. Fulbright and D. G. Hewitt, editors. Wildlife science: linking ecological theory and management applications. CRC Press, Boca Raton, Florida, USA.

INE-SEMARNAP. 1999. Proyecto de recuperación del oso negro (*Ursus americanus*). Serie PREP. INE-SEMARNAP, México, D. F., México.

IUCN. 2009. Red List of Threatened Species. http://www.iucnredlist.org. Accessed 10 Aug 2009.

Jiménez, G. M. A. 1981. Especies en peligro y proceso de desaparición en Nuevo León. Boletín Informativo del Centro de Investigaciones Biológicas de la UANL 3:5–10.

Jonkel, C. J., and I. M. Cowan. 1971. The black bear in the spruce-fir forest. Wildlife Monographs 27:1–57.

Juárez-Casillas, L. A. 2006. Hábitos alimentarios del oso negro (*Ursus americanus*) en la Sierra de Picachos, Nuevo León, México. Thesis, UANL, Monterrey, México.

Lara, M. K. E. 2006. Modelos de distribución de los mamíferos terrestres de Tamaulipas en peligro de extinción, amenazados y sujetos a protección especial. Thesis, Instituto Tecnológico de Ciudad Victoria, Tamaulipas, México.

LeCount, A. L. 1986. Black bear field guide: a manager's manual. Special Report No 16. Arizona Game and Fish Department, Phoenix, USA.

LeCount, A. L. 1987. Characteristics of a northern Arizona black bear population. Federal Aid Project W-78-R, Work Plan 2, Job 22. Arizona Game and Fish Department, Phoenix, USA.

Leopold, A. S. 1959. Wildlife of Mexico: the game birds and mammals. University of California Press, Berkeley, USA.

Lobello, R. L. 1989. Status of the black bear in Big Bend National Park, Texas. Regional Conference of the US/Mexico Border States on Parks and Wildlife 3:6.

Martínez, A., J. A. Delgadillo, D. Herrera, and J. González. 2014. Nutrición del oso negro (*Ursus americanus eremicus*) en las Serranías del Carmen, Coahuila. Ciencia UANL 68:39–48.

McDaniel, J. 1974. Habitat of the black bear in Florida. Eastern Black Bear Workshop 2:157–162.

McDonald, J. E., and T. K. Fuller. 2005. Effects of spring acorn availability on black bear diet, milk composition, and cub survival. Journal of Mammalogy 86:1022–1028.

McKinney, B. R., and J. A. Delgadillo. 2004. Manual para el manejo del oso negro mexicano: guía para manejadores. CEMEX, Monterrey, Nuevo León, México.

McKinney, B. R., and M. T. Pittman. 2001. Habitat, diet, home range, and seasonal movement of resident and relocated black bears in West Texas. Wildlife Diversity Program, Research Project WER57-state. Texas Parks and Wildlife, Austin, USA.

McKinney, B. R., P. Robles Gil, R. Skiles, V. J. Delgadillo, and J. Rojo. 2006. El Carmen–Big Bend: un innovador modelo de alianzas públicas y privadas para la conservación transfronteriza. Pages 101–112 *in* R. Mittermeier, C. F. Kormos, C. G. Mittermeier, P. Robles Gil, T. Sandwith, and C. Besançon, editors. Conservación transfronteriza: una nueva visión para áreas protegidas. CEMEX, Monterrey, Nuevo León, México.

McLellan, B., and D. C. Reiner. 1994. A review of bear evolution. International Conference on Bear Research and Management 9:85–96.

Medellín, R. A., C. Manterola, M. Valdez, D. G. Hewitt, D. Doan-Crider, and T. E. Fulbright. 2005. History, ecology, and conservation of the pronghorn antelope, bighorn sheep, and black bear in Mexico. Pages 397–400 *in* J. L. E. Cartron, G. Ceballos, and R. S. Felger, editors. Biodiversity, ecosystems and conservation in northern Mexico. Oxford University Press, New York, New York, USA.

Mijangos, J. L. A. 2009. Estimación poblacional del oso negro americano (*Ursus americanus eremicus*) con microsatélites de ADN a partir de muestras no invasivas en Lampazos, Nuevo León. Thesis, UANL, Monterrey, México.

Moreno, A. C. N. 2008. Ecología conductual del oso negro (*Ursus americanus*) en la Sierra Madre Occidental. Thesis, Instituto de Ecología, Jalapa, Veracruz, México.

Moreno-Valdez, A. 1991. Distribución del oso negro (*Ursus americanus*, Pallas) en Tamaulipas, su dieta y factores socioeconómicos que afectan sus poblaciones. SEDUE, Delegación Federal Tamaulipas, Ciudad Victoria, México.

Moreno-Valdez, A. 1998. Mamíferos del Cañón del Huajuco, Municipio de Santiago, Nuevo León, México. Revista Mexicana de Mastozoología 3:5–25.

Moreno-Valdez, A., and E. P. Vázquez-Farías. 2005. Los mamíferos terrestres de Tamaulipas. Pages 213–219 *in* L. L. Barrientos, A. S. Correa, J. V. V. Horta, and J. J. García, editors. Biodiversidad

Tamaulipeca. Volume 1. Instituto Tecnológico de Ciudad Victoria, Tamaulipas, México.

Niño-Ramírez, J. A. 1989. Análisis preliminar de la dieta de verano del oso negro (*Ursus americanus*), en la Sierra Los Picachos, Higueras, Nuevo León, México. Thesis, UANL, Monterrey, México.

Onorato, D. P., and E. C. Hellgren. 2001. Black bear at the border: natural recolonization of the Trans-Pecos. Pages 245–259 *in* D. S. Maher, R. F. Noss, and J. L. Larkin, editors. Large mammal restoration: ecological and sociological challenges in the 21st century. Island Press, Washington, DC, USA.

Onorato, D. P., E. C. Hellgren, R. A. Van Den Bussche, and D. L. Doan-Crider. 2004. Phylogeographic patterns within a metapopulation of black bear (*Ursus americanus*) in the American southwest. Journal of Mammalogy 85:140–147.

Onorato, D. P., E. C. Hellgren, R. A. Van Den Bussche, D. L. Doan-Crider, and R. Skiles Jr. 2006. Genetic structure of American black bear in the desert southwest of North America: conservation implications for recolonization. Conservation and Genetics 8:565–576.

Pelton, M. R., A. B. Coley, T. H. Eason, D. L. Doan Martínez, J. A. Pederson, F. T. van Manen, and K. M. Weaver. 1998. American black bear plan. Pages 144–156 *in* C. Servheen, S. Herrero, and B. Peyton, compilers. Bears: status survey and conservation action plan. IUCN/SSC Bear and Polar Bear Specialist Groups, IUCN, Gland, Switzerland, and Cambridge, United Kingdom.

Powell, R. A., J. W. Zimmerman, and D. E. Seaman. 1997. Ecology and behavior of North American black bears: home ranges, habitat and social organization. Chapman and Hall, London, United Kingdom.

PROFEPA. 2008. Informe anual. http://www.profepa.gob.mx. Accessed 15 Oct 2009.

Rodríguez-Martínez, A., C. N. Moreno-Arzate, E. R. González-Sierra, and C. A. López-González. 2006. Spatial ecology and population structure of endangered black bear (*Ursus americanus*) in northwestern Mexico. Western Black Bear Workshop 9:8–19.

Rogers, L. L. 1987. Factors influencing dispersal in the black bear. Pages 75–84 *in* B. D. Chepko Sade and Z. T. Halpin, editors. Mammalian dispersal patterns. University of Chicago Press, Illinois, USA.

Rogers, L. L. 1993. The role of quality in the natural regulation of a black bear population. Western Black Bear Workshop 4:95–100.

Rogers, L. L., and A. W. Allen. 1987. Habitat suitability index models: black bear, Upper Great Lakes Region. Biological Report 82 (10.144). USFWS, Washington, DC, USA.

Rojas-Martínez, A., and L. A. Juárez-Casillas. 2013. Primer registro de oso negro americano (*Ursus americanus*) para el estado de Hidalgo, México. Revista Mexicana de Biodiversidad 84:1018–1021.

Rudis, V. A. 1995. Regional assessment of remote forest and black bear habitat from forest resource surveys. Journal of Wildlife Management 59:170–180.

Ryan, C. W., J. C. Pack, W. K. Igo, and A. Billings. 2007. Influence of mast production on black bear nonhunting mortalities in West Virginia. Ursus 8:46–53.

Salgado, G., J. F. González, C. P. V. Moreno, and G. J. G. Villarreal. 2006. Preliminary study of a black bear population in the Sierra de Gomas and Sierra de Milpillas in Nuevo Leon, Mexico. International Bear News 15:24.

Salgado-De Los Santos, G. 2007. Programa de recuperación, conservación y manejo del oso negro (*Ursus americanus eremicus*) en el noroeste de Nuevo León, México. Boletín Informativo 3:3–10.

Salgado-De Los Santos, G., F. González-Jiménez, P. V. Moreno-Chacón, and J. G. Villarreal-González. 2005. Situación actual de la población de oso negro (*Ursus americanus eremicus*) en la Sierra de Gomas y Sierra de Milpillas (Cuenca Palo Blanco), Salinas Victoria, Nuevo León, México. Pages 138–147 *in* Memorias del 22 Simposio sobre Fauna Silvestre "Gral. MV. Manuel Cabrera Valtierra." Facultad de Medicina Veterinaria y Zootecnia, UNAM, México, D. F., México.

Sáyago, M. del C. Silva, and C. López. 2005. Oso negro, crónica de una falsa acusación. Naturalia 14:20–24.

SEMARNAT. 2008a. Informe de la situación del medio ambiente en México. Compendio de estadísticas ambientales. http://www.semarnat.gob.mx. Accessed 20 Oct 2009.

SEMARNAT. 2008b. Especies prioritarias de fauna registradas en Unidades de Manejo para la Conservación de la Vida Silvestre (UMA) extensivas. http://www.semarnat.gob.mx. Accessed 20 Oct 2009.

SEMARNAT. 2010. Norma Oficial Mexicana NOM-059-SEMARNAT-2010, Protección ambiental—Especies nativas de México de flora y fauna silvestres—Categorías de riesgo especificaciones para su inclusión, exclusión o cambio—lista de especies en riesgo. DOF. Segunda Sección. 30 de diciembre de 2010. México, D. F., México.

SEMARNAT-CONANP. 2005. Estudio previo justificativo para el establecimiento del Área Natural Protegida "Sierra Gorda de Guanajuato." Secretaría de Recursos Naturales, CONANP, México, D. F., México.

SEMARNAT-CONANP. 2009. Programa de acción para la conservación de la especie: lobo gris mexicano (*Canis lupus baileyi*). SEMARNAT and CONANP, México, D. F., México.

SEMARNAT-CONANP. 2012. Programa de acción para la conservación de la especie: oso negro americano (*Ursus americanus*). SEMARNAT and CONANP, México, D. F., México.

SEMARNAP-INE. 1997. Programa de Conservación de la Vida Silvestre y Diversificación Productiva en el Sector Rural. 1997–2000. SEMARNAT and INE, México, D. F., México.

Servín, J. I. M., J. E. Chacon, M. Huxley, and N. Alfonso Pérez. 1997. Los mamíferos del estado de Durango. INE, Centro Regional-Durango. Informe técnico, Proyecto P064. http://www.conabio.gob.mx. Accessed 20 Aug 2009.

Shaw, J. H. 1985. Introduction to wildlife management. McGraw-Hill, New York, New York, USA.

Sierra-Corona, R., I. A. Sáyago-Vázquez, M. Silva-Hurtado, and C. A. López-González. 2005. Black bear abundance, habitat use and food habits in the Sierra San Luis, Sonora. Pages 444–448 *in* G. J. Gottfried, B. S. Gebow, L. G. Eskew, and C. Edminster, compilers. Connecting mountain islands and desert seas: biodiversity and management of the Madrean Archipelago 2. RMRS-P-36. US Forest Service, Rocky Mountain Research Station, Fort Collins, Colorado, USA.

Sosa, F. V. 1998. Inventario de los mamíferos de las Reservas de la Biósfera Mapimí, La Michilía, El Cielo y Calakmul. Informe final SNIB-CONABIO, Proyecto No. P027. INE, México, D. F., México.

Stringham, S. F. 1983. Roles of adult males in grizzly bear population biology. Ursus 5:140–151.

Stringham, S. F. 1984. Responses by grizzly bear population dynamics to certain environmental and biosocial factors. Dissertation, University of Tennessee, Knoxville, USA.

Stringham, S. F. 1990. Black bear reproductive rate relative to body weight in hunted populations. Ursus 8:425–432.

Stringham, S. F. 2010. Ghost grizzlies and other rare bruins. Wildwatch, Soldotna, Alaska, USA.

Treviño, J. C., and C. Jonkel. 1986. Do grizzly bears still live in Mexico? Ursus 6:11–13.

Van Den Bussche, R. A., J. B. Lack, D. P. Onorato, L. C. Gardner, B. R. McKinney, J. Delgadillo. M. J. Chamberlain, D. White Jr., and E. C. Hellgren. 2009. Mitochondrial DNA phylogeography of black bear (*Ursus americanus*) in central and southern North America: conservation implications. Journal of Mammalogy 90:1075–1082.

Varas, C. 2007. Black bears blocked by the border. Pages 87–92 *in* A. Córdova and C. A. de la Parra, editors. A barrier to our shared environment: the border fence between the United States and Mexico. Secretariat of Environment and Natural Resources, National Institute of Ecology, El Colegio de la Frontera Norte, National Institute of Ecology, and Southwest Consortium for Environmental Research and Policy, México, D. F., México.

Vargas, F. M. 1997. Parques nacionales de México. Volume 2. Zonas norte y sur. SEMARNAT-INE, México, D. F., México.

Vargas, J. A., and A. H. Hernández. 2001. Distribución altitudinal de la mastofauna en la Reserva de la Biósfera "El Cielo," Tamaulipas, México. Acta Zoológica Mexicana 82:83–109.

Villa, R. B. 1977. Major game mammals and their habitats in the Chihuahuan Desert region. Transactions of the Symposium on the Biological Resources of the Chihuahuan Desert Region 3:155–167.

Williamson, D. F. 2002. In the black: status, management, and trade of the American black bear (*Ursus americanus*) in North America. TRAFFIC North America and World Wildlife Fund, Washington, DC, USA.

Wooding, S., and R. Ward. 1997. Phylogeography and Pleistocene evolution in the North American black bear. Molecular Biology and Evolution 14:1096–1105.

Zepeda-González, J., J. Arroyo-Cabrales, O. J. Polaco, and A. Jiménez-Guzmán. 1997. Notas acerca de la distribución de algunos mamíferos del sur de Nuevo León, México. Revista Mexicana de Mastozoología 2:101–112.

13 Jaguar and Puma in Mexico

Raul Valdez, Octavio César Rosas Rosas, and Rodrigo Núñez Pérez

The jaguar (*Panthera onca*) and puma (*Puma concolor*) are the largest felids in the Americas and are sympatric over the entire range of the jaguar in Mexico (fig. 13.1). In tropical and subtropical areas, jaguars and pumas share habitats with smaller wild cats: ocelot (*Leopardus pardalis*), margay (*L. wiedii*), and jaguarundi (*Puma yagouaroundi*). Pumas occupy a wider range of habitats than jaguars. Pumas are distributed throughout Mexico along both coasts, in mountainous areas, deserts, tropical habitats, and deciduous dry, conifer, and oak forests. They have one of the widest distributions of any mammal in the Western Hemisphere (Leopold 1959, Chávez Tovar and Ceballos 2014a). Jaguars, on the other hand, are distributed mainly in tropical forests with adequate vegetation cover and low human population density. In tropical and subtropical Mexico, they occur in deciduous and perennial forests, thorny forests, xeric shrublands, and cloud forests (Sanderson et al. 2002, Chávez Tovar and Ceballos 2014b). In both cases, these cats are under pressure because of habitat fragmentation, prey depletion, and illegal hunting (Valdez 1999, Rosas-Rosas et al. 2003, 2008). Additionally, widespread cattle ranching, agriculture expansion, and unlawful human settlement in protected natural areas threaten their survival in the long term (Galindo et al. 2016). The extensive past and present expansion of human-dominated landscapes

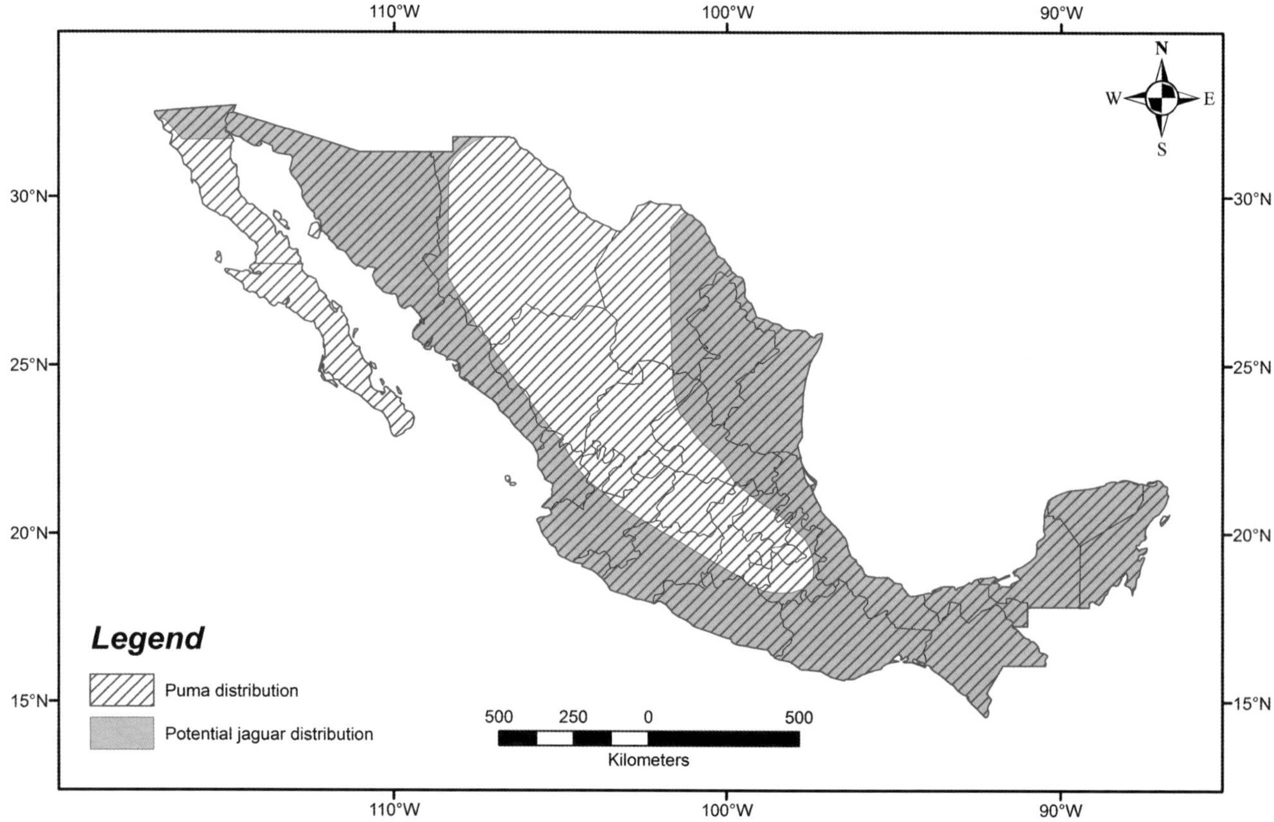

Figure 13.1. Distribution of the puma and potential distribution of the jaguar in Mexico.

and the concomitant habitat conversion are the major threats to these felid species.

The jaguar is highly threatened throughout its range from Mexico to South America because of habitat degradation, modification, and fragmentation, in addition to predator control practices and reduction of its prey (Sunquist and Sunquist 2009a). The value of the jaguar's fur was the main reason that about 13,000 jaguar skins were exported from Latin America to the US and Europe in 1973 (Koford 1975). Jaguars are officially protected by law, and sport hunting has been banned since 1987 (SEMARNAP 1997). Currently, jaguars are a priority species in Mexico (SEMARNAT and CONANP 2009, SEMARNAT 2010). However, illegal hunting of jaguars continues on federal, communal, and private lands, mainly because of livestock predation (López-González and Brown 2002, Rosas-Rosas et al. 2008). When conflict exists between carnivores and humans because of livestock predation, negative attitudes toward large predators increase. Reduced prey availability is another major threat to the puma.

In Mexico, the puma is a major management concern where there are conflicts with livestock interests and where UMAs have fenced in prey species for sport hunting (Rosas-Rosas et al. 2003, 2008; Ávila-Nájera et al. 2011). Because cattle ranching is the main economic activity in areas where pumas and jaguars occur, livestock losses from predation by these felid species are one of the most important economic factors affecting ranch owners (Rosas-Rosas and Valdez 2010). Predator control has been a common management practice. Ranch owners usually hire professional or local hunters (*tigreros*) to remove jaguars through direct hunting or by poisoning the remains of native prey or livestock. In the states of Nuevo León, Chihuahua, Sonora, and San Luis Potosí, predator control is applied on ranches where pumas and jaguars prey on cattle. In most regions where trophy hunting is practiced on game ranches, owners believe pumas to be a threat to their investments in raising and maintaining game species (Rosas-Rosas et al. 2003), and predator control is considered necessary. However, trapping and poisoning large cats has not been common, even in areas with large livestock enterprises, at least in the last decade, but poisoning does occur (Rosas-Rosas, personal observation; Zarco-González et al. 2012; Rodríguez-Soto et al. 2013).

Historical Perspective

In pre-Hispanic times, the jaguar was culturally important to communities in Mexico. In Mexican native cultures, the jaguar was a controversial species. For some, the jaguar was a divine entity, while for others it was the representation of evil, symbolizing night and nocturnal power. In ancient cultures, jaguars possessed magical and religious qualities and were associated with natural elements such as thunder, the sun, the moon, mountains, caves, and fire. Along with other animals, the jaguar is the protector of hunting and hunting rituals (González Torres 2001). It has several ethnic names. However, the most common name, "jaguar," originated from the Tupi-Guarani word *yaguará*, which means "animal that catches its prey in a single bound" (Gamero-Idiáquez 1978), referring to how jaguars capture prey. Spaniards named it *tigre* because of its similarity to the Asian tiger (*Panthera tigris*). The name "tigre" has spread throughout the continent, but in Mexico this name is more rooted in rural communities.

The word "puma" is of Inca origin. The puma is known by several names in rural communities in Mexico: *león*, mountain lion, brown lion, deer lion, and *onza*. "León" is the most common name used in rural Mexico (Young and Goldman 1946). Predators such as the puma usually evoked power and were generally considered benefactors and friends (Gill 2010).

Description

Jaguar

The jaguar is the largest feline in the Americas. Its body color varies from pale yellow to reddish brown, with black spots and white on the chest and inside the limbs. There is also a melanistic phase (black or dark, with rosettes visible in oblique light). Melanistic jaguars have been observed in Central and South America but have not been recorded in Mexico (Hoogesteijn and Mondolfi 1993, SEMARNAT 2006, Sunquist and Sunquist 2009b). The jaguar's coat is short, compact, and slightly bristly, and the hair is slightly longer on the neck and abdomen than on the rest of the body. The head is broad, robust, and bulky and the ears are short and wide (Leopold 1959). Large males weigh up to 80 kg (176 lb) in northeastern Mexico (Moreno Valdéz 1987). There is distinct sexual dimorphism; males are larger and more robust. In western Mexico, jaguar males weigh 40 to 65 kg (88 to 145 lb), and females between 25 and 40 kg (55 and 88 lb) (Núñez Pérez 2006). In general, male jaguars weigh between 50 and 70 kg (110 and 154 lb), and females are 35–50 kg (77–110 lb) (SEMARNAT 2006). In the Greater Lacandona Ecosystem (GLE) in southeastern Mexico, female jaguars were 1.60 times heavier than female pumas, and male jaguars were 1.68 times heavier than male pumas (de la Torre et al. 2017b). Tracks of the forefoot and hind foot of jaguars exhibit four toes and a main pad. The forefoot is wider than the hind foot. Jaguar forefoot tracks are 6–10 cm (2.3–3.9 in) long and 7–10 cm (2.7–3.9 in) wide; hind foot tracks are 6.5–10 cm (2.5–3.9 in) long and 6–9 cm (2.3–3.5 in) wide (Aranda 2000).

Puma

Among North American felids, the puma is surpassed in mass only by the jaguar. The puma is generally yellowish brown on the head and back, varying from reddish brown to sandy (fig. 13.2) (Leopold 1959, Chávez Tovar and Ceballos 2014a). The belly is whitish, and the hair is short and dense. The dorsal surface of the ears and the tip of the tail are black. Cubs have a spotted pattern, which disappears when they are six months old. There is well-marked sexual dimorphism in the puma, with females being smaller and less robust than males. The puma is larger at higher latitudes, and smaller toward the equator (Iriarte et al. 1990). Two adult puma males in Sonora weighed 60 kg (132 lb) and 37 kg (82 lb), and an adult female weighed 25 kg (55 lb) (Rosas-Rosas, unpublished data). In the Calakmul region of the Yucatán Peninsula, the average weight of eight males was 42 kg (93 lb). Puma forefoot tracks are 6.5–8.5 cm (2.5–3.3 in) long and 7–10 cm (2.7–3.3 in) wide. Hind foot tracks are 6.5–9 cm (2.5–3.5 in) long and 6–8.5 cm (2.3–3.3 in) wide (Aranda 2000).

Taxonomy

Jaguar

A detailed review of the evolutionary relationships of the Felidae led to the reclassification of the jaguar,

Figure 13.2. Puma in the Sierra del Abra in Tanchipa Biosphere Reserve, San Luis Potosí, Mexico. Photo by Octavio C. Rosas Rosas.

formerly in the genus *Felis*, into the genus *Panthera*, which includes the large cats from the Old World (Wozencraft 1993). Roaring is characteristic of cats in the genus *Panthera* and is made possible by a sensitive throat ligament (Sunquist and Sunquist 2009a). The jaguar is the only representative of the genus *Panthera* occurring in the Americas.

Four jaguar subspecies were formerly recognized in Mexico, based on anatomical characters (Hall 1981, Wozencraft 1993). However, a review of cranial measurements did not reveal significant differences to differentiate subspecies (Larson 1997). A genetic analysis of 40 individuals from Mexico to southern Brazil showed no detectable geographical differences, and the most significant differences existed between the populations north and south of the Amazon River (Eizirik et al. 2001). Culver and Hein (2016), based on a molecular genetic study, showed no evidence to support subspecies differentiation in Mexico, but the Sonoran subpopulation was an evolutionarily significant unit.

Puma

The puma lineage possibly originated from an American cheetah-like cat (*Miracinonyx inexpectatus*) that inhabited North America approximately 3.2–0.6 million years ago (mya) during the Pliocene-Pleistocene. This felid may be the link between the present puma and the North American cheetah (*Acinonyx trumani*) (Kurtén 1976, Van Valkenburgh et al. 1990). Fossils from the La Brea tar pits in California suggest that there was a puma species with physical characteristics intermediate between those of the North American cheetah and the current puma (Kurtén 1976). Puma fossils found in the US and Mexico are from 0.2–0.01 mya (Kurtén and Anderson 1980). Fossil evidence suggests that the puma evolved in North America and then spread to South America about 4–2 mya (Kurtén 1976, Van Valkenburgh et al. 1990). However, molecular analysis indicates that the puma has its own distinct lineage within the same group as the cheetah (*Acinonyx jubatus*) and the jaguarundi; hence it is not closely related to small wild cats of the genus *Felis* (Johnson and O'Brien 1997, Culver 2010).

The puma is classified into 32 subspecies and has a continuous distribution from North to South America. The subspecies classification is based on anatomical features of the skull, body size, coat color, and geographical distribution (Young and Goldman 1946). However, using molecular genetic data, Culver (2010) recommended recognition of six subspecies, of which only one (*cougar*) occurs in Mexico and most of North America.

Distribution

Jaguar

The jaguar was originally distributed continuously from the southern US to Patagonia, but the current distribution in Mexico is discontinuous and fragmented (Sanderson et al. 2002, Rabinowitz and Zeller 2010). They have been extirpated from more than 50% of their total historical range in the last 100 years (Rabinowitz and Zeller 2010, de la Torre et al. 2017a). In Mexico, much of their habitat has been fragmented (Chávez Tovar and Ceballos 2014b). Previously undetected populations have been reported in the Huasteca Potosina region in San Luis Potosí in northwestern coastal Mexico (Villordo-Galván et al. 2010), in western Mexico in Nayarit, Colima, Jalisco, and Michoacán (Núñez Pérez 2007), and in Hidalgo (Aguilar-Lopez et al. 2015) and Morelos (Valenzuela-Galvan et al. 2015). Distribution models for Mexico suggest that unoccupied habitats for the jaguar are widespread. In the southwestern US, jaguars have been recently recorded in Arizona and New Mexico (Glenn 1996, McCain and Childs 2008, Culver 2016). However, females have not been reported since the early 1900s in this region, except for a female killed in the White Mountains in Arizona in 1963, so the viability of a resident population remains questionable. Recorded individuals may have originated from populations in northeastern Sonora.

In northwestern Mexico, the jaguar occurs as far north as the municipality of Huásabas, Sonora, in the Sierra Madre Occidental, where it inhabits subtropical scrub habitats and oak forest ecotones (Valdez et al. 2002, López González and Brown 2002). It is distributed continuously southward to Nayarit and Jalisco, and in two small coastal regions of Michoacán and Oaxaca (Sanderson et al. 2002). In Michoacán and Guerrero, its occurrence is uncertain, but it could be present in well-conserved areas of the Sierra Madre del Sur. Along the coast of the Gulf of Mexico, the jaguar occurs in a subprovince of the Sierra Madre Oriental, the Gran Sierra Plegada, extending from Allende in Nuevo León to the Huasteca in San Luis Potosí, and to the Sierra Gorda in Querétaro. It is found in tropical oak forest, tropical deciduous forest, cloud forest, tropical rainforest, piedmont scrub, and pine-oak forest and ecotones from coastal areas to elevations above 2,400 m (7,874 ft) (Rosas-Rosas and López Soto 2002, Villordo Galván et al. 2010). There may be two isolated populations in the Sierra de Tamaulipas and the Sierra de Maratines in Tamaulipas (Caso 2007); however, it has been suggested that the population in the Sierra Madre Oriental could be isolated from other populations (Sanderson et al. 2002, Rabinowitz and Zeller 2010). In the region of Hidalgo, a continuous southern habitat does not exist because of habitat fragmentation in the state of Veracruz.

The jaguar is occasionally reported in areas uncommon for the species. In 2006, an individual was recorded in the Sierra de Nanchititla, in the state of Mexico (Monroy-Vilchis et al. 2008). Jaguars may disperse along vegetated corridors, even through human-dominated landscapes. A jaguar corridor between the Sierra Madre Occidental and Oriental through the Chihuahuan Desert has been modeled based on environmental features but is unlikely (Rabinowitz and Zeller 2010). Current fragmentation resulting from roads, intensive agricultural development, and lack of continuous vegetation cover represents a barrier for individual dispersal. In southeastern Mexico, the most important jaguar populations are in Pantanos de Centla in Tabasco and Laguna de Términos in Campeche. The most important populations in Mexico and Mesoamerica occur on the Yucatán Peninsula in Campeche and Quintana Roo (Sanderson et al. 2002, Ceballos et al. 2016). Areas with priority jaguar populations in Mexico are the Calakmul region in Campeche and Quintana Roo, Sierra Madre Occidental (Sonora and Sinaloa), Sierra Madre Oriental (Tamaulipas, Nuevo León, and San Luis Potosí), Marismas Nacionales wetlands in Sinaloa and Nayarit, tropical forests in southern Nayarit and Jalisco, and tropical forests from Oaxaca and Chiapas to Campeche and Quintana Roo (SEMARNAT 2006) (fig. 13.3).

Figure 13.3. Jaguar in the state of Sonora, Mexico. Photo by Primero Conservation.org.

Puma

The puma is the most widely distributed large felid in North America, but its distribution has been reduced by as much as two-thirds of its historical range (Pierce and Bleich 2003). The puma occurs from British Columbia, Canada, and the northern US continuously to Chile and Argentina in Patagonia (Currier 1983, Culver 2010). In Mexico, the puma is found in most states except in areas with large human populations, including central Mexico and much of the Mexican Altiplano (Chávez Tovar and Ceballos 2014a). The puma can live near human settlements and has been seen near large cities including Guadalajara, Mexico City, and Monterrey (del Castillo 2009). Currently there is not a detailed distribution map of the species. The great plasticity of this species, specifically in its diet (Núñez et al. 2000, Rosas-Rosas et al. 2003, Rosas-Rosas et al. 2008, Monroy-Vilchis et al. 2009a), has facilitated its survival in almost all habitat types in Mexico. However, it is being extirpated regionally, mainly in those areas that have been transformed by agricultural development and urbanization.

Life History

Food Habits

Taber et al. (1997) emphasized the importance of determining the availability of prey and its habitat use, as well as jaguar and puma densities, to more effectively understand the role of both species in the partitioning of resources. Where jaguars and pumas are sympatric and medium-sized prey is reduced, cattle are a diet item (Cassaigne et al. 2016). In the southern part of the state of Mexico, livestock predation risk was positively associated with elevation and distance to a road or human settlement, but negatively associated with distance to cliffs and dense vegetation (Zarco-González et al. 2012). In areas where medium-sized wild prey is available, both species may select prey with a high energy content, such as deer and peccaries. In habitats with high prey densities, the puma may select a greater variety of prey than the jaguar. The degree of habitat fragmentation can also affect the number of prey species consumed. For example, in minimally fragmented habitats, both species tend to consume fewer prey species, but in more fragmented landscapes, diet diversity can be greater, and consumption of domestic animals can increase. However, in fragmented landscapes with limited prey availability, diet overlap between the two species will be higher. Also, in habitats with abundant wild prey, they tend to avoid feeding on livestock (Cassaigne et al. 2016). In general, jaguars tend to feed on larger prey than pumas, but both species also feed on medium- and small-sized prey (Oliveira 2002, Gómez-Ortiz et al. 2015). However, Moreno et al. (2006) hypothesized that in areas where pumas and jaguars are sympatric, the local extinction of jaguars may cause a shift in puma prey in response to competitive release; on Barro Colorado Island in Panama, pumas shifted to larger prey (peccaries and deer) in the absence of jaguars. Food habit studies should compare DNA to differentiate between fecal samples of pumas and jaguars; species identification based on signs and the size and shape of feces can be inaccurate (Rueda et al. 2013).

In areas where both species occur, diet composition can vary greatly depending on prey availability, distribution, vulnerability, and abundance. In the Tamasopo region of tropical San Luis Potosí where there were 14 prey species, pumas and jaguars shared 7 of them. Collared peccary was the main prey of both felids (jaguar 39.6% and puma 45.8%), and there was a diet overlap of 93.4%. Both felid species also preyed on white-tailed deer, but in small quantities (jaguar 7.64% and puma 5.88% of biomass consumed). Brocket deer (*Mazama temama*) (10.38%) and domestic goat (7.72%) appeared only in the puma diet. Most of the prey in the jaguar and puma diets consisted of medium-sized mammals (1–15 kg, 2.2–33 lb), such as the common opossum (*Didelphis virginianus*), raccoon (*Procyon lotor*), and coati (*Nasua narica*) (Rueda et al. 2013). In the same region, in a study conducted in a protected area and adjacent fragmented landscape, collared peccary contributed 35.2% to the biomass of jaguar diets, and white-tailed deer contributed 51.9% to puma diets. Both felids fed almost exclusively on wild species, indicating an adequate prey base (Hernández-SaintMartín et al. 2015). In a study in northeastern Sonora based on two jaguars and seven pumas, white-tailed deer was the most frequent prey found at puma kill sites (67%). However, deer represented 45% and lagomorphs 20% of all individuals eaten, based on relative numbers of prey consumed. Jaguars and pumas preyed on calves (*Bos taurus*) at approximately the same rate (jaguars 3.7 calves per year, pumas 4.9 calves per year). However, in contrast to the findings of Rosas-Rosas et al. (2008), jaguars killed calves at a lower frequency, and pumas fed on calves at a higher frequency (Cassaigne et al. 2016). Refer to Zarco-González et al. (2013) for a puma and jaguar livestock predation risk model based on environmental and livestock management variables.

Jaguar

The jaguar is considered an opportunistic carnivore, and its diet is also based on the abundance and availability of prey (Seymour 1989). More than 85 species have been reported in jaguar diets, including mammals, birds, reptiles, fish, and invertebrates. Although the jaguar has a diverse diet, it prefers mammals with a mass of over 1 kg (2.2 lb) (Emmons 1987, Seymour 1989, López-González and Miller 2001). In the Calakmul Reserve in Campeche, the most important prey were collared peccary (*Pecari tajacu*, 42%), coati (*Nasua narica*, 18%) and nine-banded armadillo (*Dasypus novemcinctus*, 12%) (Aranda 1994). In the Chamela-Cuixmala Reserve on the Jalisco coast, the most-consumed prey species were white-tailed deer (*Odocoileus virginianus*, 41%), coati (17%), and armadillo (14%) (Núñez et al. 2000). In this last case, the jaguar showed a high degree of dietary overlap with the puma (84%); however, the puma had a wider food preference (11 vs. 7 prey species) and fed on smaller prey. In a study of four communities adjacent to the Calakmul and Montes Azules Biosphere Reserves in Chiapas, jaguars preyed on domestic sheep, perhaps because the sheep's docility rendered them more vulnerable to predation, and on cattle, goats, and equines. Jaguars also preyed on dogs, perhaps because dogs were abundant and opportunistically available (Amador-Alcalá et al. 2013). In a scat analysis study in northeastern Sonora, the northernmost distributional range of the jaguar in Mexico, the most-consumed species were cow calves (57%), white-tailed deer (24%), lagomorphs (5%), and collared peccary (5%) (Rosas-Rosas et al. 2008). Predation of cattle by jaguars in this region was associated with individuals who developed a preference for cattle. However, in a study using molecular analysis of scat and GPS kill site clusters in the same area, the frequency of calves in the jaguar diet (based on two animals) was 16%. Jaguars fed on nine prey species, of which 52% weighed less than 15 kg (33 lb) (Cassaigne et al. 2016).

Puma

Pumas are generalist predators, and food habits vary with latitude (Pierce and Bleich 2003). In temperate regions of North America, pumas prey on larger mammals (ungulates), while in tropical habitats, they feed on smaller prey and a larger variety. Iriarte et al. (1990) suggested that pumas may select pray based on its availability and vulnerability, as well as on habitat characteristics and potential competition with jaguars. In arid habitats of northwestern Mexico, pumas preyed on bighorn sheep without adversely affecting the sheep population. The bighorn sheep in this area were monitored before and after the study was conducted, and the prey population was stable during this period. Pumas also fed on mule deer (*Odocoileus hemionus*), collared peccary, white-tailed deer, and lagomorphs (Rosas-Rosas et al. 2003). However, pumas can induce changes in desert bighorn populations (Wakeling et al. 2009). In the Sierra Madre Occidental in Sonora, pumas fed mainly on white-tailed deer, then on collared peccary, coatis, raccoons, cattle, and lagomorphs (Rosas-Rosas et al. 2008). In Aguascalientes, a semiarid region in central Mexico, the puma's diet at two study sites consisted of large and medium-sized mammals such as white-tailed deer, collared peccary, raccoon, and cattle (de la Torre and de la Riva 2009). In central Mexico, pumas fed on 17 species, but their main prey was nine-banded armadillo, followed by white-nosed coati and white-tailed deer, which represented 95% of the total prey consumed (Monroy-Vilchis et al. 2009a). In the same area, a small number of jaguars fed on nine prey species, with armadillo and domestic goat representing 62% of the total occurrence frequency. The armadillo had the highest energy content and was probably also selected because it was the most vulnerable (Gómez-Ortiz and Monroy-Vilchis 2013). In Chiapas, around communities contiguous with nature reserves, pumas fed on cattle, domestic sheep, goats, and dogs (Amador-Alcalá et al. 2013); this study supports the theory that in disturbed tropical habitats, pumas feed on larger prey.

Reproduction

The gestation period of jaguars is 93–105 days, and of pumas 92–96 days, and both have one to four cubs, although the jaguar more commonly produces one or two young, and pumas generally two. In both species, kittens are born blind and have spotted coats. Newborn jaguars weigh approximately 800 g (1.8 lb), and pumas about 400–500 g (0.9–1.1 lb); both species are altricial (Sunquist and Sunquist 2009b, 2009c; Logan and Sweanor 2001; Pierce and Bleich 2003). Cubs of both species begin to follow the mother when they are approximately two months old and remain with her for 15 to 24 months for food and security. Jaguars and pumas reach sexual maturity at the age of 2–3 years and live 10–12 years, but in captivity they can live up to 22 years (Logan and Sweanor 2001). A jaguar recently recorded in Arizona was estimated to be 15–16 years old, making it the oldest-known jaguar in the wild (Arizona Game and Fish Department, personal communication).

Jaguar

Jaguar reproduction varies regionally. In South America, kittens have been observed during June, August, September, and December. In Mexico, births and seasonality coincide with prey abundance (Seymour 1989). Leopold (1959) noted that jaguar births in Mexico occur in July and September, with some regional variation. In northeastern Sonora, tracks of a female with kittens were found during July and August. Female jaguars with cubs have been observed in Sonora and San Luis Potosí during the summer rainy season, which is also the peak of births of whitetails, collared peccaries, and other medium-sized mammals (Rosas-Rosas, unpublished data). In western Mexico, jaguar reproduction occurs throughout the year, but births tend to concentrate in April and August (R. Núñez, unpublished data).

Puma

Pumas in Mexico probably breed throughout the year (Chávez Tovar and Ceballos 2014a). Pumas, like jaguars, are polygamous; males can mate with several females within their territories. Females may also mate with males other than territorial males (Logan and Sweanor 2001). Births generally tend to coincide with births of white-tailed deer and other small and medium-sized mammals. In Jalisco and Nayarit, the birth season is similar to that of the jaguar, at the end of the dry season (R. Núñez, unpublished data). Adult males and females associate only during the mating season. Pumas mate frequently; copulation can occur 50 times a day. During estrus, females visit scrapes (claw scratches) made by males and vocalize, possibly indicating their reproductive status. Males must often compete with other males for access to females, which sometimes leads to fatal encounters (Logan and Sweanor 2001).

Behavior

Jaguars and pumas are solitary, and their social organization is based on activity areas and home ranges. Felids communicate indirectly with conspecifics to advertise social status, territorial claims, and mating status, and to defend resources (Sunquist and Sunquist 2002). Olfactory cues are especially important. Feces are cylindrical, dark, and thick, measuring 2.5–4 cm (1–5 in) in diameter and 10–30 cm (3.9–11.8 in) in length. Jaguars and pumas may patrol their territory on trails made by other wildlife and domestic animals, and along riverbanks and ocean beaches (Aranda 2000). Water availability for jaguars and pumas and their prey is a critical habitat component (Astete et al. 2016).

In Sonora and San Luis Potosí, jaguar feces and scrapes are common in areas of overlap between males along creeks and canyons, and sometimes on trails. In Chamela-Cuixmala, it is common to find feces and scrapes at intersections of highland creeks. Communication among individual jaguars includes grunting, and other vocalizations that generally occur at dawn and dusk. Although vocalizations can be heard year-round, they occur mostly during spring and summer in western Mexico. Jaguars and pumas also communicate by scratching the soil, as well as the soft bark of trees such as palo blanco (*Piscidia mollis*) in Sonora, and clavellina (*Pseudobombax* spp.) in western Mexico. Jaguars and pumas scratch bushes to scent mark, and also rub their scent on rocks (McCain and Childs 2008).

In a study in the Calakmul and Sian Ka'an Biosphere Reserves on the Yucatán Peninsula, jaguars used habitats between forested and degraded areas. They used tall and medium forests more than expected, based on availability when compared with disturbed habitats. The less-used habitats were areas devoted to livestock and agricultural production, and areas around human settlements. Human settlements have an especially significant effect on jaguar spatial distribution. The highest percentage (80%) of jaguar locations were more than 6.5 km (4.0 mi) from settlements, and jaguars avoided areas less than 1 km (0.6 mi) from settlements. As expected, areas near settlements were the most disturbed. Hunting by traditional and subsistence hunters occurred within a minimum radius of 6 km (3.7 mi) from settlements, and there was considerable overlap between the prey favored by hunters and jaguars (Zarza et al. 2011). In the Maya Forest, male and female jaguars used habitats differently; males avoided short forest, which was preferred by females, and males did not avoid roads, but females did. Males also occurred in low-intensity livestock ranching and farming areas, but females made greater use of intact forests, and consequently female habitats were more fragmented that those of males (Conde et al. 2010). Studies in central Mexico (Monroy-Vilchis et al. 2009b, Hernández-SaintMartín et al. 2013) revealed that jaguars are primarily nocturnal and pumas diurnal. However, in northeastern Sonora, both species were active throughout the day. Also, jaguar ranges overlapped more frequently in areas with deer and calves, and those of pumas overlapped in areas with calves more than in areas with other prey (López González and Gutiérrez-González 2017). In the tropical rainforest habitat of the Greater Lacandona Ecosystem (GLE) in southeastern Mexico, habitat destruction and fragmentation seemed to affect pumas less severely than jaguars (de la Torre et al. 2017b).

Territoriality and Home Range

Like many wild felids, jaguars and pumas occur in habitats that provide resources. Only the females take care of the litter, and they have smaller territories than males. Generally, a male territory includes at least one female territory. The home range size of both species is variable but is influenced by their body mass, abundance and availability of prey, habitat suitability, and social interactions between the

Figure 13.4. Jaguar habitat, the Río Aros in the state of Sonora, Mexico. Photo by Primero Conservation.org.

two species, which can also determine the patterns of habitat use. Where sympatric, they may coexist through different habitat use, spatiotemporal segregation, mutual avoidance, or different prey preferences (Rueda et al. 2013, de la Torre et al. 2017b). Studies of home ranges in Mexico have not considered the distribution and abundance of prey (fig. 13.4).

Jaguar

In Calakmul, Campeche, the home range of a male jaguar was estimated to be 56 km² (22 mi²); those of females varied from 25 to 133 km² (9.6 to 51.4 mi²) (Chávez et al. 2007). In western Mexico in Jalisco, female jaguar home ranges averaged 24 km² (9.3 mi²), while those of males averaged 36 km² (13.9 mi²) (Núñez et al. 2002, Núñez Pérez 2006). In a study in a tropical area in San Luis Potosí, individual jaguars and pumas used the same trails and had overlapping activity patterns, indicating that the main ecological segregation strategy of these felids was the differential use of prey (Hernández-SaintMartín et al. 2015). However, in the GLE, jaguars and pumas with overlapping home ranges had little overlap of core areas and were spatially and temporally separated because they avoided using the same sites at the same time. Jaguars were primarily nocturnal, and pumas were diurnal. Jaguar home ranges were two to six times larger than those of pumas. Female jaguars had a mean annual home range of 181.4 ± 4.0 km² (69.9 ± 2.5 mi²), and that of males was 431.6 ± 152.6 km² (166.6 ± 58.9 mi²) (de la Torre et al. 2017b). In the Maya Forest of southeastern Mexico, three breeding patches had areas between 281 and 801 km² (108.5 and 309.3 mi²). Functional corridors were in areas in which forest cover was within 240 m (262 yd) and the slope was moderate to medium or flat; stepping-stone patches are critical for linking habitats (de la Torre et al. 2016).

Puma

Activity patterns of pumas have been extensively studied in the US (Pierce and Bleich 2003). In general, male home ranges are larger than those of females and can be as large as 797 km² (308 mi²), while the range of an adult female in the southwestern US can be as small as 18 km² (7 mi²). High variation in puma home range size could be a product of the techniques used to determine them, reproductive status, habitat quality, and population density (Logan and Sweanor 2001). In northern Sonora, Mexico, in a subtropical thorn scrub and oak forest, two pumas were monitored for one year; the home range of an adult male during the dry season was 87 km² (36 mi²) and included riparian habitat, and an adult female had a home range of 50 km² (19 mi²) that included riparian habitat frequently traveled by the male (Rosas-Rosas, unpublished data). The home ranges of two female pumas in a tropical forest in western Mexico during the dry season were 25 km² (9.7 mi²), compared to 60 km² (23 mi²) for two male pumas. In contrast, during the rainy season, home ranges increased to approximately 60 km² (23 mi²) for females and 90 km² (34.7mi²) for males (Núñez Pérez 2006). In western Sonora, pumas utilized riparian areas and riverbeds during the dry season. In Calakmul, Campeche, in southern Mexico, home ranges averaged 133 km² (51.4 mi²) (Chávez et al. 2007). In the GLE, the home range of one female puma was 34.3 km² (13.2 mi²), and that of males was 72.0 ± 85.2 km² (27.8 ± 32.9 mi²) (de la Torre et al. 2017b).

Population Dynamics

Density

Jaguar

Jaguar density varies regionally, probably because of habitat quality, prey abundance, and degree of human disturbance. The highest densities have been recorded in undisturbed tropical forests; in general, jaguar densities diminish at higher latitudes. Estimated jaguar densities in Central and South America are 2.36–8.8 individuals/100 km² (39 mi²) (Silver et al. 2004, Salom Pérez et al. 2007). In Chiapas, Mexico, density was 1 jaguar/13–20 km² (5–7.7 mi²) (Aranda 1996), and 1 jaguar/15–30 km² (5.8–11.6 mi²) in Calakmul, Campeche (Aranda 1998). Furthermore, as determined by camera trapping in this same region, the density was 3.33–6.67 individuals/100 km² (39 mi²) (Chávez et al. 2007). In Jalisco, densities ranged from 2.8 to 5.3 individuals/100 km² (39 mi²) in deciduous forest (Núñez Pérez 2011a, 2014a), and 2.1 jaguars/100 km² (39 mi²) at higher elevations (Núñez Pérez 2014b). In Nayarit, densities ranged from 4.5 to 6 jaguars/100 km² (39 mi²) in Sierra de Vallejo and Marismas Nacionales (Núñez et al. 2010a, 2010b). On the northeastern Yucatán Peninsula, density was 3–5 jaguars/100 km² (39 mi²) (Faller et al. 2007), and 0.7 to 3.6/100 km² (39 mi²) in another study in the same region (Ávila-Nájera et al. 2015). The density was 3–5 jaguars/100 km² (39 mi²) in the Sierra Madre Oriental in San Luis Potosí, where a jaguar population was not thought to exist given the high number of local communities (Ávila Nájera 2009). In the Meseta de Cacaxtla Protected Area in Sinaloa, density was 1.59 jaguars/100 km² (39 mi²) (Coronel-Arellano et al. 2017). During a 13-year study (2000–2012) in northeastern Sonora in a reserve where cattle were removed in 2003, 48 jaguars were recorded based on camera-trap data, and mean jaguar density was 1.87 ± 0.47 jaguars/100 km² (39 mi²); changes in survival or density were not detected during the study period (Gutiérrez-González et al. 2015). However, jaguars in northeastern Sonora are now on the verge of extinction (Cassaigne et al. 2016).

Puma

Puma densities can be expected to vary in Mexico because pumas occupy temperate and tropical habitats where prey abundance and diversity vary. Estimates of puma densities in Mexico are few. It is unlikely that puma distribution and abundance depend on large prey (Laundré and Hernández 2010), because of the great diversity and availability of small prey in tropical habitats with low human disturbance. In Central and South America, estimated densities were 0.5–19.4 individuals/100 km² (39 mi²) (Noss et al. 2004, Kelly et al. 2008). In western Mexico in the Chamela-Cuixmala Reserve in Jalisco, puma density was 3.5 animals/100 km² (39 mi²) (Núñez et al. 2002), and 5 pumas/100 km² (39 mi²) (Núñez-Pérez 2011b). In the state of Mexico, densities ranged from 2.74 to 6.86 pumas/100 km² (39 mi²) (Soria Díaz et al. 2010). On the northeastern Yucatán Peninsula in a protected area, density was 1.8 to 5.2 pumas/100 km^2 (39 mi^2) (Ávila-Nájera et al. 2015). Rosas-Rosas et al. (2003) estimated a density of 3 pumas/100 km^2 (39 mi^2) in the Altar Desert in northern Sonora. In northeastern Sonora, the density was 2 pumas/100 km^2 (39 mi^2) (Rosas-Rosas 2006). Luna and López (2005) found a density of 4.19–5.57 pumas/100 km^2 (39 mi^2) in the Sierra de San Luis, Sonora. In the Colima volcanic region, density was 0.9 adults/100 km^2 (39 mi^2) (Burton 2006), based on a small sample. In the Chihuahuan Desert, puma density was related to the intensity of habitat use by humans; pumas had a lower density in human-dominated landscapes (Laundré et al. 2009).

Present Status

Jaguar

The jaguar is listed as Near Threatened in the IUCN Red List (Caso et al. 2008), with a declining trend in abundance and distribution, and it is listed in Appendix I of CITES. It is listed as endangered in Mexico (SEMARNAT 2010). There are about 5,430 jaguars in Mexico, based on the density and population size in Jaguar Conservation Management Areas (Ceballos et al. 2016). The jaguar lacks a scientifically reliable national ecological assessment, and an updated population status in Mexico is needed to adequately develop regional long-term management programs. Although the jaguar was once widely distributed in Mexico, it is present in only half of its historically occupied area because of habitat destruction and fragmentation, illegal hunting, and decreases in prey animals (Sanderson et al. 2002, Rosas-Rosas et al. 2010). The largest population in Mexico is in the Maya Forest in southeastern Mexico and, along with the jaguar habitat in northwestern Guatemala, encompasses an area of 45,000 km^2 (17,375 mi^2). In this and other regions of Mexico, they are most abundant in nature reserves and in relatively undisturbed habitats (de la Torre et al. 2016). Refer to de la Torre et al. (2017a) for a review of the status of jaguar subpopulations throughout their distributional range.

One of the least abundant jaguar populations in Mexico is in northeastern Sonora. Rosas-Rosas and Bender (2012) reported a jaguar density of 1.1/100 km^2 (39 mi^2), and Gutiérrez-González et al. (2012) recorded a density of 1.05 ± 0.40/100 km^2 (39 mi^2), based on a 16-month study ending in May 2009 in northeastern Sonora. Rosas-Rosas et al. (2010) reported that more than 11 jaguars were killed from 1999 to 2004 in the same general area. In another study in the same region, Cassaigne et al. (2016) recorded only two live jaguars during a three-year period (2011–2013), and three illegally killed during the same period. This population is of great conservation concern because it is the probable source of jaguars recorded in the US and is on the verge of local extinction.

The construction of roads, dams, and other infrastructure, agricultural expansion, and decreases in prey are major challenges to jaguar conservation. Ecological studies of the jaguar are few, and detailed life history information is lacking. For example, we have no basic knowledge of female productivity, juvenile recruitment, or cause-specific mortality. Also, there is limited information relative to habitat degradation, spatial requirements, and movement behavior (de la Torre et al. 2017b). Of special concern is genetic connectivity of subpopulations. Jaguar populations in Mexico have some of the lowest genetic diversity values, probably because of habitat deterioration and isolation (Roques et al. 2016). Populations in the Sierra Madre Oriental in northeastern Mexico and in the Maya Forest in southeastern Mexico exhibit a genetic division that is probably a result of separation by human-dominated landscapes (Wultsch et al. 2016). Culver and Hein (2016) and Zanin et al. (2016) detected a lack of genetic connectivity between populations in northwestern Mexico and southern Mexico. Rodríguez-Soto et al. (2013) and Dueñas-López et al. (2015) identified several potential biological corridors that could connect jaguars in the northern Sierra Madre Oriental with populations in southern Mexico, but habitat restoration would be required in and among the "stepping stones."

Puma

The IUCN lists the puma as a species of Least Concern, and it is included in Appendix II of CITES, but with a decreasing population trend. However, the Central American (*cougar*) population is listed in Appendix I of CITES (Nielsen et al. 2015). The status of the puma in Mexico is critical because persistent predator control programs are indiscriminately removing individuals (Chávez Tovar and Ceballos 2014a). In most states where predation on livestock is severe, ranchers resort to predator control without following a scientific management plan, because there is a lack of federal and state conservation intervention programs (Rosas-Rosas et al. 2008, 2010; Rosas-Rosas and Valdez 2010). In addition, the mismanagement of livestock enterprises and the lack of wildlife professionals and funding for wildlife programs hamper effective efforts to conserve wild felids in Mexico. Fortunately for the puma, it can tolerate areas of greater disturbance than the jaguar and can establish viable populations in desert environments.

Management

Jaguar

Jaguar conservation in Mexico is being planned by the National Jaguar Conservation Strategy (NJCS), a collaborative effort among federal and state governments, academic institutions, and NGOs based on ecological research (Ceballos et al. 2016). This strategy consists of acquiring ecological, behavioral, population density, and size data for jaguars, following a standardized, systematic camera-trapping protocol focused on 16 national jaguar conservation sites from Sonora and Tamaulipas to the Yucatán Peninsula. These sampling sites are in four priority regions (North Pacific, Central Pacific, South Pacific, and Yucatán Peninsula), and in eight Jaguar Conservation Units (JCUs). Each JCU encompasses a viable jaguar population and is designed based on habitat availability and connectivity, prey base, and level of threat from human activity. Hence, regional efforts are essential in planning JCUs (Zarza et al. 2013). Based on camera trapping, the highest jaguar densities (7.4 individuals/100 km^2) (39mi^2) occur in the Yum Balam region of Quintana Roo, on the Yucatán Peninsula (Ceballos et al. 2016). The JCUs emphasize several strategic components, including monitoring jaguar populations and prey, mitigating the effects of human-induced infrastructure, resolving human-jaguar conflict, and providing communication and education programs. Of major concern in human-dominated landscapes with associated infrastructure such as paved roads is the loss of connectivity among populations. Refer to Ceballos et al. (2016) for a review of the national jaguar conservation strategy.

In the southwestern US and northwestern Mexico, the Technical Subgroup of the Jaguar Recovery Team (TSJRT) has been entrusted with the goal of recovering and delisting the jaguar, using a long-term framework for recovery similar to that used by JCUs. A major recovery strategy for the TSJRT is the assessment, protection, and restoration of the quantity, quality, and connectivity of habitats needed to support viable populations of jaguars. Restoring jaguars in the southwestern US will require travel corridors in order to facilitate connectivity with Mexican jaguar populations (TSJRT 2016). Grigione et al. (2009) identified 21 Cat Conservation Units and 7 Cat Conservation Corridors in the US Mexico border region in which to concentrate efforts to protect felids, but only 8.9% of these corridors and 1.1% of the units have protected status. In the southern Maya Forest of southeastern Mexico and northwestern Guatemala, the largest tract of tropical forest in Mesoamerica, de la Torre et al. (2016) found only one habitat patch with a viable jaguar. Hence, habitat maintenance and restoration are key components in conserving jaguar populations, but the associated human-related social and economic challenges must be considered in the planning process. Jaguar livestock predation produces negative attitudes toward predators and is compounded by entrenched mismanagement of rangelands (Rosas-Rosas et al. 2008; Rosas-Rosas and Valdez 2010). In addition, without an understanding of the cultural and socioeconomic roots of poverty and the effects of habitat conversion, ecosystem maintenance and restoration cannot be accomplished.

Puma

The puma is subject to sport hunting in 11 Mexican states (Baja California, Baja California Sur, Campeche, Coahuila, Chihuahua, Durango, Jalisco, Nuevo León, Sinaloa, Sonora, and Tamaulipas) (SEMARNAT 2010). In the recent past, the federal government has permitted predator control programs to regionally eliminate pumas in Sonora, Chihuahua, and Nuevo León, among other areas. In Sonora and Nuevo León, from 1993 to 2012, 55 pumas were killed (Rosas Rosas 2006; Rosas-Rosas et al. 2003, unpublished data). In Jalisco in 2005, an estimated eight jaguars were illegally killed because of purported livestock predations (R. Núñez Pérez, unpublished data). A national database of large felids killed each year by state should be established to determine areas of mortality so that these data can be incorporated into management plans. Predator control is also intensively used as a regional management tool to increase bighorn sheep, white-tailed deer, and mule deer populations. However, the problem of predation in many cases ends only temporarily after removing a problem individual. The removal of a specific individual that regularly consumes domestic animals requires its identification; this could be possible with camera traps and with the assistance of trained personnel with predator control experience, but these are often not available. This also requires the knowledgeable management of domestic animals and game ranch wildlife to minimize predation, although in many cases, predation is not a major cause of mortality. Large wild cats such as the puma and jaguar require large areas of living space, connectivity, cover, and an adequate prey base to maintain sustainable populations. Basic studies on the productivity of females, activity patterns, and predator-prey relationships are necessary to develop a scientifically based management plan for pumas and jaguars in Mexico.

To initiate appropriate management plans, detailed studies of population dynamics, productivity, and survival are of primary importance. In addition, land managers of livestock and wildlife enterprises need to know how to sustainably manage rangelands. Cattle ranching without sustainable planning results in habitat loss for wildlife. Support programs for sustainable livestock production that minimizes conflicts where large carnivores are present are urgently needed. Basic livestock management practices that help alleviate predation should include monitoring movements of cows and their calves, creating alternative water sources away from riparian areas and areas with dense vegetative cover to distribute cattle during the dry season, providing food and water during droughts, synchronizing calving, and keeping cattle in sturdy enclosures (Rosas-Rosas et al. 2008, 2010; Peña-Mondragón et al. 2016). Additional management practices should include reducing the overabundance of dogs, which can be a food and disease source for jaguars; training efficient guard dogs; keeping livestock far away from cultivated fields; and increasing prey abundance to improve availability of preferred prey for predators by regulating subsistence hunting (Amador-Alcalá 2013). Additionally, innovative techniques such as conditioned taste aversion to treat offending animals need to be tested under controlled conditions (R. Thompson, personal communication). All these methods and tools, if used properly, will empower livestock owners to reduce livestock predation and increase their income in the presence of apex carnivores.

Challenges and Future Perspectives

Although Mexican federal government agencies such as SEMARNAT have made efforts to initiate puma and jaguar conservation initiatives, funding for field studies and management programs is inadequate (Valdez et al. 2006). The jaguar, despite being a protected priority species, has an uncertain future because of continuing regional extirpations. Jaguars and pumas, like other large predators, require landscape-level conservation because of their wide-ranging habitat needs, including well-linked habitat corridors that afford opportunities for reproduction and survival, enhanced gene flow, avoidance of inbreeding depression, and dispersal and colonization (Gubbi et al. 2017). The construction of a wall along the US-Mexico border, the increasing militarization of the border region (exacerbated by confrontational political agendas), and the upgrading and expansion of transportation infrastructure are major obstacles to animal movements, denying access to resources and reducing the effective population size of predators and their prey. This highlights the need for real partnerships between neighboring countries; jaguar populations in Mexico are ecologically intertwined with populations in Guatemala and Belize, and the Mexico-US populations are similarly interrelated. Transboundary conservation projects can help protect critical habitats through the efforts of governments, NGOs, and the private sector (King and Wilcox 2008).

Concerted efforts in large felid ecological research were not initiated in Mexico until the early 2000s. Mexican biologists have made significant research contributions, as is evident from the refereed journals cited herein, and intensive studies continue. New technologies are driving this effort to heightened and more accurate levels not foreseen a few decades ago. However, developing scientifically based initiatives to manage predator populations and their habitats over their wide distributions, and acquiring the necessary funding to accomplish ecological research goals, are daunting challenges to wildlife biologists. An even greater challenge is integrating such efforts with the human needs of the rural populace (Valdez et al. 2006). Most jaguars and pumas occur outside protected areas, and hence there is a need to develop tolerance toward large predators. The extensive past and continuing expansion of the human population and the insatiable demand for livestock meat protein are the major threats to jaguars and pumas. It is crucial that conservation and management plans include the participation of communities, ejidos, and private landowners at all levels of planning, and that economically viable opportunities be generated in rural communities that coexist with large predators (Valdez et al. 2006, Figel et al. 2011, Clark et al. 2016). Also, a basic goal should be to establish efficient livestock management programs that allow owners to prevent predation and enhance livestock production while sustaining jaguar and puma populations (Thompson and Cassaigne 2017). Environmental education is another crucial component of any conservation program. Educational endeavors should include familiarizing rural communities with the advantages of retaining biodiversity, the diversification of commercial opportunities deriving from wildlife enterprises, and the sustainable use of all natural resources (Rosas-Rosas et al. 2015). Also, private protected areas can be effective and should be encouraged (Núñez-Pérez 2011a). An innovative approach is to enact policies that support inheritance and income tax relief in exchange for conservation agreements that support healthy ecosystems, including predators and prey.

Top predators have conservation value, especially in preventing trophic cascades. They can serve as umbrella species for animal communities and reserve networks, thus conserving connectivity for sympatric mammals and providing complementary ecosystem conservation strategies (Sergio et al. 2006, Thornton et al. 2015). Large carnivores are charismatic and symbolize Mexico's spectacular biodiversity and wild landscapes. The jaguar and puma are primary examples of the decreasing distribution and regional extirpation of Mexico's large mammals, and of the wildlife management challenges that must be confronted and resolved to conserve Mexico's biodiversity. Mexico must commit to ensuring that predators and their habitats are sustainably managed, and to initiating effective conservation efforts to halt the continued squandering of its biodiversity, especially of two of its most impressive fauna species, the jaguar and puma.

Literature Cited

Aguilar-Lopez, M., J. Ramos-Frías, A. E. Rojas-Martínez, and C. Cornejo-Latorre. 2015. First record of jaguar (*Panthera onca*) from the state of Hidalgo, Mexico. Western North American Naturalist 75:520–525.

Amador-Alcalá, S., E. J. Naranjo, and G. Jiménez-Ferrer. 2013. Wildlife predation on livestock and poultry: implications for predator conservation in the rainforest of south-east Mexico. Oryx 47:243–250.

Aranda, M. 1994. Importancia de las pecaríes en la alimentación del jaguar. Acta Zoológica Mexicana 62:11–22.

Aranda, M. 1996. Distribución y abundancia del jaguar (*Panthera onca*) (Carnivora: Felidae) en el estado de Chiapas, México. Acta Zoológica Mexicana 68:45–52.

Aranda, M. 1998. Densidad y estructura poblacional del jaguar (*Panthera onca*) en la Reserva de la Biósfera Calakmul, Campeche, México. Acta Zoológica Mexicana 75:199–201.

Aranda, M. 2000. Huellas y otros rastros de los mamíferos grandes y medianos de México. CONABIO and INE, Xalapa, Veracruz, México.

Aranda, M., and V. Sánchez Cordero. 1996. Prey spectra of jaguar (*Panthera onca*) and puma (*Puma concolor*) in tropical forests of Mexico. Studies on Neotropical Fauna and Environment 31:65–67.

Astete, S., et al. 2016. Living in extreme environments: modeling habitat suitability for jaguars, pumas, and their prey in a semiarid habitat. Journal of Mammalogy 98:464–474.

Ávila Nájera, D. M. 2009. Estimación de la abundancia del jaguar y sus presas en el municipio de San Nicolás de los Montes, San Luis Potosí. Thesis, Colegio de Postgraduados, Campus Montecillo, Texcoco, México.

Ávila Nájera, D. M., C. Chávez, M. A. Lazcano-Barrero, S. S. Pérez-Elizalde, and L. Calcantar-Carbajal. 2015. Estimación poblacional y conservación de felinos (Carnivora: Felidae) en el norte de Quintana Roo, México. Revista de Biología Tropical 63:799–813.

Ávila Nájera, D. M., O. C. Rosas-Rosas, L. A. Tarango-Arámbula, J. F. Martínez-Montoya, and E. Santoyo-Brito. 2011. Conocimiento, uso y valor cultural de seis presas del jaguar (*Panthera onca*) y su relación con éste, en San Nicolás de los Montes, San Luis Potosí, México. Revista Mexicana de Biodiversidad 82:1020–1028.

Burton, A. M. 2006. Puma abundance on the Colima volcanic complex, western Mexico. Revista Mexicana de Mastozoología 10:92–99.

Caso, A. 2007. Situación del jaguar en el estado de Tamaulipas. Pages 19–24 *in* G. Ceballos, C. Chávez, R. List, and H. Zarza, editors. Conservación y manejo del jaguar en México: estudios de caso y perspectiva. CONABIO-Alianza WWF/Telcel-UNAM, México, D. F., México.

Caso, A., et al. 2008. *Panthera onca*. The IUCN Red List of Threatened Species 2008: e.T15953A5327466. http://dx.doi.org/10.2305/IUCN.UK.2008.RLTS.T15953A5327466.en. Accessed 18 February 2017.

Cassaigne, I., et al. 2016. Diet of pumas (*Puma concolor*) in Sonora, Mexico, as determined by GPS kill sites and molecular identified scat, with comments on jaguar (*Panthera onca*) diet. Southwestern Naturalist 61:125–132.

Ceballos, G., H. Zarza, C. Chávez, and J. F. González-Maya. 2016. Ecology and conservation of jaguars in Mexico: state of knowledge and future challenges. Pages 273–289 *in* A. A. Aguirre and R. Sukumar, editors. Tropical conservation: perspectives on local and global priorities. Oxford University Press, New York, New York, USA.

Chávez, C., G. Ceballos, and M. Amín. 2007. Ecología poblacional del jaguar y sus implicaciones para la conservación en la Península de Yucatán. Pages 91–100 *in* G. Ceballos, C. Chávez, R. List, and H. Zarza, editors. Conservación y manejo del jaguar en México: estudios de caso y perspectivas. CONABIO-Alianza WWF/Telcel-UNAM, México, D. F., México.

Chávez Tovar, C., and G. Ceballos. 2014a. Puma. Pages 505–507 *in* G. Ceballos, editor. Mammals of Mexico. Johns Hopkins University Press, Baltimore, Maryland, USA.

Chávez Tovar, C., and G. Ceballos. 2014b. Jaguar. Pages 507–509 *in* G. Ceballos, editor. Mammals of Mexico. Johns Hopkins University Press, Baltimore, Maryland, USA.

Clark, S. G., M. B. Rutherford, and D. J. Mattson. 2016. Introduction: large carnivores, people and governance. Pages 1–28 *in* S. G. Clark and M. B. Rutherford, editors. Large carnivore conservation: integrating science and policy in the North American west. University of Chicago Press, Illinois, USA.

Conde, D. A., et al. 2010. Sex matters: modeling male and female habitat differences for jaguar conservation. Biological Conservation 143:1980–1988.

Coronel-Arellano, H., N. E. Lara-Díaz, and C. A. López-González. 2017. Abundancia y densidad de jaguar (*Panthera onca*) en el APFF Meseta de Cacaxtla, Sinaloa, México. Acta Zoológica Mexicana (n.s.): 33:116–119.

Culver, M. 2010. Lessons and insights from evolution, taxonomy, and conservation genetics. Pages 27–40 *in* M. Hornocker and S. Negri, editors. Cougar: ecology and conservation. University of Chicago Press, Illinois, USA.

Culver, M. 2016. Jaguar surveying and monitoring in the United States (ver. 1.1, November 2016). Open-File Report 2016–1095. US Geological Survey, Reston, Virginia, USA.

Culver, M., and A. O. Hein. 2016. Jaguar taxonomy and genetic diversity of southern Arizona, United States, and Sonora, Mexico. Open-File Report 2016–1109. US Geological Survey, Reston, Virginia, USA.

Currier, M. P. J. 1983. *Felis concolor*. Mammalian Species 200:1–7.

de la Torre, J. A., and G. de la Riva. 2009. Food habits of pumas (*Puma concolor*) in a semiarid region of central Mexico. Mastozoología Neotropical 16:211–216.

de la Torre, J. A., J. F. González-Maya, H. Zarza, G. Ceballos, and R. A. Medellín. 2017a. The jaguar's spots are darker than they appear: assessing the global conservation status of the jaguar *Panthera onca*. Oryx: doi:1017/S0030605316001046.

de la Torre, J. A., J. M. Núñez, and R. A. Medellín. 2016. Habitat availability and connectivity for jaguars (*Panthera onca*) in the southern Mayan forest: conservation priorities for a fragmented landscape. Biological Conservation 206:270–282.

de la Torre, J. A., J. M. Núñez, and R. A. Medellín. 2017b. Spatial requirements of jaguars and pumas in southern Mexico. Mammalian Biology 84:52–60.

del Castillo, A. 2009. Diario Milenio. http://www.milenio.com/node/317738. Accessed 15 December 2010.

Dueñas-López, G., et al. 2015. Connectivity among jaguar populations in the Sierra Madre Oriental, Mexico. Therya 6:449–468.

Eizirik, E., J. H. Kim, M. Menotti-Raymond, P. G. Crawshaw, S. J. O'Brien, and W. Johnson. 2001. Phylogeography, population history and conservation genetics of jaguars (*Panthera onca*, Mammalia, Felidae). Molecular Ecology 10:65–79.

Emmons, L. H. 1987. Comparative feeding ecology

of felids in a Neotropical rainforest. Behavioral Ecology and Sociobiology 20:271–283.

Faller, J. C., C. Chávez, S. Johnson, and G. Ceballos. 2007. Densidad y tamaño de la población de jaguar en el noreste de la Península de Yucatán. Pages 111–122 *in* G. Ceballos, C. Chávez, R. List, and H. Zarza, editors. Conservación y manejo del jaguar en México: estudios de caso y perspectivas. CONABIO-Alianza WWF/Telcel-UNAM, México, D. F., México.

Figel, J. J., E. Duran, and D. B. Bray. 2011. Conservation of the jaguar *Panthera onca* in a community-dominated landscape in montane forests in Oaxaca, Mexico. Oryx 45:554–560.

Galindo, F., et al. 2016. Conservation and livestock production in tropical Mexico: the need for sustainable options. Pages 371–380 *in* A. A. Aguirre and R. Sukumar, editors. Tropical ecology: perspectives on local and global priorities. Oxford University Press, New York, New York, USA.

Gamero-Idiáquez, I. 1978. Mamíferos de mi tierra. Volume 1. Banco Central de Honduras, Tegucigalpa, Honduras.

Gill, R. B. 2010. To save the mountain lion: evolving philosophy of nature and cougars. Pages 5–16 *in* M. Hornocker and S. Negri, editors. Cougar: ecology and conservation. University of Chicago Press, Illinois, USA.

Glenn, W. 1996. Eyes of fire: close encounters with a borderland jaguar. Texas Western Press, El Paso, USA.

Gómez-Ortiz, Y., and O. Monroy-Vilchis. 2013. Feeding ecology of puma *Puma concolor* in Mexican montane forests with comments about jaguar *Panthera onca*. Wildlife Biology 19:179–187.

Gómez-Ortiz, Y., O. Monroy-Vilchis, and G. D. Mendoza-Martínez. 2015. Feeding interactions in an assemblage of terrestrial carnivores in central Mexico. Zoological Studies 54:1–8.

González Torres, Y. 2001. Animales y plantas en la cosmovisión mesoamericana. Plaza y Valdés, INAH, and CONACULTA, México, D. F., México.

Grigione, M. M., et al. 2009. Identifying potential conservation areas for felids in the USA and Mexico: integrating reliable knowledge across an international border. Oryx 43:78–86.

Gubbi, S., et al. 2017. From intent to action: a case study for the expansion of tiger conservation from southern India. Global Ecology and Conservation 9:11–20.

Gutiérrez-González, C. E., M. A. Gómez-Ramírez, and C. A. López González. 2012. Estimation of the density of the near threatened jaguar *Panthera onca* in Sonora, Mexico, using camera trapping and an open population model. Oryx 46:431–437.

Gutiérrez-González, C. E., M. A. Gómez-Ramírez, C. A. López González, and P. F. Doherty Jr. 2015. Are private reserves effective for jaguar conservation? PLoS ONE 10(9):e0137541.

Hall, E. R. 1981. The mammals of North America. Second edition. John Wiley and Sons, New York, New York, USA.

Hernández-SaintMartín, A. D., et al. 2013. Activity patterns of jaguar, puma and their potential prey in San Luis Potosi, Mexico. Acta Zoológica Mexicana (n.s.) 29:520–533.

Hernández-SaintMartín, A. D., O. C. Rosas-Rosas, J. Palacio-Nuñez, L. A. Tarango-Arámbula, F. Clemente-Sánchez, and A. L. Hoogesteijn. 2015. Food habits of jaguar and puma in a protected area and adjacent fragmented landscape of northeastern Mexico. Natural Areas Journal 35:308–317.

Hoogesteijn, R., and E. Modolfi. 1993. The jaguar. Armitrano, Caracas, Venezuela.

Iriarte, J. A., W. L. Franklin, W. E. Johnson, and K. H. Redford. 1990. Biogeographic variation of food habits and body size of the American puma. Oecologia 85:185–190.

Johnson, W. E., and S. J. O'Brien. 1997. Phylogenetic reconstruction of the Felidae using I6s RNA and NADH-5 mitochondrial genes. Journal of Molecular Evolution 44:S98–S116.

Kelly, M. J., et al. 2008. Estimating puma densities from camera trapping across their study sites: Bolivia, Argentina, and Belize. Journal of Mammalogy 89:408–418.

King, B., and S. Wilcox. 2008. Peace parks and jaguar trails: transboundary conservation in a globalizing world. GeoJournal 71:221–231.

Koford, C. B. 1975. Felids of Latin America: importance and future prospects. Publicaciones Biológicas, Instituto de Investigaciones Científicas UANL 1:131–141.

Kurtén, B. 1976. Fossil puma (Mammalia: Felidae) in North America. Netherlands Journal of Zoology 26:502–534.

Kurtén, B., and E. Anderson. 1980. Pleistocene mammals of North America. Columbia University Press, New York, New York, USA.

Larson, S. E. 1997. Taxonomic re-evaluation of the jaguar. Zoo Biology 16:107–120.

Laundré, J. W., and L. Hernández. 2010. What we know about pumas in Latin America. Pages 76–90 *in* M. Hornocker and S. Negri, editors. Cougar: ecology and conservation. University of Chicago Press, Illinois, USA.

Laundré, J. W., J. Loredo Salazar, L. Hernández, and D. Núñez López. 2009. Evaluating potential factors affecting puma abundance in the Mexican Chihuahuan Desert. Wildlife Biology 15:207–212.

Leopold, A. S. 1959. Wildlife of Mexico: the game birds and mammals. University of California Press, Berkeley, USA.

Logan, K. A., and L. L. Sweanor. 2001. Desert puma: evolutionary ecology and conservation of an enduring carnivore. Island Press, Washington, DC, USA.

López González, C., and D. E. Brown. 2002. Status and distribution of the jaguar in Sonora, Mexico. Pages 379–392 *in* R. A. Medellín et al., editors. El jaguar en el nuevo milenio. UNAM, Wildlife Conservation Society, and Fondo de Cultura Económica, México, D. F., México.

López González, C., and C. E. Gutiérrez-González. 2017. Jaguar interactions with pumas and prey at the northern edge of jaguars' range. PeerJ 5:e2886.

López González, C. A., and B. J. Miller. 2001. Do jaguars (*Panthera onca*) depend on large prey? Western North American Naturalist 62:218–222.

Luna, S. H., and C. A. López. 2005. Abundance and food habits of cougars and bobcats in the Sierra San Luis, Sonora, Mexico. Pages 416–420 *in* G. J. Gottfried et al., compilers. Connecting mountain islands and desert seas. RMR-P-36. USDA Forest Service, Rocky Mountain Research Station, Fort Collins, Colorado, USA.

McCain, E. B., and J. L. Childs. 2008. Evidence of resident jaguars (*Panthera onca*) in the southwestern United States and the implications for conservation. Journal of Mammalogy 89:1–10.

Monroy-Vilchis, O., Y. Gómez, M. Janczur, and V. Urios. 2009a. Food niche of *Puma concolor* in central Mexico. Wildlife Biology 15:1–9.

Monroy-Vilchis, O., C. Rodríguez-Soto, M. Zarco-González, and V. Urios. 2009b. Cougar and jaguar habitat use and activity patterns in central Mexico. Animal Biology 59:145–157.

Monroy-Vilchis, O., O. Sánchez, U. Aguilera Reyes, and P. Suárez. 2008. Jaguar (*Panthera onca*) in the state of Mexico. Southwestern Naturalist 53:533–537.

Moreno, R. S., R. W. Kays, and R. Samudio Jr. 2006. Competitive release in diets of ocelot (*Leopardus pardalis*) and puma (*Puma concolor*) after jaguar (*Panthera onca*) decline. Journal of Mammalogy 87:808–816.

Moreno Valdéz, A. 1987. Determinación y distribución de mamíferos nativos del Cañón del Huajuco, Santiago, N. L. Thesis, UANL, Monterrey, Nuevo León, México.

Nielsen, C., D. Thompson, M. Kelly, and C. A. Lopez-Gonzalez. 2015. *Puma concolor* (errata version published in 2016). The IUCN Red List of Threatened Species 2015: e.T18868A97216466. Ver. 3.1. Accessed 18 February 2017.

Noss, A. J., M. J. Kelly, H. B. Camblos, and D. I. Rumiz. 2004. Pumas y jaguares simpátricos: datos de trampas-cámara en Bolivia y Belice. Pages 238–246 *in* E. Cabrera, C. Mercolli, and R. Resquín, editors. VI Congreso Manejo de Fauna Silvestre en Amazonia y Latinoamérica, Wildlife Conservation Society, Lima, Peru.

Núñez, R., J. Gómez, E. Saracho, and V. H. Vázquez. 2010a. Abundancia del jaguar (*Panthera onca*) y del ocelote (*Leopardus pardalis*) en la Reserva de la Biosfera de Marismas Nacionales, Nayarit. Memorias del Congreso de la Sociedad Mastozoológica de México, León, Guanajuato, México.

Núñez, R., B. Miller, and F. Lindzey. 2000. Food habits of jaguars and pumas in Jalisco, Mexico. Journal of Zoology 252:373–379.

Núñez, R., B. Miller, and F. Lindzey. 2002. Ecology of jaguars in the Chamela-Cuixmala Biosphere Reserve, Jalisco, Mexico. Pages 107–126 *in* R. A. Medellín et al., editors. El jaguar en el nuevo milenio. UNAM, Wildlife Conservation Society, and Fondo de Cultura Económica, México, D. F., México.

Núñez, R., E. Saracho, D. C. Anguiano, and E. Corona. 2010b. Censo nacional del jaguar y sus presas en Sierra de Vallejo Nayarit. Reporte final CENJAGUAR. Instituto de Ecología, UNAM, México, D. F., México.

Núñez Pérez, R. 2006. Área de actividad, patrones

de actividad y movimiento del jaguar (*Panthera onca*) y del puma (*Puma concolor*), en la Reserva de la Biósfera "Chamela-Cuixmala," Jalisco. Thesis, UNAM, México, D. F., México.

Núñez Pérez, R. 2007. Distribución y situación del jaguar en el occidente de México. Pages 25–40 *in* G. Ceballos, C. Chávez, R. List, and H. Zarza, editors. Conservación y manejo del jaguar en México: estudios de caso y perspectivas. CONABIO-Alianza WWF/Telcel-UNAM, México, D. F., México.

Núñez Pérez, R. 2011a. Estimating jaguar population density with camera-traps: a comparison with radio-telemetry estimates. Journal of Zoology 285:39–45.

Núñez Pérez, R. 2011b. Evaluación y monitoreo participativo de las poblaciones de felinos silvestres y sus conflictos en la región del Bajo Balsa-Costa Sierra. Reporte técnico. CEPLADE-Gobierno del Estado de Michoacán, México.

Núñez Pérez, R. 2014a. Efecto de actividad humana en la distribución y abundancia del jaguar en la costa de Jalisco. Thesis, UNAM, México, D. F., México.

Núñez Pérez, R. 2014b. Estrategia para la conservación del jaguar (*Panthera onca*) en la Reserva de la Biosfera Sierra de Manantlán. Fase 2. Reporte técnico. Fondo Mexicano para la Conservación de la Naturaleza, México, D. F., México.

Oliveira, T. G. 2002. Ecología comparativa de la alimentación del jaguar y el puma en el Neotrópico. Pages 265–288 *in* R. A. Medellín et al., editors. El jaguar en el nuevo milenio. Fondo de Cultura Económica, UNAM, and Wildlife Conservation Society, México, D. F., México.

Peña-Mondragón, J. L., A. Castillo, A. Hoogesteijn, and E. Martínez-Meyer. 2016. Livestock predation by jaguars *Panthera onca* in south-east Mexico: the role of local peoples' practices. Oryx 50:1–9.

Pierce, B. M., and V. C. Bleich. 2003. Mountain lion (*Puma concolor*). Pages 744–757 *in* G. A. Feldhamer, B. C. Thompson, and J. A. Chapman, editors. Wild mammals of North America: biology, management, and conservation. Second edition. Johns Hopkins University Press, Baltimore, Maryland, USA.

Rabinowitz, A., and K. A. Zeller. 2010. A range-wide model of landscape connectivity and conservation for the jaguar, *Panthera onca*. Biological Conservation 143:939–945.

Rodríguez-Soto, C., O. Monroy-Vilchis, and M. Zarco-González. 2013. Corridor for jaguar (*Panthera onca*) in Mexico: conservation strategies. Journal for Nature Conservation 21:438–443.

Roques, S., et al. 2016. Effects of habitat deterioration on the population genetics and conservation of the jaguar. Conservation Genetics 17:129–139.

Rosas Rosas, O. C. 2006. Ecological status and conservation of jaguars in northeastern Sonora, Mexico. Dissertation, New Mexico State University, Las Cruces, USA.

Rosas-Rosas, O. C., and L. C. Bender. 2012. Population status of jaguars (*Panthera onca*) and pumas (*Puma concolor*) in northeastern Sonora, Mexico. Acta Zoológica Mexicana 28:86–101.

Rosas-Rosas, O. C., L. C. Bender, and R. Valdez. 2008. Jaguar and puma predation on cattle calves in northeastern Sonora, Mexico. Rangeland Ecology and Management 61:554–560.

Rosas-Rosas, O. C., L. C. Bender, and R. Valdez. 2010. Habitat correlates of jaguar kill sites of cattle in northeastern Sonora, Mexico. Human Wildlife Conflicts 4:113–121.

Rosas-Rosas, O. C., J. de D. Guerrero Rodríguez, and A. D. Hernández-SaintMartín. 2015. Manual de prácticas ganaderas para regiones con grandes carnívoros en la Sierra Madre Oriental. Colegio de Postgraduados and Comisión Nacional de Áreas Naturales Protegidas, Puebla, México.

Rosas-Rosas, O. C., and J. H. López Soto. 2002. Status and distribution of the jaguar in Nuevo León, Mexico. Pages 393–401 *in* R. A. Medellín et al., editors. El jaguar en el nuevo milenio. UNAM, Wildlife Conservation Society, and Fondo de Cultura Económica, México, D. F., México.

Rosas-Rosas, O. C., and R. Valdez. 2010. The role of landowners in jaguar conservation in Sonora, Mexico. Conservation Biology 24:366–371.

Rosas-Rosas, O. C., R. Valdez, L. C. Bender, and D. Daniel. 2003. Food habits of pumas in northwestern Sonora, Mexico. Wildlife Society Bulletin 31:528–535.

Rueda, P., G. D. Mendoza, D. Martinez, and O. C. Rosas-Rosas. 2013. Determination of the jaguar (*Panthera onca*) and puma (*Puma concolor*) diet in a tropical forest in San Luis Potosi, Mexico. Journal of Applied Animal Research 41:484–489.

Salom Pérez, R., E. Carillo, J. C. Sáenz, and J. Mora. 2007. Critical condition of jaguar *Panthera onca* population in Cocovado National Park, Costa Rica. Oryx 41:51–56.

Sanderson, E. W., et al. 2002. Prioridades geográficas para la conservación del jaguar. Pages 601–628 *in* R. A. Medellín et al., editors. El jaguar en el nuevo milenio. UNAM, Wildlife Conservation Society, and Fondo de Cultura Económica, México, D. F., México.

SEMARNAP. 1997. Programa de conservación de la vida silvestre y diversificación productiva en el sector rural 1997–2000. SEMARNAP, México, D. F., México.

SEMARNAT. 2006. Proyecto para la conservación y manejo del jaguar en México. Serie proyectos de recuperación de especies prioritarias. Número 14. SEMARNAT, México, D. F., México.

SEMARNAT. 2010. Norma Oficial Mexicana NOM-059-ECOL-2001, Protección ambiental, especies nativas de México, flora y fauna silvestre—Categorías de riesgo y especificación por su inclusión, exclusión o cambio—Lista de especies en riesgo. DOF. Segunda sección. SEMARNAT, México, D. F., México.

SEMARNAT and CONANP. 2009. Programa de acción para la conservación de la especie: jaguar. Gobierno Federal, México, D. F., México.

Sergio, F., I. Newton, L. Marchesi, and P. Pedrini. 2006. Ecologically justified charisma: preservation of top predators delivers biodiversity conservation. Journal of Applied Ecology 43:1049–1055.

Seymour, K. L. 1989. *Panthera onca*. Mammalian Species 340:1–12.

Silver, S., L. Ostro, L. Marsh, L. Maffei, A. Noss, M. Kelly, R. Wallace, H. Gómez, and G. Ayala. 2004. The use of camera-traps for estimating jaguar *Panthera onca* abundance and density using capture/recapture analysis. Oryx 38:148–154.

Soria Díaz, L., O. Monroy Vilchis, M. Rodríguez Soto, M. Zarco González, and V. Urios. 2010. Variation of abundance and density of *Puma concolor* in zones of high and low concentration of camera traps in central Mexico. Animal Biology 60:361–371.

Sunquist, M. E., and F. C. Sunquist. 2002. Wild cats of the world. University of Chicago Press, Illinois, USA.

Sunquist, M. E., and F. C. Sunquist. 2009a. Family Felidae. Pages 54–130 *in* D. E. Wilson and R. A. Mittermeier, editors. Handbook of the mammals of the world. Volume 1. Carnivores. Lynx Edicions, Barcelona, Spain.

Sunquist, M. E., and F. C. Sunquist. 2009b. Genus *Panthera*. Pages 127–139 *in* D. E. Wilson and R. A. Mittermeier, editors. Handbook of the mammals of the world. Volume 1. Carnivores. Lynx Edicions, Barcelona, Spain.

Sunquist, M. E., and F. C. Sunquist. 2009c. Genus *Puma*. Pages 156–158 *in* D. E. Wilson and R. A. Mittermeier, editors. Handbook of the mammals of the world. Volume 1. Carnivores. Lynx Edicions, Barcelona, Spain.

Taber, A. B., A. J. Novaro, N. Neris, and F. H. Colman. 1997. The food habits of the sympatric jaguar and puma in the Paraguayan Chaco. Biotropica 29:204–213.

Thompson, R., and I. Cassaigne. 2017. The empowerment of livestock owners and the education of future generations to reduce human-feline carnivore conflicts. Pages 413–422 *in* C. Castaño-Uribe, C. A. Lasso, R. Hoogesteijn, and E. Payán-Garrido, editors. Conflicto entre felinos y humanos en América Latina. Instituto de Investigación de Recursos Biológicos Alexander von Humboldt, Bogotá, DC, Colombia.

Thornton, D., K. Zeller, C. Rondinini, L. Boitani, K. Crooks, C. Burdett, A. Rabinowitz, and H. Quigley. 2015. Assessing the umbrella value of a range-wide conservation network for jaguars (*Panthera onca*). Ecological Applications 26:1112–1124.

TSJRT (Technical Subgroup of the Jaguar Recovery Team). 2016. Jaguar draft recovery plan (*Panthera onca*). US Fish and Wildlife Service, Albuquerque, New Mexico, USA.

Valdez, R. 1999. Jaguar. Pages 378–388 *in* S. Demarais and P. R. Krausman, editors. Ecology and management of large mammals in North America. Prentice Hall, Upper Saddle River, New Jersey, USA.

Valdez, R., J. C. Guzmán-Aranda, F. J. Abarca, L. A. Tarango-Arámbula, and F. Clemente Sánchez. 2006. Wildlife conservation and management in Mexico. Wildlife Society Bulletin 34:270–282.

Valdez, R., A. Martínez Mendoza, and O. C. Rosas Rosas. 2002. Historical and present jaguar habitat components in northeastern Sonora, Mexico. Pages 367–378 *in* R. A. Medellín et al., editors.

El jaguar en el nuevo milenio. UNAM, Wildlife Conservation Society, and Fondo de Cultura Económica, México, D. F., México.

Valenzuela-Galvan, D., F. Castro-Campos, J. Servín, M. Martínez-Barona, and J. C. Martínez-Montes. 2015. First contemporary record of jaguar in Morelos State and the Sierra de Huautla Biosphere Reserve. Western North American Naturalist 75:370–373.

Van Valkenburgh, B., F. Grady, and B. Kurtén. 1990. The Plio-Pleistocene cheetah-like cat *Miracinonyx inexpectatus* of North America. Journal of Vertebrate Paleobiology 10:434–454.

Villordo Galván, J. A., et al. 2010. The jaguar (*Panthera onca*) in San Luis Potosí, México. Southwestern Naturalist 55:394–402.

Wakeling, B. F., R. Lee, D. Brown, R. Thompson, M. Tluczek, and M. Weisenberger. 2009. The restoration of desert bighorn sheep in the Southwest, 1951–2007: factors influencing success. Desert Bighorn Council Transactions 50:1–17.

Wozencraft, C. 1993. Order Carnivora. Pages 279–384 *in* D. E. Wilson and D. M. Reeder, editors. Mammal species of the world. Smithsonian Institution Press, Washington, DC, USA.

Wultsch, C., et al. 2016. Genetic diversity and population structure of Mesoamerican jaguars (*Panthera onca*): implications for conservation and management. PloS ONE 11 (10): e0162377.

Young, S. P., and E. A. Goldman. 1946. The puma: mysterious American cat. American Wildlife Institute, Washington, DC, USA.

Zanin, M., et al. 2016. Gene flow and genetic structure of the puma and jaguar in Mexico. European Journal of Wildlife Research 62:461–469.

Zarco-González, M. M., O. Monroy-Vilchis, and J. Alaníz. 2013. Spatial model of livestock predation by jaguar and puma in Mexico: conservation planning. Biological Conservation 159:80–87.

Zarco-González, M. M., O. Monroy-Vilchis, C. Rodríguez-Soto, and V. Urios. 2012. Spatial factors and management associated with livestock predations by *Puma concolor* in central Mexico. Human Ecology 40:631–638.

Zarza, H., C. Chávez, and G. Ceballos. 2011. Uso de hábitat del jaguar a escala regional en un paisaje con actividades humanas en el sur de la Península de Yucatán. Pages 101–110 *in* G. Ceballos, C. Chávez, R. List, and H. Zarza, editors. Conservación y manejo del jaguar en México: estudios de caso y perspectivas. CONABIO-Alianza WWF/Telcel-UNAM, México, D. F., México.

Zarza, H., C. Chávez, and G. Ceballos. 2013. Importancia de los iniciativos regionales para la consolidación de una estrategia nacional de conservación de jaguar: el caso México. Pages 191–201 *in* C. Castaño-Uribe, J. F. González-Maya, D. Zárrate-Charry, C. Ange-Jaramillo, and I. M. Vela-Vargas, editors. Plan de conservación de felinos del Caribe Colombiano. Fundación Herencia Ambiental Caribe, ProCat Colombia, and Sierra to Sea Institute, Santa Marta, Colombia.

14 Peccaries in Mexico

Rafael Angel Reyna-Hurtado, Eduardo J. Naranjo, Salvador Mandujano, and Ignacio March Mifsut

Peccaries are restricted to the New World. Of the three recognized species, two, the white-lipped peccary (*Tayassu pecari*) and the collared peccary (*Pecari tajacu*), occur in Mexico. The white-lipped peccary occurs from northern Argentina through southeastern Mexico, and the collared peccary occurs from northern Argentina to the southwestern US. The third species, the Chacoan peccary (*Catagonus wagneri*), known only from fossil specimens until the 1970s, was documented as an extant species in the dry forest of the Chaco region in Paraguay, Argentina, and Bolivia (Wetzel et al. 1975). A fourth species, the giant peccary (*Pecari maximus*), from the Amazon region, was proposed by Van Roosmalen et al. (2007); however, it is still undetermined whether it is a subspecies of collared peccary (Gongora et al. 2007).

Peccaries perform several ecological functions: they are seed predators, seed dispersers, and prey of large predators, and they can alter the structure and composition of plant communities. In addition, they are among the preferred prey species of subsistence and sport hunters throughout their distributional range (Sowls 1997). For many indigenous communities, peccaries are the most important source of animal protein (Donkin 1985, Stearman 1992, Sowls 1997). In addition, peccaries have the potential to provide a significant source of income; some communities in the northern Amazon Basin that practice sustainable population management are becoming nationally and internationally certified to commercialize peccary skins (Fang et al. 2008).

Historical Perspective

Peccaries were important in the diets and cultures of pre-Columbian peoples; early explorers mentioned that the Maya, Aztec, and Yaqui, among others, hunted peccaries and kept them captive as a source of food. *Díaz* del Castillo (2005) in 1632 described the presence of "abundant wild boar with the belly button on their back." Presumably, Montezuma also kept peccaries in his amazing zoological park, described by Cortés in the mid-1520s.

Taxonomy

Peccaries are piglike animals classified in the family Tayassuidae; the collared peccary is the only species in the Nearctic, and the other species are restricted to the Neotropics. Of the five subspecies of white-lipped peccary, only the Mexican subspecies (*ringens*), which extends into northern Nicaragua (March 1993), occurs in Mexico. Vernacular names include *pecarí labios blancos*, *jabalí*, *tropero*, *huangana*, *baquiro*, *coche de monte*, and *chancho cariblanco* in Mexico and other Latin American countries; *senso* and *hach kekén* in Chiapas; *jabalí* and *hauilla* in Campeche; and *chancho de monte* on the Yucatán Peninsula.

Hall (1981) recognized 14 subspecies of collared peccaries in North America, of which 3 occur in Mexico: *angulatus* occurs along the Gulf of Mexico in Tamaulipas and South Texas; *sonorensis* occurs in central Sonora and in the US in Arizona and New Mexico; and *yucatanensis* occurs on the Yucatán Peninsula (Hellgren and Bissonette 2003). Genetic studies have revealed two major groups of collared peccaries throughout their distributional range that could merit subspecies status, or perhaps species status (Gongora et al. 2006). Álvarez-Castañeda et al. (2015) divided collared peccaries from Mexico into two species (*Dicotyles angulatus* and *D. crassus*) based on skull morphometry and color differences. Efforts to classify collared peccaries show that this species is highly diverse and that more research is needed to clearly elucidate its taxonomic status throughout its distributional range. In this chapter, we will consider the collared peccary as a single species (*P. tajacu*).

Distribution

White-lipped

The white-lipped peccary originally occurred from Veracruz through most of southeastern Mexico (Leopold 1959, March 1990); however, its distributional range has been reduced by 84% in the last 50 years (Altrichter et al. 2012). This species has been extirpated from Veracruz and Tabasco, and single populations exist in Oaxaca and Yucatán (Altrichter et al. 2012, Naranjo et al. 2015, C. Alcerreca, personal communication); the only remaining large populations occur in Chiapas, Campeche, and Quintana Roo (fig. 14.1).

Collared

The collared peccary is widely distributed; it extends from the southern states and the Yucatán Peninsula to the north along both coasts and the two cordilleras (western and eastern Sierra Madre) to South and West Texas, southern Arizona, and New Mexico in the US (fig. 14.2). In central Mexico it has been recorded in Morelos, which represents its northernmost distribution in south-central Mexico (Mason-Romo et al. 2008).

Description

White-lipped

The white-lipped peccary is a small to medium-sized ungulate with relatively long legs and small hooves. The tail is reduced and almost absent; the head is massive and has a round nasal disk that enables foraging in the soil. The jaws are powerful and can break some of the hardest nuts (i.e., seeds of palm trees: *Mauritia flexuosa*, *Astrocaryum* sp., *Jessenia* sp., *Iriartea ventricosa*, and *Socratea exorrhiza*) (Kiltie 1981a). The white-lipped peccary has a pair of lower and upper tusks that are used mainly for defense, and to make a clacking sound that is characteristic of this species when alarmed or irritated. All peccaries have a unique dorsal gland; the substance excreted has a characteristic smell that probably has a communication function (Sowls 1997). Most of the body is black except for a white line around the lips that extends into the cheeks (from which the species gets it common name), and

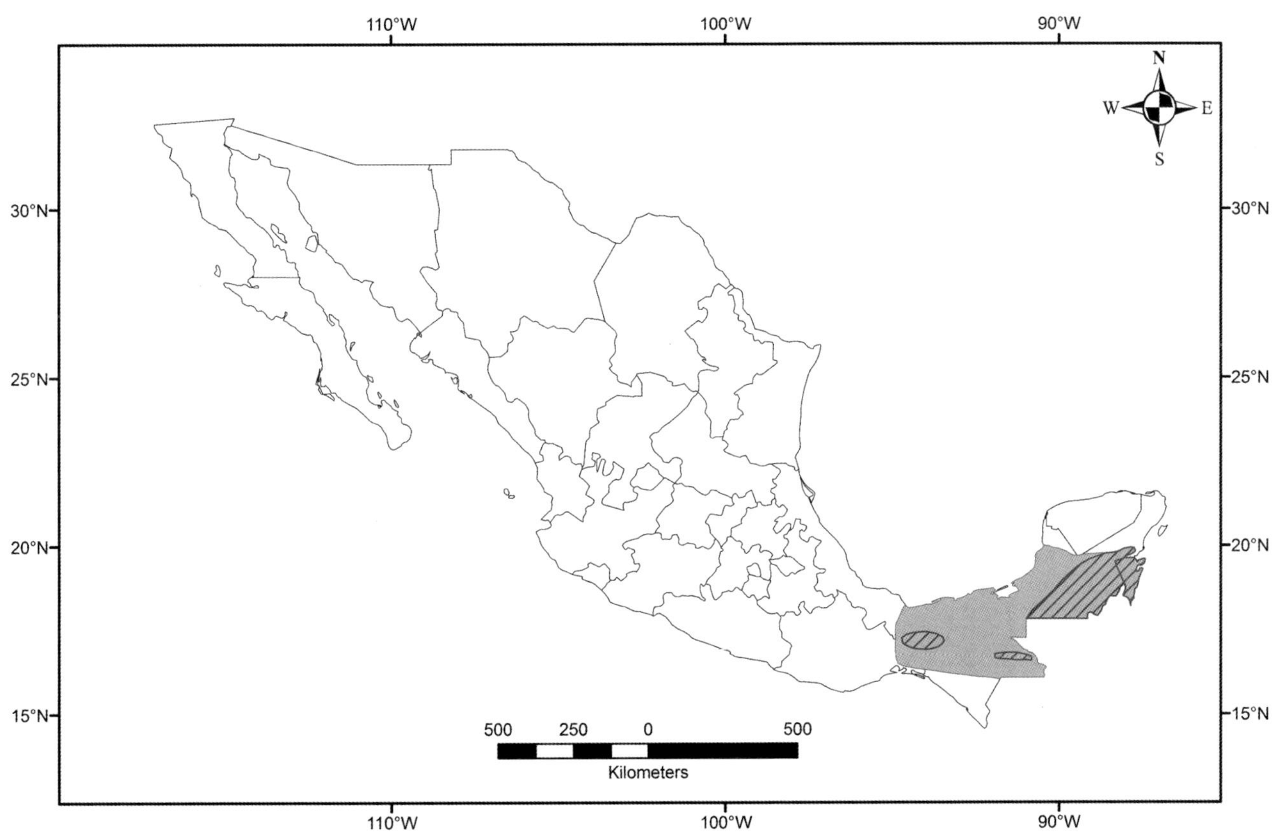

Figure 14.1. Potential (shaded) and current (barred) distribution of the white-lipped peccary in Mexico.

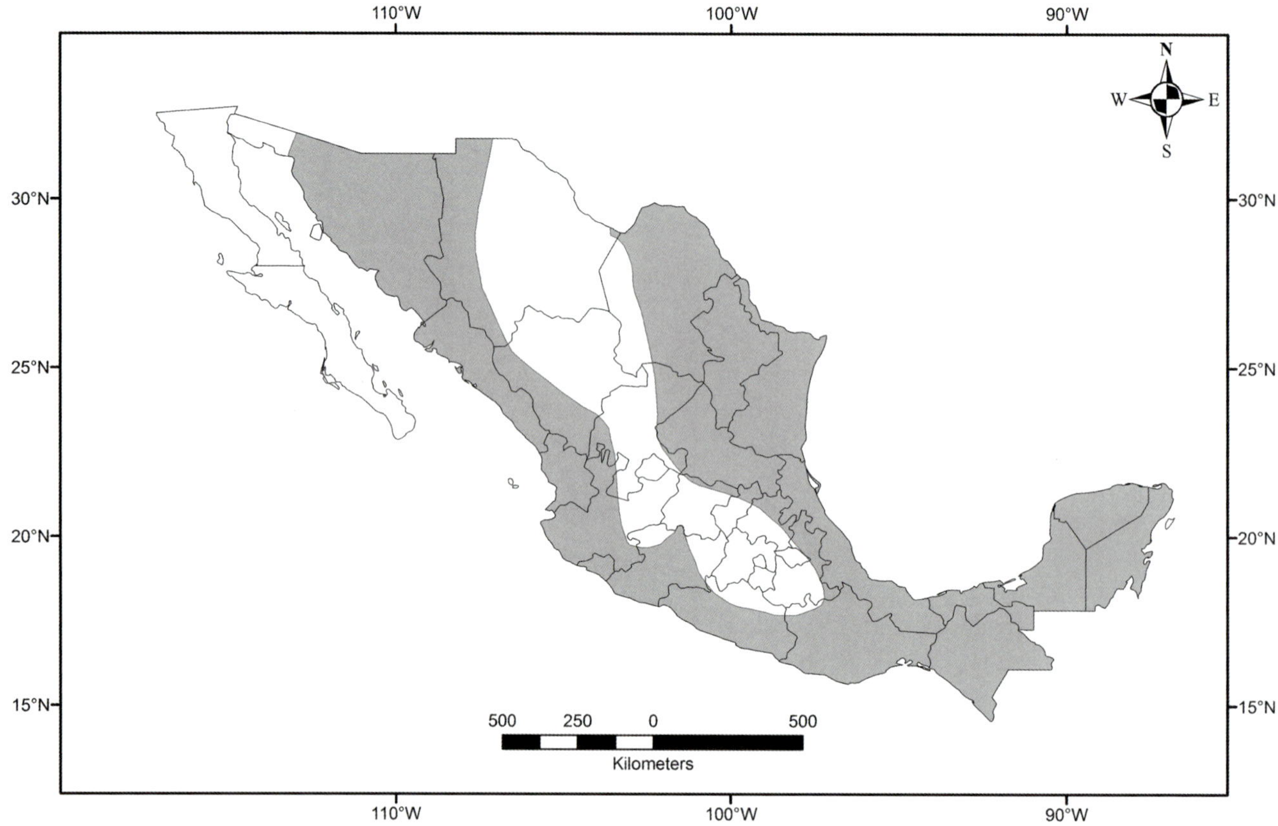

Figure 14.2. Potential distribution of the collared peccary in Mexico.

reddish areas around the neck (fig. 14.3). Neonates are a mixture of red-brown, black, and cream stripes; by the third year of age, the color changes from reddish to black. Adults attain body lengths of between 93 and 130 cm (3.0 and 4.3 ft). Individuals measured in the Calakmul Biosphere Reserve (CBR) in Campeche ranged from 104 to 124 cm (3.4 to 4.0 ft) in length. Body masses of individuals from throughout their range vary from 25 to 42 kg (55 to 93 lb). Fifteen individuals from Calakmul weighed between 25 and 38 kg (55 to 84 lb), with a mean of 30 kg (66 lb) (E. Rojas-Flores, personal communication). The sexes are not distinct anatomically.

Collared

The collared peccary, the smallest tayassuid, ranges in total length between 93.2 and 103 cm (3.0 and 3.4 ft), and body weights of adults range from 15 to 28 kg (33 to 62 lb) (Sowls 1997). The head is larger in relationship to the body than it is in the white-lipped peccary; the legs are shorter, and they have a rounder appearance than those of white-lipped peccaries. The hair is a mix of black, white, and yellow, which gives the species a grayish appearance, highlighting a dark line of hair along the back, while white hairs form a collar around the neck, from which the common name is derived (fig. 14.4). The difference in jaw force and access to certain fruits has been proposed as a niche separation between the two species; only white-lipped peccaries can break very hard nuts (Kiltie 1981b). Collared peccaries also have a pair of long tusks that are used to dig wallows, defend against predators, and create an alarm sound (clacking) when stressed (Sowls 1997). Newborns are reddish, but after the age of three months, the body color is similar to that of adults (March 2005).

Life History

Diet

White-lipped

The white-lipped peccary is omnivorous but prefers a greater percentage of fruits and seeds than vegetative parts or animal material. Kiltie (1981a) analyzed the contents of 34 stomachs; reproductive plant parts (flowers, seeds, and fruits) constituted 61% of the

Figure 14.3. White-lipped peccary in Calakmul Biosphere Reserve in Campeche, Mexico. Photo by Rafael Reyna.

Figure 14.4. Collared peccaries in Calakmul Biosphere Reserve in Campeche, Mexico. Photo by Rafael Reyna.

contents, while 39% consisted of vegetative parts (leaves, stems, and roots). The stomach contents of nine white-lipped peccaries from the CBR consisted of 81% fruit, while the rest comprised leaves, fibrous material, and animal matter (Pérez-Cortéz and Reyna-Hurtado 2008). It is possible that animal material has been underestimated in the diet of this species. Bodmer (1991) found 11% animal material in 46 samples, and R. Reyna-Hurtado (personal observation) documented groups of white-lipped peccaries in the CBR that spent long periods, even days, foraging in flooded forest during a season when earthworms and other larvae were abundant and fruit was not available. Herds also spent long periods foraging around pond edges capturing eels (*Ophisternum aenigmaticum*) (fig. 14.5) (Pérez-Cortéz and Reyna-Hurtado, 2008).

Collared

Collared peccaries are omnivorous as well but have a more diversified diet than white-lipped peccaries. They can survive eating cactus (*Opuntia* spp.) and

Figure 14.5. White-lipped peccaries searching for eels in a pond in Calakmul Biosphere Reserve in Campeche, Mexico. Photo by Rafael Reyna.

roots in xeric habitats of Sonora, Arizona, and Texas, while a preference for fruits is evident in tropical forests (Sowls 1997). In the deciduous tropical forest of Jalisco, diets consisted of between 46% and 50% roots, between 39% and 43% leaves, and only 10% to 11% fruits (Martínez Romero and Mandujano 1995). In the Calakmul region, Pérez-Cortéz and Reyna-Hurtado (2008) found that stomach contents of collared peccary consisted of 56% fruits, 33% leaves, 8.5% fibrous material (probably from roots), and only 0.5% animal matter. The diverse food items in the diets of collared peccaries and the ability to switch from one food item to another probably explain the persistence of this species in highly disturbed areas (Leopold 1959, Sowls 1997).

Reproduction

March (1993) reported that white-lipped peccaries can reproduce year-round, but a marked breeding season probably occurs between April and June. In the CBR and surrounding areas, Reyna-Hurtado (2009) reported that 18 of 19 births recorded occurred from late December through March. The gestation period lasted between 156 and 172 days, and the average litter size was two, with a range of one to three. Naranjo and Bodmer (2007) reported that sex ratios in both species conformed to the expected 1:1 ratio in the Lacandon Forest in Chiapas, although there was a slightly higher percentage of males than females in both species. The maximum longevity reported for the collared peccary in captivity is 13 years (Mayer and Wetzel 1987). The longevity of white-lipped peccaries is unknown, but animals older than six years have been recorded in the Peruvian Amazon (Fang et al. 2008).

The gestation period of collared peccaries is between 141 and 151 days; females usually produce two litters and give birth any time of year (March 2005). Naranjo and Bodmer (2007) reported a sex ratio of about 1:1 for animals hunted in communities of the Lacandon Forest in Chiapas. The maximum longevity is 24 years and 8 months in captivity (Crandall 1964, March and Mandujano 2005); in the wild, average longevity is 7 years (Sowls 1997). However, a female recorded in the wild in Arizona was estimated to be 15 years old based on dental annuli (Day 1985).

Growth

Newborn white-lipped peccaries weigh 1.1–1.35 kg (2.4–2.9 lb) and reach maturity at two years of age. Collared peccary newborns weigh around 0.5 kg (1.1 lb) and reach sexual maturity at the age of 11 months; females are probably receptive shortly thereafter (March 2005).

Energy requirements

White-lipped

White-lipped peccaries feed on high-energy food items, such as fruits and large seeds. These are usually dispersed in tropical forests, but their spatial and temporal availability are unpredictable (Reyna-Hurtado et al. 2012). Thus the white-lipped peccary is a mobile species that continually searches for resources over large areas (Reyna-Hurtado et al. 2009). In addition, they live in large and cohesive groups that require large quantities of food to fulfill their energy requirements. The movements of white-lipped peccary herds have been documented on Maracá Island, Brazil (Fragoso 1998); in Corcovado National Park, Costa Rica (Carrillo et al. 2002); in the Atlantic Forest (Keuroghlian et al. 2004) and Cerrado areas of Brazil (Almeida-Jácomo 2004); and in Mexico in the CBR, where herds traveled up to 17 km (10.5 mi) (Reyna-Hurtado et al. 2009).

Collared

Collared peccaries have a different strategy to cope with their daily energy requirements. They feed on less-energetic food items like roots, herbs, and cactus, but they also live in smaller groups than white-lipped peccaries and frequently split into subgroups. This species is considerably less mobile than the white-lipped peccary, presumably because it requires less energy (Sowls 1997).

Water requirements and thermal relationships

White-lipped

White-lipped peccaries require drinking water frequently. Because they lack sweat glands, they wallow at the edges of ponds and in drying streams and rivers, and they take mud baths to avoid overheating when temperatures are high, and to help shield their bodies from external parasites. White-lipped peccaries also drink water almost daily. Movements of more than 17 km (10.5 mi) to reach watering sites were documented during the dry season in both Calakmul (Reyna-Hurtado 2007) and Montes Azules (Naranjo 2002) Biosphere Reserves; these sites became centers of foraging movements (Reyna-Hurtado et al. 2009). This behavior is very well known among subsistence and sport hunters, who hunt this species around ponds and streams (Naranjo 2002, Reyna-Hurtado et al. 2010).

Collared

Collared peccaries use several strategies to cope with the drier habitats they inhabit (Zervanos 1975, Zervanos and Hadley 1973, Zervanos and Day 1977, Sowls 1997). In Chamela, Jalisco, collared peccaries feed on water-rich fruits of *Spondias* trees that fall to the ground after being shaken loose by chachalacas (*Ortalis poliocephala*) (Mandujano and Martínez Romero 1997). In the southwestern US, collared peccaries live in desert habitats without standing water. They fulfill their water requirements by eating succulent cactus (*Opuntia* spp.) and roots, and by spending long periods in the shade or in caves (Day 1985, Sowls 1997).

Behavior

White-lipped

White-lipped peccaries are a social species that forms larger herds than any other ungulate in dense tropical forests (Kiltie and Terborgh 1983). They form herds of 10 to 300 animals, although there are confirmed reports of herds of 400 and 700 in the Amazon Basin (J. Fragoso and R. Bodmer, personal communication), and there are historical reports of groups of 1,000 animals (Mayer and Wetzel 1987). In Mexico, group size varies from 5 to 80 in the Lacandon Forest (Naranjo 2002; I. March, personal observation), and from 11 to 44 in the CBR (fig. 14.6) (Reyna-Hurtado et al. 2009; R. Reyna-Hurtado, personal observation). In the Petén Forest of northern Guatemala, groups of 10 to 60 were observed (Moreira 2009). Herds are usually cohesive (Fragoso 1998, Reyna-Hurtado et al. 2009), but groups may split into subgroups temporarily (Carrillo et al. 2002, Keuroghlian et al. 2004). In the CBR, Reyna-Hurtado et al. (2009) followed radio-marked animals and found high group fidelity despite several occasions when different groups remained close and even traveled together for several days without forming a single herd.

Figure 14.6. White-lipped peccaries wallowing in mud in a pond in Calakmul Biosphere Reserve in Campeche, Mexico. Photo by Rafael Reyna.

Collared

Collared peccaries form smaller groups than white-lipped peccaries, numbering up to 50 animals in desert areas of the US, and even smaller groups (<20) in tropical forests (Robinson and Eisenberg 1985, March 2005). In the deciduous tropical forests of Chamela, groups of up to 12 peccaries were observed, although the most frequent group size was 1 to 4 (Mandujano 1999). In El Eden, a semideciduous forest of the Yucatán Peninsula, González-Marín et al. (2008) observed herds of 2 to 15. In the Calakmul region, the largest group observed consisted of 9 individuals (R. Reyna-Hurtado, personal observation), while in the Lacandon Forest, there were groups of 1 to 15, with a mean of 2.3 per group (Naranjo 2002).

Movements

White-lipped

White-lipped peccaries are very mobile and can travel as far as 17 km (10.5 mi) in a single day (Reyna-Hurtado et al. 2012). The presence of this species in a certain area is unpredictable, and its movements have been described as migratory, nomadic, and episodic (Sowls 1997, Fragoso 1999, Altrichter et al. 2001). Groups can appear and remain in one area for months or even years, and then disappear for several years (Fragoso 1994, Reyna-Hurtado et al. 2009). Several indigenous groups persistently hunt white-lipped peccaries when they are in a localized area (Donkin 1985, Stearman 1992, Reyna-Hurtado 2007). In the CBR, four groups that were followed for 7–18 months traveled long distances (7–17 km, 4.3–10.5 mi) in an irregular pattern, except for one group that stayed in the same area during the dry season and moved to another area during the rainy season. The other three groups moved unpredictably, but movements could have been related to watering sources (Reyna-Hurtado et al. 2009, 2012).

Collared

The movements of collared peccary have been described primarily for arid habitats of the southwestern US, where visibility is higher than in dense tropical forests. Collared peccary groups moved mainly at dawn and dusk, with individual herd members spreading over 100 m (328 ft) (Sowls 1997); this contrasted with white-lipped peccaries, which moved in tight groups usually within visual and auditory distance of each other (Fragoso 1998, Reyna-Hurtado et al. 2009). Collared peccaries have been found as far as 8 km (4.9 mi) from where they were marked; however, these could be dispersal movements, because this species usually restricts its movements within 5 km² (1.9 mi²) (Ellisor and Harwell 1969, Schweinsburg 1971).

Communication

White-lipped

Both peccary species maintain communication through physical, olfactory, and acoustic signals. Usually they greet each other by rubbing their nose against the dorsal gland of a mate to acquire a group odor identifiable to all herd members (Sowls 1997). The white-lipped peccary also has a broad vocal repertoire, and herd members are usually in contact acoustically, which is especially important when traveling through dense vegetation (March 1993). The contact call is a low growl or moan, but white-lipped peccaries also clatter their teeth and bark sharply during conflict, and they will use a deep bark as an alarm call. Adults and juveniles also make a raspy bleat (Mayer and Wetzel 1987). White-lipped peccaries frequently clack their tusks, usually during stressful periods, accompanied by barking at different intensities. When they detect a potential danger such as human or jaguar odor, they respond with loud, explosive clacking of their tusks as they flee. This explosion of sound and movement ends in seconds, and then they usually stop and remain quiet, a strategy believed to enable them to listen and smell to better detect the origin of the threat. If the danger continues, they again initiate clacking, barking, and running. On some occasions, the running/clacking and stopping is repeated more than five times as the group flees (R. Reyna-Hurtado, personal observation).

Collared

Collared peccaries also depend on odor as well as auditory communication. Sowls (1997) categorized vocalizations into nine types used to maintain group cohesion, locate other members, establish dominance, warn others, and react to a threatening situation. Collared peccaries also rub each other's backs, which spreads the product of the dorsal gland for within-group identification (Sowls 1997).

Home Range

White-lipped

White-lipped peccaries have home ranges several times larger than those of any other tropical forest ungulate. In the only estimate of white-lipped peccary home ranges in Mexico, which were recorded in the CBR, home ranges measured from 77.2 to 121.8 km² (29.8 to 47 mi²) (determined using the Minimum Convex Polygon method for two groups of 25 and 18 individuals) (Reyna-Hurtado et al. 2009). In other countries, similar or larger home ranges were recorded for groups on Maracá Island, Roraima, Brazil: 109 and 200 km² (42 and 77 mi²) for groups of 124–134, and more than 200 individuals, respectively (Fragoso 1998, 2004); and for groups living in a mosaic of open and closed habitats in the Cerrado ecosystem in Goiás State, Brazil: 288, 137, and 114 km² (111, 52, and 44 mi²) for groups of 65, 75, and 45 individuals, respectively (Almeida-Jácomo 2004). Home range estimates for white-lipped peccaries from closed habitats in other countries are smaller than those from Mexico: 21.8 km² (8.4 mi²) for a 39-member herd on Maracá Island (Fragoso 1998); 37.2 km² (14.3 mi²) for a 40–70-member herd in Corcovado National Park, Costa Rica (Carrillo et al. 2002); 18.71 km² (7.22 mi²) for a 42-member herd in the Atlantic Forest region of Brazil (Keuroghlian et al. 2004); and 14.3–84.8 km² (5.5–32.7 mi²) for other herds ranging from 15 to 130 animals in the Cerrado ecosystem of Goiás, Brazil (Almeida-Jácomo 2004).

Collared

The only estimate of collared peccary home range in Mexico was 7.3 km² (2.8 mi²) for a herd in the Lacandon Forest, Chiapas (Naranjo 2002). Annual home ranges of collared peccaries in Arizona varied from 0.5 to 3.1 km² (0.2 to 1.2 mi²) (Gabor and Hellgren 2000). In fragments of the Brazilian Atlantic Forest, home ranges consisted of 123 and 305 ha (303 and 753 ac) (Keuroghlian et al. 2004). Fragoso (1999) reported home ranges of 10.5 to 11.7 km² (4 to 4.5 mi²) on Maracá Island, Brazil. In a tropical forest in French Guiana, home ranges varied from 157 to 243 ha (387 to 600 ac) for herds of 5 to 11 animals (Judas and Henry 1999).

Density

White-lipped

White-lipped peccary densities are variable. Naranjo and Bodmer (2007) reported densities of 7.93 and 1.08 ind/km² (20.86 and 2.84 ind/mi²) in slightly hunted and persistently hunted sites, respectively, in the Lacandon rainforest, while Reyna-Hurtado et al. (2010) recorded a density of 0.43 ind/km² (1.13 ind/mi²) in the semidry forests of the CBR. In the Petén Forest of northern Guatemala, which forms a continuous forest with Calakmul, Novack et al. (2005) reported a density of 9.6 ind/km² (25.2 ind/mi²) in a nonhunted site using the line transect method. These differences, however, could be influenced by the method used, or the higher-quality habitat in the Petén Forest compared to the drier forest in Calakmul (Reyna-Hurtado et al. 2009).

Keuroghlian et al. (2004) and Cullen et al. (2000), both working in southern Brazil, reported densities between 3.9 and 10.2 ind/km² (10.2 and 26.8 ind/mi²) in fragments of the Atlantic Forest. Fragoso (1994, 2004) reported densities of 139–542 ind/km² (365–1,426 ind/mi²) in 1988, which declined to 1.4–8.3 ind/km² (3.6–21.8 ind/mi²) in 1992 after an apparent disease-driven population crash on Maracá Island, Brazil. Desbiez (2007) found densities of 13.7 and 3.0 ind/km² (36 and 7.9 ind/mi²) in forested and open Cerrado habitat of the Brazilian Pantanal, respectively. Altrichter (2005) estimated 0.33 ind/km² (0.86 ind/mi²) in hunted sites, and 1.04 ind/km² (2.73 ind/mi²) in nonhunted sites of the dry forest in the Argentine Gran Chaco.

Collared

Population density of collared peccary in Mexico in Chamela dry forest in Jalisco was 4.9 ind/km² (12.8 ind/mi²) (Martínez Romero and Mandujano 1995) and 7.5 ind/km² (19.7 ind/mi²) (Mandujano 2007). Naranjo and Bodmer (2007) estimated 1.53 and 1.15 ind/km² (4.02 and 3.02 ind/mi²) in slightly hunted and persistently hunted sites of the Lacandon Forest, respectively. On the Yucatán Peninsula, a density of 1.9 ind/km² (5 ind/mi²) was reported in El Eden, a small, protected, semidry tropical forest (González-Marín et al. 2008). In other areas of the distributional range, collared peccaries have been estimated to have the following densities: Peres (1996) found densities ranging from 0.4 to 11.6 ind/km² (1.1 to 30.5 ind/mi²) in Amazon forest sites subjected to different degrees of hunting pressure. In Atlantic Forest fragments, Keuroghlian et al. (2004) found 5.9 ind/km² (15.5 ind/mi²), with a range between 2.8 and 8.9 ind/km² (7.3 and 23.4 ind/mi²), while Cullen et al. (2000) reported a similar density in the same area of 6.4 ind/km² (16.8 ind/mi²), with a range of 3.7 to 11.2 ind/km² (9.7 to 129.4 ind/mi²). Altrichter (2005) found a very low density of this species in the hot and dry environment of the Chaco region in Argentina, with 0.62 ind/km² (1.63 ind/mi²) in hunted areas, and 0.91 ind/km² (2.39 ind/mi²) in nonhunted areas. In two national parks in eastern Colombia, Gómez and Montenegro (2012) estimated a density between 0.05 and 0.24 ind/km² (0.13 and 0.63 ind/mi²). Sowls (1997) summarized several density estimations for the collared peccary as follows: south-central Arizona, 3.0–4.7 ind/km² (7.9–12.3 ind/mi²) (Schweinsburg 1971) and 2.1–4.5 ind/km² (5.5–11.8 ind/mi²) (Supple 1983); southern Arizona, 4.5–11.5 ind/km² (11.8–30.2 ind/mi²) (Day 1985); West Texas, 1.2–2.5 (3.1–6.5) and 3.3–11.0 ind/km² (8.6–29 ind/mi²) (Bissonette 1982); South Texas, 3.8–8.8 ind/km² (10–23.1 ind/mi²) (Low 1970); Barro Colorado, Panama, 16.0 ind/km² (42.1 ind/mi²) (Eisenberg 1989) and 9.3 ind/km² (24.4 ind/mi²) (Glanz 1982); southern Peru, 5.6 ind/km² (14.7 ind/mi²) (Emmons 1987); Masaguaral, Venezuela, 8.5 ind/km² (22.3 ind/mi²) (Eisenberg et al. 1979); Guatopo National Park, Venezuela, 1.9 ind/km² (5 ind/mi²) (Schaller 1983); Peruvian Amazon, 3.3 ind/km² (8.6 ind/mi²) (Bodmer 1989); and the Neotropics in general, 12.0 ind/km² (31.5 ind/mi²) (Robinson and Redford 1986).

Productivity and recruitment

The high productivity of both peccary species—females can produce one to three young per litter from two to ten years of age—qualifies them as candidates for sustainable harvest programs (Robinson and Redford 1986, Bodmer et al. 1997, Fang et al. 2008). Both species represent an important source of income for subsistence hunters, and a commercial hunting program is being planned in the Peruvian Amazon (Fang et al. 2008). However, sustainable rates of extraction observed in the Amazon rainforest cannot be attained in other areas of the Americas, particularly in Mexico and Central America, where forested areas are considerably smaller.

Survival and mortality factors

Both peccary species are significant prey of jaguars and pumas, occasionally of ocelots, and probably of crocodiles. There are several reports of jaguar predation of both peccary species in Mexico, including historical reports by Leopold (1959) and Álvarez del Toro (1991), among others. Aranda (1990) reported the presence of peccary hairs in 45% of jaguar feces examined in the CBR. In this same area, Amín (2004) reported that both peccary species are important components of jaguar diets, and R. Reyna-Hurtado (personal observation) witnessed three incidents in which jaguars killed and ate white-lipped peccaries in 2005 and 2006.

Human-caused mortality by hunting is probably the biggest threat to peccary populations in Mexico. However, the population responses of the two species can differ: the collared peccary can resist relatively high rates of extraction and still remain abundant in some areas, while the white-lipped peccary can be extirpated under intense hunting, especially when combined with forest loss or fragmentation (Weber 2000, Naranjo and Bodmer 2007, Reyna-Hurtado and Tanner 2007, Altrichter et al. 2012). Part of this difference can be explained by their social behavior. White-lipped peccaries remain in cohesive herds when confronted by hunters, making them easy prey, while collared peccary herds immediately flee from hunters and split into subgroups. However, collared peccaries can also be easy prey for hunters because they tend to seek shelter in caves or logs where they can be cornered, especially when hunted with dogs. Once cornered, they can be approached and killed, even with a pointed stick (R. Reyna-Hurtado, personal observation).

Another difference is their ranging behavior and feeding habits. White-lipped peccaries require larger areas of forest than collared peccaries, because white-lipped peccaries feed in larger groups and focus on energy-rich foods such as fruits that usually have a patchy distribution (Kiltie and Terborgh 1983, Reyna-Hurtado et al. 2012). In contrast, collared peccary herds frequently split into subgroups and feed on a wider array of resources (fruits, herbs, roots, and animal matter) occurring in smaller patches (Sowls 1997). White-lipped peccaries are more vulnerable to habitat loss and fragmentation because they require larger forested areas. Also, their dependence on aquatic habitats makes them vulnerable during the dry season, because they usually require a daily watering source (Sowls 1997, Reyna-Hurtado 2007). Finally, although we lack long-term data in Mexico, populations of white-lipped peccaries could be subject to fluctuations similar to those recorded in the northern Amazon, where cycles of around 15 years have been observed during which populations attain high densities and then crash to lower densities, apparently as a result of disease or depleted resources (Fragoso 1997, 2004).

Emigration and Immigration

In Arizona, five collared peccaries were observed joining different groups (Sowls 1997). Fragoso (1994) also observed emigration of white-lipped peccaries from one group to another. In the Brazilian Atlantic Forest and in Corcovado National Park in Costa Rica, Keuroghlian et al. (2004) and Carrillo et al. (2002), respectively, reported that groups of white-lipped peccaries often share members and exchange individuals. In the only specific study on white-lipped peccaries in Mexico, Reyna-Hurtado et al. (2009) did not find evidence of movements between groups; individuals remained within a single herd during the 18 months of the study, even when herds came within 5 m (16.4 ft) of each other.

Population Trends and Current Status

In Mexico, both peccary species are subject to heavy hunting pressure. In an area surrounding the CBR, collared peccaries were the most frequently hunted species in one community forest (ejido) (Weber 2000), and both species were among the top five prey species in three other ejidos (Escamilla et al. 2000). The white-lipped peccary has disappeared from most ejidos because of hunting pressure and increasing forest fragmentation. In areas where white-lipped peccaries still survive, they are pursued intensively, especially during the dry season when they are most vulnerable; Reyna-Hurtado et al. (2010) documented the extirpation of an entire herd by hunters during one dry season.

White-lipped

The distributional range of white-lipped peccaries has been reduced by overhunting and forest fragmentation (Altrichter et al. 2012). The species is listed in Appendix II of CITES, but its status is poorly known. The IUCN Red List raised its status from Lower Risk/Least Concern to Near Threatened in 2008 because of its rapid decline (Reyna-Hurtado et al. 2008), and to Vulnerable in 2013 (Keuroghlian et al. 2013). Sixty years ago, Leopold (1959) observed a decrease in its numbers and distribution in Mexico because of habitat loss and excessive hunting. Since then, white-lipped peccaries have disappeared from almost all of Veracruz and Tabasco, with isolated populations remaining in Oaxaca and Yucatán, and relatively large and stable populations occurring only in Chiapas, Quintana Roo, and Campeche (March 1993, 2005). However, the white-lipped peccary was only recently listed as an endangered species in Mexico (SEMARNAT 2010), and sport hunting was permitted in Campeche and Quintana Roo until 2012. Its current status is even more precarious, because subsistence hunting and forest fragmentation continue outside protected areas, and a conservation plan at a regional or national scale has yet to be developed (Reyna-Hurtado et al. 2009).

Collared

The collared peccary is listed as a species of Least Concern by the IUCN (Gongora et al. 2011) but is not included on the endangered species list of Mexico (SEMARNAT 2010). It is still relatively common in large forested areas, and where large forest fragments are interspersed with croplands. However, it is one of the principal prey species of rural subsistence hunters and is subjected to heavy hunting pressure, especially in large protected areas such as Calakmul, El Triunfo, La Sepultura, Montes Azules, and Sian Ka'an Biosphere Reserves (Weber 2000, Naranjo 2002, March and Mandujano 2005, Ramírez-Barajas and Naranjo 2007). The killing of up to six collared peccaries in a single day by organized groups of subsistence hunters with trained dogs was observed in the Calakmul region (R. Reyna-Hurtado, personal observation), and recent evidence shows a decline in their relative abundance in some hunted areas (M. Briceño, unpublished data). The status of the collared peccary is unknown in several areas of Mexico, especially along the Pacific and Gulf coasts.

Population management

White-lipped

White-lipped peccaries require specific management attention because of their complex social behavior, space requirements, susceptibility to disease, and vulnerability to human impacts resulting from forest fragmentation and hunting pressure. The formulation of management plans requires knowledge of population densities, group sizes, herd home ranges, areas shared by groups, age structure, breeding season, and sex ratios. White-lipped peccaries require an area up to 10 times larger than that needed by other tropical ungulates (Reyna-Hurtado et al. 2009); they live in groups of variable size; and they share space with other groups. This species has special habitat requirements because of its need for watering sources and intact tropical forests. Density estimation is challenging because the animals are difficult to observe when conducting transects, which is perhaps the most common method of density estimation for large tropical mammals (R. Reyna-Hurtado, personal observation).

Another factor to consider when determining hunting quotas for white-lipped peccaries is their social behavior and herd composition. No more than 5–10% of any single group should be allowed to be hunted. For example, in an area with a mean group size of 20, no more than 2 individuals per season should be hunted from any given group to maintain a sustainable population (Reyna-Hurtado et al. 2010).

Collared

Collared peccaries have the potential to be sustainably hunted given their relatively high productivity, rapid sexual maturation, and ability to survive in disturbed areas. In two studies comparing hunted and nonhunted or slightly hunted populations in Calakmul and the Lacandon Forest in Mexico, collared peccary densities did not vary between sites, and they were among the most abundant ungulates in disturbed areas; their harvest rate appeared to be sustainable in a few communities (Naranjo and Bodmer 2007, Reyna-Hurtado and Tanner 2007). Indeed, in some areas, subsistence hunting of collared peccaries at reasonable levels could be recommended as a way to limit damage to field crops. In Peru, both peccary species are extracted at a rate of 56,000 pelts/year by subsistence hunters who hunt over a vast area of the Amazon forest (Fang et al. 2008). However, recent studies (M. Briceño, unpublished data) have found decreasing populations of collared peccaries in some community forests of the Calakmul region in southern Mexico, where this species has been hunted intensively for many years.

Habitat Requirements

White-lipped

White-lipped peccaries occur principally in humid tropical forests; however, they can live in dry or semidry forests, depending on water availability. In the CBR, water was the most important landscape component (Reyna-Hurtado et al. 2012). The movements of four groups were influenced by the availability and distribution of watering sites (*aguadas*). During the dry season, white-lipped peccaries also depended on temporary watering sources, such as water that collected in depressions in rocks (*sartenejas*), where water can persist after large ponds dry up. These were an important source of water for peccary groups, which sometimes traveled 2–4 km (1.2–2.4 mi) in a straight line to reach them (Reyna-Hurtado et al. 2012). An analogous situation occurs at Montes Azules Biosphere Reserve in Chiapas, where peccary herds remain close to watering sources (rivers, streams, and lagoons) during the driest months, February through May (Naranjo 2002). White-lipped peccaries depended heavily on two types of forest in the CBR: medium-height semiperennial forests, and flooded forests. They fed on fruits, especially ramon (*Brosimum alicastrum*) and zapote (*Manilkara zapota*) in the medium-height semiperennial forest, and they visited the flooded forest in the rainy season to feed on invertebrates such as earthworms and larvae, which were especially available at that time (Reyna-Hurtado et al. 2009). In the Lacandon Forest, white-lipped peccaries also used flooded areas frequently and relied heavily on fruits of the same species, as well as those of *Licania platypus*, *Pouteria sapota*, *Attalea butyracea*, and especially *Bactris balanoidea*. In general, across the entire distributional range, the largest populations of white-lipped peccaries occur in undisturbed tall humid tropical forests where water, fruits, and nuts are abundant. In these types of forests, white-lipped peccaries can represent a significant proportion of the mammalian biomass (Peres 1996, Kiltie and Terborgh 1983).

Collared

In contrast, collared peccaries are adapted to live in diverse environments from xeric subtropical areas such as deserts in southern Arizona, Texas, and Sonora, to humid tropical perennial forest (Sowls 1997). In tropical areas, collared peccaries do well in disturbed habitats, such as secondary forest near farm fields in tropical regions (Reyna-Hurtado and Tanner 2005), and they can be a nuisance to agricultural producers, because they can consume large quantities of corn, squash, manioc, and other crops (N. Arias, Nuevo Becal ejido, personal communication). Collared peccaries can occur in landscapes consisting of interspersed primary forest, disturbed native habitats, and croplands—places where white-lipped peccaries would be extirpated.

Habitat Management

Direct habitat improvement

To conserve the white-lipped peccary, it is necessary to preserve extensive areas of primary tropical forest with permanent water sources in Mexico. In particular, communal forests around large protected areas (i.e., Calakmul, Montes Azules, and Sian Ka'an Biosphere Reserves) should be retained to improve connectivity among peccary populations threatened with habitat fragmentation. Evergreen, subdeciduous, and flooded forest patches are especially important in southern Mexico, because they usually contain key water sources for both wildlife and people during the dry season. Land managers need to ensure that peccaries and other wildlife have access to standing water year-round. Sport and subsistence hunters frequently take advantage of the dry season to hunt white-lipped peccaries near water bodies. This is when this species is more vulnerable to hunters and wild predators.

Collared peccary populations can benefit by the presence of small cultivated clearings. Many hunters in some ejidos around the CBR used to allocate half a field for their own consumption and leave the other half to attract brocket deer, white-tailed deer, and collared peccaries (R. Reyna-Hurtado, personal observation). Artificial water and food sources could benefit peccary populations in some instances.

Retaining patches of fruiting trees and conserving undisturbed patches of tropical forest will benefit both collared and white-lipped peccary populations.

For both species, the protection of core habitat areas where hunting is prohibited is essential for the long-term survival of local populations.

Coordination with livestock management

Extensive cattle ranches and white-lipped peccary populations are incompatible. Conventional cattle ranching in the Neotropics requires extensive areas cleared of native vegetation and replanted with exotic grasses. White-lipped peccaries need extensive areas of mature forest with minimal fragmentation. Disease transmission is another potential problem that could have disastrous consequences for white-lipped peccary populations if they come in close contact with livestock, pigs, or other domestic animals (Fragoso 1997, Reyna-Hurtado et al. 2014).

Coordination with timber management

White-lipped peccary and collared peccary management can be compatible with logging if the area is large enough to allow rotation of cut and uncut plots in cycles of several years. White-lipped peccaries usually avoid human-disturbed areas, although in some ejidos in both the Calakmul and Lacandon regions, this species occurs in spite of moderate and seasonal timber extraction. Another problem is that loggers and other workers, when living in isolated forests, tend to engage in excessive hunting and can cause significant harm to wildlife populations, even resulting in local extirpations.

Collared peccary populations can tolerate timber extraction activities better than white-lipped peccaries because they are more resilient and take advantage of small forest clearings for foraging. However, under unrestricted hunting pressure, collared peccary populations can be extirpated (M. Briceño, unpublished data).

Coordination with other uses

For both peccaries, population management can be compatible with minimal environmental disturbances, such as ecotourism, wildlife photography, or low-impact extractive practices such the harvest of *xate* palms, chicle gum, mahogany seeds, or wild honey, which are common in southern Mexico's tropical forests. Sustainable hunting can be practiced in coordination with community-based conservation programs in areas where large tracts of disturbed and undisturbed habitat mosaics remain. Peccary population size, herd movement patterns, group size, and availability of water and food resources must be taken into consideration when developing a management plan. Because collared peccary populations still occur in human-disturbed landscapes, this species is an ideal candidate for implementing pilot sustainable harvest programs where hunting is restricted and population parameters are closely monitored (Fang et al. 2008).

Management Challenges in Mexico

Managing white-lipped peccaries in Mexico is a major challenge, considering that their distributional range has been reduced by 84% (Altrichter et al. 2012), and only a few potentially viable populations remain. Efforts to conserve populations (or subpopulations) of this species, particularly outside extant protected areas, such as Calakmul, Montes Azules, Chimalapas, and Sian Ka'an Biosphere Reserves, are urgently needed. These subpopulations will ensure connectivity and dispersal between protected areas and will preclude the isolation of populations within protected areas. Subsistence and sport hunting are urgent issues, because even populations inhabiting high-quality habitats could be extirpated. Because of their precarious status and the lack of ecological data, sport hunting of white-lipped peccaries should be prohibited. This species has the potential to be a game animal, but not until communities develop sustainable management programs that include solid monitoring protocols (Reyna-Hurtado et al. 2010).

Because of its wide distribution and presence in several climatic zones and ecoregions, and its high tolerance of human disturbance as well as its potential economic worth, the collared peccary is an excellent candidate for pilot management programs in rural communities (Fang et al. 2008). Rural communities in southern Mexico currently lack the wildlife management expertise of private landowners in northern areas, but the adaptability of the collared peccary lends itself to adaptive management schemes (Weber et al. 2006). Rural communities are having the greatest negative impacts on peccary populations; it is urgent that these communities adopt sustainable wildlife management programs.

The destruction and fragmentation of tropical forests is the major issue confronting wildlife management in the tropics. The sustainable management of native plant communities and ecosystems is essential to the survival of wildlife populations. Native mature forests serve as a reservoir of animal and plant populations, but they can also be incorporated into community protected areas that link existing wildlife reserves with movement and dispersal corridors. These areas are ideal for low-impact activities that can provide economic incentives for rural communities to practice sustainable management of natural resources. The continued existence of peccaries will depend on the successful implementation of wildlife conservation and management programs that will benefit both wildlife and human communities.

Literature Cited

Almeida-Jácomo, A. T. 2004. Ecologia manejo e conservação do Queixada *Tayassu pecari* in Parque Nacional das Emas e em propiedades rurais de seu entorno. Dissertation, University of Brasilia, Brazil.

Altrichter, M. 2005. The sustainability of subsistence hunting of peccaries in the Argentine Chaco. Biological Conservation 126:351–362.

Altrichter, M., E. Carrillo, J. Saenz, and T. K. Fuller. 2001. White-lipped peccary (*Tayassu pecari*, Artiodactyla: Tayassuidae) diet and fruit availability in a Costa Rican rain forest. Revista de Biología Tropical 49:1183–1192.

Altrichter, M., A. Taber, H. Beck, R. Reyna-Hurtado, and A. Keuroghlian. 2012. Range-wide declines of a key Neotropical ecosystem architect, the Near Threatened white-lipped peccary *Tayassu pecari*. Oryx 46:87–98.

Álvarez-Castañeda, S. T., T. Álvarez, and N. González-Ruiz. 2015. Keys for identifying Mexican mammals in the field and in the laboratory. CIBNOR and UAM, La Paz, Baja California, México.

Álvarez del Toro, M. 1991. Los mamíferos de Chiapas. Second edition. Gobierno del Estado, Tuxtla Gutiérrez, Chiapas, México.

Amín, M. 2004. Patrones de alimentación y disponibilidad de presas del jaguar (*Panthera onca*) y del puma (*Puma concolor*) en la Reserva de la Biósfera Calakmul, Campeche, México. Dissertation, Instituto de Ecología, UNAM, México, D. F., México.

Aranda, M. 1990. El jaguar (*Panthera onca*) en la reserva de la Biosfera de Calakmul: morfometría, hábitos alimentarios y densidad de población. Thesis, Universidad Nacional de Costa Rica, Heredia, Costa Rica.

Bissonette, J. A. 1982. Ecology and social behavior of the collared peccary in Big Bend National Park, Texas. US Department of the Interior, National Park Service, Scientific Monograph Series No. 16. 85 pp.

Bodmer, R. 1989. Frugivory in Amazonian Artiodactyla: evidence for the evolution of the ruminant stomach. Journal of Zoology 219:457–467.

Bodmer, R. 1991. Influence of digestive morphology on resource partitioning in Amazonian ungulates. Oecologia 85:361–365.

Bodmer, R., J. F. Eisenberg, and K. H. Redford. 1997. Hunting and the likelihood of extinction of Amazonian mammals. Conservation Biology 2:460–466.

Carrillo, E., J. C. Saenz, and T. K. Fuller. 2002. Movements and activities of white-lipped peccaries in Corcovado National Park, Costa Rica. Biological Conservation 108:317–324.

Crandall, L. S. 1964. The management of wild animals in captivity. University of Chicago Press, Illinois, USA.

Cullen, L., Jr., R. E. Bodmer, and C. Valladares-Padua. 2000. Effects of hunting in habitat fragments of the Atlantic forests, Brazil. Biological Conservation 95:49–56.

Day, G. I. 1985. Javelina research and management in Arizona. Arizona Game and Fish Department, Phoenix, USA.

Desbiez, A. L. J. 2007. Wildlife conservation in the Pantanal: habitat alteration, invasive species and bushmeat hunting. Dissertation, University of Kent, Canterbury, United Kingdom.

Díaz del Castillo, B. 2005. Historia verdadera de la conquista de la Nueva España. Mexicanos Unidos, México, D. F., México.

Donkin, R. A. 1985. The peccary—with observations on the introduction of pigs to the New World. American Philosophical Society, Philadelphia, Pennsylvania, USA.

Eisenberg, J. F. 1989. Mammals of the Neotropics: the northern Neotropics. University of Chicago Press, Illinois, USA.

Eisenberg, J. F., M. A. O'Connell, and P. V. August. 1979. Density, productivity and distribution of mammals in two Venezuelan habitats. Pages 187–207 *in* J. F. Eisenberg, editor. Vertebrate ecology in the northern Neotropics. Smithsonian Institute Press, Washington, DC, USA.

Ellisor, J. E., and W. F. Harwell. 1969. Mobility and home range of collared peccary in southern Texas. Journal of Wildlife Management 33:425–427.

Emmons, L. H. 1987. Comparative feeding ecology of felids in a Neotropical rainforest. Behavioral Ecology and Sociobiology 20:271–283.

Escamilla, A., M. Sanvicente, M. Sosa, and C. Galindo-Leal. 2000. Habitat mosaic, wildlife availability, and hunting in the tropical forest of Calakmul, Mexico. Conservation Biology 14:1592–1601.

Fang, T. R., E. Bodmer, P. E. Puertas, P. Mayor-Aparicio, P. Pérez-Peña, R. Acero-Villanes, and D. T. S. Hayman. 2008. Certificación de pieles de pecaríes en la Amazonia Peruana. Wust Editions, Lima, Peru.

Fragoso, J. M. V. 1994. Large mammals and the community dynamics of an Amazonian rain forest. Dissertation, University of Florida, Gainesville, USA.

Fragoso, J. M. V. 1997. Desapariciones locales del baquiro labiado (*Tayassu pecari*) en la Amazonía: migración, sobre-cosecha o epidemia? Pages 309–312 *in* T. G. Fang, R. Bodmer, R. Aquino, and M. Valqui, editors. Manejo de fauna silvestre en la Amazonía. Universidad Nacional de la Amazonía Peruana, University of Florida, UNDP/GEF, and Instituto de Ecología, La Paz, Bolivia.

Fragoso, J. M. V. 1998. Home range and movement patterns of white-lipped peccary (*Tayassu pecari*) herds in the northern Brazilian Amazon. Biotropica 30:458–469.

Fragoso, J. M. V. 1999. Perception of scale and resource partitioning by peccaries: behavioral causes and ecological implications. Journal of Mammalogy 80:993–1003.

Fragoso, J. M. V. 2004. A long-term study of white-lipped peccary (*Tayassu pecari*) population fluctuation in northern Amazonia. Pages 286–296 *in* K. Silvius, R. E. Bodmer, and J. M. V. Fragoso, editors. People in nature: wildlife conservation in South and Central America. Columbia University Press, New York, New York, USA.

Gabor, T. M., and E. C. Hellgren. 2000. Variation in peccary populations: landscape composition or competition by an invader? Ecology 81:2509–2524.

Glanz, W. E. 1982. The terrestrial mammal fauna of Barro Colorado Island: censuses and long-term changes. Pages 455–468 *in* E. G. Leigh Jr., A. S. Rand, and D. M. Windsor, editors. The ecology of a tropical forest: seasonal rhythms and long-term changes. Smithsonian Institute Press, Washington, DC, USA.

Gómez, B., and O. Montenegro. 2012. Abundancia del pecarí de collar (*Pecari tajacu*) en dos áreas protegidas de la Guayana colombiana. Mastozoología Neotropical 19:163–178.

Gongora, J., S. Morales, J. E. Bernal, and C. Moran. 2006. Phylogenetic divisions among collared peccaries (*Pecari tajacu*) detected using mitochondrial and nuclear sequences. Molecular Phylogenetics and Evolution 41:1–11.

Gongora, J., R. Reyna-Hurtado, H. Beck, A. Taber, M. Altrichter, and A. Keuroghlian. 2011. *Pecari tajacu.* IUCN Red List of Threatened Species 2011:e.T41777A10562361. http://dx.doi.org/10.2305/IUCN.UK.2011–2.RLTS.T41777A10562361. Accessed 23 Nov 2015.

Gongora, J., A. Taber, A. Keuroghlian, M. Altrichter, R. E. Bodmer, P. Mayor, C. Moran, C. S. Damayanti, and S. González. 2007. Re-examining the evidence for a "new" peccary species, "*Pecari maximus*," from the Brazilian Amazon. PPHSG Newsletter 7:19–26.

González-Marín, R. M., S. Gallina, S. Mandujano, and M. Weber. 2008. Densidad y distribución de ungulados silvestres en la Reserva Ecológica El Edén, Quintana Roo, México. Acta Zoológica Mexicana 24:73–93.

Hall, E. R. 1981. The mammals of North America. Second edition. Volume 2. John Wiley and Sons, New York, New York, USA.

Hellgren, E. C., and J. A. Bissonette. 2003. Collared peccary. Pages 867–876 *in* G. A. Feldhamer, B. C. Thompson, and J. A. Chapman, editors. Wild mammals of North America: biology, management and conservation. Second edition. Johns Hopkins University Press, Baltimore, Maryland, USA.

Judas, J., and O. Henry. 1999. Seasonal variation of home range of collared peccary in tropical rain forest of French Guiana. Journal of Wildlife Management 63:546–552.

Keuroghlian, A., A. Desbiez, R. Reyna Hurtado, M. Altrichter, H. Beck, A. Taber, and J. M. V. Fragoso. 2013. *Tayassu pecari.* IUCN Red List of Threatened Species. Version 2013.1. www.iucnredlist.org. Accessed 4 Sep 2015.

Keuroghlian, A., D. P. Eaton, and W. S. Longland. 2004. Area use by white-lipped and collared peccaries (*Tayassu pecari* and *Tayassu tajacu*) in a tropical forest fragment. Biological Conservation 120:411–425.

Kiltie, R. A. 1981a. Stomach content of rain forest peccaries (*Tayassu tajacu* and *T. pecari*). Biotropica 13:234–236.

Kiltie, R. A. 1981b. The significance of interlocking canines in the forest peccaries (Tayassuidae). Journal of Mammalogy 62:459–469.

Kiltie, R. A., and J. Terborgh. 1983. Observations on the behavior of rain forest peccaries in Peru: why do white-lipped peccaries form herds? Zeitschrift für Tierpsychologie 62:241–255.

Leopold, A. S. 1959. Wildlife of Mexico: the game birds and mammals. University of California Press, Berkeley, USA.

Low, W. A. 1970. The influence of aridity on reproduction of the collared peccary (*Dicotyles tajacu* Linn.) in Texas. Dissertation, University of British Columbia, Vancouver, Canada.

Mandujano, S. 1999. Variation in herd size of collared peccaries in a Mexican tropical forest. Southwestern Naturalist 44:199–204.

Mandujano, S. 2007. Carrying capacity and potential production of ungulates for human use in a Mexican tropical dry forest. Biotropica 39:519–524.

Mandujano, S., and L. E. Martínez Romero. 1997. Fruit fall caused by chachalacas (*Ortalis poliocephala*) on red mombin trees (*Spondias purpurea*): impact on terrestrial fruit consumers, especially the white-tailed deer (*Odocoileus virginianus*). Studies on Neotropical Fauna and Environment 32:1–3.

March, I. J. 1990. Evaluación de hábitat y situación actual del pecarí de labios blancos *Tayassu pecari* en México. Thesis, Universidad Nacional de Costa Rica, Heredia, Costa Rica.

March, I. J. 1993. The white-lipped peccary (*Tayassu pecari*). Pages 13–22 *in* W. L. R. Oliver, editor. Pigs, peccaries and hippos. IUCN, Gland, Switzerland.

March, I. 2005. Pecarí de labios blancos. Pages 522–524 *in* G. Ceballos and G. Oliva, editors. Los mamíferos silvestres de México. CONABIO and Fondo de Cultura Económica, México, D. F., México.

March, I., and S. Mandujano. 2005. Pecarí de collar. Pages 524–527 *in* G. Ceballos and G. Oliva, editors. Los mamíferos silvestres de México. CONABIO and Fondo de Cultura Económica, México, D. F., México.

Martínez Romero, L. E., and S. Mandujano. 1995. Habitos alimentarios del pecari de collar (*Pecari tacaju*) en un bosque tropical caducifolio de Jalisco, México. Acta Zoológica Mexicana 64:1–20.

Mason-Romo, E. D., E. P. Villa-Mendoza, G. Rendon-Alquicira, and D. Valenzuela-Galván. 2008. Primer registro del pecarí de collar (*Pecari tajacu*) en el estado de Morelos. Revista Mexicana de Mastozoología 12:170–175.

Mayer, J. J., and R. M. Wetzel. 1987. *Tayassu pecari*. Mammalian Species 293:1–7.

Moreira, J. 2009. Patrones diarios de actividad, tamaño y abundancia relativa de manadas de jabalí *Tayassu pecari* (Link, 1795), en el Parque Nacional Mirador-Rio Azul, Petén, Guatemala. Thesis, Universidad San Carlos, Guatemala, Guatemala.

Naranjo, E. J. 2002. Population ecology and conservation of ungulates in the Lacandon Forest, Mexico. Dissertation, University of Florida, Gainesville, USA.

Naranjo, E. J., S. A. Amador-Alcala, F. A. Falconi-Briones, and R. Reyna-Hurtado. 2015. Distribución, abundancia y amenazas a las poblaciones de tapir centroamericano (*Tapirus bairdii*) y pecarí de labios blancos (*Tayassu pecari*) en México. Therya 6:227–249.

Naranjo, E. J., and R. E. Bodmer. 2007. Source-sink systems and conservation of hunted ungulates in the Lacandon Forest, Mexico. Biological Conservation 138:412–420.

Novack, A., M. Main, M. Sunquist, and R. Labinsky. 2005. Foraging ecology of jaguar (*Panthera onca*) and puma (*Puma concolor*) in hunted and non-hunted sites within the Maya Biosphere Reserve, Guatemala. Journal of Zoology 267:167–178.

Peres, C. A. 1996. Population status of white-lipped *Tayassu pecari* and collared peccaries *T. tajacu* in hunted and unhunted Amazonian forest. Biological Conservation 77:115–123.

Pérez-Cortéz, S., and R. Reyna-Hurtado. 2008. La dieta de los pecaríes (*Pecari tajacu* y *Tayassu pecari*) en la región de Calakmul, Campeche, México. Revista Mexicana de Mastozoología 12:17–42.

Ramírez-Barajas, P. J., and E. J. Naranjo. 2007. La cacería de subsistencia en una comunidad de la zona maya, Quintana Roo, Mexico. Etnobiología 5:65–85.

Reyna-Hurtado, R. 2007. Social ecology of the white-lipped peccary (*Tayassu pecari*) in Calakmul Forest, Campeche, Mexico. Dissertation, University of Florida. Gainesville, USA.

Reyna-Hurtado, R. 2009. Conservation status of the white-lipped peccary (*Tayassu pecari*) outside the Calakmul Biosphere Reserve in Campeche, Mexico: a synthesis. Tropical Conservation Science 2:159–172.

Reyna-Hurtado, R., C. A. Chapman, S. Calme, and E. Pedersen. 2012. Searching in heterogeneous environments: foraging strategies in the white-lipped peccary (*Tayassu pecari*). Journal of Mammalogy 93:124–133.

Reyna-Hurtado, R., J. Moreira, M. Briceño, M. Sanvicente, R. McNab, R. García, M. Mérida, E. Sandoval, G. Ponce, and D. Hyeroba. 2014. White-lipped peccaries with skin problems in the Maya Forest. Suiform Soundings 13:28.

Reyna-Hurtado, R., E. Naranjo, C. A. Chapman, and G. W. Tanner. 2010. Hunting and the conservation of a social ungulate: the white-lipped peccary *Tayassu pecari* in Calakmul, Mexico. Oryx 44:88–96.

Reyna-Hurtado, R., E. Rojas-Flores, and G. W. Tanner. 2009. Home range and habitat preferences of white-lipped peccary groups (*Tayassu pecari*) in a seasonal tropical forest of the Yucatan Peninsula, Mexico. Journal of Mammalogy 90:1199–1209.

Reyna-Hurtado, R., A. Taber, M. Altrichter, J. Fragoso, A. Keuroghlian, and H. Beck. 2008. *Tayassu pecari*. 2008 IUCN Red List of Threatened Species. www.iucnredlist.org. Accessed 13 Nov 2012.

Reyna-Hurtado, R., and G. Tanner. 2005. Habitat preferences of an ungulate community in Calakmul Forest, Campeche, Mexico. Biotropica 37:676–685.

Reyna-Hurtado, R., and G. W. Tanner. 2007. Ungulate relative abundance in hunted and non-hunted sites in Calakmul Forest (southern Mexico). Biodiversity and Conservation 16:743–757.

Robinson, J. G., and J. F. Eisenberg. 1985. Group size and foraging habits of collared peccary (*Tayassu tajacu*). Journal of Mammalogy 66:153–155.

Robinson, J. G., and K. H. Redford. 1986. Body size, diet and population density of Neotropical forest mammals. American Naturalist 128:665–680.

Schaller, G. 1983. Mammals and their biomass on a Brazilian ranch. Arquivos de Zoologia 31:1–36.

Schweinsburg, R. E. 1971. Home range, movements and herd integrity of the collared peccary. Journal of Wildlife Management 35:455–460.

SEMARNAT. 2010. Norma Oficial Mexicana NOM-059-SEMARNAT- 2010, Protección ambiental—Especies nativas de México de flora y fauna silvestres—Categorías de riesgo y especificaciones para su inclusión, exclusión o cambio—Lista de especies en riesgo. Segunda Sección. DOF.

Sowls, L. K. 1997. Javelinas and the other peccaries: their biology, management and use. Second edition. Texas A&M University Press, College Station, USA.

Stearman, A. M. 1992. Neotropical hunters and their neighbors: Sirionó, Chimane and Yuquí hunting on the Bolivian frontier. Pages 108–128 *in* K. Redford and C. Padoch, editors. Conservation of Neotropical forests (working from traditional use). Columbia University Press, New York, New York, USA.

Supple, V. C. 1983. The dynamics of collared peccary dispersion into available range. Federal Aid in Wildlife Restoration, Project W-78. Arizona Game and Fish Department, Phoenix, USA.

Van Roosmalen, M. G. M., L. Frenz, P. van Hooft, H. H. de Iong, and H. Leirs. 2007. A new species of living peccary (Mammalia: Tayassuidae) from the Brazilian Amazon. PPHSG Newsletter 7:15–19.

Weber, M. 2000. Effects of hunting on tropical deer populations in southeastern Mexico. Thesis, University of London, United Kingdom.

Weber, M., G. García-Marmolejo, and R. Reyna-Hurtado. 2006. The tragedy of the commons Mexican style: a critique to the Mexican UMAs concept as applied to wildlife management and use in southeastern Mexico. Wildlife Society Bulletin 34:1480–1488.

Wetzel, R. M., R. E. Dubos, R. L. Martin, and P. Myers. 1975. *Catagonus*, an "extinct" peccary, alive in Paraguay. Science 189:379–381.

Zervanos, S. M. 1975. Seasonal effects of temperature on the respiratory metabolism of the collared peccary (*Tayassu tajacu*). Comparative Biochemistry and Physiology 50:365–371.

Zervanos, S. M., and G. I. Day. 1977. Water and energy requirements of captive and free-living collared peccaries. Journal of Wildlife Management 41:527–532.

Zervanos, S. M., and N. F. Hadley. 1973. Adaptational biology and energy relationships of the collared peccary (*Tayassu tajacu*). Ecology 54:759–774.

Tapir in Mexico

Eduardo J. Naranjo

15

Baird's tapir (*Tapirus bairdii*), the only species in the order Perissodactyla and family Tapiridae that occurs in Mexico, is commonly known as *danta* or *anteburro* in rural areas. Subspecies have not been described. It is the largest terrestrial native mammal in southern tropical Mexico (March and Naranjo 2005, Lira et al. 2014b). It is listed as endangered by the Mexican government (SEMARNAT 2010, IUCN 2015). This species is relatively unstudied compared to many other Mexican mammals. Tapirs are difficult to study because they occur in low densities, and their secretive behavior makes them difficult to observe (Naranjo 2009). Many aspects of their life history and behavior have not been studied. The aim of this chapter is to provide a preliminary synthesis of the status, distribution, and life history of Baird's tapir, based on the few studies conducted in southern Mexico and Central America.

Historical Perspective and Value

Living tapirs resemble ancestral forms of perissodactyls (odd-toed ungulates that include tapirs, horses, and rhinoceroses). Fossil records suggest that these mammals originated in the Early Eocene of Asia and spread into the Americas (Hershkovitz 1954, Eisenberg 1981). Tapirs occupied large portions of North America and Eurasia by the Pliocene, when they entered South America (Eisenberg 1989, 1997). They had disappeared from most of that range by the Late Pleistocene (Feldhamer et al. 1999). Since then, the global distribution of tapirs has been diminishing, largely because of human activities such as deforestation and overhunting (Brooks et al. 1997). Current distribution of tapirs includes tropical rainforests of Southeast Asia (Malayan tapir, *T. indicus*); tropical and subtropical forests of Central America (Baird's tapir) and South America (Brazilian tapir, *T. terrestris*); and paramo ecosystems of the northern Andes (Mountain tapir, *T. pinchaque*). Baird's tapir plays an important role in the dynamics of tropical ecosystems as seed dispersers and predators, and they have been used as a food source in rural communities of Mesoamerica since pre-Columbian times (Eisenberg 1989, O'Farrill et al. 2006, Naranjo 2009).

Distribution

Baird's tapir was formerly distributed throughout the tropical, subtropical evergreen, and subdeciduous forests of southern and eastern Mexico, from sea level to about 2,000 m (6,560 ft) (Naranjo 2009). Deforestation and habitat fragmentation have reduced current tapir distribution in Mexico to extensive tropical and montane forests remaining in the states of Campeche, Chiapas, Oaxaca, Quintana Roo, and Veracruz (March and Naranjo 2005, Naranjo et al. 2015). Areas with verified tapir presence in southeastern Mexico are the Calakmul, Balam-Ku, and Balam-Kin reserves in eastern Campeche; the Sian Ka'an Biosphere Reserve and the tropical forests of central and southern Quintana Roo; the El Ocote, El Triunfo, La Sepultura, Lacantún, and Montes Azules Biosphere Reserves, the Lacandon Forest, and Sierra Madre in Chiapas; Uxpanapa in Veracruz; and Chimalapas, Lagunas de Chacahua National Park, Sierra Veinte Cerros, and Sierra Mixe in Oaxaca (Naranjo 2009, Naranjo et al. 2015) (fig. 15.1). Virtually all areas in Mexico where tapirs are present have at least 100 km² (39 mi²) of habitat in good condition. The areas with the highest potential to shelter large (but not necessarily viable) tapir populations are Calakmul–Balam-Ku–Balam-Kin in Campeche; Sian Ka'an in central Quintana Roo; Montes Azules–Lacantún in Chiapas; and Chimalapas in Oaxaca (Naranjo et al. 2015). The existence of small tapir populations is also possible in the Pantanos de Centla Biosphere Reserve along the border between Tabasco and Guatemala, on the western coast of Oaxaca and eastern coast of Guerrero, in Los Petenes and Laguna de Terminos

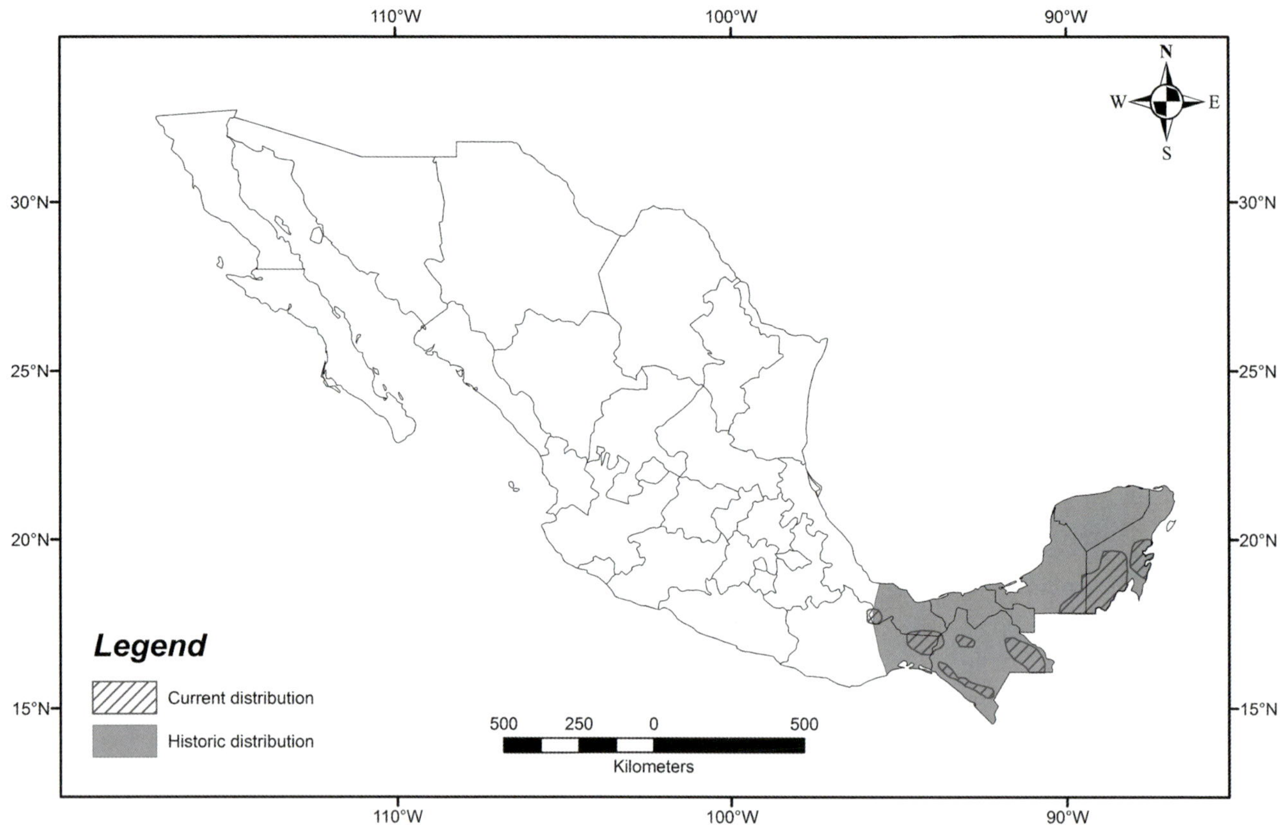

Figure 15.1. Distribution of Baird's tapir in Mexico. Historical distribution is shown in gray and current distribution in black hatched lines.

coastal forests in Campeche, on the southern tip of Yucatán, and in the Yum Balam and El Eden reserves in northern Quintana Roo (Naranjo 2009).

Description

The body of Baird's tapir is robust and cylindrical, its tail is short, and its head is noticeably large, with a long upper lip forming a flexible proboscis (Eisenberg 1989). Limbs are thick and short, with four toes on the forefeet and three toes on the hind feet (fig. 15.2). All toes end in wide, thick black nails that form hooves (Reid 1997). The thick, short body hair is dark brown to dark gray in adults, with light gray tones on the chest, throat, and tips of the ears. Young tapirs are reddish brown, with a pattern of white stripes and spots on their back and sides. Adult length may reach up to 2 m (6.5 ft), and biomass usually ranges between 150 and 300 kg (330 and 660 lb) (March and Naranjo 2005, Lira et al. 2014b). At a distance, both females and males appear to be about the same size and weight. However, available records show that female tapirs are slightly heavier than males (Padilla and Dowler 1994, Naranjo 2002).

Figure 15.2. Baird's tapir exhibiting alert behavior in the rainforest border of Montes Azules Biosphere Reserve, Chiapas, Mexico. Photo by Eduardo Naranjo.

Life History

Diet

In Mexico, tapirs are known to feed on parts of at least 98 plant species classified in 50 families (Naranjo 2009). Plant families frequently represented in tapir feces are Moraceae, Rubiaceae, Fabaceae,

Asteraceae, and Solanaceae (Cruz 2001, Rivadeneyra 2007). Plant parts consumed by this ungulate include fruits, leaves, shoots, bark, and flowers (Naranjo 1995, 2009; Olmos 1997). Depending on the availability of food items, tapirs can shift their foraging strategy among habitat types and seasons. The most noticeable changes in proportions of food items ingested by Baird's tapir throughout the year are those related to fruit consumption. Fruits (2–18%) usually constitute a smaller proportion of tapir food than leaves and other fiber sources (82–98%) (Naranjo and Cruz 1998, Rivadeneyra 2007). Tapirs are important seed dispersers or seed predators, depending on the plant species. Bulky seeds such as those of *Licania platypus* and *Pouteria sapota* are chewed and spat, but not ingested. In contrast, smaller seeds such as those of *Bactris balanoidea*, *Brosimum alicastrum*, *Ficus* spp., *Manilkara zapota*, and *Spondias mombin* are either ingested and later defecated without damage, or destroyed by chewing or digestion (Janzen 1981, Olmos 1997, Naranjo 2009). Some of the seeds consumed by tapirs maintain or increase their germinative power after passing through the tapir's digestive system. Cruz (2001) demonstrated this for *Acacia milleriana*, *Spondias mombin*, and *Ficus* spp. in Chiapas, as did O'Farrill et al. (2006) for *Manilkara zapota* in Campeche. In southeastern Mexico, oily fruits and seeds of trees such as *Attalea butyracea*, *Bactris balanoidea*, *Licania platypus*, *Manilkara zapota*, and *Pouteria sapota* are among the most frequently consumed by tapirs (O'Farrill et al. 2006, Rivadeneyra 2007, Naranjo 2009).

Reproduction

The mating behavior of Baird's tapir is still poorly known in the wild. Gestation lasts between 390 and 400 days (about 13 months) in captivity (Álvarez del Toro 1966). Females are able to mate when they are three to four years old, and they have short estrous periods (about 48 hours) every 50–80 days (Barongi 1993). The minimum period between litters is 17 months. They give birth to one young weighing around 10 kg (22 lb), which stays with its mother for about a year (Leopold 1959, Eisenberg 1989, Álvarez del Toro 1991). During the rut, males engage in intensive active behaviors such as walking, running, swimming, biting females on their feet and back, and repeatedly trying to copulate with them. Once a female has accepted a male, she may copulate up to five times (Epigmenio Cruz, Tuxtla Gutierrez Zoo, Mexico, personal communication). Sex ratios are not uniformly even (1:1). In Montes Azules Biosphere Reserve (MABR) in Chiapas, females and males represented 66.7% and 33.3% of all identified tapirs, respectively (Naranjo 2002).

Physiology

Longevity

Well-maintained captive tapirs can live up to 30 years (Matola et al. 1997). However, factors such as disease, parasites, predators, and stress reduce the average longevity of Baird's tapirs to probably less than 20 years in the wild.

Growth

Captive newborn tapirs gain over 2 kg (4.4 lb) per week (Wilson and Wilson 1973). Juveniles reach their adult size at around 18 months of age. All the milk teeth erupt within the first month, and all the permanent teeth erupt after about 29 months (Matola et al. 1997).

Energy and water requirements

The large body mass of tapirs, coupled with their habitat preferences, requires them to spend most of their active hours (about 90%) searching for food. Although their main food items are shoots and leaves, fruits constitute their most important source of calories, and these are actively sought throughout the forest floor (O'Farrill et al. 2006, Rivadeneyra 2007, Naranjo 2009). High-quality tapir habitat always includes rivers, streams, lagoons, ponds, and swamps, where they may spend several hours a day for thermal regulation as well as for relief from biting and stinging insects (March and Naranjo 2005). Aquatic habitats are also important as escape cover from potential predators such as humans and large felines.

Behavior

Baird's tapirs tend to be solitary and may be active both nocturnally and diurnally. Nonetheless, they travel much longer distances during the first hours of darkness (Foerster 1998). In areas close to human settlements, and where they are subjected to hunting pressure, tapirs may become almost completely nocturnal to avoid contact with humans and domestic animals (Naranjo and Cruz 1998). Tapirs are excellent swimmers, and they often travel along watercourses. On dry land, they frequently move through a well-defined trail system in search of food or water. Although these ungulates are shy and try to avoid encounters with humans, in extreme situations they may attack, especially when females with offspring are chased (March and Naranjo 2005).

The primary predators of tapirs are humans, jaguars, pumas, and large crocodiles (Eisenberg 1989, Álvarez del Toro 1991, Naranjo 2009).

Marking

An atypical aspect of habitat use by Baird's tapir relates to its defecation habits. Tapirs frequently defecate in shallow water; however, they sometimes use particular spots on dry land, forming latrines or defecation sites (Fragoso 1997, Naranjo and Cruz 1998). In areas with abundant streams and ponds, most defecation (> 90%) occurs in permanent or seasonal water bodies (Naranjo 2009). Most of the observed defecation sites at MABR were in isolated shallow pools along clear streams and rivers surrounded by dense vegetation (Naranjo 2002). In areas where water is scarce, a significant proportion of latrines are usually observed on dry land, especially along mountain crests. Tapirs probably use these high places as both marking sites and corridors to move between different habitat types (Naranjo and Cruz 1998). Urine is often also used for territorial marking, especially by adult males, which are capable of spraying urine as far as 5 m (16.4 ft) along their trails (Kuehn 1986).

Communication

Senses of smell and hearing, which facilitate nonvisual intraspecific communication, are well developed in tapirs. Acoustic communication consists primarily of high-pitched vocalizations similar to whistles, as well as clicking noises and nasal snorts, depending on the situation (Hunsaker and Hahn 1969). Tapirs also produce olfactory signals by spraying urine over their territories and defecating in either water bodies or dry latrines (Naranjo 2009).

Spacing and movements

The only estimate of tapir home range in Mexico (0.67 km^2, 0.26 mi^2) was reported by Naranjo and Bodmer (2002), who monitored two radio-collared females in MABR for about two months. In Costa Rica, home ranges measured between 0.75 and 1.8 km^2 (0.29 and 0.7 mi^2) (Williams 1984, Foerster and Vaughan 2002). Variations in tapir movements and home range size may be related to differences in habitat quality (e.g., food and water availability) (Naranjo 2009, Lira et al. 2014a). Aquatic habitats are an important component of their home ranges. Foerster and Vaughan (2002) found variations among individual movements: 379–720 m (1,243–2,362 ft) per night, and home ranges of 1.6 km^2 (0.6 mi^2) for males and 1.02 km^2 (0.39 mi^2) for females. There was considerable overlap in home ranges among individuals.

Population Dynamics

Density

Available densities obtained from line transect sampling in Mexico ranged from 0.07 to 0.24 tapirs/km^2 (0.18 to 0.63 tapirs/mi^2) (Lira et al. 2004, Naranjo and Bodmer 2007). Track-based indices of abundance varied between 0.24 and 1.07 tracks/km (0.38 and 1.72 tracks/mi) (Naranjo 2009). These variations may result from differences in topography, moisture, dominant vegetation type, food and water availability, presence of domestic animals, hunting, and human presence, among other factors (Naranjo 2009). In general, density tends to be higher in unhunted sites, as shown by Naranjo and Bodmer (2007). These authors found densities of 0.24 ± 0.09 tapirs/km^2 (0.63 ± 0.23 tapirs/mi^2) in unhunted sites of Montes Azules Biosphere Reserve, in contrast to densities of only 0.05 ± 0.04 tapirs/km^2 (0.13 ± 0.11 tapirs/mi^2) in persistently hunted sites around protected areas in the Lacandon Forest, Chiapas.

Population Trends and Current Status

The low density of tapirs in Mexico implies that few large protected areas could potentially shelter viable populations (Naranjo et al. 2015). Available density and habitat assessments indicate that a population size of 2,600 tapirs has been estimated in Mexico (Naranjo 2009). If this estimate is credible, the largest remaining tapir populations in Mexico occur in the following sites: Calakmul–Balam-Ku–Balam-Kin (*n* = 800); Montes Azules–Lacantún (*n* = 600); Sian Ka'an–central Quintana Roo (*n* = 450); and Chimalapas (*n* = 450). Based on the criteria of Traill et al. (2007), none of these populations are viable. However, some of these populations may still have connectivity with others, as is the case with the corridors in Calakmul–Sian Ka'an, Chimalapas–Uxpanapa–El Ocote, and Montes Azules–Sierra del Lacandon, Guatemala (Naranjo 2009).

Productivity and Survival

The low productivity of tapirs is probably a result of their (1) long period of sexual immaturity; (2) small litter size (one young in most cases) (fig. 15.3); and (3) long gestation (13 months) and lactating (8–10 months) periods (Álvarez del Toro 1966, March and Naranjo 2005). Survival rates are high for tapir populations inhabiting high-quality habitats without human disturbance. Besides humans (their main predator), jaguars, pumas, and crocodiles are significant predators, especially of young and juvenile tapirs (Eisenberg 1989, Álvarez del Toro 1991).

Unhunted tapir populations generally have a large proportion of adults, while juveniles and young occur in small percentages (Montenegro 1999). Proportions of adult, juvenile, and young individuals from an unhunted site in MABR were 85.8%, 7.1%, and 7.1%, respectively (Naranjo 2009). Reasons for the low numbers of juveniles and young in tapir populations may be that (1) the species has a very low growth rate; (2) individuals are long-lived (they can live up to 30 years); and (3) natural mortality rates tend to be very low in the absence of human disturbance (Bodmer et al. 1997, Matola et al. 1997).

Population Management

Because tapirs occur at low densities under natural conditions, and populations recover slowly from heavy hunting pressure, population management often consists of strategies to maintain or increase their numbers, either by protecting or improving extant habitat (see next section), or by reintroducing the species into areas where it has been extirpated. Line transect sampling, track counts, camera traps, and interviews of local people are the recommended monitoring methods for this species (Naranjo 2000, 2009; Lira et al. 2014a) (fig. 15.4).

Current socioeconomic conditions in southern Mexico (e.g., rapid economic development and expanding human population) present obstacles to reintroducing tapirs in areas where they have been extirpated. However, a few recently protected areas may offer an opportunity for their reestablishment. Captive-breeding programs in public and private zoos can be an important source of individuals for reestablishment programs. These programs can benefit from the protocols for tapir husbandry and reintroduction that have been developed by the IUCN Tapir Specialist Group (Médici et al. 2008; protocols available at www.tapirspecialistgroup.org).

Habitat Requirements

Tapir habitat preference studies show that tapirs prefer (1) a high density of permanent water sources;

Figure 15.3. Baird's tapir newborn at Tuxtla Gutierrez Zoo, Chiapas, Mexico. Photo by Eduardo J. Naranjo.

Figure 15.4. Baird's tapir in the Uxpanapa rainforest, Veracruz, Mexico. Photo by Eduardo J. Naranjo.

(2) a diverse and dense understory (which implies abundant food) (fig. 15.5); (3) large extensions of riparian vegetation; (4) low fire incidence; and (5) low hunting pressure and minimal human presence (Muench 2001, Naranjo 2002, 2009, Tobler et al. 2006, Lira et al. 2014a). Tapirs frequently use lowland, mature tropical evergreen forests as well as palm swamps, which offer abundant high-quality food and aquatic habitats, in addition to resting cover (Naranjo 2002).

Most tapir habitats in Mexico currently occur within protected areas, and good-quality habitat remaining outside protected areas is increasingly being transformed for human use because of its potential for agriculture, cattle grazing, and timber extraction (Naranjo et al. 2015). These habitat transformations are currently occurring near Los Chimalapas, Calakmul, El Triunfo, Lacantún, La Sepultura, Montes Azules, Selva El Ocote, and Sian Ka'an Biosphere Reserves (Lira et al. 2014a, Naranjo et al. 2015). Loggers, farmers, and ranchers continue to receive government subsidies to expand habitat-destroying activities and increase the removal of native vegetation (Naranjo 2009). To reduce these impacts, managers of protected areas should invest

Figure 15.5. Baird's tapir sniffing banana leaves (Musa sp.) in the rainforest border of Montes Azules Biosphere Reserve, Chiapas, Mexico. Photo by Eduardo Naranjo.

time and resources in outreach and cooperative programs focused on avoiding further deforestation and overhunting by neighboring communities.

Habitat Management

In southern Mexico, a high proportion of the original cover (primarily tropical forest) has been destroyed or fragmented as well as defaunated during the last 50 years (March and Flamenco 1996). As a rule, tapir populations decline as their habitats are progressively fragmented (Naranjo and Bodmer 2007). This has occurred during the last several decades in forested areas of the Mexican states currently or formerly sheltering tapir populations, such as Oaxaca, Quintana Roo, Tabasco, Veracruz, and Yucatán. In these areas, major threats include forest clearings to create livestock pastures, slash-and-burn agriculture, soil erosion by rain and wind on steep slopes, pollution of water bodies by pesticides and fertilizers, and competition for forage and space with domestic animals such as cows, horses, and pigs (Naranjo 2009, Naranjo et al. 2015).

Effective habitat protection is the most important management practice to maintain sustainable tapir populations. The creation of preserve networks and of corridors between existing protected areas would also help improve habitat availability for tapirs in Mexico. Law enforcement is needed to curb environmentally destructive practices (e.g., uncontrolled hunting, extensive cattle grazing, and burning) within and around protected areas, as these practices constitute a continuing threat in southeastern Mexico and are causing tapir populations to become increasingly isolated (Naranjo 2009). Isolation of tapir populations is especially prevalent in the Sierra Madre of Chiapas, the forests of Oaxaca and Veracruz, and the northern Yucatán Peninsula. Further isolation of these populations can be mitigated by encouraging connectivity among forest fragments through cooperative programs to establish community reserves, and promoting environmentally friendly agricultural and land-use practices such as agroforestry systems (e.g., organic cacao and shade coffee plantations), sustainable cattle grazing, and sustainable harvest of native plant species around protected areas (Naranjo 2001, 2009).

Baird's tapir can be a focal species for ecotourism. However, excessive numbers of tourists in protected areas may be counterproductive for tapirs and other wildlife. But when properly planned and managed, ecotourism can be a nonconsumptive alternative for promoting the conservation and sustainable management of tapirs and other vulnerable species.

Management Challenges in Mexico

Habitat and population management practices as well as ecological research are necessary for sound tapir conservation and management. These measures will not be sufficient, however, if local communities are not involved and do not gain some benefit from them. Many residents of communities within or around protected areas where tapirs occur ignore the existence of such preserves (Naranjo 2002). It is essential to promote the participation of these communities in managing and conserving wildlife populations and their habitats. Baird's tapir populations cannot sustain commercial hunting, but the sustainable harvest of species with higher productivity (e.g., armadillos, peccaries, and white-tailed deer) can fulfill the needs of communities for subsistence hunting in a comanagement system (Bodmer and Puertas 2000).

Threats to Baird's tapir and its habitat in Mexico are complex and challenging. Local economic development programs, such as the training and hiring of workers as park guards, environmental educators, or field assistants in research and monitoring projects, are essential. Other residents may be helped by local, state, or national government programs encouraging agroforestry or ecotourism projects around protected areas.

Finally, an essential element in fostering wise management and conservation of tapirs and their habitats in Mexico is the training of professionals. In this sense, the growing number of conservation-oriented graduate and undergraduate programs and courses available in academic institutions throughout southern Mexico is encouraging. Environmental education and communication programs should also be strongly supported and encouraged by federal and state governments in the elementary and high school systems in both urban and rural areas, particularly those near tapir habitats.

Literature Cited

Álvarez del Toro, M. 1966. A note on the breeding of Baird's tapir at Tuxtla Gutiérrez Zoo. International Zoo Yearbook 6:196–197.

Álvarez del Toro, M. 1991. Los mamíferos de Chiapas. Second edition. Gobierno del Estado de Chiapas, Tuxtla Gutiérrez, México.

Barongi, R. A. 1993. Husbandry and conservation of tapirs. International Zoo Yearbook 32:7–15.

Bodmer, R. E., J. F. Eisenberg, and K. H. Redford. 1997. Hunting and the likelihood of extinction of Amazonian mammals. Conservation Biology 11:460–466.

Bodmer, R. E., and P. E. Puertas. 2000. Community-based co-management of wildlife in the Peruvian Amazon. Pages 395–412 *in* J. G. Robinson and E. L. Bennett, editors. Hunting for sustainability in tropical forests. Columbia University Press, New York, New York, USA.

Brooks, D. M., R. E. Bodmer, and S. Matola, editors. 1997. Tapirs: status survey and conservation action plan. IUCN/SSC Tapir Specialist Group, Gland, Switzerland, and Cambridge, United Kingdom.

Cruz, E. 2001. Hábitos de alimentación e impacto de la actividad humana sobre el tapir (*Tapirus bairdii*) en la Reserva de la Biósfera La Sepultura, Chiapas, México. Thesis, El Colegio de la Frontera Sur, San Cristóbal de Las Casas, Chiapas, México.

Eisenberg, J. F. 1981. The mammalian radiations: an analysis of trends in evolution, adaptation, and behavior. University of Chicago Press, Illinois, USA.

Eisenberg, J. F. 1989. Mammals of the Neotropics. Volume 1. The northern Neotropics. University of Chicago Press, Illinois, USA.

Eisenberg, J. F. 1997. Introduction. Pages 1–2 *in* D. M. Brooks, R. E. Bodmer, and S. Matola, editors. Tapirs: status survey and conservation action plan. IUCN/SSC Tapir Specialist Group, Gland, Switzerland, and Cambridge, United Kingdom.

Feldhamer, G. A., L. C. Drickamer, S. H. Vessey, and J. F. Merritt. 1999. Mammalogy: adaptation, diversity, and ecology. Second edition. WCB/McGraw-Hill, New York, New York, USA.

Foerster, C. R. 1998. Ecología de la danta centroamericana (*Tapirus bairdii*) en un bosque lluvioso tropical de Costa Rica. Thesis, Universidad Nacional de Costa Rica, Heredia, Costa Rica.

Foerster, C. R., and C. Vaughan. 2002. Home range, habitat use, and activity of Baird's tapir in Costa Rica. Biotropica 34:423–437.

Fragoso, J. M. 1997. Tapir-generated seed shadows: scale-dependent patchiness in the Amazon rain forest. Journal of Ecology 85:519–529.

Hershkovitz, P. 1954. Mammals of northern Colombia. Preliminary report no. 7. Tapirs (genus *Tapirus*), with a systematic review of American species. Proceedings of the United States National Museum 103:465–496.

Hunsaker, D., and T. C. Hahn. 1969. Vocalization of the South American tapir (*Tapirus terrestris*). Animal Behavior 13:69–74.

IUCN. 2015. The IUCN Red List of Threatened Species. Version 2015–3. www.iucnredlist.org. Accessed 7 Nov 2015.

Janzen, D. H. 1981. Digestive seed predation by a Costa Rican Baird's tapir (*Tapirus bairdii*). Biotropica 13:59–63.

Kuehn, G. 1986. Tapiridae. Pages 931–934 *in* M. E. Fowler, editor. Zoo and wild animal medicine. W. B. Saunders, London, United Kingdom.

Leopold, A. S. 1959. Wildlife of Mexico: the game birds and mammals. University of California Press, Berkeley, USA.

Lira, I., M. Briones, and G. Sánchez. 2014a. Abundancia relativa, estructura poblacional, preferencia de hábitat y patrones de actividad del tapir centroamericano *Tapirus bairdii* (Perissodactyla: Tapiridae), en la Selva de Los Chimalapas, Oaxaca, México. Revista de Biología Tropical 62:1407–1419.

Lira, I., I. J. March, and E. J. Naranjo. 2014b. Baird's tapir. Pages 578–579 *in* G. Ceballos, editor. Mammals of Mexico. Johns Hopkins University Press, Baltimore, Maryland, USA.

Lira, I., E. J. Naranjo, D. M. Güiris, and E. Cruz. 2004. Ecología de *Tapirus bairdii* (Perissodactyla: Tapiridae) en la Reserva de la Biósfera El Triunfo (Polígono I), Chiapas, México. Acta Zoológica Mexicana 20:1–21.

March, I. J., and A. Flamenco. 1996. Evaluación rápida de la deforestación en las áreas naturales protegidas de Chiapas (1970–1993). El Colegio de la Frontera Sur, San Cristóbal de Las Casas, Chiapas, México.

March, I. J., and E. J. Naranjo. 2005. Tapir (*Tapirus bairdii*). Pages 496–497 *in* G. Ceballos and G. Oliva, editors. Los mamíferos silvestres de México. CONABIO and Fondo de Cultura Económica, México, D. F., México.

Matola, S., A. D. Cuarón, and H. Rubio-Torgler. 1997. Status and action plan of Baird's tapir (*Tapirus bairdii*). Pages 29–45 *in* D. M. Brooks, R. E. Bodmer, and S. Matola, editors. Tapirs: status survey and conservation action plan. IUCN/SSC Tapir Specialist Group, Gland, Switzerland, and Cambridge, United Kingdom.

Médici, P., P. R. Mangini, A. Goncalves, and L. A. Salas. 2008. Guidelines for tapir reintroductions and translocations. IUCN/SSC Tapir Specialist Group, Gland, Switzerland. www.tapirspecialistgroup.org. Accessed 7 Nov 2015.

Montenegro, O. L. 1999. Observaciones sobre la

estructura de una población de tapires (*Tapirus terrestris*) en el sureste de la Amazonía peruana. Pages 437–442 *in* T. G. Fang, O. L. Montenegro, and R. E. Bodmer, editors. Manejo y conservación de fauna silvestre en América Latina. Instituto de Ecología, La Paz, Bolivia.

Muench, C. E. 2001. Patrones de uso del hábitat del tapir (*Tapirus bairdii*) en dos localidades de la Selva Lacandona, Chiapas. Thesis, UNAM, México, D. F., México.

Naranjo, E. J. 1995. Hábitos de alimentación del tapir (*Tapirus bairdii*) en un bosque lluvioso tropical de Costa Rica. Vida Silvestre Neotropical 4:32–37.

Naranjo, E. J. 2000. Estimación de abundancia y densidad en poblaciones de fauna silvestre tropical. Pages 37–46 *in* E. Cabrera, C. Mercolli, and R. Resquín, editors. Manejo de fauna silvestre en la Amazonia y Latinoamérica. Fundación Moisés Bertoni and CITES, Asunción, Paraguay.

Naranjo, E. J. 2001. El tapir en México. Biodiversitas 36:9–11.

Naranjo, E. J. 2002. Population ecology and conservation of ungulates in the Lacandon Forest, Mexico. Dissertation, University of Florida, Gainesville, USA.

Naranjo, E. J. 2009. Ecology and conservation of Baird's tapir in Mexico. Tropical Conservation Science 4:140–158.

Naranjo, E. J., S. A. Amador, F. A. Falconi, and R. A. Reyna-Hurtado. 2015. Distribución, abundancia y amenazas a las poblaciones de tapir centroamericano (*Tapirus bairdii*) y pecarí de labios blancos (*Tayassu pecari*) en México. Therya 6:227–249.

Naranjo, E. J., and R. E. Bodmer. 2002. Population ecology and conservation of Baird's tapir (*Tapirus bairdii*) in the Lacandon Forest, Mexico. Tapir Conservation 11:25–33.

Naranjo, E. J., and R. E. Bodmer. 2007. Source-sink systems of hunted ungulates in the Lacandon Forest, Mexico. Biological Conservation 138:412–420.

Naranjo, E. J., and E. Cruz. 1998. Ecología del tapir en la Reserva de la Biósfera La Sepultura. Acta Zoológica Mexicana 73:111–125.

O'Farrill, G., S. Calme, and A. González. 2006. *Manilkara zapota*: a new record of a species dispersed by tapirs. Tapir Conservation 15:32–35.

Olmos, F. 1997. Tapirs as seed dispersers and predators. Pages 3–9 *in* D. M. Brooks, R. E. Bodmer, and S. Matola, editors. Tapirs: status survey and conservation action plan. IUCN/SSC Tapir Specialist Group, Gland, Switzerland, and Cambridge, United Kingdom.

Padilla, M., and R. C. Dowler. 1994. *Tapirus terrestris*. Mammalian Species 481:1–8.

Reid, F. A. 1997. A field guide to the mammals of Central America and southeast Mexico. Oxford University Press, New York, New York, USA.

Rivadeneyra, E. 2007. Hábitos alimentarios del tapir (*Tapirus bairdii*) en el sector sur de la Reserva de la Biósfera Montes Azules, Chiapas, México. Thesis, Universidad Autónoma de Yucatán, Mérida, México.

SEMARNAT. 2010. Norma Oficial Mexicana NOM-059-SEMARNAT-2010. Protección ambiental, especies nativas de flora y fauna silvestres de México, categorías de riesgo y especificaciones para su inclusión, exclusión o cambio, y lista de especies en riesgo. DOF, 30 de diciembre de 2010:1–78.

Tobler, M. W., E. J. Naranjo, and I. Lira-Torres. 2006. Habitat preferences, feeding habits and conservation of Baird's tapir in Neotropical montane oak forests. Pages 347–361 *in* M. Kappelle, editor. Ecology and conservation of Neotropical montane oak forests. Springer, Berlin, Germany.

Traill, L. W., C. J. Bradshaw, and B. W. Brook. 2007. Minimum viable population size: a meta-analysis of 30 years of published estimates. Biological Conservation 139:159–166.

Williams, K. D. 1984. The Central American tapir (*Tapirus bairdii*) in northwestern Costa Rica. Dissertation, Michigan State University, East Lansing, USA.

Wilson, R. A., and S. Wilson. 1973. Diet of captive tapirs. International Zoo Yearbook 13:213–217.

16 Mule Deer in Mexico

Louis A. Harveson, Alfonso Ortega-Sánchez, Reagan Gage, and Carlos H. Alcalá-Galván

Introduction

Mule deer (*Odocoileus hemionus*) are large (30 to 150 kg, 66 to 330 lb) ungulates in the desert mountains, grasslands, and coastal regions of northern and western Mexico. The common name *bura* in Mexico and the names "mule deer" and "black-tailed deer" are derived from their distinct anatomy. Our knowledge of the life history and ecology of mule deer is based primarily on studies conducted in the US (Swank 1958, Anthony and Smith 1977, Lawrence et al. 2004, Marshal et al. 2006); relatively few studies have been conducted in Mexico (Martínez Muñoz et al. 2003, Sánchez Rojas and Gallina 2003). Refer to Heffelfinger (2006) for a comprehensive review of desert mule deer ecology and management. This chapter provides an overview of the life history and ecology of mule deer in Mexico, based principally on published studies conducted in Mexico and the southwestern US.

Historical Perspective

Mule deer evolved from white-tailed deer about 11,000–9,000 years ago (Geist 1998). In Mexico, mule deer have experienced abundance, local extirpation, constant exploitation, and more recently, active conservation and management.

Extirpation and recovery

Leopold (1959) stipulated that the socioeconomic conditions in Mexico would limit most mule deer populations from achieving ecological carrying capacity because of continual subsistence hunting, primarily by rural populations. Mule deer have been subjected to many land-use practices. One major positive change has been the transfer of ejidos to the private sector (Stoleson et al. 2005), which has benefited wildlife (see chapters 1 and 21). Specifically, many large landowners now realize the value and economic importance of mule deer. Private landowners have been instrumental in protecting mule deer populations from illegal hunting, thus allowing them to increase in many regions of northern Mexico. Big game hunters prize desert mule deer for their majestic antlers, and they are willing to pay thousands of dollars for the opportunity to hunt them.

Since the late 1990s, there has been renewed interest in restoring desert mule deer populations. The New Mexico Department of Game and Fish, Arizona Game and Fish Department, and Texas Parks and Wildlife Department have played an active role in restoration efforts by providing surplus deer for Mexico. Private landowners and corporations have been the primary recipients of mule deer transplants in Mexico. More than 700 mule deer from Texas, New Mexico, and Arizona have been translocated to the states of Coahuila, Nuevo León, and Zacatecas (table 16.1).

Value

Heffelfinger (2006) provided a historical review of aboriginal human relationships with large mammals in the Desert Southwest. Mule deer were probably rare and are noticeably absent from archaeological sites as a food item. Tinkler (1978) noted that mule deer in the Sonoran Desert were never numerous and were hunted to the point that few isolated populations remained in the early 1900s. Intensive commercial hunting of mule deer and other large mammals in the Sonoran Desert supported the demand for meat by restaurants in Tijuana, Mexicali, and Ensenada. In the Chihuahuan Desert of Chihuahua and Coahuila, mule deer were also subjected to intensive hunting pressure (Tinkler 1978) to supply game meat for sale in the cities of Juárez, Chihuahua, Parral, Torreón, and Monterrey. In 1921, Tinkler (1978) observed a flatcar train at Sabinal with 78 mule deer carcasses destined for markets in Ciudad Juárez. Most mule deer herds in Chihuahua and Coahuila were

Table 16.1. Efforts to restore mule deer through translocation from the southwestern US to Mexico, 1990–2010.

Reintroduction						
Year	Species	Source	Locality	Male	Female	Total
1993–1994	*Odocoileus spp.*[a]	Unknown	Nuevo León	66	215	281
1997–1998	*O. hemionus*	Jeff Davis, Pecos, and Brewster Counties, Texas	Nuevo León, Coahuila	Unknown	Unknown	30
1999–2000	*Odocoileus spp.*[b]	Unknown	Unknown	20	41	61
2005–2006	*O. hemionus*	Unknown	Coahuila, Zacatecas	19	91	110
2006–2007	*O. hemionus*	Pecos and Brewster Counties, Texas	Coahuila	10	70	80
2007–2008	*O. hemionus*	Pecos and Brewster Counties, Texas	Coahuila	0	197	197
2008–2009	*O. hemionus*	Pecos and Brewster Counties, Texas	Coahuila	0	60	60

[a] Reported as O. virginianus.
[b] Species unknown.

historically abundant but were dramatically reduced, especially near populated areas. Other naturalists have also reported the overexploitation of mule deer in Mexico (Leopold 1959, Baker 1977, Challenger 1998). In Mexico, mule deer are a game species regulated by SEMARNAT. Through a permitting process, landowners registered with UMAs are allowed to legally harvest mule deer. However, because of the lack of financial resources to enforce game laws, mule deer continue to be illegally hunted.

Livestock production is the primary land use in mule deer habitats. Unfortunately, these regions face social and economic challenges that outweigh ecological considerations (Valdez et al. 2006), and most mule deer habitat in Mexico has been overgrazed. Only recently have landowners begun to prioritize mule deer (Rosas-Rosas et al. 2003, Valdez et al. 2006). In Sonora, some private landowners have focused their efforts on big game ranching operations, including mule deer, white-tailed deer, and bighorn sheep, and in some instances they have removed livestock to benefit wildlife populations (Valdez et al. 2006). In portions of Chihuahua and Coahuila, corporate landowners have also destocked rangelands to restore ecosystems, including restoration of mule deer herds (Martínez García 2009).

Taxonomy

Naturalists have recognized from 9 to 11 subspecies of mule deer in North America (Cowan 1956, Anderson and Wallmo 1984). Subspecies occurring in Mexico include the desert (*crooki*) mule deer (with the widest distribution occurring in northern Mexico in Coahuila, Chihuahua, and Sonora and extending into the US); Tiburón Island (*sheldoni*) mule deer (occurring exclusively on Tiburón Island); southern (*fulginata*) mule deer (occurring in southwestern California and northwestern Baja California); Cedros Island (*cerrosensis*) mule deer (occurring exclusively on Cedros Island); and peninsular (*peninsulae*) mule deer (occurring in southern Baja California). Molecular genetics studies have questioned the validity of most subspecies delineations. Latch et al. (2009) provided insight into the evolutionary history of mule deer and recognized several preliminary subspecies groupings based on haplotypes, including *hemionus*, *fulginata*, *crooki*, *eremicus*, *inyoensis*, *californicus*, *peninsulae*, and *sheldoni*, but subspecies designations remain unclear.

Distribution

Mule deer occur from Alaska to the southern tip of Baja California and as far east as the Great Plains of Kansas and Nebraska in the US. Desert mule deer in Mexico are associated with arid environments, including portions of the Chihuahuan and Sonoran Deserts (fig. 16.1). Previously published distribution maps for desert mule deer in Mexico generally include the states of Coahuila, Chihuahua, Sonora, and Baja California, and portions of northeastern Durango, northern Zacatecas, extreme western Nuevo León, extreme northwestern San Luis Potosí, and eastern Baja California Sur.

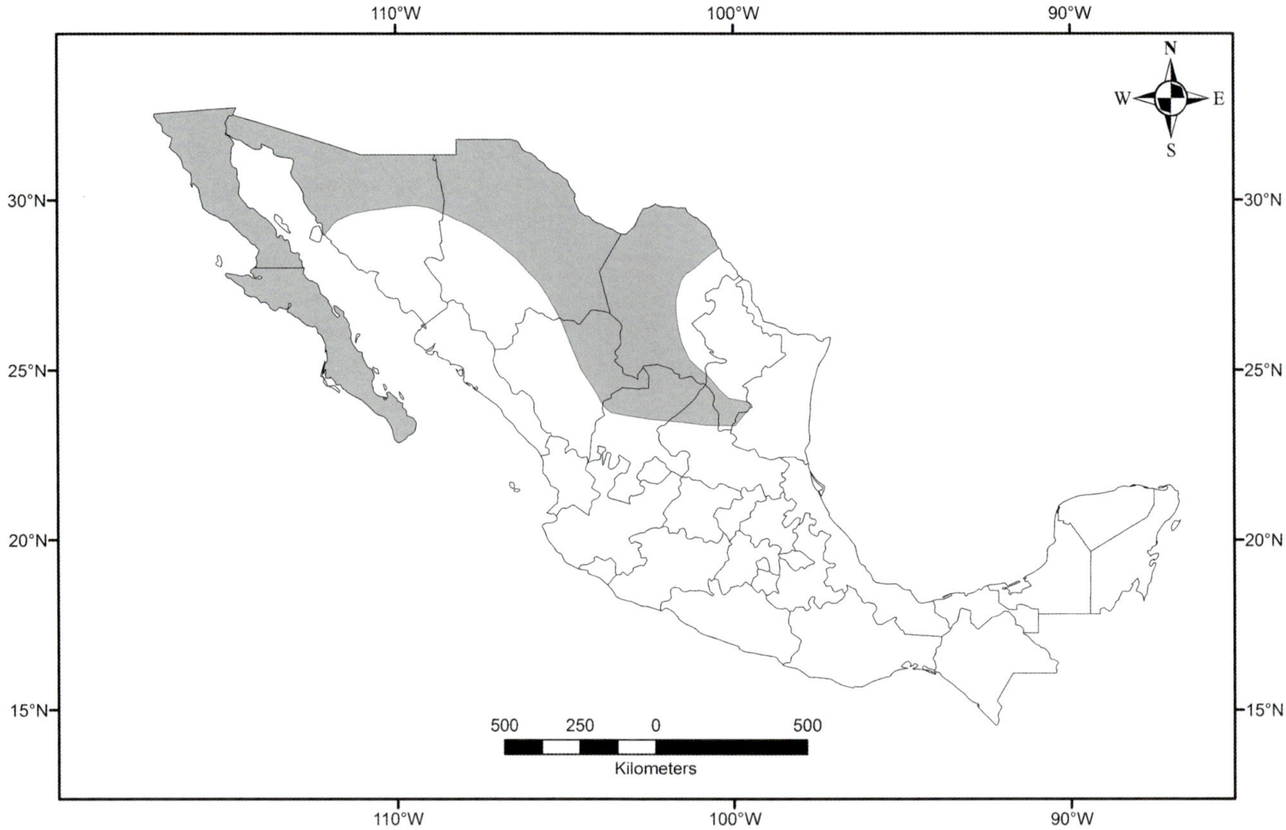

Figure 16.1. Distribution of mule deer in Mexico. Adapted from Geist (1998), Heffelfinger (2006), and Latch et al. (2009).

Description

Mule deer are sexually dimorphic, with adult bucks having larger body mass (90–115 kg, 198–253 lb) than does (60–75 kg, 132–165 lb). Females are antlerless, but males grow antlers that attain lengths of 100 cm (39.3 in). Mule deer does reach maximum weight at two years, and bucks at nine years (Anderson et al. 1974). Fawns average 3 kg (6.6 lb) at birth and grow rapidly during their first six months (Geist 1998).

Antlers begin growing in late spring (May) (fig. 16.2) and are hardened by fall (September). They are shed annually, usually in early spring (March). Antler characteristics vary greatly among individuals, but three general patterns of mule deer antler morphology have been proposed: (1) antlers with small or absent brow tines, (2) antlers that generally grow out and up, and (3) mature antlers in which the beams fork, and each fork develops an additional fork. Typical mule deer antlers include 8 points (4 points × 4 points without brow tines), or 10 points (5 points × 5 points with brow tines (fig. 16.3).

Figure 16.2. Juvenile mule deer in velvet at Rancho Tres Papalotes in Chihuahua, Mexico. Photo by Enrique Perez.

Except for the first six to nine months of life when

Figure 16.3. This radio-collared 10-point (5 × 5) buck exhibits the classic antler morphology of mule deer, including the double fork and small brow tines. Photo by Louis A. Harveson.

fawns have spotted coats, mule deer coloration varies from reddish (spring and summer) to grayish brown (fall and winter). Northern mule deer are generally darker than those in deserts. Mule deer in Baja California have a distinct black dorsal stripe (Heffelfinger 2006).

Mule deer have a black nose, a white or gray muzzle, and a black area on their forehead. Ears of mule deer are considerably larger than those of white-tailed deer (*O. virginianus*) or brocket deer (*Mazama* spp.), and their name is derived from their mule-like ears. The tails of mule deer are described as white and ropelike, with a black tip. Mule deer are also known for their white rump patch. Their strong legs allow them to cross rough and obstacle-filled landscapes.

Natural History

Diet

Mule deer have adapted their diet to many diverse habitats (Kufeld et al. 1973, Krausman et al. 1997), and they can use a broad variety of forage species and other diverse food items according to natural and human-influenced differences in the quality and quantity of vegetation available. Variation in diet composition should correspond to the dynamics of resource availability in the different ecological regions that mule deer inhabit.

Desert mule deer diets have been widely studied in the southwestern US (Krausman et al. 1997), including northern subdivisions of the Sonoran Desert (Urness et al. 1971, McCulloch 1973, Anthony 1976, Anthony and Smith 1977, Short 1977, Krausman et al. 1989, Marshal et al. 2004) and the Chihuahuan Desert (Anderson et al. 1965, Boeker et al. 1972, Krausman 1978, Leopold and Krausman 1987a). Few studies on mule deer diets have been conducted in Sonora (Alcalá Galván and Krausman 2012). References on the plant species mule deer have been observed to consume include Leopold (1959) for northern Mexico and northern Baja California, and Pérez Gil (1981) for Cedros Island.

Studies conducted in four different areas of central and western Sonora, Mexico, identified 96 plant species consumed by mule deer. Sixty-nine (72%) of those species had not previously been reported in desert mule deer diets (Alcalá Galván and Krausman 2012). Mule deer diet composition varies by area, season, and year. In the southwestern US (McCulloch 1973, Krausman et al. 1989, Krausman et al. 1997, Marshal et al. 2004) and Mexico (Alcalá Galván and Krausman 2012), browse is the dominant forage consumed by desert mule deer; browse species make up 77–88% of mule deer diets in Sonora, Mexico.

Desert mule deer consume more forbs during spring and summer. Anderson et al. (1965) found that forbs may become the most important forage in all seasons during wet years. Forbs made up an average of 5–10% of mule deer diets in central and western Sonora (Alcalá Galván and Krausman 2012). This fluctuation was related to the geographical and seasonal distribution of rainfall (Peek and Krausman 1996, Krausman et al. 1997). Use of cacti was highly variable among areas, but chainfruit cholla (*Opuntia fulgida*) was consumed consistently in all areas, and in most seasons.

As in desert mule deer diets in the southwestern US (Krausman et al. 1997), grasses appeared as the lowest forage class in mule deer diets throughout central and western Sonora in most seasons: however, higher consumption of native grasses (32% of the diet) occurred in one area of central Sonora during summer (Alcalá Galván and Krausman 2012). In that study, five native grasses accounted for more than 65% of the total grass consumption throughout the year. Nonnative buffelgrass (*Cenchrus ciliaris*) made up less than 5% of the summer diet of desert mule deer and accounted for less than 1% of the annual diet.

According to Alcalá Galván and Krausman (2012), this specific study area included good cover and well-established native grasses. Most of central Sonora was probably once covered by native perennial grasses and was classified as Sonoran savanna grassland (Brown 1994). Higher availability of grasses likely resulted in the increased consumption of grasses reported by Alcalá Galván and Krausman (2012). Other researchers have also reported high consumption of grass by mule deer. For example, Hungerford (1974) reported that grasses contributed up to 47% of the summer diets of mule deer on the Kaibab Plateau in Arizona, and Neff (1974) reported that grasses accounted for 32% or less of mule deer diets in the Beaver Creek area of north-central Arizona. The high use of grasses by mule deer is somewhat contrary to the findings of previous research on the physiological adaptations of mule deer; their small rumen-to-body-weight ratio was thought to limit their ability to digest grasses (Hanley 1982). Mubanga et al. (1985) stated that mule deer experienced digestive problems and limited their forage intake when fed with high proportions of grass under confined conditions.

Reproduction

Mating and breeding

When the mating season (rut) begins, dominant mule deer bucks begin to overlap the home ranges of doe-fawn groups. Bucks begin to monitor does to determine whether they are in estrus by using ritualized courtship behaviors (Geist 1981). Mule deer bucks are physically ready to reproduce starting in December, although most does do not come into estrus until February. After copulation, bucks may move over large distances to find another doe that is ready to breed (Weber and Galindo Leal 2005).

The timing of the mule deer breeding season varies across their range in Mexico. Leopold (1959) reported that the breeding season occurred from December to February in portions of Mexico, but it occurred from November to December in Baja California. However, in southern Baja California, Gallina-Tessaro et al. (1988) and Gallina et al. (2000) reported that mating occurred between December and late February. In the Colorado River delta, the rut occurs in February (Stone and Rhoads 1905). The primary time for mating on Cedros Island is September to November (Pérez Gil 1981). Gallina et al. (1992) reported that the rut for the subspecies *peninsulae* occurred from December to February.

Gestation

Gestation lasts about 200–207 days (Robinette et al. 1977) throughout the range of mule deer. Leopold (1959) reported that fawning occurs from late June to July in Baja California and from late July to August in Sonora. Stone and Rhoads (1905) reported fawning in August in the Colorado River delta. Similarly, Gallina (1989) found that in the Sierra de la Laguna Biosphere Reserve of Baja California, most fawning occurs in August, which coincides with the period when forage is most available. In southern Baja California, fawning occurs from July to August

(Gallina-Tessaro et al. 1988, Gallina et al. 2000). Pérez Gil (1981) documented fawning dates as early as April on Cedros Island.

Sex ratios

Sex ratios also vary, and adult sex ratios are directly related to harvest pressure. As with most ungulates, mule deer sex ratios approximate 1:1 at birth, as found by Gallina et al. (1992). Adult sex ratios are more variable. Gallina (1990) found a 54:100 buck to doe ratio in adults, whereas Álvarez Cárdenas (1995) calculated a buck to doe ratio of 75:100 using the same method. Both Gallina (1990) and Álvarez Cárdenas (1995) used fecal pellet morphometry to ascertain sex ratios. Most research has shown that sex ratios have little effect on fawn recruitment (Horejsi et al. 1988, McCulloch and Smith 1991).

Physiology

Longevity

The oldest age recorded for mule deer is 13 years (Heffelfinger 2006). In Montana, no mule deer older than 20 years were recorded (Matson 1988).

Growth

Mule deer fawns are weaned at two months of age (Dixon 1934, Heffelfinger 2006). While Nichol (1938) found that healthy mule deer fawns weigh 3.2–3.6 kg (7–7.9 lb) at birth, he also reported that weights doubled in two weeks and quadrupled in 30–40 days. He also reported that the spots on fawns disappeared in about 2.5 months.

Body mass data for mule deer on the Kaibab Plateau include the following average weights: yearlings = 46.3 kg (102 lb); 2.5 yr = 64.4 kg (142 lb); 3.5 yr = 75.3 kg (166 lb); 4.5–5.5 yr = 83.9 kg (185 lb); and 6 yr or more = 88 kg (194 lb). Yearling does averaged 37.7 kg (83 lb), those 2.5 years or older averaged 41.3 kg (91 lb), and those 3.5–5.5 years old averaged 44 kg (97 lb). Body weights began to plateau at 6.5 years (McCulloch and Smith 1991, Heffelfinger 2006).

Antler growth usually begins in May in the southwestern US (Hanson 1955, Truett 1971, Hoffmeister 1986, Heffelfinger 2006). Increasing testosterone levels cause antlers to mineralize, and the velvet is shed in September (Clark 1953, Swank 1958, Cantú and Richardson 1997). Usually by mid-March or early April, antlers are shed because of the decrease in testosterone following the rut (Truett 1971). Older deer usually have larger antlers than deer five to seven years old (Heffelfinger 2006) because nutrient allocation to body development and maintenance has priority over antler growth. Harvest data from the Chihuahuan Desert of Texas have shown that overall antler growth was maximized at 7.5 years, with an average of 9.6 antler points. Body weights also peaked at 7.5 years. Little difference was observed between 6.5- and 7.5-year-olds in either antler growth or body weight (Gray and Richardson 2008). Antler size is reduced without proper levels of protein, energy, and calcium (Ullrey 1983).

Energy requirements

The determining nutritional factors for mule deer are rainfall and the intensity of grazing by domestic livestock (Heffelfinger 2006). Nutritional stress is the greatest for desert mule deer during June or early July if rains are delayed. This is also the period when does need to sustain growth for fetus development and lactation after giving birth (Heffelfinger 2006). Many environmental factors influence energy requirements, and they change according to season and the different physiological stages of a deer's life. Ullrey et al. (1970) reported that mature does require 25 kcal of digestible energy (DE) per lb of body weight per day. During peak lactation, this number can increase to 33 kcal DE/lb/day. Kirkpatrick et al. (1975) found 70 kcal DE/lb/day to be the daily energy requirement of fawns. While the energy intake of mature bucks may decrease about 50% during the rut, 50 kcal DE/lb/day should allow body stores to replenish and maximum growth to occur.

Water requirements

In times of high metabolic demand, sufficient water intake is critical. For deer in West Texas, a spacing of dependable water sources every 1.6–2.4 km (1–1.5 mi) was considered adequate (Varner 2002). Mule deer in southwestern Arizona visited water catchments more frequently than any other species of wildlife. Visitation peaked around sunset and decreased throughout the night. Individual deer visited more often than groups (O'Brien et al. 2006). Mule deer in the state of Sonora travel a mean distance of 1.5–2.0 km (0.9–1.2 mi) in the central plains, and 2.1–3.5 km (1.3–2.1 mi) in the Gulf Coast area (Alcalá Galván and Krausman 2013). Krausman et al. (2006) summarized various desert mule deer studies and concluded that deer are dependent on water, especially in dry periods. Movements relative to known water sources are 0.8–5.0 km (0.5–3.1 mi), and the optimal spacing of water developments is less than 3.2–4.8 km (2–3 mi). A study in the Trans-Pecos region of Texas found that females used sites closer to water sources than males, but water sources did not appear to be an important factor in the scale of home range selection for either gender (Lawrence 1995). Nichol (1938) reported that water consumption by mule deer changed depending on the forage consumed and the time of year. When deer were fed dry forage, water consumption increased 25–65%. In summer, 2.2 liters (0.6 gal) were consumed per day, while winter consumption dropped to 1.1 liters (0.3 gal) per day. An estimated 3.8–9.5 liters (1.0–2.5 gal) of water may be consumed per day when forage does not provide much moisture (Elder 1954, Hervert and Krausman 1986). With higher temperatures and lower humidity (May–July), mule deer visit water more frequently (Rosenstock et al. 2004). Desert mule deer drank water mostly between 1900 and 2200 hours (Hervert and Krausman 1986). In the summer, Rodgers (1977) recorded visits to water mostly during the night, and Hazam and Krausman (1988) recorded desert mule deer visits to watering sites once a day. Deer concentrate around water sources in dry months (Brownlee 1979, Wood et al. 1970). Pregnant does drink water up to four times per day and remain within 0.4 km (0.25 mi) of water sources (Clark 1953). According to Bowyer (1986), the distribution of *O. h. fuliginatus* is strongly affected by the availability of drinking water. Hervert and Krausman (1986) reported that to survive periods of water stress, deer generally require a source of standing water. In the Chihuahuan Desert in Durango as well as in Baja California, studies have reported deer concentrations around watering sources (Sánchez Rojas and Gallina 2000a, Gallina et al. 1991). During the dry season, deer concentrated within 0.8–1.6 km (0.5–1 mi) of water sources (Ordway and Krausman 1986, Bowyer 1986).

Thermal relationships

Cain et al. (2006) described the ways in which desert ungulates thermoregulate, including behavioral modifications, changes in regional blood flow, evaporative cooling, and morphology. In hot and dry periods, mule deer can adjust the timing of their activity, becoming more crepuscular or nocturnal (Hayes and Krausman 1993, Cain et al. 2006). Desert species usually possess relatively longer, thinner appendages and a higher surface-area-to-volume ratio to assist with heat dissipation (Cain et al. 2006). Fox and Krausman (1994) found that desert mule deer does sought hiding cover and shade on the upper portions of steep slopes in mountains. Truett (1971) reported that mule deer used slopes for thermal regulation by staying on sunny slopes in cooler mornings, and shady slopes as the day warmed. Similarly, Tull et al. (2001) reported that desert mule deer used shady bed sites on hotter days, and sunny bed sites on cooler days. Studies by Leopold and Krausmann (1987b) and Hayes and Krausman (1993) reported more deer activity during the night and less in the daytime when summer temperatures were high.

Behavior

Social structure

In the Mapimí Biosphere Reserve in the state of Durango, Sánchez Rojas and Gallina (1998) reported that mule deer have an aggregated distribution and are social animals. Weber and Galindo Leal (2005) found mule deer to be relatively more gregarious than white-tailed deer. They reported doe-fawn-yearling social units of 2–8 individuals. While mature bucks were solitary, juvenile bucks formed social groups of 4–10 individuals. Mandujano and Gallina (1996) found that larger herds were prominent in open habitats, probably because of the low availability of water (Gallina et al. 1991). Álvarez Cárdenas et al. (1994) reported the smallest aggregations of deer in December and the largest in March, based on fecal pellet groupings in Baja California Sur. Gallina et al. (1992) also reported the smallest social groups during December and the largest in March, using the same method.

Movements

In the southwestern US, mule deer do not migrate but have been reported to make seasonal and daily movements motivated by weather, food and water distribution, fawning, segregation of sexes, and disturbances caused by humans or livestock (Heffelfinger 2006). Little is known about local movements of mule deer in Baja California (Weber and Galindo Leal 2005). For most of the year in Baja California, mule deer remain at higher elevations, but after snow falls, they move to lower elevations (Leopold 1977, Weber and Galindo Leal 2005). Gallina et al. (2000) and Velázquez and Reyes (1976) mentioned the possibility of local movements in the Sierra de la Laguna of Baja California. Several studies have reported mule deer moving from their normal home range to access water sources. Rautenstrauch and Krausman (1989) found that desert mule deer temporarily moved up to 32.2 km (20 mi) in search of freestanding water.

Bucks make large seasonal movements during the rut (Rodgers 1977, Koerth and Bryant 1982, Relyea and Demarais 1994). Does tend to move to higher elevations that provide more fawning cover (Fox and Krausman 1994, Heffelfinger 2006). Mule deer have been recorded to move from 0.1 to 1.2 km (0.06 to 0.75 mi) in one day (Rodgers 1977, Dickinson and Garner 1979). Daily mule deer activities reported by Koenen and Krausman (2002) averaged 26% of daylight hours standing, 6% bedding down, 29% traveling, and 39% foraging. Mellink (2005) observed mule deer on the Baja California Peninsula using areas near the tops of mountain ranges for most of the year, until winter snowfall caused a shift to lower elevations. In West Texas, James (2013) studied the movements of mature bucks between fed and nonfed study sites. He documented obvious crepuscular activity in movement rates (1 km/5 hr, 0.62 mi/5 hr), where bucks moved at higher rates in the morning (0430–0830 hr) and evening (1630–2030 hr). Additionally, he documented a difference in movement rates (1 m/5 hr, 3.3 ft/5 hr) between fed (501.7 m, 661.7 ft) and nonfed sites (419.2 m, 1,375.3 ft).

Spacing

Ordway and Krausman (1986) reported home ranges of 14–45 km^2 (5.4–17 mi^2) for bucks, and 2–18 km^2 (0.7–7 mi^2) for does in southern Arizona. Although most studies in the southwestern US have reported home ranges of 48.3–80.5 km^2 (18.6–31 mi^2), Rautenstrauch and Krausman (1989) reported home ranges of up to 351 km^2 (136 mi^2). James (2013) reported that home ranges of mature mule deer bucks in West Texas on nonfed sites were larger (45 km^2, 17 mi^2) than on fed sites (35 km^2, 14 mi^2). Geist (1981) reported that home ranges of dominant mule deer bucks overlapped with those of fawn-doe groups during the rut. Home ranges in the Sonoran Desert are much larger than those in the Chihuahuan Desert (Heffelfinger 2006). Mule deer were discontinuously distributed in the Mapimí Biosphere Reserve in the Chihuahuan Desert (Sánchez Rojas and Gallina 2000a). A study in northern Coahuila comparing site fidelity of hard-released (no acclimatization time) versus soft-released desert mule deer showed average movements of hard-released deer to be 4–11 km (2.4–6.8 mi) from the release site, while soft-released deer averaged 0.9–12 km (0.6–7.4 mi) (Martínez García 2009). Of the hard-released deer, 60% remained less than 5 km (3.1 mi) from the release site, and 75% of the soft-released individuals moved more than 5 km (3.1 mi) away. Average home ranges were 2,880.1 ha (7,116.9 ac) and 3,455.18 ha (8,537.9 ac) for soft- and hard-released deer, respectively (Martínez García 2009). Relyea (1992) observed that desert mule deer bucks were more active during postrut than during prerut or peak rut, while movements were high during both the prerut and postrut in the Trans-Pecos region of Texas during the mating season.

Population Dynamics

The most influential factor for mule deer population dynamics in arid environments is rainfall (Álvarez Cárdenas et al. 1994, Heffelfinger 2006, Walser 2006). Seasonality and weather constantly affect the demographics of mule deer habitats and influence population dynamics (McKinney 2003, Mellink 2005, Heffelfinger 2006, Walser 2006). Mule deer populations in the southwestern US and northern Mexico experience the effects of drought conditions, rather than the harsh winters that affect northern mule deer populations (Anthony 1976, Smith and LeCount 1979, Leopold and Krausman 1991, Cantú and Richardson 1997). Walser (2006) reported that drought intensity directly affects population abundance and fawn production. This information is useful when we observe changes through time, and it gives us a general idea of how populations fluctuate in relation to different factors. It is important for wildlife managers to understand the effects of climatic variation on population dynamics in order to make sound management decisions. For example, late spring and early summer precipitation was most influential on survival of desert mule deer fawns, whereas winter precipitation had the most influence on population abundance (Walser 2006).

Density

Deer densities (number of deer/area) vary, and it is important to define density for a specific time because of seasonal fluctuations (Heffelfinger 2006). Despite the difficulties inherent in density estimation, several authors have evaluated different areas across the mule deer's distribution in Mexico (Gallina et al. 1992, Galindo Leal 1993, Álvarez Cárdenas et al. 1994, Sánchez Rojas and Gallina 1998, Álvarez Cárdenas et al. 1999b, Ahumada Cervantes 2000, Sánchez Rojas and Gallina 2000b, Lozano Cavazos 2003). Mule deer densities in Mexico vary from 0.70 deer/km^2 (1.8 deer/mi^2) in the Chihuahuan Desert in Durango, to 42 deer/km^2 (110 deer/mi^2) in the state of Baja California. Sánchez Rojas and Gallina (2000a, 2007) suggested that lower densities occur in the Chihuahuan Desert because of the limited carrying capacity of this ecosystem.

Population Trends and Current Status

Unlike state agencies that regulate wildlife in the US, agencies in Mexico do not currently monitor long-term population trends of mule deer. Therefore, little is known about the trends of mule deer in Mexico. Historically, populations ranged from the Baja California Peninsula through the entire Sonoran and Chihuahuan Deserts, including parts of the states of Zacatecas, San Luis Potosí, Nuevo León, and Tamaulipas (Weber and Galindo Leal 2005). With increasing ecotourism (Valdez et al. 2006), many ranchers and landowners have started to promote wildlife populations by initiating activities ranging from hunting restrictions to species reintroductions. Mule deer have been reintroduced in the states of Nuevo León and Coahuila. There is a limited population in the state of Zacatecas.

Productivity and recruitment

The constant changes in deer populations can be considered a result of additions (reproduction and immigration) and losses (death and emigration). When mortality exceeds recruitment, a natural decline will occur. In contrast, when recruitment exceeds mortality, populations will increase. Recruitment is expressed as a rate or ratio (Heffelfinger 2006). A population can be considered productive when it has a 134:100 fawn to doe ratio, which has been reported for extreme western populations in Mexico (Álvarez Cárdenas et al. 1994, 1999a). In the Chihuahuan Desert in West Texas, fawn to doe ratios varied from 24:100 to 48:100 (Gray and Richardson 2008). This great variability was attributed to the decline of desert mule deer in West Texas that occurred in the late 1970s.

Survival and mortality factors

Mountain lions have been reported as the main predator of mule deer (Leopold 1959, Álvarez Cárdenas et al. 1994, Lawrence et al. 2004, Mellink 2005); however, coyotes (*Canis latrans*), bobcats (*Lynx rufus*), and golden eagles (*Aquila chrysaetos*) will also prey on mule deer fawns (Mellink 2005). Predation does not necessarily regulate mule deer populations (Heffelfinger 2006). Natural mortality is typically higher in males than in females. Mortality can also be expressed as a ratio or rate. Under stressful conditions such as extended drought, mortality in most cases increases (Mellink 2005, Heffelfinger 2006). Hunting may have a more dramatic impact on mule deer populations in Mexico than on those in the US. Although harvest is typically biased toward adult deer and males (Lawrence et al. 2004), harvest pressure in portions of Mexico may be more evenly distributed. Predation usually accounts for mortalities of young deer (Lawrence et al. 2004). The most stressful period for mule deer is the postrut. A combination of factors (increased activity associated with searching for mates, reduced nutrient reserves, and seasonal lows in forage quality) can cause severe nutritional stress, especially for adult males (Lawrence et al. 2004).

Evidence of bluetongue and anaplasmosis was found in a deer population in Baja California (Contreras et al. 2007). This suggests biological continuity within the same biogeographic region, and other diseases not tested for (e.g., Lyme disease, bartonellosis) could be present. Further studies are needed to determine whether these diseases are affecting deer populations (Contreras et al. 2007).

Population Management

The initial step in managing a deer population is to determine the objective (Fulbright and Ortega Santos 2006); different management strategies may be used to accomplish different objectives. Management programs should be based on quality scientific investigations. Research improves management strategies and provides objective information that helps managers make informed decisions (Garton et al. 2005). It is important to collect current information from the population to be managed, as well as information that will aid understanding of long-term population trends (Heffelfinger 2006). Deer populations are constantly changing, and in arid environments, local habitat and environmental conditions have a great influence on such changes (Álvarez Cárdenas 1995, Mellink 2005, Heffelfinger 2006). Management decisions use data from surveys and harvest records (Heffelfinger 2006). Surveys usually estimate population characteristics based on representative samples of the area to be surveyed.

To establish a harvest management program, resource managers should follow the steps outlined by Carpenter (2000) and ask the following questions: (1) Where are we now? (inventory the population for a baseline); (2) How do we get there? (establish population objectives); (3) Did we get there? (implement harvest strategy); and (4) Did it work? (evaluate strategy compared to population goals). Ultimately, management decisions regarding mule deer populations must be based on strong scientific evidence.

Habitat Requirements

Habitat is the area or space selected by a species where conditions and resources (e.g., food, cover, and water) allow survival and reproduction (Hall et al. 1997). The distribution of mule deer in Mexico encompasses a complex of conditions, from chaparral and coniferous forest on the Baja California Peninsula, to desert scrublands in the Sonoran Desert in central and northwestern Sonora, to desert shrublands and grasslands in the Chihuahuan Desert of the Mexican High Plateau (e.g., the northern half of Chihuahua, most of Coahuila, northeastern Durango, northernmost Zacatecas, and small portions of western Nuevo León). Every large ecoregion presents a variety of topographic characteristics and vegetation communities with a variety of foods, cover types, and water availability for mule deer.

The size of individual home ranges varies according to the spatial arrangement of resources, and how they can meet the daily and seasonal requirements of mule deer. Although intra- and interspecific relationships affect movements and activities, the size of home ranges is influenced mainly by the amount and quality of forage, the presence of cover for thermal and escape protection, and water accessibility. Esparza-Carlos et al. (2011) demonstrated that habitat suitability can change drastically among years as precipitation affects forage production. Mule deer movements and activity patterns are structured to meet an average daily demand of 1.5–2.0 kg (3.3–4.4 lb) of forage, 4–6 liters (0.9–1.2 gal) of water, and adequate vegetation structure to provide thermal cover and concealment. The following is a brief description of the habitat requirements of mule deer that have been quantified in various ecological regions of Mexico.

Baja California Peninsula

Mediterranean chaparral on the northwestern Baja California Peninsula contains the highest densities of mule deer in Mexico (Galindo Leal 1993). Mule deer in that region prefer open, wet areas with higher availability of forbs, and they avoid dense woody vegetation (Mellink 2005). Mule deer habitat preferences in temperate forests appear to contrast at both extremes of the peninsula. In the coniferous forest of the Sierra de San Pedro Mártir in northern Baja California, mule deer prefer plains with 20–40% canopy cover, abundant forbs, relatively nonrocky soil, and proximity to water (Ahumada-Cervantes 2000). Mule deer occupy mountaintops during most of the year; winter snowfall initiates movements to lower elevations (Mellink 2005). In the pine-oak forest of the Sierra de la Laguna on the southern extreme of the peninsula, mule deer prefer sites with slopes of more than 30%, stony soils, and dense shrubs (Gallina et al. 1991, Álvarez Cárdenas et al. 1999b). The driest areas of the peninsula support the lowest densities of mule deer, and habitat is characterized by desert scrub of the Lower Colorado Valley and Vizcaíno subdivisions of the Sonoran Desert (Gallina et al. 1991). Mule deer on Cedros Island use open desert scrub, chaparral, and closed pine forest (McCarthy et al. 1998).

Sonora

Mule deer are distributed over approximately 8 million ha (19.7 million ac) in Sonora. The western portions of their range correspond with the Sonoran Desert ecoregion. Some populations of mule deer are also found in a small portion of northeastern Sonora, which is considered a transitional zone between the Sonoran and Chihuahuan Deserts (Burgess 1995), and in the Apache Highlands ecoregion (Nature Conservancy 2009).

Habitat relationships and requirements in the

Sonoran Desert have been extensively studied in Arizona and California (Mackie et al. 1982, Anderson and Wallmo 1984, Krausman 1985, Rautenstrauch and Krausman 1989, Heffelfinger 2006). Climate and vegetation in these areas are described as within the Arizona Upland and Lower Colorado Valley subdivisions of the Sonoran Desert. Conditions in these subdivisions continue into the north-central and northwestern portions of Sonora. In these areas, mule deer require home ranges of up to 172 km^2 (453 mi^2) (Krausman 1985, Rautenstrauch and Krausman 1989). The two southern subdivisions of the Sonoran Desert—the plains of Sonora in the central portion of the state, and the central Gulf Coast on the western extreme—have distinctive biotic characteristics (Brown 1994). Simultaneous studies of habitat preferences on two representative sites in these subdivisions showed that mule deer have an annual home range of about 14 km^2 (37 mi^2) in the plains of Sonora, where vegetation is dominated by shrubs and medium-sized trees with sparse cacti and abundant grasses and forbs (Alcalá Galván and Krausman 2013). Mule deer require a home range of double that size (27 km^2, 71 mi^2) in the more arid conditions of the central Gulf Coast, where vegetation is dominated by sarcocaulescent desert scrub with shrubs, abundant tall cacti, and few understory plants (Alcalá Galván and Krausman 2013).

In general, studies on both sides of the Arizona-Sonora border support the notion that mule deer prefer xeroriparian vegetation and adjacent plains for daily activities throughout the year (Fox and Krausman 1994, Krausman and Etchberger 1995, Alcalá Galván and Krausman 2013). Habitats preferred by mule deer have higher cover of shrubs and trees as well as better availability of browse and forbs. These studies also confirm that mule deer prefer sites close to sources of free water. In the southwestern US, mule deer are usually found within 2.4 km (1.5 mi) of free water (Hanson and McCulloch 1955, Swank 1958, Boroski and Mossman 1996). In Sonora, mean distances to the nearest water source were 1.5–2.0 km (0.9–1.2 mi) in the central plains and 2.1–3.5 km (1.3–2.1 mi) in the Gulf Coast area (Alcalá Galván and Krausman 2013). Populations and habitat relationships of mule deer in northeastern Sonora have not been studied, although they appear to use desert scrub and oak-grassland savanna.

Mexican High Plateau

The distribution of mule deer on the Mexican High Plateau matches the limits of the Chihuahuan Desert. Three main subdivisions have been proposed for this ecoregion: (1) the Trans-Pecos in the northern portion, from the area along the Pecos River in Texas and New Mexico to northeastern Chihuahua (52% of the region); (2) the Mapimian or central portion, embracing eastern Chihuahua and western Coahuila (35%); and (3) the Saladan in the south, occupying southern Coahuila, northeastern Durango, and a small portion of northern Zacatecas (13%) (Morafka 1977, 1991).

There is a notable difference in the number of scientific studies conducted on mule deer in the US compared to studies conducted in Mexico. Heffelfinger (2006) summarized several studies of mule deer from West Texas that consistently estimated home ranges of 5–13 km^2 (13–34 mi^2), although James (2013) reported significantly larger home ranges for West Texas. Mule deer movements and densities were influenced by availability of free water in West Texas (Wood et al. 1970, Brownlee 1979, Dickinson and Garner 1979). In the Mapimí Biosphere Reserve in the border area of Chihuahua, Coahuila, and Durango, mule deer preferred areas near water, with higher diversity and density of woody plants (Sánchez Rojas and Gallina 2000a). Another factor associated with the density of mule deer was the unevenness of the terrain. The importance of this factor as a habitat variable is thought to be associated with providing more visibility and thus avoiding predators (Sánchez Rojas and Gallina 2000b). Ortega-Sanchez (2013) found a strong habitat preference by translocated mule deer for higher browse diversity in xeroriparian habitats in northern Coahuila.

Habitat Management

Direct habitat improvement

Broad-scale habitat improvements for mule deer in Mexico are not commonly implemented. Federal and state wildlife management agencies do not have programs to ensure the improvement and conservation of mule deer habitat. Nonetheless, some ranchers are aware of the economic benefits of mule deer hunting and have decided to protect mule deer populations and assign specific areas for their benefit of in order to increase deer numbers and the availability of trophy deer. Direct habitat improvement in localized areas includes creation of artificial sources of water, enhancement of native forbs and shrubs (as food), predator control, and reduction or total exclusion of cattle. Documentation is needed regarding the intensity and effectiveness of these practices for mule deer population performance, and the effects on other wildlife species.

Coordination with livestock management

Most mule deer habitat in Mexico consists of rangelands grazed by livestock, with minimal consideration of wildlife. Therefore, extensive habitat alterations have occurred in a large part of the mule deer's range. These alterations include the reduction of forage productivity because of the detrimental effects of livestock overgrazing, the addition of artificial sources of water, and the transformation of desert scrublands into nonnative grass pastures. In Sonora, about 20% of mule deer range has been altered by clear-cutting and the seeding of buffelgrass pastures (Alcalá Galván and Krausman 2013). Mule deer habitat improvement should focus on mitigating impacts caused by livestock (fig. 16.4), restoring cover and forage plants as natural components of shrublands, and enhancing the quality and availability of water.

Comprehensive guidelines for mule deer habitat management in the southwestern desert ecoregion (US) are provided by Heffelfinger et al. (2006). Several points need to be considered in developing livestock management plans compatible with conservation of mule deer habitat. An appropriate plan that involves common use of rangelands and maximizes the benefits for mule deer and livestock should be established. Livestock may browse important deer shrubs excessively (Swank 1958, Knipe 1977). However, with proper stocking rates, correct timing, and adequate levels of use, livestock grazing can be compatible with conservation of mule deer resources (Lyons and Wright 2003).

A management plan should also consider protection of xeroriparian areas, in which the habitat provides mature trees for thermal and screening cover, and higher forb availability. Xeroriparian habitats are

Figure 16.4. Habitat management for mule deer in the Sonoran Desert includes managing cattle stocking rates to avoid competition. Photo by Fernando A. Ibarra Flores.

preferred in the Sonoran and Chihuahuan Deserts (Krausman et al. 1985, Alcalá Galván and Krausman 2013, Ortega-Sanchez 2013). Grazing plans should include effective procedures to evaluate trends in vegetation and key soil variables. Monitoring should be conducted consistently using reliable and efficient methods (Holechek and Gault 2000).

Management Challenges in Mexico

Despite a century or more of overexploitation in Mexico, the future of mule deer is promising. Recent land-use changes, spurred by hunting operations for mule deer and bighorn sheep (Rosas-Rosas et al. 2003, Valdez et al. 2006), have laid the foundation for mule deer conservation in Mexico. Two simple requirements must be met for mule deer populations to flourish in Mexico: harvest regulation and habitat management (fig. 16.5).

Paramount to any long-term conservation strategy for mule deer in Mexico is protection. Leopold (1959), Baker (1977), and others have documented the decline of mule deer in Mexico via subsistence hunting and other forms of poaching. However, regulation alone cannot protect mule deer and other wildlife species. Enforcement, education, and a host of socioeconomic issues must be addressed at a national level before mule deer conservation will be a reality. Several groups of private landowners have successfully curtailed the illegal take of wildlife in northern Coahuila and Sonora (Valdez et al. 2006). In these instances, land use has changed because landowners have prioritized their wildlife enterprise above cattle operations, or else corporate and nongovernmental interests have purchased land for restoration purposes.

The second requirement for mule deer to flourish in Mexico is quality habitat (fig. 16.6) Mule deer are adaptable and can survive in many habitats in

Figure 16.5. The future of mule deer conservation in Mexico lies with the potential hunting revenues generated by trophy hunting. Many landowners in Sonora have initiated hunting operations that have resulted in the destocking of livestock and the implementation of habitat improvements for mule deer and other wildlife. Photo by Carlos Alcalá-Galván.

Figure 16.6. Mule deer in the Chihuahuan Desert at Rancho Tres Papalotes in Chihuahua, Mexico. Photo by Enrique Perez.

Mexico. The underlying factor for quality mule deer habitat is healthy range conditions. Mule deer consume little grass, but grass is essential for fawning cover. Likewise, mule deer rely heavily on brush and succulents for consumption and for cover. Most of the former range of mule deer has been overgrazed by livestock (Martínez Muñoz et al. 2003). In many instances, destocking rangelands alone will not allow the recovery of degraded rangelands; many will require active habitat management to stimulate recovery and restoration. Techniques such as seeding, contouring, and other forms of erosion control will be required in some areas. Private landowners, corporations, nongovernmental organizations, and federal agencies must work cooperatively if mule deer are to be managed in a sustainable manner.

Literature Cited

Ahumada Cervantes, R. 2000. Propuesta de plan de manejo para el venado bura (*Odocoileus hemionus fuliginatus*) en la Sierra de San Pedro Mártir. Thesis, UABC, Ensenada, México.

Alcalá Galván, C. H., and P. R. Krausman. 2012. Diets of desert mule deer in altered habitats in the lower Sonoran Desert. California Game and Fish Journal 98:81–103.

Alcalá Galván, C. H., and P. R. Krausman. 2013. Home range and habitat use by desert mule deer in altered habitats. California Game and Fish Journal 99:65–69.

Álvarez Cárdenas, S. 1995. Estudio poblacional y hábitat del venado bura, *Odocoileus hemionus peninsulae*, en la Sierra de la Laguna, Baja California Sur. Thesis, UNAM, México, D. F., México.

Álvarez Cárdenas, S., S. Gallina, and P. Gallina Tessaro. 1994. Dinámica poblacional del venado bura de la Sierra de la Laguna, Baja California Sur, México. Pages 114–135 *in* IV Simposio sobre Venados en México. UNAM, México, D. F., México.

Álvarez Cárdenas, S., S. Gallina, P. Gallina Tessaro, and S. Díaz Castro. 1999a. Mule deer population dynamics in a relictual oak-pine forest in Baja California Sur, Mexico. Pages 197–210 *in* Proceedings of ecology and management of forests, woodlands, and shrublands in the dryland regions of United States and Mexico: perspectives of the 21st Century. University of Arizona Press, Tucson, USA.

Álvarez Cárdenas, S., S. Gallina, P. Gallina Tessaro, and R. Domínguez Cadena. 1999b. Habitat availability for the mule deer (Cervidae) population in a relictual oak-pine forest in Baja California Sur, Mexico. Tropical Zoology 12:67–78.

Anderson, A. E., L. G. D. E. Medin, and D. C. Bowden. 1974. Growth and morphometry of the carcass, selected bones, organs, and glands of mule deer. Wildlife Monographs 39:1–122.

Anderson, A. E., W. A. Snyder, and G. W. Brown. 1965. Stomach content analyses related to condition in mule deer, Guadalupe Mountains, New Mexico. Journal of Wildlife Management 29:352–366.

Anderson, A. E., and O. C. Wallmo. 1984. *Odocoileus hemionus.* Mammalian Species 219:1–9.

Anthony, R. G. 1976. Influence of drought on diets and numbers of desert deer. Journal of Wildlife Management 40:140–144.

Anthony, R. G., and N. S. Smith.1977. Ecological relationships between mule deer and white-tailed deer in southeastern Arizona. Ecological Monographs 47:255–277.

Baker, R. H. 1977. Mammals of the Chihuahuan Desert region: future prospects. Pages 221–225 *in* R. H. Wauer and D. H. Riskind, editors. Transactions of the symposium on the biological resources of the Chihuahuan Desert region, United States and Mexico. National Park Service Transactions and Proceedings Number 3. US Department of Interior, Washington, DC, USA.

Boeker, E. L., V. E. Scott, H. G. Reynolds, and B. A. Donaldson.1972. Seasonal habits of mule deer in southwestern New Mexico. Journal of Wildlife Management 36:56–63.

Boroski, B. B., and A. S. Mossman. 1996. Distribution of mule deer in relation to water sources in northern California. Journal of Wildlife Management 60:770–776.

Bowyer, T. 1986. Habitat selection by southern mule deer. California Fish and Game Journal 72:153–169.

Brown, R. T. 1994. Biotic communities: southwestern United States and Mexico. University of Utah Press, Salt Lake City, USA.

Brownlee, S. L. 1979. Water development for desert mule deer. Booklet 7000–32. Texas Parks and Wildlife Department, Austin, USA.

Burgess, T. L. 1995. Desert grassland, mixed shrub savanna, shrub steppe, or semidesert scrub: the dilemma of coexisting growth forms. Pages 31–67 *in* M. P. McClaran, V. DeVender, and R. Thomas, editors. The desert grassland. University of Arizona Press, Tucson, USA.

Cain, J. W., P. R. Krausman, S. S. Rosenstock, and J. C. Turner. 2006. Mechanisms of thermoregulation and water balance in desert ungulates. Wildlife Society Bulletin 34:582–591.

Cantú, R., and C. Richardson. 1997. Mule deer management in Texas. Texas Parks and Wildlife Department, Austin, USA.

Carpenter, L. H. 2000. Harvest management goals. Pages 193–213 *in* S. Demarais and P. R. Krausman, editors. Ecology and management of large mammals in North America. Prentice Hall, Upper Saddle River, New Jersey, USA.

Challenger, A. 1998. Utilización y conservación de los ecosistemas terrestres de México: pasado, presente, y futuro. CONABIO, México, D. F., México.

Clark, E. D. 1953. A study of the behavior and movements of the Tucson Mountain mule deer. Thesis, University of Arizona, Tucson, USA.

Contreras, J., E. Mellink, R. Martínez, and G. Medina. 2007. Parasitos y enfermedades del venado bura (*Odocoileus hemionus fuliginatus*) en la parte norte de la Sierra de San Pedro Mártir, Baja California, México. Revista Mexicana de Mastozoología 11:8–20.

Cowan, I. 1956. What and where are the mule and black-tailed deer? Pages 334–339 *in* W. P. Taylor, editor. The deer of North America. Stackpole, Harrisburg, Pennsylvania, USA.

Dickinson, T. G., and G. W. Garner. 1979. Home range use and movements of desert mule deer in southwestern Texas. Proceedings of the Annual Conference of Southeastern Fish and Wildlife Agencies 33:267–278.

Dixon, J. S. 1934. A study of the life history and food habits of the mule deer in California. California Fish and Game 20:1–142.

Elder, J. B. 1954. Notes on summer water consumption by desert mule deer. Journal of Wildlife Management 18:540–541.

Esparza-Carlos, J. P., J. W. Laundré, and V. J. Sosa. 2011. Precipitation impacts on mule deer habitat use in the Chihuahuan Desert of Mexico. Journal of Arid Environments 75:1008–1015.

Fox, K. B., and P. R. Krausman. 1994. Fawning habitat of desert mule deer. Southwestern Naturalist 39:269–275.

Fulbright, T. E., and J. A. Ortega Santos. 2006. White-tailed deer habitat: ecology and management on rangelands. Texas A&M University Press, College Station, USA.

Galindo Leal, C. 1993. Densidades poblacionales de los venados cola blanca, cola negra y bura en Norteamérica. Pages 371–391 *in* R. A. Medellín and G. Ceballos, editors. Avances en el estudio de mamíferos en México. Asociación Mexicana de Mastozoología, México, D. F., México.

Gallina, S. 1989. El hábitat del venado bura en la Sierra de la Laguna. Memorias VI Simposio sobre Fauna Silvestre. UNAM, México, D. F., México.

Gallina, S. 1990. La población del venado bura de la Sierra de la Laguna, Baja California Sur. Memorias II Simposio Internacional de la Vida Silvestre. UNAM, México, D. F., México.

Gallina, S., S. Álvarez Cárdenas, and P. Gallina Tessaro. 2000. Familia Cervidae. Pages 793–815 *in* S. T. Álvarez Castañeda and J. L. Patton, editors. Mamíferos del noroeste de México II. Centro de Investigaciones Biológicas del Noroeste, La Paz, Baja California Sur, México.

Gallina, S., P. Gallina Tessaro, and S. Álvarez Cárdenas. 1991. Mule deer density and pattern distribution in the pine-oak forest at the Sierra de la Laguna in Baja California Sur, Mexico. Ethology, Ecology and Evolution 3:27–33.

Gallina, S., P. Gallina Tessaro, and S. Álvarez Cárdenas. 1992. Hábitat y dinámica poblacional del venado bura. Pages 297–327 *in* A. Ortega, editor. Uso y manejo de los recursos naturales de la Sierra de la Laguna, Baja California Sur. Centro de Investigaciones Biológicas and World Wildlife Foundation, México, D. F., México.

Gallina Tessaro, P., A. González Romero, G. Arnaud Franco, S. Gallina, and S. Álvarez Cárdenas. 1988. Mastofauna. Pages 209–228 *in* L. Arriaga and A. Ortega, editors. La Sierra de la Laguna de Baja California Sur. Centro de Investigaciones Biológicas de Baja California Sur, La Paz, México.

Garton, E. O., J. T. Ratti, and J. H. Giudice. 2005. Research and experimental design. Pages 43–71 *in* C. E. Braun, editor. Techniques for wildlife investigations and management. The Wildlife Society, Bethesda, Maryland, USA.

Geist, V. 1981. Behavior: adaptive strategies in mule deer. Pages 157–224 in O. C. Wallmo, editor. Mule and black-tailed deer of North America. Wildlife Management Institute and University of Nebraska Press, Lincoln, USA.

Geist, V. 1998. Deer of the world: their evolution, behavior, and ecology. Stackpole, Mechanicsburg, Pennsylvania, USA.

Gray, S. S., and C. L. Richardson. 2008. TPWD mule deer buck harvest and antler development data in the Trans-Pecos. Pages 7–76 *in* B. J. Warnock, P. H. Dickerson, and L. A. Harveson, editors. Proceedings of the Trans-Pecos Wildlife Conference—2008. Sul Ross State University, Alpine, Texas, USA.

Hall, L. S., P. R. Krausman, and M. L. Morrison. 1997. The habitat concept and a plea for standard terminology. Wildlife Society Bulletin 25:173–182.

Hanley, T. A. 1982. The nutritional basis for food selection by ungulates. Journal of Range Management 35:146–152.

Hanson, W. R. 1955. Field observations of deer in the Prescott study area. Federal Aid Project W-71-R, WP3, Job 2. Arizona Game and Fish Department, Phoenix, USA.

Hanson, W. R., and C. Y. McCulloch. 1955. Factors influencing the distribution of mule deer on Arizona rangelands. Transactions of the North American Wildlife Conference 20:568–588.

Hayes, C. L., and P. R. Krausman. 1993. Nocturnal activity of female desert mule deer. Journal of Wildlife Management 57:897–904.

Hazam, J. E., and P. R. Krausman. 1988. Measuring water consumption of desert mule deer. Journal of Wildlife Management 52:528–534.

Heffelfinger, J. 2006. Deer of the Southwest: a complete guide to natural history, biology, and management of the southwestern mule deer and white-tailed deer. Texas A&M University Press, College Station, USA.

Heffelfinger, J. R., C. Brewer, C. H. Alcalá Galván, B. Hale, D. L. Weybright, B. F. Wakeling, L. H. Carpenter, and N. L. Dodd. 2006. Habitat guidelines for mule deer: Southwest Deserts Ecoregion. Mule Deer Working Group, Western Association of Fish and Wildlife Agencies, Boise, Idaho, USA.

Hervert, J. J., and P. R. Krausman. 1986. Desert mule deer use of water developments in Arizona. Journal of Wildlife Management 50:670–676.

Hoffmeister, D. F. 1986. Mammals of Arizona. University of Arizona Press, Tucson, USA.
Holechek, J. L., and D. Gault. 2000. Grazing intensity guidelines. Rangelands 22:11–14.
Horejsi, R. G., D. D. Haywood, and R. H. Smith. 1988. The effects of hunting on a desert mule deer population. Federal Aid Project W-78-R, WP2, Job 27. Arizona Game and Fish Department, Phoenix, USA.
Hungerford, C. R. 1974. Response of Kaibab mule deer to management in summer range. Journal of Wildlife Management 34:852–862.
James, A. S. 2013. The effects of supplemental feed on seasonal home ranges and habitat use of mature mule deer bucks in Trans-Pecos, Texas. Thesis, Sul Ross State University, Alpine, Texas, USA.
Kirkpatrick, R. L., G. F. Buckland, W. A. Abler, P. F. Scanlon, J. B. Whelan, and H. E. Burkhart. 1975. Energy and protein influences on blood urea nitrogen of white-tailed deer fawns. Journal of Wildlife Management 39:692–698.
Knipe, T. 1977. The Arizona whitetail deer. Special Report 6. Arizona Game and Fish Department, Phoenix, USA.
Koenen, K. K., and P. R. Krausman. 2002. Habitat use by desert mule deer in a semidesert grassland. Southwestern Naturalist 47:353–362.
Koerth, B. H., and F. C. Bryant. 1982. Home ranges of mule deer bucks in the Texas Panhandle. Prairie Naturalist 14:122–124.
Krausman, P. R. 1978. Forage relationship between two deer species in Big Bend National Park, Texas. Journal of Wildlife Management 42:101–107.
Krausman, P. R. 1985. Impacts of the Central Arizona Project on desert mule deer and desert bighorn sheep. Contract No. 9–07–30-X069. US Bureau of Reclamation, Phoenix, Arizona, USA.
Krausman, P. R., and R. C. Etchberger. 1995. Response of desert ungulates to a water project in Arizona. Journal of Wildlife Management 59:292–300.
Krausman, P. R., A. J. Kuenzi, R. C. Etchberger, K. R. Rautenstrauch, L. L. Ordway, and J. J. Hervert. 1997. Diets of desert mule deer. Journal of Range Management 50:513–522.
Krausman, P. R., B. D. Leopold, R. F. Seegmiller, and S. G. Torres. 1989. Relationships between desert bighorn sheep and habitat in western Arizona. Wildlife Monographs 102:1–66.
Krausman, P. R., K. R. Rautenstrauch, and B. D. Leopold. 1985. Xeroriparian systems used by desert mule deer in Texas and Arizona. Pages 144–149 *in* R. Johnson et al., editors. Riparian ecosystems and their management: reconciling conflicting uses. General Technical Report RM-120. US Forest Service, Fort Collins, Colorado, USA.
Krausman, P. R., S. S. Rosenstock, and J. W. Cain. 2006. Developed waters for wildlife: science, perception, values, and controversy. Wildlife Society Bulletin 34:582–591.
Kufeld, R. C., O. C. Wallmo, and C. Feddema. 1973. Foods of the Rocky Mountain mule deer. Research Paper RM-111. US Forest Service, Washington, DC, USA.
Latch, E. K., J. A. Heffelfinger, J. A. Fike, and O. E. Rhodes. 2009. Species-wide phylogeography of North American mule deer: cryptic glacial refugia and postglacial recolonization. Molecular Ecology 18:1730–1745.
Lawrence, R. K. 1995. Population dynamics and habitat use of desert mule deer in the Trans- Pecos region of Texas. Thesis, Sul Ross State University, Alpine, Texas, USA.
Lawrence, R. K., S. Demarais, R. A. Reylea, S. P. Haskell, W. B. Ballard, and T. L. Clark. 2004. Desert mule deer survival in southwest Texas. Journal of Wildlife Management 68:561–569.
Leopold, A. S. 1959. Wildlife of Mexico: the game birds and mammals. University of California Press, Berkeley, USA.
Leopold, A. S. 1977. Fauna silvestre de México. IMERNAR, México, D. F., México.
Leopold, B. D., and P. R. Krausman. 1987a. Diets of two desert mule deer herds in Big Bend National Park, Texas. Southwestern Naturalist 32:449–455.
Leopold, B. D., and P. R. Krausman. 1987b. Diurnal activity patterns of desert mule deer in relation to temperature. Texas Journal of Science 39:49–53.
Leopold, B. D., and P. R. Krausman. 1991. Factors influencing desert mule deer distribution and productivity in southwestern Texas. Southwestern Naturalist 36:67–74.
Lozano Cavazos, E. A. 2003. Factores que afectan el uso de hábitat del venado bura (*Odocoileus hemionus crooki* Mearns) en el noroeste de Coahuila. Thesis, UAAAN, Saltillo, Coahuila, México.
Lyons, R. K., and B. D. Wright. 2003. Using livestock to manage wildlife habitat. Publication

B-6136. Texas Cooperative Extension Service, College Station, USA.

Mackie, R. J., K. L. Hamlin, and D. F. Pac. 1982. Mule deer. Pages 862–877 *in* J. A. Chapman and G. A. Feldhamer, editors. Wild mammals of North America: biology, management, economics. Johns Hopkins University Press, Baltimore, Maryland, USA.

Mandujano, S., and S. Gallina. 1996. Size and composition of white-tailed deer group in a tropical dry forest in Mexico. Ethology, Ecology and Evolution 8:125–131.

Marshal, J. P., V. C. Bleich, N. G. Andrew, and P. R. Krausman. 2004. Seasonal forage use by desert mule deer in southeastern California. Southwestern Naturalist 49:501–505.

Marshal, J. P., V. C. Bleich, P. R. Krausman, M. L. Reed, and N. G. Andrew. 2006. Factors affecting habitat use and distribution of desert mule deer in an arid environment. Wildlife Society Bulletin 34:609–619.

Martínez García, J. L. 2009. Site fidelity and post release movements of translocated mule deer in northern Coahuila, Mexico. Thesis, Sul Ross State University, Alpine, Texas, USA.

Martínez Muñoz, A., D. G. Hewitt, S. Valenzuela, J. I. Uvalle, A. E. Estrada, J. J. Avendano, and R. Aranda. 2003. Habitat and population status of desert mule deer in Mexico. Zeitschrift für Jagdwissenschaft 49:14–24.

Matson, G. 1988. Matson's Laboratory Progress Report No. 10. Winter 1988.

McCarthy, A., R. Blouch, D. Moore, and C. M. Wemmer. 1998. Deer: status survey and conservation action plan. IUCN, Species Survival Commission, Deer Specialist Group, Cambridge, United Kingdom.

McCulloch, C. Y. 1973. Seasonal diets of mule and white-tailed deer. Pages 1–37 *in* Deer nutrition in Arizona chaparral and desert habitats. Special Report No. 3. Arizona Game and Fish Department, Phoenix, USA.

McCulloch, C. Y., and R. H. Smith. 1991. Relationship of weather and other variables to the condition of the Kaibab deer herd. Federal Aid Project W-78-R, WP2, Job 12. Arizona Game and Fish Department, Phoenix, USA.

McKinney, T. 2003. Precipitation, weather and mule deer. Pages 219–238 *in* J. C. deVos, editor. Mule deer conservation: issues and management strategies. Berryman Press, Logan, Utah, USA.

Mellink, E. 2005. El venado bura de Baja California. Pages 363–372 *in* V. Sánchez Cordero and R. A. Medellín, editors. Contribuciones mastozoológicas en homenaje a Bernardo Villa. Instituto de Ecología, UNAM, and CONABIO, México, D. F., México.

Morafka, D. J. 1977. A biogeographical analysis of the Chihuahuan Desert through its herpetofauna. Dr. W. Junk B. V., The Hague, Netherlands.

Morafka, D. J. 1991. An interdisciplinary definition of the Chihuahuan Desert. Is it obtainable or desirable? Pages 23–34 *in* R. Kent and V. Harnapp, editors. 1989 Yearbook of the Conference of Latin American Geographers. Louisiana State University, Baton Rouge, USA.

Mubanga, G., J. L. Holechek, R. Valdez, and S. D. Schemnitz. 1985. Relationships between diet and fecal nutritive quality in mule deer. Southwestern Naturalist 30:573–578.

Nature Conservancy. 2009. The Apache Highlands Ecoregion. http://www.nature.org/wherewework/northamerica/states/arizona/preserves/art1942.html. Accessed 27 Jul 2009.

Neff, D. J. 1974. Forage preferences of trained mule deer on the Beaver Creek watersheds. Special Report No. 4. Arizona Game and Fish Department, Phoenix, USA.

Nichol, A. A. 1938. Experimental feeding of deer. University of Arizona Technical Bulletin 75:1–39.

O'Brien, C. S., R. B. Waddell, S. S. Rosenstock, and M. J. Rabe. 2006. Wildlife use of water catchments in southwestern Arizona. Wildlife Society Bulletin 34:582–591.

Ordway, L. L., and P. R. Krausman. 1986. Habitat use by desert mule deer. Journal of Wildlife Management 32:334–349.

Ortega-Sanchez, A. 2013. Evaluation of a translocated population of desert mule deer in the Chihuahuan Desert of northern Coahuila, Mexico. Dissertation, Texas A&M University, College Station, USA.

Peek, J. M., and P. R. Krausman. 1996. Grazing and mule deer. Pages 183–192 *in* P. R. Krausman, editor. Rangeland wildlife. Society for Range Management, Denver, Colorado, USA.

Pérez Gil, R. 1981. A preliminary study of deer from

Cedros Island, Baja California, Mexico. Thesis, School of Natural Resources, University of Michigan, Ann Arbor, USA.

Rautenstrauch, K. R., and P. R. Krausman. 1989. Influence of water availability and rainfall on movements of desert mule deer. Journal of Mammalogy 70:197–201.

Relyea, R. A. 1992. Desert mule deer movement and behavior in southwest Texas. Thesis, Texas Tech University, Lubbock, USA.

Relyea, R. A., and S. Demarais. 1994. Activity of desert mule deer during the breeding season. Journal of Mammalogy 75:940–949.

Robinette, W. L., N. V. Hancock, and D. A. Jones. 1977. The Oak Creek mule deer herd in Utah. Publication No. 77–15. Utah Division of Wildlife Resources, Salt Lake City, USA.

Rodgers, K. J. 1977. Seasonal movement of mule deer on the Santa Rita Experimental Range. Thesis, University of Arizona, Tucson, USA.

Rosas-Rosas, O. C., R. Valdez, L. C. Bender, and D. Daniel. 2003. Food habits of pumas in northwestern Sonora, Mexico. Wildlife Society Bulletin 31:528–535.

Rosenstock, S. S., M. J. Rabe, C. S. O'Brien, and R. B. Waddell. 2004. Studies of wildlife water developments in southwestern Arizona: wildlife use, water quality, wildlife diseases, wildlife mortalities, and influences on native pollinators. Technical Guidance Bulletin No. 8. Arizona Game and Fish Department, Phoenix, USA.

Sánchez Rojas, G., and S. Gallina. 1998. Densidad y patrones de agregación del venado bura (*Odocoileus hemionus*) en la Reserva de la Biósfera de Mapimí, Durango, México. VI Simposio sobre Venados en México. UNAM, Xalapa, Veracruz, México.

Sánchez Rojas, G., and S. Gallina. 2000a. Mule deer (*Odocoileus hemionus*) density in a landscape element of the Chihuahuan Desert, Mexico. Journal of Arid Environments 44:357–368.

Sánchez Rojas, G., and S. Gallina. 2000b. Factors affecting habitat use by mule deer (*Odocoileus hemionus*) in the central part of the Chihuahuan Desert, Mexico: an assessment with univariate and multivariate methods. Ethology, Ecology and Evolution 12:405–417.

Sánchez Rojas, G., and S. Gallina. 2003. Mule deer (*Odocoileus hemionus*) density in a landscape element of the Chihuahuan Desert, Mexico. Journal of Arid Environments 44:357–368.

Sánchez Rojas, G., and S. Gallina. 2007. La metapoblación del venado bura en la Reserva de la Biósfera de Mapimí, México: consideraciones para su conservación. Cuadernos de la biodiversidad: publicación cuatrimestral del Centro Iberoamericano de la Biodiversidad 22:7–15.

Short, H. L. 1977. Food habits of mule deer in a semidesert grass-shrub habitat. Journal of Range Management 30:206–209.

Smith, R. H., and A. LeCount. 1979. Some factors affecting survival of mule deer fawns. Journal of Wildlife Management 43:657–665.

Stoleson, S. H., R. S. Felger, G. Ceballos, C. Raish, M. F. Wilson, and A. Burquez. 2005. Recent history of natural resource use and population growth in northern Mexico. Pages 52–86 *in* J. L. E. Cartron, G. Ceballos, and R. S. Felger, editors. Biodiversity, ecosystems, and conservation in northern Mexico. Oxford University Press, New York, New York, USA.

Stone, W., and S. N. Rhoads. 1905. On a collection of birds and mammals from the Colorado delta, lower California. Proceedings of the Academy of Natural Sciences of Philadelphia 1905:676–690.

Swank, W. G. 1958. The mule deer in Arizona chaparral. Game Bulletin No. 3. Arizona Game and Fish Department, Phoenix, USA.

Tinkler, B. 1978. Mexican wilderness and wildlife. University of Texas Press, Austin, USA.

Truett, J. C. 1971. Ecology of the desert mule deer, *Odocoileus hemionus crooki* Mearns, in southeastern Arizona. Dissertation, University of Arizona, Tucson, USA.

Tull, J. C., P. R. Krausman, and R. J. Steidl. 2001. Bed-site selection by desert mule deer in southern Arizona. Southwestern Naturalist 46:354–357.

Ullrey, D. E. 1983. Nutrition and antler development in white-tailed deer. Pages 49–60 *in* R. D. Brown, editor. Antler development in Cervidae. Caesar Kleberg Wildlife Research Institute, Kingsville, Texas, USA.

Ullrey, D. E., W. G. Youatt, H. E. Johnson, L. D. Fay, and B. L. Bradley. 1970. Digestible and metabolizable energy requirements for winter maintenance of Michigan white-tailed does. Journal of Wildlife Management 34:863–869.

Urness, P. J., W. Green, and R. K. Watkins. 1971. Nutrient intake of deer in Arizona chaparral and desert habitats. Journal of Wildlife Management 35:469–475.

Valdez, R., J. C. Guzmán Aranda, F. J. Abarca, L. A. Tarango Arámbula, and F. Clemente Sánchez. 2006. Wildlife conservation and management in Mexico. Wildlife Society Bulletin 34:270–282.

Varner, W. L. 2002. Basic deer nutrition. Pages 46–54 *in* L. A. Harveson, P. M. Harveson, and C. Richardson, editors. Proceedings of the Trans-Pecos Wildlife Conference. Sul Ross State University, Alpine, Texas, USA.

Velázquez, V., and S. Reyes. 1976. El venado bura (*Odocoileus hemionus peninsulae*) en el sur de la Península de Baja California, México. Boletín de Fauna 7:1–31.

Walser, R. E. 2006. Influence of precipitation and the effects of season change on desert mule deer populations in Trans-Pecos, Texas. Thesis, Sul Ross State University, Alpine, Texas, USA.

Weber, M., and C. Galindo Leal. 2005. Venado bura. Pages 515–517 *in* G. Ceballos and G. Oliva, editors. Los mamíferos silvestres de México. CONABIO, México, D. F., México.

Wood, J. E., T. S. Bickle, W. Evans, J. C. Germany, and V. W. Howard Jr. 1970. The Fort Stanton mule deer herd: some ecological and life history characteristics with special emphasis on the use of water. Agricultural Experiment Station Bulletin 567. New Mexico State University, Las Cruces, USA.

17 White-Tailed Deer in Northern Mexico

Salvador Mandujano, Sonia Antonieta Gallina Tessaro, and J. Alfonso Ortega-S.

The two extant deer species of the genus *Odocoileus*, white-tailed deer (*O. virginianus*) and mule deer (*O. hemionus*), occur in Mexico and other regions of North America. However, during the Pleistocene, at least one species that is now extinct (*O. halli*) inhabited Mexico (Arroyo-Cabrales et al. 2007). White-tailed deer have been valued by many cultures since the pre-Hispanic period (Serra and Valdez 1989, Téllez 1996). This species has several names in indigenous languages: *mazatl* (Aztec); *ceh*, *uac*, *nac*, *keej*, *yalem keej* (fawn, Maya), *tankelem keej* (yearling, Maya), *nojoch keej* (adult, Maya); *axuni* (Tarasco); *guej* (Lacandon); *macha* (Huichol); *muxati* (Coras); and *phatehe* (Otomi) (Gallina et al. 2010b). The importance of white-tailed deer in Mexico is undeniable; they are a source of food, sport hunting, and handicrafts (see chapter 3). The use of white-tailed deer in religious ceremonies by indigenous groups has been documented by Mandujano and Rico Gray (1991), Montiel Ortega et al. (2000), González-Pérez and Briones-Salas (2000), and Naranjo et al. (2004). One of the days of the Aztec calendar was dedicated to deer, which reveals their symbolism as a deity since pre-Hispanic times. The importance of subsistence hunting in the present has been documented by M. Weber (see chapter 18), who reported that the largest volume of game meat harvested in the Calakmul region of Campeche was venison.

In northern Mexico, the economic value of white-tailed deer as the principal big game trophy species has become increasingly important since the 1990s. Ranchers have made land management changes to enhance habitat for white tailed-deer and forage production for domestic livestock; deer are an important component of ranch income (Ortega-S. et al. 2013). Although the white-tailed deer was close to extirpation in the 1960s, its economic value in northern Mexico provided an impetus for population recovery by triggering ranch management and conservation actions to increase populations to obtain additional income (Villarreal 1999). Presently, deer densities vary from 4.3 to 25 deer/km^2 (6.9 to 40.3 deer/mi^2) in the northern states of Tamaulipas, Nuevo León, and Coahuila (Ortega-S. et al. 2011). In this chapter, we review the ecology and management of white-tailed deer in diverse environments in Mexico; see chapter 18 for a review of white-tailed deer in tropical Mexico.

Distribution

The white-tailed deer is the most widely distributed cervid in the New World. It occurs from latitude 60° N in southern Canada through most of the US, throughout Central America, and in South America as far south as northern Brazil and southern Peru at latitude 15° S (Smith 1991, Gallina et al. 2010b). White-tailed deer occur in temperate pine, oak, and fir forests, mixed oak-pine forests, shrublands, tropical dry forests, semievergreen and evergreen tropical forests, and secondary vegetation communities. Of the 38 subspecies of white-tailed deer, 14 occur in Mexico, and only 2 (*texanus* and *carminis*) do not inhabit tropical forests; the other 12 subspecies include some tropical forest within their distributional range (Mandujano et al. 2010). This species is widespread in Mexico (figs. 17.1 and 17.2), except on the Baja California Peninsula and in some areas of northern Chihuahua and Sonora (Leopold 1959). The total potential distributional area in Mexico is estimated at 1,750,000–1,800,000 km^2 (2,822,580–2,903,225 mi^2) (Hall 1981, Villarreal 1999). However, this may be an overestimation since in some regions of Mexico deer may have been locally extirpated by overharvesting and habitat loss. For example, Heffelfinger (2006) included a map that indicated a wide area of central and northern Mexico where this species may be absent.

Classification of white-tailed deer subspecies is based on the anatomical characteristics (e.g., size,

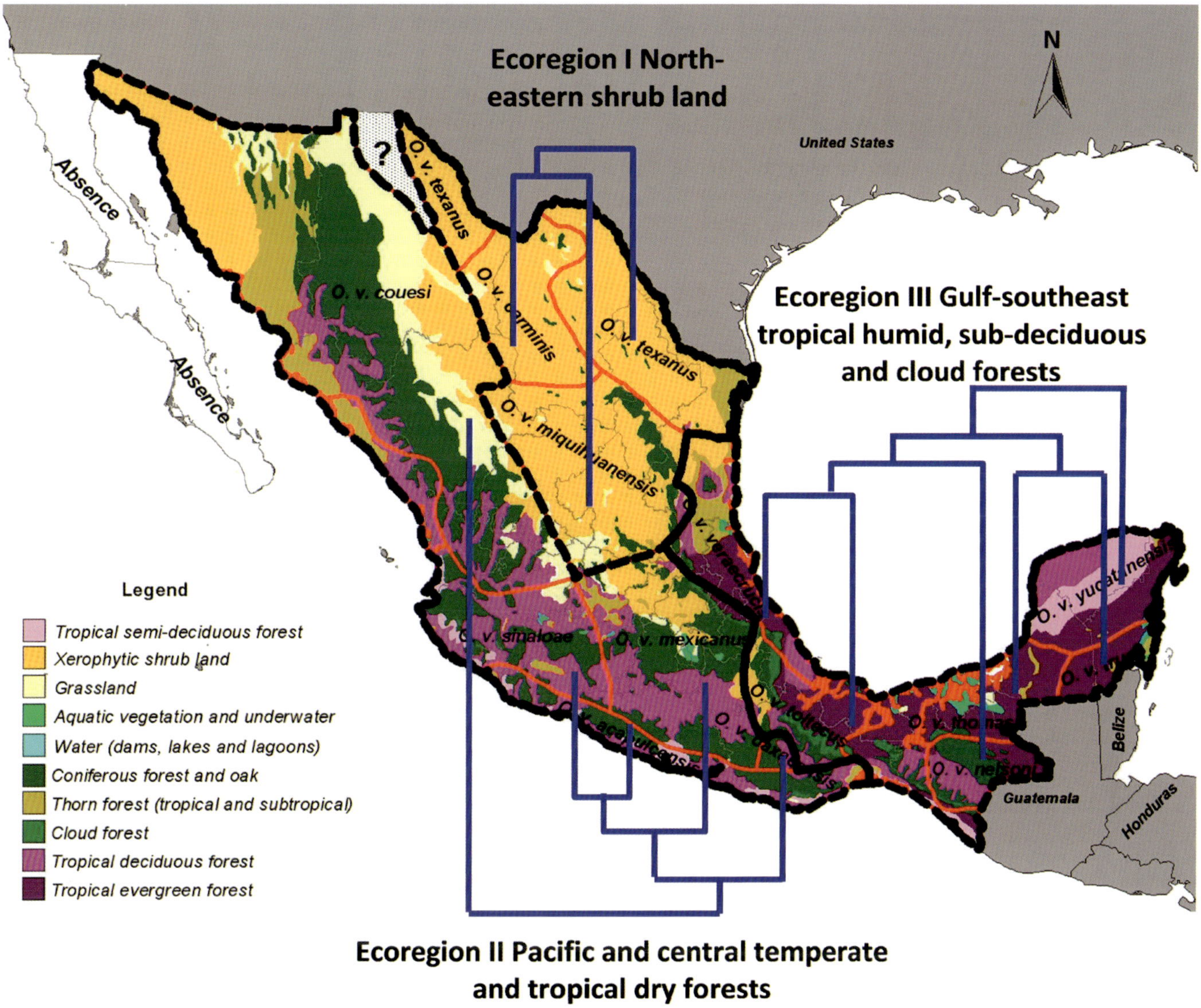

Fig. 17.1. Ecoregional distribution of white-tailed deer subspecies in Mexico, based on the distribution map of Hall (1981) and floristic provinces of Rzedowski and Reyna-Trujillo (1990).

Fig. 17.2. Excellent habitat for white-tailed deer in the Tamaulipan Biotic Province in the state of Nuevo León, northeastern Mexico. Photo by Jorge Villarreal González.

pelage color, and antler size and shape) of limited museum specimens (Kellogg 1956). There are few subspecies studies based on detailed or quantitative anatomical or genetic data in Mexico (Logan-López et al. 2006, 2007). The subspecies with the greatest distribution is *couesi* (fig. 17.3), which inhabits more than 500,000 km² (193,000 mi²), followed by *miquihuanensis* (fig. 17.4) (~200,000 km², 322,580 mi²); *mexicanus* (figs. 17.5 and 17.6); *texanus* (fig. 17.7), *sinaloae*, and *thomasi* (~106,000–172,000 km², 170,960–277,400 mi²). The subspecies *carminis*, *acapulcensis*, *veraecrucis*, and *yucatanensis* occur in an area of between 60,000 and 106,000 km² (96,700 and 170,960 mi²). The remaining subspecies, *toltecus*, *nelsoni*, *truei*, and *oaxacensis*, have the most restricted distributions (< 62,000 km², 100,000 mi²) (fig. 17.1). The state with the greatest number of subspecies is Oaxaca, with five or six, followed by Veracruz, Guanajuato, Durango, Jalisco, and Guerrero, with three or four. States with only one subspecies include Distrito Federal, Estado de México, Morelos, Tabasco, Tlaxcala, and Yucatán. The subspecies inhabiting the greatest number of states is *mexicanus* (13 of 30, including Distrito Federal), followed by *couesi*, *miquihuanensis* and *sinaloae*. The only subspecies limited to one state is *nelsoni*, while *oaxacensis* and *truei* are limited to two states (Mandujano et al. 2010). The most-studied subspecies are *texanus*, *couesi*, *sinaloae*, *mexicanus*, and *yucatanensis*. The least studied are *carminis*, *miquihuanensis*, *veraecrucis*, *toltecus*, *truei*, *nelsoni*, *oaxacensis*, *acapulcensis*, and *thomasi* (Mandujano 2004).

The subspecies *miquihuanensis* and *couesi* are found mainly in shrublands and temperate forests, respectively; however, they are also found marginally in dry tropical forests. The subspecies *sinaloae*, *mexicanus*, *acapulcensis*, and *yucatanensis* occur mostly in deciduous or tropical dry forest. However, *veraecrucis*, *toltecus*, *thomasi*, *nelsoni*, and *truei* inhabit mainly tropical evergreen, tropical semideciduous, and cloud forests (Mandujano et al. 2010).

De la Rosa-Reyna et al. (2012) collected samples from 13 sites within the distribution of five subspecies: *texanus*, *carminis*, *veraecrucis*, *sinaloae*, and *yucatanensis*. Diversity and genetic structure were estimated with 12 microsatellite markers. Heterozygosity was comparable to what was observed in other North American white-tailed deer except for *yucatanensis*, and the genetic structure was different among the

Fig. 17.3. The Coues subspecies of white-tailed deer has the greatest distribution in Mexico. Photo by Marco Aurelio Montaño Ayala.

Fig. 17.4 White-tailed deer (subspecies *miquihuanensis*) in the state of Nuevo León in northern Mexico. Photo by Jorge Villarreal González.

Fig. 17.5. White-tailed deer (subspecies *mexicanus*) from central Mexico. Photo by Salvador Mandujano.

Fig. 17.6. White-tailed deer habitat in the state of Puebla, Mexico. During the dry season, almost all plants drop their leaves, on which the deer in this region feed. Photo by Poncho Ortega.

Fig. 17.7. White-tailed deer (subspecies *texanus*) in northeastern Mexico. Photo by Jorge Villarreal González.

13 sampling sites. Molecular analysis of variance revealed that sampling sites grouped by subspecies or geographic region explained a moderate portion of the genetic variation. However, population differentiation was not minimized to a major grouping level, such as within subspecies or regional groups. Genetic distance among sampling sites was correlated with geographical distance, but some sites that were geographically close were genetically different, as in the state of Yucatán. The Yucatán white-tailed deer were genetically different from other subspecies and had the lowest genetic diversity, which is consistent with the biogeographic history of the region. The three subspecies of white-tailed deer in Michoacán

(*mexicanus*, *sinaloae*, and *acapulcencis*) exhibited moderate diversity among subspecies and low values of genetic differentiation (López-Cervantes 2010). White-tailed deer populations in Mexico pose diverse management challenges, and additional information is required to determine regional diversity and genetic structure.

Life History

Food habits

Food habits and diet composition of white-tailed deer and the nutritional value of plants consumed have been extensively studied in the thorn-scrub areas of northern Mexico, mainly within the distribution of *texanus* in Nuevo León, Coahuila, and Tamaulipas (Ramírez-Lozano 2004). These studies indicate that white-tailed deer select diverse plants and plant parts depending on nutritional requirements, which vary with the sex and age of animals, season of the year, and other factors. Some of the most common plants with high energy and protein content in deer diets include *Zephyranthes arenicola*, *Heliotropium angiospermum*, *Cynanchum barbigerum*, *Ruellia corzoi*, *Pithecellobium pallens*, *Prosopis glandulosa*, *Celtis pallida*, species of *Acacia* and *Opuntia*, *Polianthes maculosa*, *Sida filicaulis*, *Forestiera angustifolia*, *Leucaena leucocephala*, *Helietta parvifolia*, *Castela texana*, *Caesalpinia mexicana*, *Panicum hallii*, *Lantana macropoda*, *Larrea tridentata*, *Condalia lycioides*, *Aristida longiseta*, *Eragrostis mexicana*, *Cercidium macrum*, and *Solanum elaeagnifolium*. Refer to Ramírez-Lozano (2004) for an extensive list of plant species in white-tailed deer diets.

In the oak-pine forests of La Michilía Biosphere Reserve in the Sierra Madre Occidental in Durango, deer consumed 135 species from 41 families and preferred shrubs (51%) and trees (32%) (Gallina et al. 1981). The most important shrub species were mistletoe (*Phoradendron bolleanum* and *P. villosum*) (Gallina 1988), *Arctostaphylos* spp., *Pithecellobium leptophyllum*, and *Condalia hookeri*. Eighteen tree species were important in the diets, including *Juniperus deppeana*, *J. durangensis*, *Quercus* spp., *Pinus* spp., and *Arbutus glandulosa*. A seasonal difference in food habits was found during the wet season in September and early October, when the percentage of forbs increased nearly 30%. Brush and tree species with high protein, low cell wall content, high digestibility, and low seasonal variation were consumed by deer (Gallina 1993a). Deer dietary preferences did not differ among age categories. In the Sierra de San Blas in Pabellón, Aguascalientes, Clemente (1984) found that deer consumed 31 plant species from 17 families. Diets consisted of 49% forbs and 45% shrubs in summer; 15% forbs, 35% fruits, and 39% shrubs in autumn; and 20% forbs, 61% shrubs, and 18% trees in winter. Crude protein in plants consumed was 9%, 8%, and 6% in summer, autumn, and winter, respectively. Dry matter consumption estimated for a doe (weighing 45 kg, 99.2 lb), was 2.3 kg (5 lb) in summer, 1.9 kg (4.1 lb) in autumn, and 1.2 kg (2.6 lb) in winter. In the mountains of San Luis Potosí, Luévano-Esparza (1991) found that the most-consumed species were *Atriplex canescens*, *Dalea bicolor*, *Rhus virens*, *Opuntia* spp., *Solanum elaeagnifolium*, *Sphaeralcea angustifolia*, and *Lepidium* spp.

According to Arceo et al. (2005), white-tailed deer selected 82 plant species from 20 families in the Chamela tropical dry forest in Jalisco, with 12 species representing 50% of the annual diet. The main families were Euphorbiaceae, Leguminosae, Convolvulaceae, and Sapindaceae, which accounted for 80%, 64%, and 37% of the diet during the rainy, transition, and dry seasons, respectively. The most important species were *Acalypha langiana*, *Coursetia caribaea*, *Tephrosia leiocarpa*, *Cardiospermum halicacabum*, *Thouinia paucidentata*, *Ayenia micrantha*, *Spondias purpurea*, and *Rourea glabra*. Deer were more selective during the rainy season and became generalists during the dry season, when resources were scarce. Vegetative plant parts (leaves and twigs) constituted the basis of their annual diet, but during the dry season deer consumed significantly more flowers and fruits. Some of the most important fruits were those of *Spondias purpurea*, *Ficus* spp., *Brosimum alicastrum*, *Opuntia excelsa*, and several species of legumes. Consumption of grasses was low. In this region, Silva-Villalobos et al. (1999) found that plants consumed by deer had more nitrogen-free extract, protein, and polyphenols and less fiber than did plants that were not consumed. Tropical dry forest plants had higher nutritional value during the rainy season than during the dry season. Tropical semievergreen forest plants were higher in fiber and lower in protein and nitrogen-free extract throughout the year. Fiber was negatively correlated with nitrogen-free extract and protein content.

In the tropical dry forest in the Mixteca region of southern Puebla, Villarreal-Espino (2006) recorded 133 plant species in 50 families in white-tailed deer diets; legumes represented 20%, followed by Cactaceae (14%), Gramineae (7%), and Asteraceae and Agavaceae (6%). The main species in deer diets were *Agave kerchovei, Quercus glaucoides, Acacia pennatula, Acacia subangulata, Leucaena leucocephala, Eysenhardtia polystachya, Pithecellobium dulce, Haematoxylum brasiletto, Harpalyce loeseneriana, Mimosa luciana, Cercidium praecox, Pachyrhizus* spp., *Waltheria americana*, and *Montanoa* spp. Less important plants included *Quercus castanea, Castela tortuosa, Ceiba parvifolia*, and *Lippia graveolens*.

Percentage of fecal nitrogen (FN) has been used as an index for habitat quality and use. In the dry tropical forests of the Sierra de Huautla, Morelos, Buenrostro-Silva et al. (2008) reported higher concentrations of FN for females than for males. They suggested sexual segregation was responsible for the difference in diet quality because does used habitats with higher-quality plants.

Reproduction

Compared with other aspects of the ecology of this species, the reproductive biology of white-tailed deer has been little studied in Mexico. Weber et al. (1994) studied the reproductive biology of *couesi* in La Michilía Biosphere Reserve in Durango, Mexico. Average mating date was February 12, and the reproductive season lasted approximately 95 days, beginning in January and ending in early April. Estrus (heat) lasted 18 ± 1.5 hours. There were no differences between the mating dates of wild and captive deer. Fawns were born in August and September after an average gestation period of 199 days. There was a significant difference between the length of gestation periods for single fawns (197.3 days) and twin fawns (200.3 days). The greatest testicular diameter occurred in February, while antlers were dropped at the end of May. Velvet was shed in October, but juvenile males retained the velvet two weeks longer than adult males. Studies conducted in captivity showed that the breeding season in *couesi* began in December and ended in March (Rosas-Becerril 1992). In the Chapultepec Zoo in Mexico City, the deer breeding season was from October to December (Rosas-Alvarado 1994).

Martínez-Romero (2004) analyzed estradiol and progesterone levels in captive females and testosterone in males of the *mexicanus* subspecies in the Flor del Bosque Protected Area in Puebla and found that the reproductive season started with hormonal changes in November, but copulation did not take place until January. Changes in olfactory frequency and pursuit coincided with hormonal changes and the fertility cycle. Testosterone content in feces was lost after 9.6 days, progesterone after 5.9 days, and estradiol after 1 month. The ratio of estradiol to testosterone in fecal pellets can be used to determine sex ratios in free-ranging deer populations.

In tropical regions of Mexico, reproduction in white-tailed deer tends to be much less seasonal than it is in temperate environments (Galindo-Leal and Weber 1997). In the tropical dry forest of Chamela, Jalisco, *sinaloae* breeds between November and January, and fawns are born between June and August (Mandujano and Gallina 1996). At the Sierra de Huautla Biosphere Reserve in Morelos, *mexicanus* fawns are born during June and July, and newborn twins are commonly observed in autumn (Buenrostro-Silva et al. 2008). In this region, at the beginning of the reproductive season males rub their antlers against shrubs and small trees to remove the velvet in October and November; bucks select shrubs and small trees such as *Lysiloma divaricatum* and *Conzattia multiflora*. In the Calakmul region of Campeche, young fawns and/or fetuses in harvested females, as well as males with velvet antlers, have been observed almost year-round, suggesting a lack of reproductive seasonality at this latitude (Weber 2005).

Physiology

According to Weber and Hidalgo (1999), at La Michilía, Durango, the rate of growth of captive *couesi* male fawns was 225.5 g/day (7.9 oz/day), with an average of 112 days of nursing and an average weaning weight of 14.5 kg (31.9 lb). Mature females reached their maximum growth at three years of age, while males continue growing and may increase in weight until they reach five years. In this study, the measurements that best differentiated male and female growth were total body length and hind leg length.

Energy requirements depend on basal metabolism, daily activity, and physiological stage. Few studies have been conducted on this topic in Mexico. According to Gallina and Bello (2006), males of *texanus* spend more energy (1,726 ± 38 kcal/day/

animal) than females (1,556 ± 35 kcal/day/animal). Energy expenditure during feeding is less than during rest or motion. Apparently, deer adjust their energy expenditure depending on environmental conditions.

Water demand at the population level was estimated at 12 liters/ha (3.17 gal/2.4 ac) at the end of the dry season in the tropical dry forest of Jalisco (Mandujano and Gallina 1995). An important food resource in this habitat was the fruits of *Spondias purpurea* (Anacardiaceae) (Mandujano et al. 1994). These fruits constituted 13% of the diet during the dry season (Arceo et al. 2005) and could contribute 2–30 liters/ha (0.53–7.93 gal/2.4 ac) of water depending on annual fruit production (Mandujano and Gallina 1995). In the Mixteca tropical dry forest of Puebla, Villarreal-Espino and Marín (2005) reported that *mexicanus* consumed 65% fruits, 39% flowers, and 8% *Opuntia* spp. Flowers of *Ceiba parvifolia* had the greatest water content, followed by the flowers of *Pachycereus weberi* and fruits of *Ficus* spp. and *Spondias* spp.

Information on thermal relationships of white-tailed deer in central and southern Mexico is limited. Deer select specific sites in which to lie down, ruminate, and rest, and sites that provide protective cover for their offspring. These sites need to offer protection from predators, as well as thermal cover that allows deer to minimize both the absorption of heat from exposure to the sun and loss of water as a consequence of high temperatures. Cover is more important in arid and semiarid zones, where it is needed to help prevent water loss during times of high temperature (more than 40°C, 104°F) and limited water availability, which prevail for at least five months of the year (February to June). Females of the *texanus* subspecies selected significantly different resting sites than bucks, which preferred sites with higher thermic cover, volume, and vegetation height (Gallina et al. 2010a). The most common brush species found in loafing sites were *Prosopis glandulosa*, *Acacia farnesiana*, and *A. rigidula*. These sites were close to water tanks and were used between 1100 and 1400 hours, when temperatures were generally higher.

Behavior

White-tailed deer behavior is very complex and may change seasonally and from year to year. Deer displayed 106 behavioral patterns at the Chapultepec Zoo (Rosas-Alvarado 1994). In the Bosque La Primavera Laboratory in Guadalajara, Jalisco, deer displayed 24 behavioral patterns, but only 3 (loafing, feeding, and traveling) could be identified with radio tracking (Mandujano et al. 1996). These activities are important for understanding energy budgets (Gallina and Bello 2006).

White-tailed deer form two basic social groups: family groups centered on a matriarch (the mother and female fawns of previous generations); and fraternal groups (adults and occasionally yearling males) (Marchinton and Hirth 1984). Mixed groups occur only during the rut. It is possible to find mixed feeding groups, but these assemblages are temporary. Dominance hierarchies minimize conflicts and avoid aggression within groups, reducing energy expenditure and risk of injury. Social organization and behavior vary in relation to habitat characteristics, and group size is inversely related to the density of vegetative cover (Hirth 1977). In central Mexico, it is common to find only does with their fawns. In the tropical dry forest of Chamela, Jalisco, small groups and solitary individuals were frequently observed. Mean size of groups was 1.5 individuals (range = 1.0 to 1.9). Solitary individuals were frequent throughout the year. Females with fawns were the most frequent social group (Mandujano and Gallina 1996).

Movements and home range

Usable space required by deer varies depending on sex and age, season of the year, habitat productivity, and juxtaposition and interspersion of habitat components (Fulbright and Ortega-S. 2013). Home ranges of adult deer remain fairly constant from year to year (Tierson et al. 1985, Inglis et al. 1986); however, home range fidelity may vary from season to season (Hellickson et al. 2008).

Home ranges vary seasonally. Gallina and Bello (1998) reported an increase in the home ranges of males and females during the reproductive season; lactating females had the smallest home ranges. However, in areas with low population density, males may travel longer distances to find females in estrus; under these conditions, home ranges would increase (Fulbright and Ortega-S. 2013). In the tropical dry forest on the Pacific coast in Mexico, home ranges of *sinaloae* females and males during the dry season were 11 and 26 ha (27 and 64 ac), respectively. In another study of the same subspecies in a cloud forest of the Sierra de Manantlán Biosphere Reserve, Jalisco, home range size for females was 14.5 ha (35.8 ac) during the wet season and 16.9 ha (41.7 ac) in the dry season (González-Pérez 2003). Activity centers are defined as areas within the home range where deer spend time feeding, loafing, or bedding; these areas are therefore visited at different times (Inglis et al. 1986). The size of activity centers for white-tailed deer in northern Mexico varied from 59 to 63 ha (145 to 155 ac) (Gallina and Bello 1998). Those in South Texas varied from 0.8 to 15 ha (1.9 to 37 ac) (Inglis et al. 1986). Home ranges for white-tailed deer bucks in Nuevo León, northern Mexico, were 255 ha (630 ac) in a dry year and 198 ha (489 ac) in a wet year, while for does these values were 171 and 212 ha (422 and 523 ac) for dry and wet years, respectively.

Population Dynamics

Density

Estimation of population density has been one of the main objectives of white-tailed deer studies in Mexico. Considerable site-to-site variation is evident from available estimations of population density among regions and vegetation types. Sampling error, hunting intensity, and differences in sampling and analytical techniques contribute to the variation. Ecological factors that determine carrying capacity and human impact are fundamental in explaining density patterns. For example, shrub habitats of northern Mexico, temperate oak-pine forests, and tropical dry forests could support higher densities than tropical semideciduous forests and rainforests. On the other hand, protected areas and some managed areas may maintain higher densities than heavily disturbed areas or areas where hunting is not regulated (Gallina 2012). Using rainfall as a predictive variable to estimate ungulate biomass in Neotropical forests, Mandujano and Naranjo (2010) estimated that temperate forests and tropical dry forests could support a biomass three times greater than that in tropical rainforests.

Recruitment, survival, and mortality factors

In the pine-oak forest of La Michilía Biosphere Reserve in Durango, Gallina (1994) suggested that white-tailed deer density varied yearly according to the previous year's rainfall. Age structure was 28% fawns, 51% yearlings, and 21% adults. Survival rates were 0.87 for fawns to yearlings, 0.50 for female subadults to adults, and 0.30 for male subadults to adults. Fecundity in adult females was 1.27, and in subadults 0.64, with an average population growth of 0.12. In the tropical dry forest in Jalisco, age structure was 44% adults, 25% juveniles, and 31% fawns, with a sex ratio of three to four females per one male. Fecundity rate was 1.03 fawns per adult female; females did not fawn until their second year. Survival rate for fawns to juveniles was 0.79, and for juveniles to adults 0.63. The first dry season for fawns is critical in determining their survival; therefore, years with low precipitation could increase mortality and be a limiting factor for population increase. Predation was important for fawn survival at the end of the dry season, when fawns were more vulnerable because of food scarcity and less secure cover (Mandujano and Gallina 2005).

Ungulate population dynamics are affected by a combination of environmental and density-dependent factors. Environmental factors include rainfall, temperature, food availability, competition, predation, and hunting. These factors determine the physical condition of females before the rut and during gestation, the birth date and birth weight of fawns, vulnerability to diseases and depredation, and nutritional needs for growth and fat accumulation in juveniles. Other factors influencing vulnerability to depredation of dispersing young bucks include the age at which they reach sexual maturity, weight, antler growth, and the general condition of the deer herd. Also, these factors may be influenced by the social behavior of individuals, as well as density-dependent mechanisms that function when the population reaches carrying capacity. In environments with well-defined seasons, population dynamics are influenced by density-dependent factors determined by climatic conditions affecting food availability. In temperate habitats in northern Mexico, the critical period occurs in the dry season when food availability is low. In semiarid environments and in some types of tropical habitats, the critical season occurs during the dry season, when water is scarce and forage availability and nutritive values are low.

Diseases

Cantú et al. (2007) reported the presence of *Babesia bovis* and *B. bigemina*, disease agents responsible for bovine piroplasmosis, or tick fever, which is lethal in cattle in northern Mexico. In another study, 427 blood samples were collected from white-tailed deer in Nuevo León, Tamaulipas, and Coahuila, 4.2% of which were positive for *B. bigemina* and 1.7% for *B. bovis*. Additionally, 87% of deer samples were positive for *Babesia* spp. antibodies (Cantú et al. 2009). This suggests that white-tailed deer may act as a reservoir for *B. bigemina* and *B. bovis*, but it is not known whether *Babesia* spp. may be able to complete the transmission cycle in white-tailed deer.

In another study in northern Mexico, Martinez et al. (1999) collected blood samples from 350 white-tailed deer to determine the prevalence of antibodies against five infectious agents affecting ruminants. They reported a prevalence of 87% for the different types of bluetongue virus, 72% for the epizootic hemorrhagic disease virus, 3% for *Borrelia burgdorferi*, and 69% for *Anaplasma marginale*; none were positive for *Brucella abortus*, *B. melitensis*, or *B. ovis*. Cantú et al. (2008) reported that 5.6% of white-tailed deer samples were positive for antibodies against leptospirosis, 41.1% for bovine infectious *rhinotracheitis*, 63.5% for bovine viral diarrhea, and 0% for brucellosis.

Romero-Castañón et al. (2008) surveyed gastrointestinal parasites and ectoparasites in wild and domestic ungulates in the Selva Lacandona, Chiapas, including those of Baird's tapir (*Tapirus bairdii*), collared peccary (*Pecari tajacu*), white-lipped peccary (*Tayassu pecari*), white-tailed deer, red brocket deer (*Mazama americana*), horses (*Equus caballus*), cattle (*Bos taurus*), and pigs (*Sus scrofa*). Ectoparasites included 14 species of Ixodidae and one hippoboscid; of the endoparasites, 10 were new host records and 15 were new geographical records for Mexico.

Montes et al. (1998) studied gastrointestinal parasites and egg and oocyst output in the feces of captive white-tailed deer (*yucatanensis*) in Yucatán. Seven genera were determined (*Haemonchus* spp., *Cooperia* spp., *Isospora* spp., *Eimeria* spp., *Trichuris* spp., *Strongyloides* spp., and *Moniezia* spp.), which represented five orders. The most frequent genera were *Haemonchus*, *Isospora*, and *Eimeria*. This was the first report of the genus *Isospora* in this region. Frequency and level of fecal egg and oocyst outputs were variable during the year and increased during the rainy season. Coccidia and strongylida were positively correlated with relative humidity, environmental temperature, and rainfall.

Predation

The effect of predators on white-tailed deer is controversial because of complex interactions with other factors. Main predators of white-tailed deer in Mexico include mountain lions (*Puma concolor*), coyotes (*Canis latrans*), and bobcats (*Lynx rufus*) (Villarreal 1999). In general, white-tailed deer fawns usually require brushy vegetation cover for bed sites in areas with medium to tall grasses. Wildlife operations that eliminate cattle grazing often observe an increase in fawn survival along with an increasing canopy cover of grasses. The reduction in fawn mortality is sometimes attributed to the increase in hiding cover provided by medium to tall grasses (Fulbright and Ortega-S. 2013). Grazing strategies involving a high percentage of herbaceous vegetation use, such as short-duration grazing, experienced increased fawn predation by coyotes as compared to continuous stocking (Hyde et al. 1987). Information on predator control in Mexico is scarce. Ezcurra and Gallina (1981) found that predator removal in La Michilía, Durango, did not reverse the declining trend in the deer population. In North America, Ballard et al. (2001) indicated that predator control was successful when predation was identified as a limiting factor, deer populations were below carrying capacity, and the local coyote population was reduced by about 70%, among other factors.

Current Status and Population Trends

Current status and trends of white-tailed deer in Mexico can be considered positive, especially in northern Mexico. In the 1970s, the white-tailed deer was almost extinct in northern Mexico (Villarreal 1999). It is encouraging that densities can reach 25 deer/km^2 (40.3 deer/mi^2) in brushlands of Coahuila and Nuevo León (Ortega-S. et al. 2011). However, in tropical forests, densities may be less than 1 deer/km^2 (1.61 deer/mi^2). Among other factors, subsistence hunting has a marked effect on white-tailed deer populations (see chapter 18).

Habitat Requirements

Habitat requirements for white-tailed deer in northern Mexico have been discussed by Fulbright and Ortega-S. (2013). Optimal white-tailed deer habitats include mottes of dense, diverse brush vegetation interspersed across the landscape to provide good-quality cover for bucks and does. In more arid environments, the presence of prickly pear (*Opuntia* spp.) and other succulents is very important. Cover of medium and tall grasses close to brush cover is important to provide resting and hiding cover for fawns. Proper grazing management of domestic animals is essential to maintain this condition. Brush vegetation interspersed with clearings dominated by diverse herbaceous plant communities is indispensable in providing feeding areas. Drainage areas in the landscape are important because moisture accumulation is needed for diverse vegetation, and these areas can also serve as movement corridors. The juxtaposition and interspersion of superficial water sources, along with the other habitat characteristics mentioned, are necessary to produce the classes of forage needed to fulfill the nutritional requirements of deer. Maximum and minimum values for water consumption of fawns, juveniles, and adults varied from 0.8 to 3.9 liters (0.21 to 1.03 gal) per day in Jalisco (Mandujano 2004).

Habitat use

In mixed pine-oak forests in Durango, *couesi* preferred flat sites such as mesas or elevated areas, where plant diversity and available biomass were greater (Gallina 1994). Habitat use by the same subspecies in the Sierra del Laurel in Aguascalientes was associated with vegetation, slope, altitude, and distance to water (Medina-Torres et al. 2008). In this region, high habitat use occurred in subtropical shrubland less than 2 m (6.5 ft) in height, and in oak forest with secondary vegetation. Deer selected sites with slopes of 30% or less, and distance to a watering source was between 537 and 846 m (1,761 and 2,775 ft). In a temperate forest of the Sierra Norte of Oaxaca, *oaxacensis* used four vegetation associations (Ortíz-Martínez et al. 2005), but the highest densities (1.7 deer/km^2, 2.7 deer/mi^2) were in *Abies-Pinus* and *Quercus-Pinus* vegetation associations, in which shrub density, height, cover, and volume, as well as the ruggedness of the topography, were significantly greater than in *Pinus-Quercus* (0.34 deer/km^2, 0.54 deer/mi^2) and *Pinus-Abies* (0.71 deer/km^2, 1.1 deer/mi^2) associations. The *Abies-Pinus* vegetation association was characterized by a dense canopy, high plant diversity, and short distance to a permanent watering source (average 153 m, 501 ft). Among the four vegetation associations, the *Abies-Pinus* and *Quercus-Pinus* associations had food availability, horizontal and vertical cover protection, and favorable escape cover. In Aguascalientes, Mexico, areas with higher deer densities (4–11 deer/km^2, 6.4–17.7 deer/mi^2) were associated with pine-oak forests with a developed understory, north exposure, higher humidity, water availability, higher species diversity, low cattle density, and forest management (Villalobos 1998), as well as more protection from predators (Nuñez et al. 2000).

In Bajo Balsas, Michoacán, Yañez (2009) found that slope (12 to 20%), vegetation type (pine-oak forest and tropical dry forest), and human activities and density were the main variables that affected the distribution of white-tailed deer at the regional scale (700,000 km^2, 2,702,715 mi^2). At the local scale, deer densities were associated with slope, tree species composition, and higher protective cover. González-Marín et al. (2008) found that *yucatanensis* at El Eden Ecological Reserve in Quintana Roo were concentrated in secondary vegetation (acahual) that represented 37.2% of the total surface area, where four other types occurred: tropical forest (26.4%), tintal (18.2%), palm savanna (10.8%), and savanna with aquatic vegetation (7.4%). In the acahual, deer found abundant food species, including *Metopium brownei*, *Lysiloma latisiliqua*, and *Gymnopodium floribundum*, as well as dense vegetation for protection from predators and unfavorable weather.

Habitat improvement

Habitat management in the US and parts of northern Mexico has been common for several decades (Villarreal 1999, Fulbright and Ortega-S. 2013), but it is less common in central and southern Mexico, where habitat management for white-tailed deer has been implemented only in specific sites. For example, in the tropical dry forests of the Mixteca Poblana (Villarreal-Espino 2006) and Sierra Costa of Michoacán (Mendoza et al. 2006), some of the practices applied include identification of areas for different activities, and planting of browse species such as red mombin (*Spondias purpurea*), prickly pear (*Opuntia* spp.), leucaena (*Leucaena leucocephala*), and huizache (*Acacia farnesiana*), among others. Watershed management includes reforestation for water recharge, protection of watering springs, use of water troughs to improve water distribution, food plots with plants that may serve as water sources during the dry season, small areas planted with corn (*Zea mays*) and beans (*Phaseolus vulgaris*), and reforestation and soil stabilization in riparian areas.

Other practices include improving forage availability for white-tailed deer by shredding native vegetation to promote regrowth of browse species, and planting valuable browse species. Recovering degraded areas, using prescribed fire, and avoiding the establishment of introduced pastures for cattle or poorly planned clearings are also used for vegetation management. Thousands of hectares are integrated in these activities (Mendoza et al. 2006). Elsewhere, a preliminary analysis suggests that precipitation of less than 120 mm (4.7 in) from October to May did not satisfy water requirements for plants and deer in a tropical dry forest in Jalisco; therefore, it would be necessary to improve water availability through the end of the dry season (Mandujano and Gallina 1995).

Deer management in cattle operations

Deer and cattle can coexist because of their preference for different plant species. Deer prefer shrub and tree species and cattle prefer grasses, and they maintain these patterns almost throughout the year (Gallina 1993b). Although this remains a controversial topic, some evidence suggests that proper cattle grazing may be beneficial for deer habitat. In tropical regions, Mandujano and Naranjo (2010) suggested that tropical dry forest can support a higher wild ungulate and cattle biomass than tropical wet forest. Several studies have demonstrated that trees and shrubs in tropical dry forests are better food sources for livestock than those in tropical wet forests (Carranza Montaño et al. 2003). Nonetheless, cattle management practices differ between these tropical forests. For example, the original rainforest landscape has been largely transformed in the region of Montes Azules, Chiapas, to maintain livestock on pastures planted with introduced grasses. As a consequence, cattle densities are commonly 100–150 cows/km^2 (161–241/mi^2) (45,000–67,500 kg/km^2 [160,011–240,017 lb/mi^2], based on a mean weight of 450 kg [992 lb] per head), which often leads to overgrazing (Mandujano and Naranjo 2010). In contrast, in regions dominated by tropical dry forests, cattle are frequently set free to graze, which causes changes in the original forest cover and supports lower cattle densities (e.g., 10–14 cows/km^2 [16.2–22.5 cows/mi^2], or 4,500–6,300 kg/km^2 [16,000–22,400 lb/mi^2]). This is similar to the ungulate biomass values reported in other Neotropical and Palaeotropical dry forests. Cattle can influence deer distribution and abundance, because deer tend to avoid human and dog encounters (Corona 2003). Additional research is needed to document possible parasite transmission between white-tailed deer and domestic animals, especially goats and sheep, which may compete for food with white-tailed deer when introduced to tropical dry forests (Villarreal-Espino 2006).

Productivity and carrying capacity

Few estimations of carrying capacity (K) exist in Mexico. In La Michilía Biosphere Reserve, Durango, Gallina (1993b) estimated K based on available plant biomass given that both deer and cattle consume forbs, cattle consume mainly grasses, and deer consume shrubs and tree leaves. In mixed pine-oak forests, densities of 0.22 deer/ha and 0.16 cows/ha (0.09 deer/ac and 0.06 cows/ac) can coexist, whereas grasslands can support only 0.0025 deer/ha and 0.22 cows/ha (0.001 deer/ac and 0.09 cows/ac). The exclusive use of habitat by one ruminant species means it should be possible to maintain 0.41 deer/ha (0.17 deer/ac) in mixed pine-oak forests and 0.22 cows/ha (0.09 cows/ac) in grasslands. In tropical dry forests in Jalisco, Mandujano (2007) estimated K between 16 and 18 deer/km^2 (9.9 and 11.1 deer/mi^2), while in the Mixteca region in southern Mexico, Villarreal-Espino (2006) estimated a K of 10 deer/km^2 (16.1 deer/mi^2). In both regions, population density is below carrying capacity. Mortality from predators and hunting, and food scarcity during the dry season, seem to be limiting factors. Additionally, possible density-dependent effects between population density and fecundity rate have been suggested in Chamela, Jalisco (Mandujano and Gallina 2005, Mandujano 2007).

Sustainable Use

White-tailed deer are highly valued in Mexico as a source of protein and artisanal crafts, and they are an important component in the rituals of indigenous cultures (Mandujano and Rico-Gray 1991, Greenberg 1992, González-Pérez and Briones-Salas 2000, Naranjo et al. 2004, León and Montiel 2008). Hunting of this species continues to be important for subsistence and sport. The white-tailed deer is the most important big game species in Mexico, and sport hunting has increased considerably in UMAs (wildlife management units for conservation) (González-Marín et al. 2003, Segovia-Castillo and Hernández-Betancourt 2003), providing an important source of income (Villarreal-Espino 2006).

In Mexico there are few estimates of maximum sustainable yields (MSY). A comparison of hunting practices and preferences of Lacandon, Tzeltal, and Mestizo hunters from communities adjacent to Montes Azules Biosphere Reserve in the Lacandon Forest, Chiapas, revealed that wildlife was hunted primarily for food and to reduce crop damage. Total biomass extracted from 32 wildlife species was 8,160 kg/yr (17,989 lb/yr). Ungulates and rodents made up 87% of the total biomass harvested. Harvest rates were positively correlated with the intrinsic rate of natural increase of species. Species that had higher reproductive rates were hunted more frequently. There were no correlations between harvest rates and body mass, standing biomass, density, or local economic value of the game species. Therefore, natural rates of increase should be considered when managing subsistence hunting, and hunting should be regulated, preferably through community-based management, for the benefit of resident communities and local wildlife populations (Naranjo et al. 2004).

In a protected tropical dry forest in Chamela, Jalisco, on the Pacific coast, Mandujano (2008) estimated the number of deer and peccaries that could be sustainably hunted. The population equilibrium point, estimated from the density-dependent relationship between the finite population growth rate and the current density, suggested an estimated carrying capacity of 16.5–17.2 deer/km^2 (25.9–27.7 deer/mi^2) and 9.3–9.5 peccaries/km^2 (15–15.3 peccaries/mi^2). The estimated potential production for human use was 2.1 deer/km^2 (3.3 deer/mi^2) and 4.4 peccaries/km^2 (7 peccaries/mi^2) based on the model of Robinson and Redford (1991).

White-Tailed Deer Sport Hunting

Only 5 of the 14 white-tailed deer subspecies in Mexico are included in current trophy record books, such as those of the Boone and Crockett Club and Safari Club International; these subspecies are the focus of management because of their economic value. They are *texanus*, *couesi*, *carminis*, *miquihuanensis*, and in recent years, *mexicanus* (Villarreal-Espino 2002). The remaining subspecies are not formally recognized as trophies by national or international sport hunters because of their smaller antler size and the lack of an individual category in the trophy record books. This situation is beginning to change, however, and greater regional importance is being placed on the hunting of each of the 14 individual subspecies (Villarreal 2009). As a result, one worrying aspect of white-tailed deer management is the translocation of subspecies in Mexico into areas where they were not historically found. Because of this possibility, SEMARNAT requires that any animal that is translocated is of the same subspecies as that found in the area of translocation. This legislation is part of Article 81 of the General Wildlife Law of Mexico.

This is particularly important in large UMAs, where animals are often released with little regard for hybridization. For this reason, there is an urgent need for a geographical information database that would allow authorities and ranch owners to gain a better understanding of the subspecies they manage, and to maintain the genetic variability of the species and the integrity of the subspecies. The Comisión Estatal de Flora y Fauna Silvestre de Nuevo León (CEFFSNL) recently proposed the inclusion of nine hunting regions, with the objective of recognizing the 14 different subspecies of white-tailed deer in Mexico as trophy categories (Villarreal 2009).

Translocation Implications

The lack of data supporting the validity of the biogeographical limits of white-tailed deer subspecies means that there should be strict control over the deliberate, or even accidental, translocation of subspecies to areas where they have not been historically reported. One important challenge is to define the geographic limits of each subspecies. The translocation of animal species has resulted in potential interspecific competition and the introduction of diseases and parasites (Kock et al. 2007). In Mexico, this has become a problem at the intraspecific level. For example, Galindo-Leal and Weber (1994) reported incidences of dystocia (difficulty in parturition), probably caused by the translocation of animals of one subspecies, *texanus*, into the distribution range of *couesi* in La Michilía, Durango. Unfortunately, the indiscriminate translocation of *texanus* is common, with the intention of "improving the breed," primarily because of its larger antlers. There is growing opposition to this practice. Although CEFFSNL had always supported sport hunting because of its economic benefit to ranches, it advocated a policy that opposed the translocation of *texanus* from northern Mexico (Coahuila, Nuevo León, and Tamaulipas) to other regions of Mexico, even under regulated management schemes such as those in UMAs, to avoid the hybridization of subspecies that are adapted to different ecological conditions (Villarreal 2009).

Reintroduction or release of deer in areas where they have been extirpated has been practiced mainly in northern Mexico, but this is poorly documented. To measure the success of introducing captive-bred deer into the wild, 30 captive white-tailed deer fitted with radio collars were released in the Sierra Madre Occidental (Laundré et al. 2007). Deer experienced high mortality; 21 days after their release, 28 out of the 30 animals had died, and the remaining animals did not survive for more than 150 days. The main cause of death was predation, with coyotes and domestic dogs accounting for 67% of the mortality. Suggestions for future releases included increasing the number of deer released to a minimum of 40 animals and controlling the level of predation.

Management Challenges in Mexico

One of the most important challenges for conservation and management of white-tailed deer in Mexico is education of the general public, especially in rural areas, about the importance and ecological and economic value of conserving white-tailed deer. The economic value of the white-tailed deer for sport hunting in northern Mexico has facilitated its conservation, but similar management programs are still needed in central and southern Mexico. Maintaining the integrity of white-tailed deer subspecies is another crucial challenge for wildlife conservation in Mexico. Translocation of white-tailed deer individuals from captive facilities for repopulation purposes should be limited to areas within the distributional range of each subspecies to avoid hybridization.

Research efforts on white-tailed deer in the last decade have generated a great amount of information, especially in northern Mexico, but specific information about the ecology and biology of individual subspecies is needed to develop appropriate management programs in other regions of Mexico. Ecotourism and sport hunting represent an opportunity to promote a culture of conservation for the species, which should be promoted especially in central and southern Mexico where the economic value of this species has yet to be realized.

Additional Sources of Information

This chapter integrates the basic biological and management information available for white-tailed deer in Mexico. Additional sources of information include Leopold (1959), Galindo-Leal and Weber (2005), Gallina et al. (2010b), and Ortega-S.et al. (2011). Additionally, books devoted to white-tailed deer ecology and management include Ffolliott and Gallina (1981), Yah (1983), Vaughan and Rodríguez (1994), Galindo-Leal and Weber (1997), Villarreal (1999), Ramírez-Lozano (2004, 2012), Villarreal-Espino (2006), Fulbright and Ortega-S. (2007, 2013), Rojo-Curiel et al. (2007), Mandujano et al. (2010), and Mandujano (2011).

Literature Cited

Arceo, G., S. Mandujano, S. Gallina, and L. A. Perez-Jimenez. 2005. Diet diversity of white-tailed deer in a Mexican tropical forest. Mammalia 69:159–168.

Arroyo-Cabrales, J., O. J. Polaco, and E. Johnson. 2007. An overview of the Quaternary mammals from Mexico. Courier Forschungsinstitut Senckenberg 259:191–203.

Ballard, W. B., D. Lutz, T. W. Keegan, L. H. Carpenter, and J. C. deVos. 2001. Deer-predator relationships: a review of recent North American studies with emphasis on mule and black-tailed deer. Wildlife Society Bulletin 29:99–115.

Buenrostro-Silva, A., S. Gallina, and G. Sanchez-Rojas. 2008. Los talladeros de machos de venado cola blanca *Odocoileus virginianus mexicanus* (Gmelin, 1788) y su ubicación para definir los sitios reproductivos. Pages 219–233 *in* C. Lorenzo, E. Espinoza, and J. Ortega, editors. Avances en el estudio de los mamíferos de México II. Asociación Mexicana de Mastozoología, México, D. F., México.

Cantú, A., J. A. Ortega-S., J. Mosqueda, Z. Garcia-Vasquez, S. E. Henke, and J. E. George. 2007. Immunological and molecular identification of *Babesia bovis* and *Babesia bigemina* in free-ranging deer in northern Mexico. Journal of Wildlife Diseases 43:504–507.

Cantú, A., J. A. Ortega-S., J. Mosqueda, Z. Garcia-Vasquez, S. E. Henke, and J. E. George. 2008. Prevalence of infectious agents in free-ranging white-tailed deer in northeastern Mexico. Journal of Wildlife Diseases 44:1002–1007.

Cantú, A., J. A. Ortega-S., J. Mosqueda, Z. Garcia-Vasquez, S. E. Henke, and J. E. George. 2009. Epizootiology of Babesia bovis and Babesia bigemina in free-ranging white-tailed deer in northeastern Mexico. Journal of Parasitology 95:536–542.

Carranza-Montaño, M. A., L. R. Sánchez Vázquez, M. R. Pineda López, and R. Cuevas Guzmán. 2003. Calidad y potencial forrajero de especies del bosque tropical caducifolio de la Sierra de Manantlán, México. Agrociencia 37:203–210.

Clemente, S. 1984. Utilización de la vegetación nativa en la alimentación del venado cola blanca (*Odocoileus virginianus* Hays) en el estado de Aguascalientes. Thesis, Colegio de Postgraduados, Chapingo, Estado de México, México.

Corona, P. 2003. Bases biológicas para el aprovechamiento del venado cola blanca en el Ejido El Limón de Cuachichinola, Municipio de Tepalcingo, Morelos. Thesis, Instituto de Ecología, Xalapa, Veracruz, México.

De la Rosa-Reyna, X. F., R. D. Calderón-Lobato, G. M. Parra-Bracamonte, A. M. Sifuentes-Rincón, R. W. DeYoung, F. J. García-de León, and W. Arellano-Vera. 2012. Genetic diversity and structure among subspecies of white-tailed deer in Mexico. Journal of Mammalogy 93:1158–1168.

Ezcurra, E., and S. Gallina. 1981. Biology and population dynamics of white-tailed deer in northwestern Mexico. Pages 79–108 *in* P. F. Ffolliott and S. Gallina, editors. Deer biology, habitat requirements and management in western North America. Instituto de Ecología, México, D. F., México.

Ffolliott, P., and S. Gallina, editors. 1981. Deer biology, habitat requirements and management in western North America. Instituto de Ecología, México, D. F., México.

Fulbright, T. E., and J. A. Ortega-S. 2007. Ecología y manejo de venado cola blanca. Texas A&M University Press, College Station, USA.

Fulbright, T. E., and J. A. Ortega-S. 2013. White-tailed deer habitat: ecology and management on rangelands. Texas A&M University Press, College Station, USA.

Galindo-Leal, C., and M. Weber. 1994. Translocation of deer subspecies: reproductive implications. Wildlife Society Bulletin 22:117–120.

Galindo-Leal, C., and M. Weber. 1997. El venado de la Sierra Madre Occidental: ecología, manejo y conservación. Ediciones Culturales S.A. de C.V.–CONABIO and Ediciones Culturales, México, D. F., México.

Galindo-Leal, C., and M. Weber. 2005. Venado cola blanca, *Odocoileus virginianus* (Zimmermann, 1708). Pages 517–521 *in* G. Ceballos and G. Oliva, editors. Los mamíferos silvestres de México. CONABIO and Fondo de Cultura Económica, México, D. F., México.

Gallina, S. 1988. Importancia del injerto (*Phoradendron* spp.) para el venado. Southwestern Naturalist 33:21–25.

Gallina, S. 1993a. White-tailed deer and cattle diets at La Michilía, Durango, Mexico. Journal of Range Management 46:487–492.

Gallina, S. 1993b. Biomasa disponible y capacidad de carga en la reserva La Michilía, Durango. Pages 437–453 *in* R. A. Medellín and G. Ceballos, editors. Avances en el estudio de los mamíferos de México. Volume 1. Asociación Mexicana de Mastozoología, México, D. F., Mexico.

Gallina, S. 1994. Dinámica poblacional del venado cola blanca en la Reserva de la Biósfera La Michilía, México. Pages 207–234 *in* C. Vaughan and M. Rodríguez, editors. Ecología y manejo del venado cola blanca en México y Costa Rica. EUNA, Universidad Nacional, Heredia, Costa Rica.

Gallina, S. 2012. Is sustainability possible in protected areas in Mexico? Deer as an example of a renewable resource. Sustainability 4:2366–2376.

Gallina, S., and J. Bello. 2006. Costo energético para venados en una zona semiárida de México. Pages 163–168 *in* R. Bodmer, editor. Memorias del Sexto Congreso Internacional de Manejo de Fauna Silvestre en Amazonía y Latinoamérica. Wildlife Conservation Society, Lima, Peru.

Gallina, S., J. Bello, C. Contreras, and C. Delfin-Alfonso. 2010a. Daytime bedsite selection by the Texan white-tailed deer in xerophyllous brushland, north-eastern Mexico. Journal of Arid Environments 74:373–377.

Gallina, S., and J. Bello. 1998. Variación individual en el ámbito hogareño y areas de mayor actividad del venado en un matorral xerofilo del noreste de México. Pages 89–94 *in* Memorias del VI Simposio sobre Venados de México, Facultad de Medicina Veterinaria y Zootecnia División de Educación Continua, UNAM, México, D. F., México.

Gallina, S., S. Mandujano, J. Bello, H. López-Fernandez, and M. Weber. 2010b. White-tailed deer

Odocoileus virginianus (Zimmermann, 1780). Pages 101–118 *in* J. M. B. Duarte and S. González, editors. Neotropical cervidology: biology and medicine of Latin American deer. FUNEP, Jaboticabal, Brazil, and IUCN, Gland, Switzerland.

Gallina, S., E. Maury, and V. Serrano. 1981. Food habits of white-tailed deer. Pages 135–148 *in* P. F. Ffolliott and S. Gallina, editors. Deer biology, habitat requirements and management in western North America. Instituto de Ecología, México, D. F., México.

González-Marín, R., S. Gallina, S. Mandujano, and M. Weber. 2008. Densidad y distribución de ungulados silvestres en la Reserva Ecológica El Edén, Quintana Roo, México. Acta Zoológica Mexicana (n.s.) 24:73–93.

González-Marín, R. M., E. Montes, and J. Santos. 2003. Caracterización de las unidades de manejo para la conservación, manejo y aprovechamiento sustentable de la fauna silvestre en Yucatán, México. Tropical and Subtropical Agroecosystems 2:13–21.

González-Pérez, G. E. 2003. Uso del hábitat y área de actividad del venado cola blanca (*Odocoileus virginianus sinaloae* J. Allen) en la Estación Científica Las Joyas, Reserva de la Biosfera de Manantlán, Jalisco. Thesis, UNAM, México, D. F., México.

González-Pérez, G., and M. Briones-Salas. 2000. Venado cola blanca (*Odocoileus virginianus*) en comunidades indígenas de Oaxaca. Investigación Hoy 94:20–27.

Greenberg, L. S. Z. 1992. Garden hunting among the Yucatecan Maya: a coevolutionary history of wildlife and culture. Etnoecológica 1:23–33.

Hall, E. R. 1981. The mammals of North America. Volume 2. Second edition. John Wiley and Sons, New York, New York, USA.

Heffelfinger, J. 2006. Deer of the Southwest: a complete guide to the natural history, biology, and management of southwestern mule deer and white-tailed deer. Texas A&M University Press, College Station, USA.

Hellickson, M. W., T. A. Campbell, K. V. Miller, R. L. Marchinton, and C. A. DeYoung. 2008. Seasonal ranges and site fidelity of adult male white-tailed deer (*Odocoileus virginianus*) in southern Texas. Southwestern Naturalist 53:1–8.

Hirth, D. H. 1977. Social behavior of white-tailed deer in relation to habitat. Wildlife Monographs 53:1–55.

Hyde, K. J., C. A. DeYoung, and A. Garza Jr. 1987. Bed sites of white-tailed deer fawns in South Texas. Proceedings of the Annual Conference of the Southeastern Association of Fish and Wildlife Agencies 41:288–293.

Inglis, J. M., B. A. Brown, C. A. McMahan, and R. E. Hood. 1986. Deer-brush relationships on the Rio Grande Plain, Texas. Publication RM14/KS6. Texas Agricultural Experiment Station, College Station, USA.

Kellogg, R. 1956. What and where are the whitetails? Pages 31–55 *in* W. P. Taylor, editor. The deer of North America. Stackpole, Harrisburg, Pennsylvania, USA.

Kock, R. A., P. S. Soorae, and O. B. Mohammed. 2007. Role of veterinarians in re-introductions. International Zoo Yearbook 41:24–37.

Laundré, J. W., M. Durán, L. Hernández, K. M. Grajales-Tam, F. Rosales, and E. Martínez. 2007. Liberación de venado cola blanca (*Odocoileus virginianus couesi* Coues and Yarrow, 1875) en la Reserva de la Biosfera La Michilía, Durango. Pages 125–130 *in* G. Halffter, S. Guevara, and A. Melic, editors. Hacia una cultura de conservación de la diversidad biológica. Monografías Tercer Milenio, Zaragoza, Spain.

León, P., and S. Montiel. 2008. Wild meat use and traditional hunting practices in a rural Mayan community of the Yucatan Peninsula, Mexico. Human Ecology 36:249–257.

Leopold, A. S. 1959. Wildlife of Mexico: the game birds and mammals. University of California Press, Berkeley, USA.

Logan-López, K., E. Cienfuegos-Rivas, F. C. Sánchez, G. Mendoza Martínez, A. M. Sifuentes Rincón, and L. A. Tarango Arámbula. 2006. Caracterización morfométrica de cuatro subespecies de venado cola blanca (*Odocoileus virginianus*) en la zona noreste de México. Revista Científica 1:14–22.

Logan-López, K., E. Cienfuegos-Rivas, A. M. Sifuentes-Rincón, M. González-Paz, F. Clemente-Sánchez, G. Mendoza Martínez, and L. A. Tarango Arámbula. 2007. Patrones de variación genética en cuatro subespecies de venado cola blanca del noreste de México. Agrociencia 41:13–21.

López-Cervantes, J. A. 2010. Diversidad y diferenciación genética en tres subespecies de venado cola blanca *Odocoileus virginianus* del Estado de Michoacán, México analizada con microsatélites nucleares. Thesis, Universidad Michoacana de San Nicolás de Hidalgo, Morelia, Michoacán, México.

Luévano-Esparza, J. 1991. Dietas veraniegas del jabalí, venado, cabra y caballo en la Sierra de La Mojonera, Vanegas, S. L. P. Colegio de Postgraduados, Montecillo, Estado de México, México.

Mandujano, S. 2004. Análisis bibliográfico de los estudios de venados en México. Acta Zoológica Mexicana (n.s.) 20:211–251.

Mandujano, S. 2007. Carrying capacity and potential production of ungulates for human use in a Mexican tropical dry forest. Biotropica 39:519–524.

Mandujano, S. 2008. Precipitación, capacidad de carga y potencial de uso de los ungulados en un bosque tropical seco. Pages 637–660 *in* E. Espinoza Medinilla and C. Lorenzo, editors. Avances Mastozoológicos. Volume 2. México, D. F., México.

Mandujano, S. 2011. Bibliografía de estudios de venados en México. Instituto Literario de Veracruz, Xalapa, México.

Mandujano, S., C. A. Delfín-Alfonso, and S. Gallina. 2010. Comparison of geographic distribution models of white-tailed deer *Odocoileus virginianus* (Zimmermann, 1780) subspecies in Mexico: biological and management implications. Therya 1:1–68.

Mandujano, S., and S. Gallina. 1995. Disponibilidad de agua para el venado cola blanca en un bosque tropical seco de México. Vida Silvestre Neotropical 4:107–118.

Mandujano, S., and S. Gallina. 1996. Size and composition of white-tailed deer groups in a tropical dry forest in Mexico. Ethology, Ecology and Evolution 8:255–263.

Mandujano, S., and S. Gallina. 2005. Dinamica poblacional del venado cola blanca (*Odocoileus virginianus*) en un bosque tropical seco. Pages 317–330 *in* V. Sánchez-Cordero and R. Medellín, editors. Contribuciones mastozoológicas en homenaje a Bernardo Villa. Instituto de Biología and Instituto de Ecología, UNAM, México, D. F., México.

Mandujano, S., S. Gallina, and S. H. Bullock. 1994. Frugivory and dispersal of *Spondias purpurea* (Anacardiaceae) in a tropical dry forest of Mexico. Revista de Biología Tropical 42:105–112.

Mandujano, S., and E. Naranjo. 2010. Ungulate biomass across a rainfall gradient: a comparison of data from Neotropical and Palaeotropical forests and local analysis in Mexico. Journal of Tropical Ecology 26:13–23.

Mandujano, S., A. Pérez-Arteaga, R. E. Sánchez-Mantilla, and S. Gallina. 1996. Diferenciación de pautas de actividad del venado con ayuda de radiotransmisores con sensor de movimiento. Acta Zoológica Mexicana (n.s.) 67:65–78.

Mandujano, S., and V. Rico-Gray. 1991. Hunting, use, and knowledge of the biology of the white-tailed deer, *Odocoileus virginianus* (Hays), by the Maya of central Yucatan, Mexico. Journal of Ethnobiology 11:175–183.

Marchinton, R. L., and D. H. Hirth. 1984. Behavior. Pages 129–168 *in* L. K. Halls, editor. White-tailed deer: ecology and management. Stackpole, Harrisburg, Pennsylvania, USA.

Martinez, A., A. Salinas, F. Martinez, A. Cantu, and D. K. Miller. 1999. Serosurvey for selected disease agents in white-tailed deer from Mexico. Journal of Wildlife Diseases 35:799–803.

Martínez-Romero, L. E. 2004. Determinación de fechas de aprovechamiento del venado cola blanca (*Odocoileus virginianus*) a través de hormonas sexuales y comportamiento. Thesis, Instituto de Ecología, Xalapa, Veracruz, México.

Medina-Torres, S. M., E. García, M. Marquez, H. Vaquera, A. Romero, and M. Martínez. 2008. Factores que influyen en el uso del hábitat por el venado cola blanca (*Odocoileus virginianus couesi*) en la Sierra del Laurel, Aguascalientes, México. Acta Zoológica Mexicana (n.s.) 24:191–212.

Mendoza, J., A. López, V. Tapia, and G. Ramos. 2006. Rescate, conservación, manejo y aprovechamiento sustentable del venado cola blanca, *Odocoileus virginianus acapulcensis*, en la Sierra-Costa de Michoacán, una experiencia de la UCCOFF A. C. X Simposio sobre Venados en México. Facultad de Medicina, Veterinaria y Zootecnia, UNAM, México, D. F., Mexico.

Montes, R. C., R. I. Rodríguez, and J. F. Torres. 1998. Seguimiento anual de la parasitosis gastrointestinal de venados cola blanca *Odocoileus virginianus* (Artiodactyla: Cervidae) en cautiverio en Yucatán, México. Revista de Biología Tropical 46:821–827.

Montiel Ortega, S., L. M. Arias Reyes, and F. Dickinson. 2000. La cacería tradicional en el norte de Yucatán: una práctica comunitaria. Revista de Geografía Agrícola 29:43–51.

Naranjo, E. J., M. M. Guerra, R. E. Bodmer, and J. E. Bolaños. 2004. Subsistence hunting by three ethnic groups of the Lacandon forest, Mexico. Journal of Ethnobiology 24:233–253.

Nuñez, R., B. Miller, and F. Lindzey. 2000. Food habits of jaguars and pumas in Jalisco, Mexico. Journal of Zoology 252:373–379.

Ortega-S., J. A., S. Mandujano, J. Villarreal, M. I. Di Mare, H. López-Arévalo, M. Correa, and M. Molina. 2011. White-tailed deer in Latin America. Pages 565–597 *in* D. Hewitt, editor. Biology and management of white-tailed deer. CRC Press, Boca Raton, Florida, USA.

Ortega-S., J. A., A. Melgoza, F. A. Ibarra-Flores, E. A. Gonzalez-Valenzuela, M. H. Martin-Rivera, J. M. Avila-Curiel, F. Ayala-Alvarez, C. Pinedo, and O. Rivero. 2013. Exotic grasses and wildlife in northern Mexico. Wildlife Society Bulletin 37:537–545.

Ortíz-Martínez, T., S. Gallina, M. Briones-Salas, and G. González-Pérez. 2005. Densidad poblacional y caracterización del hábitat del venado cola blanca (*Odocoileus virginianus oaxacensis*, Goldman and Kellog, 1940) en un bosque templado de la Sierra Norte de Oaxaca, México. Acta Zoológica Mexicana (n.s.) 21:65–78.

Ramírez-Lozano, R. G. 2004. Nutrición del venado cola blanca. UANL, Unión Ganadera Regional de Nuevo León and Fundación Produce, Nuevo Léon, Monterrey, México.

Ramírez-Lozano, R. G. 2012. Alimentación del venado cola blanca: biología y ecología nutricional. Palibrio, Bloomington, Indiana, USA.

Robinson, J. G., and K. H. Redford. 1991. Sustainable harvest of Neotropical forest mammals. Pages 415–429 *in* J. G. Robinson and K. H. Redford, editors. Neotropical wildlife use and conservation. University of Chicago Press, Illinois, USA.

Rojo-Curiel, A., J. L. Cruz, G. Solano, and R. Hernández. 2007. Plan de manejo tipo de venado cola blanca en zonas templadas y tropicales en México. DGVS and SEMARNAT, México, D. F., México.

Romero-Castañón, S., B. G. Ferguson, D. Guiris, D. González, S. López, A. Paredes, and M. Weber. 2008. Comparative parasitology of wild and domestic ungulates in the Selva Lacandona, Chiapas, Mexico. Comparative Parasitology 75:115–126.

Rosas-Alvarado, A. 1994. Etograma del venado cola blanca en cautiverio en México. Pages 181–203 *in* C. Vaughan and M. Rodríguez, editors. Ecología y manejo del venado cola blanca en México y Costa Rica. EUNA, Universidad Nacional, Heredia, Costa Rica.

Rosas-Becerril, P. 1992. Patrones reproductivos del venado cola blanca (*Odocoileus virginianus*) durante su estación reproductiva en La Michilía, Durango. Thesis, UNAM, México, D. F., México.

Rzedowski, J., and T. Reyna-Trujillo. 1990. Tópicos biogeográficos. Mapa IV.8.3. *In* Atlas nacional de México. Volume 3. Instituto de Geografía, UNAM, México, D. F., México.

Segovia-Castillo, A., and S. Hernández-Betancourt. 2003. La cacería de subsistencia en Tzucacab, Yucatán, México. Tropical and Subtropical Agroecosystems 2:49.

Serra, M., and R. Valdez. 1989. Importancia de los venados en Teremote-Tlaltenco. Ciencia y Desarrollo 15:63–72.

Silva-Villalobos, M. A., S. Mandujano, G. Arceo, L. A. Pérez-Jiménez, and S. Gallina. 1999. Nutritional value of plants consumed by the white-tailed deer in a tropical forest of Mexico. Vida Silvestre Neotropical 8:38–42.

Smith, W. P. 1991. *Odocoileus virginianus*. Mammalian Species 388:1–13.

Téllez, E. 1996. El venado en las culturas pristinas Mesoamericanas. Pages 16–20 *in* V Simposio sobre Venados en México. UNAM, Quintana Roo, México.

Tierson, W. C., G. F. Mattfeld, R. W. Sage Jr., and D. F. Behrend. 1985. Seasonal movements and home ranges of white-tailed deer in the Adirondacks. Journal of Wildlife Management 49:760–769.

Vaughan, C., and M. Rodríguez, editors. 1994. Ecología y manejo del venado cola blanca en México y Costa Rica. EUNA, Universidad Nacional, Heredia, Costa Rica.

Villalobos, V. 1998. El venado cola blanca en la Sierra Fría, Aguascalientes. Cuaderno de Trabajo (Agricultura y Recursos Naturales) 89:1–33.

Villarreal, J. G. 1999. Venado cola blanca: manejo y aprovechamiento cinegético. Unión Ganadera Regional de Nuevo León, Monterrey, México.

Villarreal, J. 2009. Propone el Consejo Estatal de Flora y Fauna de Nuevo León la regionalización cinegética de venados cola blanca mexicanos para libro récords internacionales del Safari Club International. Boletín Cuenca Palo Blanco, Consejo Estatal de Flora y Fauna de Nuevo León 4:3–7.

Villarreal-Espino, O. A. 2002. El *grand slam* del venado cola blanca mexicano: una alternativa sostenible. Archivos de Zootecnia 51:187–193.

Villarreal-Espino, O. 2006. El venado cola blanca en la Mixteca Poblana: conceptos y métodos para su conservación y manejo. Fundación Produce, Puebla, México.

Villarreal-Espino, O. A., and M. Marín. 2005. Agua de origen vegetal para el venado cola blanca mexicano. Archivos de Zootecnia 54:191–196.

Weber, M. 2005. Ecology and conservation of tropical deer populations in the Greater Calakmul Region, Mexico. Dissertation, University of Durham, Durham, United Kingdom.

Weber, M., and R. Hidalgo. 1999. Morfometría, patrones de crecimiento y ganancia de peso en venados cola blanca (*Odocoileus virginianus*) en semi-cautiverio en Durango y Toluca, México. Veterinaria México 30:183–188.

Weber, M., P. Rosas-Becerril, A. Morales, and C. Galindo-Leal. 1994. Biología reproductiva del venado cola blanca en Durango, México. Pages 111–127 *in* C. Vaughan and M. Rodríguez, editors. Ecología y manejo del venado cola blanca en México y Costa Rica. EUNA, Universidad Nacional, Heredia, Costa Rica.

Yah, D. 1983. La cacería del venado. Secretaría de Educación Pública, Yucatán, México.

Yañez, C. 2009. Distribución y densidad poblacional del venado cola blanca en el Bajo Balsas, Michoacán. Thesis, Instituto de Ecología, Xalapa, Veracruz, México.

18 Brocket Deer and Tropical White-Tailed Deer in Mexico

Manuel Weber

Brocket deer (*Mazama* spp.) and white-tailed deer or whitetails (*Odocoileus virginianus*) have been by far the most important big game mammals in Mexico since the pre-Hispanic period. The reason is threefold: (1) deer are the most abundant and widespread large game mammals; (2) they are strongly preferred by sport and subsistence hunters as a source of food and recreation; and (3) deer may be the only large game mammals available to hunters in tropical ecosystems of Mexico because of the natural absence or extirpation of larger or equivalent-sized mammals, such as the tapir (*Tapiris bairdii*) and white-lipped peccary (*Tayassu pecari*). This chapter outlines the natural history, ecology, and management of the three species of native tropical deer: red brocket (*M. americana*), Yucatán brown brocket (*M. pandora*), and white-tailed deer. These deer continue to have cultural significance for native and rural societies in Mexico.

Historical Perspective

Brocket and white-tailed deer provide up to 80% of the total subsistence hunting biomass in local communities (Escamilla et al. 2000, Weber 2000, Naranjo 2002) and are the most important game trophies for sport hunters. Leopold (1959) emphasized the importance of deer as a wildlife asset for Mexico, but it was not until 20 years later that Mexican deer biologists began documenting detailed aspects of the natural history and ecology of white-tailed deer in temperate Mexico (see chapter 17). It took even longer to initiate research on tropical deer species (Mandujano 2004, Weber 2005).

Extirpation and Recovery

With the probable exception of deer in natural protected areas and on some private lands, deer populations have been undergoing a dynamic process of extirpation and recovery in many regions of Mexico. These trends in many regions of the Neotropics were likely the consequence of a combination of subsistence and sport-hunting overharvest, habitat conversion and loss (mostly from slash-and-burn agriculture and conversion of forests to livestock pastures), predation by mammalian carnivores in some regions, and interspecific competition, in particular with livestock.

Value

The economic value of deer in rural tropical communities is grossly underestimated by modern society. In the consumer-driven societies of large cities, where many processed goods and foods are readily available in supermarkets, the existence of semi-hunter-gatherer societies has been almost forgotten or relegated to the past. However, these societies continue to exist in tropical Mexico, and deer are the primary source of meat protein. By far the highest value of deer in rural communities is as a source of meat protein (Escamilla et al. 2000, Weber 2000, Naranjo 2002, Weber and Gonzalez 2003). Moreover, deer have other values; they are cultural heritage and totemic animals in Mesoamerican cultures, including the peninsular Mayans, Lacandons, Tojolabals, Tzotzils, and Triques (Galindo-Leal and Weber 1998).

Taxonomy

It was only recently realized that two species of brocket deer exist in Mexico. Merriam first used the name *Mazama pandora* when he described the species in 1901. In 1915, Allen kept the name and aligned it with other *Mazama* species. Several authors treated this species inconsistently, assigning it to the brown or red brocket deer groups. Goldman and Moore (1945) included the name *pandora* as a red brocket subspecies of *M. sartorii*. Hershkovitz (1951) classified *pandora* as a subspecies of the South American brown brocket, *M. gouazoubira*, but subsequently (1966) considered the Yucatán brown brocket as a color variant of the red brocket and classified it as *M. americana pandora*. Other authors classified it as a disjunct subspecies of the South American brown brocket (Czernay 1987, Bisbal 1994). Medellín et al. (1998) examined over 70 specimens of brocket deer, including 24 specimens from the Yucatán Peninsula, and determined there were two species: red and brown. Weber and Medellín (2010) also concluded that the Yucatán brown brocket deer was a distinct species.

The red brocket has had a similarly complicated taxonomic history. Its taxonomy in Mexico is less clear than that of the white-tailed deer and the Yucatán brown brocket. The red brocket is classified as *M. americana* or *M. temama* (Geist 1998, Smith et al. 1986). However, the recent designation of the red brocket as *M. temama* (previously recognized as a subspecies of *M. americana*) is not based on a study of systematics or genetics (Weber and Gonzalez 2003, Geist 1998). In this chapter, I use the original name *M. americana* for the Mexican red brocket.

All karyotypes determined from captive white-tailed deer in Mexico have a diploid chromosome number of 2n = 70 (Barragán 2002). The fundamental number (FN) of this species is 74 ($66 \times 1 + 2 \times 2$ + sex chromosomes $\times$ 2). The validity of white-tailed deer subspecies in tropical ecosystems of Mexico is controversial (Villarreal 1999). Of the originally recognized 14 subspecies described in Mexico, perhaps only *texanus* and *carminis* do not occur in a tropical or subtropical region (see chapter 17). The other 12 subspecies—whether they are valid taxa or ecotypes remains in dispute—occur in one or several tropical regions.

Distribution

Brown brocket

The Yucatán brown brocket deer occurs on the Yucatán Peninsula of southeastern Mexico. Although this region has undergone moderate habitat fragmentation and land conversion, the historical and current distributions may not differ substantially. There are records from the three states of the Yucatán Peninsula: Yucatán, Campeche, and Quintana Roo. Despite the vegetation heterogeneity and habitat gradient in the Yucatán from very wet and seasonally flooded in the south to drier and rarely flooded in the north, the species is widespread (Weber and Gonzalez 2003, Weber and Medellín 2010). It is sympatric with the red brocket only in southern areas, and mostly in tropical rainforests. The species likely occurs in northern Guatemala and Belize, but there are no published records or specimens (fig. 18.1).

Red brocket

The red brocket deer occurs in humid tropical forests of southeastern Mexico as far north as the Sierra Madre Occidental in Tamaulipas, including true rainforests, cloud forests, and savannas with interspersed patches of tropical rainforest. Its distribution includes San Luis Potosí, Tamaulipas, Veracruz, Hidalgo, Puebla, Tabasco, Campeche, Yucatán, Chiapas, and Oaxaca (fig. 18.2) (García-Marmolejo et al. 2013, 2015).

White-tailed deer

In southern latitudes, white-tailed deer range from tropical lowlands at sea level to mixed forest-brush areas. In tropical environments at lower elevations, they are more common in arid and shrub habitats than in humid forests (Baker 1984). However, although their population density is now low, white-tailed deer occur in true tropical rainforests such as the Selva Lacandona, Chiapas (Naranjo 2002).

The distribution of white-tailed deer in tropical and subtropical Latin America covers approximately 5.5 million km^2 (2.1 million mi^2) (Brokx 1984) and seems to be expanding (Weber and Gonzalez 2003, Barrera-Salazar et al. 2015). Historically, they were probably not as common as they are today but were nearly as widely distributed. Numbers increased after land clearing and forest exploitation, but then they were reduced because of subsistence overhunting. In many regions with historical environmental abuse, such as Veracruz, Tabasco, and Chiapas, their current distribution can be patchy; in some areas there are remnant populations, and in others they are extirpated (fig. 18.3).

Description

The three species of tropical deer in Mexico are relatively easily distinguished by anatomical differences. The largest is the white-tailed deer, followed by the Yucatán brown brocket; the red brocket is the smallest of the three (fig. 18.4).

White-tailed deer

The white-tailed deer is a polytypic species adapted to varied tropical environments. Its anatomical diversity

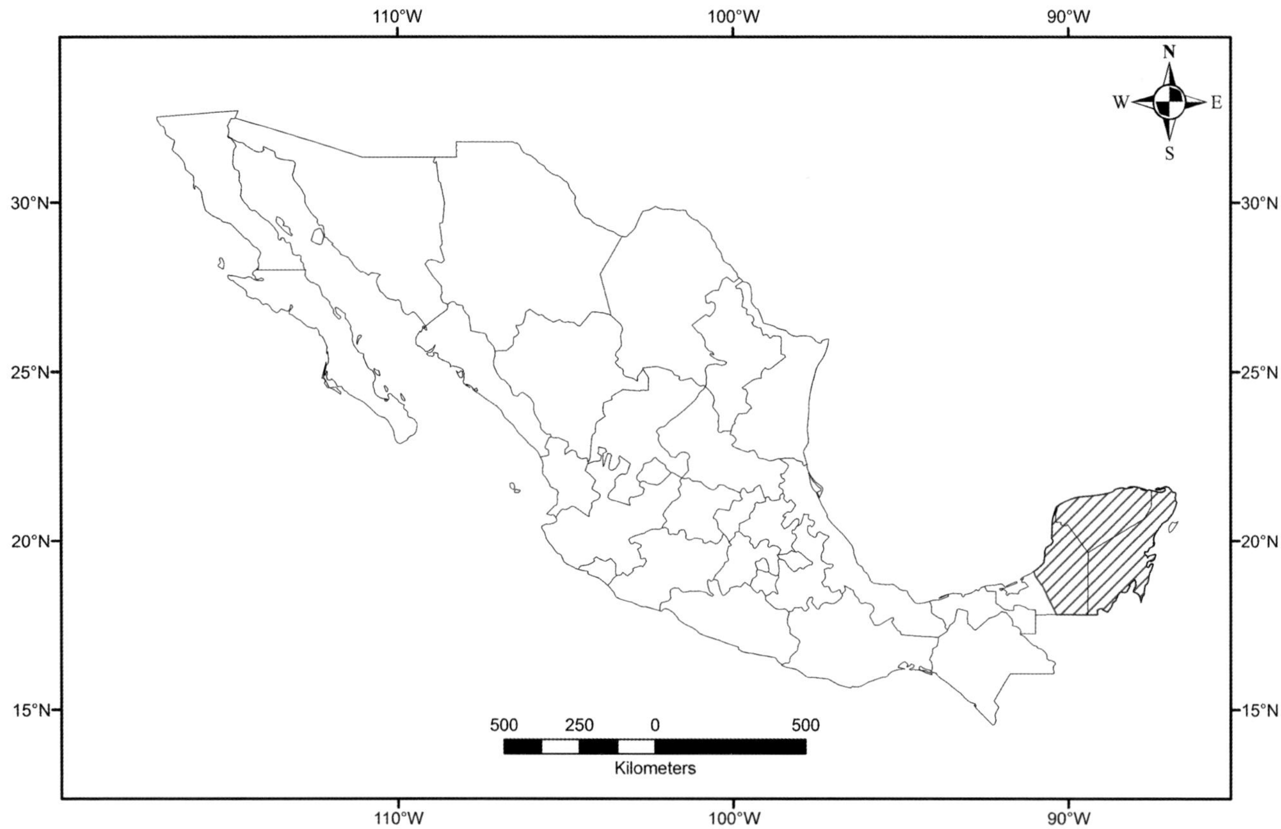

Figure 18.1. The distribution of the Yucatán brown brocket deer in Mexico includes the entire Yucatán Peninsula. Modified from IUCN (2009).

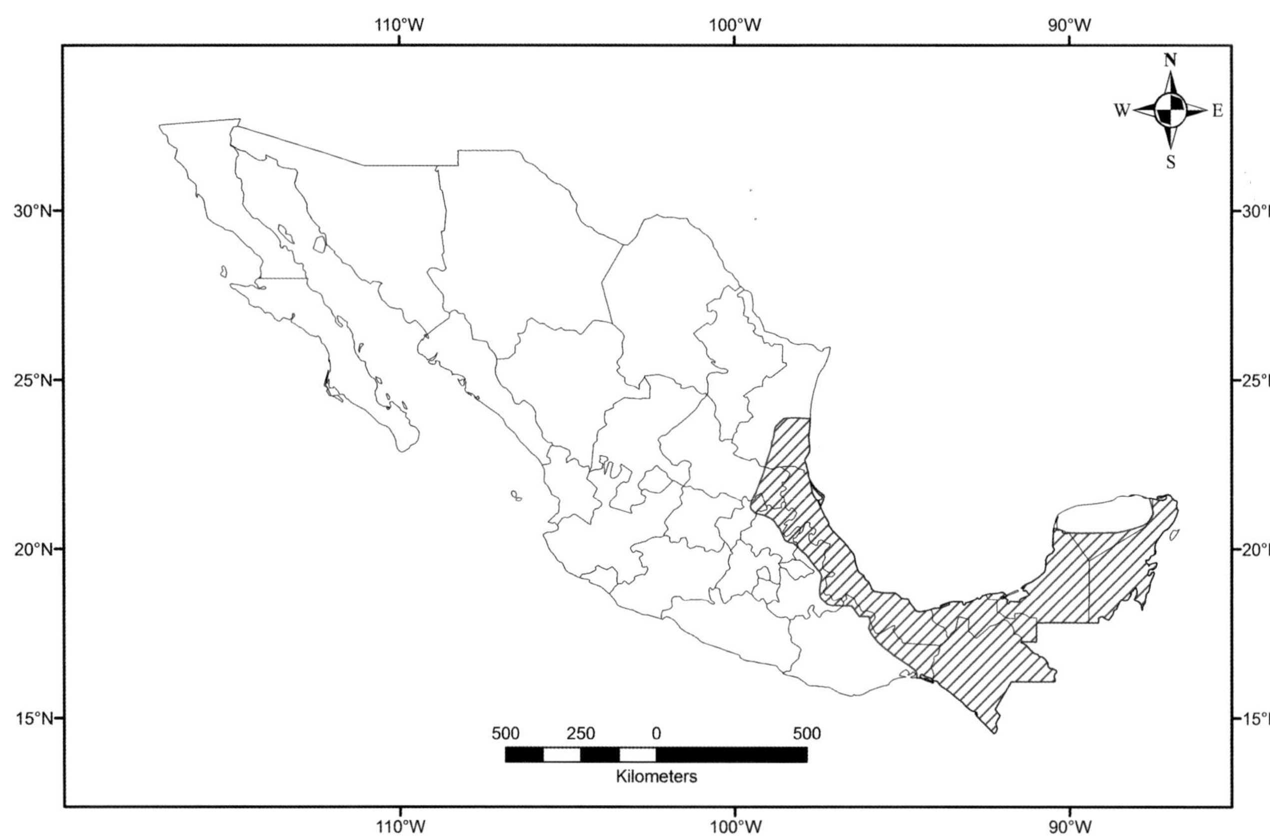

Figure 18.2. Distribution of the red brocket deer in Mexico. Modified after IUCN (2009).

Figure 18.3. Distribution map of the white-tailed deer (Odocoileus virginianus) based on Hall (1981) and Smith (1991). Modified from IUCN (2009).

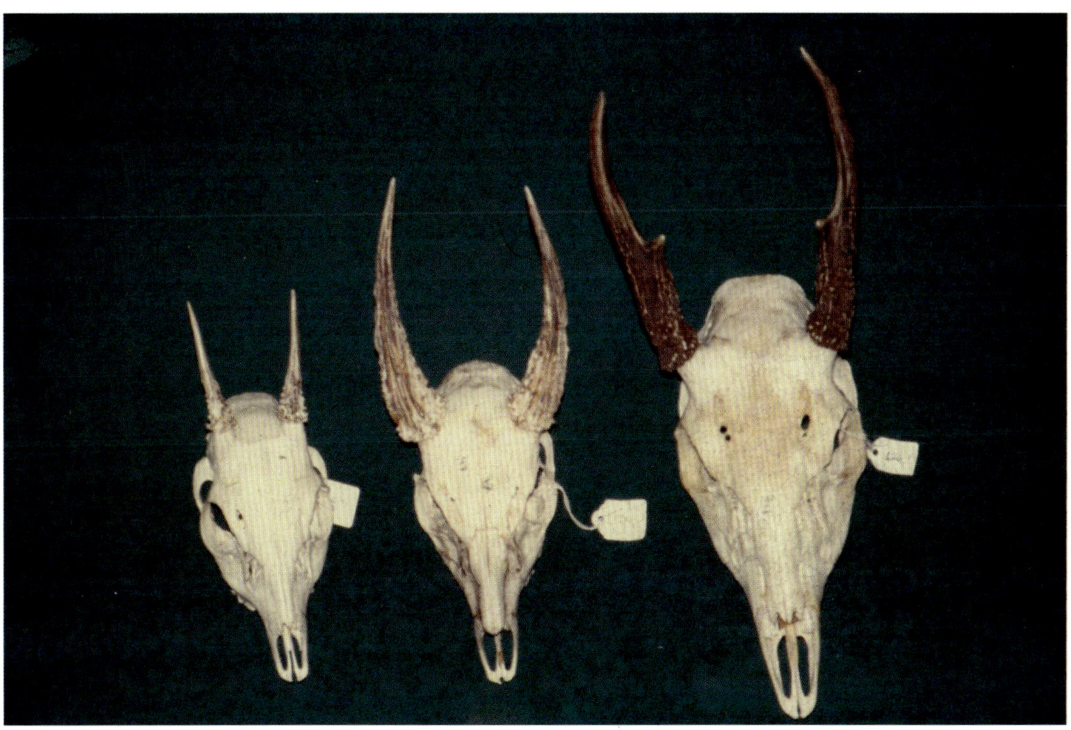

Figure 18.4. Skulls of a red brocket (left), Yucatán brown brocket (center), and tropical white-tailed deer (right) from Campeche, Mexico. Photo by Manuel Weber.

is reflected in its body weight, external dimensions, coat color, antler growth, and assorted physiological and behavioral adaptations. There is wide variation in size and coat color among tropical subspecies. In general, coat color is darker in humid areas, paler in drier and open brushland, and reddish in subtropical and rainforest environments (Baker 1984). Fawns are reddish brown, with white dorsal spots that disappear at three to four months of age (Hesselton and Hesselton 1982).

Most whitetails in subtropical and tropical climates are smaller, with males weighing less than 60 kg (132 lb) and does less than 35 kg (77 lb). However, heavier deer have been recorded in Campeche and Yucatán; some bucks harvested by subsistence hunters in Calakmul, in southeastern Mexico, weighed more than 65 kg (143 lb) (fig. 18.5). Antlered males have been recorded from April through February, but periods of antler presence change with latitude.

Yucatán brown brocket

The Yucatán brown brocket deer is medium sized, slightly larger than the Mexican red brocket, and smaller than the sympatric white-tailed deer. The dorsal pelage is uniformly brown to gray brown, becoming paler to whitish ventrally. The hair is relatively short and sparse. The rostrum is relatively broad, short, and tapered, and the ears are relatively short for the genus. There is a tuft of stiff, bristlelike hairs between the eyes, and the forehead is wide. The antlers of this species are distinctive, with widely separated bases and massive pedicels that are deeply furrowed unless the antlers are worn. The antlers are long, and certainly longer than those of any red brocket from the Yucatán or elsewhere. The antlers diverge, often curving so the tips are convergent, but with a wide space between them. In the Greater Calakmul Region (GCR) of Campeche, at the base of the Yucatán Peninsula, average body mass of 13 adult males (10 shot and 3 live) was 20.5 kg (45.1 lb) (sd = 3.8 kg, 8.7 lb), while that of 8 adult females (6 shot and 2 live) was 17.3 kg (38.1 lb) (sd = 4.2 kg, 9.2 lb) (fig. 18.6).

Figure 18.5. This adult female white-tailed deer was killed by subsistence hunters in Campeche, Mexico, and is being transported, as is typical, on the back of a bicycle. Photo by Manuel Weber.

Red brocket

The red brocket is the smallest deer species in Mexico. Average weight of 5 adult males (4 shot and 1 live) was 16.3 kg (35.9 lb) (sd = 3.8 kg, 8.3 lb), and that of 4 females (3 shot and 1 live) was 13.2 kg (29.1 lb) (sd = 2.4 kg, 5.2 lb). It has a much slenderer body than the Yucatán brown brocket, with a less massive

Figure 18.6. Red brocket female and fawn (top) and adult buck (bottom). Photos by Rafael Reyna and Manuel Weber.

head and shorter ears. It has a deep red coat, and the legs and lower parts of the body are lighter than the rest of the body. The head is also narrower than that of the brown brocket. Both males and females tend to have very dark hair in the frontal bone area of the head, and this hair tends to vary in length regionally in the Mexican tropics. The tail has abundant white hair ventrally, and dark red hair dorsally. However, the tail is not used for flagging (exposing the hair on the underside of the tail as an alarm signal), as in white-tailed deer females; it is inconspicuous even when deer are running. Antlers tend to be short and pointed and normally do not grow more than 5 cm (2 in) in length (fig. 18.7).

Life History

Diet

White-tailed deer

White-tailed and brocket deer are selective feeders that choose plants and plant parts with considerable discrimination. Whitetails, which are opportunistic concentrate feeders, are browsers that prefer the leaves and limbs of shrubs and trees, although they also consume a wide variety of other foods including sedges, fruits, nuts, forbs, mushrooms, and grasses (table 18.1). Diets change regionally and seasonally (Gallina et al. 1981, Gallina 1993, Di Mare 1994, Mandujano et al. 2004, Weber 2005).

In the tropical dry forest of Chamela, Jalisco, the main plants consumed by whitetails were in the families Euphorbiaceae, Leguminosae, Convolvulaceae, and Sapindaceae, which made up 80%, 64%, and 37% of diets during the rainy, transition, and dry seasons, respectively (Arceo et al. 2005). The estimated mean percentage of protein for plants consumed was 14%. The percentage of free nitrogen extract was 49%, and mean annual fiber percentage was 24%. An important food resource in this habitat was fruits of *Spondias purpurea* (Anacardiaceae) (Mandujano et al. 1994). These fruits constituted 13% of the diet during the dry season (Arceo et al. 2005) and could contribute 2–30 liters/ha/yr (0.8–12.5 gal/ac/yr) of water depending on annual fruit production (Mandujano and Gallina 1995).

Comparative food habits in a sympatric habitat

Food habits of the three deer species in the tropical semideciduous rainforest of southeastern Mexico (Calakmul region) were studied using stomach samples systematically collected over a three-year period (1997–1999), and opportunistically from 2001 to 2002. Brocket deer, like whitetails, are selective feeders, but there were major differences in the diets of the three species. The zapote tree (*Manilkara zapota*) is a keystone plant resource for deer and other wildlife in this region. Whitetails fed on the fruits, leaves, and stems of 26 tree species, palms, and shrubs. They also ate large amounts of fruits during the wet season (June to November). There

Figure 18.7. Female red brocket deer. Photo by Manuel Weber.

Table 18.1. Comparison of white-tailed deer diets in different habitats and countries.

Lat.	Habitats (Country)	Plant families	Species	Main food items		Species
				Rainy season[a]	Dry season	
3° N	Coastal marshes (Suriname)[b]	10	14			*Avicennia germinans, Sesuvium portulacastrum*
6° N	Tropical forests, plains, and cultivated areas (Venezuela)[c]	33	125	forbs grasses	leaves fruits	*Copernicia tectorum, Acacia macracantha, Caesalpinia coriaria, Guazuma ulmifolia, Guttardia divaricata, Platymiscium pinnatum, Randia hebecarpa*
9° N	Tropical dry forest (Costa Rica)[d]	23	45	forbs grasses	fruits leaves	*Brachiaria spp., Cyperus spp., Oplismenus spp., Dalechampia scandens, Ipomoea spp., Pithecellobium saman, Guazuma ulmifolia, Pisonia aculeata, Lonchocarpus spp., Waltheria indica, Spondias spp., Tabebuia spp.*
17° N	Tropical semideciduous rainforest (Mexico)[e]		26	leaves stems	fruits	*Brosimum alicastrum, Manilkara sapota*
19° N	Tropical dry forest (Mexico)[f]	20	82	leaves vines forbs	leaves vines grasses fruits	*Spondias purpurea, Rourea glabra, Acalypha langiana, Coursetia caribaea, Tephrosia leiocarpa, Cardiospermum halicacabum, Thouinia paucidentata, Ayenia micrantha*
22° N	Temperate deciduous and mixed forest (Mexico)[g]	17	31	leaves forbs	leaves forbs	*Arctostaphylos pungens, Quercus rugosa, Castilleja canescens, Viola barroetana, Juniperus deppeana*
23° N	Temperate deciduous and mixed forest (Mexico)[h]	41	135	forbs leaves	leaves	*Phoradendron bolleanum, P. villosum, Arctostaphylos pungens, A. longifolia, A. polifolia, Pithecellobium leptophyllum, Condalia hookeri, Juniperus durangensis, Quercus spp., Arbutus glandulosa*
28° N	Mesquite and chaparral–mixed grass, (Mexico)[i]		79	forbs grasses	leaves	*Acacia wrightii, Acacia farnesiana, Castela texana, Celtis pallida, Porlieria angustifolia, Cercidium macrum, Desmanthus virgatus, Eysenhardtia polystachya, Karwinskia humboldtiana, Leucaena leucocephala, Pithecellobium ebano, Prosopis glandulosa, Zanthoxylum fagara, Opuntia engelmannii*

Source: Modified from Gallina et al. (2010).

[a]Rainy season includes spring, summer, and fall in temperate regions, and dry season includes winter.

[b]Branan et al. (1985).

[c]Granado (1989), Danields (1987).

[d]Di Mare (1994).

[e]Weber (2005).

[f]Arceo et al. (2005).

[g]Clemente (1984).

[h]Gallina et al. (1981), Gallina (1993).

[i]Ramírez-Lozano (2004), Villarreal (1999).

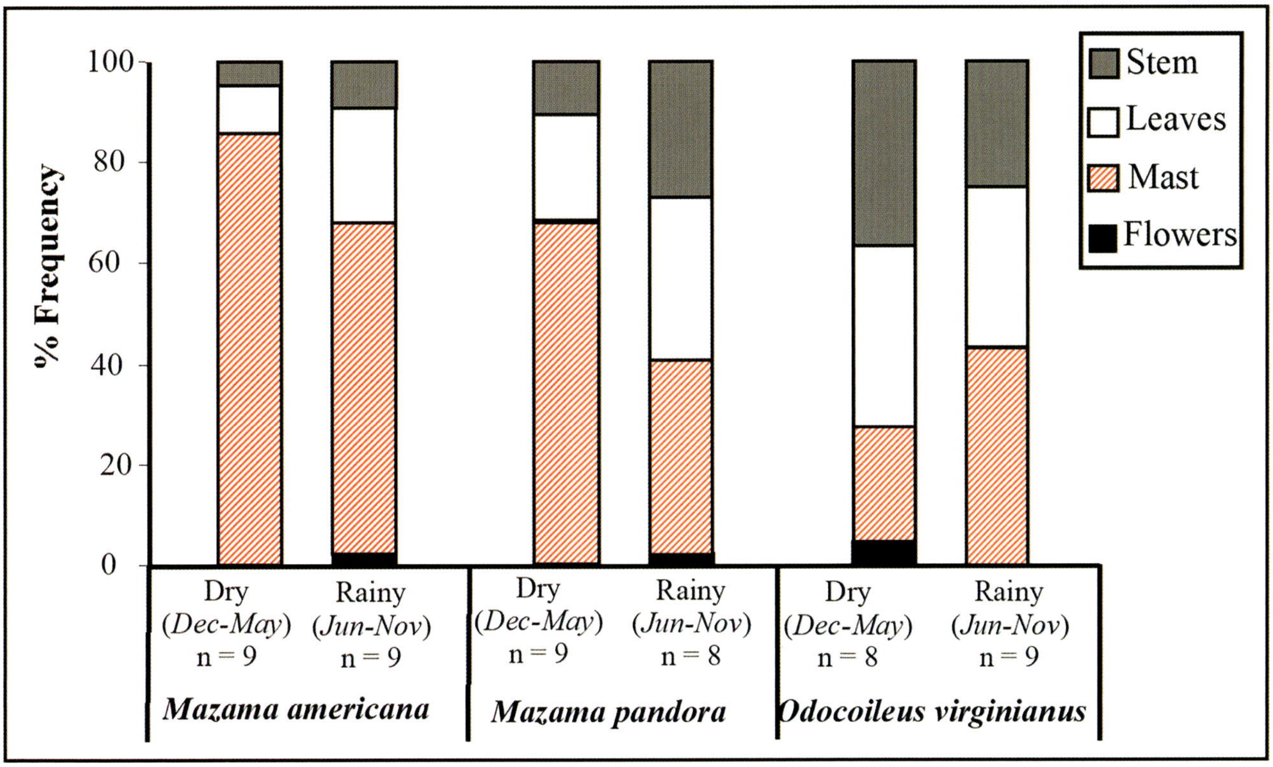

Figure 18.8. Gross seasonal diet (frequency of plant vegetative and reproductive parts, by percentage) determined from analysis of stomach contents (n = 52: brown brocket, n = 18; red brocket, n = 17; whitetail, n = 17) in the Greater Calakmul Region, southeastern Mexico.

was considerable dietary overlap between whitetails and brown brockets during this season. Whitetails consumed a wider array of plant resources year-round (26 species from 36 available in the entire deer diet spectrum), but brown brockets consumed the largest array during the wet season (Weber 2005).

The diet of red brockets was the most specialized of the three species. They consumed plants from fewer than 10 classes (at any taxonomic level) during the dry season and from 10 to 15 classes during the wet season, or a total of 20 species from 46 available in the entire deer diet spectrum (Weber 2005). Their diet consisted of up to 80% fruits and seeds year-round. In contrast, brown brockets showed strong seasonal variations, being mainly frugivores during the dry season (63% fruits), and largely browsers during the wet season, with 60% of the diet consisting of leaves and stems. Whitetail diets were closer to those of a year-round browser, with up to 70% made up of leaves and stems year-round, although there was some seasonal variation, and more fruit consumption during the wet season. Consumption of flowers by the three species was minimal and occurred only during the wet season by both brocket species and during the dry season by whitetails (fig. 18.8).

Analysis of the richness of brown brocket deer diets at the plant family, genus, and species levels revealed strong preferences for species in the families Moraceae (*Brosimum alicastrum*), Sapotaceae (*Manilkara zapota*), Leguminosae (*Eugenia* spp., *Haematoxylum* spp., and other species), Lauraceae (*Nectandra salicifolia*), Myrtaceae (mostly *Thevetia* spp.), and Arecaceae (mostly *Chamaedorea* spp. and *Sabal* spp.). The brown brocket also consumed large amounts of tree fruits of *Manilkara zapota* and *Brosimum alicastrum* year-round, and in the red brocket, more than 80% of the total annual aggregated percent volume (APV) per stomach consisted of fruits of these two tree species.

The three deer species are clearly separated into different ecological niches by differences in the vegetative and reproductive portions of the plants consumed. This difference is evident between the red brocket and the other two deer species, but less obvious between the brown brocket and whitetail. The red brocket is clearly a specialist, as evidenced by its consumption of fruits in tall and medium semievergreen forests, while the red brocket and whitetail exhibit a wider range of plant uses and preferences from the available spectrum (Weber 2005, Weber and Medellín 2010).

Summary of food habits

In general, the brown brocket diet was richer in plant species during the wet season, and much more restricted during the dry season. This suggests nutritional stress during the dry season. While the red brocket diet is composed year-round of no more than 15 plant species, diets of the brown brocket and whitetail can be nearly twice as rich at almost any taxonomic plant level, particularly during the wet season (Weber 2005).

The degree of dietary overlap is greater between the brown brocket and whitetail than between the red brocket and the other two species. Again, this suggests that when resource availability is limiting (e.g., during the dry season), some competition might occur between the former two species more often than competition with the red brocket. The separation of deer by different habitat use patterns supports this hypothesis (Weber 2005). No other study before this had ever measured the degree of dietary overlap for small, forest-dwelling ungulates in Neotropical forests (Weber and Medellín 2010).

Overall, in the semideciduous rainforest, white-tailed deer consumed the widest array of plant resources available year-round (26 species from 36 available in the whole deer diet spectrum), but the brown brocket consumed the largest array during the wet season. The diet of the red brocket was the most specialized, including fewer than 10 classes of plants (at any taxonomic level) during the dry season, and 10 to 15 plant classes during the wet season, or a total of 20 species from 46 available in the whole deer diet spectrum (Weber 2005). These data suggest that the red brocket is a frugivore that depends heavily on the fruits of *Manilkara zapota* and *Brosimum alicastrum*. The brown brocket is intermediate between a browser and a frugivore and also depends heavily on these two plant species, but it uses a wider range of plant resources than the red brocket. The white-tailed deer is mainly a browser and an opportunistic frugivore, consuming the widest array of plant resources. Surprisingly, the brown brocket uses a wider range of plant resources than the whitetail during the wet season.

Results of this deer diet study in the Calakmul region contrast with those obtained in southern latitudes in other Neotropical forests. Branan et al. (1985) studied diets of white-tailed deer and two brocket deer species (brown and *M. gouazoubira*) in Suriname and found no dominance of a single plant species in their diets. Fungi were an important component of red brocket diets, but the three deer species were nonselective feeders in Suriname. Bodmer (1989, 1991) found that 80% of the diet of these same species was fruits in the Peruvian Amazon. Similarly, Gayot et al. (2004) found that the diet of these species in French Guiana was composed mainly of fruits (56% in red brockets and 68% in gray brockets) and classified them as nonselective feeders.

Reproduction

White-tailed deer

The mean gestation period in white-tailed deer is about 202 days (Halls 1978) but differs among subspecies, ranging from 187 to 222 days. Litter size ranges from one to three (twins are common in northern latitudes but uncommon in the tropics) and is related to genetic factors and nutrition (Verme and Ullrey 1984). Neonates have spotted pelage that is lost three months after birth. Females usually breed at the age of 1.5 years, and males also attain sexual maturity at 1.5 years. The ultimate determinant of the mating season is adequate nutrition, which is cued by photoperiod (Verme and Ullrey 1984). In northern subspecies, mating occurs in autumn and winter, changing with latitude, but reproductive cycles of South American white-tailed deer are inadequately documented (Brokx 1984). In the tropics, breeding can occur year-round, with peaks in certain seasons (Geist 1998). The fawning period for whitetails in Venezuela occurs mainly from July to November (rainy season), but a second peak occurs during February and March (dry season). The fawning peak in the llanos of northwestern South America occurs mainly from November to February, perhaps an adaptation to heavy rains and widespread flooding on the lower llanos. In Peru, fawning varies regionally but generally occurs from January to March (Brokx 1984).

In a tropical dry forest in Chamela, whitetails bred between November and January, and fawns were born between June and August (Mandujano et al. 2004). In Puebla, Martínez-Romero (2004) found that the reproductive season started in November, but copulation did not take place until January. In the Chapultepec Zoo in Mexico City, deer bred from October to December (Rosas-Alvarado 1990).

The reproductive biology of the white-tailed deer in Mexico has been little studied compared to other ecological aspects of the species. Many aspects of the reproduction of the Coues white-tailed deer were studied in Durango, Mexico, in captivity and in the wild (Weber 1992, Galindo-Leal and Weber 1998). The average fawning date was February 12. The reproductive season lasted approximately 95 days, starting in January and terminating at the beginning of April. The interval between estrus occurrences was 26 days (± 3.5 days). Behavioral estrus lasted 18 hours (± 1.5 hours) at this latitude. Fawns were born in August and September after a gestation period of 198 to 200 days (Rosas-Becerril 1992, Weber and Galindo-Leal 1992). Males began the rut in November, with some behavioral and physiological changes occurring at this time, and the rut terminated in May with the shedding of antlers. Antler velvet was shed during the first two weeks of October. The largest testicular diameters were recorded in February, at the same time that most females were in estrus (Weber and Galindo-Leal 1995). Growth and weight gain patterns of fawns have also been studied in central and northern Mexico; average weight gain was 225 g/day (7.93 oz/day) for naturally reared fawns (Weber and Hidalgo 1999).

In tropical regions of Mexico, white-tailed deer reproduction tends to be much less seasonal than it is in temperate environments (Galindo-Leal and Weber 1998). However, little is known about seasonality and other aspects of the reproduction of this species in the tropics. In the Calakmul region of Campeche, very young fawns, fetuses, and males with velvet antlers have been observed almost year-round, suggesting a lack of reproductive seasonality at this latitude. McCoy-Colton and Vaughan-Dickhaut (1985) reported fawning in April and May in Guanacaste, Costa Rica, at the beginning of the rainy season.

Brocket deer

No systematic study of the reproductive biology of the Yucatán brocket has been conducted. Anecdotal and opportunistic observations in the GCR suggest that the species is not a seasonal breeder. Overall, eight newborn fawns have been observed in the wild (W. Weber, personal observation) in March (two), May (one), July (two), August (two), and October (one) in this area. All were single fawns, and no twins have been recorded based on extensive interviews of hunters and fieldwork from 1996 to date.

We have only anecdotal information on red brocket deer reproduction in Mexico. In southeastern Campeche, interviews of hunters and a few records of harvested females reveal that pregnant females can occur year-round, suggesting aseasonal breeding, but detailed studies are needed to determine whether there is a peak season for red brocket rutting, mating, and births in this and other tropical and montane areas, where reproductive patterns might be different.

Data on the shedding of velvet and antlers by the two brocket deer in Mexico are scarce. However, they do shed their antlers. In a collection of more than 120 brocket deer skulls collected in the GCR from 1996 to 2008, the great majority (87%) had completed antler growth, and the rest were either antlerless (3%) or in some state of antler development or velveted (10%). No data on antler phenology exist for brocket deer in Mexico, and little for the genus *Mazama* in general.

Longevity

The only study that addresses the survival and longevity of deer in tropical Mexico is that of Weber (2005). In this study, ages were determined from tooth wear and counts of the cementum annuli on teeth extracted from more than 150 deer skulls collected by subsistence hunters. This enabled the estimation of survival rates, age structure, and maximum longevity of the three species of sympatric deer in Calakmul, Campeche (fig. 18.9).

In the GCR, sympatric white-tailed deer and the two brockets attained a well-structured age population, with most harvested deer belonging to the one- to two-year-old age classes. However, three brocket specimens (two brown and one red) were 14 years old.

Behavior

In the tropics of Mexico, both white-tailed and brocket deer are solitary, and small groupings (mostly a male and female or a female and fawns) are the only social structure known. Aggregations of males, such as those commonly found in temperate environments, are uncommon.

White-tailed deer occupy a well-defined home

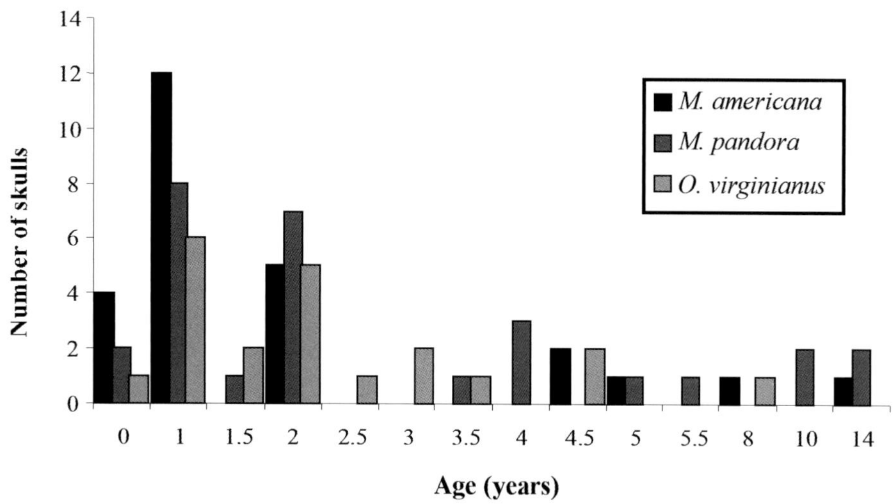

Figure 18.9. Population structure based on age classes of red brocket (n = 27), brown brocket (n = 28) and white-tailed deer (n = 23) in the Greater Calakmul Region, southeastern Mexico.

range, but they are not territorial. Home ranges are influenced by age, sex, density, social interactions, and latitude, as well as season and habitat characteristics. Size of home ranges varies inversely with population density and vegetative cover. Annual home ranges vary from 59 to 520 ha (145 to 1,284 ac) (Marchinton and Hirth 1984).

In northeastern Mexico, whitetail home ranges averaged 193 ha (476 ac) for females and 234 ha (578 ac) for males (1995–1998) in xerophyllous brushland. The differences noted were between the fawning and reproductive seasons in some years and were caused by differences in the quantity and distribution of precipitation (Bello-Gutiérrez et al. 2004). In the same area, Soto-Werschitz (2000) compared home ranges of females with and without fawns; females without fawns reduced their home range size during the fawning period, but females with fawns maintained their home range size, probably because of their energetic requirements during the lactating period.

Whitetails were radio tracked in a tropical dry forest in the Chamela Reserve in Jalisco. The mean distance traveled by a female was 1.5 and 2.6 km/day (0.93 and 1.6 mi/day) during the dry and wet season, respectively. Home ranges encompassed 11 ha (27 ac) and 24–44 ha (59–108 ac), respectively. During the dry season, a male traveled 2.5 km/day (1.5 mi/day), and the home range size in the dry season was 26 ha (64 ac) (Sánchez-Rojas et al. 1997). In a cloud forest in the Manantlán Biosphere Reserve in Jalisco, six female *sinaloae* had a mean home range of 14.5 ha (35.8 ac) during the wet season, and 16.9 ha (41.7 ac) in the dry season (González-Pérez 2003).

In Chamela, whitetails preferred to forage in tropical dry forest year-round, especially during the wet season, probably because the available palatable plant species (foliage density and biomass) were highly nutritious. Also, this habitat had low predation risk compared with that in tropical semievergreen forest along streams and rivers in this area (Núñez and Miller 1997, López-González et al. 1998, Mandujano et al. 2004).

No data on home range size or other spatial ecology exist for the Yucatán brown brocket or red brocket deer in Mexico. Only four animals were trapped in two years of discontinuous trapping effort in the GCR. Only one survived long enough to provide data on home range size; the other three were killed by predators or poachers or died from postcapture myopathy. The surviving subadult brown brocket doe exhibited a small home range of less than 50 ha (124 ac) during the three months of radio monitoring; it inhabited a mixed patch of semideciduous tropical rainforest and flooded tropical forest.

In a behavioral study of 26 female and 13 male captive whitetails in the Chapultepec Zoo in Mexico City, Rosas-Alvarado (1990) identified 106 behavioral patterns. White-tailed deer exposed to humans (presence or absence) in two kinds of enclosed plots in Banderilla and Xalapa, Veracruz, suggested changes in feeding behavior (Serio-Silva 1999).

Whitetails are crepuscular, but activity varies according to several environmental variables, mainly the distribution and amount of rain (Gallina et al. 1998, Gallina et al. 2003, Marchinton and Hirth 1984). White-tailed deer form two basic social groups in temperate environments: family groups centered around an adult female (the mother and her female fawns of previous years); and fraternal groups of adult males and occasionally yearling males (Marchinton and Hirth 1984). Mixed groups occur only during the mating period. It is possible to find mixed feeding groups, but these assemblages are temporary, and they do not form social groups. Dominance hierarchies minimize conflicts and avoid aggression within groups, reducing energy expenditure and risk of injury (Marchinton and Hirth 1984). Social organization and behavior vary in relation to habitat characteristics; group size is inversely related to density cover (Hirth 1977, Mandujano and Gallina 1996). However, whether these patterns also occur in the tropics is unknown. In a tropical dry forest in the Sierra de Huautla in Morelos, females consumed a higher-quality diet (with higher fecal nitrogen values) than males (Buenrostro Silva 2005).

In a cloud forest in the Manantlán Biosphere Reserve in Jalisco, six radio-tracked female whitetails exhibited bimodal daily activity budgets: 0900–1200 hr and 1900–0000 hr (González-Pérez 2003). Deer in arid zones were more active during crepuscular hours (1700–2100 hr) and early in the morning (0500–0900 hr). During the breeding season, deer were active at all hours of the day. A negative correlation was found between activity and temperature (Gallina et al. 2003). Distances traveled per day by whitetails in northeastern Mexico were significantly different among years (mean = 7,016 ± 354 m, 4.35 ± 0.21 mi).

Significant differences in season-year interactions were detected, probably because of significant differences in seasonal precipitation (Bello et al. 2003, Bello-Gutiérrez et al. 2004). Local hunters in the GCR harvest white-tailed and brocket deer mostly at night (using spotlights), suggesting that deer are active mostly at night in this tropical area. However, many white-tailed deer (but few brockets) are also harvested during the day in agricultural areas and gardens.

On San Lucas Island, Costa Rica, deer family groups were composed of a doe and fawns of that year; adult males were solitary. Adult males and females formed mixed groups during the fawning and breeding seasons, but only in the latter season were there bonds between the sexes (Rodríguez-Ramírez 1987).

Habitat and vegetation use by the three sympatric deer species was also studied in the GCR using sightings and hunter kill-site locations for six years (1996–2001) (table 18.2) (Weber 2005, 2008). No other information on the behavior of the two Mexican brocket deer species has been published.

Population Dynamics

Abundance

In Mexico, the white-tailed deer is the most widely studied deer (Weber and Galindo-Leal 1995); 75% of 502 deer-related publications between 1850 and 2001 were on whitetails (Mandujano 2004). Of the 14 subspecies, the most studied are *texanus*, *couesi*, and *sinaloae*. There are also some studies on *mexicanus* and *yucatanensis*, but 7 subspecies (*veraecrucis*, *toltecus*, *truei*, *nelsoni*, *oaxacensis*, *acapulcensis*, and *thomasi*), most from tropical regions, are less well known. This is critical because all subspecies inhabit tropical areas (Mandujano and Bello 1998) where rural inhabitants commonly practice subsistence hunting.

Table 18.2. Habitat type inside buffers (mean values are ha within a 200-m diameter buffer) based on deer sightings and culling site GPS positions on habitat vegetation map (30 × 30 m pixels).

Vegetation type (Two degrees of freedom)	Kruskal-Wallis Mean rank values (ha)			Test statistics		
	Red brocket (n = 27)	Brown brocket (n = 30)	Whitetail (n = 35)	H	p value	Sig.
Agriculture	35.4[a]	44.37[a]	56.81[b]	17.8	$p < 0.001$	***
Secondary growth	40.0	47.08	51.10	5.02	$p > 0.05$	ns
Deforested	34.70[a]	46.23[a]	55.83[b]	14.0	$p < 0.001$	***
Tall evergreen forest	65.83[b]	36.20[a]	40.41[a]	21.6	$p < 0.001$	***
Flooded forest	37.50[a]	57.77[b]	43.79[a]	11.1	$p < 0.005$	**
Short deciduous forest	44.30	45.70	48.89	2.05	$p > 0.1$	ns
Unclassified pixels	0.202	0.088	0.000			

Note: Kruskal-Wallis p values were ranked as follows: *** = highly significant when $p < 0.001$, ** = marginally significant when $p < 0.01$, and ns = nonsignificant when $p > 0.05$. A multiple-comparisons (Dunn's) ranks test (not adjusted for ties) was significant when $p < 0.05$; different letters within rows indicate a statistically significant difference. This test was not performed if the Kruskal-Wallis test was not significant.

In Los Tuxtlas, Veracruz, a tropical area, the abundance index was low for the white-tailed deer (0.025) compared with the red brocket deer (0.36) (Bello 1993). In Tabasco, in Parque Estatal Sierra de Tabasco and the Sierra de Tenosique, the index was 1.0 and 0.5, respectively (Jesús De La Cruz and Bello 2004).

Estimating the abundance of brocket deer within populations is a challenge. The solitary and wary nature of these two deer species, and the fact that they live in habitats where the use of standard survey methodology (i.e., distance sampling) is at best problematic, makes indirect techniques (abundance indices such as track and pathway counts, and subsistence hunting harvest data) the most common methods used (Weber 2005, García-Marmolejo et al. 2013, 2015).

Density

Densities estimated using permanent transects in the GCR (1996–2001) also indicate a higher density of brocket (0.09 deer/km^2, 0.14 deer/mi^2) than white-tailed deer (0.02 deer/km^2, 0.03 deer/mi^2). Density estimates were almost twice as low in temporary transects for both species (table 18.3). Although comparisons of density estimates with those obtained from the sign encounter rate (SER) technique are not possible because SER cannot be translated to density estimates (deer/km^2), densities are proportionally similar for brockets and whitetails. The mean group size of brocket species and whitetails was similar, and most observations were of solitary individuals (table 18.3). In Mexico, white-tailed deer densities varied from 1 to 20 deer/km^2 (1.6 to 32.2 deer/mi^2) in the different geographic locations studied (Gallina et al. 2010).

Density estimates of the two brocket deer species inhabiting the Yucatán Peninsula range from 0.90 to 1.5 deer/km^2 (1.4 to 2.4 deer/mi^2). However, these are probably underestimates since subsistence hunting harvest data suggest much higher densities in certain habitat types. The relative abundance of red and brown brockets is from three to five times that of whitetails, suggesting that the tropical habitats of brocket deer in the Yucatán are suboptimal habitat for whitetails (Weber 2005).

Status, survivorship, and mortality

Populations of the Yucatán brown brocket deer are well represented in most of its historical distribution. One area extensively studied is the GCR in southeastern Campeche, where this species is still common and widespread. Studies are needed in the northern

Table 18.3. Deer densities and mean deer group sizes for red and brown brocket and whitetails in permanent transects (used for monitoring deer population trends from 1996 to 1999), and temporary transects (for comparison of abundance in different habitat types during 2001) in the Greater Calakmul Region, southeastern Mexico.

Transect type	Abundance	Method	Two brocket species	Whitetail
Fixed permanent transects, 1996–1999	Absolute	Distance 3.5	Deer density (± se)/km^2 0.90 ± 0.72	Deer density/km^2 0.021
	Relative	SER*	Deer sign/km 2.78 ± 0.68	Deer sign/km 0.009 ± 0.005
	Relative	Deer sighting rate	Deer/km 0.030	Deer/km 0.01
		Mean deer group size (range)	1.02 (1–2)	1.42 (1–3)
Stratified random temporary transects, 2001	Absolute	Distance 3.5	Deer density (± se)/km^2 0.085 ± 0.063	Deer density (± se)/km^2 0.009 ± 0.004
	Relative	SER	Deer sign/km 1.03 ± 0.82	Deer sign/km 0.018
	Relative	Deer sighting rate	Deer/km 0.021	Deer/km 0.008
		Mean deer group size (range)	1.02 (1–2)	1.12 (1–2)

*SER = sign encounter rate.

(Yucatán state) and western extremes of its distribution (i.e., along the Tabasco-Campeche border), as well as thorough surveys in Quintana Roo along the eastern coast of the Yucatán Peninsula, where hunting and habitat destruction might be affecting populations. Detailed studies are also needed in the Petén region in Guatemala and Belize. Productivity and recruitment data are unavailable for tropical deer in Mexico.

A life table based on skull ages of the three tropical deer species collected in the GCR shows a typical ungulate mortality pattern characterized by high mortality of juvenile deer at one and two years of age, and a more stable mortality rate in all other cohorts. The presence of mature individuals of 10, 12, and even 14 years of age suggests that deer are probably not being overharvested at current harvest rates (table 18.4).

The life table and estimation of survival curves (fig. 18.10) are based on broad assumptions. Data gathered from hunters can have age-class biases, and the rate of increase is not taken into consideration (r = 0). To calculate *lx*, it is assumed that all individuals at age one survived (therefore *px* = 1000), and that all individuals within the last age class died (therefore *px* = 0). These assumptions limit the usefulness of the life table and survival curves for management purposes because true changes in mortality patterns at the species level cannot be detected. More field data are needed to understand these mortality patterns (Caughley and Sinclair 1994, Leeuwenberg and Robinson 2000). However,

Table 18.4. Life table of brown and red brocket and white-tailed deer collected from 1996 to 2001 in the Greater Calakmul Region, southeastern Mexico (n = 152).

Age	Frequency (rounded)	Survival	Mortality	Mortality rate	Survival rate
X	fx	lx	dx	qx	px
1	67	1	0.418	0.418	0.582
2	39	0.582	0.329	0.565	0.435
3	17	0.253	0.056	0.220	0.780
4	13	0.197	0.108	0.548	0.452
5	6	0.089	0.030	0.330	0.670
8	4	0.059	0.015	0.254	0.746
10	3	0.044	0	0	1
14	3	0.044	0	0	1

Note: Calculations after Caughley and Sinclair (1994).

several researchers have used data from subsistence hunters to estimate age structures in deer and other wildlife populations (Townsend 1995, Bodmer et al. 1997, Leeuwenberg and Robinson 2000, Hurtado-Gonzalez and Bodmer 2004).

In general, there were more brown brocket males harvested than females (t = 3.24, df = 2, $p < 0.005$). No statistical differences were found between the sex ratios of the other two species. There were no differences in sex ratios in the age classes assessed within one-year intervals in the three deer species harvested. In contrast, differences were observed in sex ratios for fawn, juvenile, and adult age classes of both brocket species. There were significantly more juveniles of red brocket harvested than fawns and adults (Mann-Whitney test; U = 67.4, df = 2, $p < 0.05$) (fig. 18.9). No differences in sex ratios (in this age class classification) were detected in the other two species (Mann-Whitney test; red brocket: U = 12.34, df = 2, $p > 0.1$; whitetail: U = 17.42, df = 2, $p > 0.5$).

If we compare the age classification method used by Hurtado-Gonzalez and Bodmer (2004), which used four age classes divided by two-year intervals, with the more precise aging technique of tooth cementum annuli, the resulting survival curves (fig. 18.10) are smoother and closer to a normal distribution. Data obtained using this method did not require logarithmic transformation.

While differences in survival rates among juvenile cohorts of both brocket species (second age class) were detected with the first method (Kolmogorov-Smirnov test; D = 2.24, $p < 0.05$), there were no differences using the smoothed distribution obtained with the more accurate aging technique (Kolmogorov-Smirnov test; D = 0.62, $p > 0.1$). It was not possible to compare these two techniques for age structure classification in whitetails because of the small number of skulls available by cohort (figs. 18.9 and 18.10).

Habitat

The Yucatán brown brocket is a habitat generalist throughout its range. The only information available on habitat use patterns is from Weber (2005) and Reyna-Hurtado and Tanner (2005). Both studies found that the species used habitats according to the availability of vegetation types, without clear habitat preferences. In comparison with the red brocket and whitetail, with which the brown brocket is sympatric in many areas of its distribution, the Yucatán brown

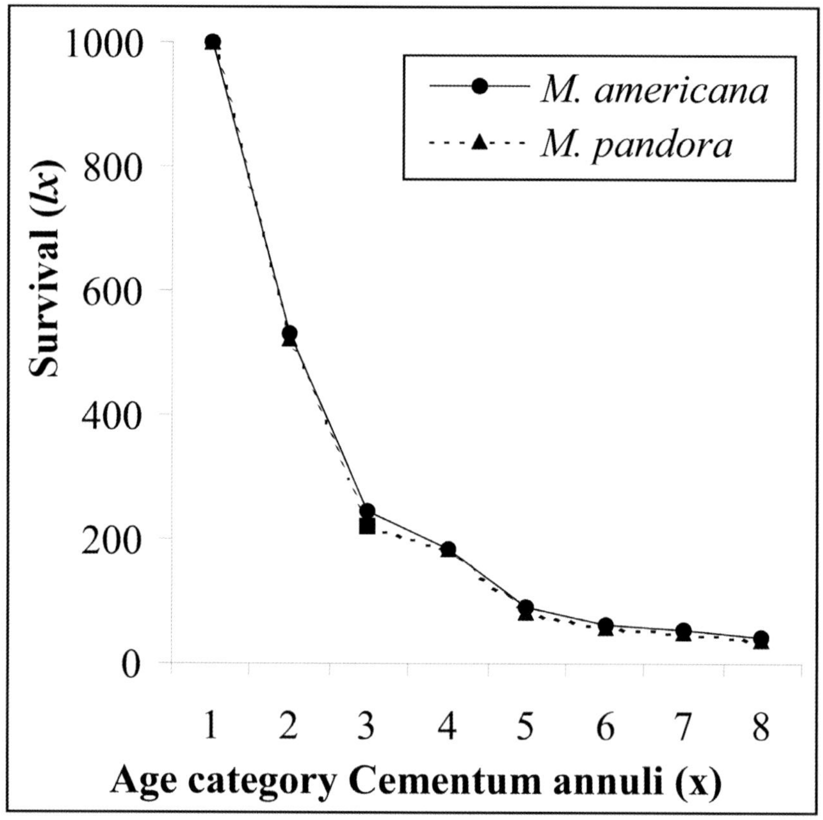

Figure 18.10. Survivorship curves by age classes determined using the cementum annuli histological technique (*n* = *78)* from red and brown brocket deer in the Greater Calakmul Region, southeastern Mexico. Calculations of survivorship are after Caughley and Sinclair (1994, table 3).

brocket is clearly a habitat generalist. However, the red brocket strongly prefers tall and medium semi-evergreen forests, and whitetails prefer deforested (second-growth and agricultural) areas on the Yucatán Peninsula (Weber 2008, García-Marmolejo et al. 2015).

White-tailed deer live in a wide range of habitats from temperate to subtropical, including semiarid environments in North America, as well as rainforests and other equatorial plant associations such as deciduous forests and savannas in Central America and northern South America (Brokx 1984, Danields 1987, 1991, Smith 1991, Flores-Armilla et al. 2011). Whitetails are abundant in mixed pine-oak forests in Mexico (Ffolliott and Gallina 1981) and prefer level sites, such as mesas or elevated areas, where plant diversity and available biomass are greater (Gallina 1994). They are also found in second-growth forests and in thickets and forest-savanna ecotones in Guatemala, Honduras, Belize, El Salvador, Costa Rica, and Panama (Mendez 1984). The amount and temporal distribution of precipitation is an ecological limitation in more-southern latitudes and at lower elevations (Ffolliott and Gallina 1981, Mendez 1984, Villarreal 1999). Whitetails favor more-mesic climates and vegetation in arid regions. In Andean South American countries, their distribution is not limited by elevation but rather by steep, arid habitat and by rainforest on mountain slopes (Brokx 1984).

White-tailed deer are extremely adaptable. They thrive in close association with humans and with agricultural and industrial land conversions. Their habitat requirements are met in practically every ecosystem, but they reach their highest densities in hardwood forests and brushlands (Teer 1991). Whitetails inhabit tropical dry forest in Chamela, on the central Pacific coast of Mexico, which is characterized by temporal and spatial availability of forage, nutrients, and water (Mandujano and Gallina 1995, Silva-Villalobos et al. 1999, Mandujano et al. 2004). In the Manantlán Biosphere Reserve in Jalisco, they prefer cloud forest (González-Pérez 2003).

In tropical evergreen forest, whitetails have benefited from converted habitats (*acahuales*), peasant gardens (milpas), and other farmed lands (Jesús De La Cruz and Bello 2004). They occur in secondary forests with trees 6 m (16.4 ft) high and gentle slopes (11°), with dense shrub cover of 90%

(Bello-Gutiérrez et al. 2004). Tropical rainforest has long been considered suboptimal habitat for the white-tailed deer (Leopold 1959). Recent detailed studies of habitat use in tropical rainforests in Mexico (García-Marmolejo et al. 2015, Naranjo 2002, Reyna-Hurtado 2002, Weber 2005) support Leopold's original hypothesis.

In the arid region of northeastern Mexico, dominated by xerophyllous brushland, whitetail habitat preferences were influenced by habitat type, sex, and year. Females used vegetation associations with dense cover (*Acacia-Castela* brushland). Males, however, selected most habitat types according to availability, showing preferences for open habitats dominated by *Flourensia*. In arid zones, when precipitation conditions are favorable (about 400 mm [15.7 in] annually), the quality and quantity of available forage are key factors in determining habitat use by deer. However, if drought or predation increases, thermal and hiding cover become relatively more important than forage availability (Gallina and Bello 2000, Bello et al. 2001).

Human Impacts

The effects and impacts of subsistence hunting on ecosystem structure and functioning have been studied by Redford (1992), Peres (2000, 2001), and Robinson and Bennett (2000). The traditional belief that local and indigenous societies live in harmony with nature has been rejected during the past 10 years, with overwhelming evidence of overharvest derived from field research in Africa, Asia, and Latin America (Alvard and Kaplan 1991, Alvard 1995, Robinson and Bennett 2000). Humans will not harvest wildlife populations in a sustainable fashion when their individual or communal needs are threatened or placed at risk by doing so (Redford 1992, Alvard et al. 1996). Weber (2005) documented a local quasi extirpation of white-tailed deer and stable brocket deer populations in two large rural communities in southern Campeche, most likely because of overhunting (fig. 18.11).

On the frontiers of human colonization in Latin America, subsistence hunting is one of the most important activities affecting wildlife populations (Bodmer et al. 1994, 1997). Frontier regions may sustain a large proportion of global biodiversity and, paradoxically, are usually located in areas where human populations are experiencing extreme poverty and high population growth (Bodmer 1995a, Alvard et al. 1996, Ericson 1996). Subsistence and commercial hunting can clearly overexploit wildlife

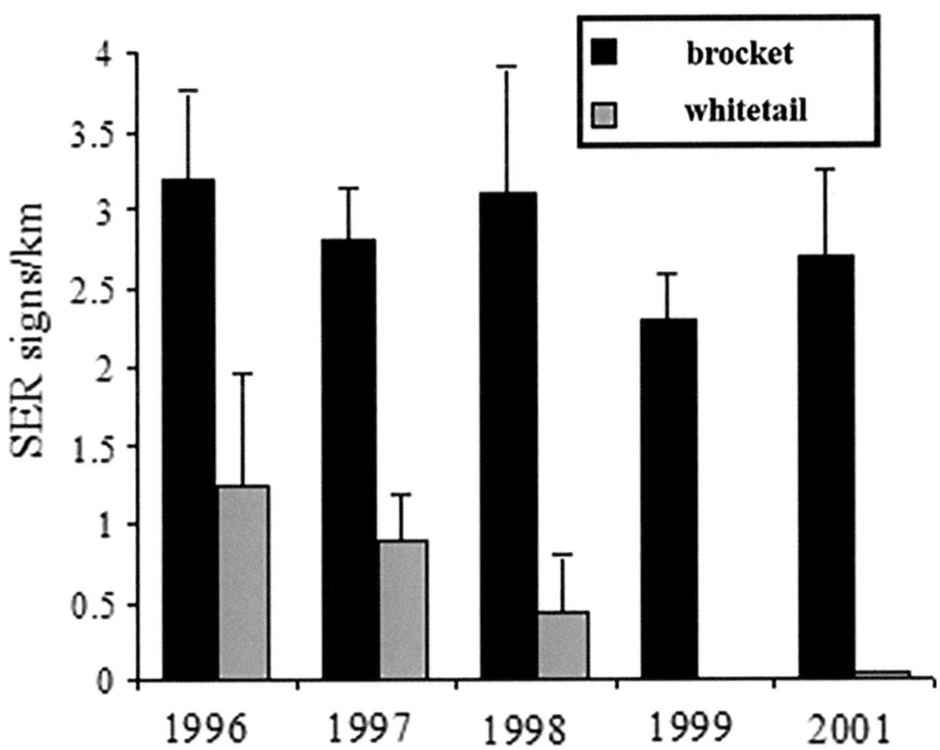

Figure 18.11. Trend in relative abundance of red and brown brocket and white-tailed deer monitored monthly in permanent transects from 1996 to 1999 in the Greater Calakmul Region, southeastern Mexico. Error bars represent one standard error of SER (sign encounter rate) for nonindependent temporal replicates on permanent transects.

populations in these regions (Bodmer et al. 1994). For example, conservative estimates of game yields indicate that as many as 23.5 million vertebrate game species (the equivalent of 84,224 tons of bush meat) are consumed each year by the rural population of the Brazilian Amazon (Peres 2000).

In a study that monitored subsistence hunting in the GCR, hunters from eight indigenous and mestizo communities reported a total harvest of 1,028 animals killed in three months. Deer (brocket deer and whitetails combined) made up almost 14% of the total harvest, and 52% of the total animal biomass, with 65 white-tailed and 78 brocket deer harvested during this period (table 18.5). Each hunter killed an average of 0.76 white-tailed deer and 0.81 brocket deer (1.5 deer/trimester), with a total biomass of 2,925 and 1,520 kg (6,448 and 3,351 lb), respectively. White-tailed and brocket deer ranked number 1 and 3, respectively, as meat providers (biomass) to the hunters surveyed in the Calakmul region (table 18.5).

In the GCR, hunters harvested deer within a mean distance of 13.5 km (8.3 mi) from their hometowns. Overall, hunters traveled an average of 2.5 km (1.5 mi) farther during the dry season. However, this difference was strongly influenced by the fact that they traveled an average of 4 km (2.4 mi) farther to hunt brown brocket than to hunt the other two deer species (fig. 18.12). This difference was even larger (9 km [5.5 mi] farther) during the wet season for brown brocket and is statistically significant (Kruskal-Wallis test, $x^2 = 4.75$, df = 1, $p < 0.005$).

Ungulates are by far the most important meat

Table 18.5. Biomass of prey species collected by subsistence hunters, and meat consumption by eight indigenous and mestizo communities from June to October 1997 in the Calakmul region, southeastern Mexico.

Common name	Scientific name	Reported harvest, no. and (%)[a]	Harvest per capita, kg[b]	Meat consumption per capita, kg[c]	Average weight, kg[d]	Total biomass, kg	Proportion of total biomass
White-tailed deer	*Odocoileus virginianus*	65 (6.3)	0.77	1.93	45	2,925	0.29
Collared peccary	*Pecari tajacu*	158 (15.4)	1.80	1.56	15	2,370	0.23
Brocket deer	*Mazama spp.*	78 (7.5)	0.92	1.00	20	1,520	0.15
Paca	*Agouti paca*	125 (12.4)	1.40	0.57	7	875	0.085
White-lipped peccary	*Tayassu pecari*	20 (1.9)	0.23	0.33	25*	500	0.049
Armadillo	*Dasypus novemcintus*	137 (13.5)	1.63	0.31	3.5	479	0.047
Ocellated turkey	*Meleagris ocellata*	79 (7.6)	0.94	0.27	5	396	0.039
Great curassow	*Crax rubra*	99 (9.5)	1.17	0.26	4*	395	0.035
Coati	*Nasua narica*	141 (13.7)	1.67	0.18	2	282	0.030
Agouti	*Dasyprocta punctata*	86 (8.3)	1.02	0.11	2*	172	0.018
Crested guan	*Penelope purpurascens*	40 (3.8)	0.47	0.05	2	120	0.015
Other small mammals and birds	Varied	Varied	0.05	0.001	< 1	< 100	0.009
Total		1,028	12.02	6.54	130.5	10,034	1
Arithmetic Mean		93.45	1.09	0.59	11.86	912.22	
Standard Error of the Mean		13.11	0.151	0.193	4.091	286.21	

Note: Harvest data were collected from August to October 1997. Species are listed in order of their total biomass.
[a]Includes the reported species harvested in the three months prior to the survey.
[b]Per capita species harvest of the subsample of 83 hunters included in the analysis.
[c]Per capita monthly meat consumption based on an average of 6.07 (± 2.4 sd) family members in the subsample of 84 hunters surveyed.
[d]Average weight is an absolute value and was obtained from harvested samples in this study when $n > 10$, unless otherwise noted (*). The weight is the mean of the values reported by Robinson and Redford (1991), Jorgenson (1993), and MacNab (1998).

source for local residents in the Neotropics (Robinson and Redford 1991, Bodmer et al. 1994, Bodmer 1995a, 1995b, Cullen et al. 2000, Peres 2000, 2001, Jerozolimski and Peres 2003). Peccaries, tapir, and deer top the list of ungulate game species most sought after by subsistence hunters (Leopold 1959, Robinson and Redford 1991, Bodmer et al. 1994, Clayton et al. 1997, Peres 2000, Hurtado-Gonzalez and Bodmer 2004).

Subsistence hunting studies in the Neotropics suggest that deer can withstand considerable hunting pressure compared with other ungulate species, without an apparent effect on population size (Robinson and Redford 1991, Bodmer et al. 1994, 1997, Bodmer 1995a, Cullen et al. 2000). Deer resilience to hunting has been attributed to two main factors: high reproductive rates and adaptability (Robinson and Redford 1991, Robinson and Bodmer 1999, Robinson and Bennett 2000). However, high levels of poaching by villagers near Barro Colorado National Monument, Panama, were strongly correlated with low population densities of white-tailed and brocket deer (Wright et al. 2000).

Captive deer are common in Mexico, and many UMAs harbor captive white-tailed deer. In the state of Yucatán, there are 12 UMAs with captive deer (González-Marín 2002), and Tabasco had more than 5 (Mendoza-Castillo and Gamboa-González 1993). Fifteen UMAs currently have captive whitetails (Bello-Gutiérrez et al. 2009), of which 3 are zoos (León-Castro 2000). Most zoos in Mexico have white-tailed deer, but few have red or brown brocket deer.

Management Challenges in Mexico

The main problem in managing tropical deer populations in Mexico is the paucity of scientific information generated in the last 40 years. Although great advances have been made in documenting many aspects of the biology and ecology of the three tropical deer species, there are still many gaps in our knowledge of fundamental aspects necessary for their management. Much information is still anecdotal or unreliable, either because of the small number of animals studied and the short period of the studies, or because it cannot be extrapolated to other areas given the enormous habitat heterogeneity and differences in climate, elevation, and human dimensions where deer are still present in the Mexican tropics (Weber and Gonzalez 2003).

Another major management challenge is the regulation, control, and biological and social understanding of subsistence hunting and the illegal harvest of deer. Subsistence deer hunting is currently a basic activity for many inhabitants of tropical Mexico, but its regulation and control is mandatory if deer populations are to persist and thrive outside protected areas. Biological and social understanding of subsistence hunting includes the transfer of current scientific knowledge to different levels of decision makers, including politicians, environmental bureau-

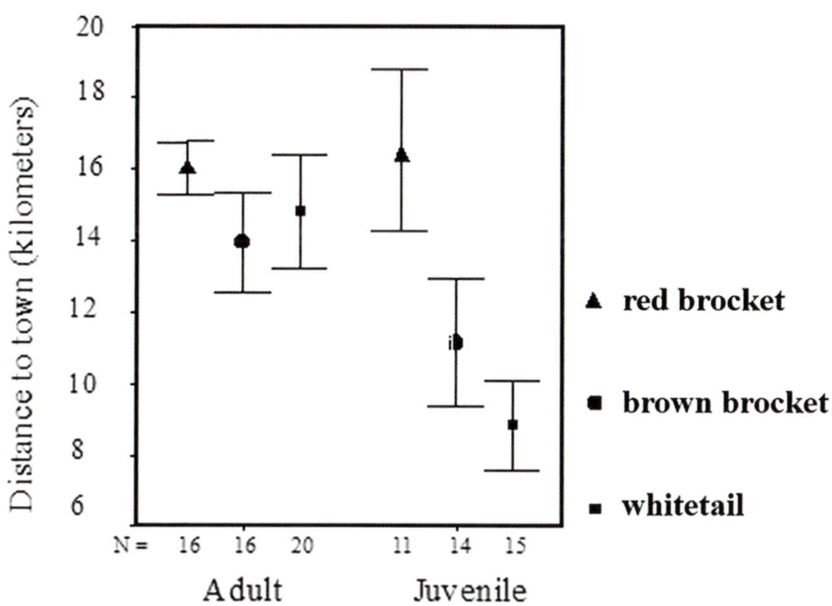

Figure 18.12. Mean distance (and one confidence interval) from the home town of hunters of red and brown brocket and white-tailed deer (adults and juveniles) based on hunting and sighting data in the Greater Calakmul Region, southeastern Mexico

crats, and environmental NGOs. This, however, has proved extremely difficult to attain; a good example is the poor performance of some UMAs in the tropics of Mexico (Weber et al. 2006, García-Marmolejo et al. 2013, 2015). In central Panama, high levels of poaching in several forests were correlated with the local extirpation of white-tailed deer, and with very low population densities of red brocket deer (Wright et al. 2000), which supports the data from the Calakmul studies.

If the Calakmul Biosphere Reserve (CBR) is to succeed in its long-term goal of maintaining the biodiversity of one of the biologically richest regions in Mesoamerica, the impact of subsistence hunting in the last frontier of colonization in Mexico must be included as a high priority in the conservation and development plans for that region. Failing to do so will probably lead to the depletion of wildlife populations in villages around the CBR, and to increased pressure on natural resources in this protected area.

Conservation Status

All subspecies of white-tailed deer in Mexico are listed in the category of Least Concern, and none are listed in CITES (Gallina and Lopez Arevalo 2008, IUCN 2009, Weber et al. 2009, Duarte and Vogliotti 2015).

In tropical dry forest, perennial dry years and the hunting of females could result in a rapid decrease in the deer population. In the Chamela region, jaguars (*Panthera onca*), pumas (*Puma concolor*), and ocelots (*Leopardus pardalis*) prey on deer (López-González et al. 1998, Núñez and Miller 1997). Predation plays a key role in deer populations, especially during fawning and at the end of the dry season, when individuals might be more vulnerable. The effects of predation on tropical deer have not been studied. An alarming mortality factor is the indiscriminate killing of both sexes, and all ages, throughout the year by subsistence hunters (Contreras-Moreno 2008).

Although white-tailed deer are highly resilient and adaptable, local extinctions from overhunting have been documented in suboptimal habitats with high hunting pressure, such as the Calakmul tropical rainforest region (Weber 2000, 2005; Weber and Gonzalez 2003). Together with peccaries, brocket deer, and some caviomorph rodents (i.e., pacas [*Agouti paca*] and agoutis [*Dasyprocta spp.*]), white-tailed deer are the main game species of subsistence and sport hunters in Mesoamerica south of the Tropic of Cancer. This species has been continually subjected to persistently high hunting pressure, yet deer still occur in most habitats throughout this region. The maximum annual harvest should maintain a minimum density of 0.6 deer/km^2 (0.37 deer/mi^2) to achieve source-sink dynamics in a heavily hunted subsistence scenario in southeastern Mexico (Weber 2008).

The IUCN (2009) lists the Yucatán brown brocket deer as Data Deficient, meaning that current information is inadequate to make a direct or indirect assessment of its risk of extinction based on its distribution or population status. This species is not included in the Mexican List of Species at Risk. More research is needed throughout its distribution, in particular in the thorn-shrubland and short-deciduous tropical forests in northern Yucatán, and in those areas where subsistence hunting and habitat fragmentation might be having negative impacts on this endemic deer species. The IUCN also lists the red brocket as Data Deficient because of insufficient data to evaluate its status. This species is also in critical need of research to establish a thorough knowledge of its natural history on which to base regional management plans. Other basic needs are regional socioeconomic studies to determine the effects of subsistence hunting, and sustainable hunting programs.

Opinions differ about the role of UMAs in the conservation of white-tailed deer in Mexico. Sánchez-Rojas et al. (2009) emphasized the importance of UMAs as a complementary strategy for the conservation and sustainable use of this species in forested areas in central Mexico. Most UMAs do not have the critical size to support minimum viable populations (MVP) of white-tailed deer, while protected areas (biosphere reserves, natural resource protected areas, and flora and fauna protected areas) could potentially support the MVP of this species (Mandujano and González-Zamora 2009). A recent evaluation suggests a minimum critical area of 1,667 to 50,000 ha (4,119 to 123,552 ac) to support an MVP of 500 deer, or 16,670 to 500,000 ha (41,192 to 1.2 million ac) for long-term viability of 5,000 deer, depending on regional deer density (Mandujano and González-Zamora 2009). These authors suggest a system of conservation at a regional level in which Protected Natural Areas (ANPs) and UMAs are incorporated, assuming source-sink and archipelago reserve models, where connectivity can play an important role in the movement of individuals between populations.

The white-tailed deer is one of the most important game mammals in Latin America, especially in Mexico, Nicaragua, Costa Rica, Panama, Colombia, Venezuela, Ecuador, and Peru (Ojasti 1996). Deer in Mexico are much less abundant now than they were in the 1940s, and the general decline appears to be the result of indiscriminate hunting, changes in land use, and competition with domestic animals (Brokx 1984). These factors need to be further understood and incorporated into conservation and management actions in order for the three species of deer to thrive in tropical ecosystems in Mexico.

Acknowledgments

Yucatán and red brocket deer research in Calakmul was funded by the Mexican Fund for Nature Conservation, Stanford University's Center for Conservation Biology, and CONACyT. The cooperation of many local and indigenous hunters from the Greater Calakmul Region, in particular Nicolas Árias Domínguez, has been essential for many years and is deeply appreciated. Rafael Reyna assisted with fieldwork from 2007 to 2009. Students Hilario Garrido-Chim and Mario Ramírez-Alcocer provided assistance with the literature review.

Literature Cited

Alvard, M. 1995. Shotguns and subsistence hunting in the Neotropics. Oryx 29:58–65.

Alvard, M., and H. Kaplan. 1991. Procurement technology and prey mortality among indigenous Neotropical hunters. Pages 79–104 *in* M. Stiner, editor. Human predators and prey mortality. Westview Press, Boulder, Colorado, USA.

Alvard, M., J. G. Robinson, K. H. Redford, and H. Kaplan. 1996. The sustainability of subsistence hunting in the Neotropics. Conservation Biology 11:977–982.

Arceo, G., S. Mandujano, S. Gallina, and L. A. Perez-Jimenez. 2005. Diversity of diet of white-tailed deer in a Mexican tropical forest. Mammalia 69:159–168.

Baker, R. H. 1984. Origin, classification and distribution. Pages 1–18 *in* L. K. Halls, editor. White-tailed deer: ecology and management. Stackpole, Harrisburg, Pennsylvania, USA.

Barragán, K. 2002. Caracterización cromosómica de venados cola blanca (*Odocoileus virginianus*) (Zimmermann, 1780). Thesis, Universidad Nacional de Colombia, Bogotá, Colombia.

Barrera-Salazar, A., S. Mandujano, O. A. Villarreal Espino-Barros, and D. Jiménez-García. 2015. Classification of vegetation types in the habitat of white-tailed deer in a location of the Tehuacán-Cuicatlán Biosphere Reserve, Mexico. Tropical Conservation Science 8:547–563.

Bello, J. 1993. Situación actual del orden Artiodactyla en la región de Los Tuxtlas, Veracruz. Thesis, Universidad Veracruzana, Xalapa, México.

Bello, J., S. Gallina, and M. Equihua. 2001. Characterization and habitat preferences by white-tailed deer (*Odocoileus virginianus*) in Mexico with high drinking water availability. Journal of Range Management 54:537–545.

Bello, J., S. Gallina, and M. Equihua. 2003. Comparación de los movimientos del venado cola blanca en dos sitios con diferente disponibilidad de agua del noreste de México. Pages 59–66 *in* R. Polanco, editor. V Congreso Internacional de Manejo de Fauna Silvestre en la Amazonía y Latinoamérica. CITES and Fundación Natura, Bogotá, Colombia.

Bello-Gutiérrez, J., F. M. Contreras-Moreno, and C. E. Pedrero-Hernández. 2009. Análisis de las unidades para la conservación, manejo y aprovechamiento de la fauna silvestre, en el estado de Tabasco. Pages 194–199 *in* Memorias IV Simposio sobre Fauna Cinegética en México. Benemérita Universidad Autónoma de Puebla, México.

Bello-Gutiérrez, J., C. C. Guzmán-Aguirre, and J. Santos-Zuñiga. 2004. Aspectos ecológicos del venado cola blanca en la región sierra del estado de Tabasco. *In* Memorias IX Simposio de Venados en México. UNAM, ANGADI, and UAEH, Pachuca, Hidalgo, México.

Bisbal, F. J. E. 1994. Biología poblacional del venado matacán (*Mazama* spp.) (Artiodactyla: Cervidae) en Venezuela. Revista de Biología Tropical 42:305–313.

Bodmer, R. E. 1989. Ungulate biomass in relation to feeding strategy within Amazonian forests. Oecologia 81:547–550.

Bodmer, R. E. 1991. Influence of digestive morphology on resource partitioning in Amazonian ungulates. Oecologia 85:361–365.

Bodmer, R. E. 1995a. Managing Amazonian wildlife: biological correlates of game choice by detribalized hunters. Ecological Applications 5:872–877.

Bodmer, R. E. 1995b. Priorities for the conservation of mammals in the Peruvian Amazon. Oryx 29:23–28.

Bodmer, R. E., J. F. Eisenberg, and K. Redford. 1997. Hunting and the likelihood of extinction of Amazonian mammals. Conservation Biology 11:460–466.

Bodmer, R. E., T. G. Fang, L. Moya, and R. Gill. 1994. Managing wildlife to conserve Amazonian forests: population biology and economic considerations of game hunting. Biological Conservation 67:29–35.

Branan, W. V., M. C. M. Werkhoven, and R. L. Marchinton. 1985. Food habits of brocket and white-tailed deer in Suriname. Journal of Wildlife Management 49:972–976.

Brokx, P. A. 1984. White-tailed deer of South America. Pages 525–546 *in* L. K. Halls, editor. Ecology and management of the white-tailed deer. Stackpole, Harrisburg, Pennsylvania, USA.

Buenrostro Silva, A. 2005. Segregación sexual y su relación con la calidad de la dieta del venado cola blanca (*Odocoileus virginianus mexicanus*) en el Ejido El Limón, Tepalcingo, Morelos. Thesis, Instituto de Ecología, Xalapa, Veracruz, México.

Caughley, G., and A. R. E. Sinclair. 1994. Wildlife ecology and management. Blackwell Scientific, Boston, Massachusetts, USA.

Clayton, L., K. M. Keeling, and E. J. Milner-Gulland. 1997. Bringing home the bacon: a spatial model of wild pig hunting in Sulawesi, Indonesia. Ecological Applications 7:642–652.

Clemente, F. 1984. Utilización de la vegetación nativa en la alimentación del venado cola blanca (*Odocoileus virginianus* Hays) en el estado de Aguascalientes. Thesis, Colegio de Postgraduados, Chapingo, Estado de México, México.

Contreras-Moreno, F. M. 2008. Ecología poblacional del venado cola blanca (*Odocoileus virginianus thomasi*) en la r/a San Joaquín municipio de Balancán, Tabasco, México. Thesis, Universidad Juárez Autónoma de Tabasco, Villahermosa, México.

Cullen, L., Jr., R. E. Bodmer, and C. V. Pádua. 2000. Effects of hunting in habitat fragments of the Atlantic forests, Brazil. Biological Conservation 95:49–56.

Czernay, S. 1987. Die Spiesshirsche und Pudus. Die neue Brehm-Bücherei. Ziemsen Wittenberg, Lutherstadt, Germany.

Danields, H. 1987. Ecología nutricional del venado caramerudo (*Odocoileus virginianus gymnotis*) en los llanos centrales. Dissertation, Universidad Central, Caracas, Venezuela.

Danields, H. 1991. Biologia y hábitat del venado caramerudo. Pages 59–66 *in* El venado en Venezuela: conservación, manejo, aspectos biológicos y legales. FUDECI, PROFAUNA, and FEDECAVE, Caracas, Venezuela.

Delfin, C. 2002. Clasificación y evaluación del hábitat como primera fase para el establecimiento y operación de una UMA con fines de aprovechamiento del venado cola blanca mexicano en la Mixteca Poblana. Thesis, Instituto de Ecología, Xalapa, Veracruz, México.

Di Mare, M. I. 1994. Hábitos alimentarios del venado cola blanca en la Isla San Lucas, Puntarenas, Costa Rica. Pages 63–90 *in* C. Vaughan and M. Rodríguez, editors. Ecología y manejo del venado cola blanca en México y Costa Rica. EUNA, Heredia, Costa Rica.

Duarte, J. M. B., and A. Vogliotti. 2015. *Mazama americana*. IUCN Red List of Threatened Species 2015. http://www.iucnredlist.org. Accessed 29 Jun 2016.

Ericson, J. 1996. Conservation and development on the border of the Calakmul Biosphere Reserve, Campeche, Mexico. Thesis, Humboldt State University, Arcata, California, USA.

Escamilla, A., M. Sanvicente, M. Sosa, and C. Galindo-Leal. 2000. Habitat mosaic, wildlife availability and hunting in the tropical forest of Calakmul, Mexico. Conservation Biology 14:1562–1601.

Ffolliott, P. F., and S. Gallina, editors. 1981. Deer biology, habitat requirements and management in western North America. Instituto de Ecología, México, D. F., México.

Flores-Armilla, V., S. Gallina, J. R. García, V. Sánchez-Cordero, and M. Jaramillo. 2011. Selección del hábitat por el venado cola blanca *Odocoileus virginianus mexicanus* (Gmelin, 1788) y su densidad poblacional en dos localidades de la región centro del Corredor Biológico Chichinautzin, Morelos, México. Therya 2:3–12.

Galindo-Leal, C., and M. Weber. 1998. El venado de la Sierra Madre Occidental: ecología, manejo y conservación. EDICEUA and CONABIO, México, D. F., México.

Gallina, S. 1993. White-tailed deer and cattle diets at La Michilía, Durango, Mexico. Journal of Range Management 46:487–492.

Gallina, S. 1994. Uso del hábitat por el venado cola blanca en la Reserva de la Biosfera La Michilía, Mexico. Pages 299–314 *in* C. Vaughan and M. Rodríguez, editors. Ecología y manejo del venado cola blanca en México y Costa Rica. EUNA, Heredia, Costa Rica.

Gallina, S., and J. Bello. 2000. Comportamiento del venado cola blanca y su hábitat en el noreste de México. Pages 69–81 *in* Memorias VII Simposio sobre Venados de México. UNAM, México, D. F., México.

Gallina, S., P. Corona-Zárate, and J. Bello. 2003. El venado cola blanca: comportamiento en zonas semiáridas del noreste de México. Pages 165–173 *in* R. Polanco, editor. Selección de trabajos. V Congreso Internacional de Manejo de Fauna Silvestre en la Amazonía y Latinoamérica. CITES and Fundación Natura, Bogotá, Colombia.

Gallina, S., and H. Lopez Arevalo. 2008. *Odocoileus virginianus.* IUCN Red List of Threatened Species. Version 2009.1. http://www.iucnredlist.org. Accessed 12 Nov 2011.

Gallina, S., S. Mandujano, J. Bello, H. F. L. Arévalo, and M. Weber. 2010. White-tailed deer *Odocoileus virginianus* (Zimmermann 1780). Pages 101–118 *in* J. M. B. Duarte and S. Gonzalez, editors. Neotropical cervidology: biology and medicine of Latin American deer. IUCN and FUNEP, Gland, Switzerland.

Gallina, S., M. E. Maury, and V. Serrano. 1981. Food habits of white-tailed deer. Pages 133–148 *in* P. F. Ffolliott and S. Gallina, editors. Deer biology, habitat requirements and management in western North America. Instituto de Ecología, México, D. F., México.

Gallina, S., A. Pérez-Arteaga, and S. Mandujano. 1998. Patrones de actividad del venado cola blanca (*Odocoileus virginianus texanus*) en un matorral xerófilo de México. Boletín de la Sociedad Biológica 69:221–228.

García-Marmolejo, G., L. Chapa-Vargas, E. Huber-Sanwald, M. Weber, O. C. Rosas-Rosas, and J. Martinez-Calderas. 2013. Potential distributional patterns of three wild ungulate species in a fragmented tropical region of northeastern Mexico. Tropical Conservation Science 6:539–557.

García-Marmolejo, G., L. Chapa-Vargas, M. Weber, and E. Huber-Sanwald. 2015. Landscape composition influences abundance patterns and habitat use of three ungulate species in fragmented secondary deciduous tropical forests, Mexico. Global Ecology and Conservation 3:744–755.

Gayot, M., O. Henry, G. Dubost, and D. Sabatier. 2004. Comparative diet of the two forest cervids of the genus *Mazama* in French Guiana. Journal of Tropical Ecology 20:31–43.

Geist, V. 1998. Deer of the world: their evolution, behavior and ecology. Stackpole, Harrisburg, Pennsylvania, USA.

Goldman, E. A., and R. T. Moore. 1945. The biotic provinces of Mexico. Journal of Mammalogy 26:347–360.

González-Marín, R. M. 2002. Diagnóstico de situación de las unidades para la conservación, manejo y aprovechamiento sustentable de fauna silvestre (UMAS), en el estado de Yucatán, México. Thesis, Universidad Autónoma de Yucatán, Mérida, México.

González-Pérez, G. E. 2003. Uso del hábitat y área de actividad del venado cola blanca (*Odocoileus virginianus sinaloae* J. Allen) en la Estación Científica Las Joyas, Reserva de la Biosfera de Manantlán, Jalisco. Thesis, UNAM, México, D. F, México.

Granado, A. 1989. Dieta del venado caramerudo (*Odocoileus virginianus gymnotis*) en el Socorro, estado Guarico. Thesis, Universidad Central de Venezuela, Caracas, Venezuela.

Hall, E. R. 1981. The mammals of North America. Volume 2. Second edition. John Wiley and Sons, New York, New York, USA.

Halls, L. K. 1978. The white-tailed deer. Pages 43–65 *in* J. L. Schmidt and D. L. Gilbert, editors. Big game of North America: ecology and management. Stackpole, Harrisburg, Pennsylvania, USA.

Hershkovitz, P. 1951. Mammals from British Honduras, Mexico, Jamaica, and Haiti. Fieldiana (Zoology) 31:547–569.

Hershkovitz, P. 1966. Mice, land bridges and Latin American faunal interchange. Pages 725–751 *in* R. L. Wenzel and V. J. Tipton, editors. Ectoparasites of Panama. Field Museum of Natural History, Chicago, Illinois, USA.

Hesselton, W. T., and R. M. Hesselton. 1982. White-tailed deer. Pages 878–901 *in* J. A. Chapman and G. A. Feldhamer, editors. Wild mammals of North America. Johns Hopkins University Press, Baltimore, Maryland, USA.

Hirth, D. H. 1977. Social behavior of white-tailed deer in relation to habitat. Wildlife Monographs 53:1–53.

Hurtado-Gonzalez, J. L., and R. E. Bodmer. 2004. Assessing the sustainability of brocket deer hunting in the Tamshiyacu-Tahuayo Communal Reserve, northern Peru. Biological Conservation 116:1–7.

IUCN. 2009. IUCN Red List of Threatened Species. http://www.iucnredlist.org. Accessed 12 Nov 2011.

Jerozolimski, A., and C. A. Peres. 2003. Bringing home the biggest bacon: a cross-site analysis of the structure of hunter-kill profiles in Neotropical forests. Biological Conservation 111:415–425.

Jesús De La Cruz, A., and J. Bello. 2004. Estado actual de las poblaciones de venados (Mammalia: Cervidae) en el ejido Oxolotán, Tabasco. Memoria IX Simposio de Venados en México. UNAM, ANGADI, and UAEH, Pachuca, Hidalgo, México.

Jorgenson, J. P. 1993. Gardens, wildlife densities and subsistence hunting by Maya Indians in Quintana Roo, Mexico. Dissertation, University of Florida, Gainesville, USA.

Leeuwenberg, F. J., and J. G. Robinson. 2000. Traditional management of hunting in a Xavanté community in central Brazil: the search for sustainability. Pages 375–394 *in* J. G. Robinson and E. L. Bennett, editors. Hunting for sustainability in Neotropical forests. Columbia University Press, New York, New York, USA.

León-Castro, R. 2000. Criaderos intensivos de venado cola blanca (*Odocoileus virginianus*) en Tabasco: situación actual y perspectivas. Thesis, Universidad Juárez Autónoma de Tabasco, Villahermosa, México.

Leopold, A. S. 1959. Wildlife of Mexico: the game birds and mammals. University of California Press, Berkeley, USA.

López-González, C., M. Hidalgo, L. Cantu, and A. González-Romero. 1998. Ocelotes y coyotes como depredadores de venado en un bosque tropical seco. Pages 102–105 *in* VI Simposio sobre Venados en México. UNAM, México, D. F., México.

MacNab, R. 1998. Comparative impacts of chicle and xiate harvest on wildlife of the Maya Biosphere Reserve, Guatemala. Thesis, University of Florida, Gainesville, USA.

Mandujano, S. 2004. Análisis bibliográfico de los estudios de venados en México. Acta Zoológica Mexicana (n.s.) 20:211–251.

Mandujano, S., and J. Bello. 1998. Asociación de las subespecies de venado cola blanca con los tipos de vegetación en México. Pages 187–195 *in* Simposio de Venado en México. UNAM, México, D. F., México.

Mandujano, S., and S. Gallina. 1995. Comparison of deer censusing methods in a tropical dry forest. Wildlife Society Bulletin 23:180–186.

Mandujano, S., and S. Gallina. 1996. Size and composition of white-tailed deer groups in a tropical dry forest in Mexico. Ethology, Ecology and Evolution 8:255–263.

Mandujano, S., S. Gallina, G. Arceo, and L. A. Pérez-Jiménez. 2004. Variación estacional del uso y preferencia de los tipos vegetacionales por el venado cola blanca en un bosque tropical de Jalisco. Acta Zoológica Mexicana (n.s.) 20:45–67.

Mandujano, S., S. Gallina, and S. H. Bullock. 1994. Frugivory and dispersal of *Spondias purpurea* (Anacardiaceae) in a tropical dry forest of Mexico. Revista de Biología Tropical 42:105–112.

Mandujano, S., and A. González-Zamora. 2009. Evaluation of natural conservation areas and wildlife management units to support minimum viable populations of white-tailed deer in Mexico. Tropical Conservation Science 2:237–250.

Marchinton, R. L., and D. H. Hirth. 1984. Behavior. Pages 129–168 *in* L. K. Halls, editor. White-tailed deer: ecology and management. Stackpole, Harrisburg, Pennsylvania, USA.

Martínez-Romero, L. E. 2004. Determinación de fechas de aprovechamiento del venado cola blanca (*Odocoileus virginianus*) a través de hormonas sexuales y comportamiento. Thesis, Instituto de Ecología, Xalapa, Veracruz, México.

McCoy-Colton, M. B., and C. Vaughan-Dickhaut. 1985. Resultados preliminares del estudio del venado cola blanca (*Odocoileus virginianus*) en Costa Rica. Pages 25–79 *in* Investigaciones sobre fauna silvestre de Costa Rica. Universidad Nacional de Costa Rica, San José, Costa Rica.

Medellín, A. R., A. L. Gardner, and J. M. Aranda.

1998. The taxonomic status of the Yucatán brown brocket *Mazama pandora* (Mammalia: Cervidae). Proceedings of the Biological Society of Washington 111:1–14.

Mendez, E. 1984. White-tailed deer populations and habitats of Mexico and Central America. Pages 513–524 *in* L. K. Halls, editor. White-tailed deer: ecology and management. Stackpole, Harrisburg, Pennsylvania, USA.

Mendoza-Castillo, H., and J. Gamboa-Gonzáles. 1993. Establecimiento de un criadero de venado cola blanca (*Odocoileus virginianus yucatanensis*) en cautiverio. Technical report. Universidad Autónoma de Chapingo, Estado de México, México.

Naranjo, J. E. 2002. Population ecology and conservation of ungulates in the Lacandon Forest, Mexico. Dissertation, University of Florida, Gainesville, USA.

Núñez, R., and B. Miller. 1997. Food habits of jaguars and pumas. Proceedings of the Seventh International Theriological Congress, Acapulco, Guerrero, México.

Ojasti, J. 1996. Wildlife utilization in Latin America: current situation and prospects for sustainable management. Conservation Guide No. 25. FAO, Rome, Italy.

Peres, C. A. 2000. Effects of subsistence hunting and habitat fragmentation on Amazonian forest vertebrates. Conservation Biology 14:240–253.

Peres, C. A. 2001. Synergistic effects of subsistence hunting and habitat fragmentation on Amazonian forest vertebrates. Conservation Biology 15:1490–1505.

Ramírez Lozano, R. G. 2004. Nutrición del venado cola blanca. UANL, Unión Ganadera Regional de Nuevo León, and Fundación Produce, Monterrey, México.

Redford, K. H. 1992. The empty forest. Bioscience 42:412–422.

Reyna-Hurtado, R. A. 2002. Hunting effects on ungulate populations in Calakmul, Mexico. Thesis, University of Florida, Gainesville, USA.

Reyna-Hurtado, R. A., and G. W. Tanner. 2005. Habitat preferences of ungulates in hunted and non-hunted areas in the Calakmul forest, Campeche, Yucatán, Mexico. Biotropica 37:676–685.

Robinson, J. G., and E. L. Bennett. 2000. Carrying capacity limits for sustainable hunting in tropical forests. Pages 13–30 *in* J. G. Robinson and E. L. Bennett, editors. Hunting for sustainability in Neotropical forests. Columbia University Press, New York, New York, USA.

Robinson, J. G., and R. E. Bodmer. 1999. Towards wildlife management in tropical forests. Journal of Wildlife Management 63:1–13.

Robinson, J. G., and K. H. Redford, editors. 1991. Neotropical wildlife use and conservation. University of Chicago Press, Illinois, USA.

Rodríguez-Ramírez, M. A. 1987. Tamaño y composición de los grupos sociales del hato de venado cola blanca (*Odocoileus virginianus*) de la Isla San Lucas, Costa Rica. Thesis, Universidad Nacional de Costa Rica, Heredia, Costa Rica.

Rosas Alvarado, A. 1990. Comportamiento del venado en cautiverio. Thesis, Escuela Nacional de Estudios Superiores-Iztacala, and UNAM, Tlalnepantla, Estado de México, México.

Rosas Becerril, P. 1992. Patrones reproductivos del venado cola blanca (*Odocoileus virginianus couesi*), durante su estación reproductiva en la Reserva de la Biosfera La Michilía, Durango, México. Thesis, Escuela Nacional de Estudios Profesionales Iztacala, Los Reyes Iztacala, Estado de México, México.

Sánchez-Rojas, G., C. Aguilar Miguel, and E. Hernández Cid. 2009. Estudio poblacional y uso de hábitat por el venado cola blanca (*Odocoileus virginianus*) en un bosque templado de la Sierra de Pachuca, Hidalgo, México. Tropical Conservation Science 2:204–214.

Sánchez-Rojas, G., S. Gallina, and S. Mandujano. 1997. Área de actividad y uso del hábitat de dos venados cola blanca (*Odocoileus virginianus*) en un bosque tropical de la costa de Jalisco, México. Acta Zoológica Mexicana (n.s.) 72:39–54.

Serio-Silva, J. C. 1999. Conducta en cautiverio de dos grupos de venado cola blanca (*Odocoileus virginianus*) con diferente grado de exposición al humano. Veterinaria México 30:323–328.

Silva-Villalobos, G., S. Mandujano, G. Arceo, S. Gallina, and L. A. Pérez-Jiménez. 1999. Nutritional value of plants consumed by the white-tailed deer in a tropical forest of Mexico. Vida Silvestre Neotropical 8:1–6.

Smith, W. P. 1991. *Odocoileus virginianus.* Mammalian Species 388:1–13.

Soto-Werschitz, A. 2000. Estrategias de comportamiento de las hembras del venado cola blanca texano durante la época de crianza. Thesis, Instituto de Ecología, Xalapa, Veracruz, México.

Taylor, W. P., editor. 1956. The deer of North America. Stackpole, Harrisburg, Pennsylvania, USA.

Teer, J. 1991. The white-tailed deer: natural history and management principles. Pages 17–27 *in* El venado en Venezuela: conservación, manejo, aspectos biológicos y legales. FUDECI, PROFAUNA, and FEDECAVE, Caracas, Venezuela.

Townsend, W. R. 1995. Living on the edge: Siriono hunting and fishing in lowland Bolivia. Dissertation, University of Florida, Gainesville, USA.

Verme, L. J., and D. E. Ullrey. 1984. Physiology and nutrition. Pages 91–118 *in* L. K. Halls, editor. White-tailed deer: ecology and management. Stackpole, Harrisburg, Pennsylvania, USA.

Villarreal, J. 1999. Venado cola blanca: manejo y aprovechamiento cinegético. Unión Ganadera Regional de Nuevo León, Monterrey, México.

Weber, M. 1992. Un arnés marcador para la detección de estros en ciervos, basado en observaciones de comportamiento reproductivo. Veterinaria México 23:353–356.

Weber, M. 2000. Effects of hunting on tropical deer populations in south-eastern Mexico. Thesis, University of London, United Kingdom.

Weber, M. 2005. Ecology and conservation of tropical deer populations in the Greater Calakmul Region, Mexico. Dissertation, University of Durham, United Kingdom.

Weber, M. 2008. Un especialista, un generalista y un oportunista: uso de tipos de vegetación por tres especies de venados en Calakmul, Campeche. Pages 483–496 *in* C. Lorenzo, E. Medinilla, and J. Ortega, editors. Avances en el estudio de los mamíferos de México. Publicaciones especiales. Volume 2. AMMAC, ECOSUR, and UNAM, México, D. F., México.

Weber, M., P. C. De Grammont, and A. D. Cuarón. 2009. *Mazama pandora*. IUCN Red List Version 2009.1. http://www.iucnredlist.org. Accessed 11 Nov 2011.

Weber, M., and C. Galindo-Leal. 1992. Distocia en venado cola blanca: informe de un caso reincidente. Veterinaria México 23:69–72.

Weber, M., and C. Galindo-Leal. 1995. History, needs and perspectives of deer research and conservation in Mexico. Pages 40–41 *in* J. Milne, editor. Recent advances in deer biology. Moredun Research Institute, Edinburgh, Scotland, United Kingdom.

Weber, M., G. García-Marmolejo, and R. Reyna-Hurtado. 2006. The tragedy of the commons: wildlife management units in southeastern Mexico. Wildlife Society Bulletin 34:1480–1488.

Weber, M., and S. Gonzalez. 2003. Latin American deer diversity and conservation: a review of status and distribution. Ecoscience 10:443–454.

Weber, M., and R. Hidalgo. 1999. Morfometría, patrones de crecimiento y ganancia de peso de venados cola blanca en cautiverio en Durango y Toluca, México. Veterinaria México 30: 183–188.

Weber, M., and R. Medellín. 2010. Yucatán brown brocket deer *Mazama pandora* (Merriam 1901). Pages 211–216 *in* J. M. B. Duarte and S. Gonzalez, editors. Neotropical cervidology: biology and medicine of Latin American deer. IUCN and FUNEP, Gland, Switzerland.

Wright, J. S., H. Zeballos, I. Dominguez, M. M. Gallardo, M. C. Moreno, and R. Ibanez. 2000. Poachers alter mammal abundance, seed dispersal and seed predation in a Neotropical forest. Conservation Biology 14:227–239.

Pronghorn in Mexico

19

Whitney J. Gann, Louis A. Harveson, Carlos E. Gonzalez-Gonzalez, and Raul Valdez

The American pronghorn (*Antilocapra americana*), referred to as *berrendo* or *antílope* in Spanish, is endemic to western North America, with a distribution extending from southwestern Canada to northern Mexico. The pronghorn is the fastest land mammal in North America. Herds have been recorded running at 64–72 km/hr (39–44 mi/hr), with maximum speeds of 86.5 km/hr (53.7 mi/hr) (Kitchen and O'Gara 1974, Nowak and Paradiso 1983). The pronghorn's success on North American grasslands is the result of anatomical, physiological, and behavioral adaptations that were shaped by selection in a highly seasonal climate (Maslin 2009, de las Heras and Islas-Espinosa 2015). They were an evolutionarily successful group and evolved into several genera and species (McKenna and Bell 1977, Heffelfinger et al. 2004). Pronghorn running speed is thought to have evolved in response to avoiding predation by the American cheetah (*Miracinonyx trumani*) during the Pleistocene epoch 2.6 million years ago (mya). Antilocaprids first appeared in North America 20–18 mya. The grasslands of North America appeared during the Pleistocene epoch, 1.8 mya to 10,000 years ago: the present species, based on fossil evidence, evolved less than 1 mya (O'Gara and Janis 2004).

There is abundant information on pronghorn biology and management methodologies (McCabe et al. 2004, O'Gara and Yoakum 2004, Autenrieth et al. 2006, Valdés et al. 2006). However, our knowledge is based primarily on studies conducted in the US, where—except for the Sonoran subspecies (*sonoriensis*), which is protected under the Endangered Species Act—the pronghorn is a species of least concern (USFWS 1967, Hoffmann et al. 2008). In contrast, the three native subspecies in Mexico (*mexicana*, *sonoriensis*, and *peninsularis*) are listed as endangered in Appendix 1 of CITES (2014).

The rapid decrease in pronghorn populations in Mexico has prompted governmental and nongovernmental entities to develop action plans to aid pronghorn conservation. Initial conservation efforts began in 1922, when Mexican president Álvaro Obregón banned the hunting of pronghorn (SEMARNAT 2009, Cancino et al. 2014). The Federal Hunting Law of 1952 continued the ban on pronghorn hunting, and Mexican Official Rule NOM-059-ECOL-1994 reiterated legal protection in 2001 and 2010 and listed the pronghorn as endangered (SEMARNAT 2010).

Description

Pronghorn are proportionately long-legged, small-bodied artiodactyls distinguished by large white areas of hair on the rump, on the sides of the face, in two bands on the throat, on the underparts, and partially on the sides of the body. The slightly curved horns have a straight bony core and are shed annually. A single prong projects forward in males. Pronghorn have a woolly undercoat covered by long, straight, coarse, brittle guard hairs (Brown and Ockenfels 2007). A subcutaneous muscle layer causes the hollow hairs to stand erect, allowing for thermoregulation (O'Gara and Yoakum 2004). Hair color varies from yellowish to tan, except for blackish coloring on top of the nose (Hoffmeister 1986). Weight of adult males ranges from 43 to 51 kg (94 to 112 lb), and of females from 40 to 50 kg (88 to 110 lb); however, Sonoran pronghorn tend to be slightly smaller and paler than the Mexican and peninsular subspecies (O'Gara and Yoakum 2004, Brown et al. 2006a). The pronghorn's reduced stomach capacity, wide trachea, and large heart and lungs (relative to animals of similar size) are among many physiological adaptations for speed (McKean and Walker 1974, O'Gara and Yoakum 2004).

Taxonomy

The pronghorn is the sole living representative of the family Antilocapridae. There are five recognized subspecies: American (*americana*), Oregon (*oregona*),

Mexican (*mexicana*), Sonoran (*sonoriensis*), and peninsular pronghorn (*peninsularis*); however, only the Mexican, Sonoran, and peninsular occur in Mexico. Relative to the genetic diversity in peninsular and Sonoran populations, peninsular pronghorn exhibit a lower mean heterozygosity (0.31 and 0.48, respectively), and a lower mean number of alleles per locus (2.05 and 4.86, respectively), indicating significant genetic divergence (Munguia-Vega et al. 2013, Klimova et al. 2014).

Historical Distribution and Conservation Efforts

Before European settlement, an estimated 40–60 million pronghorn inhabited North America; however, there were fewer than 20,000 at the beginning of the twentieth century (O'Gara and Yoakum 2004, SEMARNAT 2009). Pronghorn populations have been extirpated from much of their historical range as a result of habitat loss and fragmentation. Land-use changes in Mexico during the early twentieth century contributed to an estimated 70% loss of pronghorn habitat (Laliberte and Ripple 2004), with some subspecies losing up to 90% (Cancino et al. 1998). Agricultural and industrial development, fences, roads, canals, and railroad tracks also contributed to declines by creating movement barriers (Christensen et al. 1995, Weber and González 2003). Furthermore, overhunting, land-use changes, and long-term livestock grazing have contributed to population declines and habitat loss. Domestic livestock grazing has been extensive and severe in southeastern Arizona and northeastern Sonora, Mexico, causing soil erosion and compaction, changes in plant species composition and vegetation structure, and an increase in shrubs. Cattle numbers were at their peak in the 1890s, when severe drought exacerbated the effects of overgrazing and caused complete loss of habitat over a substantial portion of pronghorn range (Bahr 1991, Brown and Ockenfels 2007).

Pronghorn historically inhabited the Baja California Peninsula and the northern Mexican Plateau in Sonora, Chihuahua, Coahuila, Nuevo León, and Tamaulipas, and south through Durango, Zacatecas, San Luis Potosí, Guanajuato, Querétaro, and Hidalgo (fig. 19.1). In the early twentieth century, Nelson

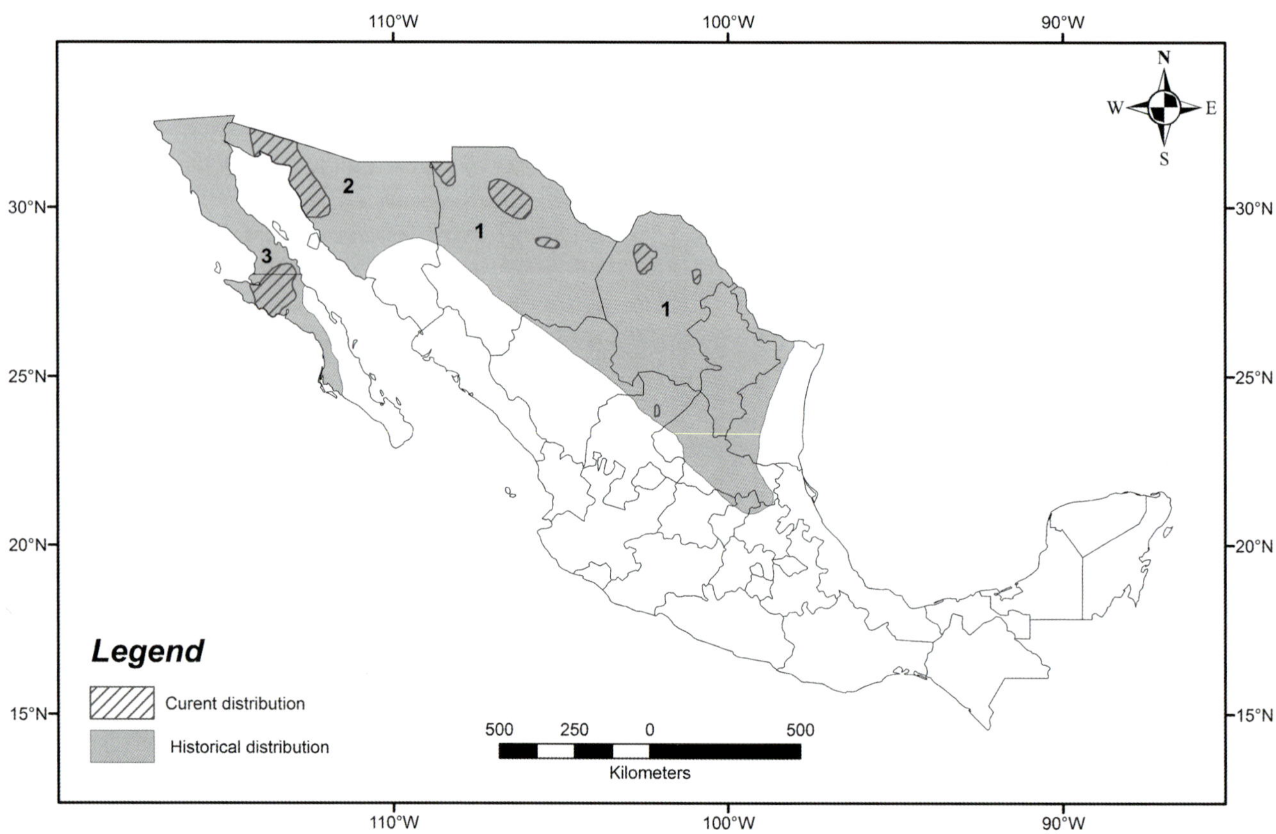

Figure 19.1. Current and historical distribution of pronghorn subspecies in Mexico: (1) Mexican (*mexicana*); (2) Sonoran (*sonoriensis*); and (3) peninsular (*peninsularis*).

(1925) estimated that the Mexican population consisted of about 2,400 individuals. Sonoran and peninsular pronghorn probably overlapped in a small area of northeastern Baja California, near San Felipe. Currently, peninsular and Sonoran pronghorn are allopatric; populations in Baja California represent the present southernmost distribution (Medellín et al. 2005). Furthermore, the peninsular subspecies is geographically isolated and has received little research attention (Nelson 1925, Leopold 1959, Monson 1968, Cancino 1994).

Mexican pronghorn

Mexican pronghorn historically occurred in southeastern Arizona, southern New Mexico, and east to the Big Bend region of West Texas and northern Mexico. In Mexico, the range included Chihuahua, eastern Sonora, Durango, Coahuila, parts of Nuevo León, and Tamaulipas and extended south through Zacatecas and San Luis Potosí to Hidalgo (Cancino et al. 2014). They were extirpated from Coahuila in the early 1960s. Over the past 50 years, populations have declined significantly in Chihuahua and have been extirpated on the central plateau (Cancino et al. 2014). Currently, the largest populations occur in north-central Chihuahua, and pronghorn have been reestablished in western Coahuila in Valle Colombia (fig. 19.2).

Several recovery efforts in the historical range of the Mexican pronghorn have not been successful. In 1967, 22 pronghorn from Colorado were translocated to Isla Tiburón in the Gulf of California. Drought and predation eradicated the population in three years (SEMARNAT 2009). In 1996, 65 pronghorn were translocated from New Mexico and released on El Cimarrón Ranch in Valle Colombia, Coahuila (Guadarrama and Valdés 2001). During the first year, emigration reduced the size of the entire herd by 54%. To supplement the first introduction, a second translocation in 1998 added an additional 85 individuals from New Mexico (Martinez et al. 2000). There were an estimated 45 pronghorn in 2009 (USFWS 2015). Despite successful reintroduction, fawn survival has been low. In 1997 and 1999, there were 0.3 and 0.2 surviving fawns per female, respectively. Unfortunately, insufficient habitat limits potential population growth (Medellín et al. 2005). Translocations in 2005 (47 individuals) and 2006 (57 individuals) to a private ranch in Coahuila led to the establishment of two reproductive groups; however, in 2008, only one group remained.

Relative to reestablishment efforts in their former

Figure 19.2. Small group of pronghorn in the Chihuahuan Desert in the Trans-Pecos region of Texas. Photo by James Weaver.

southern range, 52 pronghorn (19 males and 33 females) were moved from New Mexico in 1972 to Ramos, San Luis Potosí; however, prolonged drought, emigration, and predation led to the translocated herd's disappearance by 1987. In 2000, 12 pronghorn (3 males and 9 females) were translocated from Wyoming to Bioparque Estrella in Nuevo León, Mexico; however, the population was extirpated by 2004 (Cancino and Guenzel 2002). In 2006, 130 pronghorn fawns (70 males and 60 females) were transported from Wyoming to captive-breeding pens in Zacatecas to initiate reintroduction programs, but the population has greatly decreased (SEMARNAT 2009, Cancino et al. 2014).

Peninsular pronghorn

Peninsular pronghorn were historically distributed widely on the Baja California Peninsula, from Monterey Bay to Magdalena Bay, Baja California Sur, and on the Gulf of California side of the Baja Peninsula south of San Felipe (Nelson 1925, Hall 1981, Brown et al. 2006a, USFWS 2010). Peninsular pronghorn currently have a narrow distribution; they are isolated in the midportion of the Baja California Peninsula in an area of approximately 5,000 km^2 (1,930 mi^2) (Cancino et al. 2010), within the 25,467 km^2 (9,832 mi^2) El Vizcaíno Biosphere Reserve in the western Vizcaíno Desert. Their status beyond the Vizcaíno Desert, which represents a small portion of their original range, is unknown. Their restricted distribution and low numbers prompted the federal government to initiate a recovery plan in 1983. By the early 1990s, the area of peninsular pronghorn distribution had decreased from 40,000 km^2 (15,400 mi^2) to 5,000 km^2 (1,900 mi^2) (Cancino et al. 1994).

Sonoran pronghorn

Sonoran pronghorn historically ranged over more than 90,000 km^2 (34,750 mi^2) in the US and Mexico. They occurred throughout southwestern Arizona, northwestern Sonora, and the extreme northern end of the Baja California Peninsula. By the 1920s, Sonoran pronghorn had declined to around 100 animals in the US (Nelson 1925) and oscillated between 50 and 100 animals from the 1920s to the mid-1980s (Halloran 1957, Monson 1968, Carr 1971). The population in Sonora, Mexico, consisted of about 600 animals in 1925 but had declined by almost half by 1993 (Snow 1994, USFWS 2015), while the US population had rebounded to an estimated 280 animals by 1994 (USFWS 2015). Unfortunately, few individuals in northwestern Sonora move across the Arizona border, and isolation between the US and Mexican populations has made natural recovery improbable.

Presently, Sonoran pronghorn occupy approximately 7.6% of their historical range (USFWS 2010). Their current range is limited to approximately 10,903 km^2 (4,209 mi^2), of which 3,781 km^2 (1,459 mi^2) is in Mexico, and 7,122 km^2 (2,749 mi^2) in the US (USFWS 2010). The Pinacate Biosphere Reserve was created in 1993 because of an awareness of pronghorn declines, and increased management efforts. Additionally, four conservation units have been designated to facilitate transboundary conservation efforts: two in the US and two in Mexico. Two of these units, Kofa (19,179 km^2, 7,405 mi^2) and Cabeza Prieta (6,498 km^2, 2,508 mi^2), are in southwestern Arizona (fig. 19.3). Because of successful management efforts, the endangered population in Arizona has increased to 202 animals (as of December 2014), and there were 58 pronghorn in the Kofa National Wildlife Refuge (NWR) (USFWS 2015). The other two units, Pinacate (1,513 km^2, 584 mi^2) and Quitovac (1,671 km^2, 645 mi^2), are in northwestern Sonora, Mexico. Quitovac supports nearly 80% of the Sonoran pronghorn population in Mexico. The Pinacate unit is within the El Pinacate and Gran Desierto de Altar Biosphere Reserve (7,146 km^2, 2,759 mi^2) in northwestern Sonora, and on other federally owned and protected lands, as well as ejido and private lands. Fortunately, recent declines in the Quitovac and Pinacate populations have not been as severe as the declines in Arizona (Valdés and Manterola 2001). The most recent estimates for Quitovac and Pinacate were 434 in 2013 (Sonoran Pronghorn Recovery Team 2014) and 122 in 2014 (USFWS 2015), respectively.

Historically genetic and demographic interchange between pronghorn in Sonora and Arizona likely occurred; however, the four Sonoran populations are now geographically isolated because of barriers, including roads and fences. Mexico's Highway 2 and the international boundary fence act as barriers to movement between the Pinacate and US subpopulations (USFWS 2002), while Sonoran pronghorn habitat in Mexico is bisected by Highway 8 and associated fences (USFWS 2002).

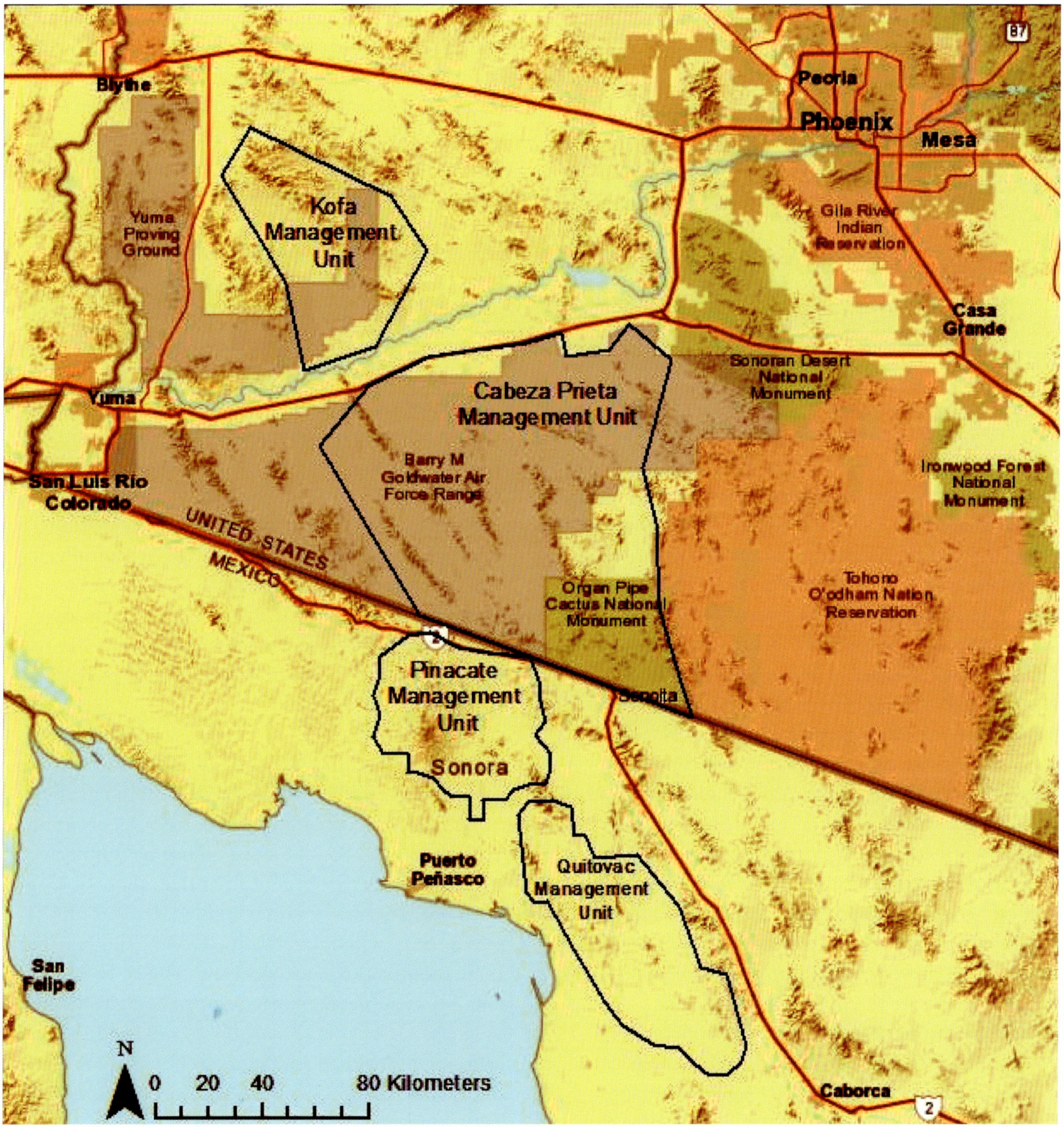

Figure 19.3. Pronghorn population management units in southwestern Arizona (Kofa and Cabeza Prieta), and in Sonora, northwestern Mexico (Pinacate and Quitovac).

Life History

Diet

Pronghorn graze on grasses, forbs, and moss and move continuously through the landscape while foraging (Cancino et al. 2014). They are considered concentrate selectors; their primary diet is forbs. However, their diets are elastic and vary by biome (grassland vs. shrubland), season, and food availability (Autenrieth et al. 2006). Their facial structure and long, narrow jaw allow targeted selection of vegetation. An adult pronghorn requires an average of 1.1–1.4 kg (2.4–3.1 lb) of air-dried, quality forage per day (fig. 19.4) (O'Gara and Yoakum 2004).

In a herd of pronghorn reintroduced to northern Coahuila, 59 plant species were identified in diets, including the poisonous buffalobur (*Solanum ros-*

Figure 19.4. Female pronghorn grazing in the Chihuahuan Desert. Photo by J. Alfonso Ortega-S.

tratum) and silverleaf (*S. elaeagnifolium*) nightshades. Forbs made up 70% of the diet, and grasses and shrubs only 12.5% each (Martinez et al. 2000). In central Chihuahua, shrubs made up a majority of the diet (53%), forbs were 43%, and grasses 4% (de Peña and Neira 1982). Common forbs include dwarf copper globemallow (*Sphaeralcea angustifolia*), prairie clover (*Dalea nana*), croton (*Croton* spp.), euphorbia (*Euphorbia* spp.), white sagebrush (*Artemisia ludoviciana*), common mallow (*Malva neglecta*), and copperleaf (*Acalypha* spp.); common shrubs include Rio Grande stickpea (*Calliandra conferta*), spreading asphead (*Aspicarpa humilis*), Gregg's prairie clover (*Dalea greggii*), yellowspine thistle (*Cirsium ochrocentrum*), and cactus (*Opuntia* spp.); grasses include feather fingergrass (*Chloris virgata*), muhly (*Muhlenbergia* spp.), and grama (*Bouteloua* spp.) (de Peña and Neira 1982, Martinez et al. 2000).

Sonoran pronghorn in southwestern Arizona eat chainfruit cholla (*Cylindropuntia* spp.), mesquite (*Prosopis glandulosa*), mistletoe (*Phoradendron* spp.), wild carrot (*Daucus pusillus*), buckwheat (*Eriogonum* spp.), ragweed (*Ambrosia* spp.), milkvetch (*Astragalus* spp.), careless weed (*Amaranthus palmeri*), brome (*Bromus* spp.), broom snakeweed (*Gutierrezia sarothrae*), jumping cholla (*Opuntia fulgida*), and Arizona blanket-flower (*Gaillardia arizonica*) (Hailey 1979, Hughes and Smith 1990, Hughes 1991, Hervert et al. 1997, 2000). In southwestern Arizona, the major diet components were cacti (44%), mostly cholla fruits (70%), followed by forbs (33%), woody plants (22%), and grasses (0.4%) (Hughes and Smith 1990). Conversely, browse predominated (43–53%) in all seasons except in wet summers, when browse decreased to 28% and forbs increased to 42% of the diet (Hervert et al. 2000). In Sonora, Mexico, diet consisted of 69% forbs, 22% woody plants, 7% cacti, and 2% grasses (Castillo 1993). Diet of the peninsular pronghorn consisted of 44% shrubs, 22% forbs, 4% grasses, and 30% unidentified material (Cancino 1994).

Reproduction

Pronghorn are polygamous. Females become sexually mature at 16 months of age but occasionally conceive at approximately 5 months (Mitchell 1967, O'Gara 1978, Byers 2003); males become sexually mature at one year of age (O'Gara 1978). Both males and females continue to reproduce for 10 years, but approximately 25% of females of reproductive age do not give birth every year (Mitchell 1967). Pronghorn populations can recover quickly from population crashes because of high productivity. Specifically, in years with above-normal rainfall, pronghorn have been known to average 1.9 fawns/doe (Harveson 2006). In the Trans-Pecos, fawn production (fawns/100 does) ranges from 10% to 75% and averages 38% (Simpson et al. 2007).

Bucks congregate in summer to search for estrous females for breeding; bucks of all subspecies typically form harems, either directly by defending females or indirectly by defending a resource (O'Gara and Yoakum 2004). Southern pronghorn have longer rutting periods that begin earlier than those in northern latitudes. In Texas and Mexico, breeding seasons vary with rainfall and elevation and may last three months (Lehman and Davis 1942). In Texas and Arizona, breeding may begin as early as July or as late as October (Lehman and Davis 1942, Phelps 1981). Chihuahuan and Sonoran populations mate in September and October (Treviño 1978, Castillo 1993) and typically give birth from mid-February to April (Bright and Hervert 2005). In a captive population, mating of Sonoran pronghorn occurred from 16 to 30 June (Wilson et al. 2008). In this pen, Sonoran pronghorn captured in Arizona birthed from mid-February to early March, while those captured in Sonora, Mexico, gave birth in mid to late March (Wilson et al. 2008). Parturition in Sonoran pronghorn coincides with spring forage abundance from February through May (USFWS 1998). Mexican and peninsular pronghorn populations breed in November–December and June–August, respectively.

Mexican pronghorn birthed in June–July, and peninsular in January–March (Treviño 1978, Cancino et al. 1995, 2014). Individual does leave herds to search for fawning areas.

Gestation is about 240 days, and twins are more common than single fawns (O'Gara 1978). Does occasionally gather with fawns and form nursery groups. Females initiate the weaning process as early as 4 weeks, and by 12 weeks, fawns are fully weaned; nursing has been observed as late as September (Byers 1997). Pronghorn fawns grow rapidly in the presence of nutritious forage and adequate moisture; by about 45 days of age, fawns can outrun predators (Byers 1997). In Arizona, estimates of Sonoran pronghorn fawn recruitment per 100 females varied from 0 to 78 fawns in 1995–2002 (Bright and Hervert 2005), whereas in the Trans-Pecos, the long-term (1977–2014) fawn to doe ratio was 0.35 (Texas Parks and Wildlife Department [TPWD], unpublished data). Furthermore, reproductive rates of pronghorn does in the Chihuahuan Desert of the Trans-Pecos were 0.22, 0.11, and 0.16 in 2010, 2011, and 2012, respectively (TPWD, unpublished data). These reproductive rates are well below those documented elsewhere (Danvir 2000, Barnowe-Meyer et al. 2010).

Longevity and mortality factors

Longevity for pronghorn is 10 years in the wild, and 12 years in captivity (Carey and Judge 2000), but average lifespan is 9 years or less (Yoakum 1978). Fawn survival (fawns/100 does) up to five months old can vary by as much as 7–82% (O'Gara and Yoakum 2004). For example, fawn survival in Hudspeth County, Texas, fluctuated between 6% and 18% during 1990–1993 (Canon 1993), and between 8% and 50% during 2011–2013 in the Trans-Pecos. Average annual adult mortality rate was 15% in the Trans-Pecos during 1990–1993 (Canon 1993) and 20% in Arizona during 1995–2002 (Bright and Hervert 2005). In New Mexico during 2006–2010, adult female mortality ranged between 22% and 67% and male mortality between 11% and 37% (Bender et al. 2013).

In their southern distribution, a positive relationship between pronghorn abundance and increasing precipitation suggests a strong influence of long-term precipitation trends. Fawn production in this region is also influenced by immediate moisture conditions (Simpson et al. 2007). Total winter precipitation, in addition to the length of time between winter and summer rains, is the most important factor determining fawn survival in Arizona deserts (Bright and Hervert 2005), New Mexico (Brown et al. 2006b), and West Texas (Simpson et al. 2007). Specifically, delayed onset of summer rains results in scarce forage and increases fawn mortality rates. Additionally, annual mortality rates of adult Sonoran pronghorn in Arizona averaged 7–13% in wet years (deVos and Miller 2005), and 83% during drought (2002) (Bright and Hervert 2005). In the Chihuahuan Desert, large losses in isolated herds have been attributed to drought and severe winters (Buechner 1950). Severe weather, in combination with barriers (i.e., net-wire fences) that prohibit movement, has resulted in large die-offs from starvation (Hailey et al. 1966). Research in other arid regions has also suggested similar relationships between pronghorn population dynamics and precipitation, indicating that precipitation may be a limiting factor in the southwestern portion of their range (Buechner 1950, Hailey et al. 1966, Yoakum and O'Gara 2000, deVos and Miller 2005, Simpson et al. 2007).

Direct mortality is frequently caused by predation by coyotes (*Canis latrans*), golden eagles (*Aquila chrysaetos*), bobcats (*Lynx rufus*), mountain lions (*Puma concolor*), and disease, primarily clostridiosis (Cancino et al. 2014). Predation on pronghorn fawns is most often associated with coyotes; however, bobcats and golden eagles can also be effective predators. In a fawn mortality study in Utah, Beale and Smith (1973) found that 27 of 44 fawns that died within their first four months of life were killed by bobcats. Coyotes were the primary predator in West Texas, accounting for 82% of fawn predation during 1990–1993 (Canon 1993). In Arizona, of 32 radio-collared pronghorn mortalities during 1995–2002, 12 resulted from predation, 5 from capture myopathy, 4 from drought-related factors (i.e., malnutrition, starvation, or dehydration), and 11 from unknown causes (Bright and Hervert 2005). Threats also include reduced forage quality, altered habitat structure, climate change, reduced access to and availability of water, disease, loss of genetic diversity, human disturbance, and accidental death or poaching.

Water requirements

Variation in location, season, and forage quality contributes to differing water needs across pronghorn range. However, apparent water needs are further confounded by the pronghorn's ability to conserve water through physiological mechanisms. Consumption of succulent plants, plus various physical and physiological adaptations that conserve water, allows pronghorn to undergo long periods without drinking (Reynolds 1984). Hervert et al. (2005) attributed the preference for associations of paloverde, cacti, mixed scrub, and chainfruit cholla to the availability of chainfruit cholla fruits and the preformed water they provide. However, preformed water and metabolic water measured from some Sonoran pronghorn diets revealed that water intake from forage is not adequate to meet minimum water requirements (Fox 1997, Fox et al. 2000, Tluczek 2012). Rather, water consumption by pronghorn varies inversely with the quantity and succulence of vegetation consumed. Pronghorn did not drink water, even if available, when plant moisture content was 75% or more. Conversely, during dry conditions, animals drank about 3.3 liters (0.9 gal) per day, consuming variable amounts between temperature extremes. Pronghorn have been observed at human-made watering troughs, and tracks have been documented at seasonal potholes (Beale and Smith 1970, Wright and deVos 1986, Morgart et al. 2005).

Historically, rivers that flowed in Sonoran pronghorn habitat in Mexico, such as the Río Sonoyta and the Río Sonora, were potentially important for survival of Sonoran pronghorn (USFWS 1998). Historical descriptions suggest the existence of a riparian greenbelt that could have provided green forage when food and water resources were limited (USFWS 1998). In fact, drying of rivers in Sonora may have contributed to Sonoran pronghorn population declines (Carr 1971).

Social structure

Pronghorn live in herds of mixed sexes comprising females with fawns and young males. Adult males are generally solitary or live in small groups. Males will occasionally form small male-only herds. Peninsular pronghorn exhibit highly variable group sizes, ranging from one male-female pair, to groups of 15 females, to one male (Cancino et al 2014). Group sizes vary throughout the year with environmental conditions but tend to be largest in winter (Byers 1997). In years when succulent forage is readily available, pronghorn are generally in smaller but more numerous and widespread groups. In years of poor and limited forage, pronghorn are in fewer, larger groups, concentrated in productive forage areas (Bright et al. 2011). Average group sizes of Sonoran pronghorn in winter in Arizona were 5.1 in 2006, 7.3 in 2008, and 5.7 in 2010 (Bright and Hervert 2011). In Mexico, average group sizes of Sonoran pronghorn during winter surveys ranged from 3.4 to 12.0 animals (Bright et al. 2011). Treviño (1978) observed a herd of 52 pronghorn in Chihuahua.

During the reproductive period, dominant males mark territories with urine, excrement, and secretions from subauricular glands. They defend territories using antagonistic behaviors such as staring, vocalizing, and approaching intruders. Male aggression and territory defense may also be a strategy to prevent young males from reproducing. Territory size varies according to the availability of food, number of animals per group, and environmental conditions. Young males one to two years old and nonreproductive adults form groups that do not defend territories. Reproductive-age females establish their own groups in and around territories already established by dominant males. Descendants of these groups become active members of the same groups six weeks after birth and join a preexisting hierarchy in the group (Byers 1997).

Home range and spacing

Movement information on pronghorn in Mexico is lacking. In southern Arizona, home range size for individual Sonoran pronghorn varied from 43 to 2,873 km^2 (16 to 1,109 mi^2), with an average of 511 km^2 (197 mi^2) (Hervert et al. 2005). However, these were much larger than the home ranges of other subspecies in desert communities in Hudspeth County (42.4 km^2, 16 mi^2) (Canon 1993) and Presidio County (83.8 km^2, 32 mi^2) (Hoffman 2015), Texas; Utah (60.4 km^2, 23.3 mi^2) (Bates 2000); central Arizona (88 km^2, 34 mi^2) (Ockenfels et al. 1994); and New Mexico (22.6 km^2, 8.7 mi^2) (Clemente et al. 1995).

Home ranges of pronghorn reflect habitat quality (Yoakum 1978); the large home ranges of Sonoran pronghorn likely indicate widely dispersed resources (Hervert et al. 2005). Pronghorn move in response to forage conditions and water availability that can change with sporadic rainfall (fig. 19.5). As a result, they require large expanses of contiguous habitat with suitable forage. Sonoran pronghorn in northwestern

Figure 19.5. Male and female pronghorn in the Chihuahuan Desert in the Trans-Pecos region of Texas. Photo by Borderlands Research Institute for Natural Resource Management at Sul Ross State University.

Arizona moved up to 130 km (81 mi) each year between summer and winter habitats (Hervert et al. 2000). Drought-induced movement has been observed in the Trans-Pecos of Texas, where a female pronghorn traveled 48 km (29.8 mi) before settling on a fawning area (Garrison 2014).

Some movement from Mexico to Arizona was historically common in Sonoran pronghorn populations, but transboundary migration is now limited because of barriers. For example, Mexico's Highway 2 and the international boundary fence restrict movement between the Pinacate and US subpopulations (USFWS 2002). Habitat in Mexico is bisected by Highway 8 and associated fences (fig. 19.6), but the extent of movement restriction is unclear.

Figure 19.6. Highways and fences limit the movement of pronghorn; the extent of the restriction is unclear. Photo by J. Alfonso Ortega-S.

Population Trends and Current Status

Inconsistencies among pronghorn population data remain a problem for understanding population trends in Mexico (tables 19.1–19.3). Variation in dates surveyed, methods, goals, sampling effort, and sample size has led to differences in population estimates and has complicated management efforts.

Mexican pronghorn

By the 1970s, Mexican pronghorn occurred only in Chihuahua, in 10 isolated populations with an estimated 533 animals (Treviño 1978). The 8 currently remaining populations occur in the three main areas of Zona Norte–Casas Grandes, Zona Central, and Zona Sur–La Perla, and specifically in the regions of La Perla, La Gregoria, San Luis, Terraceño, El Sueco-Moctezuma, Janos-Ascención, and Coyame. In 1993, there were between 214 and 307 pronghorn in Chihuahua (González-Romero and Lafon 1993). There were 282–564 in 2001, and 318 in 2007 (table 19.1) (Azuara et al. 2001, Colchero et al. 2007, CONANP 2010). Continued declines prompted the establishment of the Janos Biosphere Reserve in 2009 in northwestern Chihuahua; in 2010, it contained an estimated 106 individuals (CONANP 2010). The current population status is unknown.

In Coahuila, pronghorn occur in Valle Colombia and Rancho El Novillo (Pallares 1999). Aerial surveys from 2001 to 2003 over Valle Colombia showed a stable average population of 41, with a ratio of 0.38 fawns per female (Valdés and Manterola 2007, SEMARNAT 2009). In 2007, there were 80 pronghorn in Coahuila; more recent data are not available. The present status of pronghorn in Rancho El Novillo is unknown. In Zacatecas, a herd of probably no more than 20 is the remnant of a population introduced in 2006 (table 19.1).

Peninsular pronghorn

There were an estimated 500 peninsular pronghorn in Baja California in the 1920s (table 19.2). By the 1970s, there were fewer than 100 (Jaramillo 1989), and only about 75 remain in the wild (Klimova et al. 2014). Despite protection from hunting since 1922, populations continued to decline until most peninsular pronghorn were isolated in El Vizcaíno Biosphere Reserve. Population estimates are variable within the reserve's core area and have fluctuated, ranging from a low of 22 animals in 1990 to 175 in 1993 (Cancino et al. 1996, 2010). In 2000, there were an estimated 250 pronghorn in the area, of which most were in the reserve (Cancino et al. 2010).

In 1998, 1999, and 2000, a total of 22 pronghorn were captured from El Vizcaíno Biosphere Reserve to establish a captive-breeding program. In 2000, the first captive births occurred. In 2008, there were 468:200 in the wild and 268 in captivity (table 19.2) (AZA Antelope and Giraffe Advisory Group 2009, de las Heras and Islas-Espinosa 2015). By 2009, there were 600, consisting of both wild and captive animals (SEMARNAT 2009), and in 2012 the population estimate had decreased to 450 (Klimova et al. 2014).

Table 19.1. Population estimates of Mexican pronghorn in Valle Colombia, Coahuila, and in Chihuahua and Zacatecas, Mexico.

Location and year	Estimate	Source
Valle Colombia, Coahuila		
1999	19	Valdés and Manterola (2007), SEMARNAT (2009)
2001	37	Valdés and Manterola (2007), SEMARNAT (2009)
2002	43	Valdés and Manterola (2007), SEMARNAT (2009)
2003	42	Valdés and Manterola (2007), SEMARNAT (2009)
2005	43–50	Valdés and Manterola (2007), SEMARNAT (2009)
2007	80	Colchero et al. (2007)
Chihuahua		
1977	533	Treviño (1978)
1993	214–307	González-Romero and Lafon (1993)
1996	318	Valdés et al. (2006)
2001	282–564	Azuara et al. (2001)
2007	318	Colchero et al. (2007)
Zacatecas		
2014	20	Cancino et al. (2014)

Sonoran pronghorn

In 2000, there were an estimated 445 Sonoran pronghorn in three populations: two in Mexico and one in the US. Sonoran populations in Mexico declined from 683 in December 2004 to 634 in 2006, and there were 482 in 2009, and 241 in 2011 (table

Table 19.2. Population estimates of peninsular pronghorn in El Vizcaíno Biosphere Reserve, Baja California Sur, Mexico.

Year	Estimate	Source
1920	500	Nelson (1925)
1970	75	Klimova et al. (2014)
1977	83[a]	Cancino et al. (2010)
1980	26[a]	Cancino et al. (2010)
1985	97[a]	Cancino et al. (2010)
1990	22[a]	Cancino et al. (2010)
1993	175[a]	Cancino et al. (2010)
1995	149[a]	Cancino et al. (2010)
1996	200	Valdés et al. (2006)
2000	250[a]	Ticer and deVos (2001)
2003	41[a]	Cancino et al. (2010)
2007	391[b]	Colchero et al. (2007)
2008	468[b]	de las Heras and Islas-Espinosa (2015)
2009	600[b]	SEMARNAT (2009)
2012	450[b]	Kilmova et al. (2014)

[a]Within largest core zone of El Vizcaíno Biosphere Reserve.
[b]Estimates include captive populations.

19.3) (USFWS 2010). This decline was primarily a result of changes in the Quitovac population, which declined from 624 in 2004 to 189 in 2011, while the Pinacate population remained relatively stable. The specific cause of this decline in Quitovac is unknown; however, the Quitovac population increased to 434 in 2013 (Sonoran Pronghorn Recovery Team 2014).

Habitats

Pronghorn inhabit grasslands, sagebrush plains, and deserts. Elevations range from near sea level to 3,350 m (10,990 ft) (O'Gara 1999). Peninsular and Sonoran pronghorn occur between sea level and 200 m (656 ft), while Mexican pronghorn occur between 1,400 and 1,600 m (4,590 and 5,250 ft) (González-Romero and Lafon 1993). Little information is available for pronghorn habitats in Mexico, but pronghorn are typically found in subclimax vegetation communities historically maintained by wildfire and grazing herbivores (Yoakum 1979). Undulating topography, a mixture of forage types, and annual rainfall of 230–350 mm (9.05–13.77 in) are typical of habitats across most pronghorn ranges (Autenrieth 1978). Extreme variation also exists in fawn bed-site selection, suggesting that fawns select an optimal combination of local microhabitat features based on regional and temporal availability. Important characteristics for fawns in the Trans-Pecos region of Texas in the Chihuahuan Desert include low shrub cover and density, high grass cover, lower bed slope, limited rock cover, and high visibility (50–150 m, 164–492 ft) (Canon and Bryant 1997).

Table 19.3. Population estimates of Sonoran pronghorn in Sonora, Mexico.

Year	Estimate	Source
1920	100	Nelson (1925)
1925	600	Snow (1994)
1993	280 and 313	Castillo (1993), Snow (1994)
1996	433	Valdés et al. (2006)
2000	346	USFWS (2015)
2001	200–350	Valdés and Manterola (2001)
2002	285	USFWS (2015)
2004	683	Bright and Hervert (2006)
2006	634	Bright and Hervert (2006), USFWS (2015)
2007	433 and 360	Colchero et al. (2007), Klimova et al. (2014)
2009	482	Bright and Hervert (2011), USFWS (2015)
2011	241	USFWS (2015)
2013	434[a]	Sonoran Pronghorn Recovery Team (2014)

[a]Quitovac population only.

Pronghorn depend on vigilance and speed to escape predators; these adaptations are most suited to terrain that is relatively flat and open, without rugged slopes and mountains. Location data collected from 3,219 radio-collared Sonoran pronghorn in the US from 1994 to 2002 showed that 10% were in areas with slopes greater than 20% (O'Brien et al. 2005). Sonoran pronghorn prefer slopes and hills in the paloverde–chainfruit cholla association (Hervert et al. 2005). Densely vegetated washes and communities with paloverde, cacti, mixed scrub, and chainfruit cholla are preferred by some populations during the hottest, driest periods of the year because they provide thermal cover as well as forage with higher preformed water content. Thus, as with vegetation communities, preference for vegetation structure varies seasonally as conditions change (USFWS 2015).

Mexican pronghorn

Historically, Mexican pronghorn inhabited plains and open valleys from the Sierra Madre Occidental to the Valley of Mexico (Treviño 1978), a dry-temperate and semiarid region. Currently, animals are confined to the lowest section of the Valle Colombia, Coahuila. The Coahuila region consists of four vegetation types: natural grassland (0.66% forb cover; 12% total cover); halophyte grassland (3.6% forb cover; 22% total cover); rosetophyllous scrubland (0.72% forb cover; 33% total cover); and microphyllous scrubland (0.22% forb cover; 31% total cover). After the 1995 and 1998 reintroductions in Coahuila, pronghorn were observed in halophyte grassland communities (Martinez et al. 2000).

Peninsular pronghorn

The peninsular population occurs in plains and areas characterized by xeric shrubs and desert scrub microphyllous vegetation within the El Vizcaíno Biosphere Reserve. In the central Baja California Peninsula, terrain consists of low, rolling hills. Vegetation is dominated by shrubs, including alkali heath (*Frankenia palmeri*), brittlebush (*Encelia* spp.), and Adam's tree (*Fouquieria diguetii*), which are low in stature and widely spaced. The inland dune area contains rolling hills with taller vegetation such as pincushion (*Chaenactis* spp.) and a substrate of sand, stabilized dunes, and sandy plains (Brown and Webb 1979). Ground cover is typically less than 50% (Cancino et al. 2002).

Sonoran pronghorn

Sonoran pronghorn are found exclusively in the Sonoran desert scrub biome, which has a bimodal rainfall pattern that allows greater diversity in vegetation structure than in the Great Basin, Mojave, or Chihuahuan Deserts. The Sonoran Desert occupies the western half of Sonora, Mexico, and large areas of southeastern California, southwestern Arizona, and the Baja California Peninsula (Turner and Brown 1994). Sonoran pronghorn are found in two of the five subdivisions of the Sonoran Desert: the Lower Colorado River Valley (97% of the population) and Arizona Upland subdivisions (deVos and Miller 2005, Hervert et al. 2005). Specifically, they inhabit low dunes, sandy meadows, low hills, and basaltic areas (SEMARNAT 2009). Pronghorn in the Pinacate area occur in extensive sand flats and volcanic cinder flats, and in loose soil patches interspersed with lava fields (Bright et al. 2011). In the Quitovac area, pronghorn use semistabilized dunes, or *medanos* (fixed dunes with cholla) (Bright et al. 2011). Carr (1971) reported seeing Sonoran pronghorn frequently in the Pinta Sands area and suggested that dunes are important in the spring, when annuals are present. Some factors limiting pronghorn occupancy are occurrence and continuance of drought, which could predispose animals to predation; lack of available succulent cacti such as jumping cholla for forage; and possibly, lack of available freestanding water (Beale and Smith 1970, Brown 1982, Nowak and Paradiso 1983).

Habitat Management

Habitat improvements such as water development, supplemental feeding, and forage enhancement are uncommon in Mexico (Yoakum 1979), and as a consequence, a large portion (possibly up to 40%) of the population remains unaffected by management practices. Fences have been repeatedly documented as a serious mortality factor for pronghorn. Proper fence construction incorporates designs that facilitate movement and minimize entanglement.

Direct habitat management

Pronghorn need high-quality forage to meet nutritional requirements. Habitat disturbance may increase forage quality, at least temporarily (Yoakum 2004). Fire is an integral component in maintaining grassland habitats on which pronghorn depend (Hulbert 1969). However, in regions prone to periodic drought, low precipitation can hinder recovery of vegetation following burns for multiple years. French (2015) suggests that pronghorn habitat in the Trans-Pecos could be enhanced by using prescribed fire to achieve a patchwork effect. This provides interspersion of relatively undisturbed areas, which may retain productivity should drought conditions cause heavily disturbed patches to recover slowly.

Forage enhancement plots and supplemental water may also prove useful; providing these in key areas of pronghorn habitat may promote fawn survival by increasing availability of limiting resources to lactating females and foraging fawns during drought (Hervert et al. 2001, deVos and Miller 2005). Since 2002, five forage plots have been established in Arizona (USFWS 2015) and are used extensively by Sonoran pronghorn (Morgart et al. 2005). Access to supplemental forage may increase survival and recruitment by allowing individuals to meet nutritional needs quickly and easily during drought, pregnancy, and lactation (Fox et al. 2000, Hervert et al. 2001).

Captive-breeding programs have also been highly effective in the recovery of pronghorn in Arizona and Mexico (Cancino et al. 2005, USFWS 2015). On the Baja California Peninsula, captive-breeding programs were initiated in 1998. As of 2003, there were 90 captive pronghorn in the program, and the recovery plan for peninsular pronghorn now includes captive breeding as a main component (Cancino et al. 1995, 2005). The captive-breeding facility for peninsular pronghorn (Cancino et al. 2005) has been successful in rearing large numbers of individuals for eventual release into historically occupied habitat. The success of the peninsular breeding program prompted the establishment of a similar breeding facility for Sonoran pronghorn in the US (Cancino et al. 2005). Captive-breeding programs established on Cabeza Prieta National Wildlife Refuge were successful in producing 115 Sonoran pronghorn for release in Kofa NWR in

2013 (USFWS 2015). In 2014, there were 87 animals in the breeding program, and from 2006 to 2014, 128 pronghorn were released into the wild (Sonoran Pronghorn Recovery Team 2014). The captive-breeding program at Cabeza Prieta NWR is ongoing and could be a useful tool for increasing genetic diversity in Arizona, especially after an 80% decline in 2002. Captive studies have also been useful in understanding pronghorn behavior. Human disturbance (e.g., a person walking or running past pronghorn in an enclosed pen, or a truck, or a person entering a holding pen) caused increased heart rates in American pronghorn in captivity (Workman et al. 1992).

Additional management strategies may help increase fawn survival and recruitment. For example, controlled grazing regimes will ensure that fawns have access to hiding cover and will decrease potential livestock competition with adult pronghorn for space. Time-specific predator-control measures can increase pronghorn fawn survival (Spencer and Beach 2006, Brown 2009). Coyotes are the most effective predator of pronghorn fawns (Barrett 1984, Gregg et al. 2001); most mortalities occur before fawns reach one month of age (Barrett 1984). However, predator control data are lacking in Mexico. The pronghorn population on Anderson Mesa in north-central Arizona has exhibited a long-term correlation between coyote suppression and pronghorn fawn survival (Neff and Woolsey 1979, Smith et al. 1986). Similar relationships have been reported in Utah (Udy 1953) and Texas (Hailey 1979). However, Larsen (1970) found no conclusive evidence that coyotes limited antelope populations in New Mexico.

Management challenges

Pronghorn numbers, primarily in Sonora, have varied greatly in the last two decades. Increasing occurrence of drought in southwestern North America throughout the last decade has resulted in climatic changes that negatively impact pronghorn in northern Mexico (deVos and Miller 2005). Habitat loss and fragmentation are greatest in Sonora, Mexico, where much of the habitat is privately or communally owned. Long-term Sonoran pronghorn survival depends on improving habitat conditions and reducing factors that limit range expansion and localized movement, primarily during drought.

Current and future habitat loss is only a minor threat to Sonoran pronghorn populations in the Pinacate Biosphere Reserve; however, land-use practices such as farming are allowed in the reserve (USFWS 2015). Agriculture is prohibited in the nucleus of the reserve, but protection declines outside this area, and agricultural activities occur on ejidos and private lands (CONANP 1995). In contrast, the Quitovac population exists entirely on land with little protection from habitat alteration. Habitat loss from mining operations is the greatest stressor (USFWS 2015). Two mining operations are located in a natural movement corridor for Sonoran pronghorn. La Herradura Mine has expanded to the southeast and northwest and currently occupies approximately 500 km^2 (193 mi^2) of Sonoran pronghorn habitat, while continuing to expand rapidly (USFWS 2015). More recently, a new mining project was initiated approximately 15 km (9.3 mi) southeast of La Herradura. The two mines are connected by an access road 10 m (32.8 ft) wide that causes habitat disturbance up to 20 m (65 ft) on either side of the road and cuts pronghorn habitat into two sections. The road also affects drainage patterns, which may alter hydrological processes enough to impact native plant survival (USFWS 2015).

Pronghorn habitat across Mexico is further fragmented by habitat conversion, physical barriers, and human disturbances such as border infrastructure, fences, roads, railroads, canals, transmission corridors, and mines. The deep, fertile soil of grasslands, flat topography, and well-defined seasonality favor the development of agriculture and livestock, subjecting pronghorn to conflict with human interests. In the last five years, conversion of grasslands to agricultural land has increased markedly in various regions of Chihuahua occupied by pronghorn, as well as in the Valle Colombia, Coahuila. Human disturbance may further prevent pronghorn from using an area, and areas with extensive human use may function as a barrier to pronghorn despite otherwise suitable habitat conditions. Future conservation efforts should be centered around development of a culture of conservation based on recognition of the cultural and biological value of wildlife in Mexico. Efforts should be made to publicize the importance of society's role in the protection, conservation, recovery, and sustainable use of pronghorn antelope and their habitats.

Literature Cited

Autenrieth, R. 1978. Guidelines for the management of pronghorn antelope. Proceedings of the Biennial Pronghorn Antelope Workshop 8:472–526.

Autenrieth, R. E., D. E. Brown, J. Cancino, R. M. Lee, R. A. Ockenfels, B. W. O'Gara, T. M. Pojar, and J. D. Yoakum. 2006. Pronghorn management guide: 2006. Biological and management principles and practices designed to sustain pronghorn populations from Canada to Mexico. Pronghorn Workshop and North Dakota Game and Fish Department, Bismarck, USA.

AZA Antelope and Giraffe Advisory Group. 2009. Antelope and Giraffe Advisory Group regional collection plan. Fifth edition. Association of Zoos and Aquariums, Silver Spring, Maryland, USA.

Azuara, D., R. A. Medellín, C. Manterola, and M. Valdés. 2001. Pronghorn (*Antilocapra americana mexicana*) population surveys in Chihuahua, estimated by aerial surveys. Pages 95–106 *in* J. Cancino, editor. Proceedings of the Nineteenth Biennial Pronghorn Workshop. Centro de Investigaciones Biológicas del Noroeste, La Paz, Baja California Sur, México.

Bahr, C. J. 1991. A legacy of change: historic human impact on vegetation in the Arizona borderlands. University of Arizona Press, Tucson, USA.

Barnowe-Meyer, K. K., P. J. White, T. L. Davis, D. W. Smith, R. L. Crabtree, and J. A. Byers. 2010. Influences of wolves and high-elevation dispersion on reproductive success of pronghorn (*Antilocapra americana*). Journal of Mammalogy 91:712–721.

Barrett, M. W. 1984. Movements, habitat use, and predation on pronghorn fawns in Alberta. Journal of Wildlife Management 48:542–550.

Bates, S. 2000. Preliminary results from a radiotelemetry study of pronghorn antelope at Dugway Proving Ground, Utah. Pronghorn Antelope Workshop Proceedings 17:42–44.

Beale, D. M., and A. D. Smith. 1970. Forage use, water consumption, and productivity of pronghorn antelope in western Utah. Journal of Wildlife Management 34:570–582.

Beale, D. M., and A. D. Smith. 1973. Mortality of pronghorn antelope fawns in western Utah. Journal of Wildlife Management 37:343–352.

Bender, L. C., J. C. Boren, H. Halbritter, and S. Cox. 2013. Factors influencing survival and productivity of pronghorn in a semiarid grass-woodland in east-central New Mexico. Human-Wildlife Interactions 7:313–324.

Bright, J. L., and J. J. Hervert. 2005. Adult and fawn mortality of Sonoran pronghorn. Wildlife Society Bulletin 33:43–50.

Bright, J. L., and J. J. Hervert. 2006. Sonoran pronghorn 2004: Mexico aerial survey summary. Technical Report 241. Arizona Game and Fish Department, Phoenix, USA.

Bright, J. L., and J. J. Hervert. 2011. Sonoran pronghorn aerial survey summary: 2006–2010. Technical Report 263. Arizona Game and Fish Department, Tucson, USA.

Bright, J. L., J. J. Hervert, and M. C. M. Torres. 2011. Sonoran pronghorn 2009: Mexico aerial survey summary. Technical Report 265. Arizona Game and Fish Department, Tucson, USA.

Brown, D. E., editor. 1982. Biotic communities of the American Southwest: United States and Mexico. Desert Plants (Special Issue) 4:1–342.

Brown, D. E. 2009. Effects of coyote removal on pronghorn and mule deer populations in Wyoming. Thesis, Utah State University, Logan, USA.

Brown, D. E., J. Cancino, K. B. Clark, M. Smith, and J. Yoakum. 2006a. An annotated bibliography of references to historical distributions of pronghorn in southern and Baja California. Bulletin of the Southern California Academy of Sciences 105:1–16.

Brown, D. E., and R. A. Ockenfels. 2007. Arizona's pronghorn antelope: a conservation legacy. Arizona Antelope Foundation, Phoenix, USA.

Brown, D. E., D. Warnecke, and T. McKinney. 2006b. Effects of midsummer drought on mortality of doe pronghorn (*Antilocapra americana*). Southwestern Naturalist 51:220–225.

Brown, D. E., and P. M. Webb. 1979. A preliminary reconnaissance of the habitat of the peninsular pronghorn. Journal of the Arizona-Nevada Academy of Science 14:232–234.

Buechner, H. K. 1950. Life history, ecology, and range use of the pronghorn antelope in Trans Pecos, Texas. American Midland Naturalist 43:257–354.

Byers, J. A. 1997. American pronghorn: social adaptations and the ghosts of predators past. University of Chicago Press, Illinois, USA.

Byers, J. A. 2003. Pronghorn (*Antilocapra americana*). Pages 998–1008 *in* G. A. Feldhamer, B.

C. Thompson, and J. A. Chapman, editors. Wild mammals of North America: biology, management and conservation. Second edition. Johns Hopkins University Press, Baltimore, Maryland, USA.

Cancino, J. 1994. Food habits of the peninsular pronghorn. Proceedings of the Biennial Pronghorn Antelope Workshop 16:176–185.

Cancino, J., and R. J. Guenzel. 2002. Aero transfer of American pronghorn fawns from Wyoming to Mexico: another step toward the assisted reproduction of pronghorn. Proceedings of the Biennial Pronghorn Antelope Workshop 20:84–86.

Cancino, J., R. Lee, and F. Ramírez. 2014. Berrendo. Pages 467–487 *in* R. Valdez and J. A. Ortega-S, editors. Ecología y manejo de fauna silvestre en México. Editorial Colegio de Postgraduados, Texcoco, Estado de México, México.

Cancino, J., P. Miller, J. Bernal Stoopen, and J. Lewis. 1995. Population and habitat viability assessment for the peninsular pronghorn (*Antilocapra americana peninsularis*). IUCN Species Survival Commission and Conservation Breeding Specialist Group, Apple Valley, Minnesota, USA.

Cancino, J., A. Ortega-Rubio, and R. Rodriguez-Estrella. 1998. Population size of the peninsular pronghorn in Baja California Sur, Mexico. California Fish and Game 84:25–30.

Cancino, J., A. Ortega-Rubio, and J. A. Sanchez-Pacheco. 1996. Status of an endangered subspecies: the peninsular pronghorn at Baja California. Journal of Arid Environments 32:463–467.

Cancino, J., R. Rodríguez-Estrella, and P. Miller. 2010. Using a population viability analysis for management recommendations of the endangered endemic peninsular pronghorn. Acta Zoológica Mexicana 26:173–189.

Cancino, J., R. Rodríguez-Estrella, and A. Ortega. 1994. First aerial survey of historical range for peninsular pronghorn of Baja California, Mexico. Journal of the Arizona-Nevada Academy of Science 28:46–50.

Cancino, J., V. Sanchez-Sotomayor, and R. Castellanos. 2002. Alternative capture technique for the peninsular pronghorn. Wildlife Society Bulletin 30:256–258.

Cancino, J., V. Sanchez-Sotomayor, and R. Castellanos. 2005. Capture, hand raising, and captive management of peninsular pronghorn. Wildlife Society Bulletin 33:61–65.

Canon, S. K. 1993. Fawn survival and bed-site characteristics of Trans-Pecos pronghorn. Dissertation, Texas Tech University, Lubbock, USA.

Canon, S. K., and F. C. Bryant. 1997. Bed-site characteristics of pronghorn fawns. Journal of Wildlife Management 61:1134–1141.

Carey, J. R., and D. S. Judge. 2000. Longevity records: life spans of mammals, birds, amphibians, reptiles, and fish. Odense Monographs on Population Aging. Odense University Press, Denmark.

Carr, J. N. 1971. Sonoran pronghorn endangered species investigation. Progress report. Arizona Game and Fish Department, Phoenix, USA.

Castillo, C. 1993. Informe técnico y programa de manejo para el berrendo sonorense (*Antilocapra americana sonoriensis*) en Sonora, México. Gobierno del Estado de Sonora and Centro Ecológico de Sonora, Hermosillo, México.

Christensen, A. G., L. J. Lyon, R. Pendersen, P. Harrington, W. L. Bodie, R. Johnson, and B. W. O'Gara. 1995. Ungulate assessment in the Columbia River Basin. USFS, USDI, and BLM, Washington, DC, USA.

CITES. 2014. CITES Appendix I. http://www.cites.org/eng/app/appendices.php. Accessed 22 Feb 2016.

Clemente, F., R. Valdez, J. L. Holechek, P. J. Zwank, and M. Cardenas. 1995. Pronghorn home range relative to permanent water in southern New Mexico. Southwestern Naturalist 40:38–41.

Colchero, F., L. L. León-Paniagua, and A. G. Navarro. 2007. Hábitat del berrendo en México. Instituto Nacional de Ecología, SEMARNAT, México D. F., México. http://www2.inecc.gob.mx/publicaciones/libros/494/habitat.html. Accessed 21 Apr 2016.

CONANP. 1995. Programa de manejo: Biósfera del Pinacate y Gran Desierto de Altar. CONANP, México, D. F., México.

CONANP. 2010. Monitoreo de las poblaciones de berrendo (*Antilocapra americana*) en la Reserva de la Biosfera Janos. CONANP, México, D. F., México. http://www.conanp.gob.mx/acciones/fichas/janos/info/info.pdf. Accessed 4 Apr 2016.

Danvir, R. E. 2000. Environmental and density dependent effects on a northern Utah pronghorn population. Proceedings of the 1996 Pronghorn Antelope Workshop 17:36–41.

de las Heras, A., and M. Islas-Espinosa. 2015. Peninsular pronghorn conservation: too many paradigms, too few indicators. Pages 126–145 *in* A. Latawiec and D. Agol, editors. Sustainability indicators in practice. Walter de Gruyter, Warsaw, Poland.

de Peña, R. H., and J. M. P. Neira. 1982. Hábitos alimentícios del berrendo (*Antilocapra americana*) en la región central de Chihuahua. Pastizales 6:2–9.

deVos, J. C., and W. H. Miller. 2005. Habitat use and survival of Sonoran pronghorn in years with above-average rainfall. Wildlife Society Bulletin 33:35–42.

Fox, L. M. 1997. Nutritional content of forage in Sonoran pronghorn habitat, Arizona. Thesis, University of Arizona, Tucson, USA.

Fox, L. M., P. R. Krausman, M. L. Morrison, and R. M. Katting. 2000. Water and nutrient content of forage in Sonoran pronghorn habitat, Arizona. California Fish and Game 86:216–232.

French, J. T. 2015. Pronghorn diet, nutrition, and habitat assessment in Trans-Pecos, Texas. Thesis, Sul Ross State University, Alpine, Texas, USA.

Garrison, T. O. 2014. Post-release survival and movements of translocated pronghorn in Trans-Pecos, Texas. Thesis, Sul Ross State University, Alpine, Texas, USA.

González-Romero, A., and A. Lafon. 1993. Distribución y estado actual del berrendo (*Antilocapra americana*) en México. Pages 409–420 *in* R. Medellín and G. Ceballos, editors. Avances en el estudio de los mamíferos de México. Asociación Mexicana de Mastozoología, La Paz, Baja California Sur, México.

Gregg, M. A., M. Bray, K. M. Kilbride, and M. R. Dunbar. 2001. Birth synchrony and survival of pronghorn fawns. Journal of Wildlife Management 65:19–24.

Guadarrama, E., and M. Valdés. 2001. Reintroducción y monitoreo de berrendos en el estado de Coahuila. Proceedings of the Pronghorn Antelope Workshop 19:110.

Hailey, T. L. 1979. A handbook for pronghorn antelope management in Texas. Series 20. Texas Parks and Wildlife Department, Austin, USA.

Hailey, T. L., J. W. Thomas, and R. M. Robinson. 1966. Pronghorn die-off in Trans-Pecos Texas. Journal of Wildlife Management 30:488–496.

Hall, E. R. 1981. The mammals of North America. Second edition. John Wiley and Sons, New York, New York, USA.

Halloran, A. F. 1957. A note on the Sonoran pronghorn. Journal of Mammalogy 38:423.

Harveson, L. A. 2006. Life history and ecology of pronghorn. Pages 1–4 *in* K. A. Cearley and S. Nelle, editors. Proceedings of the 2006 Pronghorn Symposium. Texas Cooperative Extension Service, Amarillo, USA.

Heffelfinger, J. R., B. W. O'Gara, C. M. Janis, and R. Babb. 2004. A bestiary of ancestral antilocaprids. Proceedings of the Biennial Pronghorn Antelope Workshop 20:87–111.

Hervert, J. J., J. L. Bright, M. T. Brown, L. A. Piest, and R. S. Henry. 2000. Sonoran pronghorn population monitoring: 1994–1998. Technical Report 162. Arizona Game and Fish Department, Phoenix, USA.

Hervert, J. J., J. L. Bright, R. S. Henry, L. A. Piest, and M. T. Brown. 2005. Home-range and habitat-use patterns of Sonoran pronghorn in Arizona. Wildlife Society Bulletin 33:8–15.

Hervert, J. J., L. A. Piest, W. Ballard, R. S. Henry, M. T. Brown, and S. Boe. 1997. Sonoran pronghorn population monitoring: progress report. Arizona Game and Fish Department, Phoenix, USA.

Hervert, J. J., L. A. Piest, M. T. Brown, and R. S. Henry. 2001. Sonoran pronghorn recovery: habitat enhancements to increase fawn survival. Proceedings of the Pronghorn Antelope Workshop 19:19–127.

Hoffman, J. K. 2015. Post-release survival and movements of translocated pronghorn in Trans-Pecos, Texas. Thesis, Sul Ross State University, Alpine, Texas, USA.

Hoffmann, M., J. Byers, and J. Beckmann. 2008. *Antilocapra americana.* IUCN Red List of Threatened Species. http://www.iucnredlist.org/details/1677/0. Accessed 22 Feb 2016.

Hoffmeister, D. F. 1986. Mammals of Arizona. University of Arizona Press, Tucson, USA.

Hughes, K. S. 1991. Sonoran pronghorn use of habitat in southwest Arizona. Thesis, University of Arizona, Tucson, USA.

Hughes, K. S., and N. S. Smith. 1990. Sonoran pronghorn use of habitat in southwest Arizona. Final Report 14–16–009–1564 RWO#6. Arizona Cooperative Fish and Wildlife Research Unit, Tucson, USA.

Hulbert, L. C. 1969. Fire and litter effects in undisturbed bluestem prairie in Kansas. Ecology 50:874–877.

Jaramillo, F. 1989. Contribución al conocimiento y

conservación del berrendo de Baja California (*Antilocapra americana peninsularis*, Nelson 1912: Antilocapridae, Mammalia) en el Desierto del Vizcaíno, México. Thesis, UNAM, México, D. F., México.

Kitchen, D. W., and B. W. O'Gara. 1974. Social behavior and ecology of the pronghorn. Wildlife Monographs 38:1–96.

Klimova, A., A. Munguia-Vega, J. I. Hoffman, and M. Culver. 2014. Genetic diversity and demography of two endangered captive pronghorn subspecies from the Sonoran Desert. Journal of Mammalogy 95:1263–1277.

Laliberte, A. S., and W. J. Ripple. 2004. Range contractions of North American carnivores and ungulates. BioScience 54:123–138.

Larsen, P. 1970. A six-year study of antelope productivity and survival in southern New Mexico. Proceedings of the Antelope States Workshop 4:97–103.

Lehman, V. W., and J. B. Davis. 1942. Experimental wildlife management in the South Texas chaparral. Program 1-R Quarterly Report. Texas Game, Fish, and Oyster Commission, Austin, USA.

Leopold, A. S. 1959. Wildlife of Mexico: the game birds and mammals. University of California Press, Berkeley, USA.

Martinez, M. A., R. E. P. Miranda, S. J. I. Uvalle, R. R. Aranda, S. Chakeredza, and U. Meulen. 2000. Monitoring a pronghorn (*Antilocapra americana mexicana*) population reintroduced to the northeast of Mexico. Journal of Agriculture in the Tropics and Subtropics 101:141–161.

Maslin, M. 2009. Global warming: a very short introduction. Oxford University Press, Oxford, United Kingdom.

McCabe, R. E., B. W. O'Gara, and H. M. Reeves. 2004. Prairie ghost: pronghorn and human interactions in early America. University Press of Colorado, Boulder, USA.

McKean, T., and B. Walker. 1974. Comparison of selected cardiopulmonary parameters between the pronghorn and the goat. Respiratory Physiology 21:365–370.

McKenna, M. C., and S. K. Bell. 1997. Classification of mammals above the species level. Columbia University Press, New York, New York, USA.

Medellín, R. A., C. Manterola, M. Valdés, D. Hewitt, D. Doan-Crider, and T. E. Fulbright. 2005. History, ecology, and conservation of the pronghorn antelope, bighorn sheep, and black bear in Mexico. Pages 387–404 *in* J. E. Cartron, G. Ceballos, and R. E. Felger, editors. Biodiversity, ecosystems, and conservation in northern Mexico. Oxford University Press, Oxford, United Kingdom.

Mitchell, G. 1967. Minimum breeding age of female pronghorn antelope. Journal of Mammalogy 48:489–490.

Monson, G. 1968. The desert pronghorn. Desert Bighorn Council Transactions 12:63–69.

Morgart, J. R., J. J. Hervert, P. R. Krausman, J. L. Bright, and R. S. Henry. 2005. Sonoran pronghorn use of anthropogenic and natural water sources. Wildlife Society Bulletin 33:51–60.

Munguia-Vega, A., A. Klimova, and M. Culver. 2013. New microsatellite loci isolated via next-generation sequencing for two endangered pronghorn from the Sonoran Desert. Conservation Genetics Resources 5:125–127.

Neff, D. J., and N. W. Woolsey. 1979. Effect of predation by coyotes on antelope fawn survival on Anderson Mesa. Special Report 8. Arizona Game and Fish Department, Phoenix, USA.

Nelson, F. W. 1925. Status of the pronghorn antelope, 1922–1924. Agriculture Bulletin 1346. USDA, Washington, DC, USA.

Nowak, R. M., and J. L. Paradiso. 1983. Walker's mammals of the world. Volume 2. Johns Hopkins University Press, Baltimore, Maryland, USA.

O'Brien, C. S., S. S. Rosenstock, J. J. Hervert, J. L. Bright, and S. R. Bol. 2005. Landscape- level models of potential habitat for Sonoran pronghorn. Wildlife Society Bulletin 33:24–34.

Ockenfels, R. A., A. A. Alexander, C. L. D. Ticer, and W. K. Carrel. 1994. Home ranges, movement patterns and habitat selection of pronghorn in central Arizona. Technical Report 13. Arizona Game and Fish Department, Phoenix, USA.

O'Gara, B. W. 1978. *Antilocapra americana.* Mammalian Species 90:1–7.

O'Gara, B. W. 1999. *Antilocapra americana.* Pages 339–341 *in* D. E. Wilson and S. Ruff, editors. The Smithsonian book of North American mammals. Smithsonian Institution Press, Washington, DC, USA.

O'Gara, B. W., and C. M. Janis. 2004. The fossil record. Pages 27–39 *in* B. W. O'Gara and J. D. Yoakum, editors. Pronghorn ecology and management. University Press of Colorado, Boulder, USA.

O'Gara, B. W., and J. D. Yoakum, editors. 2004. Pronghorn ecology and management. University Press of Colorado, Boulder, USA.

Pallares, E. 1999. Jaguares, tucanes y otros animales de la fauna mexicana. Subsecretaría de Educación Básica y Normal, México, D. F., México.

Phelps, J. S. 1981. Biological observations on the Sonoran pronghorn. Pages 28–33 *in* J. S. Phelps, editor. The Sonoran pronghorn. Special Report 10. Arizona Game and Fish Department, Phoenix, USA.

Reynolds, T. D. 1984. Daily summer movements, activity patterns, and home range of pronghorn. Northwest Science 58:300–311.

SEMARNAT. 2009. Programa de acción para la conservación de la especie: berrendo (*Antilocapra americana*). Instituto Nacional de Ecología and SEMARNAT, México, D. F., México.

SEMARNAT. 2010. Norma Oficial Mexicana NOM-059-SEMARNAT-2010, Protección ambiental—Especies nativas de México de flora y fauna silvestres—Categorías de riesgo especificaciones para su inclusión, exclusión o cambio—Lista de especies en riesgo. Diario Oficial de la Federación. Segunda Sección. 30 de diciembre de 2010. México, D. F., México.

Simpson, D. C., L. A. Harveson, C. E. Brewer, R. E. Walser, and A. R. Sides. 2007. Influence of precipitation on pronghorn demography in Texas. Journal of Wildlife Management 71:906–910.

Smith, R. H., D. J. Neff, and N. G. Woolsey. 1986. Pronghorn response to coyote control: a benefit:cost analysis. Wildlife Society Bulletin 14:226–231.

Snow, T. K. 1994. Sonoran pronghorn aerial survey summary 1992–1994. Technical Report 51. Arizona Game and Fish Department, Phoenix, USA.

Sonoran Pronghorn Recovery Team. 2014. Sonoran pronghorn monthly update: February–March 2014. Cabeza Prieta NWR, Ajo, Arizona, USA.

Spencer, J. O., Jr., and R. H. Beach. 2006. Wildlife damage management protection efforts for a vulnerable pronghorn antelope population in northwestern Nevada: 2000–2004. Proceedings of Vertebrate Pest Conference 22:269–273.

Ticer, C. L., and J. C. deVos. 2001. Pronghorn province and state status report. Proceedings of the Biennial Pronghorn Antelope Workshop 19:7–18.

Tluczek, M. 2012. Diet, nutrients, and free water requirements of pronghorn antelope on Perry Mesa, Arizona. Thesis, Arizona State University, Tempe, USA.

Treviño, J. 1978. Number and distribution of pronghorn antelope in Chihuahua, Mexico. Thesis, New Mexico State University, Las Cruces, USA.

Turner, R. M., and D. E. Brown. 1994. Tropical-subtropical desert lands: Sonoran desert scrub. Pages 180–221 *in* D. E. Brown, editor. Biotic communities: southwestern United States and northwestern Mexico. University of Utah Press, Salt Lake City, USA.

Udy, J. R. 1953. Effects of predator control on antelope populations. Publication 5. Utah State Department of Fish and Game, Salt Lake City, USA.

USFWS. 1967. US endangered species list. Federal Register 32:4001.

USFWS. 1998. Final revised Sonoran pronghorn recovery plan. USFWS, Albuquerque, New Mexico, USA.

USFWS. 2002. Recovery criteria and estimates of time for recovery actions for the Sonoran pronghorn: a supplement and amendment to the final revised Sonoran pronghorn recovery plan. USFWS, Albuquerque, New Mexico, USA.

USFWS. 2010. Final environmental assessment for reestablishment of Sonoran pronghorn. USFWS, Albuquerque, New Mexico, USA.

USFWS. 2015. Draft recovery plan for the Sonoran pronghorn (*Antilocapra americana sonoriensis*). Second revision. USFWS, Albuquerque, New Mexico, USA.

Valdés, M., E. de la Cruz, E. Peters, and E. Pallares, compilers. 2006. El berrendo en México: acciones de conservación. Agrupación Sierra Madre and INE-SEMARNAT, México, D. F., México.

Valdés, M., and C. Manterola. 2001. La conservación del berrendo (*Antilocapra americana*) en México. Biodiversitas 35:1–6.

Valdés, M., and C. Manterola. 2007. Reintroducción del berrendo en Coahuila. INE-SEMARNAT, México, D.F., México. http://www2.inecc.gob.mx/publicaciones/libros/494/reintro.html. Accessed 21 Apr 2016.

Weber, M., and S. González. 2003. Latin America deer diversity and conservation: a review of status and distribution. Ecoscience 10:443–454.

Wilson, R. R., P. R. Krausman, and J. R. Morgart. 2008. Behavior and timing of life-history events in a semi-captive population of the Sonoran pronghorn (*Antilocapra americana sonoriensis*). Southwestern Naturalist 53:389–393.

Workman, G. D., T. D. Bunch, J. W. Call, F. C. Evans, L. S. Neilson, and E. M. Rawlings. 1992. Sonic boom and other disturbance impacts on pronghorn antelope (*Antilocapra americana*). Report. US Air Force, Hill Air Force Base, Utah, USA.

Wright, R. L., and J. C. deVos Jr. 1986. Final report on Sonoran pronghorn status in Arizona. Arizona Game and Fish Department, Phoenix, USA.

Yoakum, J. D. 1978. Pronghorn. Pages 103–121 *in* J. L. Schmidt and D. L. Gilbert, editors. Big game of North America: ecology and management. Wildlife Management Institute, Washington, DC, and Stackpole, Harrisburg, Pennsylvania, USA.

Yoakum, J. D. 1979. Managing rangelands for pronghorns. Rangelands 1:146–148.

Yoakum, J. D. 2004. Management plans, environmental impact statements and guides. Pages 541–569 *in* B. W. O'Gara and J. D. Yoakum, editors. Pronghorn ecology and management. University Press of Colorado, Boulder, USA.

Yoakum, J. D., and B. W. O'Gara. 2000. Pronghorn. Pages 559–577 *in* S. Demarais and P. R. Krausman, editors. Ecology and management of large mammals in North America. Prentice Hall, Upper Saddle River, New Jersey, USA.

20 Desert Bighorn Sheep in Mexico

Andrew V. Sandoval, Raul Valdez, and Alejandro Espinosa-T.

Wild sheep in North America occur from Alaska to Mexico. Two species are recognized: (1) Dall's and Stone's sheep (*Ovis dalli*) in Alaska and northwestern Canada, and (2) Rocky Mountain and desert bighorn sheep (*Ovis canadensis*) in southwestern Canada, the western US, and northern Mexico. Only desert bighorn sheep occur in Mexico (fig. 20.1). Wild sheep can tolerate temperatures varying from −20°C in Alaska to more than 45°C in Mexico. Their relatively compact bodies are adapted for living in precipitous montane environments. They generally occupy habitats free of tall vegetation that could obstruct their vision. Males possess long, massive horns that curl near the side of the face. The longest recorded horn and greatest basal horn circumference of a Mexican specimen measured 112 cm (44 in) and 43.2 cm (17 in), respectively. Desert bighorn in Mexico occur in scattered populations, some of questionable viability, in desert regions of northern Mexico, and south through the Baja California Peninsula. They inhabit some of the most inhospitable and rugged desert landscapes (fig. 20.2) (Hall 1981, Valdez and Krausman 1999). Desert bighorn in Mexico are a priority species and are accorded special protection (SEMARNAT 2010).

Bighorn sheep are highly prized hunting trophies.

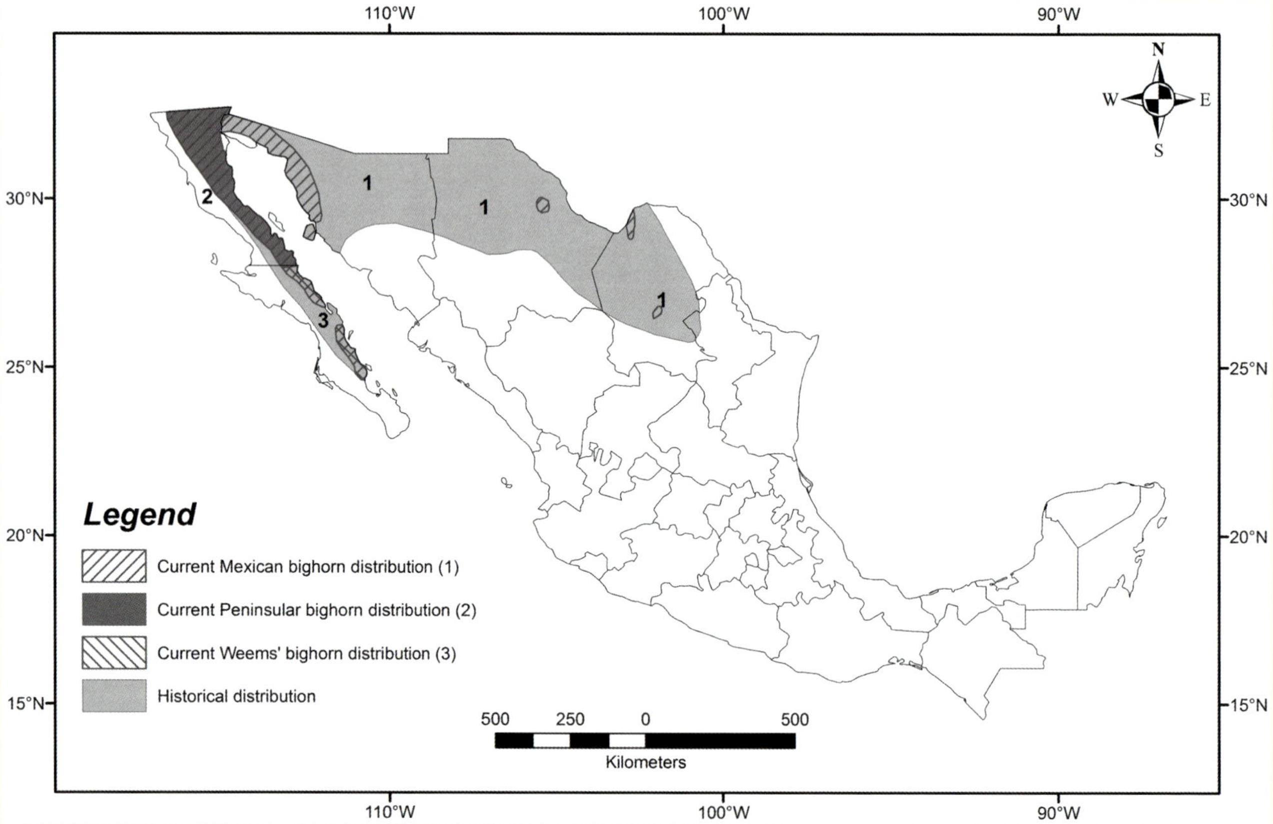

Figure 20.1. Historical and current distribution of desert bighorn subspecies in Mexico: (1) Mexican (Chihuahua, Sonora, and Coahuila); (2) peninsular (Baja California); (3) Weems' (Baja California Sur).

Figure 20.2. Bighorn sheep rely on open, steep terrain for predator avoidance since their predator-evasion strategies involve foraging diurnally in large dispersed groups in open habitat close to escape terrain. Sierra Maderas del Carmen, Coahuila, Mexico. Photo by Alejandro Espinosa-T., CEMEX/Nature.

Demand for bighorn sheep hunting permits exceeds availability. In 2012, the state of Sonora authorized 72 hunting permits, which sold for as much as US$70,000 (Valdez Aguayo 2013). Hunters have paid US$200,000 for a bighorn sheep hunting permit on Tiburón Island, Sonora. Between 1998 and 2007, the Seri Indian community earned US$3.2 million from the sale of hunting permits, and from selling live bighorn for captive-breeding purposes from Tiburón Island. In 2011, special fund-raising permits in Mexico generated close to US$1 million (Lee 2011). Sport hunting of bighorn sheep is legal in Sonora, Baja California Sur, Chihuahua, and Coahuila. The revenues generated have been used to improve social and cultural infrastructure and living standards, and to invest in bighorn sheep management programs.

Desert bighorn represented an important cultural motif of native peoples. Bighorn are widely depicted in rock art (pictographs and petroglyphs). They represent independence, tenacity, and masculinity and are used as emblems and mascots by public and private institutions as well as civil organizations in Mexico (Murray and Espinosa-T. 2006).

Description

Desert bighorn are relatively short-legged animals with a blocky appearance, but they are not as heavy bodied as Rocky Mountain bighorn. Lambs weigh 2.7–4.1 kg (5.9–9.0 lb) at birth and grow rapidly; by six months of age they weigh approximately 29.0 kg (63.9 lb) and stand 69 cm (27 in) at the shoulder. One-year-old bighorn rams appear equivalent in size to adult ewes except for a shorter rostrum, a characteristic of lambs. The massive curled, divergent horns of the males are their primary distinguishing anatomical feature. In contrast, female horns are relatively thin, sickle shaped, and usually less than 20 cm (< 8 in) long (fig. 20.3). An adult desert bighorn is 76–99 cm (30–39 in) tall at the shoulder and about 152 cm (60 in) in total length. Males are usually larger than females, although some females may be taller and have a larger body than the smallest adult males. Adult males average about 73 kg (160 lb) live weight, but they can weigh 91 kg (200 lb) or more (fig. 20.4). Adult ewes average about 48 kg (105 lb) (Hansen 1980). Considerable variation in pelage color is evident within individuals in the same population. Pelage may vary from dark tan, gray brown, and blackish in late summer to a pale buff or washed-out tan in the spring because of bleaching by the sun. Young animals tend to have lighter-colored coats than prime or older bighorn, and ewes are generally lighter colored than rams (Hansen 1980, Valdez 2011a).

Figure 20.3. Slopes of more than 60% are considered steep enough to be escape terrain. Free-ranging bighorn population in Sierra Maderas del Carmen, Coahuila, Mexico. Photo by Edgar Siller.

Figure 20.4. Mixed group of bighorn in Maderas del Carmen, Coahuila, Mexico; males possess long, massive, curving horns. Photo by Edgar Siller.

Taxonomy

Of the four subspecies of desert bighorn in North America, three occur in Mexico: peninsular (*cremnobates*) in Baja California; Weems's (*weemsi*) in Baja California Sur; and Mexican bighorn (*mexicana*) in Sonora, Chihuahua, and Coahuila. The fourth subspecies, Nelson's (*nelsoni*), occurs in the southwestern US. The classification of desert bighorn sheep is controversial (Ramey 1993, Wehausen and Ramey 1993, Gutierrez-Espeleta et al. 1998, Valdez 2011a). Mexican bighorn sheep show relatively low levels of genetic diversity and genetic differentiation, perhaps because of their recent colonization; the last populations became established in North America about 13,000 years ago (Rodríguez-Rodríguez et al. 2015). However, phylogeographic and population genetic analyses confirm the validity of two subspecies in northwestern Mexico: *mexicana* and *cremnobates* (Buchalski et al. 2016). Their taxonomy has conservation and management implications, because subspecies and genetic boundaries are important in selecting reintroduction stock. Because of their greater potential to adapt to their new environment, wild sheep from the closest population with similar habitat should be used for reintroduction purposes.

Distribution and Status

Wild sheep in Mexico have been extirpated in Chihuahua and Coahuila. The extirpation of wild sheep populations in northeastern Mexico and the southwestern US can be attributed to habitat loss, degradation, and fragmentation; diseases transmitted by domestic livestock; and illegal subsistence and sport hunting. Diseases transmitted by domestic livestock, particularly domestic sheep and goats, were the principal cause of bighorn sheep extirpation (Gross et al. 2000).

Historical distribution

Desert bighorn originally occurred in Baja California, Baja California Sur, Sonora, Chihuahua, Coahuila, and probably Nuevo León (fig. 20.5). Original desert bighorn populations now occur only in Baja California, Baja California Sur, and Sonora (Sandoval 1985, Tarango and Krausman 1997, Lee 2003a, Valdez 2011a). However, they have been reintroduced in Chihuahua and Coahuila during the past two decades. They occur in only 40% of their historical

Figure 20.5. Adult desert bighorn ram from reestablished population in the Sierra Maderas del Carmen, Coahuila, Mexico, showing typical prominent white rump patch. Photo by Santiago Gibert Isern.

range in Baja California and Baja California Sur (Jimenez-L. et al. 1996).

In northwestern Sonora, bighorn were scarce by 1894 because of excessive hunting by Papago Indians with firearms (Mearns 1907). Desert bighorn formerly occurred in almost every mountain range west of the Nogales-Hermosillo highway, but by the 1930s they had disappeared from east and north of the Sonoyta-Caborca highway. They were not known to have occurred on Tiburón Island in the Sea of Cortez prior to their successful introduction in 1975. However, morphological and DNA analyses confirmed that fossilized pellets found in a cave on the island were those of bighorn sheep. Wilder et al. (2014) concluded that native desert bighorn sheep previously colonized the island via a land bridge, most likely during the Late Pleistocene, when lower sea levels connected Tiburón Island to the mainland. They were extirpated on the island about 1,500 years ago, probably because of the inherent dynamics of isolated populations, prolonged drought, or human overkill.

In Chihuahua, 11 areas of historical bighorn occurrence have been identified, comprising 23 mountain ranges extending southeast from the New Mexico–Mexico border to the Coahuila state border (Heffelfinger and Marquez-Muñoz 2005). In Coahuila, they occurred in 14 mountain ranges as far south as latitude 25°43′02″ N (Espinosa-T. et al. 2006). In Nuevo León, bighorn probably occurred in the Sierra Pájaros Azules near the Coahuila border. Concern over the decline of desert bighorn populations prompted the federal government to establish a 10-year hunting prohibition in 1922. The prohibition was extended for another decade in 1933 and was declared permanent in 1944 (Mellink 1993). Because the public was uninformed and the prohibition was not enforced, poaching continued unabated (Sandoval 1985). Corruption at various levels of government exacerbated the problem (Dávila 1960). Hunting was eventually reopened, but the season was closed again in 1990 in the state of Baja California, and it remains closed (Valdez 2011a).

Current status

In Sonora, there are an estimated 7,560 sheep, of which 3,200 are free ranging in 27 mountain ranges, 3,600 are in 36 captive herds, and 640 are on Tiburón Island (Lee et al. 2012; Valdez Aguayo 2013; J. M. Segundo Galán, personal communication). On the Baja California Peninsula, there are 3,600. In the state of Baja California, there are an estimated 2,500 sheep (table 20.1) (Martinez-G. 2011, Lee et al. 2012). In Baja California Sur, there are two distinct bighorn sheep populations, one on the mainland with an estimated 500 animals (Lee and Segundo-Galán 2011, Lee et al. 2012), and the other on Carmen Island with 600 (Jiménez and Hernández 2010; J. D. Wehausen, personal communication) (table 20.1).

Desert bighorn were reestablished in Rancho La Guarida near Coyame in northeastern Chihuahua in 2000, and subsequent translocations were made from Sonora. Between 2007 and 2009, 14 ewes and 14 rams were removed from Rancho La Guarida to establish two additional captive-breeding populations in Chihuahua. A third captive population was established by a private landowner with sheep purchased from Sonora. By 2015, there were 300 sheep in La Guarida and 275 in the three separate captive herds, for a statewide estimate of 575 animals (Uranga-Thomas and Valdez 2011; Uranga-Thomas, personal communication) (table 20.1). In 2009, the first ram to be legally hunted in Chihuahua in more than 50 years was taken in Rancho La Guarida.

In Coahuila, there are two captive populations (Pilares UMA and Rancho San Juan), and two that are free ranging (Maderas del Carmen and Rincón de la Madera UMAs). At Pilares UMA, 48 desert bighorn (37 ewes and 11 rams) were translocated between 2000 and 2002 from Sonora to a 5,000 ha (12,355 ac) game-proof enclosure in the Sierra Pilares near Ocampo in northeastern Coahuila; in 2015, this population numbered 200 (Espinosa-T. et al. 2002; A. Espinosa-T., personal communication). In 2004, 24 bighorn (11 ewes and 13 rams) were captured in Sonora and released directly into the wild in the Sierra Maderas del Carmen; this population was augmented in 2005 with 4 ewes and 2 rams from Sonora, and subsequently with 4 transplants from the Pilares herd. In 2009, the same population was augmented with 30 animals (20 ewes and 10 rams); in 2010 with 16 (5 ewes and 11 rams); in 2012 with 18 (11 ewes and 7 rams); and in 2014 with 21 (11 ewes and 10 rams). The free-ranging herd in the Sierra Maderas del Carmen now numbers 80 (McKinney and Delgadillo-Villalobos 2005; A. Espinosa-T., personal communication).

Table 20.1. Desert bighorn sheep population estimates in Mexico.

State	Population size	Source
Sonora Free-ranging Captive	 3,200 3,600	Lee and Segundo Galán (2011), Lee et al. (2012), Valdez Aguayo (2013), J. M. Segundo Galán (pers. comm.)
Tiburón Island Other	640 120[a]	J. M. Segundo Galán (pers. comm.)
Subtotal	7,560	
Baja California	2,500	Martínez (2011), Lee et al. (2012), Escobar-Flores et al. (2015)
Baja California Sur Mainland	500[b]	Lee and Segundo Galán (2011), Lee et al. (2012)
Carmen Island	600	Lee and Segundo Galán (2011), Lee et al. (2012), J. D. Wehausen (pers. comm.)
Subtotal	3,600	
Chihuahua	575[c]	Uranga-Thomas and Valdez (2011), Uranga-Thomas (pers. comm.)
Coahuila	920[d]	Espinosa-T. et al. (2002), E. Rangel (pers. comm.), A. Espinosa-T. (pers. comm.)
Total	12,655	

[a]Includes two reestablished populations.
[b]Estimates range up to 1,000.
[c]Inlcudes 275 in captivity.
[d]Includes 340 in captivity.

At Rincón de la Madera, Coahuila, the free-ranging herd originated from a captive group of 100 sheep (60 females and 40 males) translocated from Tiburón Island, Sonora, in 2005; it had increased to 500 by 2015. At Rancho San Juan, a private preserve, there are 140 sheep. In Coahuila as a whole, there are 340 in captivity and an estimated 580 free ranging, for a total of 920 (table 20.1). The first legal harvest of a ram in Coahuila was in the Sierra Maderas del Carmen in 2010.

In general, desert bighorn are increasing in captivity and stable in wild populations in Sonora, stable in Baja California and Baja California Sur, and increasing in captivity in Chihuahua and Coahuila. Because of mountain lion predation, the one free-ranging population in the Sierra Maderas del Carmen, in extreme northern Coahuila, has not increased; it remains stable at 80 animals.

One of the most successful desert bighorn sheep transplants was on Tiburón Island in 1975. Following the release of 20 animals (16 ewes and 4 rams) from mainland Sonora (Montoya and Gates 1975), the Tiburón Island population increased rapidly to more than 500 animals. This population has served as the main source of animals to repopulate Sonora, Chihuahua, and Coahuila, and to establish numerous captive-breeding populations in Sonora. Since 1998, more than 500 animals have been removed from Tiburón Island (O'Farril 2003, Valdez Aguayo 2013).

Life History

Diet

North American wild sheep are primarily grazers; however, desert bighorn consume a considerable amount of forbs and woody plants. It is difficult to discuss seasonal food habits of desert bighorn without considering seasonal precipitation and its effects on forage availability, quality, and palatability (Krausman and Bowyer 2003). Desert bighorn are opportunists and are adapted to forage in desert habitats. Throughout southwestern North America, desert bighorn actively seek browse and forbs even when grasses are plentiful. Some browse species have a high nutritional value throughout the year. In the Sierra El Mechudo in Baja California Sur, desert bighorn were browsers; diets consisted of 73% shrubs and trees, and 27% forbs (Guerrero-Cárdenas et al. 2016). Several other studies also indicate that desert bighorn sheep in Mexico are browsers (table 20.2)

Mating and breeding

Bighorn sheep are seasonal breeders with an extended mating and breeding season, and they are serially polygynous (dominant males mate with more than one ewe during the mating season but court and guard a single female). Rams are not territorial and do not form harems. Instead, adult rams move among groups of ewes searching for those in estrus (fig. 20.6). Ewes come into estrus in late summer. During this time, ewes will accept copulation by more than one ram, but dominant rams do most of the mating. Desert bighorn are capable of mating at 18 months of age, and ewes give birth to one lamb after a gestation period of six months (Krausman et al. 1999).

An extended lambing period is an adaptive strategy that ensures lamb survival during periods of unpredictable precipitation patterns and forage production. In southern California (US), in the peninsular ranges that extend into Baja California, births occurred during seven months of the year, but 87% of lambs were born in February–April, and 55% in March. Lambs born during the peak birthing period had greater survival than those born later. Annual winter rains, the period of high plant productivity, occurred after the peak lambing months and provided ewes with nutritious forage. The highest mortality rate usually occurred during the first year. Bighorn sheep have a long life expectancy, but low fecundity (Krausman et al. 1999, Rubin et al. 2000).

Water requirements

Desert bighorn can survive on preformed water in their food, and metabolic water formed from oxidative metabolism. However, increased day length,

Table 20.2. Mean percentage of annual forage classes in diets of desert bighorn sheep in southwestern North America.

	Percentage in diet			
Location	Grasses	Forbs	Shrubs	Source
New Mexico (San Andres Mountains)	42	27	27	Sandoval (1979)
Texas (Elephant Mountain)	52	15	26	Brewer (2001)
California (Death Valley)	56	16	16	Ginnett and Douglas (1982)
Arizona (Sonoran Desert)	64	13	23	Miller and Gaud (1989)
Baja California (San Pedro Mártir)	25	15	50[a]	Martinez Gallardo and Galindo Manriquez (2002)
Baja California Sur (Sierra El Mechudo)	0	27	73	Guerrero-Cárdenas (2016)
Sonora (Sierra del Viejo)	5	33	44	Tarango et al. (2002a)
Sonora (Tiburón Island)	5	5	90	O'Farril (2003)

[a]Includes 12% cacti.

Figure 20.6. Ram exhibiting mating behavior in a reestablished population in the Sierra Maderas del Carmen, Coahuila, Mexico. Photo by Santiago Gibert Isern.

high ambient temperatures, reduced forage moisture content, and mating activities during late summer require additional water intake, resulting in dependence on surface sources of water. Desert bighorn require a minimum of 4% to 5% of their body weight in water per day during summer, and 1% to 2% of their body weight in water during winter, although the amount consumed by different individuals varies. During periods of high heat stress, bighorn restrict their activities and seek shade in caves and under overhangs and trees to avoid heat gain and thus conserve water (Turner and Weaver 1980, Cain et al. 2008). Temperature and its effect on water metabolism, and precipitation, which affects vegetative succulence and the availability of ephemeral water sources, are the two factors determining frequency of water use by desert bighorn sheep. Studies have substantiated the affinity of desert bighorn for water. In the Sierra El Mechudo in Baja California Sur, 93% of bighorn observed (n = 345) were within 500 m (1,640 ft) of permanent or ephemeral water sources (Alvarez-Cardenas et al. 2005).

Social Behavior

Herd composition

The largest social groups are nursery bands composed of ewes, lambs, and yearlings and often consist of a family group. Nursery bands are largest in spring after lambing, and smallest during fall and winter. Young ewes gradually dissociate themselves from their mothers. Young rams usually leave the nursery band at two years of age to join ram bands of mixed-age sheep. Bands of mixed-age sheep are important to bighorn sheep society because younger sheep acquire information on migration routes and home range from older sheep (Geist 1971, Krausman et al. 1999).

During the nonmating season, adult male herds are sexually segregated from female groups consisting of adults, yearlings, and lambs. In Rancho El Plomito in Sonora, males and females were generally segregated from October to June and formed common herds from July to September. Medium-sized (four to six) and large groups (more than six) consisted mostly of segregated females during winter and spring and were mixed during summer. Females formed larger groups, were closer to escape terrain, and occupied more rugged areas than males. Females minimized predation by using rugged habitat features to optimize conditions for rearing offspring (Tarango et al. 2002a, Schroeder et al. 2010).

Population Ecology

The amount of available resources such as food and water, and the effects of predation and other variables such as climate and disease, are factors that regulate bighorn sheep population growth within a given habitat. Quantity and quality of food are the main factors that limit population size and density. Food availability therefore frequently sets an upper limit on population growth. Refer to Sandoval et al. (2011) for the density and population structure (ratios of males to females, lambs to ewes, and yearlings to ewes) of wild sheep in Mexico.

Mortality factors

Most bighorn sheep populations have high lamb mortality and low postyearling mortality, followed by increasingly higher mortality at older ages. Once a bighorn reaches age 2, mortality is negligible until age 7. Mortality rates increase for both sexes after 7 years. Life expectancy after 10 years is less than 2 years. However, a few individuals may live to be 15 years of age (Krausman et al. 1999). One ewe in the San Andres Mountains in New Mexico was known to have produced a lamb at 16 years of age.

Diseases

Desert bighorn sheep are more sensitive to diseases and parasites than other big game species and are vulnerable to several diseases transmitted by domestic animals. Die-offs have occurred following contact with domestic sheep. The most serious disease vectors to bighorn sheep are domestic sheep and goats. These exotics have similar feeding habits, forage preferences, and affinity for rough terrain, and they harbor parasites and other disease agents detrimental to bighorn. Pneumonia transmission from domestic sheep to bighorn sheep usually results in high mortality, including all-age die-offs (Sandoval 1988, Besser et al. 2013). Mortalities from disease have not been documented in wild or captive sheep populations in Mexico, but this is probably because of inadequate disease monitoring.

The major mortality agents are pneumonia-causing pathogens transmitted by domestic livestock. Several bacterial species in the family Pasteurellaceae have been implicated. Bighorn sheep that survive an epizootic become carriers and transmit disease to other herds, especially by dispersing males. Lambs are especially susceptible. Disease transmission in lambs most likely occurs through contact with ewes, other lambs, and juveniles in the herd. In herds with animals that have been exposed to pneumonia, lamb recruitment is greatly reduced, sometimes for a decade (Sandoval 1988, Bunch et al. 1999, Raghavan et al. 2016). Other infectious agents in disease outbreaks in desert bighorn sheep include bluetongue, contagious ecthyma, chronic sinusitis, and psoroptic scabies; under certain circumstances these diseases can result in major population losses. Bluetongue is a viral hemorrhagic disease transmitted by biting gnats (*Culicoides* spp.) that have fed on infected hosts. Potential hosts include sympatric deer, cattle, or both. Contagious ecthyma is a parapoxvirus capable of causing significant lamb mortality during severe outbreaks. Chronic sinusitis is a bacterial infection that causes deterioration of bone in sinuses and horn cores and may be fatal (Fox et al. 2015). Chronic sinusitis is initiated by the decomposition of nasal bot fly (*Oestrus ovis*) larvae that hatch and become trapped in the sinuses of bighorn sheep. Psoroptic scabies, caused by the mite *Psoroptes ovis*, is a contagious skin disease (Bunch et al. 1999).

Predators

Although coyotes (*Canis latrans*), bobcats (*Lynx rufus*), and golden eagles (*Aquila chrysaetos*) may be seasonally significant predators in some areas, especially for lambs, they are not considered a serious threat to most bighorn sheep populations. The mountain lion (*Puma concolor*) may be the major predator of desert bighorn sheep and can be a serious limiting factor to bighorn sheep herd establishment and expansion. Lion depredation is currently the most significant mortality factor in numerous bighorn populations, and the biggest obstacle in the reintroduction of the species on historical range (Sandoval 1983, Wehausen 1996, Ross et al. 1997, Hayes et al. 2000, Rominger and Weisenberger 2000, Kamler et al. 2002).

Between 1992 and 2003, approximately 85% of all known nonhunting-related mortalities of radio-collared individuals in wild desert bighorn herds in New Mexico were caused by mountain lion predation (Rominger et al. 2004). Predation by mountain lions was the primary proximate cause of the biological extinction of the remnant San Andres population, following a significant population reduction in the late 1970s from a psoroptic mite epizootic (Rominger and Weisenberger 2000). Mountain lion predation has been identified as the primary cause of mortality among adult desert bighorns in the peninsular ranges of southern California. Between 1992 and 2003, 64 of 231 (28%) radio-collared bighorn sheep were killed by mountain lions (Rubin et al. 2005). Mountain lion predation can be a significant mortality factor in captive, recently reintroduced, or small populations of bighorn sheep. Under these circumstances, lion control is a valuable wildlife management tool.

Population Management

Hunting has been the most important management concern in Mexico (Tarango and Krausman 1997). Beginning in 1964, 100 permits per year were issued at the national level (Perez-Gil 1996). Between 1977 and 1982, the annual average was 114 permits. Between 1989 and 1999, 625 bighorn hunting permits were issued for Baja California, with a hunter success rate of 75% and a harvest of 469 bighorns, or approximately 47 per year (Lee and Mellink 1996). Concern regarding excess harvest, illegal take, and reported decline of desert bighorn in Baja California resulted in a decree by the Mexican president in 1990 that banned all hunting until accurate population estimates could be obtained and evaluated (DeForge et al. 1993). This ban currently remains in place in Baja California. In Baja California Sur, 111 sheep were harvested between 1978 and 1993. The number of permits gradually declined from 24 to 0 per year by the mid-1990s (Perez-Gil 1996). Baja California Sur reopened hunting in 1996, when four hunting permits were auctioned. In 1998–1999, there were five permits issued in three areas. By 2009–2010, 34 permits were issued for eight areas. Hunter success for 2007–2011 averaged 91% (Lee and Segundo-Galán 2011). During 2012, 72 permits were issued for Sonora (Valdez Aguayo 2013). A total of 110 permits were made available the same year for Baja California Sur, Sonora, Chihuahua, and Coahuila (Lee 2011).

Protection

The federal government is attempting to discourage the illegal taking of bighorn sheep by implementing stiff penalties and increasing surveillance in bighorn sheep areas. Limited surveillance has been conducted through a coordinated system of air and ground reconnaissance. Additional game wardens have been employed and equipped with modern equipment, such as four-wheel-drive vehicles, two-way radios, and spotting scopes. Game guards on horseback patrol areas inaccessible to vehicles (Lee 2008). There are three federally protected areas on the Baja California Peninsula inhabited by wild sheep and one in Sonora; these offer protection by restricting conflicting land-use practices and limiting illegal access. Most game guards are hired by ejidos and private landowners; federal and state governments lack funding to provide law enforcement (Valdez et al. 2006).

Captive breeding and restocking

Mexico has the most successful wild sheep captive-breeding program in North America. The UMA management system allows an individual landowner to construct and maintain an enclosed facility for the captive propagation of wildlife (see chapters 1 and 21). By 2012, 40 UMAs had been established specifically for desert bighorn sheep in Sonora; there are now 3,600 desert bighorn in captivity (Valdez Aguayo 2013). In Coahuila and Chihuahua combined, there are an estimated 575 animals in captivity.

Transplants have been used as the primary management tool for restoring bighorn sheep populations in the western US, Canada, and Mexico. Throughout North America, successful bighorn transplants have averaged 41 individuals (Singer et al. 2000). The first successful insular desert bighorn transplant on Tiberón Island in 1975 was highly successful. Likewise, sheep from Carmen Island, Baja California Sur, which increased rapidly following introductions in the 1990s, have been used to augment the population on the peninsula from which they originated; in 2005, 10 sheep (7 ewes and 3 rams) from the island were released on the mainland (Jiménez and Hernández 2010).

Habitat Requirements

Habitat components essential to bighorn sheep include food, water, escape terrain, and living space (i.e., unfragmented habitat and little or no stress or competition from other ungulates). Bighorn, especially ewe-lamb groups, rely on open, steep terrain for predator avoidance, since their predator evasion strategy involves foraging diurnally in large dispersed groups on open habitat near escape terrain. Habitat is considered sufficiently open when canopy cover is no more than 25–30%; it is characterized by broken, rugged terrain, including canyons, cliffs, rock outcrops, and talus slopes (Sandoval 1979). Slopes of 60% or more are considered steep enough to be escape terrain (Hansen 1980, Álvarez-Cárdenas et al. 2001). The value of escape terrain in providing protection from predators is positively related to the size of escape terrain patches, and inversely related to the distance between them (Berger and Wehausen 1991).

Desert bighorn are uniquely adapted to inhabit arid, remote, rugged terrain in southwestern North America. Throughout this vast region, they are sparsely scattered in isolated areas with deep canyons, rock outcrops, and numerous cliffs that provide a high degree of visibility (fig. 20.7). Use of escape terrain varies with group size, group composition, and season. Therefore, the importance of slope steepness evidently depends on a variety of factors, including availability, season, time of day, sex or age groups, and activities in which bighorn are engaged. For example, bighorn may be more vulnerable to predation when visiting watering sites. In the Sierra Isabel in Baja California, watering sites were isolated and close to escape terrain, and they had good visibility and limited tall vegetation (Escobar-Flores et al. 2016).

Habitat Management

The key to management of desert bighorn is habitat protection, maintenance, and enhancement. Desert bighorn habitats are in ecosystems with low precipitation, temperature extremes, and highly diverse landscapes. These ecosystems are easily, frequently, and permanently damaged by human encroachments. The few habitat management initiatives in Mexico have focused on water development and habitat acquisition projects.

The lack of water is a major limiting factor for any species living in the desert. Desert bighorn can satisfy most of their physiological water requirements through metabolism from vegetative succulence. Their habitat is exceedingly xeric and geologically and topographically unsuitable for natural water-storage sites. Evaluations should be made to determine the necessity of developing natural and artificial watering sites (fig. 20.8) (Cain et al. 2007).

Co-use of habitats by bighorn sheep and exotics such as burros, domestic goats and sheep, aoudad or Barbary sheep (*Ammotragus lervia*), and cattle has been attributed to declines, die-offs, and extinctions of bighorn populations (Rudolph et al. 2003). Effects

Figure 20.7. Desert bighorn occur in widely scattered, isolated areas with deep canyons, prominent rocky outcrops, and cliffs that afford high visibility. This ram is from a reestablished population in the Sierra Maderas del Carmen, Coahuila, Mexico. Photo by Santiago Gibert Isern.

Figure 20.8. Permanent natural watering site favored by desert bighorn sheep in the Santa Isabel Mountains, Baja California. Note the excellent adjacent escape terrain. Photo by Jonathan Escobar Flores.

of exotics occupying bighorn habitat include potential competition for food and water, social competition (space), disease transmission, and displacement of bighorn sheep. Negative interactions between desert bighorn and exotics continue to be a major concern throughout the occupied range of desert bighorn sheep in Mexico. DeForge et al. (1993) and Lee (2003b) documented large numbers of feral burros and goats in most of the mountain ranges surveyed in Baja California and Baja California Sur, respectively. A negative correlation was evident between bighorn sheep numbers observed and the occurrence of burros, domestic goats, and cattle.

Management Challenges

Most challenges with wild sheep management are directly attributable to human-related activities. Desert bighorn sheep may be negatively affected by human-related disturbances. Reaction to human disturbance can be affected by the type and frequency of disturbance, season of occurrence, amount of habitat affected, position of the disturbance in relation to the sheep, and degree of habituation. Dispersal of exotics, livestock grazing, oil and gas exploration and leasing, mining, off-road vehicle activity, and commercial exploitation of native plants such as candelilla (*Euphorbia antisyphilitica*) are potential land-use threats. Noise and vehicle traffic can cause abandonment of habitat. Bighorn may temporarily abandon habitat while it is being mined, which could be critical if mining occurs near lambing habitat or water sources. Roads for mining activities may provide access to previously undisturbed areas and increase the negative impacts (poaching, displacement, harassment) of people in bighorn habitat (McQuivey 1978, Valdez and Krausman 1999). Motorized off-road vehicle races continue to grow in popularity, especially in Baja California, and numerous motorized off-road vehicles have been observed in the basin between bighorn habitat in the Picacho del Diablo and the foothills of the Sierra San Felipe (DeForge et al. 1993).

Conflicts with livestock

Overgrazing, and perhaps the mere presence of cattle in desert bighorn ranges, continues to be a major reason for continuing declines of some populations of desert bighorn, and for the failure of others to increase. Presence of cattle has been negatively correlated with success of bighorn sheep transplants (Singer et al. 2000). Abundant evidence implicates domestic sheep as one of the major causes of bighorn population declines and extinctions. These domestics prefer the same habitat as bighorn, compete for forage, and carry diseases that are lethal to bighorn sheep (Krausman et al. 1996).

Illegal hunting

Illegal hunting has been recognized as one of the major factors responsible for the extirpation of desert bighorn sheep in northeastern Mexico (Espinosa-T. et al. 2006). The lack of an effective and aggressive plan for protection and law enforcement allows illegal hunting to be a continuing problem in many areas in Mexico (Sandoval 1985, Valdez et al. 2006). On Tiburón Island, the percentage of mature rams (class III and IV) declined precipitously between 1993 and 2003. Given the very restricted legal harvest (four rams per year), rams were apparently being removed from the population by other means, perhaps poaching (Lee 2003a).

Human-made barriers

Barbed and net-wire fences are detrimental to desert bighorn sheep because they restrict movements and can be a direct cause of mortality. Bighorn will crawl through or jump over fences, but they are poorly adapted for jumping because of their stocky build, and they may die if they become entangled. Four-lane divided highways may inhibit movements and cause mortality from collisions between bighorn sheep and vehicles. Housing developments also disrupt travel corridors and fragment habitat (Krausman et al. 1999). Of special concern is maintaining connectivity of wild sheep populations along the US-Mexico border. Studies of desert bighorn sheep population genetics in the Peninsular Ranges in southern California and northern Baja California revealed substantial gene flow between populations. International efforts should be directed at minimizing threats to functional connectivity associated with highway expansion, renewable energy development, and completion of a border fence (Buchalski et al. 2016.)

Exotic ungulates

The widespread occurrence of free-ranging exotic ungulates, including aoudad in northeastern Mexico, feral burros in Baja California and Sonora, and domestic sheep and goats in bighorn habitat, could pose a serious threat to desert bighorn sheep. Aoudad currently occupy potential bighorn habitat in Coahuila (Espinosa-T. et al. 2006). Their habitat requirements are like those of desert bighorn. The aoudad would likely dominate in competitive interactions with desert bighorn. Their evolutionary history and demonstrated rates of colonization indicate that these exotics are well adapted to desert environments. They have a high reproductive potential, with an annual rate of increase of approximately 75%, and they may be able to survive on a lower quantity and quality of forage. There are two known wild populations of aoudad within historical bighorn sheep habitat in Mexico. One population occurs in the Sierra Morena in Nuevo León, and the other in the Sierra Pájaros Azules, which borders the states of Nuevo León and Coahuila. Numbers and distribution of these two free-ranging herds are unknown. Both herds have probably increased and dispersed over a relatively large area of suitable habitat. Espinosa-T. et al. (2006) found evidence of free-ranging aoudad in at least three mountain ranges in Coahuila historically occupied by bighorn. The existence and unchecked dispersal of aoudad onto historical desert bighorn sheep habitat in northeastern Mexico could hinder the restoration of desert bighorn (Valdez 2011b).

In Baja California, numerous domestic and feral livestock have been reported in bighorn habitats. Burros were the most common feral animal seen during helicopter surveys; more than 200 have been observed in occupied bighorn habitat in Mexico. The largest number of burros was in the Sierra Juarez (where only four bighorn were observed) and Sierra Las Tinajas. Cattle were common in the Sierra San Pedro Mártir, Arroyo Grande, and Sierra Santa Isabel. Feral goats were observed with bighorn in the Sierra Cucapa, where only two bighorn were observed (DeForge et al. 1993). In some historical habitats in Baja California Sur, bighorn were not observed, but hundreds of goats, burros, and cows were common (Lee 2003b).

Mountain lions

Mountain lions are the principal predator of desert bighorn sheep in some areas. Approximately 85% of all known nonhunting-related mortalities in free-ranging radio-collared bighorn herds in New Mexico are caused by mountain lion predation. Predation by mountain lions was the primary proximate cause of the biological extinction of the remnant San Andres population (Rominger and Weisenberger 2000).

Predation by mountain lions is considered the principal limiting factor in all desert bighorn populations in New Mexico and Coahuila, whether they are born in the wild or raised in captivity. Predator control is a valuable wildlife management tool to protect captive, recently transplanted, or small populations. Because of their land tenure system, mountain lion population reductions would probably have to exceed 50% of the adult lions over a sustained period and encompass a large geographic area to have a positive effect on the prey population (Logan et al. 1996).

Lion control is advocated when bighorn populations are below carrying capacity, when control occurs at a focused level, and when the control effort is large enough to yield positive results. A mountain lion control program should be designed with an adaptive management approach. As bighorn sheep numbers increase and the percentage of bighorn killed by lions decreases, the number of lions removed to protect bighorn should decrease. With this approach, mountain lions may be intensively controlled at first to allow the bighorn population to establish itself to the point that it is able to sustain losses to predation. Removal becomes progressively less intensive if lion predation is not limiting bighorn population growth.

Research and Management Challenges

Basic and applied research on wild sheep is in its infancy in Mexico. State and federal wildlife agencies have not been involved in management programs except for population surveys in specific areas, and most research has been conducted by graduate students from Mexican and American universities. The successful captive-breeding programs have been funded by private landowners. Studies are lacking on movement ecology, population connectivity, habitat use, reproduction, mortality factors, and other aspects of the life histories of wild populations. Although disease outbreaks have not been documented, this is a major potential mortality factor that has not been addressed in Mexico, especially given that many population die-offs in the US have been caused by disease outbreaks. Because most wild sheep and captive herds occur on private lands and the returns from sport-hunting programs are lucrative, the private sector should assume the major responsibility for wild sheep research. Hunter conservation organizations such as the Wild Sheep Foundation and other NGOs should prioritize wild sheep research in Mexico.

Habitat management in Mexico has been neglected. Irresponsible livestock grazing has created widespread ecosystem degradation in wild sheep habitats. Also, a major effort should be initiated to increase the number of protected areas, including private and communal preserves. Community-based wildlife management programs based on sport hunting have been successful in promoting wild sheep conservation; the creation of more of these programs should be encouraged, especially in poor rural communities. Another major effort should be made to contain the spread of Barbary sheep in Coahuila, which have the potential to spread throughout northern Mexico (Valdez 2011b).

Conclusions

Public interest and legal mandates require that state and federal agencies conserve and manage desert bighorn sheep populations for their intrinsic and economic value. Desert bighorn sheep in Mexico are an important component of fragile desert ecosystems and should be sustainably managed for consumptive and nonconsumptive use. It is essential that sufficient habitat and intermountain travel corridors be provided and that exotic and domestic ungulates, human disturbance, and illegal hunting be regulated. All Mexican citizens have an obligation to ensure that Mexico's ecosystems retain their natural processes and biodiversity. Desert bighorn sheep are an integral part of the structure and heritage of desert ecosystems. They are one of the most spectacular mammals in Mexico and are a symbol of wilderness landscapes associated with rugged, montane habitats. The aesthetic value alone of desert bighorn sheep should be reason enough to ensure their survival.

Literature Cited

Álvarez-Cárdenas, S., I. Guerrero-Cárdenas, S. Díaz, P. Galina-Tessaro, and S. Gallina. 2001. The variables of physical habitat selection by the desert bighorn sheep (*Ovis canadensis weemsi*) in the Sierra del Mechudo, Baja California Sur, Mexico. Journal of Arid Environments 49:357–374.

Alvarez-Cardenas, S., R. Lee, P. Galina-Tessaro, and A. Castellanos. 2005. Desert bighorn sheep distribution, abundance, and conservation status in Sierra El Mechudo, Baja California Sur, Mexico. Abstract. Desert Bighorn Council Transactions 48:72.

Berger, J., and J. D. Wehausen. 1991. Consequences of a mammalian predator-prey disequilibrium in the Great Basin Desert. Conservation Biology 5:244–248.

Besser, T. E., E. F. Cassirer, M. A. Highland, P. Wolff, A. Justice-Allen, K. Mansfield, M. A. Davis, and W. Foreyt. 2013. Bighorn sheep pneumonia: sorting out the cause of a polymicrobial disease. Preventive Veterinary Medicine 108:85–93.

Brewer, C. 2001. Diets and seasonal forage utilization of desert bighorn sheep at Elephant Mountain Wildlife Management Area. Final Report. Project No. WBB04. Texas Parks and Wildlife Department, Austin, USA.

Buchalski, M. R., B. N. Sacks, D. A. Gille, M. C. T. Penedo, H. B. Ernest, S. A. Morrison, and W. M. Boyce. 2016. Phylogeographic and population genetic structure of bighorn sheep (*Ovis canadensis*) in North American deserts. Journal of Mammalogy 97:823–838.

Bunch, T. D., W. M. Boyce, C. P. Hibler, W. R. Lance, T. R. Spraker, and E. S. Williams. 1999. Diseases of North American wild sheep. Pages 209–237 *in* R. Valdez and P. R. Krausman, editors. Mountain sheep of North America. University of Arizona Press, Tucson, USA.

Cain, J. W., B. D. Jansen, R. R. Wilson, and P. R. Krausman. 2008. Potential thermoregulatory advantages of shade use by desert bighorn sheep. Journal of Arid Environments 72:1518–1525.

Cain, J. W., P. R. Krausman, J. R. Morgart, B. D. Jansen, and M. P. Pepper. 2007. Response of desert bighorn sheep to removal of water. Wildlife Monographs 171:1–30.

Dávila, C. 1960. Borregos y berrendos en México. Desert Bighorn Council Transactions 4:101–106.

DeForge, J. R., S. D. Ostermann, D. Toweill, P. Cyrog, and E. M. Barrett. 1993. Helicopter surveys of peninsular bighorn sheep in northern Baja California. Desert Bighorn Council Transactions 37:24–28.

Escobar-Flores, J. G., S. Álvarez-Cárdenas, R. Valdez, et al. 2015. Detección de las preferencias de hábitat del borrego cimarrón (*Ovis canadensis cremnobates*) en Baja California, mediante técnicas de teledetección satelital. Therya 6:519–534.

Escobar-Flores, J. G., R. Valdez, S. Álvarez-Cárdenas, S. Díaz-Castro, A. Castellanos Vera, J. Torres, and M. Delgado-Fernández. 2016. Utilización de aguajes por el borrego cimarrón (*Ovis canadensis cremnobates*) y análisis de calidad del agua en Sierra Santa Isabel, Baja California, México. Acta Universitaria 26:12–19.

Espinosa-T., A., A. V. Sandoval, and A. J. Contreras-B. 2006. Historical distribution of desert bighorn sheep (*Ovis canadensis mexicana*) in Coahuila, Mexico. Southwestern Naturalist 51:282–288.

Espinosa-T., A., A. V. Sandoval, and R. Valdez. 2002. Progress report: CEMEX desert bighorn sheep reestablishment program. Desert Bighorn Council Transactions 46:4–5.

Fox, K. A., N. M. Rouse, K. P. Huyvaert, K. A. Griffin, H. J. Killion, J. Jennings-Gaines, W. H. Edwards, S. L. Quackenbush, and M. W. Miller. 2015. Bighorn sheep (*Ovis canadensis*) sinus tumors are associated with coinfections by potentially pathogenic bacteria in the upper respiratory tract. Journal of Wildlife Diseases 51:19–27.

Geist, V. 1971. Mountain sheep: a study in behavior and evolution. University of Chicago Press, Illinois, USA.

Ginnett, T. F., and C. L. Douglas. 1982. Food habitats of feral burros and desert bighorn sheep in Death Valley National Monument. Desert Bighorn Council Transactions 24:81–87.

Gross, J. E., F. J. Singer, and M. E. Moses. 2000. Effects of disease, dispersal, and area on bighorn sheep restoration. Restoration Ecology 45:25–37.

Guerrero-Cárdenas, I., S. Gallina, P. Corcuera, S. Álvarez-Cárdenas, and R. Ramirez-Orduño. 2016. Diet composition and selection of the bighorn sheep (*Ovis canadensis*) in Sierra El Mechudo, Baja California Sur, Mexico. Therya 7:423–437.

Gutierrez-Espeleta, G. A., S. T. Kalinowski, W. M. Boyce, and P. W. Hedrick. 1998. Genetic variation in desert bighorn sheep. Desert Bighorn Council Transactions 42:1–10.

Hall, E. R. 1981. The mammals of North America. Second edition. Volume 2. John Wiley and Sons, New York, New York, USA.

Hansen, C. G. 1980. Habitat. Pages 64–79 *in* G. Monson and L. Sumner, editors. The desert bighorn: its life history, ecology, and management. University of Arizona Press, Tucson, USA.

Hayes, C. L., E. S. Rubin, M. C. Jorgensen, R. A. Botta, and W. M. Boyce. 2000. Mountain lion predation of bighorn sheep in the Peninsular Ranges, California. Journal of Wildlife Management 64:954–959.

Heffelfinger, J. R., and E. Marquez-Muñoz. 2005. Historical occurrence and distribution of desert bighorn sheep in Chihuahua, Mexico. Desert Bighorn Council Transactions 48:28–38.

Jiménez, S. D., and M. C. Hernández. 2010. Programa de conservación del borrego cimarrón (*Ovis canadensis weemsi*) en Baja California Sur, México. Galemys 22:447–468.

Jimenez-L., S., C. Hernandez-C., J. R. DeForge, and R. Valdez. 1996. Desert bighorn sheep (*Ovis canadensis weemsi*) recovery project in Baja California Sur, Mexico. Desert Bighorn Council Transactions 40:8–12.

Kamler, J. F., R. M. Lee, J. C. deVos, and W. B. Ballard. 2002. Survival and cougar predation of translocated bighorn sheep in Arizona. Journal of Wildlife Management 66:1267–1272.

Krausman, P. R., and R. T. Bowyer. 2003. Mountain sheep: *Ovis canadensis* and *Ovis dalli*. Pages 1095–1115 *in* G. A. Feldhamer, B. C. Thompson, and J. A. Chapman, editors. Wild mammals of North America: biology, management, and conservation. Second edition. Johns Hopkins University Press, Baltimore, Maryland, USA.

Krausman, P. R., A. V. Sandoval, and R. C. Etchberger. 1999. Natural history of desert bighorn sheep. Pages 139–208 *in* R. Valdez and P. R. Krausman, editors. Mountain sheep of North America. University of Arizona Press, Tucson, USA.

Krausman, P. R., R. Valdez, and J. A. Bissonette. 1996. Bighorn sheep and livestock. Pages 237–243 *in* P. R. Krausman, editor. Rangeland wildlife. Society for Range Management, Denver, Colorado, USA.

Lee, R. M. 2003a. A 10-year review of wild sheep management in Sonora, Mexico. Desert Bighorn Council Transactions 47:47–51.

Lee, R. M. 2003b. A review of recent wild sheep surveys in Baja California Sur, Mexico. Desert Bighorn Council Transactions 47:40–46.

Lee, R. M. 2008. Hunting as a tool for wildlife conservation: the case of sheep hunting in Mexico. Pages 53–58 *in* R. D. Baldus, G. R. Damm, and K. Wollscheid, editors. Best practices in sustainable hunting. CIC Technical Series Publication Number 1. International Council for Game and Wildlife Conservation, Budakeszi, Hungary.

Lee, R. M. 2011. Economic aspects of and the

market for desert bighorn sheep. Desert Bighorn Council Transactions 51:46–49.

Lee, R. M., G. R. Martinez, J. Zatarain, and J. G. Escobar-Flores. 2012. Observations of distribution and abundance of bighorn sheep in Baja California, Mexico. California Fish and Game 98:51–59.

Lee, R. M., and E. Mellink. 1996. Status of bighorn sheep in Mexico. Desert Bighorn Council Transactions 40:35–39.

Lee, R. M., and J. M. Segundo-Galán. 2011. Status report on desert bighorn in various states in Mexico. Desert Bighorn Council Transactions 51:80–84.

Logan, K. A., L. L. Sweanor, T. K. Ruth, and M. G. Hornocker. 1996. Cougars of the San Andres Mountains, New Mexico. Final Report. Project W-128-R. New Mexico Department of Game and Fish, Santa Fe, USA.

Martinez-G., R. 2011. Muestreo terrestre y aéreo de la población de borrego cimarrón en Baja California. Pages 24–35 *in* Estrategia estatal para la conservación y el manejo sustentable del borrego cimarrón en Baja California. Secretaría de Protección Ambiente, Gobierno de Baja California, Ensenada, México.

Martinez Gallardo, R., and E. Galindo Manriquez. 2002. Feeding habits of bighorn sheep (*Ovis canadensis cremnobates*) in San Pedro Martir Sierra, Baja California, Mexico. Desert Bighorn Council Transactions 45:111–125.

McKinney, B. R., and J. Delgadillo-Villalobos. 2005. Desert bighorn reintroduction in Maderas del Carmen, Coahuila, Mexico. Desert Bighorn Council Transactions 48:46–49.

McQuivey, R. P. 1978. The desert bighorn sheep of Nevada. Biological Bulletin No. 6. Nevada Department of Fish and Game, Las Vegas, USA.

Mearns, E. A. 1907. Mammals of the Mexican boundary. Bulletin 56. US National Museum, Washington, DC, USA.

Mellink, E. 1993. The president spoke. Pages 201–220 *in* G. P. Nabhan, editor. Counting sheep: twenty ways of seeing desert bighorn. University of Arizona Press, Tucson, USA.

Miller, G. D., and W. S. Gaud. 1989. Composition and variability of desert bighorn diets. Journal of Wildlife Management 53:597–606.

Montoya, B., and G. Gates. 1975. Bighorn capture and transplant in Mexico. Desert Bighorn Council Transactions 19:28–32.

Murray, W. B., and A. Espinosa-T. 2006. The natural setting of bighorn petroglyphs in the eastern Sierra Madre (Nuevo Leon, Coahuila, Mexico). American Indian Rock Art 32:45–51.

O'Farril, G. 2003. Dieta y uso de hábitat de borrego cimarrón (*Ovis canadensis mexicana*) y el venado bura (*Odocoileus hemionus sheldoni*) en Isla Tiburón, Sonora, México. Thesis, UNAM, México, D. F., México.

Perez-Gil, R. 1996. Importancia economica de los vertebrados silvestres de México. PG-7 Consultores and CONABIO, México, D. F., México.

Raghavan, B., K. Erickson, A. Kugadas, S. A. Batra, D. R. Call, M. A. Davis, W. J. Foreyt, and S. Srikumaran. 2016. Role of carriers in the transmission of pneumonia in bighorn sheep (*Ovis canadensis*). Biology Open: doi:10.1242/bi0.018234.

Ramey, R. R., II. 1993. Evolutionary genetics and systematics of North American mountain sheep: implications for conservation. Dissertation, Cornell University, Ithaca, New York, USA.

Rodríguez-Rodríguez, M. A., J. Gasca-Pineda, R. A. Medellín, and E. Eguarte. 2015. Analysis of genetic diversity of bighorn sheep (*Ovis canadensis*) from Mexican populations. Journal of Mammalogy 96:473–480.

Rominger, E. M., and M. E. Weisenberger. 2000. Biological extinction and a test of the "conspicuous individual hypothesis" in the San Andres Mountains, New Mexico. North American Wild Sheep Conference 2:293–307.

Rominger, E. M., H. E. Whitlaw, D. L. Weybright, W. C. Dunn, and W. B. Ballard. 2004. The influence of mountain lion predation on bighorn sheep translocations. Journal of Wildlife Management 68:993–999.

Ross, P. I., M. G. Jalkotzy, and M. Festa-Bianchet. 1997. Cougar predation on bighorn sheep in southwestern Alberta during winter. Canadian Journal of Zoology 74:771–775.

Rubin, E. S., W. M. Boyce, and V. C. Bleich. 2000. Reproductive strategies of desert bighorn sheep. Journal of Mammalogy 81:769–786.

Rubin, E. S., C. J. Stermer, C. Chun, S. G. Torres, and W. M. Boyce. 2005. Mountain lion predation on bighorn sheep: an examination of risk factors and seasonal mortality patterns. Abstract. Desert

Bighorn Council Transactions 48:71.
Rudolph, K. M., D. L. Hunter, W. J. Foreyt, E. F. Cassirer, R. B. Rimler, and A. C. S. Ward. 2003. Sharing of *Pasteurella* spp. between free-ranging bighorn sheep and feral goats. Journal of Wildlife Diseases 39:897–903.
Sandoval, A. V. 1979. Preferred habitat of desert bighorn sheep in the San Andres Mountains, New Mexico. Thesis, Colorado State University, Fort Collins, USA.
Sandoval, A. V. 1983. Mountain lions and desert bighorns: an objective assessment of the problem. Report. New Mexico Department of Game and Fish, Santa Fe, USA.
Sandoval, A. V. 1985. Status of bighorn sheep in the Republic of Mexico. Pages 86–94 *in* M. Hoefs, editor. Wild sheep: distribution, abundance, management and conservation of the sheep of the world and closely related mountain ungulates. Special Report. Northern Wild Sheep and Goat Council, Yukon, Canada.
Sandoval, A. V. 1988. Bighorn sheep die-off following association with domestic sheep: case history. Desert Bighorn Council Transactions 32:36–38.
Sandoval, A. V., R. Valdez, and A. Espinosa-T. 2011. El borrego cimarrón en México. Pages 489–518 *in* R. Valdez and J. A. Ortega-S., editors. Ecología y manejo de fauna silvestre en México. Editorial del Colegio de Postgraduados, Montecillo, Texcoco, México.
Schroeder, C. A., R. T. Bowyer, V. C. Bleich, and T. R. Stephenson. 2010. Sexual segregation in Sierra Nevada bighorn sheep, *Ovis canadensis sierrae*: ramifications for conservation. Arctic, Antarctic and Alpine Research 42:476–489.
SEMARNAT. 2010. Norma Oficial Mexicana NOM-059-ECOL-2001. Protección ambiental—Especies nativas de México de flora y fauna silvestres—Categorías de riesgo y especificaciones para su inclusión, exclusión o cambio—Lista de especies en riesgo. DOF, México, D. F., México.
Singer, F. J., C. M. Papouchis, and K. K. Symonds. 2000. Translocation as a tool for restoring populations of bighorn sheep. Restoration Ecology 8:6–13.
Tarango, L. A., and P. R. Krausman. 1997. Desert bighorn sheep in Mexico. Desert Bighorn Council Transactions 41:1–7.
Tarango, L. A., P. R. Krausman, and R. Valdez 2002a. Habitat use by desert bighorn sheep in Sonora, Mexico. Pirineos 157:219–226.
Tarango, L. A., P. R. Krausman, R. Valdez, and R. M. Kattnig. 2002b. Research observation: desert bighorn sheep diets in northwestern Sonora, Mexico. Journal of Range Management 55:530–534.
Turner, J. C., and R. A. Weaver. 1980. Water. Pages 100–112 *in* G. Monson and L. Sumner, editors. The desert bighorn: its life history, ecology, and management. University of Arizona Press, Tucson, USA.
Uranga-Thomas, R., and R. Valdez. 2011. Reintroduction of desert bighorn sheep in Chihuahua, Mexico. Desert Bighorn Council Transactions 51:32–38.
Valdez, R. 2011a. Bighorn sheep. Pages 738–739 *in* D. E. Wilson and A. Mittermeier, editors. Handbook of mammals of the world. Lynx Edicions, Barcelona, Spain.
Valdez, R. 2011b. Aoudad. Pages 714–715 *in* D. E. Wilson and A. Mittermeier, editors. Handbook of mammals of the world. Lynx Edicions, Barcelona, Spain.
Valdez, R., J. C. Guzman-Aranda, F. J. Abarca, L. A. Tarango-Arambula, and F. Clemente-Sanchez. 2006. Wildlife conservation and management in Mexico. Wildlife Society Bulletin 34:1480–1488.
Valdez, R., and P. R. Krausman. 1999. Description, distribution and abundance of mountain sheep in North America. Pages 1–22 *in* R. Valdez and P. R. Krausman, editors. Mountain sheep of North America. University of Arizona Press, Tucson, USA.
Valdez Aguayo, H. R. 2013. Borrego cimarron (*Ovis canadensis mexicana*): resultados del monitoreo aereo en el estado de Sonora, México, noviembre, 2012. Second edition. Dirección General Forestal and Fauna de Interés Cinegético de la SAGARHPA, Hermosillo, Sonora, México.
Wehausen, J. D. 1996. Effects of mountain lion predation on bighorn sheep in the Sierra Nevada and Granite Mountains of California. Wildlife Society Bulletin 24:471–479.
Wehausen, J. D., and R. Ramey II. 1993. A morphometric reevaluation of the peninsular bighorn subspecies. Desert Bighorn Council Transactions 37:1–10.
Wilder, B. T., J. L. Betancourt, C. W. Epps, R. S. Crowhurst, J. I. Mead, and E. Ezcurra. 2014. Local extinction and unintentional rewilding of bighorn sheep (*Ovis canadensis*) on a desert island. PLoS ONE 9:e91358.

21 Evolution of Wildlife Laws and Policy in Mexico

Carlos González-Rebeles Islas, Melisa Meztli Méndez Méndez, and Raul Valdez

Until recently, wildlife law, policy, and administration in Mexico were solely the prerogative of the federal government. Management and regulation of Mexican wildlife resided in the federal government and was not challenged by the states; it was not until 2000 that northern states began to take an active role in managing wildlife. This is in sharp contrast to the US, where states assumed authority over most wildlife as early as the 1700s. In Mexico, federal orders, statutes, regulations, and international treaties approved by the federal government determined national and international wildlife policy. However, the impetus to regulate wildlife did not arise until the late 1800s, when it became obvious that wildlife populations had undergone widespread extirpation because of wanton exploitation. Indeed, it was not until the 1950s that Mexico began to view wildlife as a valued resource and to enact significant regulatory legislation. Prior to this, wildlife had been relegated to a resource of least concern. Moreover, beginning in the late 1980s, Mexico's participation in wildlife treaties and agreements required its international involvement in regulating wildlife. Throughout its postcolonial history, a major concern has been Mexico's inability to enforce its wildlife and environmental laws. Also, Mexico has been unable to establish a robust wildlife program because of the continual shifting of federal agencies responsible for wildlife management, with the concomitant lack of adequate federal funding. In this chapter, we review the evolution of wildlife law, policy, and administration in Mexico since independence in 1821.

During the colonial period (1521–1821), the Spanish did not regulate hunting other than to monopolize the commercialization of animals and plants. Wildlife utilization became the exclusive right of the landed aristocracy, a policy that continued after independence. However, subsistence hunting was widespread (see chapter 1). It was not until 1870, decades after Mexico gained independence from Spain (1821), that the Civil Code (Código Civil) first established hunting regulations that limited hunting on private lands and required landowner authorization. Hunting on publicly accessible lands, including protected areas, was not restricted; wildlife was considered a public resource. The subsequent 1894 federal government decree to regulate the use of forests and rangelands included the first hunting restrictions, which imposed permits and fees, hunting seasons, and fines for noncompliance. The first federal wildlife agency, the Bureau of Hunting and Fishing, created in 1916 in the Ministry of Agriculture and Development (SAF; agency acronyms follow the Spanish names), regulated the hunting of sea lions and waterfowl. This bureau became the Department of Hunting under the Department of Forestry, Hunting, and Fishing (DAFCP) (1918–1930) in the SAF. These agencies began efforts to regulate sport hunting and, later, to protect diminishing species by not allowing them to be hunted. Big game species, such as pronghorn (*Antilocapra americana*) and bighorn sheep (*Ovis canadensis*), were particularly vulnerable to overexploitation and were some of the earliest species to receive protection (Retana 2006).

Article 27 of the 1917 postrevolution constitution officially established that natural resources (including wildlife) were owned by the nation and under federal control (Pérez Gil 1988). This was the first effort to provide a national legal basis for the protection of natural resources. From then until the 1940s, a series of federal wildlife agencies emerged, shifting wildlife conservation responsibility among them. In 1939, the then-autonomous DAFCP, a subdepartment of the SAF (1934–1939), lost influence when it became the Office of Hunting within the Ministry of Agriculture and Livestock (SAG). However, during the same period, legal regulations for the management of game species and sport hunting–related commercial businesses were enacted that relegated wildlife to a secondary forestry resource.

National Natural Protected Areas

The creation of National Natural Protected Areas (ANPs), which developed in parallel with game management, benefited wildlife. Throughout the 1930s and early 1940s, there was an awareness of the misuse of Mexico's natural resources and the need to regulate them. Under the leadership of Miguel Ángel de Quevedo, chief of DAFCP (1935–1939), up to 30% of the country was proposed to be placed in protected areas (e.g., forest reserves, national parks, and reforestation reserves). However, wildlife conservation efforts when ANPs were first being developed were limited, and ANPs did not encompass the country's biodiversity, since much of the protection focused on subhumid temperate ecological zones, especially coniferous forests. Mexico largely copied the US model of national parks and consequently protected the same types of ecosystems.

Another factor was the desirability of establishing ANPs where there were no conflicts with other land uses, such as farming and ranching. Also, nature conservation was not a government priority, and the management of ANPs was neglected. In some areas, conflicts arose because expropriated or affected villagers were not properly compensated, and many human settlements remained within ANPs, with no information about or link to conservation programs. In other cases, the lack of monitoring, regulatory enforcement, and institutional presence opened the door to land invasions by landless farmers (Vargas 1984). Thus, while the American model of parks was based on a strict protectionist principle that excluded exploitation and human interference, the Mexican model integrated communities and nature conservation, which better suited the socioeconomic reality (Carabias et al. 2008, Bezaury Creel and Gutiérrez 2009, Wakild 2012).

Hunting Regulations

The fundamental legal basis for hunting regulations in Mexico originated with the Hunting Act of 1940, soon replaced by the Federal Hunting Act (Ley Federal de Caza) of 1951. The latter regulated and established the principles for wildlife management policy until it was revoked in 2000. However, its concept of conservation remained restricted to the regulation of sport hunting, and it did not provide a legal basis for the rational use of nongame species, other than to forbid the commercialization of wildlife (Retana 2006, Anta and Carabias 2008). During the same year, the former Hunting Office became the Hunting Department within the Forest and Hunting Directorate (DFC) of the Undersecretariat of Forest and Hunting Resources (SRFC) of the SAG. These agencies promoted the development of game-breeding farms for wildlife threatened by overexploitation. In addition, programs were also developed to control wild predators that were negatively affecting livestock. By the mid-1950s, these programs had led to the near extinction of the Mexican gray wolf (*Canis lupus baileyi*) and the local extirpation of other predators (Retana 2006).

The 1960s and 1970s constituted a watershed in the history of wildlife management in Mexico. Thanks to the influence of international debates about global environmental issues, government agencies were given strong support, and a new ideology guided Mexican wildlife conservation policy. By 1960, public discourse in Mexico had adopted a comprehensive concept of wildlife (including nongame wildlife), which was viewed as an important natural resource that should be appreciated for its ecological, social, cultural, and economic values. A significant shift was the replacement of the SRFC by the Undersecretariat of Forest and Wildlife Resources (SFFS), and of the DFC by the General Directorate of Wildlife (DGVS) in 1964; the functions of these new agencies involved the management not only of sport hunting, but also of nongame wildlife use and protection.

Changes in ideology resulted in new wildlife reserves, infrastructure for research, wildlife utilization programs, and game ranches. These measures were inspired by new principles for the conservation and rational use of biological resources that spread worldwide. Agreements and resolutions enacted at international conferences (e.g., the United Nations Conference on the Human Environment in 1972) influenced and provided guiding principles for Mexican environmental policy. Furthermore, Mexico established wildlife management partnerships with the US and Canada; these partnerships would result in international cooperation agreements in the 1990s. Hemispheric agreements included the Management Plan for Aquatic Birds of North America in 1994, the Wildlife Without Borders Program, also in 1994, and the Canada-Mexico-US Trilateral Committee in 1996 (Retana 2006).

Development of the Ecosystem Approach in Wildlife Public Policy

However, there were also serious setbacks in conservation policy because of urgent new national economic development priorities, and the pressure and political influence of other government agencies related to energy, industry, and agricultural production, among others. This significantly limited the management authority of the emerging conservation agencies. In particular, agriculture received strong support through subsidies, public and private investment, and development of infrastructural and technical innovations. From 1940 to 1965, agricultural area increased 2.5 times, from 5.9 to 14.7 million ha (14.5 to 36.3 million ac). Support for Mexican agriculture was also influenced by global technological developments during the Green Revolution, including improved genetic varieties, mechanization, innovations in irrigation methods, and the use of fertilizers, pesticides, and herbicides. Many of these agricultural policies were inimical to wildlife conservation (see chapter 1).

This agricultural boom began to wane in the mid-1960s; however, it coincided with the development of livestock husbandry, a practice that synergistically increased environmental damage and lasted until the 1980s. In 1940, there were 38.8 million ha (95.8 million ac) registered for livestock production; by 1983, there were 90.4 million ha (223.3 million ac). The impetus for livestock enterprises originated in external factors, particularly funding from the World Bank and the Inter-American Development Bank. In addition, forestry became a political priority for the first time; the Forestry Law was reformed, and management approaches prioritized commercial interests. These development policies, together with administrative bodies such as the paradoxical National Commission for Deforestation (Comisión Nacional de Desmontes), promoted the unmanaged exploitation and colonization of extensive forested tracts in tropical southeastern Mexico—which, because of their isolation, had persisted virtually unexploited—with disastrous results for wildlife. Likewise, livestock husbandry expanded in northern Mexico at the expense of fragile, arid shrublands unfit for intensive grazing (Anta and Carabias 2008, Carabias et al. 2008).

The 1980s and 1990s were characterized by dynamic changes in political vision, regulatory bases, and conservation approaches for wildlife and their habitats. This internal dynamism coincided with or reflected a series of important international events; particularly relevant was the emergence of sustainable development, known worldwide through the Brundtland Report of 1987. This concept attempted to reconcile economic growth with biological conservation. Furthermore, the Convention on Biological Diversity, the United Nations Framework Convention on Climate Change, and Agenda 21, among other global initiatives that were ratified by Mexico, influenced and guided biodiversity conservation in Mexico. The result was the creation of the National Commission for the Study and Use of Biodiversity (CONABIO). During this period, Mexico also became a signatory of CITES (Retana 2006, Carabias et al. 2008).

Internal political motivations were driven by an increasing interest in conservation among various sectors of the population. This interest was in response to ecosystem degradation, particularly that resulting from excessive agricultural growth between 1940 and 1980. Scientific research on environmental issues and conservation biology advanced because of academic influence on national policy. The themes of protection and wise use of biodiversity were added to the previous focus on pollution. Of major significance was the inclusion of environmental issues in the agendas of political parties, as well as the emergence of several Mexican NGOs and the participation of international organizations with subsidiaries in Mexico such as the Nature Conservancy, Conservation International, and World Wildlife Fund (Carabias et al. 2008).

In 1982, the Federal Environmental Protection Act (LFPA) was enacted, and the Secretariat of Urban Development and Ecology (SEDUE) was established, which marked a watershed in the management of natural resources because it gave equal priority to environmental issues (Retana 2006). One of the most significant achievements of SEDUE was the creation of the National Ecology Program (1984–1988). It also integrated, under a single administrative unit, the Undersecretariat of Ecology (SE), the two most important government agencies for the conservation of biological resources in Mexico: the General Directorate of Wild Flora and Fauna (DGFFS), and the General Directorate of Parks, Reserves, and Protected Ecological Areas (DIPARES) (SEMARNAP 1997), which eventually merged with the General Directorate of Environmental Protection of Natural Resources (DGPERN) from 1985 to 1992.

Sport hunting remained a management priority for DGFFS under the regulations of the Hunting Act and its regulatory instrument, the hunting calendar. However, the agency continued to explore the idea, beginning in the 1960s, of combining other forms of wildlife use with protection plans in a comprehensive conservation strategy for all species of wildlife and their habitats. This resulted in the creation of wildlife management centers dedicated to wildlife rescue and rehabilitation, captive breeding (fig. 21.1), and restoration of endangered species, among other efforts (SEMARNAP 1997).

Under the new SE, DIPARES began an initiative to increase institutional presence and attention in protected areas. Several projects were funded to provide infrastructure, equipment, and personnel to various ANPs; in addition, some ANPs were dissolved and others created under different protection schemes. Bezaury Creel and Gutiérrez (2009) consider the period between 1988 and 1994 significant for the significant increase in land surface incorporated into the protected area system (national parks, wild fauna and flora sanctuaries, ecological reserves, etc.). The Biosphere Reserve Protected Areas program, initiated in 1975, gained momentum, and several reserves were created by presidential decree (e.g., Sian Ka'an in Quintana Roo, 1986; Manantlán in Tamaulipas, 1987; and Vizcaíno in Baja California Sur, 1988) (Carabias et al. 2008, Bezaury Creel and Gutiérrez 2009). Biosphere reserves are important for wildlife because they encompass greater ecoregional variation and increase the number of protected wildlife habitats. These reserves also became a model of comprehensive conservation, combining strict protection and the sustainable use of wildlife (consumptive and nonconsumptive) under a zoning and diversification strategy of land use.

Figure 21.1. Mule deer captive-breeding facility in Sonora, Mexico. Photo by J. Alfonso Ortega-S.

Legal and Institutional Foundations of Current Wildlife Conservation Policy

The General Law of Ecological Balance and Environmental Protection (LGEEPA) of 1988 integrated the regulation and management of natural resources, which had previously been dispersed among sectoral laws (e.g., the Fisheries Act, Forestry Act, and Federal Hunting Act). It included provisions for the possession, propagation, commercialization, and management of wildlife populations, complementing the hunting provisions. The LGEEPA established the political foundation for environmental protection in Mexico (Retana 2006, Anta and Carabias 2008, Carabias et al. 2008).

The concept of wildlife conservation and management at that time led to specialization and compartmentalization, which resulted in isolation from other fields of public policy. Activities related to agriculture, animal husbandry, forestry, and fisheries were dissociated from biodiversity conservation, even though these are integral for conservation planning and development. To this day, these aspects are treated separately from wildlife conservation (CONABIO 2006, Carabias et al. 2008).

Thus, the conservation and sustainable utilization of natural resources remained administratively disconnected. Eventually, the two general directorates (DGFFS and DIPARES) were unsuccessfully merged into one undersecretariat; management issues related to hunting and habitat protection of nongame species were never integrated. This was aggravated by the creation of the Secretariat of Social Development (SEDESOL) in 1992, and the disappearance of SEDUE and SE. This affected the DGFFS, which was subdivided, reduced to an area directorate, and transferred to the Secretariat of Agriculture and Water Resources (SARH) (SEMARNAP 1997, Carabias et al. 2008).

In 1994, after previous efforts to merge natural resource departments and agencies, there was another attempt to integrate fisheries, forestry, and water resources, the protection and use of biodiversity, and the prevention and management of pollution under one administrative body. This is the Secretariat of the Environment, Natural Resources, and Fisheries (SEMARNAP), which was intended to comprehensively coordinate policies and programs related to "promoting social development, economic growth and environmental protection" (Romero 2002, Carabias et al. 2008, Anta and Carabias 2008). However, the impact of SEMARNAP was negligible, and in 2000 its responsibilities were divided by replacing it with the Secretariat of the Environment and Natural Resources (SEMARNAT) (fig. 21.2).

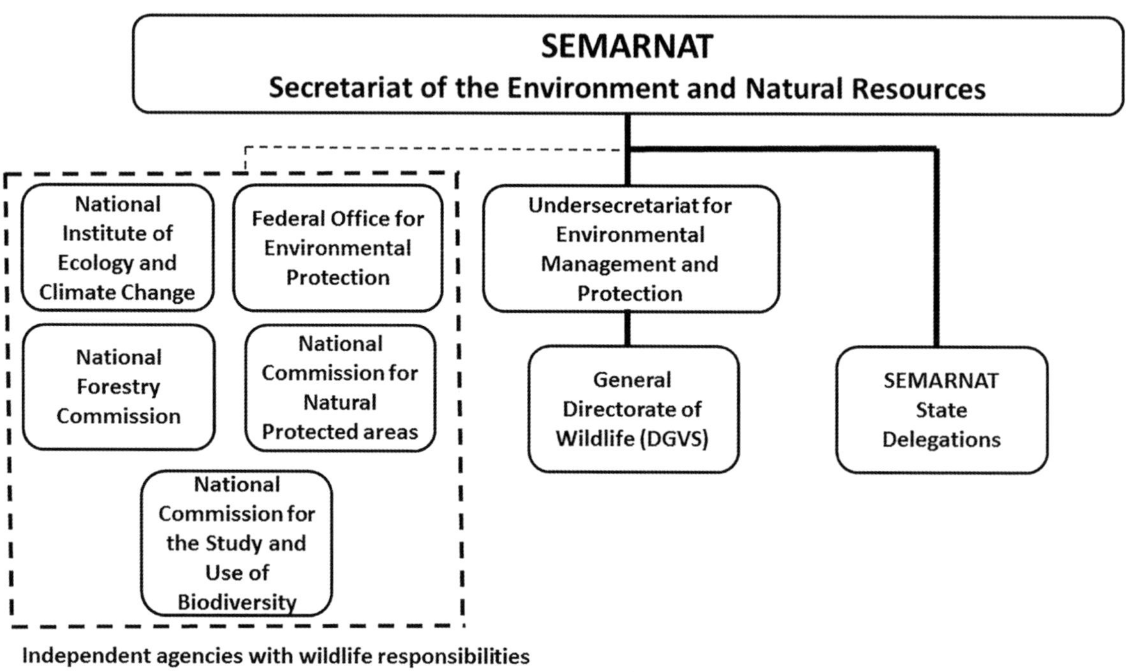

Figure 21.2. Organizational structure of Mexico's General Directorate of Wildlife (DGVS) within SEMARNAT. DGVS is directly responsible for wildlife policy, administration, and law enforcement.

Negative consequences resulted from the haste with which these institutional changes took place during the 1980s and 1990s. These included an increase in the compartmentalization of federal agencies, discontinuities in several programs and an overall decrease in their political clout, and a decrease in the bargaining power of federal agencies in charge of wildlife management, versus other government sectors (Retana 2006).

However, there were also positive aspects, such as the establishment of a regulatory basis for the modern management of natural resources in Mexico, based on a comprehensive understanding, at least in legal terms, of wild flora and fauna. The fisheries subsector was transferred to the Secretariat of Agriculture, Livestock, Rural Development, Fisheries, and Food (SAGARPA) when SEMARNAP was restructured into SEMARNAT in 2000. The public administration of the environment focused on pollution, ecosystems, biodiversity protection, and sustainable use, under the coordination of three undersecretariats. The Undersecretariat of Ecology was responsible for wildlife management policies in coordination with two decentralized government bodies: the National Institute of Ecology (a normative and regulatory body), and the Federal Office for Environmental Protection (PGPA), an agency in charge of monitoring regulatory compliance. Furthermore, the responsibilities of the Undersecretariat of Ecology were coordinated with those of CONABIO, which included the promotion and coordination of biodiversity research. This collaboration established a link between academic research and public environmental policies. The organizational structure of SEMARNAT has undergone changes over the past 10 years; the names of the undersecretariats and their functions have changed, and new general directorates have been created. In addition, some bodies that once had a central role in decision making have been reduced to a consultative status.

The creation of the current General Directorate of Wildlife (DGVS), which originated in the SARH in the mid-1990s and continued to function within SEMARNAT (fig. 21.2), marked a new era in the public management of wildlife in Mexico. This innovative administrative agency finally succeeded in consolidating an ecosystem approach to the management of biological resources. In addition, amendments made in 1996 to the LGEEPA created a legal framework that served as the foundation for this new concept of conservation. Biodiversity in general was established as the focus of regulation, including wild flora and fauna, both terrestrial and aquatic. The concept of environmentally sustainable use was also incorporated as a conservation tool (Anta and Carabias 2008, Retana 2006).

The achievements of the DGVS included the Conservation Program for Wildlife and Productive Diversification in the Rural Sector (PCVSDP) (SEMARNAP 1997), which formed the basis of the General Wildlife Law (Ley General de Vida Silvestre [LGVS]), published in 2000, and its bylaws in 2006. These instruments deserve special mention, as they are the main documents forming the legal, technical, and scientific framework under which, at present, the use and conservation of biological resources (including wildlife) are regulated in Mexico. Their structure is based on recognizing the diverse stakeholders of biological resources, establishing mechanisms that allow each to have access to wildlife resources, to manage those resources, and to legally profit from them. These new groups of recognized stakeholders, in addition to hunters, include peasants, indigenous peoples, animal and plant collectors, scientists, wildlife management technicians, and wildlife commercial dealers, none of were previously recognized by the LGEEPA (Retana 2006, Anta and Carabias 2008).

State Government Participation

Another breakthrough was the involvement of state governments in wildlife conservation. In 2000, several wildlife regulatory functions were transferred to state agencies, while the federal government maintained its authority to dictate a national wildlife policy. This provided the states with authorization to plan and act in response to local wildlife issues, which in turn could be transferred to their municipalities through local legislation and agreements. This was possible because of a consensus to support wildlife conservation policy on scientifically informed decisions; the knowledge and awareness of local institutions; the reality of the exploitation of regional biological resources at the state level; the states' ability to initiate immediate and appropriate local conservation programs; the opportunity for states to establish cooperative agreements with neighboring states; and the need to manage wildlife based on local traditions and issues (Ortiz Monserio Quintana 2011).

Wildlife conservation regulations were transferred to states through special agreements (e.g., DOF 2006a, 2006b, 2006c, 2007a, 2007b, 2009). These agreements were created at the request of states, according to each state's financial capabilities, and with the creation of an appropriate state institutional infrastructure to assume the transfer of functions. At present, only northern border states (Baja California, Chihuahua, Coahuila, Nuevo León, Sonora, and Tamaulipas) have signed wildlife agreements with the federal government (Ortiz Monasterio Quintana 2011). The regulatory functions assumed by states include authorizing, registering, and supervising UMAs and their management; mitigating damage by noxious wildlife; establishing standards for optimal management of captive wildlife; issuing permits for sport, commercial, and subsistence hunting and regulating it; and promoting wildlife cultural traditions. Nonetheless, the federal government maintains authority over international trade regulations; scientific sampling and collecting; registration of scientific collections; wildlife translocation and liberation; and wildlife use and conservation planning, among other national wildlife management priorities.

Under these special agreements, states now have the authority to administer revenues generated from wildlife use, and to fund local and rural development programs. Both federal and state governments are committed to periodically evaluating agreements. These normative instruments have been reinforced by the publication of several Mexican Official Standards or Norms (Normas Oficiales Mexicanas [NOM]) between 1994 and 2011, which have provided specifications for the regulation of whale-watching tourism, management of marine mammals in captivity, scientific collecting of biological resources, and mitigation of adverse effects on biodiversity from changes in forest land use. An important challenge still pending is the devolution of federal wildlife authority to states in central and southern Mexico (Ortiz Monasterio Quintana 2011).

Rural Productive Diversification, a New Wildlife Management Scheme

The legal use of wildlife was for many years limited to hunting. Hunting permits, regulated from 1952 to 2000 by the Federal Hunting Act, granted various intermediaries—private individuals or hunter organizations—the right to harvest wildlife. Landowners, deriving no direct economic benefits from ecosystem conservation, converted lands to agriculture (CONABIO 2006). In contrast, one of the main objectives of the rural diversification program (PCVSDP) was to integrate the utilization of wildlife with the economic development of the rural sector, thereby recognizing the economic value of wildlife as an instrument for bettering the welfare of rural communities. The PCVSDP included a detailed assessment of the comparative advantages and economic opportunities offered in markets associated with hunting, industrial inputs, nurseries, extensive and intensive breeding farms, nontimber forest products, ecotourism, and pharmacochemical research. By the end of the 1990s, the commercialization of these activities was worth 2,400 million pesos a year (SEMARNAP 1997).

The most important instruments of the PCVSDP were the Program for Conservation and Recovery of Priority Species (PREP) and the System for Wildlife Conservation Units (SUMA). The PREP was created with the goal of ensuring the protection and recovery of selected wildlife species that were at some level of risk. Indirectly, it provides protection for other species and their habitats, including charismatic species and species with cultural and economic value (SEMARNAP 1997, 2000).

The program was designed to influence habitat management and provide guidelines for sustainable use (Cervantes et al. 2008). SUMA, in turn, would integrate land ownership (private, ejidos, communal, or public) with activities focused on the strict protection and sustainable utilization of wild fauna and flora. These registered land properties were named Management Units for the Conservation of Wildlife (UMAs). This concept was innovative because it was based on multiple use of local resources, principally flora and fauna. The UMA concept promotes land management schemes among rural communities that provide social and economic development alternatives. UMAs can involve consumptive (sport or commercial hunting, captive breeding) and nonconsumptive schemes (e.g., ecotourism, research, environmental education) (see chapter 3). Extensive UMAs enable ecosystems to be sustainably maintained (through their conservation or restoration) to guarantee a sustained harvest. Intensive UMAs allow captive breeding to meet the market demand for wildlife, indirectly preventing the overexploitation of wild populations. However, some intensive UMAs also incorporate captive-breeding programs for restocking purposes (SEMARNAP 1997, 2000; Valdez et al. 2006).

Nonetheless, there has been confusion regarding the interpretation and application of the original concept, which has resulted in inappropriate management approaches and controversy about its usefulness. Furthermore, there has been excessive proliferation of intensive UMAs; 2,436 are currently registered, of which 77% are engaged in captive breeding (19% are plant nurseries, and 4% are botanical gardens) (DGVS, SEMARNAT, personal communication). The large number of intensive UMAs makes it difficult to maintain oversight, resulting in inadequate control of wildlife trafficking.

There has also been a lack of fostering and support for rural communities, as most captive-breeding farms (and plant nurseries) are private. Moreover, following the publication of the Bylaws of the General Wildlife Act in 2006, the concept of intensive UMAs was modified. Now only breeding farms (or plant nurseries) that maintain wild species for recovery and repopulation purposes are considered intensive UMAs, and these are required by the LGVS to manage native species listed in NOM-059-SEMARNAT-2001—that is, those that fall within some category of risk. All other facilities that maintain captive wildlife (mainly native species not listed in NOM-059, but also exotic species) for commercial or exhibition purposes are now considered Premises or Facilities for the Management of Wildlife (PIMVS). Because these units do not handle at-risk species, they have been deregulated; they are no longer required to submit their management plans for approval (SEMARNAT 2006), a situation that should be reconsidered. Moreover, the PREP, originally conceived to complement the conservation activities of UMAs, does not have proper monitoring or affiliations with academic research, and there is a lack of coordination, thus limiting its function as a species recovery program linking all UMAs. In 2007, management of the program was transferred to the National Commission for Natural Protected Areas (CONANP), associating it with the ANPs under the name of Conservation Program for Species at Risk (CONANP 2007).

The extensive model of wildlife management has been applied mainly on cattle ranches in northern Mexico, where livestock pastures (fig. 21.3) are interspersed with arid shrubland, which is excellent habitat for white-tailed deer (*Odocoileus virginianus*), a big game species in high demand for sport hunting. Here, local ecological and socioeconomic conditions imitate management schemes used by Texas game ranches. However, the application of this model has been limited in central and southern Mexico (González et al. 2003, Valdez et al. 2006, Weber et al. 2006) because of the different ecological, socioeconomic, and cultural environment. Its failure may also be linked to the absence of large markets, which in northern

Figure 21.3. Ranches in northern Mexico where livestock pastures are interspersed with shrublands are excellent habitat for white-tailed deer. Photo by J. Alfonso Ortega-S.

Mexico are closely linked to the US. In contrast, the states of central and southern Mexico have rural economic dynamics, and the use of wildlife resources has a predominantly cultural, ritual, subsistence, and nonleisure character (Weber et al. 2006).

In addition, corruption, management deficiencies, and lack of ecosystem management experience have interfered with the application of conservation programs. Some critics have asserted that most UMAs do not fulfill their mission of providing economic development alternatives to rural communities, and even less of contributing to conservation, since there is a negative correlation between the economic growth of UMAs and the conservation of animal populations and habitats (González et al. 2003, Valdez et al. 2006, Weber et al. 2006, García Marmolejo et al. 2008). Others have proposed revising the UMA concept and have identified priority areas for a research agenda (García Marmolejo et al. 2008, Gallina and Mandujano 2009, Gallina Tessaro et al. 2009). Some have questioned the validity of the UMA concept in the tropics (Valdez et al. 2006, Weber et al. 2006, Sisk et al. 2007). Among the concerns are the failure to include subsistence hunting in hunting harvest estimates, the undetermined ecological effects associated with the introduction of exotic species, and the proliferation of intensive breeding farms and consequent potential increase in the incidence of disease (Weber et al. 2006). Some have suggested professionalizing wildlife management by providing technical training to improve operational transparency, improving surveillance and monitoring mechanisms, and promoting educational advancement at all levels (Valdez et al. 2006, Gallina Tessaro et al. 2009). Finally, there are no statewide studies, much less nationwide research, documenting the social and economic impacts of UMAs on rural communities, even though one of the main objectives of the PCVSDP was to improve their quality of life. However, the number of UMAs continues to increase; there are currently 10,193 (69% extensive and 31% intensive), of which extensive UMAs encompass about 30 million ha (about 74 million ac) (DGVS, SEMARNAT, personal communication).

Challenges for Nature Conservation in Mexico

Most recent wildlife conservation policies in Mexico have focused on extensive UMAs, especially those based on hunting. However, the original concept involves consumptive and nonconsumptive production, and diversified utilization of flora and fauna, incorporating numerous management possibilities. A pending challenge is to integrate the rational use of resources with different management models of sustainable development that are attuned to each ecological zone (see chapter 2) and that take into account the cultural context and socioeconomic conditions of each region.

The latter is particularly important considering the new paradigms of conservation that propose to refocus biological conservation efforts toward the human landscape matrix. The goal is to make wildlife conservation economically and environmentally sustainable by minimizing environmental degradation; maintaining ecosystem structure, diversity, and functionality; and creating a diverse mosaic of land uses (Franklin 1993, Primack et al. 2001, McNeely and Scherr 2003, Amend et al. 2008, Scherr and McNeely 2008). One of the major challenges is to orient production activities to meet future demand while maintaining the natural resource base (De Haan et al. 2003, FAO 2006).

In parallel with the conservation efforts associated with UMAs, nonconventional production models have been applied in the agricultural and animal husbandry sectors (Anta and Carabias 2008). These approaches include sustainable agriculture, holistic management, agroforestry systems, organic agriculture, livestock production, and habitat restoration through traditional agricultural techniques. Various legal instruments have been developed to facilitate and promote the development of these alternative production models (e.g., the Sustainable Rural Development Act). However, as with UMAs, these models have not been successful economically because they have been localized and isolated, which has limited their potential for competing with other, more lucrative commercial land uses.

The modern concept of wildlife management in Mexico should consider an ecosystem approach. This requires an integrated strategy for natural resource management in which the planning and execution of actions for the wise use of flora and fauna should consider nonconventional (and environmentally sustainable) forms of agricultural production. Such an approach could serve to complement, or even be an alternative to, some conventional forms of production and could help control the expansion of industrialized agriculture at the expense of remaining functional wildlife habitats, in addition to mitigating environmental impacts. These measures should include nonextractive land uses (fig. 21.4) and in certain cases should be focused on the strict protection of biodiversity (González-Rebeles 2007). As mentioned previously, the particular system of land-use zoning in biosphere reserves allows, at least in theory, a combination of strict protection of some wildlife species along with diversified forms of resource utilization based on the promotion of traditional knowledge, which may assist in the regulation and control of industrialized production. These management models, developed at local or regional levels, could provide focus and direction for a comprehensive resource management strategy that could be applied in Mexico nationwide. A study of the effectiveness of ANPs in Mexico (Figueroa and Sánchez Cordero 2008, Figueroa et al. 2009) revealed that biosphere reserves have a lower rate of vegetation cover conversion: 60% of the biosphere reserves evaluated showed less than 5% conversion.

Figure 21.4. Nonextractive activities such as ecotourism should be considered in the modern concept of wildlife management in Mexico. *Falco rufigularis* in Walter Gretter Peters UMA, La Concordia, Chiapas. Mexico. Photo by Amado Ríos Valdez.

The political vision for the management of wildlife in Mexico will continue to change according to national development priorities and will depend on the international economic, social, and political context. Another remaining challenge is to merge academia with decision making, so that academics may be incorporated in developing national policy regarding the protection and sustainable use of biodiversity in Mexico. Rands et al. (2010) asserted that global conservation actions have proved insufficient, given the progressive loss of biodiversity. It will not be possible to reverse the process of environmental degradation without radical global changes in land-use strategies. In addition, public governance should consider biodiversity a "global public good" and recognize it as an integral and strategic component of global development policies of production and consumption. It remains to be seen whether Mexico will successfully attain its goal of sustainable wildlife conservation and provide for the socioeconomic and cultural needs of future generations of citizens.

Literature Cited

Amend, T., J. Brown, A. Kothari, A. Phillips, and S. Stolton, editors. 2008. Protected landscapes and agrobiodiversity values. Volume 1. Protected landscapes and seascapes. IUCN and GTZ, Kasparek, Heidelberg, Germany.

Anta, F., and J. Carabias. 2008. Consecuencias de las políticas públicas en el uso de los ecosistemas y la biodioversidad. Pages 87–153 *in* Capital natural de México. Volume 3. Políticas públicas y perspectivas de sustentabilidad. CONABIO, México, D. F., México.

Bezaury Creel, J., and D. Gutiérrez. 2009. Áreas naturales protegidas y desarrollo social en México. Pages 385–431 *in* Capital natural de México. Volume 2. Estado de conservación y tendencias de cambio. CONABIO, México, D. F., México.

Carabias, J., J. de la Maza, and E. Provencio. 2008. Evolución de enfoques y tendencias en torno a la conservación y el uso de la biodiversidad. Pages 29–42 *in* Capital natural de México. Volume 3. Políticas públicas y perspectivas de sustentabilidad. CONABIO, México, D. F., México.

Cervantes, V., J. Carabias, and V. Arriaga. 2008. Evolución de las políticas públicas de restauración ambiental. Pages 155–226 *in* Capital natural de México. Volume 3. Políticas públicas y perspectivas de sustentabilidad. CONABIO, México, D. F., México.

CONABIO. 2006. Capital natural y bienestar social. CONABIO, México, D. F., México.

CONANP. 2007. Programa de Conservación de Especies en Riesgo 2007–2012. http://www.conanp.gob.mx/pdf_especies/PROCERFinalpubmar2009.pdf. Accessed 20 Nov 2010.

De Haan, C., H. Steifeld, and H. Blackburn. 2003. Livestock and the environment: finding a balance. Commission of the European Communities, FAO, and World Bank, Suffolk, United Kingdom.

Dirzo, R., and A. Miranda. 2000. Contemporary Neotropical defaunation and forest structure, function and diversity: a sequel to John Terborgh. Conservation Biology 4:444–447.

DOF. 2006a. CONVENIO Específico para la asunción de funciones en materia de vida silvestre que celebran la Secretaría de Medio Ambiente y Recursos Naturales y el Estado de Baja California. http://dof.gob.mx/nota_to_doc.php?codnota=4932787. Accessed 20 May 2016.

DOF. 2006b. CONVENIO Específico para la asunción de funciones en materia de vida silvestre que celebran la Secretaría de Medio Ambiente y Recursos Naturales y el Estado de Sonora. http://dof.gob.mx/nota_to_doc.php?codnota=4916148. Accessed 17 May 2016.

DOF. 2006c. CONVENIO Específico para la asunción de funciones en materia de vida silvestre que celebran la Secretaría de Medio Ambiente y Recursos Naturales y el Estado de Chihuahua. http://dof.gob.mx/nota_to_doc.php?codnota=4912707. Accessed 16 May 2016.

DOF. 2007a. CONVENIO Específico para la asunción de funciones en materia de vida silvestre que celebran la Secretaría de Medio Ambiente y Recursos Naturales y el Estado de Tamaulipas. http://dof.gob.mx/nota_to_doc.php?codnota=5060622. Accessed 16 May 2016.

DOF. 2007b. CONVENIO Específico para la asunción de funciones en materia de vida silvestre que celebran la Secretaría de Medio Ambiente y Recursos Naturales y el Estado de Nuevo León. http://dof.gob.mx/nota_detalle.php?codig0=5004766&fecha=26/10/2007. Accessed 15 May 2016.

DOF. 2009. CONVENIO Modificatorio al Convenio Específico para la asunción de funciones en materia de vida silvestre que celebran la Secretaría de Medio Ambiente y Recursos Naturales y el Estado de Coahuila de Zaragoza. https://www.diariooficial.gob.mx/nota_detalle.php?codig0=5099903&fecha=23/07/2009. Accessed 18 May 2016.

FAO. 2006. World agriculture: towards 2030/2050. Interim Report. FAO, Rome, Italy.

Figueroa, F., and V. Sánchez Cordero. 2008. Effectiveness of natural protected areas to prevent land use and land cover change in Mexico. Biodiversity Conservation 17:3223–3240.

Figueroa, F., V. Sánchez Cordero, J. A. Meave, and I. Trejo. 2009. Socioeconomic context of land use and land cover change in Mexican biosphere reserves. Environmental Conservation 36:180–191.

Franklin, J. F. 1993. Preserving biodiversity: species, ecosystems, or landscapes? Ecological Applications 3:202–205.

Gallina, S., and S. Mandujano. 2009. Investigaciones sobre ecología, conservación y manejo de ungulados silvestres en México. Tropical Conservation Science 2:128–139.

Gallina Tessaro, S., A. Hernández Huerta, C. A. Delfín Alfonso, and A. González Gallina. 2009. Unidades para la conservación, manejo y aprovechamiento sustentable de la vida silvestre en México (UMA). Retos para su correcto funcionamiento. Investigación Ambiental 1:143–152.

García Marmolejo, C., G. Escalona Segura, and H. Van Der Val. 2008. Multicriteria evaluation of wildlife management units in Campeche, Mexico. Journal of Wildlife Management 72:1194–1202.

González, R. M., R. Montes, and J. Santos. 2003. Caracterización de las unidades para la conservación, manejo y aprovechamiento sustentable de fauna silvestre en Yucatán, México. Tropical and Subtropical Agroecosystems 2:13–21.

González-Rebeles, C. 2007. Zootecnia aplicada a la fauna silvestre. Pages 481–514 *in* M. E. Trujillo, editor. Introducción a la zootecnia. Facultad de Medicina Veterinaria y Zootecnia, UNAM, México, D. F., México.

Mandujano, S., and A. González Zamora. 2009. Evaluation of natural conservation areas and wildlife management units to support minimum viable populations of white-tailed deer in Mexico. Tropical Conservation Science 2:237–250.

McNeely, J., and S. Scherr. 2003. Estrategias de ecoagricultura para ayudar a alimentar al mundo y salvar la biodiversidad silvestre. Future Harvest Foundation and World Conservation Union. www.ecoagriculturepartners.org. Accessed 20 Jan 2010.

Ortiz Monasterio Quintana, A. 2011. La administración descentralizada de algunos aspectos del manejo de la vida silvestre en México: síntesis del proceso y marco jurídico. Pages 335–349 *in* O. Sánchez, O. P. Zamorano, and H. Moya, editors. Temas sobre conservación de vertebrados silvestres en México. SEMARNAT, INE, USFWS, UAT, and UAEM, México, D. F., México.

Pérez Gil, R., coordinator. 1988. Fauna silvestre y áreas naturales protegidas. Universo Veintiuno, México, D. F., México.

Primack, R., R. Rozzi, P. Feinsinger, and F. Massardo. 2001. Conservación fuera de las áreas protegidas. Pages 521–558 *in* R. Primack, R. Rozzi, P. Feinsinger, R. Dirzo, and F. Massardo, coordinators. Fundamentos de conservación biológica: perspectivas latinoamericanas. FCE, México, D. F., México.

Rands, M. R., et al. 2010. Biodiversity conservation: challenges beyond 2010. Science 329:1298–1303.

Retana, O. G. 2006. Fauna silvestre de México. Aspectos históricos de su gestión y conservación. UACAM and FCE, México, D. F., México.

Romero, L. 2002. El peso de las políticas mexicanas en la sustentabilidad de las recientes tendencias de desarrollo. Pages 30–45 *in* E. Leff et al., compilers. La transición hacia el desarrollo sustentable: perspectiva de América Latina y el Caribe. SEMARNAT, INE, UAM, and PNUMA, México, D. F., México.

Scherr, S. J., and J. A. McNeely. 2008. Biodiversity conservation and agricultural sustainability: towards a new paradigm of "ecoagriculture" landscapes. Philosophical Transactions of the Royal Society, Biological Sciences Series B 363:477–494.

SEMARNAP. 1997. Programa de la Conservación de la Vida Silvestre y Diversificación Productiva en el Sector Rural: 1997–2000. DGFS and INE, México, D. F., México.

SEMARNAP. 2000. Estrategia nacional para la vida silvestre. Logros y retos para el desarrollo sustentable 1995–2000. DGFS and INE, México, D. F., México.

SEMARNAT. 2006. La gestión ambiental en México. SEMARNAT, México, D. F., México.

Sisk, T. D, A. E. Castellanos, and G. W. Koch. 2007. Ecological impacts of wildlife conservation units policy in Mexico. Frontiers in Ecology and the Environment 5:209–212.

Valdez, R., J. C. Guzmán-Aranda, F. J. Abarca, L. A. Tarango-Arámbula, and F. C. Sánchez. 2006. Wildlife conservation and management in Mexico. Wildlife Society Bulletin 34:270–282.

Vargas, F. M. 1984. Parques nacionales de México y reservas equivalentes. Instituto Mexicano de Investigaciones Económicas, UNAM, México, D. F., México.

Wakild, E. 2012. Parables of Chapultepec: urban parks, national landscapes and contradictory conservation in modern Mexico. Pages 192–217 *in* C. R. Boyer, editor. A land between waters: environmental histories of Mexico. University of Arizona Press, Tucson, USA.

Weber, M., G. García Marmolejo, and R. Reyna Hurtado. 2006. The tragedy of the commons: wildlife management units in southeastern Mexico. Wildlife Society Bulletin 34:1480–1488.

22 Wildlife Conservation Management Challenges and Strategies in Mexico

J. Alfonso Ortega-S., Jorge G. Villarreal González, Salvador Mandujano, Sonia Antonieta Gallina Tessaro, Manuel Weber, Fernando Clemente-Sánchez, and Raul Valdez

The immense richness of Mexico's biodiversity and the associated cultural, economic, and biological benefits underlie the importance and urgency of sustainably managing its wildlife and associated habitats (Sarukhán et al. 1996). An estimated 8% to 10% of the world's plant and animal species occur in Mexico, which positions Mexico as the fifth-richest country in biodiversity (Toledo and Ordóñez 1993, Valdez et al. 2006). The most important threats to wildlife in Mexico are all the environmental effects caused by the country's population growth (Valdez et al. 2006). Humans have modified all landscapes in Mexico. By 2007, primary vegetation existed in only 98.2 million ha (243 million ac) of the potential 194 million ha (479 million ac), and secondary vegetation contributed another 14 million ha (35 million ac) (see chapter 2; INEGI 2005, 2010). Species extinctions and the extirpation and decrease of wildlife populations (see chapter 1) make evident Mexico's biodiversity crisis resulting from the legal and illegal unsustainable use of natural resources, primarily via agricultural development, deforestation, illegal trafficking of threatened species, invasive species and associated diseases and parasites, and climate change (Ehrlich and Ehrlich 1992, Ceballos et al. 2005).

Wildlife conservation in Mexico is affected by the complex interaction of political, social, and economic factors, which, because of inconsistent political and environmental policies (see chapters 1 and 21), have prevented the development of sustainable wildlife management strategies. Mexico is distinctly divided into a northern region (with low population density, arid climate, and abundant wildlife populations) and a southern region (with high population density, tropical climate, and low wildlife populations) because of biogeographical as well as ethnic, social, and economic differences. These differences contribute to the complexity of resolving regional wildlife conservation and management scenarios. Hence, challenges, opportunities, and conservation strategies for northern and southern Mexico differ considerably (Weber et al. 2006). In this chapter, we summarize the challenges and strategies outlined in this book and recommend specific actions to achieve the sustainable use of wildlife in Mexico.

Habitat Management Challenges

Habitat destruction, degradation, and fragmentation are pivotal considerations for wildlife conservation in Mexico. Because approximately 70% of the country has some degree of desertification and more than 50% of the native vegetation has been lost (SEMARNAP 1997), one of the major challenges for wildlife conservation is the recovery and maintenance of native habitats. On the other hand, the use of introduced exotic species such as buffelgrass (*Pennisetum ciliare*) will continue as long as cattle production remains economically viable. Even though introduced grasses are considered detrimental to wildlife, research data to determine the proportions of introduced grasses and native vegetation that would optimize livestock and wildlife populations are lacking. Nonetheless, in Sonora, ranches with pastures dominated by buffelgrass continue to produce trophy mule deer for sport hunting (Ortega-S. et al. 2013) (fig. 22.1).

Another important challenge is the reduction and fragmentation of habitats caused by human infrastructure, economic development such as mineral, forest, and petroleum extraction, and farming and ranching land conversions. The growth and expansion of the human population and its detrimental environmental effects will continue; that is a reality to which wildlife management must adapt. Landscape and ecosystem strategies that incorporate the concepts of matrices, patches, and corridors will have to be the basis for accomplishing sustainable management and conservation objectives (see chapter

Figure 22.1. Ranches with large tracts of introduced grasses such as buffelgrass (*Pennisetum ciliare*) in Sonora, Mexico, continue to produce trophy mule deer. Photo by Fernando A. Ibarra Flores.

3). Conservation cooperatives may be an option to conserve landscapes with agreed-upon specific sustainable wildlife use objectives; this strategy is especially important in areas where landownership is small and focused on agricultural development. The first step would be for landowners to adopt cooperative management schemes and to develop short- and long-term goals and objectives. Initial management should incorporate regulated subsistence hunting; proper water distribution; maintenance or recovery of native vegetation in riparian habitats to provide wildlife cover and movement corridors; and flora and fauna inventories, among other actions.

Mexico has case studies of successful cooperative wildlife management programs. These involve the participation of organized landowners (e.g., small landowners and *ejidatarios*) and UMAs that promote the recovery, conservation, and sustainable management of flora and fauna resources in rural communities (Ramirez F. and Mondragón 2010). An example is the Palo Blanco Basin in Nuevo León. This area consists of 53 extensive UMAs covering 130,000 ha (321,490 ac). In 1994, the area was repopulated with white-tailed deer (*Odocoileus virginianus texanus*) to initiate a sustainable sport-hunting program. The result was an increase in wildlife productivity and profitability and a change in the vision and attitude of UMA managers and landowners toward wildlife conservation and management. In this area of natural brushland and forested ecosystems, there are more than 145 species of birds, 34 mammals, 30 cacti (fig. 22.2), and over 800 species of native plants, which represent 37%, 24%, 30%, and 32% of the fauna and flora in Nuevo León, respectively. Some of the important species recovered include black bear (*Ursus americanus*), golden eagle (*Aquila chrysaetos*), black-capped vireo (*Vireo atricapillus*), century plant (*Agave ovatifolia*), and royal agave (*Agave victoriae-reginae*) (Villarreal G. 2009).

Minimum Areas to Sustain Viable Populations

Two relevant conservation questions relate to the minimum size of a viable wildlife population, and

Figure 22.2. *Echinocereus stramineus* in Cuenca Palo Blanco in Nuevo León, Mexico. Photo by Jorge Villarreal G.

the minimum size of UMAs or natural protected area (ANPs) that will sustain a viable wildlife population. This topic has been practically ignored in wildlife management in Mexico (Mandujano and González-Zamora 2009). The minimum population size for biological conservation has been debated; there is no consensus on minimum population size or how to calculate it (Soulé 1987). It has been proposed that the number of individuals required for a viable population may be hundreds, thousands, or hundreds of thousands (Traill et al. 2007). In general, the minimum number of individuals is from 500 to 5,000, but it varies depending on the life history, structure, and genetics of the population, and its habitat type. For example, the minimum area of a UMA should be between 1,667 and 50,000 ha (4,122 and 123,650 ac) to sustain a minimum white-tailed deer population of 500 individuals, or between 1,667 and 500,000 ha (4,122 and 1,236,500 ac) to sustain 5,000 deer, depending on the local population density (Mandujano and González-Zamora 2009). A lower population density would require a larger management unit.

This is a crucial point, because in those regions where UMAs are small (< 1,000 ha, or < 2,473 ac), the sustainability of a minimum viable population of the desired species is questionable, and as a consequence, a sustainable harvest would be small. In areas of small management units, a regional, landscape-level vision for wildlife management is necessary. This requires the development of a strong organizational scheme in which landowners and managers cooperatively manage resources at the landscape level with common goals and objectives, as in the Palo Blanco Management Unit in Nuevo León (Ramirez and Mondragón 2010).

Wildlife Population Management

Carnivores

Predators play an important role in ecosystems. Landowners and managers of sport-hunting operations need to accept the beneficial ecological value of predators. The presence of predators indicates the presence of prey populations. Indiscriminate control or eradication of predators to avoid domestic and game animal predation must be rejected (Gallina-Tessaro et al. 2008). Ranchers in Mexico view some predators such as coyotes (*Canis latrans*) with aversion and repugnance, an attitude initiated with the Spanish introduction of livestock (Villarreal G. 2000). Coyote predation of game species is often not significant, and predator control is not economically justifiable (see chapter 11). Villarreal G. (2013a) reported a successful translocation of white-tailed deer in areas of thorn scrubland in Cerralvo, Nuevo León. During the past 20 years, an average of 200 white-tailed deer per year are harvested without intensive predator control. Occasionally, coyotes and bobcats (*Lynx rufus*) are harvested by sport hunters.

There have been efforts to create an ecological conscience among landowners and the general public to avoid indiscriminate predator control in Nuevo León. These include a permanent public relations program (via radio, pamphlets, posters, and articles in popular publications) to educate the public about ecologically unwise coyote and mountain lion (*Puma concolor*) predator control programs. These predator conservation programs should be imitated in other states.

A positive result is an increasing black bear population trend (see chapter 12) in Coahuila. Black bears have been downlisted from an endangered species to one requiring special protection. In 2001 the DGVS solicited black bear damage complaints from landowners; legitimate livestock predation complaints were documented as well as illegal black bear predator control. In May and June 2008, at least 30 black bears were recorded in several urban areas of Monterrey. Three of these black bears (one male and two females) were captured, equipped with radio collars, and released at the Palo Blanco Management Unit (Villarreal G. 2009). In Tamaulipas, a black bear was trapped close to the city of Victoria and relocated to the zoo. These examples illustrate the rational management of black bears and the promotion of a public awareness program based on sound predator population management principles.

The future of the jaguar (*Panthera onca*) and mountain lion is uncertain because of documented regional extirpations (see chapter 13). Programs to generate resources to improve the quality of life in rural communities that coexist with these felids are needed to provide communities with incentives to coexist with predators. Programs have been initiated to compensate landowners who lose livestock when feline predation is confirmed. The National Confederation of Livestock Organizations (CNOG) and the Ministry of Agriculture, Livestock, Rural Development, Fisheries, and Food (SAGARPA) created an insurance fund in 2006 to cover Mexican ranchers against damage caused by large felid predation. The main objective of this program is to convince ranchers to avoid indiscriminate predator control (CNOG 2013).

In southern Campeche and Quintana Roo, sport hunting includes 12 species of birds and mammals, and there is also an ecotourism program for jaguar observation. In this case, a positive conservation perception was created by incorporating many diverse game species in a sport-hunting program and by recognizing the jaguar as a protected cultural and umbrella species. The legalized game bird and mammal hunting program was approved by the federal government, which was critical because the jaguar is listed as requiring special protection (Weber et al. 2006).

Ungulates

The future of big game hunting for mule deer (*Odocoileus hemionus*), white-tailed deer, and other ungulates is promising. The main requirements for the successful management of mule deer in Mexico are harvest regulations and habitat management (see chapter 16). Interest in conservation and sustainable harvest through sport hunting has in several instances triggered a reduction in cattle stocking rates (Villarreal G. 1999a), thereby promoting habitat restoration. However, a major concern for white-tailed deer conservation is the unregulated translocation of subspecies. In fact, there is great genetic differentiation between the subspecies of white-tailed deer from the central-northern and south-southeastern regions of Mexico (Ambriz-Morales et al. 2016). Enforcement of wildlife regulations that limit translocations is urgently needed (Galindo-Leal and Weber 1994), as this will help prohibit hybridization of subspecies and thus preserve the integrity and diversity of genetic resources in Mexico. The motivation to translocate mainly *texanus* white-tailed deer from northern Mexico to other regions is based on "genetic improvement," assuming that breeding *texanus* white-tailed deer with other subspecies may increase antler size and thus economic value. However, this is only partially true, because at least a segment of sport hunters would pay more to hunt other native deer subspecies in their native habitat no matter the antler size, rather than an exotic deer or hybrid. Refer to Logan-López et al. (2007) for a discussion of the "genetic contamination" caused by the translocation of *texanus* white-tailed deer in Tamaulipas and the consequences of translocation of animals to other regions. Fortunately, since 2012, seven new trophy categories of Mexican white-tailed deer have been approved, meaning that Mexican white-tailed deer subspecies will be recognized internationally as separate trophies. The new subspecies categories will also yield economic benefits (Viejo 2012).

Compiling the scientific knowledge base to manage brocket deer (*Mazama* spp.) (see chapter 18), tapir (*Tapirus bairdii*) (see chapter 15), and white-lipped peccary (*Tayassu pecari*) (see chapter 14) in southern Mexico is a major challenge. This will require the regulation of subsistence and commercial hunting. In order to develop effective management programs, subsistence and commercial hunting must be recognized as a social factor. Additionally, politicians and decision makers need to become better informed about the benefits of sustainable wildlife management. In marginalized human populations in rural areas, the people's priority is to provide meat protein for their families. Viable solutions to this challenge need to be implemented before the conservation concerns of scientists, conservationists, and politicians can be resolved.

Pronghorn (*Antilocapra americana*) populations in Mexico are small and limited in distribution (see chapter 19). The recommended minimum number for a viable population is 100 individuals (Clemente-Sanchez 1993), but this is a problem in Baja California Sur, Chihuahua, and Coahuila, where populations do not reach this size. The major challenge is to continue restoration efforts by releasing free-ranging animals and supplementing wild populations with animals from stable populations. However, captive breeding of pronghorn is an option for Baja California Sur and Sonora.

Bighorn sheep (*Ovis canadensis*) sport hunting has been incentivized because of its economic value. Even under intensive captive-breeding programs, ranchers in Sonora have sold bighorn sheep hunts for US$35,000–$60,000 or more (see chapter 20). Habitat reduction and fragmentation and the coexistence of domestic and introduced exotic species are major concerns. Bighorn sheep management challenges on the Baja California Peninsula include the reduction of livestock and feral animals. Feral donkeys (*Equus asinus*) may compete for food and watering sources with bighorn sheep. On the other hand, poaching is a great threat not only to bighorn sheep but also to mule deer, which are sympatric. Government efforts to prevent illegal hunting are not effective, and the collaboration of rural communities (ejidos) and private landowners is absolutely and urgently needed. Management of bighorn sheep on the Baja California Peninsula is focused on harvest restrictions but ignores strategies for the establishment of protected core areas, habitat improvement, and repopulation. Outreach programs from universities and state governments are needed to educate ejido communities regarding management techniques for mule deer and bighorn sheep; the economic potential of both species for sport hunting is a positive incentive.

Bighorn sheep in Sonora also produce other management challenges. Intensive captive-breeding systems may be beneficial for their conservation and sustainable use. However, two aspects need to be considered: (1) avoidance of hybridization with other subspecies, and (2) limiting the number of trophies for sport hunting in order to maintain trophy quality and high trophy prices.

Birds

Management challenges for birds are discussed in chapters 5, 7, and 8 for cracids, ducks and geese, and quails, respectively. Habitat destruction, agricultural and demographic growth, and associated rapid urban development are the main threats for game birds. Management challenges for birds may be divided in two groups: those for migratory birds and those for residents. In both groups, management should focus on those species with social and economic value and should increase abundance to attain sustainable use levels.

In the case of migratory birds, which are shared across political borders, especially in the case of ducks and geese because of their importance as game birds, efforts should focus on establishing agreements with the US and Canada to benefit populations (see chapter 7). Winter surveys of resident and overwintering populations in Mexico need to be conducted annually. This information may be used to suggest harvest quotas in the US to avoid negative effects on the number of ducks and geese that migrate to Mexico. In Mexico, the DGVS needs to analyze and make information available on the number of hunting permits issued per year based on harvest rates. Estimation of annual harvest based on the annual reports of UMAs is also needed to determine the magnitude of harvest in Mexico.

Wild turkeys (*Meleagris* spp.) involve great management challenges (see chapter 9). Populations have recovered in the past 20 years with successful reintroduction programs. A management example is the repopulation of wild turkeys in the Sierra Fria in Aguascalientes. With an initial introduction of only 84 individuals from 1992 to 1994, the entire habitat is currently occupied, and turkeys have expanded their range to adjacent mountains in Jalisco. However, the potential translocation of turkeys in other areas of Mexico without consideration of the subspecies present in the release sites may result in the genetic contamination of subspecies, which should be avoided. Habitat availability for wild turkeys is decreasing; the management challenge is to increase productivity in the available areas through habitat recovery to the point of warranting sustainable use.

The Passeriformes, another group of migratory birds, is not covered in this book but is important because of its potential economic value and harvest rates in rural communities. Unfortunately, harvest rates are undocumented and unregulated (see chapter 3). Illegal markets for these species as pets exist in all states and are more evident in large urban areas.

Mexican quail species are distributed in many diverse environments, which complicates their conservation. In general, challenges for quail conservation may focus on two groups: those in forested areas and those in grasslands and brushlands. The challenge for forest quail species is to recover and conserve habitats and maintain intact forests with a dense canopy cover of trees for protection and a thick cover of litter for foraging. This is important because the reproductive biology of forest quail species is a limiting factor, and populations as large those of quail species from brushlands and grasslands are not attainable. Tapia et al. (2002) and Zaragoza et al. (2004) report that the fragility of Montezuma quail (*Cyrtonyx montezumae*) habitats is the limiting factor in its expansion, in addition to the scarcity of the main component of its diet (bulbs and tubers) in fragmented forests (fig. 22.3). For the other quail species, the management challenge is to maintain populations during severe droughts. Habitat improvement practices including the distribution of watering sites may maintain body moisture during incubation and lead to a higher percentage of successful hatchings during drought. For example, in the Ejido Rincón de Diego Martín UMA in Salinas County, San Luis Potosí, cattle stocking rates have been reduced for eight years, and 10 exclosures have been created for conservation, management, and sustainable use in an area of 9,000 ha (22,239 ac). These practices have allowed adequate cover to be maintained, resulting in significant increases in the scaled quail population (Fernando Clemente, personal communication).

Figure 22.3. Montezuma quail at Rancho Tres Papalotes in Chihuahua, Mexico. Photo by Enrique Perez.

One of the main problems with cracid management in Mexico, in addition to the problems already discussed for birds in general, is subsistence hunting and illegal trafficking (see chapter 5). In addition to harvest restrictions and management practices, regulations must be enforced. PROFEPA is the agency responsible for enforcing laws and regulations, but it is insufficiently staffed, and rural communities therefore need to become aware of the importance of wildlife resources to reduce law violations and illegal activities.

Education Challenges

Education is a key component in successful ecosystem conservation. Environmental education needs in Mexico can be separated into several sectors: the general public, ranchers and ranch managers, and wildlife specialists. Politicians and government employees are important participants in the decision-making process for wildlife conservation. Recognition of the importance of wildlife conservation by the general public has increased considerably, but more citizen participation is needed. To accomplish this, state governments need to make a major effort to reinforce the conservation ethic. Communication media such as radio and television can play a major role in broadcasting information to increase awareness of conservation-related topics. Environmental education in elementary and high schools and hands-on programs to promote and create an ecological conscience and encourage learning about the benefits of conservation and sustainable management are basic in developing a responsive society. In this context, publication of educational materials for children such as books, videos, and other outreach tools is important, and increasing the educational level of the general public is essential.

The need for professional specialists in wildlife management is indisputable. The lack of university programs leading to professional careers in wildlife management is a major problem. Some universities offer wildlife management courses to provide basic knowledge of the discipline. The limited number of university professors with academic degrees in wildlife management is one reason for the lack of wildlife management curricula. Surprisingly, university professors in the US have not prioritized the recruitment of Mexican graduate students, despite the proximity of Mexico, the urgent need, and the country's pivotal importance for wildlife management in North America (Valdez et al. 2006).

Wildlife Research

During the past three decades, considerable ecological field research has advanced our knowledge of game species in Mexico, especially regarding the population ecology of the economically most important species. However, many of the basic questions about conservation and sustainable management discussed by Leopold in the 1950s remain unanswered (Weber et al. 2006), especially in the tropics. Wildlife management in the tropics is less advanced than that in temperate regions (Robinson and Bodmer 1999), and research has focused on a limited number of species. For example, from 1850 to 2001, there were 502 publications on Mexican deer, of which 76% were published in the last 20 years, and 75% of those were on white-tailed deer. The least-studied species in the tropics were the two tropical brocket deer species (Mandujano 2004) (fig. 22.4).

Figure 22.4. The white-tailed deer has been studied more than other wildlife species in the past 20 years, but many of the questions about conservation and management posed by Leopold in the 1950s remain unanswered. Photo by J. Alfonso Ortega-S.

Knowledge accumulation is closely linked with the number of professional specialists in wildlife management. There are professors conducting research in wildlife management at universities and research centers in Mexico, in addition to organized research groups in several regions. However, networking among researchers and institutions is not strong enough to develop well-defined research lines or national efforts to solve similar problems in different regions of the country, much less to develop multidisciplinary projects to generate solutions to specific problems across ecosystems. Two of the most important limiting factors are the reduced funding for research and the lack of political interest in conservation (Valdez et al. 2006).

Conservation Strategies

There are two official basic federal strategies for wildlife conservation and management in Mexico: ANPs and UMAs, both under the direction of SEMARNAT (see chapter 21). ANPs include terrestrial and aquatic ecosystems that have not been significantly altered by humans and provide diverse environmental services. These areas provide consumptive and nonconsumptive opportunities (see chapter 3). Many ANPs include lands with different ownership or use schemes (public lands, private properties, and communal lands), which makes it difficult to create management plans to accomplish conservation objectives. Protected areas in Mexico have increased by more than 3 million ha (7.4 million ac) and currently encompass more than 25 million ha (61.8 million ac) in 174 ANPs, or about 12% of the Mexican land area (www.conanp.gob.

mx). Information generated from research in ANPs is a very important source of ecological knowledge in addition to understanding of the social and economic factors affecting ecosystem use.

The UMA model creates supplementary economic benefits for rural populations dedicated to traditional agricultural practices (SEMARNAP 1997) (see chapters 1 and 21). There are analyses and critiques of the UMA model (Weber et al. 2006, Gallina-Tessaro et al. 2008), but it is important to recognize the innovative management of wildlife in UMAs. The economic benefits resulting from sport hunting, especially in northern Mexico, are undeniable; however, in terms of conservation, UMAs have not been as successful. The success of the UMA model may be measured by the number of registered UMAs, which by July 2012 totaled 11,839 and covered 37.4 million ha (92,490,200 ac) (www.semarnat.gob). UMAs gained popularity and acceptance, especially in northern Mexico, because sport hunting became economically viable and provided a mechanism for ranchers to diversify their income (Weber et al. 2006, Gallina and Escobedo-Morales 2009). UMAs in northern Mexico combine several factors that allow for high profitability (Villarreal G. 2008): (1) property size; most game operations occupy extensive areas of land; (2) experience and tradition to offer a variety of sport-hunting opportunities and ample customers, in combination with livestock enterprises; and (3) the location of northern UMAs close to the US border, which attracts American sport hunters with economic resources.

In contrast, the translocation of the UMA concept modeled on the UMAs in northern Mexico has negatively affected wildlife conservation and increased poverty in rural communities. An important factor to consider for southern UMAs is that species diversity is higher in tropical regions (for example, there are three species of deer, two peccaries, the tapir, and numerous game birds), but game populations are smaller. Additionally, sport hunting quotas do not consider subsistence hunting, which may lead to overharvesting. In regions where the UMA model and sport hunting are not compatible, other use schemes including nonconsumptive wildlife activities compatible with wildlife conservation need to be created. There are successful UMAs in the tropics, such as Las Mariposas in Quintana Roo (Weber et al. 2006). However, whether extensive UMAs are a viable alternative for tropical regions of Mexico remains unanswered. Also unresolved is the best wildlife conservation strategy for a variety of ecosystems with dense human populations and diverse cultures and with extremely complex social and economic problems.

Strategies for Sustainable Wildlife Use

Advantages of UMAs

Weber et al. (2006) indicate that a positive aspect of UMAs is that subsistence hunting is acknowledged, legalized, and regulated because of its importance as a food source. However, a major concern is that wildlife populations are not monitored, and wildlife harvested is not documented.

Valdez et al. (2006) agree that the UMA concept focuses on species with sport-hunting value and promotes management of individual species instead of multiple-species or ecosystem management. UMAs have promoted diversification in cattle-wildlife operations, and economic benefits have provided incentives for reduced cattle stocking rates; the value of a white-tailed deer can be equivalent to the value of 20 calves (Villarreal G. 1999). In northern Mexico, some extensive UMAs have recovered large areas of brushland habitats (*matorrales*) for game species but have also benefited nontarget wildlife (Gallina-Tessaro et al. 2008). The economic value of sport hunting, especially of white-tailed deer, has been a principal incentive to reduce cattle stocking rates, resulting in wildlife habitat recovery. On six ranches in brushland counties in Anáhuac, Lampazos, and Parás in Nuevo León, the carrying capacity for white-tailed deer varied from 1 deer/10–18 ha (25–45 ac) on ranches with cattle to 1 deer/3.5–6.5 ha (9–16 ac) on ranches without cattle (Villarreal G. 1999a).

Disadvantages of UMAs

The UMA model has not been critically evaluated (González-Marín et al. 2003). There are no data to substantiate whether wildlife populations are utilized in a sustainable manner or whether habitat recovery is widespread. Implementation of the UMA model must consider the importance of subsistence hunting for rural communities in southern Mexico (Valdez et al. 2006, Weber et al. 2006). Also, sport-hunting pressure may negatively affect some of the most vulnerable and least-studied wildlife species. Indeed, overhunting of brocket deer in the state of

Yucatán was initiated long before UMAs began to be implemented in 2000. Over 5,000 brocket deer per year have been illegally harvested, including males, females, and fawns (Betancourt et al. 1996), a harvest rate that was not sustainable.

Deficient management plans

Management plans must guarantee sustainable use of wildlife and the integrity of habitats. However, in many cases, management plans are incomplete and are based on questionable and inconsistent information (García-Marmolejo 2005). Poorly trained personnel and the lack of institutional capacity to review and regulate these plans lead to management plan plagiarism; additionally, many of these plans do not fulfill the legal requirements for UMAs (Weber et al. 2006). Many plans are based on literature reviews without field verification. This leads to a lack of confidence and limits their credibility (García-Marmolejo et al. 2008).

Wildlife population estimates

Reliable estimates of wildlife populations are required to determine harvest rates (Caughley and Sinclair 1994). Because harvest rates are based on the number of individuals per hunting season, the most appropriate parameter for this determination is population density. In northern Mexico, estimates by direct counts are easier to make because of open habitats; in the tropics, however, visibility is greatly reduced (Weber et al. 2006). Deficient methodologies, including inadequate survey methods, sampling design, and sample size in northern and southern Mexico, make it difficult to determine adequate harvest quotas and guarantee sustainable use (Weber et al. 2006).

Introduction of exotic wildlife

Many exotic species have been introduced in Mexico to make sport hunting in UMAs more attractive (Weber et al. 2006). Medellín L. (2005) reported 49 exotic species from seven mammal families and five bird families in 660 UMAs in 16 states. Potential competition among exotic species and native wildlife, as well as the danger of disease transmission, has previously been reported (Weber 1993, Caughley and Sinclair 1994, Villarreal G. 1999a, Gallina and Escobedo-Morales 2009). For example, feral hogs (*Sus scrofa*) are considered pests in the state of Nuevo León. This free-ranging species occurs in an area of 800,000 ha (1,978,400 ac) of Tamaulipan thorn scrub in 25 counties in northern, northeastern, and central Nuevo León. If drastic actions are not taken to control this exotic species, populations will continue to increase and expand throughout the state, jeopardizing public and livestock health and causing ecological deterioration. In response to this threat, an outreach campaign was initiated via magazines, posters, and television to promote their eradication (Villarreal G. 2013b).

Translocation of white-tailed deer subspecies

Because the white-tailed deer is one of the most valuable species for sport hunting, deer farming has increased considerably in northern Mexico. By 2010, 39 white-tailed deer farming operations were reported in Nuevo León alone (Ortega-S. et al. 2011). Originally, regional white-tailed deer farming for repopulation purposes was appropriate, but as the process evolved, seed stock of the subspecies *texanus* began to be sold to other regions beyond its distributional limits in order to increase antler size. This practice threatens the integrity of other white-tailed deer subspecies that are well adapted to their habitat and have great ecological and genetic value; in the future, economic value may increase considerably with the proper promotion of these subspecies for sport hunting (Galindo-Leal and Weber 1994).

Regional Management

Because of the limitations of the UMA as an independent management unit and its small size in central and southern Mexico, several authors have suggested a network of regional systems (Mandujano and González-Zamora 2009, Moya et al. 2011, Ortiz-Monasterio 2011). For example, it may be possible to connect small UMAs with large ANPs through corridors, but an important disadvantage of this would be that illegal hunters could enter ANPs through these corridors. Educational programs in neighboring rural communities and among UMA customers to create an awareness of the importance of conservation would be crucial for the success of this strategy. Landscape-level systems may allow animal dispersal movements and increase the probability of having a minimum viable population. Unimpeded gene dispersal of white-tailed deer may already be occurring in northeastern Mexico in Coahuila, Nuevo León, and Tamaulipas, given the large size of contiguous UMAs.

In order to develop a management and conservation network system at the landscape level, it is possible to use ecological models such as archipelago-type reserves, and metapopulation and source-sink models. Archipelago-type reserves have been proposed to protect beta diversity (Halffter 2007). The source-sink model involves a population with positive growth providing individuals that migrate to smaller sink populations with a negative growth rate, which persist only because of the source population. Animals from large ANPs could disperse to small UMAs. This is probably the case in the Montes Azules Biosphere Reserve in Chiapas, which may provide a source of ungulates for neighboring rural communities (Naranjo and Bodmer 2007). Therefore, the creation of a network of ANPs and UMAs connected at the landscape level may be an alternative for managing wildlife populations (Moya et al. 2011). Such a strategy would require regional UMAs to be committed to sharing information, monitoring wildlife populations, and implementing management programs in coordination with other UMAs. This is not a simple strategy; it requires the agreement and commitment of UMA owners, different levels of government, and academicians.

Despite the operational problems of UMAs in central and southern Mexico, 11,839 UMAs have been established since 2000, covering 37.4 million ha (92,490,200 ac) of natural forest, scrublands, and rainforest ecosystems. With the 22 million ha (54,406,000 ac) of ANPs, there are 59 million ha (146 million ac) with a "management plan" for sustainable wildlife use approved by SEMARNAT. Although most UMAs may be functioning inadequately, they have the flexibility to modify and improve their management strategies (Villarreal G. 2013c). Additional efforts include the creation of the Comité Nacional para el Sistema Productivo de Ganadería Diversificada (National Committee for Cattle-Wildlife Production), a committee consisting of members from different sectors of the ranching industry that coordinate and negotiate efforts to increase the competitiveness of cattle-wildlife operations (Villarreal G. 2012).

The Role of the Federal Government

Departments within federal forestry and agriculture agencies are charged with promoting wildlife as an economically important resource (Weber et al. 2006). Valdez et al. (2006) indicated that there has been growing recognition in federal agencies of the economic and ecological importance of wildlife during the past 30 years. However, the continual creation and dissolution of federal agencies responsible for wildlife administration and management, in combination with the lack of adequate funding, has not allowed the establishment of a progressive wildlife program. Until recently, wildlife management in Mexico has been based on a centralized system without state participation. Biodiversity in Mexico was acknowledged as a national priority with the creation of CONABIO in 1992. Around this same time, other agencies related to natural resource conservation were created, in most cases with limited funding and with administrators who lacked experience in biodiversity conservation and management (see chapter 21).

Federal and state governments at all levels are responsible for promoting wildlife conservation. Wildlife and its habitats are critical elements in the sustainable use of natural resources. Governments must acknowledge the economic importance of wildlife with a long-term vision, because in many cases the economic value of wildlife exceeds the value of other agroecosystem enterprises. It is necessary to develop economically viable and sustainable communal and private initiatives that include wildlife as part of planning strategies for natural resource management.

In the last 30 years, Mexico has attained important achievements in wildlife conservation, including programs for the sustainable use of wildlife by the end of the 1990s. However, challenges now are greater, more diverse, and more complex. Historically, wildlife management has developed viable alternatives to achieve the recovery and maintenance of habitats, thereby increasing wildlife populations to benefit Mexican society. This can be facilitated in species with a high reproductive potential that may quickly recover. Simultaneously, wildlife species at risk should be prioritized nationally with sufficient funding to develop ecological research and environmental education programs. These strategies have been proven to yield positive results. Ignoring the threats discussed in this book will result in losing the opportunity to achieve wildlife species conservation and sustainable use of Mexico's wildlife.

Literature Cited

Ambriz-Morales, P., X. F. De La Rosa-Reyna, A. M. Sifuentes-Rincon, G. M. Parra-Bracamonte, A. Villa-Melchor, O. Chassin-Noria, and W. Arellano-Vera. 2016. The complete mitochondrial genomes of nine white-tailed deer subspecies and their genomic differences. Journal of Mammalogy 97:234–245.

Betancourt, S. H., A. Segovia C., and F. Martínez C. 1996. Datos preliminares de la cacería de subsistencia en el sur de Yucatán, México. Pages 32–40 *in* Memorias del VI Simposio sobre Venados de México. UNAM, México, D. F., México.

Caughley, G., and A. R. E. Sinclair. 1994. Wildlife ecology and management. Blackwell Science, Oxford, United Kingdom.

Ceballos, G., P. R. Ehrlich, J. Soberón, I. Salazar, and J. P. Fay. 2005. Global mammal conservation: what must we manage? Science 309:603–607.

Clemente-Sanchez, F. 1993. Influences of range condition, cattle, and watering hole distribution of a pronghorn population in south-central New Mexico. Dissertation, New Mexico State University, Las Cruces, USA.

CNOG. 2013. Fondo de aseguramiento de la CNOG: avances y logros. Decimoseptima Reunión Anual de la CONASA. http://www.conasamexico.org.mx/conasa/docs_17a_reunion/foroaseguramientoencontingenciassanitarias/Antonio_M_Garcia_Garza.pdf. Accessed 21 Jan 2014.

Ehrlich, A. H., and P. R. Ehrlich. 1992. Causes and consequences of the disappearance of biodiversity. Pages 43–55 *in* J. Sarukhán and R. Dirzo, compilers. México ante los retos de la biodiversidad. CONABIO, México, D. F., México.

Galindo-Leal, C., and M. Weber. 1994. Translocation of deer subspecies: reproductive implications. Wildlife Society Bulletin 22:117–120.

Gallina, S., and L. A. Escobedo-Morales. 2009. Análisis sobre las unidades de manejo (UMAs) de ciervo rojo (*Cervus elaphus* Linnaeus, 1758) y wapiti (*Cervus canadensis* Erxleben, 1777) en México: problemática para la conservación de los ungulados nativos. Tropical Conservation Science 2:251–265.

Gallina-Tessaro, S. A., A. Hernández-Huerta, C. A. Delfin-Alfonso, and A. González-Gallina. 2008. Unidades para la conservación, manejo y aprovechamiento sustentable de la vida silvestre en México (UMA). Retos para su correcto funcionamiento. Investigación Ambiental 1:143–152.

García-Marmolejo, G. 2005. Caracterización y sustentabilidad de las unidades para la conservación, manejo y aprovechamiento sustentable de la vida silvestre en Campeche. Thesis, El Colegio de la Frontera Sur, Campeche, México.

García-Marmolejo, G., G. Escalona-Segura, and H. Van Der Wal. 2008. Multicriteria evaluation of wildlife management units in Campeche, Mexico. Journal of Wildlife Management 72: 1194–1202.

González-Marín, R. M., R. Montes Pérez, and J. Santos Flores. 2003. Caracterización de las unidades para conservación, manejo y aprovechamiento sustentable de fauna silvestre en Yucatán, México. Tropical and Subtropical Agroecosystems 2:13–21.

Halffter, G. 2007. Reservas archipiélago: un nuevo tipo de área protegida. Pages 281–286 *in* G. Halffter, S. Guevara, and A. Melic, editors. Hacia una cultura de conservación de la biodiversidad biológica. Volume 6. Mongrafías Tercer Milenio, Zaragoza, Spain.

INEGI. 2005. Guía para la interpretación de cartografía: uso de suelo y vegetación. Instituto Nacional de Estadística Geográfica e Informática, Aguascalientes, Aguascalientes, México.

INEGI. 2010. Carta de uso del suelo y vegetación, escala 1:25,000: Serie IV. INEGI, Aguascalientes, Aguascalientes, México.

Logan-López, K., E. Cienfuegos-Rivas, A. Sifuentes-Rincón, M. González-Paz, F. Clemente-Sánchez, G. Mendoza-Martínez, and L. Tarango-Arámbula. 2007. Patrones de variación genética en cuatro subespecies de venado cola blanca del noreste de México. Agrociencia 41:13–21.

Mandujano, S. 2004. Análisis bibliográfico de los estudios de venados en México. Acta Zoológica Mexicana (n.s.) 20:211–251.

Mandujano, S., and A. González-Zamora. 2009. Evaluation of natural conservation areas and wildlife management units to support minimum viable populations of white-tailed deer in Mexico. Tropical Conservation Science 2:237–250.

Medellín L., R. A. 2005. Vertebrados superiores exóticos en México: diversidad, distribución y efectos potenciales. INE, UNAM, and CONABIO, México, D. F., México.

Moya, H., E. Peters, and P. Zamorano. 2011. La importancia de un enfoque regional para la conservación del hábitat natural en la frontera norte de México. Pages 49–68 *in* O. Sánchez, P. Zamorano, E. Peters, and H. Moya, editors. Temas sobre conservación de vertebrados silvestres en México. SEMARNAT, México, D. F., México.

Naranjo, E. J., and R. E. Bodmer. 2007. Source-sink systems and conservation of hunted ungulates in the Lacandon forest, Mexico. Biological Conservation 138:412–420.

Ortega-S., J. A., S. Mandujano, J. Villarreal G., M. I. DiMare, H. Lopez-Arevalo, M. Molina, and M. Correa-Viana. 2011. Managing white-tailed deer: Latin America. Pages 565–597 *in* D. G. Hewitt, editor. Biology and management of *white-tailed deer*. CRC Press, Boca Raton, Florida, USA.

Ortega-S., J. A., A. Melgoza, F. A. Ibarra-Flores, E. A. Gonzalez-Valenzuela, M. H. Martin-Rivera, J. M. Avila-Curiel, F. Ayala-Alvarez, C. Pinedo, and O. Rivero. 2013. Exotic grasses and wildlife in northern Mexico. Wildlife Society Bulletin 37:537–545.

Ortiz-Monasterio, A. 2011. La administración descentralizada de algunos aspectos del manejo de la vida silvestre en México: síntesis del proceso y marco jurídico. Pages 335–371 *in* O. Sánchez, P. Zamorano, E. Peters, and H. Moya, editors. Temas sobre conservación de vertebrados silvestres en México. SEMARNAT, México, D. F., México.

Ramirez, F., and M. E. Mondragón. 2010. Conservación y aprovechamiento sustentable: La Cuenca Palo Blanco. Pages 98–99 *in* J. Carabias, J. Sarukhán, J. de la Masa, and C. Galindo Leal, editors. Patrimonio natural de México. Cien casos de éxito. CONABIO, México, D. F., México.

Robinson, J. G., and R. E. Bodmer. 1999. Towards wildlife management in tropical forests. Journal of Wildlife Management 63:1–13.

Sarukhán, J., J. Soberón, and J. Larson-Guerra. 1996. Biological conservation in a high beta diversity country. Pages 246–263 *in* E. di Castri and T. Younes, editors. Biodiversity, science and development: toward a new partnership. CAB International-IUBS, Paris, France.

SEMARNAP (Secretaría de Medio Ambiente, Recursos Naturales y Pesca). 1997. Programa de la Conservación de la Vida Silvestre y Diversificación Productiva en el Sector Rural: 1997–2000. DGFS and INE, México, D. F., México.

Soulé, M. E. 1987. Viable populations for conservation. Cambridge University Press, Cambridge, United Kingdom.

Tapia, R. J., S. Ibarra-Zimbrón, G. D. Mendoza-Martínez, C. Zaragoza-Hernández, F. Clemente-Sánchez, and L. A. Tarango-Arámbula. 2002. Densidad poblacional de la codorniz Moctezuma (*Cyrtonyx montezumae*) en la región noroeste del estado de México, México. Veterinaria México 33:255–262.

Toledo, V., and M. J. Ordóñez. 1993. The biodiversity scenario of México: a review of terrestrial habitats. Pages 757–775 *in* T. P. Ramamoorthy, R. Bye, A. Lot, and J. Fa, editors. Biological diversity of Mexico. Oxford University Press, New York, New York, USA.

Traill, L. W., C. J. A. Bradshaw, and B. W. Brook. 2007. Minimum viable population size: a meta-analysis of 30 years of published estimates. Biological Conservation 139:159–166.

Valdez, R., J. C. Guzmán-Aranda, F. J. Abarca, L. A. Tarango-Arámbula, and F. C. Sánchez. 2006. Wildlife conservation and management in Mexico. Wildlife Society Bulletin 34:270–282.

Viejo, G. J. 2012. Nuevas categorías de trofeos de caza de venados cola blanca mexicanos en el libro de records del Safari Club Internacional. Pages 192–200 *in* Memorias del XIII Simposio sobre Venados de México "Ing. Jorge G. Villarreal González." Secretaría de Educación Continua y Tecnología and UNAM, Toluca, Estado de México, México.

Villarreal G., J. G. 1999a. Venado cola blanca: manejo y aprovechamiento cinegético. Unión Ganadera Regional de Nuevo León, Monterrey, México.

Villarreal G., J. G. 1999b. Importancia del puma para la producción de venados de calidad. Revista Beefmaster (Noviembre–Diciembre) 6:14–15.

Villarreal G., J. G. 2000. Coyotefobia: un mal de la ignorancia. Revista Beefmaster (Julio–Agosto) 4:10–11.

Villarreal G., J. 2008. Historia del manejo del venado cola blanca *Odocoileus virginianus* en Nuevo León, México. Pages 131–155 *in* Memorias del XI Simposio sobre Venados en México "Jorge G. Villarreal González." UNAM, México, D. F., México.

Villarreal G., J. G. 2009. Vida silvestre de la Cuenca "Palo Blanco," Nuevo León, México. Consejo Estatal de Flora y Fauna Silvestre de Nuevo León, Parques y Vida Silvestre de Nuevo León, and UANL, Monterrey, México.

Villarreal G., J. G. 2012. Introducción al manejo y

aprovechamiento sustentable de la fauna silvestre en ranchos ganaderos diversificados. Sistema Producto Ganadería Diversificada and SAGARPA, Monterrey, México.

Villarreal G., J. G. 2013a. 20 años de experiencias en la repoblación de venado cola blanca texano *Odocoileus virginianus texanus* realizada en Cerralvo, Nuevo León, México. Pages 183–187 *in* Memorias del XXX Simposio sobre Fauna Silvestre "Gral. M. V. Manuel Cabrera Valtierra," Secretaría de Educación Continua y Tecnología and UNAM, México, D. F., México.

Villarreal G., J. G. 2013b. La plaga de jabalí europeo y marrano alzado *Sus scrofa* en Nuevo León, México. Pages 123–139 *in* Memorias de Relatorías, Conclusiones y Recomendaciones de la 21ª Reunión Anual del Consejo Técnico Consultivo Nacional de Sanidad Animal (CONASA). Monterrey, Nuevo León, México.

Villarreal G., J. G. 2013c. Mexican wildlife management synopsis: from the road to extinction to the road toward sustainability. The Wildlife Society, Southwest Section News 3:26–29.

Villarreal G., J. G. 2013d. Convocatoria para la erradicación de marrano alzado y/o jabalí europeo *Sus scrofa* en Nuevo León y México. Revista Gran Safari 4:28.

Weber, M. 1993. Ganadería de ciervos: ¿alternativa de producción animal o amenaza a la conservación de la fauna nativa? Agrociencia 3:99–113.

Weber, M., G. García-Marmolejo, and R. Reyna-Hurtado. 2006. The tragedy of the commons: wildlife management units (UMAs) in southeastern Mexico. Wildlife Society Bulletin 34:1480–1488.

Zaragoza, H. C., G. D. Mendoza-Martínez, S. Ibarra-Zimbrón, M. M. Crosby-Galván, F. Clemente-Sánchez, and B. Aguilar-Valdez. 2004. Composición de la dieta de la codorniz Moctezuma (*Cyrtonyx montezumae*) en el noroeste del estado de México, México. Veterinaria México 35:215–236.

Contributors

Carlos H. Alcalá-Galván
Congreso 365 Col. Ley 57
Hermosillo, Sonora, México 83100
sr.alcala@gmail.com

Erick H. Baur
29 Alder Lane
Burlington, VT 05401
ehbaur@hotmail.com

Leonard A. Brennan
Caesar Kleberg Wildlife Research Institute
MSC 218, 700 University Boulevard
Texas A&M University–Kingsville
Kingsville, Texas 78363
leonard.brennan@tamuk.edu

Patricia L. R. Brennan
University of Massachusetts
Psychology Department
411 Tobin Hall
Amherst, MA 01003
pbrennan@bio.umass.edu

Daniel M. Brooks
Houston Museum of Natural Science
Department of Vertebrate Zoology
5555 Hermann Park Drive
Houston, TX 77030
dbrooks@hmns.org

Antony Challenger
Lab. de Ecología de Ecosistemas
UNAM (Campus Morelia)
Antigua Carretera a Pátzcuaro 8701
Ex-Hacienda de San José de la Huerta
Morelia, Michoacán, México 58190
acbchallenger@yahoo.co.uk

Gilberto Chávez-León
Instituto Nacional de Investigaciones Forestales
Agrícolas y Pecuarias
Progreso No. 5, Barrio de Santa Catarina
Del Coyoacan, México, D. F., México 04010
chavez.gilberto@inifap.gob.mx

Felipe Chávez-Ramírez
Gulf Coast Bird Observatory
103 Hwy. 332 West
Lake Jackson, TX 77566
fchavez@gcbo.org

Fernando Clemente-Sánchez
Colegio de Postgraduados
Iturbide No. 73
Salinas de Hidalgo, San Luis Potosí, México 78620
clemente@colpos.mx

Jonás A. Delgadillo Villalobos
Naturaleza Sin Fronteras
Ave. Independencia 901-A OTE
Col. Cementos
Monterrey, Nuevo León, México 64520
jonas.villalobos@hotmail.com

Jack C. Eitniear
Center for the Study of Tropical Birds
218 Conway Drive
San Antonio, TX 78209–1716
jclintoneitniear@gmail.com

Alejandro Espinosa-T.
CEMEX
Ave. Ricardo Margáin Zozoya 325
San Pedro Garza Garcia
Nuevo León, México 66265
alejandro.espinosa@cemex.com

Reagan Gage
708 W. Main St., Suite D
Rio Grande City, TX 78582
rgage22@gmail.com

Sonia Antonieta Gallina Tessaro
Red de Biología y Conservación de Vertebrados
Instituto de Ecología, A. C.
Carretera Antigua a Coatepec No. 351
El Haya
Xalapa, Veracruz, México 91070
sonia.gallina@inecol.mx

Whitney J. Gann
Borderlands Research Institute for Natural Resource Management
Sul Ross State University
P.O. Box C-16
Alpine, TX 79832
whitney.gann@sulross.edu

Mario Alberto García Aranda
Especies, Sociedad y Hábitat, A. C.
Mar Caspio 8212 Loma Linda
Monterrey, Nuevo León, México 64120
mgarcia_20@yahoo.com.mx

David García Solórzano
Alianza por la Conservación de la Vida Silvestre
Ramón Ángel Amante No. 422
Colonia Las Torres
Hermosillo, Sonora, México 83139
cientifico@amecvis.org

Santiago Gibert Isern
Dimensión Natural
Mariano Escobedo 7, Col. 6 de Enero
Zoncuantla, Coatepec, Veracruz, México 96108
santiago@dimensionnatural.com

Fernando González-García
Red de Biología y Conservación de Vertebrados
Biblioteca de Sonidos de las Aves de México (BISAM)
Carretera Antigua a Coatepec No. 351
El Haya
Xalapa, Veracruz, México 91070
fernando.gonzalez@inecol.edu.mx

Carlos E. Gonzalez-Gonzalez
Institute of Renewable Natural Resources
Texas A&M University
1500 Research Pkwy., A110
College Station, TX 77840
lalgon@hotmail.com

Carlos González-Rebeles Islas
Facultad de Medicina Veterinaria y Zootecnia
Dpto. de Etología, Fauna Silvestre y Anim. Lab.
Universidad Nacional Autónoma de México
Ciudad Universitaria
México, D. F., México 40510
grebeles@unam.mx

Louis A. Harveson
Borderlands Research Institute for Natural Resource Management
Sul Ross State University
P.O. Box C-16
Alpine, TX 79832
harveson@sulross.edu

Fidel Hernández
Caesar Kleberg Wildlife Research Institute
Texas A&M University–Kingsville
MSC 218, 700 University Blvd.
Kingsville, TX 78363
kffh001@tamuk.edu

Lucina Hernandez (deceased)
Former director, Rice Creek Field Station
State University of New York at Oswego
Oswego, NY 13126

Alberto Lafón-Terrazas
Protección de la Fauna Mexicana A. C.
Calle 16 No. 2604, Col. Pacifico
Chihuahua, Chihuahua, México 31030
lafon@profauna.org.mx

John W. Laundré
Western Oregon University
Monmouth, OR 97361
launjohn@hotmail.com

Roel R. Lopez
Texas A&M Institute of Renewable Natural Resources
4040 Broadway, Suite 360
San Antonio, TX 78209
roel@tamu.edu

E. Alejandro Lozano-Cavazos
Universidad Autónoma Agraria Antonio Narro
Departamento de Recursos Naturales Renovables
Buenavista, Saltillo, Coahuila, México 25315
alejandro.lozano-c@uaaan.mx

Salvador Mandujano
Red de Biología y Conservación de Vertebrados
Instituto de Ecología A. C.
Km 25 Carret. Autopista Coatepec No. 351
Congregación del Haya
Xalapa, Veracruz, México
salvador.mandujano@inecol.mx

Ignacio March Mifsut
Comisión Nacional de Áreas Naturales Protegidas
Avenida Ejército Nacional No. 223, Piso 12 Ala B
1ª Sección Col. Anáhuac
Del. Miguel Hidalgo
Ciudad de México, México 11320

Melisa Meztli Méndez Méndez
Facultad de Medicina Veterinaria y Zootecnia
Dpto. de Etología, Fauna Silvestre y Anim. Lab.
Universidad Nacional Autónoma de México
México 29290
Ciudad Universitaria
México, D. F., México 40510
meztliluna@hotmail.com

Eduardo J. Naranjo
El Colegio de la Frontera Sur
Carretera Panamericana y Periférico Sur
San Cristóbal de Las Casas, Chiapas, México
enaranjo@ecosur.mx

Rodrigo Núñez Pérez
Instituto de Biología
Universidad Nacional Autónoma de México
Tercer Circuito Cultural, Ciudad Universitaria
México, D. F., México 04510
proyectojaguar@gmail.com

J. Alfonso Ortega-S.
Caesar Kleberg Wildlife Research Institute
Texas A&M University–Kingsville
MSC 218, 700 University Blvd.
Kingsville, TX 78363
poncho.ortega@tamuk.edu

Alfonso Ortega-S. Jr.
La Paloma Ranch
P.O. Box 447
Charlotte, TX 78011
poncho@pearceind.com

Markus J. Peterson
Dept. of Wildlife and Fisheries Sciences
2258 TAMU
Texas A&M University
College Station, TX 77843
mpeterson@tamu.edu

Rafael Angel Reyna-Hurtado
Conservación de la Biodiversidad
ECOSUR Campeche
Campeche, México 24500
rreyna@ecosur.mx

Octavio César Rosas Rosas
Colegio de Postgraduados
Campus San Luis Potosí
Iturbide No. 73
Salinas de Hidalgo, San Luis Potosí, México 78620
octaviocrr@colpos.mx

Andrew V. Sandoval
Borrego Cimarron Wildlife Consulting
P.O. Box 238
Chacon, NM 87713
bighornwlc@nnmt.net

Sanford D. Schemnitz (deceased)
New Mexico State University
Chacon, NM 87713–0238
Dept. of Fish, Wildlife and Conservation Ecology
P.O. Box 30003, MSC 4901
Las Cruces, NM 88003

Nova J. Silvy
Texas A&M University
Dept. of Wildlife and Fisheries Sciences
2258 TAMU
College Station, Texas 77843
n-silvy@tamu.edu

Raul Valdez
New Mexico State University
Dept. of Fish, Wildlife and Conservation Ecology
P.O. Box 30003, MSC 4901
Las Cruces, NM 88003
rvaldez@nmsu.edu

Ramiro Velázquez Rincón
GEOPROSPECT
Padre Salvador Perales Cazares 1137
Col. Misión de San Gerardo
Linares, Nuevo León, México 67778
ramiro.velazquezr@gmail.com

Jorge G. Villarreal González (deceased)
Consejo Estatal de Flora y Fauna Silvestre de Nuevo León
Padre Mier No. 545-F Poniente, Centro
Monterrey, Nuevo León, México 64000

Manuel Weber
El Colegio de la Frontera Sur
Avenida Rancho Polígono 2A
Parque Industrial Lerma
Campeche, México 24500
mweber@ecosur.mx

Lovett E. Williams Jr. (deceased)
P.O. Box 552
Hyde Park, NY 12538

Index

Other Books in the Perspectives on South Texas Series

Nesting Birds of a Tropical Frontier: The Lower Rio Grande Valley of Texas
Timothy Brush

White-Tailed Deer Habitat: Ecology and Management on Rangelands
Timothy E. Fulbright and J. Alfonso Ortega-S.

Texas Quails: Ecology and Management
Leonard A. Brennan

Ecología y Manejo de Venado Cola Blanca
Timothy E. Fulbright and J. Alfonso Ortega-S.

Petra's Legacy: The South Texas Ranching Empire of Petra Vela and Mifflin Kenedy
Jane Clements Monday and Francis Brannen Vick

Racial Borders: Black Soldiers along the Rio Grande
James N. Leiker

Plants of Deep South Texas: A Field Guide to the Woody and Flowering Species
Alfred Richardson

African Americans in South Texas History
Bruce A. Glasrud

Beef, Brush, and Bobwhites: Quail Management in Cattle Country
Fidel Hernández and Fred S. Guthery

Upland and Webless Migratory Game Birds of Texas
Leonard A. Brennan